Lecture Notes in Computer Science   12949

More information about this subseries at http://www.springer.com/series/7407

Osvaldo Gervasi · Beniamino Murgante ·
Sanjay Misra · Chiara Garau ·
Ivan Blečić · David Taniar ·
Bernady O. Apduhan · Ana Maria A. C. Rocha ·
Eufemia Tarantino · Carmelo Maria Torre (Eds.)

Computational Science and Its Applications – ICCSA 2021

21st International Conference
Cagliari, Italy, September 13–16, 2021
Proceedings, Part I

 Springer

Editors
Osvaldo Gervasi 🆔
University of Perugia
Perugia, Italy

Sanjay Misra 🆔
Covenant University
Ota, Nigeria

Ivan Blečić 🆔
University of Cagliari
Cagliari, Italy

Bernady O. Apduhan
Kyushu Sangyo University
Fukuoka, Japan

Eufemia Tarantino 🆔
Polytechnic University of Bari
Bari, Italy

Beniamino Murgante 🆔
University of Basilicata
Potenza, Potenza, Italy

Chiara Garau 🆔
University of Cagliari
Cagliari, Italy

David Taniar 🆔
Monash University
Clayton, VIC, Australia

Ana Maria A. C. Rocha 🆔
University of Minho
Braga, Portugal

Carmelo Maria Torre 🆔
Polytechnic University of Bari
Bari, Italy

ISSN 0302-9743   ISSN 1611-3349   (electronic)
Lecture Notes in Computer Science
ISBN 978-3-030-86652-5   ISBN 978-3-030-86653-2   (eBook)
https://doi.org/10.1007/978-3-030-86653-2

LNCS Sublibrary: SL1 – Theoretical Computer Science and General Issues

This Springer imprint is published by the registered company Springer Nature Switzerland AG
The registered company address is: Gewerbestrasse 11, 6330 Cham, Switzerland

Preface

These 10 volumes (LNCS volumes 12949–12958) consist of the peer-reviewed papers from the 21st International Conference on Computational Science and Its Applications (ICCSA 2021) which took place during September 13–16, 2021. By virtue of the vaccination campaign conducted in various countries around the world, we decided to try a hybrid conference, with some of the delegates attending in person at the University of Cagliari and others attending in virtual mode, reproducing the infrastructure established last year.

This year's edition was a successful continuation of the ICCSA conference series, which was also held as a virtual event in 2020, and previously held in Saint Petersburg, Russia (2019), Melbourne, Australia (2018), Trieste, Italy (2017), Beijing. China (2016), Banff, Canada (2015), Guimaraes, Portugal (2014), Ho Chi Minh City, Vietnam (2013), Salvador, Brazil (2012), Santander, Spain (2011), Fukuoka, Japan (2010), Suwon, South Korea (2009), Perugia, Italy (2008), Kuala Lumpur, Malaysia (2007), Glasgow, UK (2006), Singapore (2005), Assisi, Italy (2004), Montreal, Canada (2003), and (as ICCS) Amsterdam, The Netherlands (2002) and San Francisco, USA (2001).

Computational science is the main pillar of most of the present research on understanding and solving complex problems. It plays a unique role in exploiting innovative ICT technologies and in the development of industrial and commercial applications. The ICCSA conference series provides a venue for researchers and industry practitioners to discuss new ideas, to share complex problems and their solutions, and to shape new trends in computational science.

Apart from the six main conference tracks, ICCSA 2021 also included 52 workshops in various areas of computational sciences, ranging from computational science technologies to specific areas of computational sciences, such as software engineering, security, machine learning and artificial intelligence, blockchain technologies, and applications in many fields. In total, we accepted 494 papers, giving an acceptance rate of 30%, of which 18 papers were short papers and 6 were published open access. We would like to express our appreciation for the workshop chairs and co-chairs for their hard work and dedication.

The success of the ICCSA conference series in general, and of ICCSA 2021 in particular, vitally depends on the support of many people: authors, presenters, participants, keynote speakers, workshop chairs, session chairs, organizing committee members, student volunteers, Program Committee members, advisory committee members, international liaison chairs, reviewers, and others in various roles. We take this opportunity to wholehartedly thank them all.

We also wish to thank Springer for publishing the proceedings, for sponsoring some of the best paper awards, and for their kind assistance and cooperation during the editing process.

We cordially invite you to visit the ICCSA website https://iccsa.org where you can find all the relevant information about this interesting and exciting event.

September 2021

Osvaldo Gervasi
Beniamino Murgante
Sanjay Misra

Welcome Message from the Organizers

COVID-19 has continued to alter our plans for organizing the ICCSA 2021 conference, so although vaccination plans are progressing worldwide, the spread of virus variants still forces us into a period of profound uncertainty. Only a very limited number of participants were able to enjoy the beauty of Sardinia and Cagliari in particular, rediscovering the immense pleasure of meeting again, albeit safely spaced out. The social events, in which we rediscovered the ancient values that abound on this wonderful island and in this city, gave us even more strength and hope for the future. For the management of the virtual part of the conference, we consolidated the methods, organization, and infrastructure of ICCSA 2020.

The technological infrastructure was based on open source software, with the addition of the streaming channels on YouTube. In particular, we used Jitsi (jitsi.org) for videoconferencing, Riot (riot.im) together with Matrix (matrix.org) for chat and ansynchronous communication, and Jibri (github.com/jitsi/jibri) for streaming live sessions to YouTube.

Seven Jitsi servers were set up, one for each parallel session. The participants of the sessions were helped and assisted by eight student volunteers (from the universities of Cagliari, Florence, Perugia, and Bari), who provided technical support and ensured smooth running of the conference proceedings.

The implementation of the software infrastructure and the technical coordination of the volunteers were carried out by Damiano Perri and Marco Simonetti.

Our warmest thanks go to all the student volunteers, to the technical coordinators, and to the development communities of Jitsi, Jibri, Riot, and Matrix, who made their terrific platforms available as open source software.

A big thank you goes to all of the 450 speakers, many of whom showed an enormous collaborative spirit, sometimes participating and presenting at almost prohibitive times of the day, given that the participants of this year's conference came from 58 countries scattered over many time zones of the globe.

Finally, we would like to thank Google for letting us stream all the live events via YouTube. In addition to lightening the load of our Jitsi servers, this allowed us to record the event and to be able to review the most exciting moments of the conference.

Ivan Blečić
Chiara Garau

Organization

ICCSA 2021 was organized by the University of Cagliari (Italy), the University of Perugia (Italy), the University of Basilicata (Italy), Monash University (Australia), Kyushu Sangyo University (Japan), and the University of Minho (Portugal).

Honorary General Chairs

Norio Shiratori	Chuo University, Japan
Kenneth C. J. Tan	Sardina Systems, UK
Corrado Zoppi	University of Cagliari, Italy

General Chairs

Osvaldo Gervasi	University of Perugia, Italy
Ivan Blečić	University of Cagliari, Italy
David Taniar	Monash University, Australia

Program Committee Chairs

Beniamino Murgante	University of Basilicata, Italy
Bernady O. Apduhan	Kyushu Sangyo University, Japan
Chiara Garau	University of Cagliari, Italy
Ana Maria A. C. Rocha	University of Minho, Portugal

International Advisory Committee

Jemal Abawajy	Deakin University, Australia
Dharma P. Agarwal	University of Cincinnati, USA
Rajkumar Buyya	University of Melbourne, Australia
Claudia Bauzer Medeiros	University of Campinas, Brazil
Manfred M. Fisher	Vienna University of Economics and Business, Austria
Marina L. Gavrilova	University of Calgary, Canada
Yee Leung	Chinese University of Hong Kong, China

International Liaison Chairs

Giuseppe Borruso	University of Trieste, Italy
Elise De Donker	Western Michigan University, USA
Maria Irene Falcão	University of Minho, Portugal
Robert C. H. Hsu	Chung Hua University, Taiwan
Tai-Hoon Kim	Beijing Jaotong University, China

Vladimir Korkhov	St. Petersburg University, Russia
Sanjay Misra	Covenant University, Nigeria
Takashi Naka	Kyushu Sangyo University, Japan
Rafael D. C. Santos	National Institute for Space Research, Brazil
Maribel Yasmina Santos	University of Minho, Portugal
Elena Stankova	St. Petersburg University, Russia

Workshop and Session Chairs

Beniamino Murgante	University of Basilicata, Italy
Sanjay Misra	Covenant University, Nigeria
Jorge Gustavo Rocha	University of Minho, Portugal

Awards Chair

| Wenny Rahayu | La Trobe University, Australia |

Publicity Committee Chairs

Elmer Dadios	De La Salle University, Philippines
Nataliia Kulabukhova	St. Petersburg University, Russia
Daisuke Takahashi	Tsukuba University, Japan
Shangwang Wang	Beijing University of Posts and Telecommunications, China

Technology Chairs

| Damiano Perri | University of Florence, Italy |
| Marco Simonetti | University of Florence, Italy |

Local Arrangement Chairs

Ivan Blečić	University of Cagliari, Italy
Chiara Garau	University of Cagliari, Italy
Alfonso Annunziata	University of Cagliari, Italy
Ginevra Balletto	University of Cagliari, Italy
Giuseppe Borruso	University of Trieste, Italy
Alessandro Buccini	University of Cagliari, Italy
Michele Campagna	University of Cagliari, Italy
Mauro Coni	University of Cagliari, Italy
Anna Maria Colavitti	University of Cagliari, Italy
Giulia Desogus	University of Cagliari, Italy
Caterina Fenu	University of Cagliari, Italy
Sabrina Lai	University of Cagliari, Italy
Francesca Maltinti	University of Cagliari, Italy
Pasquale Mistretta	University of Cagliari, Italy

Augusto Montisci   University of Cagliari, Italy
Francesco Pinna   University of Cagliari, Italy
Davide Spano   University of Cagliari, Italy
Giuseppe A. Trunfio   University of Sassari, Italy
Corrado Zoppi   University of Cagliari, Italy

Program Committee

Vera Afreixo   University of Aveiro, Portugal
Filipe Alvelos   University of Minho, Portugal
Hartmut Asche   University of Potsdam, Germany
Ginevra Balletto   University of Cagliari, Italy
Michela Bertolotto   University College Dublin, Ireland
Sandro Bimonte   INRAE-TSCF, France
Rod Blais   University of Calgary, Canada
Ivan Blečić   University of Sassari, Italy
Giuseppe Borruso   University of Trieste, Italy
Ana Cristina Braga   University of Minho, Portugal
Massimo Cafaro   University of Salento, Italy
Yves Caniou   University of Lyon, France
José A. Cardoso e Cunha   Universidade Nova de Lisboa, Portugal
Rui Cardoso   University of Beira Interior, Portugal
Leocadio G. Casado   University of Almeria, Spain
Carlo Cattani   University of Salerno, Italy
Mete Celik   Erciyes University, Turkey
Maria Cerreta   University of Naples "Federico II", Italy
Hyunseung Choo   Sungkyunkwan University, South Korea
Chien-Sing Lee   Sunway University, Malaysia
Min Young Chung   Sungkyunkwan University, South Korea
Florbela Maria da Cruz   Polytechnic Institute of Viana do Castelo, Portugal
  Domingues Correia
Gilberto Corso Pereira   Federal University of Bahia, Brazil
Fernanda Costa   University of Minho, Portugal
Alessandro Costantini   INFN, Italy
Carla Dal Sasso Freitas   Universidade Federal do Rio Grande do Sul, Brazil
Pradesh Debba   The Council for Scientific and Industrial Research
                            (CSIR), South Africa
Hendrik Decker   Instituto Tecnolčgico de Informática, Spain
Robertas Damaševičius   Kausan University of Technology, Lithuania
Frank Devai   London South Bank University, UK
Rodolphe Devillers   Memorial University of Newfoundland, Canada
Joana Matos Dias   University of Coimbra, Portugal
Paolino Di Felice   University of L'Aquila, Italy
Prabu Dorairaj   NetApp, India/USA
Noelia Faginas Lago   University of Perugia, Italy
M. Irene Falcao   University of Minho, Portugal

Cherry Liu Fang	Ames Laboratory, USA
Florbela P. Fernandes	Polytechnic Institute of Bragança, Portugal
Jose-Jesus Fernandez	National Centre for Biotechnology, Spain
Paula Odete Fernandes	Polytechnic Institute of Bragança, Portugal
Adelaide de Fátima Baptista Valente Freitas	University of Aveiro, Portugal
Manuel Carlos Figueiredo	University of Minho, Portugal
Maria Celia Furtado Rocha	Universidade Federal da Bahia, Brazil
Chiara Garau	University of Cagliari, Italy
Paulino Jose Garcia Nieto	University of Oviedo, Spain
Jerome Gensel	LSR-IMAG, France
Maria Giaoutzi	National Technical University of Athens, Greece
Arminda Manuela Andrade Pereira Gonçalves	University of Minho, Portugal
Andrzej M. Goscinski	Deakin University, Australia
Eduardo Guerra	Free University of Bozen-Bolzano, Italy
Sevin Gümgüm	Izmir University of Economics, Turkey
Alex Hagen-Zanker	University of Cambridge, UK
Shanmugasundaram Hariharan	B.S. Abdur Rahman University, India
Eligius M. T. Hendrix	University of Malaga, Spain/Wageningen University, The Netherlands
Hisamoto Hiyoshi	Gunma University, Japan
Mustafa Inceoglu	EGE University, Turkey
Peter Jimack	University of Leeds, UK
Qun Jin	Waseda University, Japan
Yeliz Karaca	University of Massachusetts Medical School, USA
Farid Karimipour	Vienna University of Technology, Austria
Baris Kazar	Oracle Corp., USA
Maulana Adhinugraha Kiki	Telkom University, Indonesia
DongSeong Kim	University of Canterbury, New Zealand
Taihoon Kim	Hannam University, South Korea
Ivana Kolingerova	University of West Bohemia, Czech Republic
Nataliia Kulabukhova	St. Petersburg University, Russia
Vladimir Korkhov	St. Petersburg University, Russia
Rosa Lasaponara	National Research Council, Italy
Maurizio Lazzari	National Research Council, Italy
Cheng Siong Lee	Monash University, Australia
Sangyoun Lee	Yonsei University, South Korea
Jongchan Lee	Kunsan National University, South Korea
Chendong Li	University of Connecticut, USA
Gang Li	Deakin University, Australia
Fang Liu	Ames Laboratory, USA
Xin Liu	University of Calgary, Canada
Andrea Lombardi	University of Perugia, Italy
Savino Longo	University of Bari, Italy

Tinghuai Ma   Nanjing University of Information Science
                             and Technology, China
Ernesto Marcheggiani   Katholieke Universiteit Leuven, Belgium
Antonino Marvuglia   Research Centre Henri Tudor, Luxembourg
Nicola Masini   National Research Council, Italy
Ilaria Matteucci   National Research Council, Italy
Eric Medvet   University of Trieste, Italy
Nirvana Meratnia   University of Twente, The Netherlands
Giuseppe Modica   University of Reggio Calabria, Italy
Josè Luis Montaña   University of Cantabria, Spain
Maria Filipa Mourão   Instituto Politécnico de Viana do Castelo, Portugal
Louiza de Macedo Mourelle   State University of Rio de Janeiro, Brazil
Nadia Nedjah   State University of Rio de Janeiro, Brazil
Laszlo Neumann   University of Girona, Spain
Kok-Leong Ong   Deakin University, Australia
Belen Palop   Universidad de Valladolid, Spain
Marcin Paprzycki   Polish Academy of Sciences, Poland
Eric Pardede   La Trobe University, Australia
Kwangjin Park   Wonkwang University, South Korea
Ana Isabel Pereira   Polytechnic Institute of Bragança, Portugal
Massimiliano Petri   University of Pisa, Italy
Telmo Pinto   University of Coimbra, Portugal
Maurizio Pollino   Italian National Agency for New Technologies, Energy
                             and Sustainable Economic Development, Italy
Alenka Poplin   University of Hamburg, Germany
Vidyasagar Potdar   Curtin University of Technology, Australia
David C. Prosperi   Florida Atlantic University, USA
Wenny Rahayu   La Trobe University, Australia
Jerzy Respondek   Silesian University of Technology Poland
Humberto Rocha   INESC-Coimbra, Portugal
Jon Rokne   University of Calgary, Canada
Octavio Roncero   CSIC, Spain
Maytham Safar   Kuwait University, Kuwait
Francesco Santini   University of Perugia, Italy
Chiara Saracino   A.O. Ospedale Niguarda Ca' Granda, Italy
Haiduke Sarafian   Pennsylvania State University, USA
Marco Paulo Seabra dos   University of Coimbra, Portugal
  Reis
Jie Shen   University of Michigan, USA
Qi Shi   Liverpool John Moores University, UK
Dale Shires   U.S. Army Research Laboratory, USA
Inês Soares   University of Coimbra, Portugal
Elena Stankova   St. Petersburg University, Russia
Takuo Suganuma   Tohoku University, Japan
Eufemia Tarantino   Polytechnic University of Bari, Italy
Sergio Tasso   University of Perugia, Italy

Ana Paula Teixeira	University of Trás-os-Montes and Alto Douro, Portugal
Senhorinha Teixeira	University of Minho, Portugal
M. Filomena Teodoro	Portuguese Naval Academy/University of Lisbon, Portugal
Parimala Thulasiraman	University of Manitoba, Canada
Carmelo Torre	Polytechnic University of Bari, Italy
Javier Martinez Torres	Centro Universitario de la Defensa Zaragoza, Spain
Giuseppe A. Trunfio	University of Sassari, Italy
Pablo Vanegas	University of Cuenca, Equador
Marco Vizzari	University of Perugia, Italy
Varun Vohra	Merck Inc., USA
Koichi Wada	University of Tsukuba, Japan
Krzysztof Walkowiak	Wroclaw University of Technology, Poland
Zequn Wang	Intelligent Automation Inc, USA
Robert Weibel	University of Zurich, Switzerland
Frank Westad	Norwegian University of Science and Technology, Norway
Roland Wismüller	Universität Siegen, Germany
Mudasser Wyne	National University, USA
Chung-Huang Yang	National Kaohsiung Normal University, Taiwan
Xin-She Yang	National Physical Laboratory, UK
Salim Zabir	National Institute of Technology, Tsuruoka, Japan
Haifeng Zhao	University of California, Davis, USA
Fabiana Zollo	University of Venice "Cà Foscari", Italy
Albert Y. Zomaya	University of Sydney, Australia

Workshop Organizers

Advanced Transport Tools and Methods (A2TM 2021)

Massimiliano Petri	University of Pisa, Italy
Antonio Pratelli	University of Pisa, Italy

Advances in Artificial Intelligence Learning Technologies: Blended Learning, STEM, Computational Thinking and Coding (AAILT 2021)

Alfredo Milani	University of Perugia, Italy
Giulio Biondi	University of Florence, Italy
Sergio Tasso	University of Perugia, Italy

Workshop on Advancements in Applied Machine Learning and Data Analytics (AAMDA 2021)

Alessandro Costantini	INFN, Italy
Davide Salomoni	INFN, Italy
Doina Cristina Duma	INFN, Italy
Daniele Cesini	INFN, Italy

Automatic Landform Classification: Spatial Methods and Applications (ALCSMA 2021)

Maria Danese   ISPC, National Research Council, Italy
Dario Gioia   ISPC, National Research Council, Italy

Application of Numerical Analysis to Imaging Science (ANAIS 2021)

Caterina Fenu   University of Cagliari, Italy
Alessandro Buccini   University of Cagliari, Italy

Advances in Information Systems and Technologies for Emergency Management, Risk Assessment and Mitigation Based on the Resilience Concepts (ASTER 2021)

Maurizio Pollino   ENEA, Italy
Marco Vona   University of Basilicata, Italy
Amedeo Flora   University of Basilicata, Italy
Chiara Iacovino   University of Basilicata, Italy
Beniamino Murgante   University of Basilicata, Italy

Advances in Web Based Learning (AWBL 2021)

Birol Ciloglugil   Ege University, Turkey
Mustafa Murat Inceoglu   Ege University, Turkey

Blockchain and Distributed Ledgers: Technologies and Applications (BDLTA 2021)

Vladimir Korkhov   St. Petersburg University, Russia
Elena Stankova   St. Petersburg University, Russia
Nataliia Kulabukhova   St. Petersburg University, Russia

Bio and Neuro Inspired Computing and Applications (BIONCA 2021)

Nadia Nedjah   State University of Rio de Janeiro, Brazil
Luiza De Macedo Mourelle   State University of Rio de Janeiro, Brazil

Computational and Applied Mathematics (CAM 2021)

Maria Irene Falcão   University of Minho, Portugal
Fernando Miranda   University of Minho, Portugal

Computational and Applied Statistics (CAS 2021)

Ana Cristina Braga   University of Minho, Portugal

Computerized Evaluation of Economic Activities: Urban Spaces (CEEA 2021)

Diego Altafini   Università di Pisa, Italy
Valerio Cutini   Università di Pisa, Italy

Computational Geometry and Applications (CGA 2021)

Marina Gavrilova   University of Calgary, Canada

Collaborative Intelligence in Multimodal Applications (CIMA 2021)

Robertas Damasevicius   Kaunas University of Technology, Lithuania
Rytis Maskeliunas   Kaunas University of Technology, Lithuania

Computational Optimization and Applications (COA 2021)

Ana Rocha   University of Minho, Portugal
Humberto Rocha   University of Coimbra, Portugal

Computational Astrochemistry (CompAstro 2021)

Marzio Rosi   University of Perugia, Italy
Cecilia Ceccarelli   University of Grenoble, France
Stefano Falcinelli   University of Perugia, Italy
Dimitrios Skouteris   Master-Up, Italy

Computational Science and HPC (CSHPC 2021)

Elise de Doncker   Western Michigan University, USA
Fukuko Yuasa   High Energy Accelerator Research Organization
                               (KEK), Japan
Hideo Matsufuru   High Energy Accelerator Research Organization
                               (KEK), Japan

Cities, Technologies and Planning (CTP 2021)

Malgorzata Hanzl   University of Łódź, Poland
Beniamino Murgante   University of Basilicata, Italy
Ljiljana Zivkovic   Ministry of Construction, Transport and
                               Infrastructure/Institute of Architecture and Urban
                               and Spatial Planning of Serbia, Serbia
Anastasia Stratigea   National Technical University of Athens, Greece
Giuseppe Borruso   University of Trieste, Italy
Ginevra Balletto   University of Cagliari, Italy

Advanced Modeling E-Mobility in Urban Spaces (DEMOS 2021)

Tiziana Campisi   Kore University of Enna, Italy
Socrates Basbas   Aristotle University of Thessaloniki, Greece
Ioannis Politis   Aristotle University of Thessaloniki, Greece
Florin Nemtanu   Polytechnic University of Bucharest, Romania
Giovanna Acampa   Kore University of Enna, Italy
Wolfgang Schulz   Zeppelin University, Germany

Digital Transformation and Smart City (DIGISMART 2021)

Mauro Mazzei	National Research Council, Italy

Econometric and Multidimensional Evaluation in Urban Environment (EMEUE 2021)

Carmelo Maria Torre	Polytechnic University of Bari, Italy
Maria Cerreta	University "Federico II" of Naples, Italy
Pierluigi Morano	Polytechnic University of Bari, Italy
Simona Panaro	University of Portsmouth, UK
Francesco Tajani	Sapienza University of Rome, Italy
Marco Locurcio	Polytechnic University of Bari, Italy

The 11th International Workshop on Future Computing System Technologies and Applications (FiSTA 2021)

Bernady Apduhan	Kyushu Sangyo University, Japan
Rafael Santos	Brazilian National Institute for Space Research, Brazil

Transformational Urban Mobility: Challenges and Opportunities During and Post COVID Era (FURTHER 2021)

Tiziana Campisi	Kore University of Enna, Italy
Socrates Basbas	Aristotle University of Thessaloniki, Greece
Dilum Dissanayake	Newcastle University, UK
Kh Md Nahiduzzaman	University of British Columbia, Canada
Nurten Akgün Tanbay	Bursa Technical University, Turkey
Khaled J. Assi	King Fahd University of Petroleum and Minerals, Saudi Arabia
Giovanni Tesoriere	Kore University of Enna, Italy
Motasem Darwish	Middle East University, Jordan

Geodesign in Decision Making: Meta Planning and Collaborative Design for Sustainable and Inclusive Development (GDM 2021)

Francesco Scorza	University of Basilicata, Italy
Michele Campagna	University of Cagliari, Italy
Ana Clara Mourao Moura	Federal University of Minas Gerais, Brazil

Geomatics in Forestry and Agriculture: New Advances and Perspectives (GeoForAgr 2021)

Maurizio Pollino	ENEA, Italy
Giuseppe Modica	University of Reggio Calabria, Italy
Marco Vizzari	University of Perugia, Italy

Geographical Analysis, Urban Modeling, Spatial Statistics (GEOG-AND-MOD 2021)

Beniamino Murgante	University of Basilicata, Italy
Giuseppe Borruso	University of Trieste, Italy
Hartmut Asche	University of Potsdam, Germany

Geomatics for Resource Monitoring and Management (GRMM 2021)

Eufemia Tarantino	Polytechnic University of Bari, Italy
Enrico Borgogno Mondino	University of Turin, Italy
Alessandra Capolupo	Polytechnic University of Bari, Italy
Mirko Saponaro	Polytechnic University of Bari, Italy

12th International Symposium on Software Quality (ISSQ 2021)

Sanjay Misra	Covenant University, Nigeria

10th International Workshop on Collective, Massive and Evolutionary Systems (IWCES 2021)

Alfredo Milani	University of Perugia, Italy
Rajdeep Niyogi	Indian Institute of Technology, Roorkee, India

Land Use Monitoring for Sustainability (LUMS 2021)

Carmelo Maria Torre	Polytechnic University of Bari, Italy
Maria Cerreta	University "Federico II" of Naples, Italy
Massimiliano Bencardino	University of Salerno, Italy
Alessandro Bonifazi	Polytechnic University of Bari, Italy
Pasquale Balena	Polytechnic University of Bari, Italy
Giuliano Poli	University "Federico II" of Naples, Italy

Machine Learning for Space and Earth Observation Data (MALSEOD 2021)

Rafael Santos	Instituto Nacional de Pesquisas Espaciais, Brazil
Karine Ferreira	Instituto Nacional de Pesquisas Espaciais, Brazil

Building Multi-dimensional Models for Assessing Complex Environmental Systems (MES 2021)

Marta Dell'Ovo	Polytechnic University of Milan, Italy
Vanessa Assumma	Polytechnic University of Turin, Italy
Caterina Caprioli	Polytechnic University of Turin, Italy
Giulia Datola	Polytechnic University of Turin, Italy
Federico dell'Anna	Polytechnic University of Turin, Italy

Ecosystem Services: Nature's Contribution to People in Practice. Assessment Frameworks, Models, Mapping, and Implications (NC2P 2021)

Francesco Scorza	University of Basilicata, Italy
Sabrina Lai	University of Cagliari, Italy
Ana Clara Mourao Moura	Federal University of Minas Gerais, Brazil
Corrado Zoppi	University of Cagliari, Italy
Dani Broitman	Technion, Israel Institute of Technology, Israel

Privacy in the Cloud/Edge/IoT World (PCEIoT 2021)

Michele Mastroianni	University of Campania Luigi Vanvitelli, Italy
Lelio Campanile	University of Campania Luigi Vanvitelli, Italy
Mauro Iacono	University of Campania Luigi Vanvitelli, Italy

Processes, Methods and Tools Towards RESilient Cities and Cultural Heritage Prone to SOD and ROD Disasters (RES 2021)

Elena Cantatore	Polytechnic University of Bari, Italy
Alberico Sonnessa	Polytechnic University of Bari, Italy
Dario Esposito	Polytechnic University of Bari, Italy

Risk, Resilience and Sustainability in the Efficient Management of Water Resources: Approaches, Tools, Methodologies and Multidisciplinary Integrated Applications (RRS 2021)

Maria Macchiaroli	University of Salerno, Italy
Chiara D'Alpaos	Università degli Studi di Padova, Italy
Mirka Mobilia	Università degli Studi di Salerno, Italy
Antonia Longobardi	Università degli Studi di Salerno, Italy
Grazia Fattoruso	ENEA Research Center, Italy
Vincenzo Pellecchia	Ente Idrico Campano, Italy

Scientific Computing Infrastructure (SCI 2021)

Elena Stankova	St. Petersburg University, Russia
Vladimir Korkhov	St. Petersburg University, Russia
Natalia Kulabukhova	St. Petersburg University, Russia

Smart Cities and User Data Management (SCIDAM 2021)

Chiara Garau	University of Cagliari, Italy
Luigi Mundula	University of Cagliari, Italy
Gianni Fenu	University of Cagliari, Italy
Paolo Nesi	University of Florence, Italy
Paola Zamperlin	University of Pisa, Italy

**13th International Symposium on Software Engineering Processes
and Applications (SEPA 2021)**

Sanjay Misra	Covenant University, Nigeria

Ports of the Future - Smartness and Sustainability (SmartPorts 2021)

Patrizia Serra	University of Cagliari, Italy
Gianfranco Fancello	University of Cagliari, Italy
Ginevra Balletto	University of Cagliari, Italy
Luigi Mundula	University of Cagliari, Italy
Marco Mazzarino	University of Venice, Italy
Giuseppe Borruso	University of Trieste, Italy
Maria del Mar Munoz Leonisio	Universidad de Cádiz, Spain

Smart Tourism (SmartTourism 2021)

Giuseppe Borruso	University of Trieste, Italy
Silvia Battino	University of Sassari, Italy
Ginevra Balletto	University of Cagliari, Italy
Maria del Mar Munoz Leonisio	Universidad de Cádiz, Spain
Ainhoa Amaro Garcia	Universidad de Alcalà/Universidad de Las Palmas, Spain
Francesca Krasna	University of Trieste, Italy

**Sustainability Performance Assessment: Models, Approaches and Applications
toward Interdisciplinary and Integrated Solutions (SPA 2021)**

Francesco Scorza	University of Basilicata, Italy
Sabrina Lai	University of Cagliari, Italy
Jolanta Dvarioniene	Kaunas University of Technology, Lithuania
Valentin Grecu	Lucian Blaga University, Romania
Corrado Zoppi	University of Cagliari, Italy
Iole Cerminara	University of Basilicata, Italy

Smart and Sustainable Island Communities (SSIC 2021)

Chiara Garau	University of Cagliari, Italy
Anastasia Stratigea	National Technical University of Athens, Greece
Paola Zamperlin	University of Pisa, Italy
Francesco Scorza	University of Basilicata, Italy

Science, Technologies and Policies to Innovate Spatial Planning (STP4P 2021)

Chiara Garau	University of Cagliari, Italy
Daniele La Rosa	University of Catania, Italy
Francesco Scorza	University of Basilicata, Italy

Anna Maria Colavitti	University of Cagliari, Italy
Beniamino Murgante	University of Basilicata, Italy
Paolo La Greca	University of Catania, Italy

Sustainable Urban Energy Systems (SURENSYS 2021)

Luigi Mundula	University of Cagliari, Italy
Emilio Ghiani	University of Cagliari, Italy

Space Syntax for Cities in Theory and Practice (Syntax_City 2021)

Claudia Yamu	University of Groningen, The Netherlands
Akkelies van Nes	Western Norway University of Applied Sciences, Norway
Chiara Garau	University of Cagliari, Italy

Theoretical and Computational Chemistry and Its Applications (TCCMA 2021)

Noelia Faginas-Lago	University of Perugia, Italy

13th International Workshop on Tools and Techniques in Software Development Process (TTSDP 2021)

Sanjay Misra	Covenant University, Nigeria

Urban Form Studies (UForm 2021)

Malgorzata Hanzl	Łódź University of Technology, Poland
Beniamino Murgante	University of Basilicata, Italy
Eufemia Tarantino	Polytechnic University of Bari, Italy
Irena Itova	University of Westminster, UK

Urban Space Accessibility and Safety (USAS 2021)

Chiara Garau	University of Cagliari, Italy
Francesco Pinna	University of Cagliari, Italy
Claudia Yamu	University of Groningen, The Netherlands
Vincenza Torrisi	University of Catania, Italy
Matteo Ignaccolo	University of Catania, Italy
Michela Tiboni	University of Brescia, Italy
Silvia Rossetti	University of Parma, Italy

Virtual and Augmented Reality and Applications (VRA 2021)

Osvaldo Gervasi	University of Perugia, Italy
Damiano Perri	University of Perugia, Italy
Marco Simonetti	University of Perugia, Italy
Sergio Tasso	University of Perugia, Italy

Workshop on Advanced and Computational Methods for Earth Science Applications (WACM4ES 2021)

Luca Piroddi	University of Cagliari, Italy
Laura Foddis	University of Cagliari, Italy
Augusto Montisci	University of Cagliari, Italy
Sergio Vincenzo Calcina	University of Cagliari, Italy
Sebastiano D'Amico	University of Malta, Malta
Giovanni Martinelli	Istituto Nazionale di Geofisica e Vulcanologia, Italy/Chinese Academy of Sciences, China

Sponsoring Organizations

ICCSA 2021 would not have been possible without the tremendous support of many organizations and institutions, for which all organizers and participants of ICCSA 2021 express their sincere gratitude:

Springer International Publishing AG, Germany (https://www.springer.com)

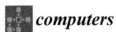

Computers Open Access Journal (https://www.mdpi.com/journal/computers)

IEEE Italy Section, Italy (https://italy.ieeer8.org/)

Centre-North Italy Chapter IEEE GRSS, Italy (https://cispio.diet.uniroma1.it/marzano/ieee-grs/index.html)

Italy Section of the Computer Society, Italy (https://site.ieee.org/italy-cs/)

University of Perugia, Italy (https://www.unipg.it)

University of Cagliari, Italy (https://unica.it/)

University of Basilicata, Italy
(http://www.unibas.it)

 MONASH University

Monash University, Australia
(https://www.monash.edu/)

Kyushu Sangyo University, Japan
(https://www.kyusan-u.ac.jp/)

Universidade do Minho
Escola de Engenharia

University of Minho, Portugal
(https://www.uminho.pt/)

Scientific Association Transport Infrastructures, Italy
(https://www.stradeeautostrade.it/associazioni-e-organizzazioni/asit-associazione-scientifica-infrastrutture-trasporto/)

REGIONE AUTÒNOMA DE SARDIGNA
REGIONE AUTONOMA DELLA SARDEGNA

Regione Sardegna, Italy
(https://regione.sardegna.it/)

COMUNE DI CAGLIARI

Comune di Cagliari, Italy
(https://www.comune.cagliari.it/)

CITTÀ METROPOLITANA DI CAGLIARI

Città Metropolitana di Cagliari

Cagliari Accessibility Lab (CAL)
(https://www.unica.it/unica/it/cagliari_accessibility_lab.page/)

Referees

Nicodemo Abate	IMAA, National Research Council, Italy
Andre Ricardo Abed Grégio	Federal University of Paraná State, Brazil
Nasser Abu Zeid	Università di Ferrara, Italy
Lidia Aceto	Università del Piemonte Orientale, Italy
Nurten Akgün Tanbay	Bursa Technical University, Turkey
Filipe Alvelos	Universidade do Minho, Portugal
Paula Amaral	Universidade Nova de Lisboa, Portugal
Federico Amato	University of Lausanne, Switzerland
Marina Alexandra Pedro Andrade	ISCTE-IUL, Portugal
Debora Anelli	Sapienza University of Rome, Italy
Alfonso Annunziata	University of Cagliari, Italy
Fahim Anzum	University of Calgary, Canada
Tatsumi Aoyama	High Energy Accelerator Research Organization, Japan
Bernady Apduhan	Kyushu Sangyo University, Japan
Jonathan Apeh	Covenant University, Nigeria
Vasilike Argyropoulos	University of West Attica, Greece
Giuseppe Aronica	Università di Messina, Italy
Daniela Ascenzi	Università degli Studi di Trento, Italy
Vanessa Assumma	Politecnico di Torino, Italy
Muhammad Attique Khan	HITEC University Taxila, Pakistan
Vecdi Aytaç	Ege University, Turkey
Alina Elena Baia	University of Perugia, Italy
Ginevra Balletto	University of Cagliari, Italy
Marialaura Bancheri	ISAFOM, National Research Council, Italy
Benedetto Barabino	University of Brescia, Italy
Simona Barbaro	Università degli Studi di Palermo, Italy
Enrico Barbierato	Università Cattolica del Sacro Cuore di Milano, Italy
Jeniffer Barreto	Istituto Superior Técnico, Lisboa, Portugal
Michele Bartalini	TAGES, Italy
Socrates Basbas	Aristotle University of Thessaloniki, Greece
Silvia Battino	University of Sassari, Italy
Marcelo Becerra Rozas	Pontificia Universidad Católica de Valparaíso, Chile
Ranjan Kumar Behera	National Institute of Technology, Rourkela, India
Emanuele Bellini	University of Campania Luigi Vanvitelli, Italy
Massimo Bilancia	University of Bari Aldo Moro, Italy
Giulio Biondi	University of Firenze, Italy
Adriano Bisello	Eurac Research, Italy
Ignacio Blanquer	Universitat Politècnica de València, Spain
Semen Bochkov	Ulyanovsk State Technical University, Russia
Alexander Bogdanov	St. Petersburg University, Russia
Silvia Bonettini	University of Modena and Reggio Emilia, Italy
Enrico Borgogno Mondino	Università di Torino, Italy
Giuseppe Borruso	University of Trieste, Italy

Michele Bottazzi	University of Trento, Italy
Rahma Bouaziz	Taibah University, Saudi Arabia
Ouafik Boulariah	University of Salerno, Italy
Tulin Boyar	Yildiz Technical University, Turkey
Ana Cristina Braga	University of Minho, Portugal
Paolo Bragolusi	University of Padova, Italy
Luca Braidotti	University of Trieste, Italy
Alessandro Buccini	University of Cagliari, Italy
Jorge Buele	Universidad Tecnológica Indoamérica, Ecuador
Andrea Buffoni	TAGES, Italy
Sergio Vincenzo Calcina	University of Cagliari, Italy
Michele Campagna	University of Cagliari, Italy
Lelio Campanile	Università degli Studi della Campania Luigi Vanvitelli, Italy
Tiziana Campisi	Kore University of Enna, Italy
Antonino Canale	Kore University of Enna, Italy
Elena Cantatore	DICATECh, Polytechnic University of Bari, Italy
Pasquale Cantiello	Istituto Nazionale di Geofisica e Vulcanologia, Italy
Alessandra Capolupo	Polytechnic University of Bari, Italy
David Michele Cappelletti	University of Perugia, Italy
Caterina Caprioli	Politecnico di Torino, Italy
Sara Carcangiu	University of Cagliari, Italy
Pedro Carrasqueira	INESC Coimbra, Portugal
Arcangelo Castiglione	University of Salerno, Italy
Giulio Cavana	Politecnico di Torino, Italy
Davide Cerati	Politecnico di Milano, Italy
Maria Cerreta	University of Naples Federico II, Italy
Daniele Cesini	INFN-CNAF, Italy
Jabed Chowdhury	La Trobe University, Australia
Gennaro Ciccarelli	Iuav University of Venice, Italy
Birol Ciloglugil	Ege University, Turkey
Elena Cocuzza	Univesity of Catania, Italy
Anna Maria Colavitt	University of Cagliari, Italy
Cecilia Coletti	Università "G. d'Annunzio" di Chieti-Pescara, Italy
Alberto Collu	Independent Researcher, Italy
Anna Concas	University of Basilicata, Italy
Mauro Coni	University of Cagliari, Italy
Melchiorre Contino	Università di Palermo, Italy
Antonella Cornelio	Università degli Studi di Brescia, Italy
Aldina Correia	Politécnico do Porto, Portugal
Elisete Correia	Universidade de Trás-os-Montes e Alto Douro, Portugal
Florbela Correia	Polytechnic Institute of Viana do Castelo, Portugal
Stefano Corsi	Università degli Studi di Milano, Italy
Alberto Cortez	Polytechnic of University Coimbra, Portugal
Lino Costa	Universidade do Minho, Portugal

Annunziata Esposito Amideo	University College Dublin, Ireland
Dario Esposito	Polytechnic University of Bari, Italy
Claudio Estatico	University of Genova, Italy
Noelia Faginas-Lago	Università di Perugia, Italy
Maria Irene Falcão	University of Minho, Portugal
Stefano Falcinelli	University of Perugia, Italy
Alessandro Farina	University of Pisa, Italy
Grazia Fattoruso	ENEA, Italy
Caterina Fenu	University of Cagliari, Italy
Luisa Fermo	University of Cagliari, Italy
Florbela Fernandes	Instituto Politecnico de Braganca, Portugal
Rosário Fernandes	University of Minho, Portugal
Luis Fernandez-Sanz	University of Alcala, Spain
Alessia Ferrari	Università di Parma, Italy
Luís Ferrás	University of Minho, Portugal
Ângela Ferreira	Instituto Politécnico de Bragança, Portugal
Flora Ferreira	University of Minho, Portugal
Manuel Carlos Figueiredo	University of Minho, Portugal
Ugo Fiore	University of Naples "Parthenope", Italy
Amedeo Flora	University of Basilicata, Italy
Hector Florez	Universidad Distrital Francisco Jose de Caldas, Colombia
Maria Laura Foddis	University of Cagliari, Italy
Valentina Franzoni	Perugia University, Italy
Adelaide Freitas	University of Aveiro, Portugal
Samuel Frimpong	Durban University of Technology, South Africa
Ioannis Fyrogenis	Aristotle University of Thessaloniki, Greece
Marika Gaballo	Politecnico di Torino, Italy
Laura Gabrielli	Iuav University of Venice, Italy
Ivan Gankevich	St. Petersburg University, Russia
Chiara Garau	University of Cagliari, Italy
Ernesto Garcia Para	Universidad del País Vasco, Spain,
Fernando Garrido	Universidad Técnica del Norte, Ecuador
Marina Gavrilova	University of Calgary, Canada
Silvia Gazzola	University of Bath, UK
Georgios Georgiadis	Aristotle University of Thessaloniki, Greece
Osvaldo Gervasi	University of Perugia, Italy
Andrea Gioia	Polytechnic University of Bari, Italy
Dario Gioia	ISPC-CNT, Italy
Raffaele Giordano	IRSS, National Research Council, Italy
Giacomo Giorgi	University of Perugia, Italy
Eleonora Giovene di Girasole	IRISS, National Research Council, Italy
Salvatore Giuffrida	Università di Catania, Italy
Marco Gola	Politecnico di Milano, Italy

A. Manuela Gonçalves   University of Minho, Portugal
Yuriy Gorbachev   Coddan Technologies LLC, Russia
Angela Gorgoglione   Universidad de la República, Uruguay
Yusuke Gotoh   Okayama University, Japan
Anestis Gourgiotis   University of Thessaly, Greece
Valery Grishkin   St. Petersburg University, Russia
Alessandro Grottesi   CINECA, Italy
Eduardo Guerra   Free University of Bozen-Bolzano, Italy
Ayse Giz Gulnerman   Ankara HBV University, Turkey
Sevin Gümgüm   Izmir University of Economics, Turkey
Himanshu Gupta   BITS Pilani, Hyderabad, India
Sandra Haddad   Arab Academy for Science, Egypt
Malgorzata Hanzl   Lodz University of Technology, Poland
Shoji Hashimoto   KEK, Japan
Peter Hegedus   University of Szeged, Hungary
Eligius M. T. Hendrix   Universidad de Málaga, Spain
Edmond Ho   Northumbria University, UK
Guan Yue Hong   Western Michigan University, USA
Vito Iacobellis   Polytechnic University of Bari, Italy
Mauro Iacono   Università degli Studi della Campania, Italy
Chiara Iacovino   University of Basilicata, Italy
Antonino Iannuzzo   ETH Zurich, Switzerland
Ali Idri   University Mohammed V, Morocco
Oana-Ramona Ilovan   Babeş-Bolyai University, Romania
Mustafa Inceoglu   Ege University, Turkey
Tadashi Ishikawa   KEK, Japan
Federica Isola   University of Cagliari, Italy
Irena Itova   University of Westminster, UK
Edgar David de Izeppi   VTTI, USA
Marija Jankovic   CERTH, Greece
Adrian Jaramillo   Universidad Tecnológica Metropolitana, Chile
Monalisa Jena   Fakir Mohan University, India
Dorota Kamrowska-Załuska   Gdansk University of Technology, Poland
Issaku Kanamori   RIKEN Center for Computational Science, Japan
Korhan Karabulut   Yasar University, Turkey
Yeliz Karaca   University of Massachusetts Medical School, USA
Vicky Katsoni   University of West Attica, Greece
Dimitris Kavroudakis   University of the Aegean, Greece
Shuhei Kimura   Okayama University, Japan
Joanna Kolozej   Cracow University of Technology, Poland
Vladimir Korkhov   St. Petersburg University, Russia
Thales Körting   INPE, Brazil
Tomonori Kouya   Shizuoka Institute of Science and Technology, Japan
Sylwia Krzysztofik   Lodz University of Technology, Poland
Nataliia Kulabukhova   St. Petersburg University, Russia
Shrinivas B. Kulkarni   SDM College of Engineering and Technology, India

Pavan Kumar	University of Calgary, Canada
Anisha Kumari	National Institute of Technology, Rourkela, India
Ludovica La Rocca	University of Naples "Federico II", Italy
Daniele La Rosa	University of Catania, Italy
Sabrina Lai	University of Cagliari, Italy
Giuseppe Francesco Cesare Lama	University of Naples "Federico II", Italy
Mariusz Lamprecht	University of Lodz, Poland
Vincenzo Laporta	National Research Council, Italy
Chien-Sing Lee	Sunway University, Malaysia
José Isaac Lemus Romani	Pontifical Catholic University of Valparaíso, Chile
Federica Leone	University of Cagliari, Italy
Alexander H. Levis	George Mason University, USA
Carola Lingua	Polytechnic University of Turin, Italy
Marco Locurcio	Polytechnic University of Bari, Italy
Andrea Lombardi	University of Perugia, Italy
Savino Longo	University of Bari, Italy
Fernando Lopez Gayarre	University of Oviedo, Spain
Yan Lu	Western Michigan University, USA
Maria Macchiaroli	University of Salerno, Italy
Helmuth Malonek	University of Aveiro, Portugal
Francesca Maltinti	University of Cagliari, Italy
Luca Mancini	University of Perugia, Italy
Marcos Mandado	University of Vigo, Spain
Ernesto Marcheggiani	Università Politecnica delle Marche, Italy
Krassimir Markov	University of Telecommunications and Post, Bulgaria
Giovanni Martinelli	INGV, Italy
Alessandro Marucci	University of L'Aquila, Italy
Fiammetta Marulli	University of Campania Luigi Vanvitelli, Italy
Gabriella Maselli	University of Salerno, Italy
Rytis Maskeliunas	Kaunas University of Technology, Lithuania
Michele Mastroianni	University of Campania Luigi Vanvitelli, Italy
Cristian Mateos	Universidad Nacional del Centro de la Provincia de Buenos Aires, Argentina
Hideo Matsufuru	High Energy Accelerator Research Organization (KEK), Japan
D'Apuzzo Mauro	University of Cassino and Southern Lazio, Italy
Chiara Mazzarella	University Federico II, Italy
Marco Mazzarino	University of Venice, Italy
Giovanni Mei	University of Cagliari, Italy
Mário Melo	Federal Institute of Rio Grande do Norte, Brazil
Francesco Mercaldo	University of Molise, Italy
Alfredo Milani	University of Perugia, Italy
Alessandra Milesi	University of Cagliari, Italy
Antonio Minervino	ISPC, National Research Council, Italy
Fernando Miranda	Universidade do Minho, Portugal

B. Mishra	University of Szeged, Hungary
Sanjay Misra	Covenant University, Nigeria
Mirka Mobilia	University of Salerno, Italy
Giuseppe Modica	Università degli Studi di Reggio Calabria, Italy
Mohammadsadegh Mohagheghi	Vali-e-Asr University of Rafsanjan, Iran
Mohamad Molaei Qelichi	University of Tehran, Iran
Mario Molinara	University of Cassino and Southern Lazio, Italy
Augusto Montisci	Università degli Studi di Cagliari, Italy
Pierluigi Morano	Polytechnic University of Bari, Italy
Ricardo Moura	Universidade Nova de Lisboa, Portugal
Ana Clara Mourao Moura	Federal University of Minas Gerais, Brazil
Maria Mourao	Polytechnic Institute of Viana do Castelo, Portugal
Daichi Mukunoki	RIKEN Center for Computational Science, Japan
Beniamino Murgante	University of Basilicata, Italy
Naohito Nakasato	University of Aizu, Japan
Grazia Napoli	Università degli Studi di Palermo, Italy
Isabel Cristina Natário	Universidade Nova de Lisboa, Portugal
Nadia Nedjah	State University of Rio de Janeiro, Brazil
Antonio Nesticò	University of Salerno, Italy
Andreas Nikiforiadis	Aristotle University of Thessaloniki, Greece
Keigo Nitadori	RIKEN Center for Computational Science, Japan
Silvio Nocera	Iuav University of Venice, Italy
Giuseppina Oliva	University of Salerno, Italy
Arogundade Oluwasefunmi	Academy of Mathematics and System Science, China
Ken-ichi Oohara	University of Tokyo, Japan
Tommaso Orusa	University of Turin, Italy
M. Fernanda P. Costa	University of Minho, Portugal
Roberta Padulano	Centro Euro-Mediterraneo sui Cambiamenti Climatici, Italy
Maria Panagiotopoulou	National Technical University of Athens, Greece
Jay Pancham	Durban University of Technology, South Africa
Gianni Pantaleo	University of Florence, Italy
Dimos Pantazis	University of West Attica, Greece
Michela Paolucci	University of Florence, Italy
Eric Pardede	La Trobe University, Australia
Olivier Parisot	Luxembourg Institute of Science and Technology, Luxembourg
Vincenzo Pellecchia	Ente Idrico Campano, Italy
Anna Pelosi	University of Salerno, Italy
Edit Pengő	University of Szeged, Hungary
Marco Pepe	University of Salerno, Italy
Paola Perchinunno	University of Cagliari, Italy
Ana Pereira	Polytechnic Institute of Bragança, Portugal
Mariano Pernetti	University of Campania, Italy
Damiano Perri	University of Perugia, Italy

Federica Pes	University of Cagliari, Italy
Marco Petrelli	Roma Tre University, Italy
Massimiliano Petri	University of Pisa, Italy
Khiem Phan	Duy Tan University, Vietnam
Alberto Ferruccio Piccinni	Polytechnic of Bari, Italy
Angela Pilogallo	University of Basilicata, Italy
Francesco Pinna	University of Cagliari, Italy
Telmo Pinto	University of Coimbra, Portugal
Luca Piroddi	University of Cagliari, Italy
Darius Plonis	Vilnius Gediminas Technical University, Lithuania
Giuliano Poli	University of Naples "Federico II", Italy
Maria João Polidoro	Polytecnic Institute of Porto, Portugal
Ioannis Politis	Aristotle University of Thessaloniki, Greece
Maurizio Pollino	ENEA, Italy
Antonio Pratelli	University of Pisa, Italy
Salvatore Praticò	Mediterranean University of Reggio Calabria, Italy
Marco Prato	University of Modena and Reggio Emilia, Italy
Carlotta Quagliolo	Polytechnic University of Turin, Italy
Emanuela Quaquero	Univesity of Cagliari, Italy
Garrisi Raffaele	Polizia postale e delle Comunicazioni, Italy
Nicoletta Rassu	University of Cagliari, Italy
Hafiz Tayyab Rauf	University of Bradford, UK
Michela Ravanelli	Sapienza University of Rome, Italy
Roberta Ravanelli	Sapienza University of Rome, Italy
Alfredo Reder	Centro Euro-Mediterraneo sui Cambiamenti Climatici, Italy
Stefania Regalbuto	University of Naples "Federico II", Italy
Rommel Regis	Saint Joseph's University, USA
Lothar Reichel	Kent State University, USA
Marco Reis	University of Coimbra, Portugal
Maria Reitano	University of Naples "Federico II", Italy
Jerzy Respondek	Silesian University of Technology, Poland
Elisa Riccietti	École Normale Supérieure de Lyon, France
Albert Rimola	Universitat Autònoma de Barcelona, Spain
Angela Rizzo	University of Bari, Italy
Ana Maria A. C. Rocha	University of Minho, Portugal
Fabio Rocha	Institute of Technology and Research, Brazil
Humberto Rocha	University of Coimbra, Portugal
Maria Clara Rocha	Polytechnic Institute of Coimbra, Portugal
Miguel Rocha	University of Minho, Portugal
Giuseppe Rodriguez	University of Cagliari, Italy
Guillermo Rodriguez	UNICEN, Argentina
Elisabetta Ronchieri	INFN, Italy
Marzio Rosi	University of Perugia, Italy
Silvia Rossetti	University of Parma, Italy
Marco Rossitti	Polytechnic University of Milan, Italy

Francesco Rotondo	Marche Polytechnic University, Italy
Irene Rubino	Polytechnic University of Turin, Italy
Agustín Salas	Pontifical Catholic University of Valparaíso, Chile
Juan Pablo Sandoval Alcocer	Universidad Católica Boliviana "San Pablo", Bolivia
Luigi Santopietro	University of Basilicata, Italy
Rafael Santos	National Institute for Space Research, Brazil
Valentino Santucci	Università per Stranieri di Perugia, Italy
Mirko Saponaro	Polytechnic University of Bari, Italy
Filippo Sarvia	University of Turin, Italy
Marco Scaioni	Polytechnic University of Milan, Italy
Rafal Scherer	Częstochowa University of Technology, Poland
Francesco Scorza	University of Basilicata, Italy
Ester Scotto di Perta	University of Napoli "Federico II", Italy
Monica Sebillo	University of Salerno, Italy
Patrizia Serra	University of Cagliari, Italy
Ricardo Severino	University of Minho, Portugal
Jie Shen	University of Michigan, USA
Huahao Shou	Zhejiang University of Technology, China
Miltiadis Siavvas	Centre for Research and Technology Hellas, Greece
Brandon Sieu	University of Calgary, Canada
Ângela Silva	Instituto Politécnico de Viana do Castelo, Portugal
Carina Silva	Polytechic Institute of Lisbon, Portugal
Joao Carlos Silva	Polytechnic Institute of Cavado and Ave, Portugal
Fabio Silveira	Federal University of Sao Paulo, Brazil
Marco Simonetti	University of Florence, Italy
Ana Jacinta Soares	University of Minho, Portugal
Maria Joana Soares	University of Minho, Portugal
Michel Soares	Federal University of Sergipe, Brazil
George Somarakis	Foundation for Research and Technology Hellas, Greece
Maria Somma	University of Naples "Federico II", Italy
Alberico Sonnessa	Polytechnic University of Bari, Italy
Elena Stankova	St. Petersburg University, Russia
Flavio Stochino	University of Cagliari, Italy
Anastasia Stratigea	National Technical University of Athens, Greece
Yasuaki Sumida	Kyushu Sangyo University, Japan
Yue Sun	European X-Ray Free-Electron Laser Facility, Germany
Kirill Sviatov	Ulyanovsk State Technical University, Russia
Daisuke Takahashi	University of Tsukuba, Japan
Aladics Tamás	University of Szeged, Hungary
David Taniar	Monash University, Australia
Rodrigo Tapia McClung	Centro de Investigación en Ciencias de Información Geoespacial, Mexico
Eufemia Tarantino	Polytechnic University of Bari, Italy

Sergio Tasso	University of Perugia, Italy
Ana Paula Teixeira	Universidade de Trás-os-Montes e Alto Douro, Portugal
Senhorinha Teixeira	University of Minho, Portugal
Tengku Adil Tengku Izhar	Universiti Teknologi MARA, Malaysia
Maria Filomena Teodoro	University of Lisbon/Portuguese Naval Academy, Portugal
Giovanni Tesoriere	Kore University of Enna, Italy
Yiota Theodora	National Technical Univeristy of Athens, Greece
Graça Tomaz	Polytechnic Institute of Guarda, Portugal
Carmelo Maria Torre	Polytechnic University of Bari, Italy
Francesca Torrieri	University of Naples "Federico II", Italy
Vincenza Torrisi	University of Catania, Italy
Vincenzo Totaro	Polytechnic University of Bari, Italy
Pham Trung	Ho Chi Minh City University of Technology, Vietnam
Dimitrios Tsoukalas	Centre of Research and Technology Hellas (CERTH), Greece
Sanjida Tumpa	University of Calgary, Canada
Iñaki Tuñon	Universidad de Valencia, Spain
Takahiro Ueda	Seikei University, Japan
Piero Ugliengo	University of Turin, Italy
Abdi Usman	Haramaya University, Ethiopia
Ettore Valente	University of Naples "Federico II", Italy
Jordi Vallverdu	Universitat Autònoma de Barcelona, Spain
Cornelis Van Der Mee	University of Cagliari, Italy
José Varela-Aldás	Universidad Tecnológica Indoamérica, Ecuador
Fanny Vazart	University of Grenoble Alpes, France
Franco Vecchiocattivi	University of Perugia, Italy
Laura Verde	University of Campania Luigi Vanvitelli, Italy
Giulia Vergerio	Polytechnic University of Turin, Italy
Jos Vermaseren	Nikhef, The Netherlands
Giacomo Viccione	University of Salerno, Italy
Marco Vizzari	University of Perugia, Italy
Corrado Vizzarri	Polytechnic University of Bari, Italy
Alexander Vodyaho	St. Petersburg State Electrotechnical University "LETI", Russia
Nikolay N. Voit	Ulyanovsk State Technical University, Russia
Marco Vona	University of Basilicata, Italy
Agustinus Borgy Waluyo	Monash University, Australia
Fernando Wanderley	Catholic University of Pernambuco, Brazil
Chao Wang	University of Science and Technology of China, China
Marcin Wozniak	Silesian University of Technology, Poland
Tiang Xian	Nathong University, China
Rekha Yadav	KL University, India
Claudia Yamu	University of Groningen, The Netherlands
Fenghui Yao	Tennessee State University, USA

Contents – Part I

General Track 2: High Performance Computing and Networks

General Track 1: Computational Methods, Algorithms and Scientific Applications

Performance Evaluation of GPS Trajectory Rasterization Methods

Necip Enes Gengeç[1]([⊠]) [iD] and Ergin Tarı[2] [iD]

[1] Graduate School of Engineering and Technology,
Istanbul Technical University, Istanbul, Turkey
`gengec@itu.edu.tr`
[2] Department of Geomatics Engineering, Istanbul Technical University,
Istanbul, Turkey
`tari@itu.edu.tr`

Abstract. The availability of the Global Positioning System (GPS) trajectory data is increasing along with the availability of different GPS receivers and with the increasing use of various mobility services. GPS trajectory is an important data source which is used in traffic density detection, transport mode detection, mapping data inferences with the use of different methods such as image processing and machine learning methods. While the data size increases, efficient representation of this type of data is becoming difficult to be used in these methods. A common approach is the representation of GPS trajectory information such as average speed, bearing, etc. in raster image form and applying analysis methods. In this study, we evaluate GPS trajectory data rasterization using the spatial join functions of QGIS, PostGIS+QGIS, and our iterative spatial structured grid aggregation implementation coded in the Python programming language. Our implementation is also parallelizable, and this parallelization is also included as the fourth method. According to the results of experiment carried out with an example GPS trajectory dataset, QGIS method and PostGIS+QGIS method showed relatively low performance with respect to our method using the metric of total processing time. PostGIS+QGIS method achieved the best results for spatial join though its total performance decreased quickly while test area size increases. On the other hand, both of our methods' performances decrease directly proportional to GPS point. And our methods' performance can be increased proportional to the increase with the number of processor cores and/or with multiple computing clusters.

Keywords: Rasterization · GPS trajectory · Data aggregation · Spatial join · Parallelization

1 Introduction

Availability of digital data is increasing with the increase of sensor device connectivity and with the decrease in data storage area costs. The availability of spatial data, the data that are having spatial components, is also increasing.

© Springer Nature Switzerland AG 2021
O. Gervasi et al. (Eds.): ICCSA 2021, LNCS 12949, pp. 3–17, 2021.
https://doi.org/10.1007/978-3-030-86653-2_1

Spatial data is collected and stored mostly with the use of Global Positioning System (GPS) receivers or other devices that are equipped with GPS units such as smart phones, navigation devices etc.

Collected GPS data with receivers is also varying. One type of data collected with GPS devices is called GPS trajectories and it is the collection of consecutive GPS locations during the travel time of a moving body [26]. In addition to GPS locations, additional information such as timestamp, speed, bearing of the movement, acceleration/deceleration can be recorded with GPS trajectory and/or can be derived from one another.

GPS trajectories are used in studies focusing on mapping data inference [1,11,14], traffic density detection [12] and transportation mode detection [3,13]. In these examples from literature GPS trajectories are used as is or represented in a generalized form such as embedding attributes into predetermined feature classes or converting GPS trajectories into raster images (Fig. 1) that are representing certain attributes (GPS point frequency, transportation mode) or their attributes' aggregation (average speed, maximum speed, average bearing). After the representation of GPS trajectories with embedding or rasterization, different data analysis methods can be applied to these derived data.

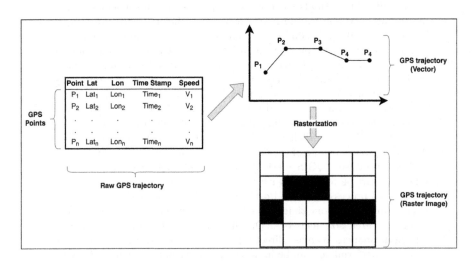

Fig. 1. A simple GPS trajectory and its rasterization.

In addition, research on GPS trajectories as GPS trajectories being the only data source, GPS trajectories fusion with satellite or aerial imagery is a new area [20]. In the context of data fusion of GPS trajectories with satellite imagery, GPS trajectories are rastered, and it is important to obtain one to one pixel match between rasterized GPS trajectories and satellite imagery to carry out the analysis accurately.

On the other hand, due to the size of GPS trajectory data, the rasterization process can be time consuming while the data size and work area increases. This issue is not limited only to GPS trajectory rasterization or aggregation, similar research domains such as spatial social media data analysis and other domains that are dealing with high volume point data.

In this contribution, we address the rasterization of GPS trajectories using open source Geographic Information System (GIS) tools and an algorithm coded using the Python programming language. These tools are evaluated according to their performances with an experiment which has the goal to rasterize multiple attributes of given GPS trajectories. Evaluation carried out only on the performance results of approaches for three tools in the same architecture is presented, no philosophical discussion is carried out yet.

To the best of our knowledge, this study is the only study comparing multiple open source tools and algorithms to understand their performance for aggregation of the big point data to structured grids and their rasterization. There has been multiple research for general performance comparison of QGIS with respect to GIS software like ArcGIS [8]. The parallelization is one of the options implemented in this study. There is various research on parallelization for GIS applications spreading from implementing big data tools into GIS software [6,7,25] to adaption of cluster based, distributed big data tools into GIS domain [5,21]. Also, there are significant research which discuss CPU and GPU acceleration, their special applications in GIS and achieved performance improvement [22–24]. Although previous researches may contribute to various future research directions combined with our research results, these researches neither focus on the point to structured grid aggregation and rasterization nor provide a performance comparison with respect to the widely used open source GIS tools.

2 Tools and Rasterization Process Flow

As in raw form, GPS trajectory data is a vector data while the aim of this research is to represent this data in raster image format. Because of these dependencies, tools that are required should be able to handle both vector and raster image data. Within the multiple open source GIS tools that are freely available, QGIS [19] and PostGIS [17] are used for the rasterization of GPS trajectories in this study since they are widely adopted in GIS field thanks to their robustness and abilities.

In Sect. 2.1 and 2.2, the rasterization abilities of QGIS and PostGIS will be examined with respect to given input data (work area boundaries, output pixel size, GPS trajectory data) and expected output rasterized GPS trajectory layers (frequency, average speed and maximum speed). In Sect. 2.3, our implementation will be explained. Finally, in Sect. 2.4, the process flow of methods will be summarized.

2.1 Rasterization with QGIS

As a GIS software, QGIS has different functions, tools and plugins for data analysis and data conversion. *Rasterize (Vector to Raster)* is one of these tools offered within QGIS. This QGIS tool acts as a user interface, collects the user inputs and runs *gdal_rasterize* tool at the background. This tool is able to get an input data and burn the pixel values that are stored in the preferred attribute field of input vector data within the predefined outer boundaries. Although QGIS has the rasterize tool, this tool is only able to rasterize given values but cannot aggregate multiple values of the same attribute in the given pixels. Though, it is possible to represent pixels in vector form (structured grid) and achieve the required aggregation with spatial join tool of QGIS. After the aggregation of GPS trajectories into the structured grid, it is possible to rasterize the aggregated attributes into raster image data.

2.2   Rasterization with PostGIS

PostGIS is the spatial database add-on for the PostgreSQL [18] database management system. PostGIS is able to store and analyze spatial data in vector and raster image form that is stored in a PostgreSQL database. PostGIS has *ST_AsRaster* tool for similar to QGIS which is accepting the input though producing only given attribute values. On the other hand, similar to QGIS, it is possible to aggregate one vector layer into another using Structured Query Language (SQL) statements. Even though PostGIS does provide rasterize functionality, it is also possible to connect QGIS to PostgreSQL database and rasterize the output data that is created with PostGIS via QGIS.

2.3   Rasterization with Python

Python is a general-purpose programming language which has many internal and external libraries such as data science libraries (Pandas) and geospatial computation (GDAL, pyproj) libraries. With the use of these libraries, it is possible to analyze GPS trajectories.

To achieve the required raster images, our own Python method was created (Algorithm 1). This method gets the GPS points, coordinates of work area boundary and pixel size as inputs and calculates the raster matrix. Unlike Post-GIS and QGIS, the Python method makes use of the structured grid definition (work area boundary and pixel size) of a raster image and carries out the calculation of outputs without creating a vector grid. The method determines the row and column of the pixel where each GPS point is contained. Following, it aggregates the required feature values and assigns the output raster image matrix values according to previously determined rows and columns.

The most computation intensive part of the Algorithm 1 is the *for* loop shown between row numbers 5 to 9. Python gives the ability to parallelize with the use of additional libraries such as Dask [4] and Swifter [2]. Swifter library uses Dask library at its backend. It is able to provide the processing time information and provides user the ability to choose parallel or normal computation options easily. Due to these features Swifter library is used for the experiments of this study.

Algorithm 1: Spatial join with Python.

Data: $P = \{p_1, p_2, p_3, ..., p_n\}$ where each p_i contains latitude $(p_{i,lat})$, longitude$(p_{i,lon})$, speed $(p_{i,speed})$.
Output raster image top-left corner coordinate (X, Y) and pixel size (px).

Result: Output images $Image_{count}$, $Image_{speed-avg}$, $Image_{speed-max}$ in matrix form.

```
1  begin
       /* Convert input coordinates into projected cartesian
          coordinate system.                                    */
2      def transformCoordinates(P_lat, P_lon):
3          | Transform WGS84 to projected WGS84
4      return P_X, P_Y
       /* Determining row and column of each GPS point within the
          output raster.                                        */
5      foreach p_i ∈ P do
6          | p_row = (p_{i,X} − (p_{i,X} mod px) − X)/px
7          | p_column = (p_{i,Y} − (p_{i,Y} mod px) − Y)/px
8          | p_i ⟵ p_row, p_column
9      end
       /* Aggregate GPS count, average and maximum speed values
          with grouping by row and column number.               */
10     def aggregateValues(P_row, P_column, P_speed):
11         | P'_count ⟵ count of records having same p_row, p_column where p ∈ P
12         | P'_speed-avg ⟵ average of p_{i,speed} having same p_row, p_column where
           |   p ∈ P
13         | P'_speed-max ⟵ maximum of p_{i,speed} having same p_row, p_column
           |   where p ∈ P
14     return P'
       /* Assign pixel values by row and column of output images.
          */
15     foreach p'_i ∈ P' do
16         | Image_count[p'_{i,row}][p'_{i,column}] ⟵ p'_{i,count}
17         | Image_speed-avg[p'_{i,row}][p'_{i,column}] ⟵ p'_{i,speed-avg}
18         | Image_speed-max[p'_{i,row}][p'_{i,column}] ⟵ p'_{i,speed-max}
19     end
20  end
```

2.4   Summary of Methods

According to the spatial data processing and rasterization abilities of QGIS, PostGIS and Python the process flow of the four methods was determined as in the Fig. 2.

The QGIS method creates a vector grid and transforms coordinates of GPS points into the coordinate system of the required output raster image. After creation of the vector grid and coordinate transformation, these are joined spatially. Finally, output of the spatial join is rasterized.

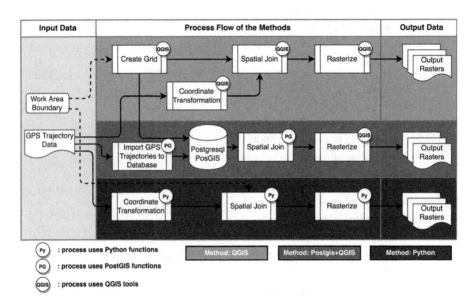

Fig. 2. Process flow of the methods; QGIS, PostGIS+QGIS and Python (Python process flow is identical for both Python and Python (Parallel) methods).

The second method is called PostGIS+QGIS because this method benefits from both PostGIS and QGIS. This method creates the vector grid with the use of QGIS. Also, GPS trajectories are required to be imported into PostgreSQL database which PostGIS is operating on. After import and grid creation, PostGIS spatially joins both data. In this method, coordinate transformation is carried out along with the spatial join. Finally, QGIS is used to rasterize the output of the spatial join.

The algorithm for the Python method is explained in Algorithm 1. This algorithm gets the GPS trajectory data directly using Python libraries and applies the coordinate transformation. After the transformation, our method calculates the output raster image matrix without the need of a vector grid and saves the output raster to the disk.

3   Setup of the Experiments

The methods defined in Sect. 2 are evaluated with the use of MTL-Trajet dataset [10]. This dataset consists of GPS trajectories collected in 2016 around Montreal, Canada. Raw data contains GPS point locations in the WGS84 Datum and timestamps. Speed of the moving objects are calculated using time difference and geodesic distances of consecutive GPS points with Geopy package [9]. These values added as an attribute to the raw GPS trajectory data. Speed value accuracy is dependent to projection and calculation method, but speed accuracy is not the main focus of this study. Because of this, achieved speed values are well enough for performance evaluation of the methods.

In the experiment, the target is to rasterize GPS point "frequency (count)", "average speed" and "maximum speed" raster images using QGIS, Post-GIS+QGIS, Python and Python (Parallel) methods. In order to understand the dependencies of performance, the experiment is carried out with varying test area size and GPS point count. Figure 3 shows the test area boundaries and Table 1 summarizes the corresponding GPS count that is used in the experiments. MTL-Trajet GPS point coverage is wider than the defined test areas. The GPS points that are outside of the test area are removed from main dataset to focus only on the performance of methods for given area. Though our Python and Python (Parallel) implementations are able to ignore the GPS points which are out of given area.

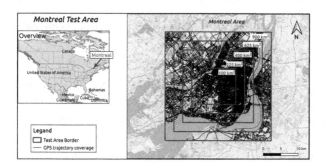

Fig. 3. MTL-Trajet dataset and test areas boundaries.

Table 1. GPS trajectory by test area that was used in the experiment.

Test Area Size (km^2)	Number of GPS points (million)
100	5
225	5, 10, 11
400	5, 10, 15
625	5, 10, 15, 18
900	5, 10, 15, 20

As explained in Sect. 1, the aim is to obtain a one to one pixel match with a given raster image. Usually the raster images that are widely used are provided in projected coordinate systems. In order to add this constraint to the experimental setup, unlike GPS trajectory data, test areas are created in projected WGS84 Datum using cartesian coordinates so that the pixel dimensions are defined in meters. According to this, GPS point coordinates must be transformed into projected WGS84 from geographic WGS84.

The aim of our research for the rasterization is to represent GPS points in raster format to use in further analysis with additional satellite imagery. Because of this, the expected output should be aligned to the satellite imagery pixel resolution. Similar studies in literature use high resolution satellite images that are having pixel resolution around 1–5 m [15,16,20]. In order to align with the literature examples the output raster image pixel size set as 5 m.

Total processing time ($t_{Totaltime}$) and spatial join time ($t_{SpatialJoin}$) are preferred as the evaluation metric. Since the process flow of each method is different, their total processing time is also varying.

Total processing time with QGIS method is calculated with

$$t_{TotalQGIS} = t_{Gridcreation}^{QGIS} + t_{CoordinateTransform}^{QGIS} + t_{SpatialJoin}^{QGIS} + t_{Rasterize}^{QGIS} \quad (1)$$

where $t_{Gridcreation}^{QGIS}$ grid creation time, $t_{CoordinateTransform}^{QGIS}$ coordinate transformation time, $t_{SpatialJoin}^{QGIS}$ spatial join time, $t_{Rasterize}^{QGIS}$ rasterization time with QGIS.

Total processing time with PostGIS+QGIS method is calculated with

$$t_{TotalPostGIS+QGIS} = t_{Gridcreation}^{QGIS} + t_{SpatialJoin}^{PG} + t_{Rasterize}^{QGIS} \quad (2)$$

where $t_{Gridcreation}^{QGIS}$ grid creation time with QGIS, $t_{SpatialJoin}^{PG}$ spatial join with PostGIS, $t_{Rasterize}^{QGIS}$ rasterization time with QGIS.

Total processing time with Python method is calculated with

$$t_{TotalPython} = t_{CoordinateTransform}^{Py} + t_{SpatialJoin}^{Py} + t_{Rasterize}^{Py} \quad (3)$$

where $t_{CoordinateTransform}^{Py}$ coordinate transformation, $t_{SpatialJoin}^{Py}$ spatial join and $t_{Rasterize}^{Py}$ rasterization using Python.

Experiments are carried out using a standard laptop which has an Intel Core I7-4600 M CPU with four 2.90 GHz cores, 16 GB RAM and Ubuntu/Linux operating system. The QGIS and PostgreSQL/PostGIS are used with their default installation settings. Spatial indexes are used for GPS point and vector grid layers in PostGIS+QGIS method. In order to avoid delays caused by memory and processor usage of another software, minimum required software was kept open while running a method.

4 Comparisons

Experiments are carried out with the methods defined in Sect. 2.4 and with the setup defined in Sect. 3. Following the experiments, output raster images are compared with the use of raster calculation. All output raster images subtracted from remaining output images one by one for given test area. This comparison aims to determine if the created raster images of each method is identical or not. If compared raster images are identical, empty raster image expected as the result of raster image subtraction. According to the comparison of the outputs of each test area, the subtraction results were empty images which proves that each method achieved the same raster images as output.

With the use of the output data from the experiments, comparison plots are created. Figures 4 and Figs. 5 show the performance comparison of methods for each test area. Firstly, as seen in Fig. 4a, Python (Parallel) and Python methods achieve best performance in terms of total processing time and followed by PostGIS+QGIS method within the 900 km² test area. On the other hand, PostGIS+QGIS achieves the best performance for spatial join time measure (Fig. 4b). QGIS method achieves very poor performance for each measure. Because of this, QGIS method is excluded from plots in the Figs. 5.

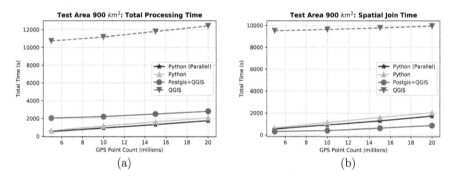

Fig. 4. (a) comparison of total processing time and (b) comparison of spatial join time for 900 km² test area.

The results for both measures are very similar for the rest of the test areas (Fig. 5). It is also visible that the Python method without parallelization is slower than PostGIS+QGIS method in small areas though performance increases with respect to the PostGIS+QGIS method while the test area size increases (Fig. 5c, 5e, 5g). In the spatial join measure, PostGIS+QGIS achieved better performance followed by the Python (Parallel) method. With the increase of GPS point count, the difference between Python methods and PostGIS+QGIS method also increases.

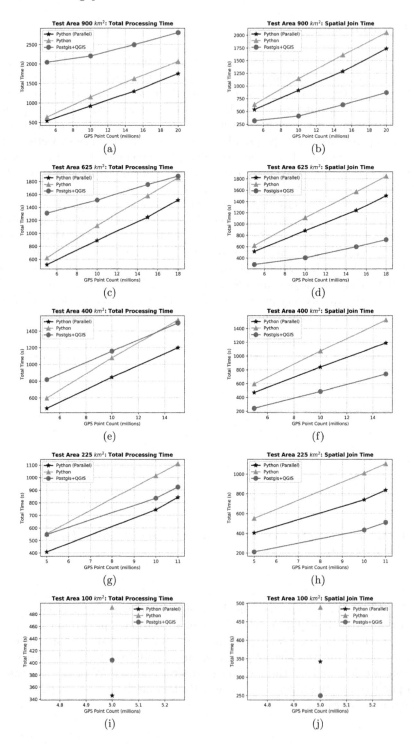

Fig. 5. Comparison of total processing time and spatial join for each test area (QGIS method's results are excluded).

There are two major reasons for the performance decrease of PostGIS+QGIS method in total processing time. The first and most important reason is the requirement of a vector grid. As summarized in Table 2, grid creation time is proportional to the test area size and increases as the test area size increases. The second reason is the importing time of the GPS trajectories to the database. Unlike the QGIS method and Python methods, GPS trajectories required to be imported to the database before starting the rest of the process for Post-GIS+QGIS method. Figure 6 shows that this process is dependent on the GPS point count and not dependent on test area size.

Table 2. Grid creation time for each test area.

Test Area Size (km^2)	Time (s)
100	78
225	164
400	299
625	466
900	706

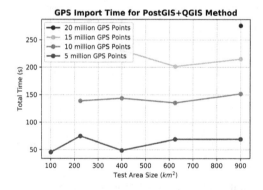

Fig. 6. Time spent for GPS trajectory data to database import for PostGIS+QGIS method.

Lastly, method results are compared internally with the performance measure by test area size when GPS count is kept constant in plots (Fig. 7). As per this comparison, total processing time of the QGIS method increases proportional to the test area size (Fig. 7d). Similar to QGIS method, PostGIS+QGIS method's total processing time also increases proportional to test area size though the increase is steeper compared to the QGIS method. On the other hand, both Python (Parallel) and Python methods show very few increases when test area size increases. Their performances are proportional to GPS point count. Python (Parallel) method is faster than Python method.

In addition to the experiments defined in Sect. 3, an additional experiment was also carried out to understand the limitations of these methods. This experiment was carried out with test area size 22120 km^2 and approximately 25 million GPS points around Montreal. Python and Python (Parallel) methods have been able to process and rasterize this area in 40 and 52 min respectively. On the other hand, QGIS and PostGIS+QGIS methods couldn't process this larger test area due to grid creation with the current hardware. Since grid creation time is proportional to test area size for both methods, in the case of a big test area, grid creation cannot be possible with the use of QGIS and at the end QGIS crashes.

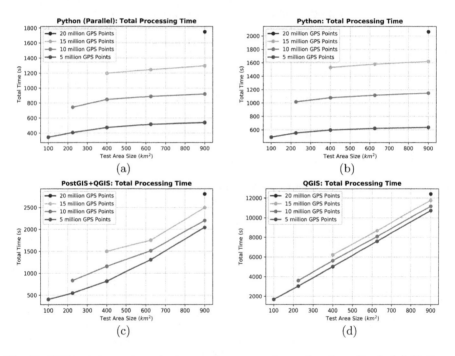

Fig. 7. GPS point count vs. test area size comparison of each method; **(a)** Python (Parallel) **(b)** Python, **(c)** PostGIS+QGIS, **(d)** QGIS.

5   Conclusions

This study evaluates the methods for rasterization of GPS trajectories. Evaluation is carried out for QGIS, PostGIS+QGIS methods and our Python and Python (Parallel) implementations. For evaluation, an experiment was carried out with varying test area and GPS trajectory size. Total processing time and spatial join time were adopted as the evaluation metric.

According to the results, the Python (Parallel) method achieves the best results among the compared methods. The Python method also showed better results with respect to QGIS and PostGIS+QGIS methods. PostGIS+QGIS method achieves the best result for spatial join. QGIS shows the worst performance for both of the metrics.

Python and Python (Parallel) methods perform slower than PostGIS+QGIS method for spatial join metric. This issue is a result of the time for indexing operation that our implementation spent which is more than the spatial join operations carried out by PostGIS. Indexing operation only dependent to GPS point data size but PostGIS+ QGIS method performance is dependent both to GPS data size and the test are size. Also, when compared to the spent time for grid creation, this delay caused by indexing is negligible. Moreover, indexing operation is more robust than grid creation. On the contrary grid creation consumes too much memory and prone to crashes. Although the PostGIS+QGIS method achieves the best spatial join performance, due to the disadvantage of grid creation and import time required for GPS points, the total performance decreased very fast while the test area size increased. Grid creation can be considered as one-time cost though it is still a disadvantage for the possible cases of different work areas in different applications domains. Similar to the PostGIS+QGIS method, in addition to weak performance of the spatial join, QGIS also performed worse while the test area size increases.

On the other hand, the Python methods' performance is proportional to point count of the GPS trajectories. This feature is proven with additional experiment which has wider test area. As per results, Python methods can work in large areas though QGIS and PostGIS+QGIS methods fail to achieve this. Because the performance is not dependent to test area size and being suitable to parallelization, it is possible to increase performance of Python methods' with distributing computation into more processor cores and/or computation clusters.

As a conclusion, our implementation performs better than QGIS and PostGIS+QGIS methods and can be used for GPS trajectory rasterization. The use of our implementation is not limited to GPS trajectory rasterization. It is also possible to use our implementation in similar problems which require rasterization and aggregation of big point-based datasets into structured grids such as spatial social media data analysis. In addition, the integration of our implementation would increase its usage in other research domains which benefit from QGIS but requiring better performance. It is possible to integrate our implementation scripts into QGIS since support Python programming language, but further research needed to determine if libraries like Swifter are compatible with QGIS Python environment.

References

1. Ahmed, M., Wenk, C.: Constructing street networks from GPS trajectories. In: Epstein, L., Ferragina, P. (eds.) ESA 2012. LNCS, vol. 7501, pp. 60–71. Springer, Heidelberg (2012). https://doi.org/10.1007/978-3-642-33090-2_7
2. Carpenter, J.: Swifter: A package which efficiently applies any function to a pandas dataframe or series in the fastest available manner (2020). https://github.com/jmcarpenter2/swifter. Accessed 15 Mar 2020

3. Dabiri, S., Heaslip, K.: Inferring transportation modes from GPS trajectories using a convolutional neural network. Transp. Res. Part C Emerg. Technol. **86**, 360–371 (2018). https://doi.org/10.1016/j.trc.2017.11.021
4. Dask: Dask: Scalable analytics in Python (2020). https://dask.org/. Accessed 15 Mar 2020
5. Eldawy, A., Mokbel, M.F.: SpatialHadoop: a mapreduce framework for spatial data. In: 31st IEEE International Conference on Data Engineering, ICDE 2015, Seoul, South Korea, April 13–17, 2015, pp. 1352–1363 (2015). https://doi.org/10.1109/ICDE.2015.7113382
6. ESRI: GIS Tools for Hadoop: Big Data Spatial Analytics for the Hadoop Framework (2020). http://esri.github.io/gis-tools-for-hadoop/. Accessed 05 May 2020
7. ESRI: Spatial Framework for Hadoop (2020). https://github.com/Esri/spatial-framework-for-hadoop. Accessed 05 May 2020
8. Friedrich, C.: Comparison of ArcGIS and QGIS for applications in sustainable spatial planning. Ph.D. thesis, uniwien (2014)
9. Geopy: Geopy: Python client for several popular geocoding web services (2020). https://geopy.readthedocs.io/en/stable/. Accessed 05 May 2020
10. Gouvernement du Québec: Déplacements MTL Trajet (2018). https://www.donneesquebec.ca/recherche/fr/dataset/vmtl-mtl-trajet. Accessed 15 Mar 2020
11. He, S., et al.: RoadRunner: improving the precision of road network inference from GPS trajectories. In: Proceedings of the 26th ACM SIGSPATIAL International Conference on Advances in Geographic Information Systems, pp. 3–12. ACM (2018)
12. Institute of Advanced Research in Artificial Intelligence: Traffic4cast Competition Track at NeurIPS 2019 (2020). https://www.iarai.ac.at/traffic4cast/traffic4cast-conference-2019/. Accessed 15 Mar 2020
13. Jiang, X., de Souza, E.N., Pesaranghader, A., Hu, B., Silver, D.L., Matwin, S.: TrajectoryNet: An Embedded GPS Trajectory Representation for Point-based Classification Using Recurrent Neural Networks, pp. 192–200 (2017)
14. Karagiorgou, S., Pfoser, D.: On vehicle tracking data-based road network generation. In: Proceedings of the 20th International Conference on Advances in Geographic Information Systems, pp. 89–98. SIGSPATIAL '12, ACM (2012). https://doi.org/10.1145/2424321.2424334
15. Mnih, V., Hinton, G.E.: Learning to detect roads in high-resolution aerial images. In: Daniilidis, K., Maragos, P., Paragios, N. (eds.) ECCV 2010. LNCS, vol. 6316, pp. 210–223. Springer, Heidelberg (2010). https://doi.org/10.1007/978-3-642-15567-3_16
16. Papadomanolaki, M., Vakalopoulou, M., Zagoruyko, S., Karantzalos, K.: Benchmarking deep learning frameworks for the classification of very high resolution satellite multispectral data. ISPRS Ann. Photogrammetry Remote Sens. Spatial Inform. Sci. **3**(7), 83–88 (2016). https://doi.org/10.5194/isprs-annals-III-7-83-2016
17. PostGIS: Spatial and Geographic objects for PostgreSQL (2020). https://postgis.net/. Accessed 15 Mar 2020
18. PostgreSQL: PostgreSQL (2020). https://www.postgresql.org/. Accessed 15 Mar 2020
19. QGIS: A Free and Open Source Geographic Information System (2020). https://qgis.org/. Accessed 15 Mar 2020
20. Sun, T., Di, Z., Che, P., Liu, C., Wang, Y.: Leveraging crowdsourced GPS data for road extraction from aerial imagery. In: Proceedings of the IEEE Conference on Computer Vision and Pattern Recognition, pp. 7509–7518 (2019)

21. Wang, F., Aji, A., Vo, H.: High performance spatial queries for spatial big data. SIGSPATIAL Special **6**(3), 11–18 (2015). https://doi.org/10.1145/2766196. 2766199
22. Zacharatou, E.T., Doraiswamy, H., Ailamaki, A., Silva, C.T., Freire, J.: GPU rasterization for realtime spatial aggregation over arbitrary polygons. Proc. VLDB Endowment **11**(3), 352–365 (2017). https://doi.org/10.14778/3157794.3157803
23. Zhang, J., You, S.: CudaGIS: report on the design and realization of a massive data parallel GIS on GPUs. In: Proceedings of the 3rd ACM SIGSPATIAL International Workshop on GeoStreaming, IWGS 2012, pp. 101–108 (2012). https://doi.org/10. 1145/2442968.2442981
24. Zhang, J., You, S., Gruenwald, L.: U2Stra: high-performance data management of ubiquitous urban sensing trajectories on gpgpus. In: International Conference on Information and Knowledge Management, Proceedings, pp. 5–12 (2012). https:// doi.org/10.1145/2390226.2390229
25. Zhao, L., Chen, L., Ranjan, R., Choo, K.K.R., He, J.: Geographical information system parallelization for spatial big data processing: a review. Cluster Comput. **19**(1), 139–152 (2016). https://doi.org/10.1007/s10586-015-0512-2
26. Zheng, Y., Zhang, L., Xie, X., Ma, W.Y.: Mining interesting locations and travel sequences from GPS trajectories. In: Proceedings of the 18th International Conference on World Wide Web - WWW '09, pp. 791–800 (2009). https://doi.org/10. 1145/1526709.1526816

An Algorithm for Polytope Overlapping Detection

Miroslav S. Petrov[1] and Todor D. Todorov[2(✉)]

[1] Department of Technical Mechanics, Technical University, 5300 Gabrovo, Bulgaria
[2] Department of Mathematics, Informatics and Natural Sciences, Technical University, 5300 Gabrovo, Bulgaria

Abstract. The intersection of polytopes is a basic problem of computational geometry with many engineering applications. Intersections of simplices or parallelotopes have been widely used in finite element grid generations. This paper is devoted to an algorithm for detecting overlapping polytopes. We present a new iterative algorithm, which is independent of the dimension of the Euclidean space. The main idea is triangulating the tested polytopes by simplicial finite elements and then investigating couples of potential simplices for an intersection. For that purpose, a method for overlapping detection of arbitrary simplices in $\mathbf{R}^n$ is developed. A detailed description of the pseudocode of the original algorithm is presented. The advantages of the proposed method are demonstrated in the twelve-dimensional case.

Keywords: Overlap detection algorithm · Overlapping domains · Intersection of simplices · Convex polytopes

AMS Subject Classifications: 52B11 · 65N30

1 Introduction

The problem of the intersection of two polytopes in Euclidean space has arisen in many Engineering activities. This includes applications in robotics, computer-aided design, computer graphics, computational mechanics, etc. There are several papers devoted to this problem [1,8,25]. This is one of the significant problems in the area of computational geometry. Most of the authors have used various iterative techniques based on numerical methods in order to establish the overlapping of polytopes [17]. Intersections of polyhedra related to the finite element method have been studied by Descantes et al. [7] and Lee et al. [12]. The main idea in [7] is a common plane to be created that separates both tested polytopes. The algorithm is restricted to the three-dimensional space and convex polyhedra. The paper by Edwards et al. [8] is devoted to the intersection of metrizable infinite-dimensional simplices. Lira et al. have applied finite element meshes in order to detect the intersection of surfaces [15].

© Springer Nature Switzerland AG 2021
O. Gervasi et al. (Eds.): ICCSA 2021, LNCS 12949, pp. 18–33, 2021.
https://doi.org/10.1007/978-3-030-86653-2_2

Brandts et al. [5] marked the beginning of a new stage in the development of the finite element method. They have discussed boundary value problems in multidimensional domains as a motivation to discover the finite element method in higher-dimensional Euclidean spaces. The authors have defined areas for further studies as: supercloseness and superconvergence; strengthened Cauchy-Schwarz inequalities; angle conditions for regularity of FEM partitions etc.

The major contribution of the paper is a feasible polytope overlapping detection algorithm. The proposed algorithm is easy for implementing. The validity of the algorithm does not depend on the complexity of the investigated polytopes. The algorithm can be successfully applied to convex polytopes in all n-dimensional Euclidean spaces. A method for identifying an overlapping of arbitrary nondegenerated k-simplex and l-simplex in n-dimensional Euclidean spaces is obtained. Intersections between polytopes of different dimensions are studied as well. The main idea is described in pseudocode. A real comparison between various iterative methods can be established in higher-dimensional spaces. For this purpose, a twelve-dimensional example is used to demonstrate the advantages of the presented algorithm for overlap detection. The new overlap detection algorithm is based on a method for solving systems of linear inequalities. Therefore, a comparison between methods for solving systems of linear inequalities is presented. The numerical tests in the twelve-dimensional Euclidean space indicate that the Motzkin method is superior among all considered methods.

This paper is organized as follows. Preliminary results on methods for solving systems of linear inequalities are presented in Sect. 2. The overlapping of simplices is considered in Sect. 3. The overlap detection algorithm in pseudocode is described in Sect. 4. Examples illustrating the new algorithm are presented in Sect. 5. The obtained results are discussed in Sect. 6.

2   Preliminaries

Our algorithm is based on solving a system of linear inequalities and the dissipation algorithm [20]. This is why we make a brief overview of the methods for solving a system of inequalities. We classify these methods into two major groups: exact methods and methods applying numerical optimizations. The exact methods for solving linear systems of inequalities have a significant advantage over iterative ones because in them the determination of a solution depends only on the round-off errors. On the other hand, any system of linear inequalities can be solved by the Fourier-Motzkin elimination [10]. The Fourier-Motzkin algorithm has been improved by Bastrakov et al. [2], Keßler [11], and Šimeček et al. [24]. But unfortunately, the classic exact methods are provided with an unacceptable computational cost. For instance, the algebraic complexity of the Fourier-Motzkin algorithm grows up to double exponentially concerning the dimensions of the linear system [10]. By reducing the large number of redundant inequalities Khachiyan [14] has obtained a computational cost equal to singly exponential time. In the second group, we mention the applications of the least square method [13], Newton's method [21], the BB-method [23], the conjugate gradient method for quadratics [28], the relaxation method [26], and the Motzkin method [9] for solving systems of inequalities.

The proposed method requires a system of linear inequalities to be solved. To this end, we choose the Motzkin iterative method. Our choice is based on a comparison between several iterative methods for solving systems of inequalities. The best performance of the Motzkin method in the twelve-dimensional space is demonstrated in Sect. 5. Our algorithm can successfully work with any convergent method for solving a system of linear inequalities.

Throughout the whole paper ε stands for a positive number, which approximates zero and the upper index indicates the dimension of the polytope. The denotations $\underline{x}$ and $\|\cdot\|$ are the radius column vector of the point $x \in \mathbf{R}^n$ and the Euclidean norm in $\mathbf{R}^n$. Let $A\left(2(n+1)\times n\right)$ and $B\left(2(n+1)\times 1\right)$ be matrices with real entries, I_n be the identity matrix of size n, and $x = (x_1, x_2, \dots x_n) \in \mathbf{R}^n$. We say that $\underline{x} < \underline{y}$, $x, y \in \mathbf{R}^n$ if $x_i < y_i$ $\forall i = 1, 2, \dots, n$. The object of interest in this section is the linear systems of inequalities

$$A\underline{x} < B \tag{1}$$

presented in a matrix form. We suppose that the block decomposition of the matrix

$$A = \begin{pmatrix} A_{11}(n \times n) \\ A_{21}((n+2) \times n) \end{pmatrix}$$

is provided with an invertible block A_{11}. The analyses of the specific system of inequalities (1) is motivated by the main problem of the paper.

Definition 1. *We say that the system (1) is consistent if (1) has at least one solution.*

We suppose that A_i is the i-th row of A, and

$$\xi = \operatorname*{argmax}_{i=1,2,\dots,2n+2} \left(A_i \underline{x} - B_i\right).$$

Additionally, we assume that all rows of the matrix A are normalized by $\|A_i\| = 1$. We define a convex functional

$$\hat{J}(x) = \frac{1}{2}\left|\hat{I}(A_\xi \underline{x} - B_\xi)\right|^2,$$

where

$$\hat{I}x = \begin{cases} x, & \text{if } x > 0 \\ 0, & \text{otherwise} \end{cases}.$$

We apply the Motzkin iterative method [9]

$$\underline{x}_{k+1} = \underline{x}_k - D\hat{J}(x_k), \quad k \geq 0. \tag{2}$$

for minimization the functional $\hat{J}(\hat{x})$. Each stationary point of $\hat{J}(\hat{x})$ is a solution of the problem (1). The choice of the initial guess will be determined further.

3   Overlapping of Two Simplices

Definition 2. *We define the n-dimensional simplex* $T = [t_1, t_2, \ldots, t_{n+1}]$ *as convex hull of affinely independent* $n + 1$ *vertices* $t_i \in \mathbf{R}^n$, $i = 1, 2, \ldots, n + 1$.

We suppose that the points t_i are affinely independent. We denote the boundary, the interior, the length of the longest edge, and the center of gravity of the n-dimensional nondegenerated simplex T by ∂T, $\mathring{T}$, $h(T)$ and $C_T = \frac{1}{n+1} \sum_{i=1}^{n+1} t_i$. We choose the cube corner $\hat{T} = [\hat{t}_1, \hat{t}_2, \ldots, \hat{t}_{n+1}]$, $\hat{t}_i = \{0, 0, \ldots, 1, \ldots, 0\}$, $i = 1, 2, \ldots, n, \hat{t}_{n+1} = \{0, 0, \ldots, 0\}$ for the reference simplex. The i-th coordinate of $\hat{t}_i$ $i = 1, 2, \ldots, n$ is equal to one, and all other coordinates are equal to zero. An arbitrary simplex T can be obtained from the reference simplex by the generic affine transformation F_T : $\underline{x} = A_T \hat{\underline{x}} + B_T$, $\hat{x} \in \hat{T}$, $x \in T$, where $A_T = \left(\underline{t}_1 - \underline{t}_{n+1} \ \underline{t}_2 - \underline{t}_{n+1} \cdots \underline{t}_n - \underline{t}_{n+1}\right)$ is the transition matrix and $B_T = \underline{t}_{n+1}$ is the translation vector.

The denotation μ_n stands for the Lebesgue measure in $\mathbf{R}^n$.

Definition 3. *The value*

$$\delta(T) = \frac{h(T)\mu_{n-1}(\partial T)}{2n\mu_n(T)},$$

is called measure of degeneracy [4, 20] of the simplex T.

Definition 4. *The measure of degeneracy related to an arbitrary triangulation* τ_h *is defined by*

$$\delta(\tau_h) = \max_{T \in \tau_h} \delta(T).$$

Our goal is to establish whether two nondegenerated simplices T and K have an intersection $T \cap K$ with $\mu_n(T \cap K) > 0$. In this case we say that both simplices intersect each other. We do not consider the case when $\mu_n(T \cap K) = 0$, $\mu_k(T \cap K) > 0$ for some k so that $1 \le k < n$.

Theorem 1. *Let* $M = A_K^{-1} A_T$ *and* $N = A_K^{-1}(B_T - B_K)$. *The n-dimensional nondegenerated simplices* T *and* K *have an intersection* $T \cap K$ *with* $\mu_n(T \cap K) > 0$, *if there are two points* $\hat{x}$, $\hat{y} \in \mathring{T}$, *not necessarily different, so that*

$$\hat{y} = M\hat{x} + N. \tag{3}$$

Proof. From (3) it follows that there are two points $\hat{x}$, $\hat{y} \in \mathring{T}$ so that $F_T(\hat{x}) = F_K(\hat{y})$. Since F_T and F_K are linear transformations, F_T maps $\hat{x}$ in $\mathring{T}$ and F_K maps $\hat{y}$ in $\mathring{K}$. The latter means that $\mathring{T} \cap \mathring{K} \neq \emptyset$, whence $\mu_n(T \cap K) > 0$. The two- and three-dimensional cases are illustrated in Fig. 1.□

By writing (3) in a scalar form

$$\hat{y}_i = \hat{\varphi}_i(\hat{x}), \quad i = 1, 2, \ldots, n \tag{4}$$

we obtain the following corollary.

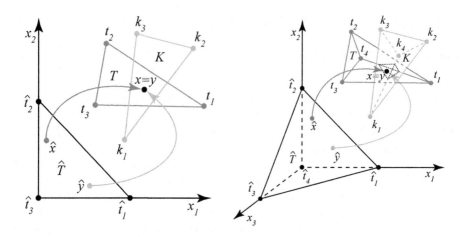

Fig. 1. Overlapping 2D and 3D simplices.

Corollary 1. *The n-dimensional simplices T and K have an intersection $T \cap K$ with $\mu_n(T \cap K) > 0$, if*

$$\hat{\varphi}_i(\hat{x}) > 0, \, i = 1, 2, \ldots, n, \, \sum_{i=1}^{n} \hat{\varphi}_i(\hat{x}) < 1 \tag{5}$$

and

$$\hat{x}_i > 0, \, i = 1, 2, \ldots, n, \, \sum_{i=1}^{n} \hat{x}_i < 1 \tag{6}$$

Proof. The relation $\hat{y} \in \overset{\circ}{\hat{T}}$ follows from (5). Additionally, $\hat{x} \in \overset{\circ}{\hat{T}}$ because of (6). Then the validity of the corollary is guaranteed by (4). □

In a matrix form the system (5)–(6) is equivalent to (1).

In Theorem 1, we consider the case when T and K are n-dimensional. Further, we solve more general problem when both simplices have different dimensions. Let $\hat{T}^l$ be a reference simplex with $\dim(\hat{T}^l) = l$.

Theorem 2. *The nondegenerated simplices T and K satisfy $\dim(T) = k$, $\dim(K) = l$, $1 \le k \le l \le n$, $k + l < 2n$,*

$$\overset{\circ}{T} \cap \overset{\circ}{K} \neq \emptyset \tag{7}$$

if there are two points $\hat{x} \in \overset{\circ}{\hat{T}}{}^k$ and $\hat{y} \in \overset{\circ}{\hat{T}}{}^l$, so that

$$\hat{\underline{y}} = M\hat{\underline{x}} + N, \tag{8}$$

where $M = Q_K^{-1} A_K^T A_T$ and $N = Q_K^{-1} A_K^T (B_T - B_K)$, $Q_K = A_K^T A_K$.

Proof. In this case, it is impossible to express $\hat{y}$ directly from the relation $F_T(\hat{x}) = F_K(\hat{y})$. That is why, we multiply the latter equality by A_K^T. Since the simplex K is nondegenerated we have $\mathrm{rank}(A_K^T A_K) = l$, i.e. Q_K is an invertible matrix. On the other hand, $\hat{x}$ and $\hat{y}$ belong to the reference simplices, which guarantees the validity of inequality (7). □

Analogously to Corollary 1, we create a system of linear inequalities to determine overlapping of two simplices.

Corollary 2. *Assume that*

$$\hat{y}_i = \hat{\varphi}_i(\hat{x}), \quad i = 1, 2, \ldots, l, \tag{9}$$

is a scalar form of (8). Then the nondegenerated simplices T and K satisfying $\dim(T) = k$, $\dim(K) = l$, $1 \le k \le l \le n$, $k + l < 2n$, *have an intersection* $T \cap K$ *with* $\mu_k(T \cap K) > 0$, *if*

$$\hat{\varphi}_i(\hat{x}) > 0, \, i = 1, 2, \ldots, l, \, \sum_{i=1}^{l} \hat{\varphi}_i(\hat{x}) < 1 \tag{10}$$

and

$$\hat{x}_i > 0, \, i = 1, 2, \ldots, k, \, \sum_{i=1}^{k} \hat{x}_i < 1. \tag{11}$$

Proof. The relations

$$\hat{x} \in \mathring{T}^k \text{ and } \hat{y} \in \mathring{T}^l$$

follows from (10) and (11). Then the validity of the corollary is guaranteed by (9).□

4   The Overlap Detection Algorithm

In this section we analyze an overlapping of two convex polytopes. The ball $\mathcal{B}(T) = \text{Ball}(C_T, r_T)$ has a radius $r_T = \max_{i=1,2,\ldots,n+1} \|C_T - t_i\|$ and a center C_T. The latter means that $\mathcal{B}(T)$ is the smallest ball with a center C_T, which covers the simplex T. Let $\mathcal{F}_n$ be the set of all nondegenerated linear maps from $\mathbf{R}^n$ to $\mathbf{R}^n$, and Ω_1 and Ω_2 are nondegenerated polytopes in $\mathbf{R}^n$. We suppose that the triangulations

$$\tau_h(\Omega_1) = \left\{ T_i = F_i(\hat{T}) \,\middle|\, F_i \in \mathcal{F}_n, \; \Omega_1 = \bigcup_{i=1}^{m} T_i, \right.$$
$$\left. \mu_n(T_i \cap T_j) = 0, \; 1 \le i < j \le m \right\}$$

and

$$\tau_H(\Omega_2) = \left\{ K_i = G_i(\hat{T}) \,\middle|\, G_i \in \mathcal{F}_n, \; \Omega_2 = \bigcup_{i=1}^{p} K_i, \right.$$
$$\left. \mu_n(K_i \cap K_j) = 0, \; 1 \le i < j \le p \right\}$$

have as small as possible $\text{Card}(\tau_h)$, $\delta(\tau_h)$, and $\text{Card}(\tau_H)$, $\delta(\tau_H)$. In the general case, the mesh parameters h and H are independent.

Petrov and Todorov [20] have obtained a dissipative algorithm for triangulating convex polytopes. For the sake of computational simplicity, they demonstrated their algorithm in the four-dimensional space. But the dissipation algorithm [20] can be applied successfully to all higher-dimensional spaces. In this paper, we obtain the triangulations τ_h and τ_H of the domains Ω_1 and Ω_2 in the following way. Each polytope Ω_i is provided with the set of nodes

$$\mathcal{N}(\Omega_i) = \{\text{all vertices of } \Omega_i \text{ and the gravity center } G(\Omega_i)\}.$$

We apply the dissipation algorithm to triangulate the polytopes Ω_i, $i = 1, 2$. We could obtain optimal results by creating Delaunay triangulations (see Cignonit et al. [6]) of Ω_1 and Ω_2 but such divisions needs much more computational work.

The proposed algorithm has four steps:

(i)   Creation of simplicial triangulations τ_h and τ_H for both polytopes;

(ii)   Generation of the cover balls $\mathcal{B}(T)$ and $\mathcal{B}(K)$ for all simplices T and K in both triangulations;

(iii)   Determination of all couples (T_i, K_j), $T_i \in \tau_h$ and $K_j \in \tau_H$ satisfying $\mu_n\left(\mathcal{B}(T_i) \bigcap \mathcal{B}(K_j)\right) > 0$;

(iv)   Overlap detection test for all couples described in step (iii).

We denote the system of linear inequalities (5) with the constrains (6) by (W). The procedure for solving the system (W) by means of an iterative numerical method is denoted by Solve. The procedure Solve indicates the consistency or inconsistency of the system (1). This procedure has two nondegenerated polytopes as input and the existence or the lack of solutions of the linear system (W) as output. The initial guess x_0 for starting the Motzkin method is chosen to be the gravity center of the reference polytope. The cardinality of both triangulations affects the number of necessary operations to make a decision about the overlapping of both polytopes. The overlap detection algorithm is presented in pseudocode below.

```
1:  Algorithm Polytopes Overlapping Detection;
2:      function Min(J : functional);
3:          var k : local;
4:          x₀ =GravityCenter(T̂);
5:          Min:=Ĵ(x₀);
6:          k=0;
7:          repeat
8:              γₖ = ‖DĴ(xₖ)‖;
9:              ξ =   argmax   (Aᵢx̲ − Bᵢ);
                   i=1,2,...,2n+2
10:             x̲ₖ₊₁ = x̲ₖ − DĴ(xₖ);
11:             if Ĵ(xₖ₊₁) <Min then Min:=Ĵ(xₖ₊₁)
12:             end if;
13:             k=k+1
14:         until γₖ ≥ ε
15:         Min:=Ĵ(xₖ);
16:     end function;
17: procedure SOLVE(T, K : simplices);
18:     var i, j : local;
```

19:    **determine** A_T, B_T **from** $T = A_T \hat{T} + B_T$;

20:    **determine** A_K, B_k **from** $K = A_K \hat{T} + B_K$;

21:    $M := A_K^{-1} A_T$; $N := A_K^{-1}\left(B_T - B_K\right)$;

22:    $\underline{\hat{\Phi}}(\hat{x}) := M\underline{\hat{x}} + N$;

23:    % The functions $\hat{\varphi}_i(\hat{x})$ are coordinates of the vector $\underline{\hat{\Phi}}(\hat{x})$.

24:    **for** $i = 1$ **to** $n + 1$ **do**

25:       **for** $j = 1$ **to** n **do**

26:          $A_{ij} = \hat{\varphi}_i(\hat{t}_{n+1}) - \hat{\varphi}_i(\hat{t}_j)$

27:       **end for**

28:    **end for**;

29:    $\left(A_{ij}\right)_{i=n+2,n+3,\ldots,2n+1, j=1,2,\ldots,n} := -I_{n+1}$;

30:    $A_{2n+2} := (1, 1, \ldots 1)$;

31:    **for** $i = 1$ **to** $n + 1$ **do**

32:       $B_i = \hat{\varphi}_i(\hat{t}_{n+1})$

33:    **end for**;

34:    $\left(B_i\right)_{i=n+2,n+3,\ldots,2n+1} := \underline{0}_n$; $B_{2n+2} := 1$;

35:    $\hat{J}(x) := \frac{1}{2}\left|\hat{I}(A_\xi \underline{x} - B_\xi)\right|^2$;

36:    **if** $\left|\text{Min}(\hat{J})\right| < \varepsilon$ **then** Solution := **exists**

37:    **else** Solution := **nonexists**

38:    **end if**;

39: **end procedure**;

40: % The Main Algorithm.

41: **begin**

42: **set** Overlap = **false**;

43: **set** ε;

44: **define** Ω_1, Ω_2;

45: **compile** $\tau_h(\Omega_1)$, $\tau_H(\Omega_2)$;

46: **for** $i = 1$ **to** m **do**

47:    **for** $j = 1$ **to** p **do**

48:       **if** $\mu_n\left(\mathcal{B}(T_i) \bigcap \mathcal{B}(K_j)\right) > 0$ **then**

49:          Solve(T_i, K_j);

50:          **if** Solution = **exists then**

51:             % An intersection is detected.

52:             Overlap = **true**;

53:             **print**('The simplices', T_i, 'and', K_j, 'are overlapping each other.')

54:          **end if**

55:       **end if**

56:    **end for**

57: **end for**

58: **if** Overlap = **true then**

59:    **print**('The polytopes have a common subset')

60: **end if**

61: **end.**

Algorithm 1 is explained in the specific case when $\dim(\Omega_1) = \dim(\Omega_2) = n$ for the sake of simplicity. The same algorithm can be extended to the case $\dim(\Omega_1) = k$, $\dim(\Omega_2) = l$, $2 \leq k + l < 2n$ by applying Corollary 2.

5   Overlap Detection Examples

In this section, we compare some low-cost numerical methods to establish overlap between polytopes.

We begin the illustration of the theory from the previous paragraph with a three-dimensional example. By the first two examples, we illustrate that the intersection between the tested polytopes is found. An axonometric drawing of the considered polyhedra in Example 1 is presented. The three-dimensional shadow of the tested polytopes in Example 2 is shown. The real advantages of the proposed algorithm are demonstrated by the overlapping of two twelve-dimensional polytopes in Example 3.

Further, we use the denotation $\mathbf{1}_k = (1, i = 1, 2, \ldots, k)$.

Example 1. *We define a transformation* L_3 : $\underline{x} = D_3 \hat{\underline{x}} + E_3$, $L_3 \in \mathcal{F}_3$ *by*

$$
D_3 = \begin{pmatrix} 1 & 0.3 & 0 \\ 0 & 1 & 0.2 \\ 0.4 & 0 & 1 \end{pmatrix} \text{ and } E_3 = 0.275598 \begin{pmatrix} 1 \\ 1 \\ 1 \end{pmatrix}.
$$

Let us consider the polyhedra:

$$
\begin{aligned}
\Omega_1 &= [t_1(0,0,0), t_2(1,0,0), t_3(0,1,0), \\
&\quad t_4(0,0,1), t_5(-1,0,0), t_6(0,-1,0), t_7(0,0,-1)] \text{ and} \\
\hat{\Omega}_2 &= [\hat{k}_8(0,0,0), \hat{k}_9(1,0,0), \hat{k}_{10}(0,1,0), \hat{k}_{11}(0,0,1), \\
&\quad \hat{k}_{12}(1,1,0), \hat{k}_{13}(0,1,1), \hat{k}_{14}(1,0,1), \hat{k}_{15}(1,1,1)].
\end{aligned}
$$

The parallelohedron Ω_2 *is obtained by*

$$
\Omega_2 = L_3(\hat{\Omega}_2). \tag{12}
$$

The main purpose in this example is to determine the presence or absence of an intersection between the polyhedra Ω_1 *and* Ω_2.

Solution. We triangulate both polyhedra as follows:
$\Omega_1 = \Big\{ T_1 = [t_1, t_2, t_3, t_4], \quad T_2 = [t_1, t_3, t_4, t_5], \quad T_3 = [t_1, t_2, t_4, t_6], \quad T_4 = [t_1, t_2, t_3, t_7], \quad T_5 = [t_1, t_4, t_5, t_6], \quad T_6 = [t_1, t_3, t_5, t_7], \quad T_7 = [t_1, t_2, t_6, t_7], \quad T_8 = [t_1, t_5, t_6, t_7] \Big\}$ and $\hat{\Omega}_2 = \Big\{ \hat{K}_1 = [\hat{k}_9, \hat{k}_{10}, \hat{k}_{11}, \hat{k}_{15}], \hat{K}_2 = [\hat{k}_8, \hat{k}_9, \hat{k}_{10}, \hat{k}_{11}], \hat{K}_3 = [\hat{k}_{10}, \hat{k}_{11}, \hat{k}_{13}, \hat{k}_{15}], \hat{K}_4 = [\hat{k}_9, \hat{k}_{11}, \hat{k}_{14}, \hat{k}_{15}], \hat{K}_5 = [\hat{k}_9, \hat{k}_{10}, \hat{k}_{12}, \hat{k}_{15}] \Big\}$. The polyhedron Ω_1 is an octahedron, which is optimally divided into eight cube corners all of them with measure $\delta(T_i) = 3.34607$ $i = 1, 2, \ldots, 8$. The polyhedron $\hat{\Omega}_2$ is a cube, which is optimally partitioned [19] into four cube corners $\hat{K}_i$ $i = 2, 3, 4, 5$ and one

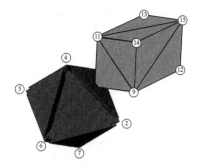

Fig. 2. An overlapping of Ω_1 and Ω_2 in the three-dimensional case.

Table 1. A comparison between the Motzkin method and the other iterative methods.

Method	CGM	BBM	MM	RM
$\epsilon \setminus$ initial guesses	$\underline{x}_0 = \underline{C}_{\hat{T}}$	$\underline{x}_0 = \underline{C}_{\hat{T}}$, $\underline{x}_1 = 5 \cdot \underline{1}_3$	$\underline{x}_0 = \underline{C}_{\hat{T}}$	$\underline{x}_0 = \underline{C}_{\hat{T}}$, $\lambda = 1.9$
10^{-6}	66	87	169	11

regular tetrahedron K_1. The cube $\hat{\Omega}_2$ is deformed by means of (12). The measures of degeneracy before and after the deformation of the cube $\hat{\Omega}_2$ are: $\delta(\hat{K}_1) = 2.44949$, $\delta(\hat{K}_i) = 3.34607$, $i = 2, 3, 4, 5$ and $\delta(K_1) = 3.04601$, $\delta(K_2) = 2.90384$, $\delta(K_3) = 4.42111$, $\delta(K_4) = 4.0847$, $\delta(K_5) = 4.32041$. The octahedron Ω_1 is not deformed. By applying Algorithm 1, we establish an intersection between T_1 and K_1, i.e. an overlapping between Ω_1 and Ω_2 is detected, see Fig. 2. The point $x_0(0.317466, 0.333466, 0.341466)$ belongs to the intersection $\overset{\circ}{\Omega}_1 \cap \overset{\circ}{\Omega}_2$. The corresponding points in the reference simplex are: $\hat{x}_0(0.00760254, 0.317466, 0.333466)$ and $\hat{y}_0(0.0277923, 0.0469174, 0.0547506)$. The vertex k_8 of Ω_2 belongs to the interior $\overset{\circ}{\Omega}_1$ of the octahedron.

We compare the Motzkin method (MM) [9] with the BB-method [23], the conjugate gradient method for quadratics (CGM) [28], and the relaxation method (RM) with various steplengths [26]. Let γ_k be the Euclidean norm of the Fréchet derivative of the objective functional. The stop criterion for all these methods is $\gamma_k < \epsilon$. An initial guess for all methods is chosen to be the gravity center of the reference finite element. We present the number of iterations for satisfying the stop criteria in Table 1. Additional initial guesses are necessary for the two-point methods.

Table 1 indicates that the Motzkin method has the best performance among other considered methods in this example. The row relaxation method strongly depends on the relaxation parameter λ. The results in Table 2 indicate that the overrelaxation is more suitable in this example.□

Table 2. The performance of the row relaxation method. The initial guess is chosen to be $\underline{x}_0 = \underline{C}_{\hat{T}}$ and $\epsilon = 10^{-6}$. The number ν_λ stands for the necessary iterations for satisfying the stop criterion.

λ	1.2	1.25	1.5	1.6	1.7	1.9
ν_λ	110	99	17	13	11	11

Example 2. *We continue with a four-dimensional example. Here, we use a deformation map* $L_4 : \underline{x} = D_4\hat{\underline{x}} + E_4,\ L_4 \in \mathcal{F}_4$ *by*

$$D_4 = \begin{pmatrix} 1 & 0.3 & 0.2 & 0.1 \\ 0.1 & 1 & 0.2 & 0.2 \\ 0.1 & 0.1 & 1 & 0.1 \\ 0.2 & 0.1 & 0.2 & 1 \end{pmatrix} \quad \text{and} \quad E_4 = 0.23 \begin{pmatrix} 1 \\ 1 \\ 1 \\ 1 \end{pmatrix}.$$

We define the polytopes:

$$\Omega_1 = \Big[t_1(0,0,0,0), t_2(1,0,0,0), t_3(0,1,0,0), t_4(0,0,1,0), t_5(0,0,0,1),$$

$$t_6(-1,0,0,0), t_7(0,-1,0,0), t_8(0,0,-1,0), t_9(0,0,0,-1) \Big]$$

and $\qquad \hat{\Omega}_2 = \Big[\hat{k}_1(0,0,0,0), \hat{k}_2(1,0,0,0), \hat{k}_3(0,1,0,0), \hat{k}_4(0,0,1,0),$
$\hat{k}_5(1,1,0,0), \hat{k}_6(0,1,1,0), \hat{k}_7(1,0,1,0), \hat{k}_8(1,1,1,0), \hat{k}_9(0,0,0,1),$
$\hat{k}_{10}(1,0,0,1), \hat{k}_{11}(0,1,0,1), \hat{k}_{12}(0,0,1,1), \hat{k}_{13}(1,1,0,1), \hat{k}_{14}(0,1,1,1),$
$\hat{k}_{15}(1,0,1,1), \hat{k}_{16}(1,1,1,1), \hat{k}_{17}\left(\frac{1}{2},\frac{1}{2},\frac{1}{2},\frac{1}{2}\right)].$
The polytope Ω_2 *is obtained by*

$$\Omega_2 = L_4(\hat{\Omega}_2). \tag{13}$$

The objects of intersection in this example are: Ω_1 *and* Ω_2.

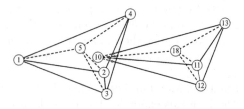

Fig. 3. The intersecting pentatopes T_1 and K_1.

Solution. The polytope Ω_1 is an orthoplex. Therefore, we divide it into sixteen tesseract corners:

$$\Omega_1 = \Big\{ T_1 = [t_1, t_2, t_3, t_4, t_5], \quad T_2 = [t_1, t_3, t_4, t_5, t_6], \quad T_3 = [t_1, t_2, t_4, t_5, t_7],$$

$T_4 = [t_1, t_2, t_3, t_5, t_8], \quad T_5 = [t_1, t_2, t_3, t_4, t_9], \quad T_6 = [t_1, t_4, t_5, t_6, t_7],$

$T_7 = [t_1, t_3, t_5, t_6, t_8], \quad T_8 = [t_1, t_3, t_4, t_6, t_9], \quad T_9 = [t_1, t_2, t_5, t_7, t_8],$

$T_{10} = [t_1, t_2, t_4, t_7, t_9], \quad T_{11} = [t_1, t_2, t_3, t_8, t_9], \quad T_{12} = [t_1, t_5, t_6, t_7, t_8],$

$T_{13} = [t_1, t_4, t_6, t_7, t_9], \quad T_{14} = [t_1, t_3, t_6, t_8, t_9], \quad T_{15} = [t_1, t_2, t_7, t_8, t_9],$

$$T_{16} = [t_1, t_6, t_7, t_8, t_9] \Big\}$$

all of them with a measure of degeneracy $\delta(T_i) = 4.24264$ $i = 1, 2, \ldots 16$. The polytope $\hat{\Omega}_2$ is a tesseract, which is optimally refined into 24 tesseract corners as it is done in [20]. All simplices K_i in

$$\hat{\tau}_H(\hat{\Omega}_2) = \left\{ \hat{K}_i \in [\hat{T}] \mid \Omega_2 = \bigcup_{i=1}^{24} \hat{K}_i, \ \mu_n \left(\hat{K}_i \cap \hat{K}_j \right) = 0, \ 1 \le i < j \le 24 \right\}$$

have the same measure of degeneracy $\delta(\hat{K}_i) = 4.24264$. After the deformation (13) the measure $\delta(\tau_H)$ of

$$\tau_H(\Omega_2) = \left\{ K_i = L_4(\hat{K}_i) \mid 1 \le i \le 24 \right\}$$

grows up to 5.85361. The latter indicates a slight deformation of Ω_2. Obviously, $\Omega_2 = \bigcup_{i=1}^{24} K_i$, $\mu_n (K_i \cap K_j) = 0$, $1 \le i < j \le 24$. The execution of Algorithm 1 indicates an intersection of the simplices $T_1[t_1, t_2, t_3, t_4, t_5]$ and $K_1[k_{10}, k_{11}, k_{12}, k_{13}, k_{18}]$, see Fig. 3. The vertex k_{10} of Ω_2 belongs to the interior of the orthoplex. The point $x_0(0.237771, 0.246095, 0.238852, 0.250095)$ belongs to the intersection between the cross polytope and the parallelotope, see Fig. 4. The preimages of x_0 in the reference simplex are: $\hat{y}_0(0.96391, 0.00147, 0.01128, 0.00583)$ and $\hat{x}_0(0.02719, 0.23777, 0.24609, 0.23885)$. □

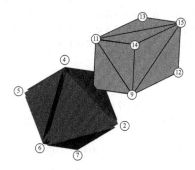

Fig. 4. An overlapping of Ω_1 and Ω_2 in the four-dimensional case.

The advantages of the Motzkin method will be demonstrated in the third example.

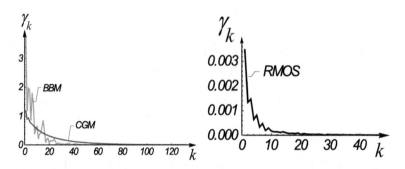

Fig. 5. The gradient descent obtained by the Barzilai-Borwein method, the conjugate gradient method, and the relaxation method with the optimal steplength.

Example 3. *Here, we consider overlapping of two twelve-dimensional simplices. Each vertex $\hat{k}_i$, $i = 1, 2, \ldots 12$ of the Freudenthal element*

$$\hat{K}[\hat{k}_1(1, 0, \ldots, 0), \hat{k}_2(1, 1, \ldots, 0, 0), \ldots, \hat{k}_{12}(1, 1, \ldots, 1, 1), \hat{k}_{13}(0, 0, \ldots, 0)]$$

has the first i coordinates equal to one and the other coordinates are equal to zero. The thirteenth node is located at the origin. The simplex $\hat{K}$ is deformed by the linear transformation $K = D_{12}\hat{K} + E_{12}$, where

$$D_{12} = \begin{pmatrix}
1 & 0.3 & 0.2 & 0.1 & 0.3 & 0.2 & 0.1 & 0.3 & 0.2 & 0.1 & 0.3 & 0.2 \\
0.1 & 1 & 0.2 & 0.2 & 0.3 & 0.2 & 0.1 & 0.3 & 0.2 & 0.1 & 0.3 & 0.3 \\
0.1 & 0.1 & 1 & 0.1 & 0.3 & 0.2 & 0.1 & 0.1 & 0.1 & 0.2 & 0.2 & 0.2 \\
0.2 & 0.1 & 0.2 & 1 & 0.3 & 0.2 & 0.1 & 0.3 & 0.2 & 0.1 & 0.1 & 0.1 \\
0.1 & 0.1 & 0.2 & 0.1 & 1 & 0.2 & 0.1 & 0.1 & 0.1 & 0.2 & 0.2 & 0.2 \\
0.3 & 0.1 & 0.3 & 0.1 & 0.3 & 1 & 0.1 & 0.3 & 0.2 & 0.1 & 0.1 & 0.1 \\
0.1 & 0.1 & 0.2 & 0.1 & 0.3 & 0.2 & 1 & 0.3 & 0.1 & 0.2 & 0.1 & 0.2 \\
0.2 & 0.1 & 0.3 & 0.1 & 0.3 & 0.2 & 0.1 & 1 & 0.2 & 0.1 & 0.1 & 0.3 \\
0.1 & 0.1 & 0.1 & 0.1 & 0.3 & 0.2 & 0.1 & 0.3 & 1 & 0.1 & 0.3 & 0.2 \\
0.2 & 0.1 & 0.2 & 0.1 & 0.3 & 0.2 & 0.1 & 0.3 & 0.2 & 1 & 0.3 & 0.2 \\
0.1 & 0.1 & 0.1 & 0.1 & 0.3 & 0.2 & 0.1 & 0.3 & 0.2 & 0.1 & 1 & 0.2 \\
0.2 & 0.1 & 0.2 & 1 & 0.3 & 0.2 & 0.1 & 0.3 & 0.2 & 0.1 & 0.3 & 1
\end{pmatrix}$$

and

$$E_{12} = (0.0744333, 0.0745333, 0.0746333, 0.0747333, 0.0748333, 0.0749333,$$
$$0.0750333, 0.0751333, 0.0752333, 0.0753333, 0.0754333, 0.0755333)^T.$$

We study the intersection between the reference cube corner $\hat{T}^{12}$ and the deformed Freudenthal element K.

Solution. Since $k_{13} \in \overset{\circ}{\hat{T}}{}^{12}$, the measure $\mu_{12}(\hat{T} \cap K)$ is positive. We calculate the transitional matrix A_K of the element K and write $A_T = I_{12}$, $B_T = \underline{0}_{12}$, $B_K = E_{12}$. The cube corners have the lowest rate of divergence [18]. Despite this $\delta(\hat{T}^{12}) = 10.93477$. To obtain a real comparison between the considered

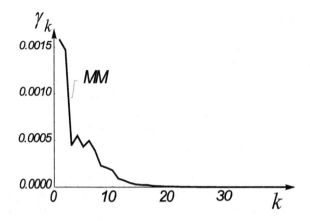

Fig. 6. The gradient descent obtained by the Motzkin method.

methods, we choose the strongly deformed element K with $\delta(K) = 132.726$. This affect the condition number of the matrix $\kappa(M_{12}) = 67.1266$. Thus we obtain ill-conditioned system (1).

We compare the Motzkin method with the BB-method, the conjugate gradient method, and the relaxation method with the optimal steplength (RMOS) [16]. An initial guess for all methods is chosen to be the gravity center of the cube corner. The relaxation method is divergent with a zero initial guess. We present the number of iterations for satisfying the stop criteria in Table 3. Additional initial guesses are necessary for the two-point methods. Table 3 indicates that the Motzkin method is superior to other considered methods in this example. Intermediate results on gradient descent are presented in Figs. 5 and 6.

Table 3. A comparison between the Motzkin method and the other iterative methods. The table present the number of necessary iterations for satisfying the stop criterion.

Method	CGM	BBM	MM	RMOS
$\epsilon \setminus$ initial guesses	$\underline{x}_0 = \underline{C}_{\hat{T}}$	$\underline{x}_0 = \underline{\hat{t}}_{13},$ $\underline{x}_1 = \underline{C}_{\hat{T}}$	$\underline{x}_0 = \underline{C}_{\hat{T}}$	$\underline{x}_0 = \underline{C}_{\hat{T}}$
10^{-6}	284	122	26	168
10^{-8}	370	164	41	287
10^{-12}	545	255	72	539

We establish good performance for the row relaxation method in the low-dimensional spaces and bad behavior of the same method in the higher-dimensional cases.

6 Conclusion

A feasible iterative algorithm for polytopes overlapping detection is developed in this paper. The algorithm is described in pseudocode. The only restriction on the polytopes tested for an intersection is both of them be convex. Concave polytopes can be studied for intersection by this algorithm if each of them is previously divided into convex subdomains. Then each part can be triangulated by simplicial finite elements. The algorithm is designed to be used in any Euclidean space. The investigation for overlapping requires triangulating of the considered polytopes. For this purpose, we use the dissipation algorithm. Thus we obtain triangulations of the tested polytopes with a small number of elements in order to reduce the number of couples of simplices. The overlappings of simplices with the same and with different dimensions are analyzed. Due to the high complexity of the Fourier-Motzkin elimination for complicated polytopes in $\mathbf{R}^n$ $n \geq 10$, the system (W) is solved iteratively. The proposed algorithm is implemented and tested in the three- and four-dimensional cases.

The numerical tests indicate that the row iterative methods have better performance than the other iterative methods. Additionally, the Motzkin method is superior to all other tested methods. The real advantages of the method are demonstrated by intersecting twelve-dimensional polytopes. Graphical presentations are shown in the three and four-dimensional spaces.

References

1. Aharoni, R., Berger, E.: The intersection of a matroid and a simplicial complex. Trans. Amer. Math. Soc. **358**(11), 4895–4917 (2006)
2. Bastrakov, S.I., Churkin, A.V., Yu, N.: Zolotykh, Accelerating Fourier-Motzkin elimination using bit pattern trees Optimization Methods & Software 2020–01-14. https://doi.org/10.1080/10556788.2020.1712600
3. Bauschke, H.H., Combettes, P.L.: Fejér monotonicity and fixed point iterations. In: Convex Analysis and Monotone Operator Theory in Hilbert Spaces. CMS Books in Mathematics (Ouvrages de mathématiques de la SMC). Springer, New York (2011)
4. Bey, J.: Simplicial grid refinement, on Freudenthal's algorithm and the optimal number of congruence classes. Numer. Math. **85**(1), 1–29 (1998)
5. Brandts, J., Korotov, S., Křížek, M.: Simplicial finite elements in higher dimensions. Appl. Math. **52**(3), 251–265 (2007)
6. Cignonit, P., Montanit, C., Scopigno, R.: DeWall: a fast divide and conquer Delaunay triangulation algorithm in E^d. Comput. Aided Des. **30**(5), 333–341 (1998)
7. Descantes, Y., Tricoire, F., Richard, P.: Classical contact detection algorithms for 3D DEM simulations: drawbacks and solutions. Comput. Geotech. **114**, 103134 (2019)
8. David, A., Edwards, K., Ondřej, F.K., Spurný, J.: A note on intersections of simplices. Bull. Soc. Math. France **139.1**, 89–95 (2011)
9. Haddock, J., Needell, D.: On Motzkin's method for inconsistent linear systems. Bit Numer. Math. **59**, 387–401 (2019)
10. Jing, R.-J., Maza, M.M., Talaashrafi, D.: Complexity estimates for Fourier-Motzkin elimination. CoRR, abs/1811.01510 (2018)

11. Keßler, C.W.: Parallel fourier-motzkin elimination. In: Bougé, L., Fraigniaud, P., Mignotte, A., Robert, Y. (eds.) Euro-Par 1996. LNCS, vol. 1124, pp. 66–71. Springer, Heidelberg (1996). https://doi.org/10.1007/BFb0024686
12. Lee, J.H., Amit, J., Eva, M.: Sevick-Muraca Fast intersections on nested tetrahedrons (FINT): an algorithm for adaptive finite element based distributed parameter estimation. J. Comput. Phys. **227**(11), 5778–5798 (2008)
13. Lei, Y.: The inexact fixed matrix iteration for solving large linear inequalities in a least squares sense. Numer. Algorithms **69**(1), 227–251 (2015)
14. Khachiyan, L.: Fourier-motzkin elimination method. In: Christodoulos, A.F., Panos, M.P. (eds) Encyclopedia of Optimization. Second Edition, pp. 1074–1077. Springer (2009)
15. Lira, W.M.: Luiz cristovao gomes coelho. In: Multiple Intersections of Finite-Element Surface Meshes. IMR, pp. 355–363. Luiz Fernando Martha (2002)
16. Mandel, J.: Convergence of the cyclical relaxation method for linear inequalities. Math. Programm. **30**, 218–228 (1984)
17. Montanari, M., Petrinic, N., Barbieri, E.: Improving the GJK algorithm for faster and more reliable distance queries between convex objects. ACM Trans. Graph. **36**(3), 30:1–30:17 (2017)
18. Petrov, M.S., Todorov, T.D.: Properties of the multidimensional finite elements. Appl. Math. Comput. **391**, 125695 (2021)
19. Petrov, M.S., Todorov, T.D.: Refinement strategies related to cubic tetrahedral meshes. Appl. Numer. Math. **137**, 169–183 (2019)
20. Petrov, M.S., Todorov, T.D.: Stable subdivision of 4D polytopes. Numer. Algorithms **79**(2), 633–656 (2018)
21. Pinar, M.Ç.: Newton's method for linear inequality systems. Eur. J. Oper. Res. **107**, 710–719 (1998)
22. Polyak, B.T.: Gradient methods for solving equations and inequalities. USSR Comput. Math. Math. Phys. **4**(6), 17–32 (1964)
23. Raydan, M.: On the Barzilai and Borwein choice of steplength for the gradient method. IMA J. Numer. Anal. **13**, 321–326 (1993)
24. Šimeček, I., Fritsch, R., Langr, D., Lórencz, R.: Parallel solver of large systems of linear inequalities using fourier-motzkin elimination. Comput. Inform. **35**, 1307–1337 (2016)
25. Talman, D.: intersection theorems on the unit simplex and the simplotope. In: Gilles, R.P., Ruys, P.H.M. (eds) Imperfections and Behavior in Economic Organizations. Theory and Decision Library (Series C: Game Theory, Mathematical Programming and Operations Research), vol. 11, pp. 257–278. Springer, Dordrecht (1994)
26. Telgen, J.: On relaxation methods for systems of linear inequalities. Eur. J. Oper. Res. **9**(2), 184–189 (1982)
27. Yen, N.D.: An introduction to vector variational inequalities and some new results. Acta Math. Vietnam **41**, 505–529 (2016)
28. Yuan, G., Hu, W.: A conjugate gradient algorithm for large-scale unconstrained optimization problems and nonlinear equations. J. Inequal. Appl. **2018**, 113 (2018)

Machine Learning-Based Numerical Dispersion Mitigation in Seismic Modelling

Kirill Gadylshin[1], Vadim Lisitsa[1]([✉])[ⓘ], Kseniia Gadylshina[2],
Dmitry Vishnevsky[2], and Mikhail Novikov[2]

[1] Sobolev Institute of Mathematics SB RAS, 4 Koptug ave.,
630090 Novosibirsk, Russia
lisitsavv@ipgg.sbras.ru
[2] Institute of Petroleum Geology and Geophysics SB RAS,
3 Koptug ave., 630090 Novosibirsk, Russia

Abstract. We present an original approach to improving seismic modelling performance by applying deep learning techniques to mitigate numerical error. In seismic modelling, a series of several thousand simulations are required to generate a typical seismic dataset. These simulations are performed for different source positions (equidistantly distributed) at the free surface. Thus, the output wavefields that corresponded to the nearby sources are relatively similar, sharing common peculiarities. Our approach suggests simulating wavefields using finite differences with coarse enough discretization to reduce the computational complexity of seismic modelling. After that, solutions for 1 to 10 percents of source positions are simulated using fine discretizations to obtain the training dataset, which is used to train the deep neural network to remove numerical error (numerical dispersion) from the coarse-grid simulated wavefields. Later the network is applied to the entire dataset. Our experiments illustrate that the suggested algorithm in the 2D case significantly (up to ten times) speeds up seismic modelling.

Keywords: Deep learning · Seismic modelling · Numerical dispersion

1 Introduction

Seismic modelling becomes a common tool to investigate peculiarities of wave propagation in realistic complex models of the Earth's interior [2,12,24] verification of the seismic processing and inversion algorithms, and as a part of the inversion methods, [21]. However, simulation of seismic wave propagation in complex media is one of the most computationally demanding problems requiring intense use of high-performance computing. In particular, if a typical seismic

K. Gadylshin and V. Lisitsa—are grateful to Mathematical Center in Akademgorodok, the agreement with Ministry of Science and High Education of the Russian Federation number 075-15-2019-1613 for the financial support. MN is supported by the Agency of the Precedent of Russian Federation, grant no. MK-3947.2021.1.5.

O. Gervasi et al. (Eds.): ICCSA 2021, LNCS 12949, pp. 34–47, 2021.
https://doi.org/10.1007/978-3-030-86653-2_3

acquisition system is considered, one has to simulate wavefields corresponding to hundreds of thousands of source positions (right-hand sides). Each simulation of a single shot gather is performed in a domain of about 10^3 km, which corresponds to 100^3 wavelength. Thus, up to $8 \cdot 10^9$ grid points are needed to obtain accurate enough numerical results. Reduction of the problem size by increasing the grid step leads to numerical error growth, which may completely destroy the solution. There are several ways to reduce the numerical dispersion, including use of high order finite-difference schemes [11], dispersion-suppression schemes [14], high-order finite element and discontinuous Galerkin methods [1,8,13]. However, the increase of the approach accuracy imminently leads to high computational intensity, including increased flops and RAM access operations.

The other option to reduce the numerical dispersion in the simulated wavefields is a post-processing [9,23]. However, the standard waveform correction procedures used in seismic processing are not efficient for numerical dispersion mitigation. The error associated with the numerical dispersion depends on the wave propagation path, velocity model etc. Thus, it can not be compensated by a single-phase shift. In this paper, we suggest an approach to post-processing based on using the deep learning technique.

Deep learning finds wide application in various fields of science. Providing a large representative training dataset, deep neural networks (DNNs) can approximate complex non-linear operators within the supervised learning workflow. These DNNs can learn about highly non-linear physics and usually provide much faster computational time than traditional simulation [6,17].

To develop an efficient algorithm for numerical dispersion mitigation, we use the following peculiarity of seismic modelling. The entire seismic dataset includes wavefields corresponding to different source positions. These positions are relatively close to each other (10 to 100 m apart). Thus, the velocity models and the simulated wavefields are similar if the source is situated nearby. It allows using a small number of sources to simulate accurate solution to be used as a training dataset. At the same time, we can simulate the entire dataset using a coarse enough grid, train the deep neural network, and then post-process the data.

The remainder of the paper has the following structure. In Sect. 2 we remind the basic concepts of seismic modelling, including the main estimates of the numerical dispersion, depending on the grid size. The description of the numerical dispersion mitigation network (NDM-net) is provided in Sect. 3. Numerical experiments illustrating the applicability of the NDM-net to the synthetic seismic data enhancement are presented in Sect. 4.

2   Seismic Modelling

Seismic wave propagation in 2D isotropic elastic media is governed by the elastic wave equation:

$$
\begin{aligned}
&\rho \frac{\partial u_1}{\partial t} = \frac{\partial \sigma_{11}}{\partial x_1} + \frac{\partial \sigma_{13}}{\partial x_3},
&&\frac{\partial \sigma_{11}}{\partial t} = (\lambda + 2\mu)\frac{\partial u_1}{\partial x_1} + \lambda\frac{\partial u_3}{\partial x_3} + f_{11}(t)\delta(\boldsymbol{x} - \boldsymbol{x}_s), \\
&\rho \frac{\partial u_3}{\partial t} = \frac{\partial \sigma_{13}}{\partial x_1} + \frac{\partial \sigma_{33}}{\partial x_3},
&&\frac{\partial \sigma_{33}}{\partial t} = \lambda\frac{\partial u_1}{\partial x_1} + (\lambda + 2\mu)\frac{\partial u_3}{\partial x_3} + f_{33}(t)\delta(\boldsymbol{x} - \boldsymbol{x}_s), \\
&&&\frac{\partial \sigma_{13}}{\partial t} = \mu\frac{\partial u_1}{\partial x_3} + \mu\frac{\partial u_3}{\partial x_1} + f_{13}(t)\delta(\boldsymbol{x} - \boldsymbol{x}_s),
\end{aligned}
\tag{1}
$$

where ρ is the mass density, λ and μ are the Lame parameters, $\boldsymbol{u} = (u_1, u_3)^T$ is the particle velocity vector, σ is the stress tensor, $f_{ij}(t)$ are the components of the source wavelet function, $\delta(\boldsymbol{x})$ is the Kroneker delta-function, $\boldsymbol{x}$ is the vector of spatial coordinates, and $\boldsymbol{x}_s$ is the source coordinate. The seismic modelling is stated in half-space $x_3 > 0$ and within bounded time interval $t \in [0, T]$.

A common way to approximate the elastic wave equation is the use of staggered grid finite differences [11,20], where the different components of the wavefield are defined at different spatial and temporal points with the use of symmetric stencils to approximate the derivatives:

$$
\begin{aligned}
\rho D_t[u_1]^{n-1/2}_{i+1/2,j} &= D_1[\sigma_{11}]^{n-1/2}_{i+1/2,j} + D_3[\sigma_{13}]^{n-1/2}_{i+1/2,j}, \\
\rho D_t[u_3]^{n-1/2}_{i,j+1/2} &= D_1[\sigma_{13}]^{n-1/2}_{i,j+1/2} + D_3[\sigma_{33}]^{n-1/2}_{i,j+1/2}, \\
D_t[\sigma_{11}]^n_{i,j} &= (\lambda + 2\mu)D_1[u_1]^n_{i,j} + \lambda D_3[u_3]^n_{i,j} + f_{11}(t^n)[\delta(\boldsymbol{x} - \boldsymbol{x}_s)]_{i,j}, \\
D_t[\sigma_{33}]^n_{i,j} &= \lambda D_1[u_1]^n_{i,j} + (\lambda + 2\mu)D_3[u_3]^n_{i,j} + f_{33}(t^n)[\delta(\boldsymbol{x} - \boldsymbol{x}_s)]_{i,j}, \\
D_t[\sigma_{13}]^n_{i+1/2,j+1/2} &= \mu D_1[u_3]^n_{i+1/2,j+1/2} + \mu D_3[u_1]^n_{i+1/2,j+1/2} + \\
&\quad + f_{13}(t^n)[\delta(\boldsymbol{x} - \boldsymbol{x}_s)]_{i+1/2,j+1/2},
\end{aligned}
\tag{2}
$$

where finite-difference operators are

$$
\begin{aligned}
D_t[g]^N_{I,J} &= \frac{g^{N+1/2}_{I,J} - g^{N-1/2}_{I,J}}{\tau}, \\
D_1[g]^N_{I,J} &= \frac{1}{h_1} \sum_{m=0}^{M} \alpha_m \left(g^N_{I+m+1/2,J} - g^N_{I-m-1/2,J} \right), \\
D_2[g]^N_{I,J} &= \frac{1}{h_2} \sum_{m=0}^{M} \alpha_m \left(g^N_{I,J+m+1/2} - g^N_{I,J-m-1/2} \right),
\end{aligned}
\tag{3}
$$

where indices I, J, N can be either integer or half-integer, and g is a smooth enough scalar function. The operator D_t approximates the temporal derivatives with the second order to the time step τ. The operators D_1 and D_2 approximate the spatial derivatives. However, depending on the choice of α_m one may construct a high-order approximation up to $2m + 2$ [11], or use theses degrees of freedom to suppress numerical dispersion [14]. Note, that we are not discussing the approximation of the right-hand sides, as it is presented in [7], and the model parameters treatment because it is studied in [16,22].

The use of symmetric stencils to approximate derivatives ensures an even order of approximation with zero coefficients of odd degrees in the differential approximation of the finite difference scheme (2). Thus the numerical error appears in the solution as a numerical dispersion without dissipation, see [19], and [3] for the details. It means that the emitted impulse will deteriorate, propagating through the media. An example of the impulse deformation due to the dispersion is presented in Fig. 1. We plot the true pulse and that travelled 30 wavelengths simulated by the second-order scheme with a spatial discretization of 10, 20, and 40 points per wavelength and Courant number equal to 0.8. Note that the maximum impulse shifts backwards in time, leading to an overestimation of the reflecting intervals depth in seismic processing and interpretation. Refining the mesh, one gets the convergence of the numerical solution to the true one, however, refining a spatial step by the factor of two leads to the increase of the problem size by the factor of 8 in 3D and 4 in 2D. Moreover, the number of flops increases by 16 in 3D and 8 in 2D because of the temporal step refinement. On the other hand, the solution obtained on the grid with 20 points per wavelength is accurate enough, and simple processing may turn it into the true one.

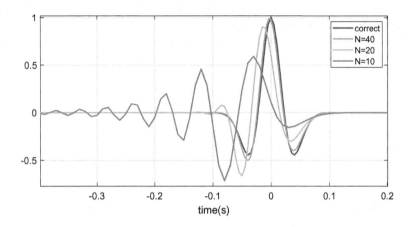

Fig. 1. An example of the pulse deformation due to numerical dispersion.

3 Numerical Dispersion Mitigation Network (NDM-net)

Convolutional Neural Networks (CNN) are usually applied to analyze visual imagery. A particular case of CNN is a U-Net [18], which was originally introduced for biomedical image segmentation. At this moment, the U-Net and its modifications have broad applications in seismic inversion, pre-stack seismic data processing and interpretation. This work suggests using the Numerical Dispersion Mitigation deep neural network (NDM-net) to learn the mapping between the synthetic seismic data modelled on a coarse grid and data modelled on a fine grid. In other words, we plan to eliminate the numerical dispersion using the Deep Learning approach.

The architecture of the network is similar to the one used by [5]. The differences are using a conventional convolutional layer instead of partial convolutions and the different input/output dimensions, see Fig. 2. These DNN contains 16 convolutional layers, eight upsampling layers, and eight concatenation layers (skip connections). The input and output tensors dimensions are $1250 \times 512 \times 2$. An activation function for the first eight convolutional layers (encoding, or feature extracting, part of the DNN) is ReLU, while the last eight convolutional layers (decoding part) have LeakyReLu activation with a negative slope coefficient equals to 0.2. We implemented NDM-net in TensorFlow. The DNN weights were randomly initialized, and Adam stochastic optimization algorithm was exploited during the training process.

In the current implementation, we consider the input/output to be regularly sampled pre-stack seismic data. For training, we used each 10-th common shot gather computed on a fine grid and its corrupted version modelled on a coarse grid. Each common shot is converted to a tensor with a dimension of $1250 \times 512 \times 2$. Here 1250 is the number of time samples in data (4 ms time discretization and 5s record time), 512 is the number of 2C receivers, and 2 is the number of recorded components (vertical and horizontal velocity components). Next, we split this dataset into training and validation datasets. Each common shot is normalized by scaling it to unit variance before being processed by the NDM-net.

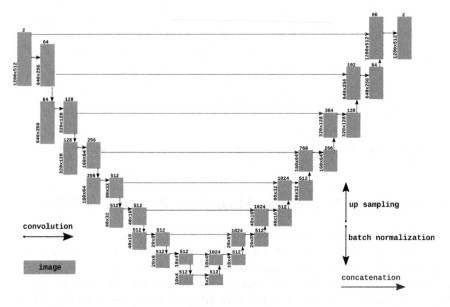

Fig. 2. The architecture of NDM-net. The Black right arrow indicates convolution operation, while the red right arrow indicates concatenation. Up and down arrows indicate upsampling and batch normalization operations correspondingly.

4   Numerical Experiments

We applied our approach to mitigate the numerical dispersion in two datasets. Both simulations were done in 2D to illustrate the applicability of the NDM-net to improve seismic modelling accuracy and efficiency.

4.1   Marmousi2 Model

First we considered the elastic Marmousi2 model [15], as presented in Fig. 3. The size of the model was 17 km in the horizontal and 3.6 km in the vertical direction. Marmousi2 is the offshore model with water at the top. To make the considerations consistent with land data acquisition, we substitute water with solid used for the ocean bottom in the model. We performed simulations of seismic waves propagation using meshes with steps equal to 1.25 m, 2.5 m, and 5 m, assuming the solution obtained on the 1.25 m grid is the exact one. Such small grid steps were chosen due to the thick low-velocity layer, that was introduced instead of water at the top of the model. Note that the original model was provided on a grid with step size 1.25 m. However, to exclude the effect of model changes when the simulation mesh is coarsening, we map the mode to the mesh with step 5 m. After that 5-meters model was used for all numerical simulations.

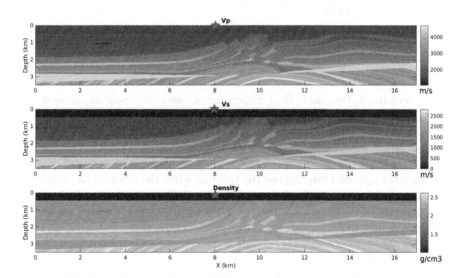

Fig. 3. Marmousi2 elastic velocity model used for synthetic data generation. The marker represents the source position at $x = 8$ km.

The acquisition included 171 sources with the distance between the sources 100 m. We recorded wavefield by 512 2C receivers for each shot with maximal source-receiver offsets equal to 6.4 km. The distance between the receivers was 25 m. Simulations were performed using the fourth-order staggered grid scheme [11]. On average, the simulation time was 5 s per shot if a 5 m grid was used; 40 s per shot for 2.5 m grid; and 4 min for 1.25 m grid, using Nvidia V100 GPU. The example of modelled seismogram on the grid 1.25 m (X = 9 km) is presented in Fig. 4.

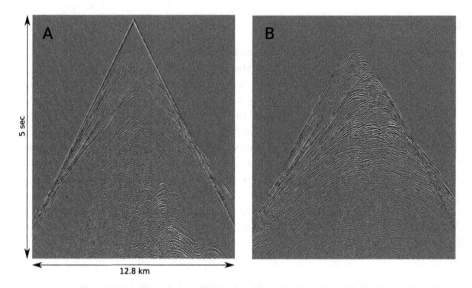

Fig. 4. Synthetic seismograms for shot positioned at $x = 9$ km: horizontal (a) and vertical (b) components calculated on a numerical grid with the spatial step 1.25 m.

We performed two numerical experiments, and for each experiment, we trained NDM-net. One was designed to map the data simulated using a 2.5 m grid to the exact solution (data acquired on the grid with steps 1.25 m). The other NDM-net was trained to map 5 m-data to the 1.25 m-data. The training was performed on the Nvidia V100 GPU. As a regularization, we used an early stopping technique and interrupted the training when the error on the validation dataset started to grow. In both cases (2.5 m to 1.25 m and 5 m to 1.25 m), the training process took about 30min. The prediction time is about 0.7 sec for one full common shot gather, while one forward modelling using FD technique on a GPU took about 40 s on 2.5 m grid and about 5 s on 5 m grid, but 5 min for the finest grid of 1.25 m.

To estimate the quality of DNN prediction, we use the normalized RMS (NRMS) as a measure of datasets similarity. NRMS is a strict sample-by-sample metric used for evaluating repeatability between two datasets in 4D seismic [10]. An acceptable level of NRMS in 4D seismic is about 20–40%. The verification of DNN predictions was performed on a testing dataset that differs from training and validation, i.e. were invisible by DNN during the training process. The NRMS plot calculated trace by trace using a sliding window of 200 ms is presented in Fig. 5. On average, the NRMS between 1.25 m data and 2.5 m-data was 30%. Application of the NDM-net reduced the NRMS down to 14%. The average NRMS between 5-m data and 1.25 m-data was about 59%, and the DNN managed to construct a prediction with the NRMS of 33%. So one may conclude that in both cases, NDM-net were able to reduce NRMS up to the acceptable level. To illustrate the effect of the NDM-net data enhancement, we provide the plots (see Figs. 6, 7) of a single seismic trace computed using different grids and then improved by the NDM-net.

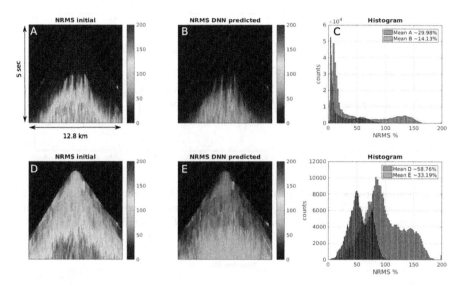

Fig. 5. NRMS plot calculated between seismograms computed on a numerical grid with the spatial steps 1.25 m and 2.5 m (a), 1.25 m and DNN predicted data using 2.5 m data as input (b), 1.25 m and 5 m (d), 1.25 m and DNN predicted data using 5 m data as input, and the corresponding histograms (c,f).

4.2   Model with Vertical Intrusion

The second set of experiments was done for a model with vertical high-contrast intrusions causing lateral heterogeneity as presented in Fig. 8. The size of the entire model was 220 km by 2.6 km. The acquisition included 1901 sources with the distance between the sources 100 m. We recorded wavefield by 512 receivers for each shot with maximal source-receiver offsets equal to 6.4 km. The distance between the receivers was 25 m. In this research, we simulated the wavefield without the surface waves by using a perfectly matched layer for $x < 0$ [4]. The source wavelet was the Ricker pulse with a central frequency 30 Hz.

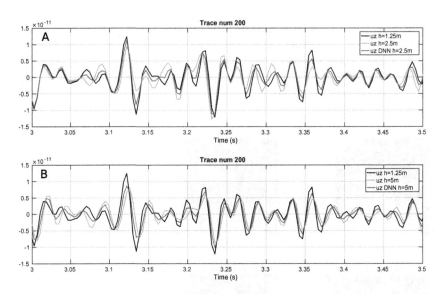

Fig. 6. Seismic traces at different positions and its DNN-predictions for the case 2.5 m-data (a) and 5 m-data (b). Black plot – vertical component on the fine grid, red plot – input data for DNN prediction and blue plot – DNN-predicted data.

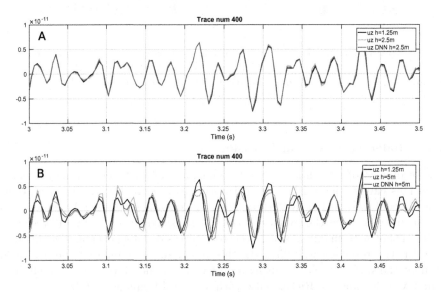

Fig. 7. Seismic traces at different positions and its DNN-predictions for the case 2.5 m-data (a) and 5 m-data (b). Black plot – vertical component on the fine grid, red plot – input data for DNN prediction and blue plot – DNN-predicted data.

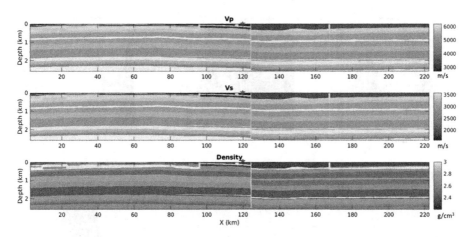

Fig. 8. Elastic velocity model used for synthetic data generation. The marker represents the source position at $x = 120$ km.

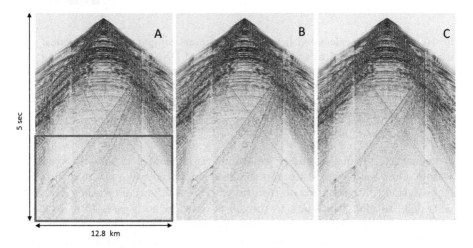

Fig. 9. Synthetic seismograms for shot positioned at $x = 120$ km: vertical component calculated on a numerical grid with the spatial steps 2.5 m (a), 5 m (b) and 10 m (c). (Color figure online)

Originally, the model was provided on a grid with the steps 50 m in horizontal and 5 m in a vertical direction. We computed three datasets using the fourth-order staggered grid scheme [11]. We considered the solution acquired at the grid with steps of 2.5 m as the accurate one, whereas two others generated using grids with 5 m and 10 m spatial steps are polluted. We provide examples of the seismograms in Fig. 9.

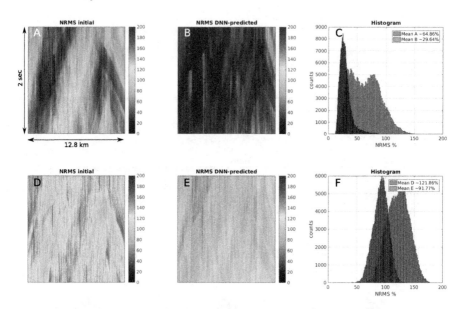

Fig. 10. NRMS plot calculated between seismograms computed on a numerical grid with the spatial steps 2.5 m and 5 m (a), 2.5 m and DNN predicted data using 5 m data as input (b), 2.5 m and 10 m (d), 2.5 m and DNN predicted data using 10 m data as input, and the corresponding histograms (c,f). NRMS were calculated in the area designated by a red rectangle on Fig. 2 (vertical component at the time from 3 s to 5 s including all receiver positions)

As in the previous example, we trained two NDM-nets for two synthetic datasets. One was designed to map the data simulated using a 5 m grid to the exact solution (data acquired on the grid with steps 2.5 m). The other NDM-net was trained to map 10 m-data to the 2.5 m-data. In both cases (5 m to 2.5 m and 10 m to 2.5 m), the training process took about 40 min. The prediction time is about 0.7 sec for one full common shot gather, while one forward modelling using FD technique on a GPU took about 40 s on 2.5 m grid and about 5 s on 5 m grid. Since the main error accumulates in the late arrivals, we calculate NRMS for the time range from 3 s to 5 s (red rectangle on the Fig. 9). The corresponding NRMS plot is presented in Fig. 10. On average, the NRMS between 2.5 m data and 5 m-data was 65%. Application of the NDM-net reduced the NRMS down to 30%. The average NRMS between 10-m data and 2.5 m-data was about 120%, which means that the 10-m data are extremely far from the true solution. As a result, the DNN managed to reduce NRMS up to the 90% level. The effect of the NDM-net data enhancement is illustrated in the Figs. 11, 12.

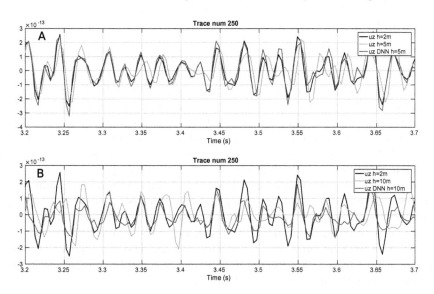

Fig. 11. Seismic traces at different positions and its DNN-predictions for the case 5 m-data (a) and 10 m-data (b). Black plot – vertical component on the fine grid, red plot – input data for DNN prediction and blue plot – DNN-predicted data.

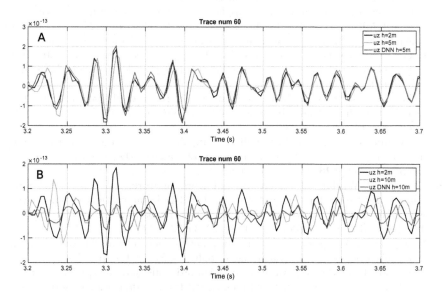

Fig. 12. Seismic traces at different positions and its DNN-predictions for the case 5 m-data (a) and 10 m-data (b). Black plot – vertical component on the fine grid, red plot – input data for DNN prediction and blue plot – DNN-predicted data.

5   Conclusions

We present an original approach to numerical simulation of seismic wavefields. The method combines conventional seismic modelling based on the finite differences with the consequent correction of the data by the DNN-based algorithm called the NDM-net. First, we generate a training dataset simulating wavefields corresponding to at most 10% of the positions of the sources using fine enough spatial discretization (up to 20 points per minimal wavelength - ppw). Second, the full dataset is generated using a coarse mesh with no more than 3–5 ppw. Note that in the 2D case, simulation of the solution using 5 ppw is 64 times faster than that with 20 ppw. Third, the NDM-net is trained to reduce the numerical error in the coarse-grid solution. Then the NDM-net is applied to correct the entire dataset. The presented results demonstrate the ability of the NDM-net to make a high-quality seismic data prediction using the synthetics generated on a coarse grid. In particular, the application of the NDM-net reduced the computational time to simulate the full dataset of 171 common shot gathers for the Marmousi2 model from 684 min to 112 min.

References

1. Baldassari, C., Barucq, H., Calandra, H., Diaz, J.: Numerical performances of a hybrid local-time stepping strategy applied to the reverse time migration. Geophys. Prospect. **59**(5), 907–919 (2011). https://doi.org/10.1111/j.1365-2478.2011.00975.x
2. Chen, G., Song, L., Liu, L.: 3D numerical simulation of elastic wave propagation in discrete fracture network rocks. Pure Appl. Geophys. **176**(12), 5377–5390 (2019)
3. Cohen, G. (ed.): Metodes numeriques d'ordre eleve pour les ondes en regime transitoire. INRIA (1994). in French
4. Collino, F., Tsogka, C.: Application of the perfectly matched layer absorbing layer model to the linear elastodynamic problem in anisotropic heterogeneous media. Geophysics **66**, 294–307 (2001)
5. Gadylshin, K., Silvestrov, I., Bakulin, A.: Inpainting of local wavefront attributes using artificial intelligence for enhancement of massive 3-D prestack seismic data. Geophys. J. Int. **223**, 1888–1898 (2020)
6. Guo, X., Li, W., Iorio, F.: Convolutional neural networks for steady flow approximation. In: Proceedings of the 22nd ACM SIGKDD International Conference on Knowledge Discovery and Data Mining - KDD '16, San Francisco, CA, USA, pp. 481–490 (2016). https://doi.org/10.1145/2939672.2939738
7. Hicks, G.: Arbitrary source and receiver positioning in finite-difference schemes using kaiser windowed sinc functions. Geophysics **67**(1), 156–165 (2002)
8. Kaser, M., Dumbser, M.: An arbitrary high-order discontinuous galerkin method for elastic waves on unstructured meshes - i. the two-dimensional isotropic case with external source terms. Geophys. J. Int. **166**(2), 855–877 (2006)
9. Koene, E., Robertsson, J.: Removing numerical dispersion artifacts from reverse time migration and full-waveform inversion, pp. 4143–4147 (2017)
10. Kragh, E., Christie, P.: Seismic repeatability, normalized rms, and predictability. Lead. Edge **21**(7), 640–647 (2002)
11. Levander, A.R.: Fourth-order finite-difference p-sv seismograms. Geophysics **53**(11), 1425–1436 (1988)

12. Lisitsa, V., Kolyukhin, D., Tcheverda, V.: Statistical analysis of free-surface variability's impact on seismic wavefield. Soil Dyn. Earthq. Eng. **116**, 86–95 (2019)
13. Lisitsa, V., Tcheverda, V., Botter, C.: Combination of the discontinuous galerkin method with finite differences for simulation of seismic wave propagation. J. Comput. Phys. **311**, 142–157 (2016)
14. Liu, Y.: Optimal staggered-grid finite-difference schemes based on least-squares for wave equation modelling. Geophys. J. Int. **197**(2), 1033–1047 (2014)
15. Martin, G.S., Wiley, R., Marfurt, K.J.: Marmousi2: an elastic upgrade for marmousi. Lead. Edge **25**(2), 156–166 (2006)
16. Moczo, P., Kristek, J., Vavrycuk, V., Archuleta, R.J., Halada, L.: 3D heterogeneous staggered-grid finite-differece modeling of seismic motion with volume harmonic and arithmetic averagigng of elastic moduli and densities. Bull. Seismol. Soc. Am. **92**(8), 3042–3066 (2002)
17. Moseley, B., Nissen-Meyer, T., Markham, A.: Deep learning for fast simulation of seismic waves in complex media. Solid Earth **11**, 1527–1549 (2020)
18. Ronneberger, O., Fischer, P., Brox, T.: U-Net: convolutional networks for biomedical image segmentation. In: Navab, N., Hornegger, J., Wells, W.M., Frangi, Al.F. (eds.) MICCAI 2015. LNCS, vol. 9351, pp. 234–241. Springer, Cham (2015). https://doi.org/10.1007/978-3-319-24574-4_28
19. Shokin, Y., Yanenko, N.: Method of Differential Approximation. Application to Gas Dynamics. Nauka, Novosibirsk (1985). in Russian
20. Virieux, J.: P-sv wave propagation in heterogeneous media: velocity-stress finite-difference method. Geophysics **51**(4), 889–901 (1986)
21. Virieux, J., Calandra, H., Plessix, R.E.: A review of the spectral, pseudo-spectral, finite-difference and finite-element modelling techniques for geophysical imaging. Geophys. Prospect. **59**(5), 794–813 (2011). https://doi.org/10.1111/j.1365-2478.2011.00967.x
22. Vishnevsky, D., Lisitsa, V., Tcheverda, V., Reshetova, G.: Numerical study of the interface errors of finite-difference simulations of seismic waves. Geophysics **79**(4), T219–T232 (2014)
23. Xu, Z., et al.: Time-dispersion filter for finite-difference modeling and reverse time migration, pp. 4448–4452 (2017)
24. Zhu, J., Ren, M., Liao, Z.: Wave propagation and diffraction through non-persistent rock joints: an analytical and numerical study. Int. J. Rock Mech. Mining Sci. **132**, 104362 (2020)

A Variant of the Nonlinear Multiscale Dynamic Diffusion Method

Andrea M. P. Valli[1], Isaac P. Santos[2], Sandra M. C. Malta[3],
Lucia Catabriga[1]([envelope]), and Regina C. Almeida[3]

[1] DI, Universidade Federal do Espírito Santo, Vitória, ES, Brazil
{avalli,luciac}@inf.ufes.br
[2] DMA, Universidade Federal do Espírito Santo, São Mateus, São Mateus, ES, Brazil
isaac.santos@ufes.br
[3] Laboratório Nacional de Computação Científica, Petrópolis, RJ, Brazil
{smcm,rcca}@lncc.br

Abstract. This paper presents a two-scale finite element formulation for a variant of the nonlinear Dynamic Diffusion (DD) method, applied to advection-diffusion-reaction problems. The approach, named here new-DD method, introduces locally and dynamically an extra stability through a nonlinear operator acting in all scales of the discretization, and it is designed to be bounded. We use bubble functions to approximate the subgrid scale space, which are locally condensed on the resolved scales. The proposed methodology is solved by an iterative procedure that uses the bubble-enriched Galerkin solution as the correspondent initial approximation, which is automatically recovered wherever stabilization is not required. Since the artificial diffusion introduced by the new-DD method relies on a problem-depend parameter, we investigate alternative choices for this parameter to keep the accuracy of the method. We numerically evaluate stability and accuracy properties of the method for problems with regular solutions and with layers, ranging from advection-dominated to reaction-dominated transport problems.

Keywords: New-dynamic diffusion method · Bubble functions · Advection-diffusion-reaction equations

1 Introduction

Stable finite element approximations to advection-diffusion-reaction problems have been the subject of intense research for the past forty years. Typical Galerkin formulations for this class of problems lack stability when advection or reaction effects are much stronger than the diffusion phenomena. Several works proposed the introduction of stabilization along the streamlines, in a linear and consistent manner [2,11,14]. To deal with instabilities along other directions, a variety of nonlinear stabilization methodologies have been developed [4,12,13,16]. Both linear

Supported by organizations CNPq, FAPERJ and FAPES.

O. Gervasi et al. (Eds.): ICCSA 2021, LNCS 12949, pp. 48–61, 2021.
https://doi.org/10.1007/978-3-030-86653-2_4

and nonlinear stabilization terms depend on user-defined parameters that ultimately yield the accuracy and stability properties of the method. More recently, the broader framework introduced by variational multiscale finite element formulations [8,9] have reformulated stabilized methods by decomposing state variables into resolved and unresolved (subgrid) scales. In the multiscale framework, stabilized methods can be seen as techniques to capture the unresolved scale variability into the resolved scale solution. Such an approach has opened new directions of research. The method proposed in [7], for example, used scale separation to introduce artificial dissipation only onto the small scales, still requiring definition of tunable parameter. This drawback was overcome in [17] by using a subgrid diffusivity stabilization that locally depends on the resolved scale solution. Using the same basis for the construction of stabilization, this method was extended in [1] by adding isotropic dissipation on all scales, in the context of non conforming two-scale methods. The so called Dynamic Diffusion (DD) method was then extended to the continuous setting in [19,20] by using bubble functions to build the subgrid space. In this way, the subgrid degrees of freedom can be condensed onto the resolved scale degrees of freedom, reducing the computational cost typical of two-scale methods. The resulting technique shows robust stability properties for advection-diffusion-reaction problems in the presence of layers [20].

   In the DD model, a nonlinear dissipation mechanism is introduced in both the resolved and unresolved scales. The artificial diffusion is dynamically evaluated depending on the resolved scale solution. Specifically, it depends on the residual and the gradient of the resolved scale solution, as well as on a characteristic local length scale. In this way, without invoking any linear stabilization, the DD method does not require any additional user-defined parameter and ultimately avoids kinetic energy accumulation at the resolved scale, precluding local and non-local oscillations in the presence of layers. Unfortunately, the numerical analysis of the DD method could not be established since the artificial diffusion operator is not upper bounded, a key issue to prove the existence of discrete solution, stability, and *a priori* error estimate. To overcome this drawback, a new-DD method was developed and analyzed in [18], and optimal convergence rates in the $L^2(\Omega)$, $H^1(\Omega)$, and energy norms were numerically shown for problems with smooth solutions.

   Here, we investigate the behavior of the new-DD method for problems with internal and external layers, ranging from advection dominated to reaction dominated transport problems. Of note, the artificial diffusion introduced by the new-DD method relies on a problem dependent parameter that plays a key role on its upper boundedness. Although it can be defined locally and depending on the problem, we perform numerical experiments to investigate alternative, and eventually simpler, choices for this parameter that does not compromise the accuracy of the method. Thus, focusing on those numerical aspects, this work is organized as follows. In Sect. 2 we present the new-DD method for solving the advection-diffusion-reaction transport problems. Section 3 contains the numerical studies and the evaluation of the method. The conclusions are summarized in Sect. 4.

2   The New Dynamic Diffusion Method

The problem of interest is modeled by the following steady-state advection-diffusion-reaction equation

$$- \epsilon \Delta u + \boldsymbol{\beta} \cdot \boldsymbol{\nabla} u + \sigma u = f, \qquad \text{in } \Omega, \tag{1}$$

where the domain $\Omega \subset \mathbb{R}^2$ has a sufficiently regular boundary Γ. It represents the transport dynamics of a scalar field u (e.g., temperature, chemical concentration, etc.), subject to the diffusion coefficient $\epsilon > 0$ and to the incompressible velocity field $\boldsymbol{\beta}$. Local dynamics are modeled by the reaction term σu, in which $\sigma \geq 0$ is the reaction coefficient; and f is the source term. To complete the problem setting, we assume for simplicity the following homogeneous Dirichlet boundary conditions

$$u = 0, \qquad \text{on } \Gamma. \tag{2}$$

Also for simplicity, and without loss of generality, we assume that both ϵ and σ are constant parameters. From now on we use $H^1(X)$ to denote the Hilbert space in X, for each $X \subseteq \Omega$, and $H^0(X) = L^2(X)$. We also denote $\| \cdot \|_{m,X}$ as the standard norm in $H^1(X)$ ($m = 1$) and in $L^2(X)$ ($m = 0$), and $(\cdot, \cdot)$ as the inner product in $L^2(X)$.

A general variational formulation of problem (1)–(2) is: find $u \in H_0^1(\Omega) = \{u \in H^1(\Omega), u = 0 \text{ on } \Gamma\}$ such that

$$B(u, v) = (f, v), \quad \forall v \in H_0^1(\Omega), \tag{3}$$

with

$$B(u, v) = \epsilon(\boldsymbol{\nabla} u, \boldsymbol{\nabla} v) + (\boldsymbol{\beta} \cdot \boldsymbol{\nabla} u, v) + \sigma(u, v). \tag{4}$$

To build the corresponding discrete multiscale formulation we first define a regular triangulation $\mathcal{T}_h = \{T\}$ of the domain Ω into regular elements, denoted by T, and we set $h = \max\{h_T : T \in \mathcal{T}_h\}$ with $h_T := diam(T)$. The Galerkin formulation associated to (3) consists of finding $u_h \in V_h^0 \subset H_0^1(\Omega)$ such that

$$B(u_h, v_h) = (f, v_h), \quad \forall v_h \in V_h^0, \tag{5}$$

where $V_h^0 = \{w \in H^1(\Omega) \mid w|_T \in \mathbb{P}_1(T), \forall T \in \mathcal{T}_h, w|_\Gamma = 0\}$, with $\mathbb{P}_1(T)$ the set of first order polynomials in elements $T \in \mathcal{T}_h$. In the multiscale framework, we consider that V_h^0 is the resolved (grid) scale space, and we define the two-scale approximation space $V_{hb}^0 = V_h^0 \oplus V_b$, which is a direct sum of V_h^0 and an unresolved subgrid (fine) scale space V_b. As in [7], we define $V_b = \{v_b \in H_0^1(\Omega); v_b|_T = span\{\psi_T\}, \forall T \in \mathcal{T}_h\}$, where $\psi_T \in H_0^1(T)$ are bubble functions. Specifically, we define $\psi_T(\boldsymbol{x}) = 27N_1(\boldsymbol{x})N_2(\boldsymbol{x})N_3(\boldsymbol{x}), \forall \boldsymbol{x} \in T$, where N_i is the basis function associated of the ith node of T. Using these definitions, the new-DD method consists of finding $u_{hb} = u_h + u_b \in V_{hb}^0$, with $u_h \in V_h^0$, $u_b \in V_b$, such that

$$B(u_{hb}, v_{hb}) + \sum_{T \in \mathcal{T}_h} D_T(u_{hb}; u_{hb}, v_{hb}) = (f, v_{hb}), \quad \forall v_{hb} \in V_{hb}^0, \tag{6}$$

in which $D_T(\cdot; \cdot, \cdot) : V_{hb}^0|_T \times V_{hb}^0|_T \times V_{hb}^0|_T \longrightarrow \mathbb{R}$ is the artificial diffusion operator given by

$$D_T(u_{hb}; u_{hb}, v_{hb}) = \int_T \xi_T\big(\kappa_h(u_{hb})\big) \boldsymbol{\nabla} u_{hb} \cdot \boldsymbol{\nabla} v_{hb} \, d\Omega, \qquad (7)$$

where $\kappa_h : V_{hb}^0 \longrightarrow V_h^0$ is a linear projection operator, such that $\kappa_h(w_{hb}) = w_h$ with $(id - \kappa_h)(w_{hb}) = w_b \in V_b$ and id is the identity operator in V_{hb}^0 [6]. The artificial diffusion at each element $T \in \mathcal{T}_h$ is given by

$$\xi_T(u_h) = \begin{cases} \mu(h_T)\dfrac{\|R(u_h)\|_{0,T}}{\eta + \tau} \,, & \text{if } Pe_T > 1 \,, \\[2ex] \quad \text{with } \eta = \|u_h\|_{1,T}, \text{ if } \sigma > 0; \quad \text{or} \quad \eta = \|\boldsymbol{\nabla} u_h\|_{0,T}, \text{ if } \sigma = 0 \,, \\[2ex] 0, \quad \text{otherwise} \,, \end{cases}$$
$$(8)$$

where $\tau > 0$ is an appropriate constant. In (8), $Pe_T = \|\boldsymbol{\beta}\|_{0,T} h_T / 2\epsilon$ is the local Péclet number, $\mu(h_T) = \mathcal{O}(h_T) > 0$ is the local subgrid characteristic length scale satisfying $\mu(h_T) \leq \mu(h) = \max\{\mu(h_T); \; T \in \mathcal{T}_h\}$, and $R(u_h) = -\epsilon \Delta u_h + \boldsymbol{\beta} \cdot \boldsymbol{\nabla} u_h + \sigma u_h - f$ is the residual of the resolved scale solution. The latter expression simplifies when $u_h|_T \in \mathbb{P}_1(T)$, which is the case considered here. Thus, henceforth we will use $R(u_h) = \boldsymbol{\beta} \cdot \boldsymbol{\nabla} u_h + \sigma u_h - f$. Note that the new-DD operator (7) introduces a nonlinearity into the formulation (6) since it depends on the residual of the resolved scale solution. The artificial diffusion (8), that depends on the $T \in \mathcal{T}_h$, regulates the amount of stabilization, being larger where the resolved solution is less accurate and vanishing otherwise. We remark that, in the latter case, the Galerkin method enriched with bubble functions is automatically recovered. It is also worth mentioning that the new-DD method can handle reaction-diffusion problems by replacing the condition $Pe_T > 1$ with $\|\boldsymbol{\nabla} u_h\|_{0,T} > 10^{-5}$, as proposed in [20].

We solve (6) by using the iterative process developed in [20] to improve convergence. The actual artificial diffusion in each $T \in \mathcal{T}_h$ in the iteration $k + 1$ is defined by combining values from two consecutive iterations in the form $\xi_T^{k+1} = \omega \hat{\xi}_T^{k+1} + (1 - \omega) \xi_T{}^k$, with $\omega = 0.5$ and $\hat{\xi}^0 = 0.0$. The quantity $\hat{\xi}_T^{k+1}$ is set equal to $\xi_T^{k+1}(u_h^k)$, which is evaluated using Eq. (8), with $\mu(h_T) = h_T = \sqrt{meas_2(T)}$, in which $meas_2(T)$ is the area of the element T. The convergence of the iterative process is reached when $\left(\left\| (u_h^{k+1} - u_h^k)/u_h^{k+1} \right\|_\infty < 10^{-6} \right)$, limited to a maximum number of 30 nonlinear iterations. Moreover, we also integrate a strategy that sets the damping factor ω equal to zero when $\|R(u_h^k)\|_{0,T}$ is close enough to $\|R(u_h^{k-1})\|_{0,T}$, with a threshold of 20% for the relative error between these residuals.

3   Numerical Results

We investigate the numerical properties of the new-DD method for 2D transport problems with smooth solutions and in the presence of layers, using both structured and unstructured mesh, and different values for τ. We numerically evaluate the convergence rates measured using different norms, and then we analyze advection-dominated and reaction-dominated transport problems whose solutions have regions with strong gradients. We analyze the quality of the solutions for three values of the parameter τ: a small constant value $\tau = 10^{-5}$, related to the experiments showed in [20]; $\tau = 1$, as proposed in [15], and a locally problem-dependent value defined as $\tau = \tau_T = \max\{\|f\|_{0,T}, 10^{-5}\}$.

3.1   Convergence Rates

First, we investigate convergence rates in the $L^2(\Omega)$, $H^1(\Omega)$, and the energy norms for transport regimes ranging from advection to diffusion dominated problems with or without reaction term, for first-order interpolating polynomials ($\mathbb{P}_1$). Simulations were performed on one structured grid (Grid1) that were uniformly refined, and on one unstructured grid with obtuse triangles (Grid2) generated and refined using Gmsh [5]. Mesh patterns are illustrated in Fig. 1. We consider the following smooth solution

$$u(x,y) = 100x^2(1-x)^2y(1-y)(1-2y) \tag{9}$$

where the source term f is defined to satisfy Eq. (1), with $\boldsymbol{\beta} = (3,2)^T$ and parameter combinations of $\sigma = 0$ or $\sigma = 1$, and $\epsilon = 10^{-6}$ or $\epsilon = 10$. Errors in $L^2(\Omega)$ ($\| \cdot \|_0$), $H^1(\Omega)$ ($\| \cdot \|_1$), and the energy ($\| \cdot \|_E$) norms (recalling that $\|v\|_E = \left(\epsilon|v|_1^2 + \sigma\|v\|_0^2\right)^{1/2}$) were evaluated for both grids and are reported in Table 1 using $\tau = 10^{-5}$, $\tau = \max\{\|f\|_{0,T}, 10^{-5}\}$ named as τ_T, and $\tau = 1$.

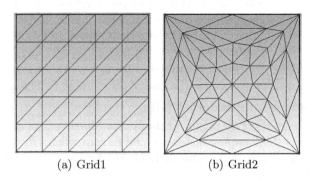

(a) Grid1   (b) Grid2

Fig. 1. Mesh patterns used in the numerical experiments.

Observe that optimal convergence rates in the $L^2(\Omega)$, $H^1(\Omega)$, and the energy norms are independently of the grid pattern, for all values of τ. Also, the order of convergence in the energy norm is similar to that in the $L^2(\Omega)$ norm when $\epsilon = 10^{-6}$ and $\sigma = 1$. This means that the zero-order term outweighs the first-order term, which does not happen in the cases with $\sigma = 0$. It is also possible to notice a decrease in the convergence rates obtained with Grid2. However, the errors associated with the unstructured Grid2 are slightly the smallest errors for all cases. In particular, we plotted the errors against mesh size on a log–log scale in Fig. 2, for $\epsilon = 10^{-6}$, $\sigma = 0$, and τ_T. Overall, the influence of τ on the convergence orders across the grids is small, and optimal orders of convergence is sustained. However, for $\tau = \tau_T$ and $\tau = 1$, we may note a slight decrease of the convergence rate for solutions obtained with Grid2 and $\sigma = 1$ (see Table 1). Such mesh-dependence on the method is still a topic under investigation.

Table 1. Convergence rates for the numerical solution of (9).

τ	σ	Grid	$\epsilon = 10^{-6}$			$\epsilon = 10$		
			$\|u - u_h\|_0$	$\|u - u_h\|_1$	$\|u - u_h\|_E$	$\|u - u_h\|_0$	$\|u - u_h\|_1$	$\|u - u_h\|_E$
10^{-5}	0	Grid1	1.99	1.04	1.04	1.99	0.99	0.99
		Grid2	1.95	1.09	1.09	1.77	0.93	0.93
	1	Grid1	2.00	1.03	1.99	1.99	0.99	0.99
		Grid2	1.96	1.09	1.95	1.77	0.93	0.93
τ_T	0	Grid1	1.94	1.01	1.01	1.99	0.99	0.99
		Grid2	1.68	0.97	0.97	1.77	0.93	0.93
	1	Grid1	1.95	1.01	1.94	1.99	0.99	0.99
		Grid2	1.70	0.97	1.70	1.77	0.93	0.93
1	0	Grid1	2.03	1.00	0.90	1.99	0.99	0.99
		Grid2	1.44	0.95	0.95	1.77	0.93	0.93
	1	Grid1	2.03	1.00	2.01	1.99	0.99	0.99
		Grid2	1.43	0.95	1.43	1.77	0.93	0.93

3.2   Solution with Two Internal Layers

We now consider an advection dominated advection-diffusion problem, first presented in [11], with new-DD artificial diffusion calculated using (8). The model is defined in the unit square domain $\Omega = (0,1)^2$, where $\boldsymbol{\beta} = (1,0)^T$, $\sigma = 0$, $\epsilon = 10^{-8}$, and $f(x,y) = 16(1 - 2x)$ if $(x,y) \in [0.25, 0.75]^2$ and zero otherwise. Homogeneous Dirichlet boundary conditions are considered on Γ. The exact solution depicted in Fig. 3(a) is equal to zero everywhere but in $(0.25, 0.75)^2$, where it is close to the parabolic function $u = (4x - 1)(3 - 4x)$ with interior layers at $(0.25, 0.75) \times \{0.25\}$ and $(0.25, 0.75) \times \{0.75\}$.

   The quality of the new-DD solutions are evaluated based on measures of spurious oscillations given by both undershoots and overshoots, as the ones shown by the SUPG (Streamline Upwind Petrov-Galerkin) solution in Fig. 3(b), for

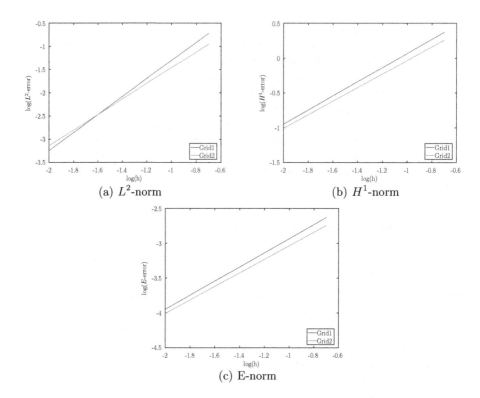

(a) L^2-norm

(b) H^1-norm

(c) E-norm

Fig. 2. Convergence rates in the $L^2(\Omega)$, $H^1(\Omega)$, and the energy norms for the problem with $\epsilon = 10^{-6}$, $\sigma = 0$, and using τ_T.

which the plane $z = 0$ was reduced to a line. Of note, the SUPG solution, implemented as in [2], was given for comparison. In Fig. 4, we compare the new-DD solutions for $\tau = 10^{-5}$ and $\tau = 1$, using a mesh with pattern Grid1 with 32×32 divisions in each direction. Both solutions are quite satisfactory, with the one obtained with $\tau = 1$ almost completely precluding spurious modes.

To observe the impact of τ in the solutions, we can accurately assess the amount of spurious oscillations in the discrete solution u_h by defining

$$min := - \min_{0.4 \le x \le 0.6} u_h(x,y), \quad diff := \max_{x \ge 0.8} u_h(x,y) - \min_{x \ge 0.8} u_h(x,y) , \quad (10)$$

where $y \in [0,1]$, and $\min u_h$ and $\max u_h$ are computed using values of u_h at the vertices of the mesh. Thus, as defined in [13], the min values measure the undershoots along the interior layers, while the $diff$ values compare the magnitude of the discrete solutions in a region downstream from the layers, where it should vanish ($x \ge 0.8$). In Table 2, we show min and $diff$ values for the new-DD method using different meshes (Grid1) with 16×16, 32×32 and 64×64 divisions, as well as different values for τ. Specifically, we set $\tau = 10^{-5}, 0.1, 1$ and $\tau = \tau_T = \max\{\|f\|_{0,T}, 10^{-5}\}$. The smallest $diff$ value was attained with

$\tau = 1$, except for the more refined mesh; in all cases, the *min* values were less than 2×10^{-5}. Independently of the τ value, the new-DD method could handle this highly complex problem and yielded quite satisfactory solutions. Of note, the solution obtained with τ_T is quite similar to that using $\tau = 10^{-5}$.

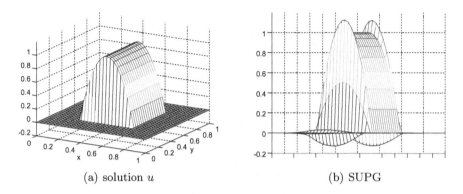

(a) solution u (b) SUPG

Fig. 3. Two internal layers problem: solution u and SUPG solution using a mesh with pattern Grid1 with 32×32 cells.

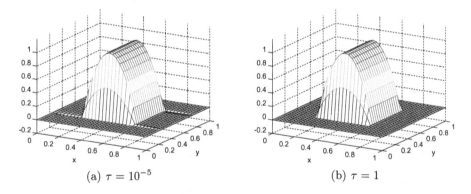

(a) $\tau = 10^{-5}$ (b) $\tau = 1$

Fig. 4. Two internal layers problem: new-DD solutions using a mesh with 32×32 divisions and $\tau = 10^{-5}$ (a) and $\tau = 1$ (b).

3.3   Reaction-Diffusion Problem

This is a reaction-diffusion problem, null velocity ($\beta = (0,0)$), characterized by dominance of reactive effects with two boundary layers. The problem is defined in the square region $\Omega = [0,1] \times [0,1]$, with $\epsilon = 10^{-6}$ and $\sigma = 1$, and source term given by

$$f(x,0) = \begin{cases} x, & for\ x \le -0.5; \\ 1-x, & for\ x > 0.5. \end{cases}$$

Table 2. The *min* and *diff* values obtained using three different meshes and different values for τ.

	τ	16×16	32×32	64×64
min	10^{-5}	8.03×10^{-11}	1.08×10^{-12}	2.18×10^{-10}
	τ_T	7.82×10^{-11}	1.02×10^{-12}	2.18×10^{-10}
	0.1	5.23×10^{-11}	1.01×10^{-9}	1.62×10^{-8}
	1	1.03×10^{-6}	4.28×10^{-6}	1.45×10^{-5}
diff	10^{-5}	1.42×10^{-2}	6.81×10^{-2}	8.51×10^{-2}
	τ_T	1.56×10^{-2}	6.23×10^{-2}	8.39×10^{-2}
	0.1	9.24×10^{-3}	3.46×10^{-3}	1.02×10^{-3}
	1	9.05×10^{-4}	1.62×10^{-4}	1.28×10^{-4}

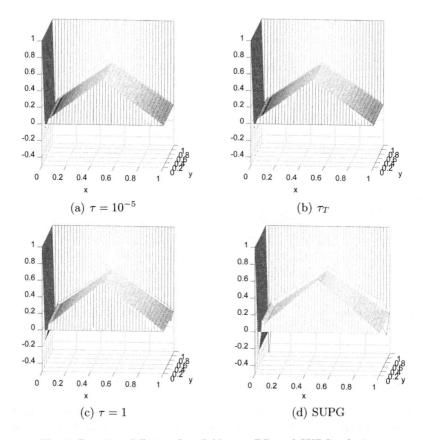

(a) $\tau = 10^{-5}$

(b) τ_T

(c) $\tau = 1$

(d) SUPG

Fig. 5. Reaction-diffusion flow field: new-DD and SUPG solutions.

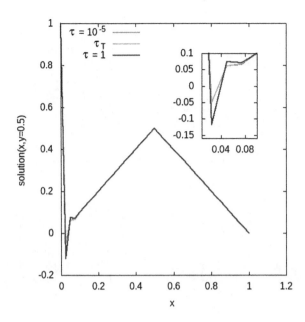

Fig. 6. Reaction-diffusion flow field: solution profiles for $0 \leq x \leq 1$ and $y = 0.5$.

Dirichlet boundary conditions are assigned to the boundary: $u(0, y) = 1$ for $0 \leq y \leq 1$, $u(x, 1) = 1$ for $0 \leq x \leq 1$ and $u(x, 0) = 0$ for $0 < x \leq 1$. Figures 5(a), 5(b) and 5(c) show the three-dimensional structure plots of new-DD solutions for a mesh (Grid1) with 40×40 division using $\tau = 10^{-5}$, $\tau = \tau_T = \max\{\|f\|_{0,T}, 10^{-5}\}$ and $\tau = 1$, respectively. Here, also for comparison, we show in Fig. 5(d) the SUPG solution as calculated in [2]. All new-DD solutions show very small smearing and almost the suppression of both overshoots and undershoots. The numerical dissipation for all τ values can be better compared in the solution profiles for $0 \leq x \leq 1$ and $y = 0.5$, see Fig. 6. Clearly, the new-DD solution with $\tau = 1$ presents the biggest amount of overshoots at the left boundary, while the other new-DD solutions produce qualitatively better solutions. As in the last example, the new-DD solutions with $\tau = 10^{-5}$ and τ_T are practically the same and they yield qualitatively excellent solutions.

3.4 2D Advection-Diffusion-Reaction Problem

Now, we examine an advection dominated problem that models a "rotating pulse" [3,10], also with known exact solution but reaction term not null ($\sigma = 2$), i.e., the artificial diffusion being defined in (8). This problem is particularly interesting since the width of the internal transition layer is very thin ($\mathcal{O}(\epsilon^{1/2})$) and its alignment with the mesh elements is not constant. The exact solution,

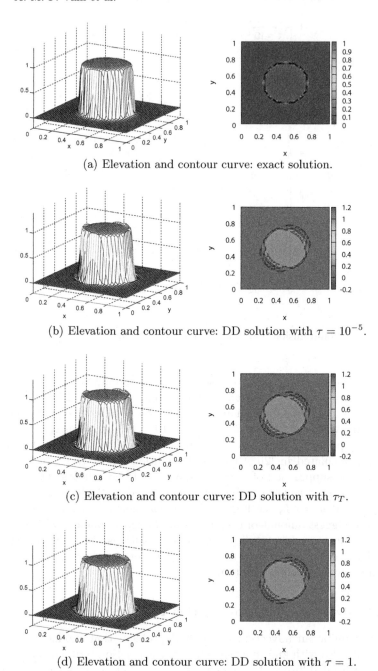

(a) Elevation and contour curve: exact solution.

(b) Elevation and contour curve: DD solution with $\tau = 10^{-5}$.

(c) Elevation and contour curve: DD solution with τ_T.

(d) Elevation and contour curve: DD solution with $\tau = 1$.

Fig. 7. Solutions of the "rotating pulse" problem for a mesh with 40×40 cells, $\tau = 10^{-5}$, τ_T and $\tau = 1$.

shown in Fig. 7(a), depicts a circular region where $u = 1$ is separated by a sharp transition from the rest of the domain, where $u = 0$. It is defined by

$$u = \frac{1}{2} + \frac{1}{\pi} \text{arctg} \left[1000(0.25^2 - (x - 0.5)^2 - (y - 0.5)^2) \right]. \tag{11}$$

The problem, in the unit square domain, uses $\epsilon = 10^{-4}$, $\sigma = 2$, and the components of the velocity field are

$$\begin{cases} \beta_x = -2(y - 1) \left(0.25^2 - (x - 0.5)^2 - (y - 0.5)^2 \right), \\ \beta_y = 2(x - 1) \left(0.25^2 - (x - 0.5)^2 - (y - 0.5)^2 \right), \end{cases}$$

if $0 \leq (x - 0.5)^2 - (y - 0.5)^2 \leq 0.25^2$, and $\beta_x = \beta_y = 0$ otherwise. The source term f and Dirichlet boundary conditions are chosen appropriately to satisfy (11). The new-DD solutions with $\tau = 10^{-5}$, $\tau = \tau_T = \max\{\|f\|_{0,T}, 10^{-5}\}$ and $\tau = 1$ are presented in Figs. 7(b)–7(c). In this example, observe that all new-DD solutions are qualitatively similar. They can represent the region of high gradient with moderate smearing and some small localized oscillations probably related to the mesh pattern. Such issues can be better observed confronting the level curves. Note that the impact of τ on the solutions is undetectable in this problem.

4   Conclusions

This work focus on the stability and convergence properties of the new-DD method developed and analyzed in [18], a consistent method that effectively adds extra stability through a nonlinear operator acting in all scales of the discretization. This variant of the Dynamic Diffusion (DD) method [20] is defined via a new design for the nonnegative, upper bounded nonlinear artificial diffusion operator. In [18], it was theoretically proved that the new-DD method has a convergence rate of $\mathcal{O}(h^{1/2})$ in the energy norm. Moreover, defining $\tau = 10^{-5}$, optimal convergence rates in the $L^2(\Omega)$, $H^1(\Omega)$, and energy norms were numerically shown for advection- and diffusion-dominated problems with smooth solutions. Of note, this artificial diffusion relies on a problem dependent parameter that plays a key role on its upper boundedness. Thus, focusing on those numerical aspects, the objective here is a preliminary evaluation of how the new diffusion operator impact the solution of 2D transport problems with smooth solutions and in the presence of layers.

Our numerical experiments have shown that the new-DD method presents robust stabilization and convergence properties, and the accuracy of the method is not very sensitive to the choice of parameter τ. Specifically, here we considered small constant values of τ, $\tau = 10^{-5}, 0.1, 1$, and also a locally defined problem-dependent value, $\tau = \tau_T = \max\{\|f\|_{0,T}, 10^{-5}\}$. The convergence rates were obtained using both structured (Grid1) and unstructured (Grid2) meshes, for the problem with smooth solution. We solved an advection dominated problem with two internal layers, a reaction-diffusion problem characterized by dominance of reactive effects with also two boundary layers, and an advection dominated problem that models a "rotating pulse", where the width of the internal transition layer is very thin and its alignment with the mesh elements is not constant. For all problems using the mesh pattern of Grid1, the new-DD solutions with $\tau = 10^{-5}$ and $\tau = \tau_T$ are practically the same and they yield qualitatively excellent solutions. In particular, we noted that the impact of τ on the new-DD solutions is undetectable in the "rotating pulse" problem. However, this is not the case for the others two problems with sharp layers, where the impact of τ is more significant. For instance, we have qualitatively better solutions using $\tau = 1$ in the two internal layers problem. However, the reaction-diffusion solution with $\tau = 1$ presents the biggest amount of overshoots at the left boundary. Overall, we consider that the new-DD method is robust for a wide scope of transport problems, showing potential for interesting future works. Possible extensions include the use of mesh adaptivity and high order approximations as well as extension for more involved problems, such as compressible and incompressible fluid flows.

Acknowledgments. This work was supported by the Foundation for Research Support of Espírito Santo (FAPES) under Grant 181/2017.

References

1. Arruda, N., Almeida, R., do Carmo, E.D.: Dynamic diffusion formulation for advection dominated transport problems. Mecánica Computacional **29**, 2011–2025 (2010)
2. Brooks, A., Hughes, T.: Streamline upwind/Petrov-Galerkin formulations for convection dominated flows with particular emphasis on the incompressible Navier-Stokes equations. Comput. Methods Appl. Mech. Eng. **32**, 199–259 (1982)
3. Cawood, M., Ervin, V., Layton, W., Maubach, J.: Adaptive defect correction methods for convection dominated, convection diffusion problems. J. Comput. Appl. Math. **116**(1), 1–21 (2000)
4. Galeão, A., do Carmo, E.D.: A consistent approximate upwind Petrov-Galerkin method for convection-dominated. Comput. Methods Appl. Mech. Eng. **10**, 83–95 (1988)
5. Geuzaine, C., Remacle, J.F.: GMSH: a 3-D finite element mesh generator with built-in pre- and post-processing facilities. Int. J. Numer. Meth. Eng. **79**(11), 1309–1331 (2009)
6. Guermond, J.L.: Stabilization of Galerkin approximations of transport equation by subgrid modeling. Math. Model. Num. Anal. **33**, 1293–1316 (1999)
7. Guermond, J.L.: Subgrid stabilization of Galerkin approximations of linear monotone operators. IMA J. Numer. Anal. **21**, 165–197 (2001)
8. Hughes, T.J.R., Scovazzi, G., Franca, L.P.: Multiscale and Stabilized Methods, pp. 1–64. American Cancer Society (2017). https://doi.org/10.1002/9781119176817.ecm2051
9. Hughes, T., Feijoo, G., Luca, M., Jean-Baptiste, Q.: The variational multiscale method - a paradigm for computational mechanics. Comput. Methods Appl. Mech. Eng. **166**, 3–24 (1998)
10. Iliescu, T.: Genuinely nonlinear models for convection-dominated problems. Comput. Math. Appl. **48**(10–11), 1677–1692 (2004)

11. John, V., Knobloch, P.: A computational comparison of methods diminishing spurious oscillations in finite element solutions of convection-diffusion equations. In: Proceedings of the International Conference Programs and Algorithms of Numerical Mathematics, vol. 13, pp. 122–136. Academy of Sciences of the Czech Republic (2006)

12. John, V., Knobloch, P.: On spurious oscillations at layers diminishing (SOLD) methods for convection-diffusion equations: part I - a review. Comput. Methods Appl. Mech. Eng. **196**(17–20), 2197–2215 (2007)

13. John, V., Knobloch, P.: On spurious oscillations at layers diminishing (SOLD) methods for convection-diffusion equations: part II - analysis for P1 and Q1 finite elements. Comput. Methods Appl. Mech. Eng. **197**(21–24), 1997–2014 (2008)

14. Johnson, C., Navert, U., Pitkäranta, J.: Finite element methods for linear hyperbolic problems. Comput. Methods Appl. Mech. Eng. **45**, 285–312 (1984)

15. Knopp, T., Lube, G., Rapin, G.: Stabilized finite element methods with shock capturing for advection-diffusion problems. Comput. Methods Appl. Mech. Eng. **191**(27), 2997–3013 (2002). https://doi.org/10.1016/S0045-7825(02)00222-0

16. Mallet, M.: A finite element method for computational fluid dynamics. Ph.D. thesis, Department of Civil Engineering, Stanford University (1985)

17. Santos, I.P., Almeida, R.C.: A nonlinear subgrid method for advection-diffusion problems. Comput. Methods Appl. Mech. Eng. **196**, 4771–4778 (2007)

18. Santos, I.P., Malta, S.M., Valli, A.M., Catabriga, L., Almeida, R.C.: Convergence analysis of a new dynamic diffusion method. Comput. Math. Appl. **98**, 1–9 (2021). https://doi.org/10.1016/j.camwa.2021.06.012

19. Valli, A., Catabriga, L., Santos, I., Coutinho, A., Almeida, R.: Multiscale dynamic diffusion method to solve advection-diffusion problems. In: XXXVI Ibero-Latin American Congress on Computational Methods in Engineering, Rio de Janeiro, RJ (2015)

20. Valli, A.M., Almeida, R.C., Santos, I.P., Catabriga, L., Malta, S.M., Coutinho, A.L.: A parameter-free dynamic diffusion method for advection-diffusion-reaction problems. Comput. Math. Appl. **75**(1), 307–321 (2018)

A Convergence Study of the 3D Dynamic Diffusion Method

Ramoni Z. S. Azevedo[1](✉) [iD], Lucia Catabriga[1,2](✉) [iD],
and Isaac P. Santos[1,3](✉) [iD]

[1] Laboratório de Otimização e Modelagem Computacional,
Programa de Pós-Graduação em Informática, Universidade Federal do Espírito Santo,
Vitória, ES, Brazil
ramoni.sedano@aluno.ufes.br
[2] Departamento de Informática, Universidade Federal do Espírito Santo,
Vitória, ES, Brazil
luciac@inf.ufes.br
[3] Departamento de Matemática Aplicada, Universidade Federal do Espírito Santo,
São Mateus, ES, Brazil
isaac.santos@ufes.br

Abstract. In this work we present a convergence study of the multi-scale Dynamic Diffusion (DD) method applied to the three-dimensional steady-state transport equation. We consider diffusion-convection and diffusion-convection-reaction problems, varying the diffusion coefficient in order to obtain an increasingly less diffusive problem. For both cases, the convergence order estimates are evaluated in the energy norm and the $L^2(\Omega)$ and $H^1(\Omega)$ Sobolev spaces norms. In order to investigate the meshes effects on the convergence, the numerical experiments were carried out on two different sets of meshes: one with structured meshes and the other with unstructured ones. The numerical results show optimal convergence rates in all norms for the dominant convection case.

Keywords: Dynamic diffusion method · Convergence estimate · 3D transport equation

1 Introduction

Let $\Omega \subset \mathbb{R}^3$ be a bounded domain with a Lipschitz boundary Γ with an outward unit normal $\boldsymbol{n}$. The steady-state diffusive–convective–reactive transport problem is described by

$$- \kappa \Delta u + \boldsymbol{\beta} \cdot \boldsymbol{\nabla} u + \sigma u = f \text{ in } \Omega, \tag{1}$$

$$u = u_D \text{ on } \Gamma, \tag{2}$$

where u represents the quantity being transported, $\kappa > 0$ is the diffusivity coefficient, $\boldsymbol{\beta} \in [L^\infty(\Omega)]^3$ is the velocity field such that $\boldsymbol{\nabla} \cdot \boldsymbol{\beta} = 0$, $\sigma \geq 0$ is the reaction coefficient, $f \in L^2(\Omega)$ is the source term and $u_D \in H^{1/2}(\Gamma)$.

© Springer Nature Switzerland AG 2021
O. Gervasi et al. (Eds.): ICCSA 2021, LNCS 12949, pp. 62–77, 2021.
https://doi.org/10.1007/978-3-030-86653-2_5

This equation plays an important role in many study fields such as dispersal of pollutants [1], chemical reaction [2], deterministic and continuous population models in ecology and economy [3], flow of large crowds of pedestrians [4], trickle-bed reactors that are employed in hydrotreatments and hydrocracking reactions in the petrochemical processing industries to the environmental detoxification of exhaust gases and polluted wastewaters from chemical plants [5], petroleum wellbore stability [6]. Due to the difficulty of solving analytically the mathematical models related to those applications, several numerical methods, such as finite differences, finite elements, finite volumes and boundary elements, have been developed to obtain approximate solutions of problem (1)–(2) [7].

In the context of finite element methods, the usual approach to cope with the instabilities caused by the convection (or reactive) term is to add extra operators to the standard Galerkin formulation with the aim of increasing the stability of the approximate solution. In general, these new formulations involves the Galerkin method coupled with two operators, one defining linear stabilized formulations and the other nonlinear, the so-called shock-capturing term. Some well-known linear stabilized formulations are the Streamline Upwind Petrov-Galerkin (SUPG) method [8], the Galerkin-Least Square (GLS) method [9], the Unusual Finite Element Method (USFEM) [10], the Continuous Interior Penalty (CIP) method [11] and the Local Projection Stabilization (LPS) [12]. Although all these linear methods present stable global solutions, non-negligible spurious oscillations are often present in the neighborhood of sharp layers. In order to recover the monotonicity properties of the continuous problem, precluding under/overshoots in the boundary layers, one can resort to the nonlinear stabilized methods. See [13,14] for an excellent review about classical nonlinear shock-capturing schemes.

The linear stabilized methods, such as SUPG, GLS, CIP and LPS, using linear interpolation, present convergence estimates with theoretical order of $\mathcal{O}(h^{3/2})$ and $\mathcal{O}(h)$ in the norms of the Sobolev spaces, $L^2(\Omega)$ and $H^1(\Omega)$, respectively, where h is the mesh size parameter [9,11,12,15]. The estimate in $H^1(\Omega)$ is optimal, whereas in $L^2(\Omega)$ is sub-optimal, leaving a gap of $\mathcal{O}(h^{1/2})$ in the order of convergence in the L^2-norm. According to [16], in general, shock-capturing schemes exhibit a priori error estimates of $\mathcal{O}(h^{1/2})$ for linear interpolants and using a mesh dependent norm, although some nonlinear methods are designed to present better convergence estimates, such as the methods presented in [17,18].

Another class of methods for solving convection dominated problems is based on the variational multiscale (VMS) framework, introduced by Hughes et al. in [19,20]. In particular, we mention the nonlinear two-scales formulations developed in [21–26]. These two-scales methods consists of adding locally and dynamically a nonlinear artificial diffusion either on the subgrid scale [24,25] or on both scales [21,26], in order to achieve stability of the numerical solution. This methodology has been applied for solving 2D problems, such as scalar transport equations, compressible and incompressible flow problems, obtaining good results [22,23,26]. In [27] the nonlinear two-scale method proposed in [21,26], called the Dynamic Diffusion (DD) method, was applied for solving 3D scalar transport equations with convection dominant.

The numerical analysis of the DD method shares the difficulties of classical nonlinear shock-capturing methods. In [28] is presented a numerical analysis of a variant of the DD method, obtaining a theoretical convergence estimate of order $\mathcal{O}(h^{1/2})$ in the energy norm, although the numerical experiments show optimal convergence in the $L^2(\Omega)$ and $H^1(\Omega)$ norms. In [26], a numerical study of the convergence estimates of the DD method, in the solution of two-dimensional problems, was carried out, obtaining optimal convergence rates on the $L^2(\Omega)$ and $H^1(\Omega)$ norms. This study is extended here for three-dimensional problems.

The aim of this work it to present a convergence study of the DD method when it is applied for solving 3D problems. We evaluated the convergence order estimates in the energy norm and the norms of the Sobolev spaces, $L^2(\Omega)$ and $H^1(\Omega)$, considering two transport equations for both, convection and diffusion dominant regimes. The numerical experiments was carried out on two different sets of meshes in order to investigate the mesh effects on the convergence order.

The remainder of this work is organized as follows. In Sect. 2 we present the numerical formulation of the DD method. In Sect. 3 the numerical experiments are conducted and the conclusions of the this paper are described in Sect. 4.

2   Dynamic Diffusion Method

The DD method is a nonlinear multiscale method that consists of adding a nonlinear artificial diffusion operator acting on both scales of the discretization [21,26]. The multiscale method is proposed by defining the enriched spaces S_E and V_E, spaces of admissible functions and test functions, as $S_E = S_h \oplus S_B$ and $V_E = V_h \oplus S_B$, where

$$S_h = \{u_h \in H^1(\Omega) : u_h|_{\Omega_e} \in \mathbb{P}_1(\Omega_e), \forall \Omega_e \in \mathcal{T}_h, u_h = g \in \Gamma\},$$
$$V_h = \{v_h \in H^1(\Omega) : v_h|_{\Omega_e} \in \mathbb{P}_1(\Omega_e), \forall \Omega_e \in \mathcal{T}_h, v_h = 0 \in \Gamma\}$$

are the discrete spaces of the Galekin method for linear tetrahedral elements (see [27] for more details), and S_B is the space spanned by bubble functions. By denoting $N_b \in H_0^1(\Omega^e)$ as the bubble function defined in each element Ω_e, we define $S_B|_{\Omega_e} = span(N_b)$ and $S_B = \oplus|_{\Omega_e} S_B|_{\Omega_e}$ for all Ω_e in $\mathcal{T}_h$. Here, we use $N_b = 256 N_1(x,y,z) N_2(x,y,z) N_3(x,y,z) N_4(x,y,z)$, where N_j is the local interpolation function associated with the nodal point j of the element Ω_e [29].

The DD method can be statement as: find $u = u_h + u_b \in S_E$, with $u_h \in S_h$ and $u_b \in S_B$, such that,

$$A(u,v) + A_{DD}(u_h; u, v) = F(v), \tag{3}$$

for all $v = v_h + v_b \in V_E$, with $v_h \in V_h$ and $v_b \in S_B$, where

$$A(u,v) = \int_{\Omega_e} \kappa \boldsymbol{\nabla} u \cdot \boldsymbol{\nabla} v + (\boldsymbol{\beta} \cdot \boldsymbol{\nabla} u) v + \sigma u v \, d\Omega_e,$$

$$A_{DD}(u_h; u, v) = \int_{\Omega_e} \tau_{DD}(u_h) \boldsymbol{\nabla} u \cdot \boldsymbol{\nabla} v \, d\Omega_e,$$

$$F(v) = \int_{\Omega_e} fv \, d\Omega_e.$$

In the nonlinear operator $A_{DD}(u_h; \cdot, \cdot)$, the function $\tau_{DD}(u_h)$ represents the amount of artificial diffusion and is defined as

$$\tau_{DD}(u_h) = \begin{cases} \mu(h) \frac{|R(u_h)|}{\|\boldsymbol{\nabla} u_h\|}, & \text{if } \|\boldsymbol{\nabla} u_h\| > 0, \\ 0, & \text{otherwise,} \end{cases} \tag{4}$$

where

$$R(u_h) = -\kappa \Delta u_h + \boldsymbol{\beta} \cdot \boldsymbol{\nabla} u_h + \sigma u_h - f \tag{5}$$

is the residue of the resolved scale of the equation, calculated in the barycenter of the element Ω_e. The symbol $|\cdot|$ represents the module of a real number and $\|\cdot\|$ is the Euclidean norm. The parameter $\mu(h)$ is given by

$$\mu(h) = \begin{cases} \frac{3}{4} \left(\sqrt[3]{V_e} + \frac{\|\boldsymbol{\beta}_B\|}{\|\boldsymbol{B}\|} \right), & \text{if } \Omega_e \cap \Gamma_+ \neq \emptyset, \\ \frac{\|\boldsymbol{\beta}_B\|}{2\|\boldsymbol{B}\|}, & \text{otherwise,} \end{cases}$$

with

$$\boldsymbol{\beta}_B = \frac{|R(u_h)|}{\|\boldsymbol{\nabla} u_h\|^2} \boldsymbol{\nabla} u_h \quad \text{and} \quad \boldsymbol{B} = \boldsymbol{\beta}_B \frac{\partial \boldsymbol{\xi}}{\partial \boldsymbol{x}},$$

where V_e is the volume of the element Ω_e, $\Gamma_+ = \{\boldsymbol{x} \in \Omega_e; \boldsymbol{\beta} \cdot \boldsymbol{n} > 0\}$ is the outflow part of Γ, $\boldsymbol{n}$ is the unit outward normal vector to the boundary Γ, $\boldsymbol{x} = (x, y, z) \in \Omega$ and $\boldsymbol{\xi} = (\xi, \eta, \zeta)$ is the vector whose components are the standard variables.

The method is solved by using an iterative procedure defined as: given u^n, we find $u^{n+1} \in S_E$ satisfying

$$A(u^{n+1}, v) + A_{DD}(u_h^n; u^{n+1}, v) = F(v), \quad \forall v \in V_E, \tag{6}$$

where the initial solution is $u^0 = 0$.

To improve convergence, we consider the weighting rule $\tau_{DD}(u_h^{n+1}) = w\tilde{\tau}_{DD}(u_h^{n+1}) + (1 - w)\tau_{DD}(u_h^n)$, with $w = 0.5$ and $\tilde{\tau}_{DD}(u_h^{n+1})$ calculated using the expression (4). The Eq. (6) results in the following local system of algebraic equations, associated with each element Ω_e,

$$\begin{bmatrix} A_{hh}^e & A_{bh}^e \\ A_{hb}^e & A_{bb}^e \end{bmatrix} \begin{bmatrix} U_h^e \\ U_b^e \end{bmatrix} = \begin{bmatrix} F_h^e \\ F_b^e \end{bmatrix}, \tag{7}$$

where the local matrices and vectors in Eq. (7) are defined as follow,

$$A_{hh}^e : \int_{\Omega_e} \left((\kappa + \tau_{DD}(u_h)) \boldsymbol{\nabla} u_h \cdot \boldsymbol{\nabla} v_h + (\boldsymbol{\beta} \cdot \boldsymbol{\nabla} u_h) v_h + \sigma u_h v_h \right) d\Omega_e;$$

$$A_{bh}^e : \int_{\Omega_e} \left((\boldsymbol{\beta} \cdot \boldsymbol{\nabla} u_b) v_h + \sigma u_b v_h \right) d\Omega_e; \qquad F_h^e : \int_{\Omega_e} fv_h \, d\Omega_e;$$

$$A^e_{hb} : \int_{\Omega_e} \left((\boldsymbol{\beta} \cdot \boldsymbol{\nabla} u_h) v_b + \sigma u_h v_b \right) d\Omega_e; \qquad F^e_b : \int_{\Omega_e} f v_b \, d\Omega_e;$$

$$A^e_{bb} : \int_{\Omega_e} \left((\kappa + \tau_{DD}(u_h)) \boldsymbol{\nabla} u_b \cdot \nabla v_b + \sigma u_b v_b \right) d\Omega_e;$$

and U^e_h and U^e_b are the vectors that store the variables $u_h|_{\Omega_e}$ and u_b, respectively. Performing a static condensation to eliminate the unknowns U^e_b at each element, the system (7) can be written in terms of the macro solution U^e_h, as follows,

$$\left(A^e_{hh} - A^e_{bh}(A^e_{bb})^{-1} A^e_{hb} \right) U^e_h = F^e_h - A^e_{bh}(A^e_{bb})^{-1} F^e_b. \tag{8}$$

Assembling all local systems (8), calculated on each element Ω_e, we obtain the global system,

$$[A_h(U^n_h)]U^{n+1}_h = F_h, \tag{9}$$

$$U^0_h = 0. \tag{10}$$

We solved the problem (9)–(10) using the nonlinear scheme described in Algorithm 1 that is based on the damping algorithm given in John and Knobloch [14]. The linear systems are solved by the Generalized Minimal Residual (GMRES) method [30].

When calculating an approximate solution, we are concerned with the accuracy of that solution obtained. The error e is defined as the difference between the exact solution and the approximate solution, $e(\boldsymbol{x}) = u_e(\boldsymbol{x}) - u_h(\boldsymbol{x})$. Normally, the a priori error estimate, measured in an appropriate norm $\|\cdot\|_*$, follows the form

$$\|e\|_* \leq Ch^p, \tag{11}$$

where h is mesh parameter, C is a constant that depends on the data of the problem and p represents the convergence rate with respect to the norm $\|\cdot\|_*$ [31]. The error estimates are calculated using the $L^2(\Omega)$, $H^1(\Omega)$ and energy norms, which are defined by

$$\|e\|^2_{L^2(\Omega)} = \int_\Omega |e|^2 d\Omega, \qquad \|e\|^2_{H^1(\Omega)} = \|e\|^2_{L^2(\Omega)} + \|\boldsymbol{\nabla} e\|^2_{L^2(\Omega)},$$

$$\|e\|^2_E = \sigma \|e\|^2_{L^2(\Omega)} + \kappa \|\boldsymbol{\nabla} e\|^2_{L^2(\Omega)}.$$

The H^1 and energy norms provide similar measures of the error under certain conditions, but as the H^1 norm does not depend on the problem, it is more convenient in some situations.

Using piece-wise linear basis function, the Galerkin finite element method presents, for $u_e \in H^2(\Omega)$, the following a priori error estimates: $\|e\|_{L^2(\Omega)} \leq Ch^2$ and $\|e\|_{H^1(\Omega)} \leq Ch$, whereas most linear stabilized methods present $\|e\|_{L^2(\Omega)} \leq Ch^{3/2}$ and $\|e\|_{H^1(\Omega)} \leq Ch$ [9,11,12,15].

Algorithm 1: Nonlinear algorithm

Data: $u_h^0 := 0$; $r^0 := \|f\|$; $w_{min} := 0.01$; $w_{max} := 1.0$;
$\quad\quad c_1 := 1.001$; $c_2 := 1.1$; $c_3 := 1.001$; $c_4 := 0.9$;
Result: Solution u_h of the system (9)-(10).

1 $w := w_{max}$; $k := 0$; $er := r^0$;
2 **while** $er > Tol$ **and** $k < MaxIter$ **do**
3 $\quad$ compute $\tilde{u}_h^{k+1}$ satisfying Eq. (9);
4 $\quad$ $firstDamp := 1$; $aux := 1$;
5 $\quad$ **while** $aux = 1$ **do**
6 $\quad\quad$ $u_h^{k+1} := u_h^k + w(\tilde{u}_h^{k+1} - u_h^k)$;
7 $\quad\quad$ compute residual r^{k+1} using Eq. (5);
8 $\quad\quad$ **if** $r^{k+1} < r^k$ **or** $w \le c_1 w_{min}$ **then**
9 $\quad\quad\quad$ **if** $r^{k+1} < r^k$ **and** $firstDamp = 1$ **then**
10 $\quad\quad\quad\quad$ $w_{max} := min(1, c_3 w_{max})$;
11 $\quad\quad\quad\quad$ $w := min(w_{max}, c_2 w)$;
12 $\quad\quad\quad$ **end**
13 $\quad\quad\quad$ $aux := 0$;
14 $\quad\quad$ **else**
15 $\quad\quad\quad$ $w := max(w_{min}, w/2)$;
16 $\quad\quad\quad$ **if** $firstDamp = 1$ **then**
17 $\quad\quad\quad\quad$ $w_{max} := max(w_{min}, c_4 w_{max})$;
18 $\quad\quad\quad\quad$ $firstDamp := 0$;
19 $\quad\quad\quad$ **end**
20 $\quad\quad$ **end**
21 $\quad$ **end**
22 $\quad$ compute absolute error $er = \|u^{k+1} - u^k\|$;
23 $\quad$ $k = k + 1$;
24 **end**

In order to evaluate numerically the convergence rate, p, given in (11), we define the function $\mathcal{E}(h) = Ch^p$ and plot the graph of $\log \mathcal{E} = p \log h + \log C$ in terms of $\log h$, resulting in a straight line whose slope is p. Thus, for h sufficiently small, a plot of $\log \|e\|_*$ versus $\log h$ will give a straight line with slope equal to p. The 3D integrals present in the definition of $\|\cdot\|_{L^2(\Omega)}$, $\|\cdot\|_{H^1(\Omega)}$ and $\|\cdot\|_E$ are solved numerically using Gaussian quadrature considering tetrahedral elements whose weights and points are described in [32].

3 Numerical Experiments

In this section, we obtain the convergence rates for a diffusion-convection problem and for a diffusion-convection-reaction problem in a unit cube, considering two set of meshes generated and refined using Gmsh [33]: GRID 1 (structured meshes), generated by dividing for 4, 8, 16, 32, 64 parts in each direction; and

GRID 2 (unstructured meshes), generated by starting with the value of the element size factor at 0.25 in the 8 points that generate the domain and dividing by 2 to obtain a new, more refined mesh. Table 1 shows the number of nodes and elements for all meshes.

Table 1. Size of the meshes using in the experiments.

(a) GRID 1			(b) GRID 2		
Meshes	Nodes	Elements	Meshes	Nodes	Elements
M1	125	384	M6	141	373
M2	729	3072	M7	685	2564
M3	4913	24576	M8	4014	18932
M4	35937	196608	M9	27458	148982
M5	274625	1572864	M10	200742	1169024

The mesh Péclet number, $Pe = \frac{\|\beta\| h}{2\kappa}$, expresses the relationship between convective and diffusive effects on a mesh with size h. It indicates if the numerical solution is convection or diffusion-dominated. When $Pe > 1$ or $\kappa < \frac{\|\beta\| h}{2}$ the numerical approximation is convection-dominated or the mesh is too coarse. In this case the Galerkin finite element method may present unphysical solution in problems with boundary and internal layers, requiring a stabilization scheme. When $Pe \leq 1$ the numerical solution is diffusion-dominated and the classical Galerkin finite element methods perform well.

Table 2. Mesh Péclet numbers in the meshes.

Meshes		Mesh Péclet numbers (Pe)		
		$\kappa = 1$	$\kappa = 10^{-3}$	$\kappa = 10^{-6}$
GRID 1	M1	1.19×10^{-1}	1.19×10^{2}	1.19×10^{5}
	M2	5.95×10^{-2}	5.95×10^{1}	5.95×10^{4}
	M3	2.97×10^{-2}	2.97×10^{1}	2.97×10^{4}
	M4	1.48×10^{-2}	1.48×10^{1}	1.48×10^{4}
	M5	7.44×10^{-3}	7.44×10^{0}	7.44×10^{3}
GRID 2	M6	1.67×10^{-1}	1.67×10^{2}	1.67×10^{5}
	M7	8.44×10^{-2}	8.44×10^{1}	8.44×10^{4}
	M8	4.64×10^{-2}	4.64×10^{1}	4.64×10^{4}
	M9	2.25×10^{-2}	2.25×10^{1}	2.25×10^{4}
	M10	1.10×10^{-2}	1.10×10^{1}	1.10×10^{4}

This error estimates study considers different values of Pe. The mesh size is defined as $h = \sqrt[3]{V}$, where $V = \max\{V_e\}$ is the volume of the largest element of the mesh. In both problems, $\boldsymbol{\beta} = (1,1,1)^T$. Therefore, the problems studied here vary the value of κ having a problem of dominant diffusion when $\kappa = 1$ and dominant convection when $\kappa = 10^{-3}$ and $\kappa = 10^{-6}$. Table 2 shows the Pe for the thickest and most refined mesh of each set of mesh types. As one can see, for dominant diffusion problem ($\kappa = 1$), $Pe < 1$ and for dominant convection problems ($\kappa = 10^{-3}$ and $\kappa = 10^{-6}$), $Pe > 1$ and the value of Pe increases when κ decreases.

Figure 1 presents the structured mesh M3 and the unstructured mesh M8. In Fig. 1a (structured mesh) the elements have the same size and are distributed in an organized way, whereas in Fig. 1b (unstructured mesh) the elements have similar sizes but are not organized.

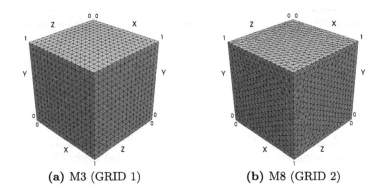

(a) M3 (GRID 1)   (b) M8 (GRID 2)

Fig. 1. Visualization of the two types of meshes used.

The computational code was developed in C language in a computer with the following characteristics: Intel Corel $i5 - 2450M$, CPU 2.50 GHz ×4 and 8 GB of RAM. The linear systems are solved with the GMRES method considering 50 Krylov vectors to restart, ILU(1) preconditioner, the maximum number of 100 cycles and a tolerance of 10^{-7}. For the nonlinear algorithm we use a tolerance of 10^{-3} and maximum number of iterations of 300. Due to the difficulty of visualizing solutions in three-dimensional domains, we present the solutions on the line passing through the cube diagonal, that is, a line that passes through points $(0,0,0)$ and $(1,1,1)$.

3.1 Diffusion-Convection (DC) Problem

Consider a DC problem in the unit cube $\Omega =]0,1[^3$ with a smooth exact solution given by $u(x,y,z) = sin(\pi x)sin(\pi y)sin(\pi z)$ and coefficients, $\boldsymbol{\beta} = (1,1,1)^T$, $\sigma = 0$, $\kappa \in \{1, 10^{-3}, 10^{-6}\}$, so that the source term and the Dirichlet boundary conditions satisfy the exact solution.

Figures 2 and 3 show the solutions obtained by the DD method on all meshes, for each $\kappa \in \{1, 10^{-3}, 10^{-6}\}$. In terms of convergence, the results obtained for the GRID 1, Fig. 2, and for the GRID 2, Fig. 3, are similar. Although the DD solutions on the coarsest meshes (M1 for GRID 1 and M6, M7 for GRID 2) are not quite accurate, when the meshes are refined, the solutions approach the exact solution, for all values of κ.

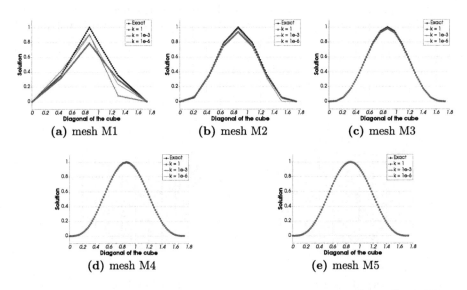

Fig. 2. Solution of DC problem at the cube diagonal – $\kappa = 1$, 10^{-3}, 10^{-6} – GRID 1

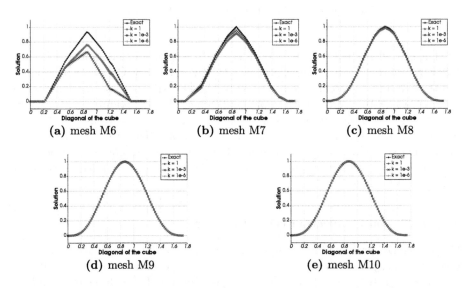

Fig. 3. Solution of DC problem at the cube diagonal – $\kappa = 1$, 10^{-3}, 10^{-6} – GRID 2.

Figures 4, 5 and 6 present the convergence rates in the $L^2(\Omega)$, $H^1(\Omega)$ and energy norms for GRID 1 and 2 considering $\kappa = 1$, 10^{-3} and 10^{-6}, respectively. In the convection-dominated case, shown in Fig. 5 ($\kappa = 10^{-3}$) and Fig. 6 ($\kappa = 10^{-6}$), optimal convergence rates are obtained for all set of meshes in both norms, $L^2(\Omega)$ and $H^1(\Omega)$. In the diffusion dominated case, Fig. 4 ($\kappa = 1$), also optimal convergence rates are obtained in the $H^1(\Omega)$ norm for both grids, whereas the $L^2(\Omega)$ norm presents a rate higher than the suboptimal one.

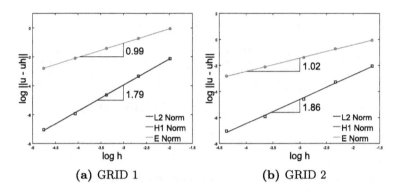

Fig. 4. Convergence rates considering the DC problem with $\kappa = 1$.

For the DC problem, $\sigma = 0$ and the energy norm satisfies $\|e\|_E = \sqrt{\kappa}\|\nabla e\|_{L^2(\Omega)} = \sqrt{\kappa}|e|_{H^1(\Omega)}$, where $|\cdot|_{H^1(\Omega)}$ is the usual semi-norm of the space $H^1(\Omega)$ and which presents the same convergence rate of the $H^1(\Omega)$ norm, for solutions $u \in H^1_0(\Omega)$, as shown in Figs. 5, 6 and 4.

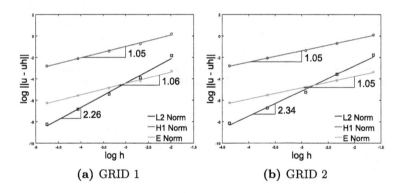

Fig. 5. Convergence rates considering the DC problem with $\kappa = 10^{-3}$.

When $\kappa = 1$, Fig. 4, the error in the $H^1(\Omega)$ norm is slightly greater than the error in energy norm, so that graphically, the two lines are superimposed and look like a single line. As κ decreases, the energy norm error also decreases, which causes its convergence rate line to move away from the $H^1(\Omega)$ norm rate line, but with the same slope (convergence rate), according to Figs. 5 and 6.

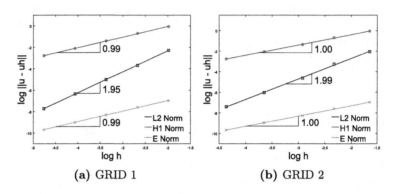

(a) GRID 1 (b) GRID 2

Fig. 6. Convergence rates considering the DC problem with $\kappa = 10^{-6}$.

3.2 Diffusion-Convection-Reaction (DCR) Problem

In this experiment, we consider a DCR problem, defined in $\Omega =]0,1[^3$, with a smooth exact solution, $u(x,y,z) = 100xyz(1-x)(1-y)(1-z)$ and coefficients, $\beta = (1,1,1)^T$, $\sigma = 1$, $\kappa \in \{1, 10^{-3}, 10^{-6}\}$, so that the source term and the Dirichlet boundary conditions satisfy the exact solution.

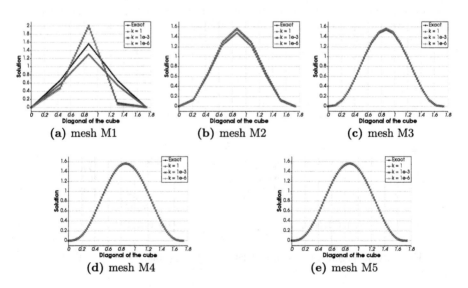

(a) mesh M1 (b) mesh M2 (c) mesh M3

(d) mesh M4 (e) mesh M5

Fig. 7. Solution of DCR problem at the cube diagonal – $\kappa = 1$, 10^{-3}, 10^{-6} – GRID 1.

Figures 7 and 8 present the DD solutions and the exact solution to the DCR problem considering $\kappa = 1$, 10^{-3}, 10^{-6}, respectively, for both set of meshes. As one can see for GRID 1 and GRID 2 the behavior of the DD solutions for DCR problem are very similar with those obtained for DC problem. In general, the DD solutions are closer to the exact solution when the mesh is refined.

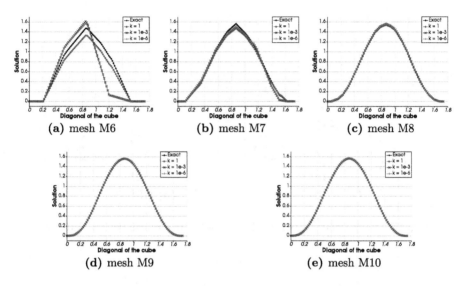

Fig. 8. Solution of DCR problem at the cube diagonal – $\kappa = 1$, 10^{-3}, 10^{-6} – GRID 2.

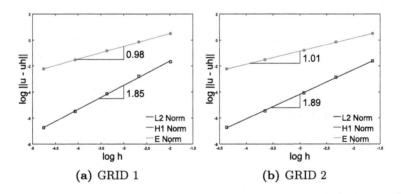

Fig. 9. Convergence rates considering the DCR problem with $\kappa = 1$.

Figures 9, 10 and 11 present the convergence rates in the $L^2(\Omega)$, $H^1(\Omega)$ and energy norms for GRID 1 and 2, respectively, considering $\kappa = 1$, 10^{-3}, 10^{-6} for the three dimensional DCR problem. The convergence rates present a behavior similar to the case of the CD problem: convection dominant problems obtained optimal rates for all set of meshes, whereas for diffusion dominated case, the $L^2(\Omega)$ norm presents a rate higher than the suboptimal one for both grids.

When $\kappa = 1$ the energy norm is precisely the $H^1(\Omega)$ norm, which can be seen in Fig. 9. When $\kappa = 10^{-6}$, the zero-order term outweighs the first-order term so that the energy norm is close to the $L^2(\Omega)$ norm, as shown in Fig. 11. When $\kappa = 10^{-3}$, the contribution of the term containing the gradient of the error affects the convergence rate. Thus, the rate presented by the energy norm is of order $\mathcal{O}(h^{1.5})$.

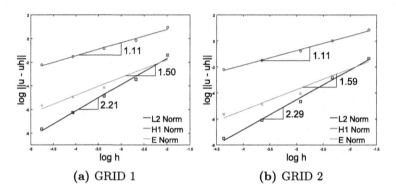

(a) GRID 1   (b) GRID 2

Fig. 10. Convergence rates considering the DCR problem with $\kappa = 10^{-3}$.

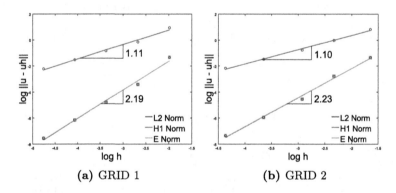

(a) GRID 1   (b) GRID 2

Fig. 11. Convergence rates considering the DCR problem with $\kappa = 10^{-6}$.

4   Conclusions

This work presents a convergence rate evaluation of the DD method applied to the three-dimensional stationary transport equation, considering diffusion-convection and diffusion-convection-reaction problems. Experiments were carried out when diffusion was dominant, $\kappa = 1$, and when convection was dominant, $\kappa = 10^{-3}$ and $\kappa = 10^{-6}$, considering two set of meshes: one structured and another unstructured.

In both problems we observed that when the convection is dominant, the method presented an optimal convergence in $L^2(\Omega)$ norm, i.e., $\mathcal{O}(h^2)$ and in $H^1(\Omega)$ norm, i.e., $\mathcal{O}(h^1)$. However, when the diffusion is dominant, the method presented a rate higher than the suboptimal one in $L^2(\Omega)$ norm, i.e., approximately $\mathcal{O}(h^{1.8})$ and an optimal convergence in $H^1(\Omega)$ norm, i.e., $\mathcal{O}(h^1)$.

The energy norm presented the behavior that was expected. In the DC problem, it presented a convergence rate equal to the rate of the $H^1(\Omega)$ norm, $\mathcal{O}(h^1)$, whereas for the DCR problem, the convergence rate varies according to the κ variation, agreeing with the $H^1(\Omega)$ and $L^2(\Omega)$ result for $\kappa = 1$ and $\kappa = 10^{-6}$. When $\kappa = 10^{-3}$, it presented a convergence rate in the order $\mathcal{O}(h^{1.5})$.

As a future work, we will apply this methodology to solve three-dimensional transient transport problems and also to investigate the convergence rates in suitable norms.

Acknowledgments. This work was supported by the Espírito Santo State Research Support Foundation (FAPES), under Grant Term 181/2017, and by the Coordenação de Aperfeiçoamento de Pessoal de Nível Superior - Brasil (CAPES) - Finance Code 001.

References

1. Friedman, A., Littman, W.: Industrial mathematics: a course in solving real-world problems. Society for Industrial and Applied Mathematics, pp. 27–45 (1994)
2. Kuttler, C.: Reaction-Diffusion equations with applications. In: Internet siminar, Technische Unversität München, pp. 3–11 (2011)
3. Jüngel, A.: Diffusive and nondiffusive population models. In: Naldi, G., Pareschi, L., Toscani, G. (eds) Mathematical Modeling of Collective Behavior in Socio-Economic and Life Sciences. Modeling and Simulation in Science, Engineering and Technology, pp 397–425. Birkhäuser Boston (2010) https://doi.org/10.1007/978-0-8176-4946-3_15
4. Colombo, R.M., Garavello, M., Lécureux-Mercier, M.: A class of nonlocal models for pedestrian traffic. Math. Models Methods Appl. Sci. **22**(04), 1150023 (2012). https://doi.org/10.1142/S0218202511500230
5. Lopes, R.J.G., Quinta-Ferreira, R.M.: Numerical assessment of diffusion-convection-reaction model for the catalytic abatement of phenolic wastewaters in packed-bed reactors under trickling flow conditions. Comput. Chem. Eng. **35**(12), 2706–2715 (2011)
6. Yin, S., Towler, B.F., Dusseault, M.B., Rothenburg, L.: Fully coupled THMC modeling of wellbore stability with thermal and solute convection considered. Transp. Porous Media **84**(3), 773–798 (2010). https://doi.org/10.1007/s11242-010-9540-9
7. Quarteroni, A.: Numerical Models for Differential Problems, vol. 16. Springer, Cham (2017). https://doi.org/10.1007/978-3-319-49316-9_21
8. Brooks, A.N., Hughes, T.J.R.: Streamline upwind Petrov-Galerkin formulations for convection dominated flows with particular emphasis on the incompressible Navier-Stokes equations. Comput. Methods Appl. Mech. Eng. **32**(1–3), 199–259 (1982)

9. Hughes, T.J.R., Franca, L.P., Hulbert, G.M.: A new finite element formulation for computational fluid dynamics: VIII. The Galerkin/Least-Squares method for advective-diffusive equations. Comput. Methods Appl. Mech. Eng. **73**(2), 173–189 (1989)
10. Franca, L., Valentin, F.: On an improved unusual stabilized finite element method for the advective-reactive-diffusive equation. Comput. Methods Appl. Mech. Eng. **190**(13), 1785–1800 (2000). https://doi.org/10.1016/S0045-7825(00)00190-0
11. Burman, E., Hansbo, P.: Edge stabilization for Galerkin approximations of convection-diffusion-reaction problems. Comput. Methods Appl. Mech. Eng. **193**(15), 1437–1453 (2004)
12. Knobloch, P., Lube, G.: Local projection stabilization for advection-diffusion-reaction problems: one-level vs. two-level approach. Appl. Numer. Math. **59**(12), 2891–2907 (2009)
13. John, V., Knobloch, P.: On spurious oscillations at layers diminishing (SOLD) methods for convection-diffusion equations: part I-a review. Comput. Methods Appl. Mech. Eng. **196**(17), 2197–2215 (2007)
14. John, V., Knobloch, P.: On spurious oscillations at layers diminishing (SOLD) methods for convection-diffusion equations: part II - analysis for P1 and Q1 finite elements. Comput. Methods Appl. Mech. Eng. **197**(21–24), 1997–2014 (2008)
15. Johnson, C., Nävert, U., Pitkäranta, J.: Finite element methods for linear hyperbolic problems. Comput. Methods Appl. Mech. Eng. **45**(1), 285–312 (1984). https://doi.org/10.1016/0045-7825(84)90158-0
16. Allendes, A., Barrenechea, G., Rankin, R.: Fully computable error estimation of a nonlinear, positivity-preserving discretization of the convection-diffusion-reaction equation. SIAM J. Sci. Comput. **39**(5), A1903–A1927 (2017). https://doi.org/10.1137/16M1092763
17. Barrenechea, G.R., John, V., Knobloch, P.: A local projection stabilization finite element method with nonlinear crosswind diffusion for convection-diffusion-reaction equations. ESAIM Math. Model. Numer. Anal. **47**(5), 1335–1366 (2013)
18. Barrenechea, G.R., Burman, E., Karakatsani, F.: Blending low-order stabilised finite element methods: A positivity-preserving local projection method for the convection-diffusion equation. Comput. Methods Appl. Mech. Eng. **317**, 1169–1193 (2017)
19. Hughes, T.J.R.: Multiscale phenomena: green's functions, the Dirichlet-to-Neumann formulation, subgrid scale models, bubbles and the origins of stabilized methods. Comput. Methods Appl. Mech. Eng. **127**(1–4), 387–401 (1995)
20. Hughes, T.J., Feijóo, G.R., Mazzei, L., Quincy, J.B.: The variational multiscale method - a paradigm for computational mechanics. Comput. Methods Appl. Mech. Eng. **166**(1–2), 3–24 (1998)
21. Arruda, N.C.B., Almeida, R.C., Dutra do Carmo, E.G.: Dynamic diffusion formulations for advection dominated transport problems. In: Dvorkin, E., Goldschmit, M., Storti, M. (eds.) Mecánica Computacional, vol. 29, pp. 2011–2025. Asociación Argentina de Mecánica Computacional, Buenos Aires (2010)
22. Baptista, R., Bento, S.S., Santos, I.P., Lima, L.M., Valli, A.M.P., Catabriga, L.: A multiscale finite element formulation for the incompressible Navier-Stokes equations. In: Gervasi, O., et al. (eds.) ICCSA 2018. LNCS, vol. 10961, pp. 253–267. Springer, Cham (2018). https://doi.org/10.1007/978-3-319-95165-2_18
23. Bento, S.S., de Lima, L.M., Sedano, R.Z., Catabriga, L., Santos, I.P.: A nonlinear multiscale viscosity method to solve compressible flow problems. In: Gervasi, O., et al. (eds.) ICCSA 2016. LNCS, vol. 9786, pp. 3–17. Springer, Cham (2016). https://doi.org/10.1007/978-3-319-42085-1_1

24. Santos, I.P., Almeida, R.C.: A nonlinear subgrid method for advection-diffusion problems. Comput. Methods Appl. Mech. Eng. **196**(45–48), 4771–4778 (2007)
25. Santos, I., Almeida, R., Malta, S.: Numerical analysis of the nonlinear subgrid scale method. Comput. Appl. Math. **31**(3), 473–503 (2012)
26. Valli, A.M.P., Almeida, R.C., Santos, I.P., Catabriga, L., Malta, S.M.C., Coutinho, A.L.G.A.: A parameter-free dynamic diffusion method for advection-diffusion-reaction problems. Comput. Math. Appl. **75**(1), 307–321 (2018)
27. Azevedo, R.Z.S., Santos, I.P.: Multiscale finite element formulation for the 3D diffusion-convection equation. In: Gervasi, O., et al. (eds.) ICCSA 2020. LNCS, vol. 12251, pp. 455–469. Springer, Cham (2020). https://doi.org/10.1007/978-3-030-58808-3_33
28. Santos, I.P., Malta, S.M., Valli, A.M., Catabriga, L., Almeida, R.C.: Convergence analysis of a new dynamic diffusion method (2021, submitted)
29. Ern, A., Guermond, J.-L.: Theory and Practice of Finite Elements. vol. 159. 1st edn. Springer, New York (2004). https://doi.org/10.1007/978-1-4757-4355-5
30. Saad, Y., Schultz, H.: GMRES: a generalized minimal residual algorithm for solving nonsymmetric linear systems. SIAM. J. Sci. Stat. Comput. **7**(3), 856–869 (1986)
31. Becker, E.B., Carey, G.F., Oden, J.T.: Finite Elements - An Introduction, vol. 1. Prentice-Hall, New Jersey (1981)
32. Zienkiewicz, O.C., Taylor, R.L.: The Finite Element Method - Volume 1: The Basis, 5th edn. Butterworth-Heinemann, Massachusetts (2000)
33. Geuzaine, C., Remacle, J.F.: Gmsh: a three-dimensional finite element mesh generator with built-in pre- and post-processing facilities. Int. J. Num. Methods Eng. **79**(11), 1309–1331 (2009). https://gmsh.info

Comparison Between Protein-Protein Interaction Networks CD4$^+$T and CD8$^+$T and a Numerical Approach for Fractional HIV Infection of CD4$^+$T Cells

Eslam Farsimadan$^{(\boxtimes)}$, Leila Moradi , Dajana Conte ,
Beatrice Paternoster , and Francesco Palmieri

University of Salerno, 84084 Fisciano, Italy
{efarsimadan,lmoradi,dajconte,beapat,fpalmieri}@unisa.it

Abstract. This research examines and compares the construction of protein-protein interaction (PPI) networks of CD4$^+$ and CD8$^+$T cells and investigates why studying these cells is critical after HIV infection. This study also examines a mathematical model of fractional HIV infection of CD4$^+$T cells and proposes a new numerical procedure for this model that focuses on a recent kind of orthogonal polynomials called discrete Chebyshev polynomials. The proposed scheme consists of reducing the problem by extending the approximated solutions and by using unknown coefficients to nonlinear algebraic equations. For calculating unknown coefficients, fractional operational matrices for orthogonal polynomials are obtained. Finally, there is an example to show the effectiveness of the recommended method. All calculations were performed using the Maple 17 computer code.

Keywords: Protein-protein interaction network · HIV infection of CD4$^+$T · Discrete Chebyshev polynomials · Fractional calculus · Absolute errors

1 Introduction

A lentivirus, the human immunodeficiency virus (HIV), causes acquired immunodeficiency syndrome (AIDS). HIV infection is distinguished by alterations in the function of T cells and homeostasis and the extreme heterogeneity between infected people and those untreated. On average, most patients infected with HIV develop AIDS in 10 to 20 years. Variations in HIV infection clinical outcomes may be due to genetic differences in HIV strains, host genetic differences, or differences in virus-specific inflammatory responses. The first HIV case was confirmed in 1980. Due to the latest count, more than 35 million people have died as a result of HIV, and over 37 million people have this virus in their bodies, posing a threat to the rest of the world. They will also convey this danger by mother-to-child transfer, unsafe sex, and other ways.

© Springer Nature Switzerland AG 2021
O. Gervasi et al. (Eds.): ICCSA 2021, LNCS 12949, pp. 78–94, 2021.
https://doi.org/10.1007/978-3-030-86653-2_6

There have already been several mathematical models developed to research the within-host dynamics of HIV infection [4,19,30]. Virus-to-cell infection was the focus of the majority of these models. Direct cell-to-cell transmitting is also a possibility for the virus to spread. An ODE model of HIV virus spread in a well-mixed compartment like the bloodstream, was proposed by Perelson [23,24]. Within the field of mathematical modeling of HIV infection, this model has had a noteworthy effect. Other several approaches have been suggested based on Perelson model [1,3,8,10,11,13,16,21,26,29,31]. However, most HIV infection modeling research has focused on integer-order ordinary differential equations [5,14,22,32].

Fractional calculus (FC) has recently been widely used in a variety of fields. Plenty of applied scientists and mathematicians have attempted to use fractional calculus to model real-world processes. Fractional kinetics in complex systems, reported in [17]. The dynamics of the fractional-order in botanical impedances were studied in [12]. A mathematical fractional-order model of human root dentin was presented in [25].

In biology, it has been determined that biological organisms' cell membranes have fractional-order electrical conductance, which is then categorized into non-integer order models. Fractional derivatives represent basic characteristics of cell-rheological conduct and have had the best achievement in the area of rheology [7]. Furthermore, it has been demonstrated that the behavior modeling by fractional ordinary differential equations (FODE) for the vestibule oculomotor neurons has more benefits than the classical integer-order modeling [2]. FODE is intrinsically linked to systems found in all biological systems. They are also associated with fractals, which are normally found in biological systems. In this study, we propose a FODE system for modeling HIV and a numerical method for solving it.

The following is the paper's structure. The primary objective of selecting FODE for modeling HIV infection of CD4$^+$T cells is described in Sect. 2. In fact, in this section, we present and compare the protein-protein interaction network of cell infection. In Sect. 3, the mathematical model will be outlined. The formulation of the DCPs and their properties are covered in Sect. 4. The numerical solution of HIV infection of CD4$^+$T cells is discussed in Sect. 5. Section 6, describes illustrative examples that show the DCP's superiority. Finally, in Sect. 7, the main concluding remark is summarized.

2   Comparison Protein-Protein Interaction Networks CD4$^+$T and CD8$^+$T

CD4$^+$T lymphocytes are a kind of white blood cell as well as a lymphocyte. A helper T cell is also known as a T cell that assists other cells. CD4$^+$T cells primarily serve as a type of T cell that helps other T cells resist virus infection, while CD8$^+$ cells are widely distributed on the surface of suppressor and cytotoxic T lymphocytes during HIV infection. The researchers concluded that HIV infection caused severe immune system problems, including the loss of CD4$^+$T cells and a reduction in the $\frac{CD4^+}{CD8^+}$T cell ratio.

In the peripheral group of healthy adults, the $\frac{CD4^+}{CD8^+}$T ratio is about 2:1, and an abnormal ratio may signify disorders related to autoimmunity or immunodeficiency. An inverted $\frac{CD4^+}{CD8^+}$T ratio (i.e., less than $\frac{1}{1}$) means that the immune system is compromised. As a result, it is critical to look into the differences between CD4$^+$ and CD8$^+$T cells at various stages of HIV infection.

AIDS is caused by the pathogen HIV, which is well-known. The researchers reasoned that analyzing the differences in mutual differentially expressed genes between CD4$^+$ and CD8$^+$Tcells at different phases of disease would reveal more about HIV.

Few studies examine the systemic features of CD4$^+$T cells at various levels of HIV infection or the differences between CD4$^+$ and CD8$^+$T cells at the same level. And this topic is beyond the scope of this article's discussion. In this study, we simply compare and contrast the construction of protein-protein interaction (PPI) networks in CD4$^+$ and CD8$^+$T cells after HIV infection. We investigate the overlapping differentially expressed genes in CD4$^+$ cells and those in CD8$^+$T cells in two PPI networks and compare two PPI networks Fig. 1, to comprehend the differentially expressed between CD4$^+$ and CD8$^+$ from a network perspective.

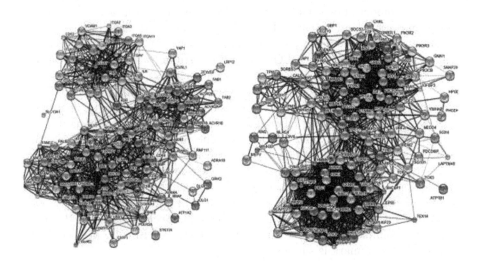

Fig. 1. PPI network constructed in CD8$^+$T cells (right) and CD4$^+$T cells (left)

The amount of overlapping of different expressed genes in each cell type is extremely high, and the two PPI networks shown in Fig. 1 are far too convoluted to provide significant network information. As a result, we compare the functional modules from each PPI network in separate networks to assess the differences between two HIV-infected cells, respectively (Figs. 2, 3). The immune responses of CD4$^+$ and CD8$^+$T cells at various stages after HIV infection are diverse, as shown in Figs. 2, 3.According to research, specific CD8$^+$T cells play a

key role in directly combating HIV infection, while CD4⁺T cells primarily serve
to support CD8⁺T cells.

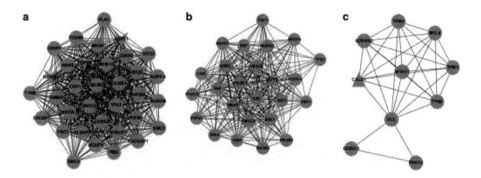

Fig. 2. (a-c) Function modules obtained by the PPI network in CD8⁺T cells.

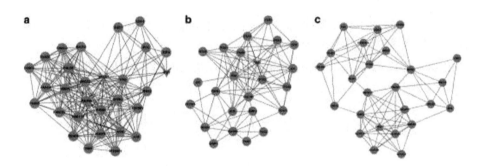

Fig. 3. (a-c) Functional modules obtained by the PPI network in CD4⁺T cells.

It has been validated by comparing networks and prior research that CD4⁺T
cells undergo gradual depletion after HIV infection and that virus reproduction
(more than 99 %) happened primarily in CD4⁺T cells in the peripheral blood and
lymphoid tissue. HIV infection can also stop CD4⁺T cells from proliferating. As
a consequence, HIV is a retrovirus that primarily infects CD4⁺T cells. CD4⁺T
cells will develop new virions after being infected, leading to more cell infection
and viral development (See Fig. 4). Therefore, for researchers, studying these
cells is critical. We want to introduce a mathematical model of fractional HIV
infection of CD4⁺T cells and find a numerical solution for it in this work.

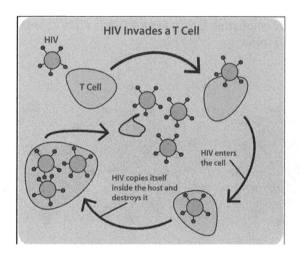

Fig. 4. HIV virus invades Tcell.

3   Mathematical Model Information

In the following, definitions of the Caputo and Riemann-Liouville fractional integral and derivative are discussed [20, 28].

Definition 1. *Let $\alpha > 0$. The Riemann-Liouville fractional integration of order α is defined as*

$$
(\mathbf{I}^\alpha \mathrm{f})\,(t) = \begin{cases} \dfrac{1}{\Gamma(\alpha)} \displaystyle\int_0^t \dfrac{\mathrm{f}(z)dz}{(t-z)^{1-\alpha}}, & \alpha > 0, \\[2mm] \mathrm{f}(t), & \alpha = 0. \end{cases}
$$

Definition 2. *Let $\alpha > 0$ and n is an integer. The Caputo fractional derivative of order α, is described as*

$$
\mathbf{D}^\alpha \mathrm{f}(t) = \begin{cases} \dfrac{1}{\Gamma(n-\alpha)} \displaystyle\int_0^t \dfrac{\mathrm{f}^{(n)}(z)}{(t-z)^{\alpha-n+1}}dz, & t>0, \quad 0 \le n-1 < \alpha < n, \\[2mm] \dfrac{d^n \mathrm{f}(t)}{dt^n}, & \alpha = n \in \mathbb{N}. \end{cases}
$$

Now we present the HIV infection model in CD4$^+$T cells with fractional-order. The following FODE is used to characterize the new system:

$$
\begin{aligned}
\mathbf{D}^\alpha T &= q - \eta T + rT(1 - \tfrac{T+1}{T_{\max}}) - kVT, \\
\mathbf{D}^\alpha I &= kVT - \beta I, \\
\mathbf{D}^\alpha V &= \mu\beta I - \gamma V,
\end{aligned} \tag{1}
$$

where $T(0) = r_1$, $I(0) = r_2$, $V(0) = r_3$, $0 \le t \le R < \infty$.

Where $V(t)$, $I(t)$, and $T(t)$ represent free HIV particles within the blood, CD4$^+$T cells contaminated by HIV, and the concentration of susceptible CD4$^+$T cells, respectively. R is any positive constant. The terms η, γ, and β indicate the normal circulation rates of non-infected T cells, virus particles, and infected T cells, respectively. $1 - \frac{T+1}{T_{\max}}$ depicts the logistic growth of healthy CD4$^+$T cells while ignoring the proliferation of infected CD4$^+$T cells. The term KVT reflects the occurrence of HIV infection of healthy CD4$^+$T cells when $k > 0$ is the infection rate. During its lifetime of CD4$^+$T cell, each infected CD4$^+$T cell is expected to generate l virus particles. The body is attempted to make CD4$^+$T cells at a constant rate q from precursors in the bone marrow and thymus. Whenever T cells are stimulated by antigen or mitogen, they multiply at a rate of r via mitosis. The maximum CD4$^+$T cell concentration in the body is denoted by $T_{\max}$.

4   Preliminary Remarks

4.1   Discrete Chebyshev Polynomials

We will focus our attention in this section on the fundamental definition, formulation, and characteristics of a large family of orthogonal polynomials namely discrete Chebyshev polynomials introduced by P.L. Chebyshev [9,18].

Definition 3. *The discrete Chebyshev polynomials $\mathcal{C}_{n,N}(x)$ are defined as*

$$\mathcal{C}_{n,N}(x) = \sum_{k=0}^{n} (-1)^k \binom{n+k}{n} \binom{N-k}{n-k} \binom{x}{k}, \quad n = 0, 1, ..., N, \qquad (2)$$

where N be a positive integer number.

The discrete Chebyshev polynomials set $\{\mathcal{C}_{n,N}, n = 0, 1, ..., N\}$ are orthogonal on $[0, N]$ according to the following discrete norm:

$$\langle \mathrm{f}, \mathrm{g} \rangle = \sum_{r=0}^{N} \mathrm{f}(r)\mathrm{g}(r). \qquad (3)$$

These polynomials have the following orthogonality property:

$$\langle \mathcal{C}_{m,N}, \mathcal{C}_{n,N} \rangle = \sum_{r=0}^{N} \mathcal{C}_{m,N}(r)\mathcal{C}_{n,N}(r) = \xi_n \delta_{mn},$$

where δ_{mn} is the Kronecker delta and ξ_n is introduced by the following relation:

$$\xi_n = \frac{(N+n+1)!}{(2n+1)(N-n)!(n!)^2}. \qquad (4)$$

By using (2), the analytical form of $\mathcal{C}_{n,N}(t)$ can be obtained as:

$$\mathcal{C}_{n,N}(t) = \sum_{k=0}^{n} \mathsf{a}_{j,n}x^k, \quad n = 0, 1, ...N, \qquad (5)$$

where

$$a_{j,n} = \sum_{r=j}^{n} \frac{(-1)^r \binom{n+r}{n} \binom{N-r}{n-r} s(r,i)}{r!}. \tag{6}$$

and $s(r,i)$ is a Stirling numbers of the first kind [27].

4.2   Shifted Discrete Chebyshev Polynomials

We construct the shifted discrete Chebyshev polynomials (SDCPs) by changing the variable $x = Nt$ for using the discrete Chebyshev polynomials over $[0,1]$. Let nth SDCPs (i.e. $\mathcal{C}_{n,N}(Nt)$) to be marked by $\mathcal{S}_{n,N}(t)$. The set of SDCPs $\{\mathcal{S}_{n,N}, n = 0,1,...,N\}$ are then orthogonal on $[0,1]$ according to the discrete norm as follow:

$$\langle f, g \rangle_* = \sum_{r=0}^{N} f\left(\frac{r}{N}\right) g\left(\frac{r}{N}\right). \tag{7}$$

The property of orthogonality for SDCPs is described by:

$$\langle \mathcal{S}_{m,N}(t), \mathcal{S}_{n,N}(t) \rangle_* = \sum_{r=0}^{N} \mathcal{S}_{m,N}\left(\frac{r}{N}\right) \mathcal{S}_{n,N}\left(\frac{r}{N}\right) = \xi_n \delta_{mn}, \tag{8}$$

where ξ_n was introduced in (4).

Let $f(t)$ defined on $[0,1]$. $f(t)$ can be expanded by the SDCPs as follows:

$$f(t) \simeq \sum_{i=0}^{N} c_i \mathcal{S}_{i,N}(t) = C^T \Psi(t), \tag{9}$$

where C and $\Psi(t)$ are $(N+1)$ vectores given by

$$C^T = [c_0, c_1, ..., c_N]^T, \tag{10}$$

$$\Psi(t) = [\mathcal{S}_{0,N}(t), \mathcal{S}_{1,N}(t), ..., \mathcal{S}_{N,N}(t)]^T, \tag{11}$$

and the coefficients c_i can be obtained by the following expression:

$$c_i = \frac{\langle f(t), \mathcal{S}_{i,N}(t) \rangle_*}{\xi_i}, \quad i = 0,1,...N. \tag{12}$$

Operational Matrices:

In the follow-up, explicit formulas for operational matrices of Riemann-Liouville fractional integration for SDCPs are provided. In addition, the SDCPs vector's product operating matrix is calculated.

Lemma 1. *[15] The inner product of the $\mathcal{S}_{n,N}(t)$ and t^r, indicated by $\lambda(n,r)$, can be calculated for a positive integer r as follows:*

$$\lambda(n,r) = \frac{1}{N^r} \sum_{j=0}^{N} \sum_{k=0}^{n} \mathsf{a}_{j,n} j^{k+r}, \tag{13}$$

where $\mathsf{a}_{j,n}$ is introduced in (6).

Theorem 1. *[6, 15] Assume that $\Psi(t)$ is the SDCPs vector specified in (11). Then, its Riemann-Liouville fractional integral of order α is*

$$\mathbf{I}^{\alpha}\Psi(t) = \mathcal{P}^{(\alpha)}\Psi(t), \tag{14}$$

where $\mathcal{P}^{(\alpha)}$ is fractional integral operational matrix and

$$\mathcal{P}_{i,j}^{(\alpha)} = \sum_{j=0}^{N} \left(\sum_{k=0}^{i} \frac{\mathsf{a}_{k,N}\lambda(k+\alpha,j)\alpha(k+1)}{\alpha(k+\alpha+1)} \right), \qquad i,j = 1,2,...,N+1. \tag{15}$$

Theorem 2. *[6, 15] Consider that $\Psi(t)$ is the SDCPs vector specified in (11) and Q be an arbitrary $(N+1)$ vector. Then*

$$\Psi(t)\Psi^T(t)Q = \tilde{Q}\Psi(t), \tag{16}$$

where $\tilde{Q}$ is the $(N+1) \times (N+1)$ product operational matrix and

$$\tilde{Q}_{i+1,j+1} = \frac{1}{\xi_j} \sum_{k=0}^{N} Q_k \langle \mathcal{S}_{k,N}(t)\mathcal{S}_{i,N}(t), \mathcal{S}_{j,N}(t) \rangle_*, \quad i,j = 0,1,...,N.$$

5   The Numerical Method

SDCPs are used to approximate the solution of fractional HIV infection of CD4$^+$T cells, In this section. Consider the FODE (1) and $\mathbf{D}^{\alpha}T(t), \mathbf{D}^{\alpha}I(t)$ and $\mathbf{D}^{\alpha}V(t)$ involved in, as follows:

$$\mathbf{D}^{\alpha}T(t) \simeq F^T\Psi(t), \quad \mathbf{D}^{\alpha}I(t) \simeq G^T\Psi(t), \quad \mathbf{D}^{\alpha}V(t) \simeq H^T\Psi(t), \tag{17}$$

where $\Psi(t)$ is the SDCPs vector specified in (11). Furthermore, F, G, H are unknown vectors that should be determined. By using of fractional Riemann-Liouville operator $\mathbf{I}^{\alpha}$, we have:

$$
\begin{cases}
T \simeq \mathbf{I}^\alpha F^T \Psi(t) + \sum_{k=0}^{m-1} T^{(k)}(0)\frac{t^k}{k!} = F^T \mathcal{P}^{(\alpha)} \Psi(t) + d_1 \Psi(t), \\[2mm]
I \simeq \mathbf{I}^\alpha G^T \Psi(t) + \sum_{k=0}^{m-1} I^{(k)}(0)\frac{t^k}{k!} = G^T \mathcal{P}^{(\alpha)} \Psi(t) + d_2 \Psi(t), \\[2mm]
V \simeq \mathbf{I}^\alpha H^T \Psi(t) + \sum_{k=0}^{m-1} V^{(k)}(0)\frac{t^k}{k!} = H^T \mathcal{P}^{(\alpha)} \Psi(t) + d_3 \Psi(t),
\end{cases}
\tag{18}
$$

where $\mathcal{P}^{(\alpha)}$ is the fractional operational matrix of SDCPs vector derived in (14).

Substituting (17)–(18) in FODE (1), we have the following residual functions as:

$$
\begin{cases}
E_1(t) \simeq (F^T \Psi(t)) - q - \eta(F^T \mathcal{P}^{(\alpha)} \Psi(t) + d_1 \Psi(t)) + r(F^T \mathcal{P}^{(\alpha)} \Psi(t) + d_1 \Psi(t)) \\[1mm]
(1 - \frac{(F^T \mathcal{P}^{(\alpha)} \Psi(t) + T(0)) + 1}{T_{\max}}) - k(H^T \mathcal{P}^{(\alpha)} \Psi(t) + d_2 \Psi(t))(F^T \mathcal{P}^{(\alpha)} \Psi(t) + d_1 \Psi(t)), \\[2mm]
E_2(t) \simeq (G^T \Psi(t)) - k(H^T \mathcal{P}^{(\alpha)} \Psi(t) + d_2 \Psi(t))(F^T \mathcal{P}^{(\alpha)} \Psi(t) + d_1 \Psi(t) + \\[1mm]
\beta(G^T \mathcal{P}^{(\alpha)} \Psi(t) + d_3 \Psi(t)), \\[2mm]
E_3(t) \simeq (H^T \Psi(t)) - \mu\beta(G^T \mathcal{P}^{(\alpha)} \Psi(t) + d_3 \Psi(t)) + \gamma(H^T \mathcal{P}^{(\alpha)} \Psi(t) + d_2 \Psi(t)).
\end{cases}
\tag{19}
$$

Now, to find the solution $T(t), I(t)$ and $V(t)$, we must first collocate the residual functions $E_i(t)$, $i = 1, 2, 3$ at the $N+1$ points. We use roots of shifted Chebyshev polynomials to find appropriate collocates, as shown below:

$$
E_i(t_j) = 0, \quad i = 1, 2, 3, \ j = 1, 2, \dots N + 1.
\tag{20}
$$

Solve the system of algebraic equations to achieve unknown coefficients of the vectors F, G, H. Finally, we obtain the numerical solution by inputting the acquired vectors F, G, H in Eq. (1).

6   Computational Results and Comparisons

The effectiveness of the shifted discrete chebyshev polynomials method for solving fractional HIV infection of CD4$^+$T cells is proved in this section. In the following, one example is given to demonstrate the properties of the new model.

– SDCPM = Shifted discrete chebyshev polynomials method
– All computations are carried out using MAPLE 17 with 16 digits precision

In the $t \in [0, 1]$, we used the described method for FODEs (1) with the initial conditions $T(0) = 0.1, I(0) = 0$, and $V(0) = 0.1$. $q = 0.1, \eta = 0.02, \beta = 0.3, r = 3, \gamma = 2.4, k = 0.0027, T_{max} = 1500, \mu = 10$ were used in this article.

Approximate solutions population of healthy CD4$^+$T cells, infected CD4$^+$T cells, and free HIV particle for $N = 8$ and different α in $[0, 1]$ are represented in Figs. 5, 6, and 7, respectively. Can be seen in Figs. 8, 9, and 10, that $T(t)$, the concentration of susceptible CD4$^+$T cells, increases rapidly, $I(t)$, the amount of

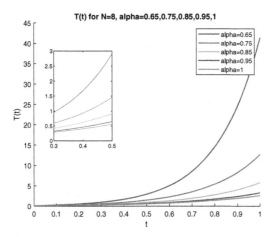

Fig. 5. Numerical results comparison $T(t)$ for $N = 8$ and different α in $[0, 1]$.

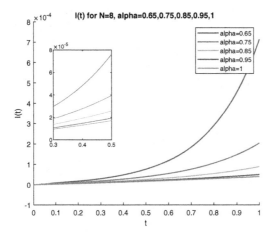

Fig. 6. Numerical results comparison $I(t)$ for $N = 8$ and different α in $[0, 1]$.

Table 1. Numerical results comparison for $T(t)$.

Method	$t = 0.2$	$t = 0.4$	$t = 0.6$	$t = 0.8$	$t = 1.0$
Runge-kutta [5]	0.208808	0.406240	0.764423	1.414046	2.591594
MVIM [14]	0.208808	0.406240	.764428	1.414094	2.208808
VIM [14]	0.208807	0.406134	0.762453	1.397880	2.506746
LADM-Pade [22]	0.208807	0.406105	0.761146	1.377319	2.329169
Bessel [32]	0.203861	0.380330	0.695462	1.275962	2.383227
SDCPM	0.208807	0.406240	0.764422	1.414045	2.591592

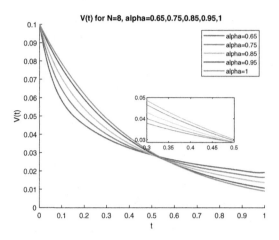

Fig. 7. Numerical results comparison $V(t)$ for $N = 8$ and different α in $[0, 1]$.

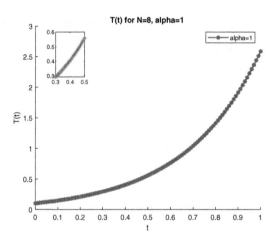

Fig. 8. Numerical result $T(t)$ for $N = 8$ and $\alpha = 1$ in $[0, 1]$.

Table 2. Numerical results comparison for $I(t)$.

Method	$t = 0.4$	$t = 0.6$	$t = 0.8$	$t = 1.0$
Runge-kutta [5]	$0.131583e - 4$	$0.212237e - 4$	$0.301774e - 4$	$0.400378e - 4$
MVIM [14]	$0.131583e - 4$	$0.212233e - 4$	$0.301745e - 4$	$0.400254e - 4$
VIM [14]	$0.131487e - 4$	$0.210141e - 4$	$0.279513e - 4$	$0.243156e - 4$
LADM-Pade [22]	$0.131591e - 4$	$0.212683e - 4$	$0.300691e - 4$	$0.398736e - 4$
Bessel [32]	$0.129355e - 4$	$0.203526e - 4$	$0.283730e - 4$	$0.369084e - 4$
SDCPM	$0.131583e - 4$	$0.212237e - 4$	$0.301773e - 4$	$0.400377e - 4$

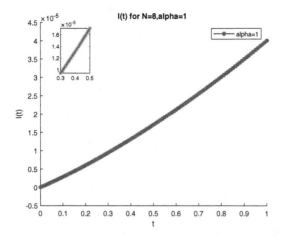

Fig. 9. Numerical result $I(t)$ for $N = 8$ and $\alpha = 1$ in $[0, 1]$.

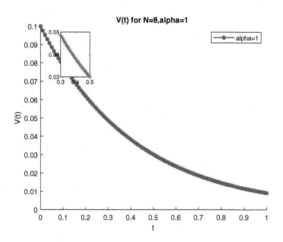

Fig. 10. Numerical result $V(t)$ for $N = 8$ and $\alpha = 1$ in $[0, 1]$.

Table 3. Numerical results comparison for $V(t)$.

Method	$t = 0.2$	$t = 0.4$	$t = 0.6$	$t = 0.8$	$t = 1.0$
Runge-kutta [5]	0.061879	0.038294	0.023704	0.014680	0.009100
MVIM [14]	0.061879	0.038295	0.023710	0.014700	0.009157
VIM [14]	0.061879	0.038308	0.023920	0.016217	0.016084
LADM-Pade [22]	0.061879	0.038313	0.024391	0.009967	0.003305
Bessel [32]	0.061879	0.038294	0.023704	0.014679	0.023704
SDCPM	0.061879	0.038294	0.023704	0.014680	0.009100

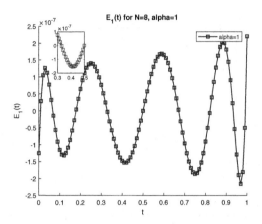

Fig. 11. Error function of $E_1(t)$ for $N = 8$ and $\alpha = 1$.

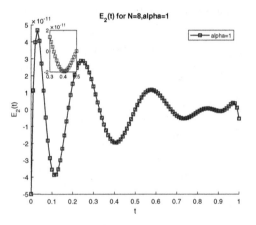

Fig. 12. Error function of $E_2(t)$ for $N = 8$ and $\alpha = 1$.

CD4$^+$T cells infected by the HIV increases significantly for $N = 8$, and $V(t)$, the number of free HIV particles in the blood decrease in a very short time after infection.

For $N = 8$ and $\alpha = 1$, Figs. 11, 12, and 13 display the error functions obtained with an accuracy of the solutions by utilizing the mentioned strategy given by Eqs. (20). The numerical values of the approximate solutions $T(t)$, $I(t)$, and $V(t)$ of the present method for $N = 8$ in the interval $[0, 1]$ are compared with the Legendre Wavelet Collocation strategy [5], the Runge-Kutta strategy [5], the variational iteration strategy [14], the modified variational iteration strategy [14], the Laplace Adomian decomposition-pade strategy [22] and the Bessel collocation [32] in Tables 1, 2, and 3. To comparison of error functions provided by Eqs. (20) for $T(t)$, $I(t)$, and $V(t)$ of the current method for $N = 8$ in the interval $[0, 1]$ are reported in Tables 4, 5, and 6. These results lead to the conclusion that the numerical solutions of the present method are better than those obtained with other methods since the absolute errors have gotten by the current strategy are superior to other methods.

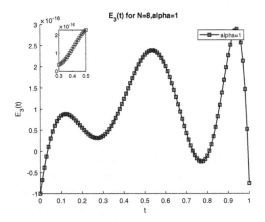

Fig. 13. Error function of $E_3(t)$ for $N = 8$ and $\alpha = 1$.

Table 4. Comparison error results for $T(t)$.

Method	$t = 0.2$	$t = 0.4$	$t = 0.6$	$t = 0.8$	$t = 1.0$
LWCM [5]	$7.50e - 06$	$2.70e - 05$	$7.34e - 05$	$1.77e - 04$	$3.98e - 04$
MVIM [14]	$7.85e - 05$	$3.00e - 04$	$8.49e - 04$	$2.14e - 03$	$5.14e - 03$
VIM [14]	$7.78e - 05$	$1.94e - 04$	$1.13e - 03$	$1.41e - 02$	$8.00e - 02$
LADM-Pade [22]	$7.77e - 05$	$1.65e - 04$	$2.43e - 03$	$3.46e - 02$	$2.58e - 01$
Bessel [32]	$4.87e - 03$	$2.56e - 02$	$6.81e - 02$	$1.36e - 01$	$2.04e - 01$
SDCPM	$6.48e - 08$	$1.47e - 07$	$1.65e - 07$	$9.09e - 08$	$2.20e - 07$

Table 5. Comparison error results for $I(t)$.

Method	$t = 0.2$	$t = 0.4$	$t = 0.6$	$t = 0.8$	$t = 1.0$
LWCM [5]	$8.36e - 10$	$1.95e - 09$	$3.20e - 09$	$4.63e - 09$	$6.44e - 09$
MVIM [14]	$1.19e - 09$	$5.29e - 09$	$1.27e - 08$	$2.27e - 08$	$3.12e - 08$
VIM [14]	$1.12e - 09$	$4.23e - 09$	$1.96e - 07$	$2.20e - 06$	$1.57e - 05$
LADM-Pade [22]	$1.20e - 09$	$6.15e - 09$	$5.78e - 08$	$8.26e - 08$	$1.21e - 07$
Bessel [32]	$2.16e - 07$	$2.17e - 07$	$8.58e - 07$	$1.78e - 06$	$3.09e - 06$
SDCPM	$1.52e - 11$	$1.94e - 11$	$1.05e - 11$	$1.42e - 12$	$5.30e - 12$

Table 6. Comparison error results for $V(t)$.

Method	$t = 0.2$	$t = 0.4$	$t = 0.6$	$t = 0.8$	$t = 1.0$
LWCM [5]	$1.00e - 10$	$1.24e - 06$	$1.15e - 06$	$9.50e - 07$	$7.61e - 07$
MVIM [14]	$5.61e - 08$	$1.06e - 06$	$5.75e - 06$	$2.01e - 05$	$5.64e - 05$
VIM [14]	$1.00e - 07$	$1.33e - 05$	$2.16e - 04$	$1.54e - 03$	$6.98e - 03$
LADM-Pade [22]	$1.08e - 07$	$1.84e - 05$	$6.87e - 04$	$4.71e - 03$	$5.80e - 03$
Bessel [32]	$6.59e - 08$	$3.79e - 08$	$2.31e - 07$	$7.87e - 07$	$1.46e - 02$
SDCPM	$2.06e - 16$	$9.64e - 17$	$6.47e - 17$	$2.12e - 16$	$8.01e - 17$

7 Conclusion

This research looked at the design of protein-protein interaction (PPI) networks and compared them to each other as well. Besides, the mathematical model of fractional HIV infection of CD4$^+$T cells was introduced, which refers to a class of nonlinear differential equation structures. The SDCPM was suggested for finding approximate solutions to the HIV infection model of CD4$^+$T cells. With the help of an example, the precision and reliability of the current procedure were illustrated. All calculations were done with the aid of a Maple 17 computer program.

References

1. Ali, N., Ahmad, S., Aziz, S., Zaman, G.: The adomian decomposition method for solving HIV infection model of latently infected cells. Matrix Sci. Math. **3**, 5–8 (2019)
2. Anastasio, T.J.: The fractional-order dynamics of bainstem vestibulo-oculomotor neurons. Biol. Cybern. **72**, 69–79 (1994)
3. Angulo, J., Cuesta, T., Menezes, E., Pedroso, C., Brites, C.: A systematic review on the influence of HLA-B polymorphisms on HIV-1 mother to child transmission. Braz. J. Infec. Dis. **23**, 53–9 (2019)
4. Asquith, B., Bangham, C.R.M.: The dynamics of T-cell fratricide: application of a robust approach to mathematical modelling in immunology. J. Theoret. Biol. **222**, 53–69 (2003)
5. Attaullah, Sohaib, M.: Mathematical modeling and numerical simulation of HIV infection model. Results Appl. Math. **7**, 100–118 (2020)
6. Conte, D., Farsimadan, E., Moradi, L., Palmieri, F., Paternoster, B.: Time-Delay Fractional Optimal Control Problems: A Survey Based on Methodology. In: Proceedings of the 8th International Conference on Fracture, Fatigue and Wear, Belgium, pp. 325–337 (2021)
7. Djordjević, V.D., Jarić, J., Fabry, B., Fredberg, J.J., Stamenović, D.: Fractional derivatives embody essential features of cell rheological behavior. Ann. Biomed. Eng. **31**, 692–699 (2003)
8. Duro, R., Pereira, N., Figueiredo, C., Pineiro, C., Caldas, C., Serrao, R.: Routine CD4 monitoring in HIV patients with viral suppression: is it really necessary? A Portuguese cohort. J. Microbiol. Immunol. Infect. **51**, 593–7 (2018)

9. Gogin, N., Hirvensalo, M.: On the generating function of discrete Chebyshev polynomials. J. Math. Sci. **2**, 224 (2017)

10. Kirschner, D.E.: Using mathematics to understand HIV immune dynamics. Not. Am. Math. Soc. **43**, 191–202 (1996)

11. Hallbergc, D., Kimariob, T., Mtuyab, C., Msuyab, M., Bjorlingc, G.: Factors affecting HIV disclosure among partners in morongo. tanzania. Int J Afr Nurs Sci 10, 49–54 (2019)

12. Jesus, I.S., Machado, J.A.T., Cunha, J.B.: Fractional electrical impedances in botanical elements. J. Vib. Control **14**, 1389–1402 (2008)

13. Li, Q., Xiao, Y.: Global dynamics of a virus immune system with virus guided therapy and saturation growth of virus. Math. Probl. Eng. 1–18 (2018)

14. Merdan, M., Gökdogan, A., Yildirim, A.: On the numerical solution of the model for HIV infection of CD4+T-cells. Comput. Math. Appl. **62**, 118–123 (2011)

15. Moradi, L., Mohammadi, F.: A comparative approach for time-delay fractional optimal control problems: discrete versus continuous Chebyshev polynomials. Asian J. Control **21**(6), 1–13 (2019)

16. Nelson, P.W., Perelson, A.S.: Mathematical analysis of delay differential equation models of HIV-1 infection. Math. Biosci. **179**, 73–94 (2002)

17. Nigmatullin, R.R., Nelson, S.O.: Recognition of the fractional kinetics in complex systems: dielectric properties of fresh fruits and vegetables form 0.01 to 1.8 GHz. Signal Process. **86**, 2744–2759 (2006)

18. Nikiforov, A.F., Suslov, S.K., Uvarov, V.B.: Classical Orthogonal Polynomials of a Discrete Variable. Springer Series in Computational Physics, Springer, Heidelberg (1991). https://doi.org/10.1007/978-3-642-74748-9_2

19. Nowak, M., May, R.: Mathematical biology of HIV infections: antigenic variation and diversity threshold. Math. Biosci. **106**, 1–21 (1991)

20. Oldham, K.B., Spanier, J.: The Fractional Calculus. Academic Press, New York (1974)

21. Omondi, E., Mbogo, W., Luboobi, L.: A mathematical modeling study of HIV infection in two heterosexual age groups in Kenya. Infec. Dis. Modell. **4**, 83–98 (2019)

22. Ongun, M.: The Laplace adomian decomposition method for solving a model for HIV infection of CD4+T-cells. Math. Comput. Modell. **63**, 597–603 (2011)

23. Perelson, A.S., Kirschner, D.E., Boer, R.D.: Dynamics of HIV infection CD4$^+$T cells. Math. Biosci. **114**, 81–125 (1993)

24. Perelson, A.S., Nelson, P.W.: Mathematical analysis of HIV-I dynamics in vivo. SIAM Rev. **41**(1), 3–44 (1999)

25. Petrovic, L.M., Spasic, D.T., Atanackovic, T.M.: On a mathematical model of a human root dentin. Dent. Mater. **21**, 125–128 (2005)

26. Ransome, Y., Thurber, K., Swen, M., Crawford, N., Germane, D., Dean, L.: Social capital and HIV/AIDS in the United States: knowledge, gaps, and future directions. SSM Popul. Health **5**, 73–85 (2018)

27. Riordan, J.: An Introduction to Combinatorial Analysis. Wiley, New York (1980)

28. Samko, S.G., Kilbas, A.A., Marichev, O.I.: Fractional Integrals and Derivatives: Theory and Applications. Gordon and Breach, Langhorne (1993)

29. Theys, K., Libin, P., Pena, A.C.P., Nowe, A., Vandamme, A.M., Abecasis, A.B.: The impact of HIV-1 within host evolution on transmission dynamics. Curr. Opin. Virol. **28**, 92–101 (2018)

30. Wang, L., Li, M.Y.: Mathematical analysis of the global dynamics of a model for HIV infection of CD4+T cells. Math. Biosci. **200**, 44–57 (2006)

31. Yddotuzbasi, S., Karacayir, M.: An exponential Galerkin method for solution of HIV infected model of CD4+t-cells. Comput. Biol. Chem. **67**, 205–12 (2017)
32. Yüzbas, S.: A numerical approach to solve the model for HIV infection of CD4+T-cells. Appl. Math. Model. **36**, 5876–5890 (2012)

A Rapid Euclidean Norm Calculation Algorithm that Reduces Overflow and Underflow

Takeyuki Harayama[1], Shuhei Kudo[2], Daichi Mukunoki[2] (iD),
Toshiyuki Imamura[2] (iD), and Daisuke Takahashi[3(✉)] (iD)

[1] Graduate School of System and Information Engineering,
University of Tsukuba, Tsukuba, Ibaraki 305-8573, Japan
[2] Center for Computational Science, RIKEN, Kobe, Hyogo 650-0047, Japan
[3] Center for Computational Sciences, University of Tsukuba,
Tsukuba, Ibaraki 305-8577, Japan
`daisuke@cs.tsukuba.ac.jp`

Abstract. The Euclidean norm is widely used in scientific and engineering calculations. However, a straightforward implementation may cause overflow/underflow and loss of accuracy. The Blue algorithm selects a scaling value from three conditional branches according to the absolute value to reduce overflow and underflow. We improve this algorithm and propose the Two-Accumulator method, which is expected to be faster by selecting it from two conditional branches. The input range of our method is slightly smaller than the Blue algorithm. However, we mitigate this problem by dynamically setting the scaling value depending on the vector size. Moreover, we combine it with double-double arithmetic to prevent rounding errors. An evaluation shows that our method combined with double-double arithmetic can be approximately 15% faster than the Blue algorithm while maintaining the same error level, but it can be approximately 46% slower, depending on the input range.

Keywords: Euclidean norm · Accurate computation · Overflow · Underflow

1 Introduction

The Euclidean norm is widely used in scientific and engineering calculations. For example, the Euclidean norm is used in the field of numerical analysis, as well as in linear calculations such as the Conjugate Gradient (CG) method and the Gram-Schmidt orthogonalization method. The Euclidean norm for the vector $x = (x_1, x_2, \ldots, x_n)$ shown in Eq. (1) is a simple calculation, where n is the vector size of the input vector.

$$\|x\| = \sqrt{x_1^2 + x_2^2 + \cdots + x_n^2} \tag{1}$$

However, the straightforward norm calculation algorithm shown as Algorithm 1 can only handle a limited range of inputs due to the effects of overflow and underflow. The Blue algorithm [4,7] and the Kahan algorithm [7] have

© Springer Nature Switzerland AG 2021
O. Gervasi et al. (Eds.): ICCSA 2021, LNCS 12949, pp. 95–110, 2021.
https://doi.org/10.1007/978-3-030-86653-2_7

been proposed to suppress overflow and underflow. However, the Blue algorithm requires at least three conditional branches, and the Kahan algorithm requires conditional branching and divisions, which are slow operations. On the other hand, there is a demand for accurate computations in scientific and engineering calculations, and MPLAPACK [13] provides high-precision BLAS and LAPACK routines. Euclidean norm calculations are prone to large errors due to the accumulation of rounding errors.

In this paper, we propose a method called the Two-Accumulator method to prevent overflow and underflow. We show that the implementation of double-word arithmetic (so-called double-double (DD) arithmetic) in the proposed method is as simple as the implementation in the Blue algorithm and can easily achieve higher accuracy. Dynamic scaling is also proposed to further reduce the overflow and underflow in the Two-Accumulator method.

Our contributions in this paper are as follows:

- The proposed method is able to suppress overflow and underflow with little change in the input range of the Blue algorithm,
- It combines well with DD arithmetic, making it easy to achieve high precision,
- It has the potential to be faster than the Blue algorithm.

In what follows, only double-precision arithmetic according to IEEE 754 is used, but application to other types of precision is possible as well.

This paper consists of the following sections. Section 2 explores related research. In Sect. 3, the Blue algorithm and the proposed method are described. Section 4 provides an overview of DD arithmetic, which is used as a precision expansion method to reduce rounding errors. The accuracy and performance are evaluated in Sect. 5. Finally, Sect. 6 summarizes the paper including suggestions for future work in this domain.

Algorithm 1. Straightforward norm calculation algorithm
___
Require: Vector x, size n
Ensure: The l_2 norm of x
 1: sum $= x_0^2$;
 2: **for** $i \leftarrow 1, n - 1$ **do**
 3:     sum$+= x_i^2$;
 4: **end for**
 5: **return** $(\sqrt{\text{sum}})$;

---

2   Related Work

To suppress overflow and underflow in Euclidean norm calculations, the Blue and Kahan algorithms bring the value of the exponential part close to zero in advance by scaling. The Kahan algorithm continuously changes the scaling value of a vector $x = (x_1, x_2, \ldots, x_n)$ according to the maximum absolute value of the input values up to x_i, which not only involves branching but also causes performance problems due to the frequent execution of slow division. The Blue algorithm is explained in the next section.

OpenBLAS [15] implements the Euclidean norm using the x87 instruction set, which uses 80-bit floating-point format. Even squares of values that overflow or underflow in double precision can be expressed entirely without overflow or underflow in 80-bit floating-point format. This makes it possible to calculate the Euclidean norm without using the Blue algorithm or the Kahan algorithm, which suppress overflow and underflow. Since the norm calculation in double-precision arithmetic has a large bytes/flop ratio and is memory-intensive, the performance may not change significantly whether the x87 instruction set or Single Instruction, Multiple Data (SIMD) instruction set is used. However, this approach cannot be used in environments where the x87 instruction set is not supported.

With the objective of improving accuracy, XBLAS [12] was proposed, which uses double-word arithmetic, like DD arithmetic, for internal operations. Further, MPLAPACK can arbitrarily select the precision from the GNU Multiple Precision Arithmetic Library (GMP) [1], MPFR [6], and the QD library [2,8].

3   Blue Algorithm and Two-Accumulator Method

3.1   Blue Algorithm

The maximum value of double-precision numbers, DBL_MAX, according to IEEE 754 is approximately 2^{1024}, and the minimum value, DBL_MIN, is approximately 2^{-1022}. Algorithm 2 shows the pseudo-code for the Blue algorithm. The Blue algorithm selects a scaling value from three conditional branches depending on the absolute value of the input. Inputs above 2^{300} are scaled by 2^{-600}. Inputs less than 2^{-300} are scaled by 2^{600}. Therefore, the value after scaling does not overflow or underflow when squared. The number of operations in the Blue algorithm is only one greater (a multiplication) than the straightforward norm calculation algorithm. However, the disadvantage of the Blue algorithm is that it requires branching for the scaling selection, making it inefficient to apply performance-enhancing methods such as vectorization.

3.2   Two-Accumulator Method with Fixed Scaling Value

The Two-Accumulator method is a simplified version of the Blue algorithm. This algorithm selects a scaling value from two conditional branches. The advantage of the Two-Accumulator method is the expected increase in speed because the number of conditional branches is smaller than in the Blue algorithm.

However, because the Two-Accumulator method only considers two conditional branches, some ranges that can be handled by the Blue algorithm cannot be handled by this simplified variant. In this section, we discuss the Two-Accumulator method with fixed scaling value 2^{498} for inputs less than 1.0 (hereinafter called *the Two-Accumulator method with fixed scaling value*). For example, the scaling value for inputs less than 1.0 in the Two-Accumulator method

---

Algorithm 2. Blue algorithm

Require: Vector $\boldsymbol{x}$, size n
Ensure: The l_2 norm of $\boldsymbol{x}$
1: $a_{\text{big}} = a_{\text{sml}} = a_{\text{med}} = 0$;
2: overFlowLimit = DBL_MAX $* s_{\text{h}}$;
3: **for** $i \leftarrow 0, n-1$ **do**
4:     absx = $|x_i|$
5:     **if** absx $\geq \gamma$ **then**
6:         a_{big}+= (absx $* s_{\text{h}})^2$;
7:     **else if** absx $\geq$ low **then**
8:         a_{med}+= absx^2;
9:     **else**
10:         a_{sml}+= (absx $* s_{\text{l}})^2$;
11:     **end if**
12: **end for**
13: **if** a_{sml} == 0 **then**
14:     **if** $\sqrt{a_{\text{big}}} \geq$ overFlowLimit **then**
15:         **return** (∞);
16:     **else if** a_{med} == 0 **then**
17:         **return** $(\sqrt{a_{\text{big}}}/s_{\text{h}})$;
18:     **else**
19:         $y_{\text{min}} = \min(\sqrt{a_{\text{med}}}, \sqrt{a_{\text{big}}}/s_{\text{h}})$;
20:         res = $\max(\sqrt{a_{\text{med}}}, \sqrt{a_{\text{big}}}/s_{\text{h}})$;
21:     **end if**
22: **else if** a_{big} == 0 **then**
23:     **if** a_{med} == 0 **then**
24:         **return** $(\sqrt{a_{\text{sml}}}/s_{\text{l}})$;
25:     **else**
26:         $y_{\text{min}} = \min(\sqrt{a_{\text{med}}}, \sqrt{a_{\text{sml}}}/s_{\text{l}})$;
27:         res = $\max(\sqrt{a_{\text{med}}}, \sqrt{a_{\text{sml}}}/s_{\text{l}})$;
28:     **end if**
29: **else**
30:     **return** $(\sqrt{a_{\text{med}}})$;
31: **end if**
32: **if** $y_{\text{min}} < \sqrt{\epsilon} *$ res **then**
33:     **return** (res);
34: **else**
35:     **return** $(\text{res} * \sqrt{1 + (y_{\text{min}}/\text{res})^2})$;
36: **end if**

---

with fixed scaling value is 2^{498}, so inputting values less than 2^{-1009} will cause underflow in the calculation of a square. Also, scaling for the input of $2^0 - 2^0 \times \epsilon$ (ϵ is the machine epsilon in double precision) results in $2^{996} \times (2^0 - \epsilon)$. Therefore, if the same input is given 2^{28} times, it will overflow. However, as will be shown in the next subsection, this can be mitigated by dynamic scaling.

3.3   Two-Accumulator Method with Dynamic Scaling

As already noted, the Two-Accumulator method with fixed scaling value cannot accommodate input values that are too large or too small. This is partly because calculating the square of a floating-point number will more than double the value of the exponent part. We should also take into account that the calculation of the Euclidean norm requires a sum of squares, which has an increasing partial sum. In other words, there is a relationship between n and the range of computable input values and the scaling values. Therefore, we propose a dynamic scaling method that sets the scaling value so that the range of computable input values is handled according to the length n of the input values. The scaling value (s_l) for inputs less than 1.0 is $s_l = 2^{512-r}$ using r in Eq. (2).

$$r \geq \lceil (\log_2 n)/2 \rceil \tag{2}$$

If n is less than or equal to 4^r, which is the vector size that can be input, it will not overflow or underflow. However, the scaling value should be as large as possible because the input range becomes smaller when the scaling value is reduced. Therefore, r is the smallest value among the possible values satisfying $(\log_2 n)/2$. The reason for $r \geq \lceil (\log_2 n)/2 \rceil$ is that n must be less than or equal to the size of the input vector. Further, the maximum value of the input range below 1.0 is $1.0 - 1.0 \times \epsilon$, which is extremely close to 2^0. If 2^0 is multiplied by a power of two and squared, the result is always an even power of two. Therefore, it is possible to add an even power of two to the maximum value $2^{1024} - 2^{1024} \times \epsilon$ that can be expressed in double precision. Algorithm 3 is the pseudo-code for the Two-Accumulator method. The constants used in Algorithms 2 and 3 are listed in Table 3.

For inputs above 1.0, the vector size that can be input changes depending on the range of input values. Therefore, the scaling value for inputs above 1.0 is fixed at 2^{-511}, and by restricting the input range, overflow and underflow can be prevented. Hereafter, when we refer simply to the Two-Accumulator method, we mean the Two-Accumulator method with dynamic scaling.

3.4   Input Range and Vector Size

In the Euclidean norm calculation, overflow or underflow is most likely to occur when calculating a square. The range and vector size that can be input for each method are different because the number of partitions for each method is different. In this subsection, we discuss the vector sizes and ranges that can be computed for the inputs that are most likely to have overflow and underflow for each method. The worst-case input value is the one that minimizes the computable vector size when all the values of the input vector consist of this value.

The worst-case input values for each method and vector sizes that can be input are shown in Table 1. Table 2 shows the maximum range of input values for which a vector of at least two elements can be computed. Since the Euclidean norm for a vector x of size 1 is $\|x\| = \sqrt{x_1^2} = |x_1|$, it can be calculated in all ranges except NaN (Not a Number). The maximum range shown in Table 2 is valid only when implemented with double-precision arithmetic according to IEEE 754, and the lower limit changes when DD arithmetic is used. The straightforward norm calculation algorithm can only calculate up to 2^0 in vector size when $2^{512} \times (2^0 - \epsilon)$ is input consecutively.

The range of possible inputs for the Two-Accumulator method and the Two-Accumulator method with fixed scaling value depends on the vector size of the input vectors. In the discussion of the worst-case input value, input values greater than 1.0 of the Two-Accumulator method are excluded. The reason for this is that, as explained in the last paragraph of the previous subsection, the worst-case input value for inputs greater than 1.0 of the Two-Accumulator method depends on the vector size of the input vector. The scaling value in the Two-Accumulator method with fixed scaling value is 2^{498}. The worst-case input value, in this case, is $2^0 - \epsilon$. Therefore, we can add 2^{28} times without going over DBL_MAX (approximately 2^{1024}). The worst-case input value in the Two-Accumulator method is also $2^0 - \epsilon$, and the minimum scaling value is 2^{496}, so if $2^0 - \epsilon$ is input consecutively, it can be input 2^{32} times. This means that inputs with a larger vector size and wider range can be used with the Two-Accumulator method compared to what is possible with the Two-Accumulator method with fixed scaling value.

In contrast, the worst-case input value for the Blue algorithm is DBL_MAX, which can be input 2^{176} times if this value is input consecutively. This is because the scaling value of the Blue algorithm for values less than 2^{300} is 2^{-600}. Therefore, the vector size that can be input is inferior to that of the Blue algorithm even for the Two-Accumulator method. However, in practical contexts where the Euclidean norm is used in scientific and engineering calculations, inputs that cannot be handled by the Two-Accumulator method would rarely occur.

Table 1. Worst-case input values for each method and vector sizes that can be input

	Worst value	Vector size
Straightforward norm calculation algorithm	$2^{512} \times (2^0 - \epsilon)$	2^0
Two-Accumulator method with fixed scaling value	$2^0 - \epsilon$	2^{28}
Two-Accumulator method	$2^0 - \epsilon$	2^{32}
Blue Algorithm	DBL_MAX	2^{176}

Algorithm 3. Two-Accumulator method

Require: Vector x, size n
Ensure: The l_2 norm of x
1: $a_{\text{big}} = a_{\text{sml}} = 0$;
2: overFlowLimit $=$ DBL_MAX $* s_{\text{h}}$;
3: $r = \lceil (\log_2 n)/2 \rceil$;
4: $s_{\text{l}} = 2^{512-r}$;
5: **for** $i \leftarrow 0, n-1$ **do**
6:    **if** $|x_i| \geq \gamma$ **then**
7:       $a_{\text{big}} += (x_i * s_{\text{h}})^2$;
8:    **else**
9:       $a_{\text{sml}} += (x_i * s_{\text{l}})^2$;
10:    **end if**
11: **end for**
12: **if** $a_{\text{sml}} == 0$ **then**
13:    **return** $(\sqrt{a_{\text{big}}}/s_{\text{h}})$;
14: **else if** $a_{\text{big}} == 0$ **then**
15:    **return** $(\sqrt{a_{\text{sml}}}/s_{\text{l}})$;
16: **else**
17:    **if** $\sqrt{a_{\text{big}}} \geq$ overFlowLimit **then**
18:       **return** (∞);
19:    **end if**
20:    $y_{\text{min}} = \min(\sqrt{a_{\text{sml}}}/s_{\text{l}}, \sqrt{a_{\text{big}}}/s_{\text{h}})$;
21:    res $= \max(\sqrt{a_{\text{sml}}}/s_{\text{l}}, \sqrt{a_{\text{big}}}/s_{\text{h}})$;
22:    **if** $y_{\text{min}} < \sqrt{\epsilon} *$ res **then**
23:       **return** (res);
24:    **else**
25:       **return** $(\text{res} * \sqrt{1 + (y_{\text{min}}/\text{res})^2})$;
26:    **end if**
27: **end if**

Table 2. Maximum range that can be input for each method

	Range
Straightforward norm calculation algorithm	$2^{-511} \leq x_i \leq 2^{511}$
Two-Accumulator method with fixed scaling value	$2^{-1009} \leq x_i \leq 2^{1009}$
Two-Accumulator method	DBL_MIN $\leq x_i \leq 2^{1022}$
Blue Algorithm	DBL_MIN $\leq x_i \leq$ DBL_MAX

Table 3. Constants in the Blue algorithm, the Two-Accumulator method with fixed scaling value, and the Two-Accumulator method

	γ	low	s_{h}	s_{l}
Blue Algorithm	2^{300}	2^{-300}	2^{-600}	2^{600}
Two-Accumulator method with fixed scaling value	1.0	—	2^{-498}	2^{498}
Two-Accumulator method	1.0	—	2^{-511}	$2^{496} \sim 2^{512}$

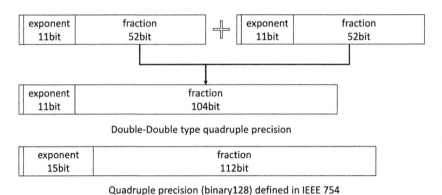

Fig. 1. Format difference between DD type and binary128

4 DD Arithmetic

4.1 Overview of DD Arithmetic

Precision expansion of floating-point arithmetic is a simple approach for reducing the accumulation of rounding errors and achieving more accurate calculations. In this paper, we leverage DD arithmetic, which is a low-cost method for achieving precision expansion. DD-type quadruple precision has 4 fewer bits in its exponent part and 8 fewer bits in its mantissa part than the quadruple precision (binary128) defined in IEEE 754 and shown in Fig. 1. DD arithmetic is faster than binary128 using software emulation. Also, the range of the exponential part that can be represented by DD-type quadruple precision is the same as that of double precision.

DD arithmetic is based on the error-free transformations by Knuth [10] and Dekker [5]. First, the addition by Knuth is shown in Algorithm 4. The result of the addition of the double-precision floating-point numbers a and b is denoted by s, and the error generated during the addition is denoted by e. The double-precision addition without error is expressed as $a + b = s + e$. Next, Dekker's addition is shown in Algorithm 5. Whilst this addition requires fewer operations compared to that of Knuth, it can only be used when $|a| \geq |b|$ is true. Next, the multiplication with error-free transformation using a Fused Multiply-Add (FMA) instruction is shown in Algorithm 6. Let p be the result of the calculation without considering the rounding error of double-precision multiplication $a \times b$ and let e be the error generated. The multiplication with error-free transformation can be expressed as $a \times b = p + e$. The multiplication with error-free transformation without using FMA instructions has also been proposed by Dekker [5].

We can calculate a DD addition using Algorithms 4 and 5, shown in Algorithm 7. The DD addition QuadAdd$(a_H, a_L, b_H, b_L, c_H, c_L)$ calculates $c = a + b$ for $a = a_H + a_L$, $b = b_H + b_L$, and $c = c_H + c_L$. QuadAdd needs to be normalized so that the calculation result $c = c_H + c_L$ satisfies $c_L \leq 0.5\text{ulp}(c_H)$, and Quick-TwoSum of Algorithm 5 is used for normalization. DD addition can be used so that calculations incur less rounding error compared to double-precision addition.

Algorithm 4. Addition with error-free transformation [10]

1: **function** $[s, e] = \text{TwoSum}(a, b)$
2:     $s = a + b$;
3:     $v = s - a$;
4:     $e = (a - (s - v)) + (b - v)$;
5: **end function**

Algorithm 5. Dekker's addition [5]

1: **function** $[s, e] = \text{Quick-TwoSum}(a, b)$
2:     $s = a + b$;
3:     $e = b - (s - a)$;
4: **end function**

DD addition can be accelerated using CPairSum by Lange and Rump [11,14]. The pseudo-code for CPairSum is shown in Algorithm 8. CPairSum omits the assumption that the calculation result satisfies $c_L \leq 0.5\text{ulp}(c_H)$. This means that Quick-TwoSum, which is executed twice in QuadAdd, will not be executed in CPairSum. Therefore, CPairSum is slightly less accurate than QuadAdd, but can be computed with fewer operations. Specifically, CPairSum performs DD addition using the difference between the results of double-precision addition and TwoSum calculation. Also, QuadAdd uses TwoSum to calculate the lower bits, while CPairSum uses double-precision operations. QuadAdd requires 20 double-precision operations, while CPairSum can perform a DD addition in 11 operations.

4.2   Euclidean Norm Calculation Using DD Arithmetic

In this paper, we use DD addition to improve the accuracy of the Euclidean norm calculation where the input and output are double precision. In the Blue algorithm and the Two-Accumulator method, the scaling and squaring of the input values do not need to be computed using DD arithmetic. Both methods multiply powers of two as scaling values, and powers of two do not cause errors even in double-precision arithmetic. The square can be calculated with TwoProd-FMA shown in Algorithm 6.

When comparing the operations for the combination with DD arithmetic, the Blue method and the Two-Accumulator method only require one more double-precision multiplication for scaling compared to the straightforward norm calculation algorithm. If DD arithmetic is used in the Kahan algorithm, DD division

---

Algorithm 6. Multiplication using the FMA instruction [9]

1: **function** $[p, e] = \text{TwoProd-FMA}(a, b)$
2:     $p = a * b$;
3:     $e = a * b - p$;   ▷ FMA
4: **end function**

Table 4. Measurement environment

CPU	Intel Xeon Gold 6126 (2.6 GHz)× 2 sockets
Memory	192GiB (DDR4)
OS	CentOS 7.7 (x86-64), kernel 3.10.0
Compiler	Intel Compiler 19.0.5.281

is required inside the loop, which complicates the implementation and makes it difficult to achieve high speed. Therefore, the Blue algorithm and the Two-Accumulator method are more effective when combined with DD arithmetic.

Algorithm 7. DD addition [2]

1: **function** $[c_H, c_L]$ = QuadAdd(a_H, a_L, b_H, b_L)
2:    $[sh, eh]$ = TwoSum(a_H, b_H);
3:    $[sl, el]$ = TwoSum(a_L, b_L);
4:    $eh = eh + sl$;
5:    $[sh, eh]$ = Quick-TwoSum(sh, eh);
6:    $eh = eh + el$;
7:    $[c_H, c_L]$ = Quick-TwoSum(sh, eh);
8: **end function**

Algorithm 8. Fast DD addition [11]

1: **function** $[c_H, c_L]$ = CPairSum(a_H, a_L, b_H, b_L)
2:    $c_H = a_H + b_H$;
3:    $[t, s]$ = TwoSum(a_H, b_H);
4:    $t = (t - c_H) + s$;
5:    $c_L = t + (a_L + b_L)$;
6: **end function**

---

5   Evaluation of Performance and Relative Error

5.1   Evaluation Method

We implemented algorithms for calculating the Euclidean norm (the straightforward norm calculation algorithm, the Blue algorithm, and the Two-Accumulator method) with double precision and combined them with precision expansion methods (QuadAdd and CPairSum). We evaluated the relative error and performance of these implementations as follows. The measurement environment is shown in Table 4. Evaluation was performed using a single core and a single thread. We specified -O3 -ipo -xSKYLAKE-AVX512 -fprotect-parens -no-vec -fma -unroll0 as the compilation options in the Intel C Compiler to evaluate the relative error, and -O3 -ipo -xSKYLAKE-AVX512 -fprotect-parens -finline -restrict to evaluate the performance.

- Straightforward norm calculation algorithm + double-precision
- Straightforward norm calculation algorithm + QuadAdd
- Straightforward norm calculation algorithm + CPairSum
- Blue algorithm + double-precision
- Blue algorithm + QuadAdd
- Blue algorithm + CPairSum
- Two-Accumulator method + double-precision
- Two-Accumulator method + QuadAdd
- Two-Accumulator method + CPairSum

First, we input uniform random numbers of $[0, 1]$ generated with double precision for the input vector size $n = 10, 100, 1000, 10000, 100000$ and compare the relative error between the value computed by each method and the value computed using MPFR with a 256-bit mantissa part. Further, we compare the relative error when random numbers of $[2^{-1014}, 2^{1014}]$ are input to each method, and evaluate whether the Two-Accumulator method can prevent overflow and underflow. We normalize the difference between the values computed using MPFR and the values computed by each method by half of the computer epsilon to highlight how much error is introduced.

In the performance evaluation, we compare the extent to which the process of preventing overflow and underflow affects the performance. We evaluate the performance difference between combinations with CPairSum or QuadAdd. The vector size at the time of measurement is 100000.

5.2   Evaluation Results

Relative Error. Table 5 shows the relative error in $[2^{-1014}, 2^{1014}]$ when each method is implemented with double-precision arithmetic. The relative errors of each Euclidean norm calculation algorithm combined with each accuracy expansion method for uniform random numbers in $[0, 1]$ are shown in Tables 6, 7 and 8. In this evaluation, auto-vectorization by the compiler is turned off to evaluate the inherent relative error of the algorithm.

First, from the relative error of each Euclidean norm calculation algorithm for the input of $[2^{-1014}, 2^{1014}]$ combined with double-precision arithmetic, the relative error is Inf because of the overflow and underflow in the straightforward norm calculation algorithm. The Blue algorithm and the Two-Accumulator method do not overflow or underflow in the calculation of the sum of squares because of the scaling when the vector size is less than $n = 100000$ in the input range $[2^{-1014}, 2^{1014}]$.

Next, we describe the relative error of each Euclidean norm calculation algorithm for the input range $[0, 1]$ combined with double-precision arithmetic. Even the straightforward norm calculation algorithm does not overflow or underflow in the input range $[0, 1]$. The Blue algorithm and the Two-Accumulator method

Table 5. Relative error of each method combined with double-precision arithmetic for the input range of $[2^{-1014}, 2^{1014}]$

n	Straightforward norm calculation method		Blue algorithm		Two-Accumulator method	
	Average	Variance	Average	Variance	Average	Variance
10	N/A	N/A	0.05	0.03	0.05	0.03
100	N/A	N/A	0.30	0.09	0.30	0.09
1000	N/A	N/A	0.55	0.19	0.55	0.19
10000	N/A	N/A	1.25	0.98	1.25	0.98
100000	N/A	N/A	5.26	13.79	5.26	13.79

Table 6. Relative error of each method combined with double-precision arithmetic for the input range of $[0, 1]$

n	Straightforward norm calculation method		Blue algorithm		Two-Accumulator method	
	Average	Variance	Average	Variance	Average	Variance
10	0.47	0.12	0.47	0.12	0.47	0.12
100	1.01	0.58	1.01	0.58	1.01	0.58
1000	3.30	6.18	3.30	6.18	3.30	6.18
10000	9.96	56.34	9.96	56.34	9.96	56.34
100000	26.51	439.52	26.51	439.52	26.51	439.52

have comparable relative errors and variances. This is because the input value is from a single region for both methods, and the computation procedure for both is the same except for the scaling by powers of two. Further, the accuracy can be improved by enabling auto-vectorization by the compiler. This is because the number of accumulators in each region changes in the Blue algorithm and the Two-Accumulator method. The accumulation of rounding errors can be mitigated by varying the number of accumulators when calculating the sum [3].

For DD-type quadruple precision, the maximum and minimum absolute values that can be represented do not change, because the exponent values are the same as in double precision. Therefore, even if the straightforward norm calculation algorithm is implemented using DD arithmetic, the conditions for the occurrence of overflow and underflow are unchanged. The use of DD arithmetic extends the number of bits in the mantissa part and prevents the accumulation of rounding errors from propagating to the double-precision output. The relative error between the Blue algorithm and the Two-Accumulator method is smaller than that combined with double-precision arithmetic.

CPairSum is a fast algorithm for DD addition (QuadAdd), so the accuracy is approximately the same as combined with QuadAdd.

Table 7. Relative error of each method combined with QuadAdd for the input range of $[0, 1]$

n	Straightforward norm calculation method		Blue algorithm		Two-Accumulator method	
	Average	Variance	Average	Variance	Average	Variance
10	0.38	0.07	0.38	0.07	0.38	0.07
100	0.39	0.07	0.39	0.07	0.39	0.07
1000	0.47	0.09	0.47	0.09	0.47	0.09
10000	0.30	0.04	0.30	0.04	0.30	0.04
100000	0.42	0.07	0.42	0.07	0.42	0.07

Table 8. Relative error of each method combined with CPairSum for the input range of $[0, 1]$

n	Straightforward norm calculation method		Blue algorithm		Two-Accumulator method	
	Average	Variance	Average	Variance	Average	Variance
10	0.38	0.07	0.38	0.07	0.38	0.07
100	0.39	0.07	0.39	0.07	0.39	0.07
1000	0.47	0.09	0.47	0.09	0.47	0.09
10000	0.30	0.04	0.30	0.04	0.30	0.04
100000	0.42	0.07	0.42	0.07	0.42	0.07

Performance. The performance of each method combined with double-precision arithmetic is shown in Fig. 2, and the performance combined with QuadAdd and CPairSum is shown in Fig. 3. GFLOPS is usually used as a measure of the performance of the Euclidean norm, but GB/s is used for the vertical axes in Figs. 2 and 3. The reason for this is that the computational amount varies slightly depending on the method and input values, and since DD arithmetic is used, GB/s, which is the unit of input vector size divided by the execution time, is more suitable for performance comparison. The straightforward norm calculation algorithm has overflow and underflow for input ranges other than $[0, 1]$ and $[2^{-300}, 2^{300}]$. Therefore, this excludes a performance comparison. First, we focus on the extent to which the process of suppressing overflow and underflow affects the performance. In the case of the implementation combined with double precision in the input range of $[0, 1]$, the performance of the straightforward norm calculation algorithm is twice that of the Two-Accumulator method. However, the performance of the implementations combined with QuadAdd and CPairSum is almost identical. This is because the Euclidean norm calculation is memory-intensive at double precision, but computation-intensive at the DD-type quadruple-precision level. The implementation combined with double precision is faster in the order of the straightforward norm calculation algorithm, the

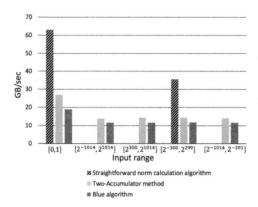

Fig. 2. Performance of implementations combined with double-precision arithmetic

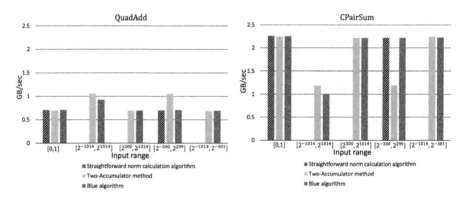

Fig. 3. Performance evaluation when combined with QuadAdd and CPairSum

Two-Accumulator method, and the Blue algorithm. The reason for this is that each method is vectorized by the compiler's automatic vectorization. Implementations that use double-precision arithmetic are vectorized by the compiler, but implementations that use DD-type quadruple-precision arithmetic are not vectorized, because there is a dependency in the process of adding the squares together. Further, the vectorization rate is higher for the straightforward norm calculation algorithm, the Two-Accumulator method, and the Blue algorithm, in that order. Therefore, the Two-Accumulator method is easier to speed up than the Blue algorithm because the number of conditional branches greatly affects the vectorization rate.

In the implementation combined with QuadAdd, the Two-Accumulator method is faster than or nearly as fast as the Blue algorithm for all input ranges. The Two-Accumulator method is faster than the Blue algorithm when all the accumulators are stored, such as $[2^{-1014}, 2^{1014}]$. For the other input ranges, the Two-Accumulator method and the Blue method are nearly identical in performance.

The performance of the implementation combined with CPairSum is up to 3.3 times faster than the implementation combined with QuadAdd. However, the Blue algorithm is faster than the Two-Accumulator method in the input range of $[2^{-300}, 2^{299}]$ only implemented using CPairSum. One possible reason for this is the relationship between the number of unpredictable conditional branches and the number of operations. The Blue algorithm processes this range of input as a single region, while the Two-Accumulator method processes it as two regions. Therefore, in this range, the Two-Accumulator method is being used under conditions such that the CPU is likely to fail in branch prediction. Therefore, the implementation combined with QuadAdd is less likely to show this difference than the implementation CPairSum because more operations are required for DD addition. On the other hand, in the implementation combined with CPairSum, the penalty for branch prediction failure is more likely to be reflected in the result, and the Blue algorithm is considered to be faster than the Two-Accumulator method.

6 Summary and Future Work

In this paper, we propose the Two-Accumulator method as an algorithm for calculating the Euclidean norm with less overflow and underflow. We combine this novel method with DD arithmetic to achieve higher accuracy. Further, we use CPairSum, which is an algorithm to speed up DD addition. The algorithms for calculating the Euclidean norm are compared in terms of input range, vector size, relative error, and performance. In terms of ease of overflow and underflow, the Two-Accumulator method is inferior to the Blue algorithm. However, in terms of accuracy, the Two-Accumulator method and the Blue algorithm are nearly identical in performance, even when combined with an accuracy expansion method. The Two-Accumulator method performs better than the Blue algorithm in the input range of $[2^{-1014}, 2^{1014}]$ combined with QuadAdd. However, the Two-Accumulator method combined with CPairSum is significantly slower than the Blue algorithm, depending on the input range. Vectorization of the Two-Accumulator method represents fruitful terrain for future research. In vectorization, the branching process changes significantly, and the number of branches has a large impact on performance. The advantage of the Two-Accumulator method with fewer conditional branches is emphasized, and the Two-Accumulator method has the potential to be faster in all ranges, as in the performance of implementations combined with double-precision arithmetic.

Acknowledgment. This research was supported in part by the Multidisciplinary Cooperative Research Program (Cygnus) in the Center for Computational Sciences (CCS), University of Tsukuba.

References

1. GMP: The GNU Multiple Precision Arithmetic Library. https://gmplib.org
2. Bailey, D.H.: QD: A double-double and quad-double package for Fortran and C++. https://www.davidhbailey.com/dhbsoftware/
3. Blanchard, P., Higham, N.J., Mary, T.: A class of fast and accurate summation algorithms. SIAM J. Sci. Comput. **42**, 1541–1557 (2020)
4. Blue, J.L.: A portable Fortran program to find the Euclidean norm of a vector. ACM Trans. Math. Softw. **4**(1), 15–23 (1978)
5. Dekker, T.J.: A floating-point technique for extending the available precision. Numer. Math. **18**(3), 224–242 (1971)
6. Hanrot, G., Lefèvre, V., Pélissier, P., Théveny, P., Zimmermann, P.: The GNU MPFR Library. https://www.mpfr.org/
7. Hanson, R.J., Hopkins, T.: Remark on algorithm 539: a modern Fortran reference implementation for carefully computing the Euclidean norm. ACM Trans. Math. Softw. **44**(3), Article 24 (2017)
8. Hida, Y., Li, X.S., Bailey, D.H.: Algorithms for quad-double precision floating point arithmetic. In: Proceedings of the 15th IEEE Symposium on Computer Arithmetic, pp. 155–162 (2001)
9. Karp, A.H., Markstein, P.: High-precision division and square root. ACM Trans. Math. Softw. **23**(4), 561–589 (1997)
10. Knuth, D.E.: The Art of Computer Programming, Volume 2: Seminumerical Algorithms, 3rd edn. Addison-Wesley, Boston (1997)
11. Lange, M., Rump, S.M.: Faithfully rounded floating-point computations. ACM Trans. Math. Softw. **46**(3), Article 21 (2020)
12. Li, X.S., et al.: Design, implementation and testing of extended and mixed precision BLAS. ACM Trans. Math. Softw. **28**(2), 152–205 (2002)
13. Nakata, M.: The MPACK (MBLAS/MLAPACK); a multiple precision arithmetic version of BLAS and LAPACK (2010). http://mplapack.sourceforge.net/
14. Rump, S.M.: Error bounds for computer arithmetics. In: 2019 IEEE 26th Symposium on Computer Arithmetic (ARITH), pp. 1–14 (2019)
15. Xianyi, Z., Qian, W., Chothia, Z.: OpenBLAS. http://www.openblas.net

A Computational Analysis of the Hopmoc Method Applied to the 2-D Advection-Diffusion and Burgers Equations

D. T. Robaina[1]([⊠]), M. Kischinhevsky[2], S. L. Gonzaga de Oliveira[3], A. C. Sena[4], and Mario João Junior[4]

[1] Escola Superior de Propaganda e Marketing, Rio de Janeiro, RJ, Brazil
professor.robaina@gmail.com
[2] UFF, Niterói, RJ, Brazil
kisch@ic.uff.br
[3] Universidade Federal de Lavras, Lavras, MG, Brazil
sanderson@ufla.br
[4] Universidade do Estado do Rio de Janeiro, Lavras, MG, Brazil
asena@ime.uerj.br, jmario@ime.uerj.br

Abstract. This paper shows the accuracy of the Hopmoc method when applied to a partial differential equation that combines both nonlinear propagation and diffusive effects. Specifically, this paper shows the numerical results yielded by the Hopmoc algorithm when applied to the 2-D advection-diffusion and Burgers equations. The results delivered by the Hopmoc method compare favorably with the Crank-Nicolson method and an alternating direction implicit scheme when applied to the advection-diffusion equation. The experiments with the 2-D Burgers equation also show that the Hopmoc algorithm provides results in agreement with several existing methods.

Keywords: Hopscotch method · Modified method of characteristics · Semi-lagrangian approach · Two-dimensional advection-diffusion equation · Two-dimensional Burgers equation

1 Introduction

Practitioners apply various approaches in the approximate solution of differential equations. Some of these approaches are based on a decomposition of operators, such as the alternating direction method [1–3], Implicit-Explicit Method (IMEX) [4,5], and the Odd-Even Hopscotch method [6–8].

The Hopmoc algorithm (see [9] and references therein) joins concepts of the Odd-Even Hopscotch method and the modified method of characteristics (MMOC for short) [10]. The Hopmoc algorithm is similar to the Odd-Even Hopscotch method because both approaches decompose the grid points into two

© Springer Nature Switzerland AG 2021
O. Gervasi et al. (Eds.): ICCSA 2021, LNCS 12949, pp. 111–120, 2021.
https://doi.org/10.1007/978-3-030-86653-2_8

subsets. The two subsets have their unknowns separately updated within a one-time semi-step. Moreover, each subset performs one explicit update and one implicit update of its unknowns. The Hopmoc algorithm evaluates time semi-steps along characteristic lines in a semi-Lagrangian approach as the MMOC operates. Recently, we showed numerical simulations of the Hopmoc method applied to the 1-D modified Burgers equation [11, 12].

This paper shows the computational results yielded by the Hopmoc method when applied to the 2-D advection-diffusion equation. This paper also compares the results provided by the Hopmoc algorithm with six recent approaches when applied to the 2-D Burgers equation.

The remainder of this paper is structured as follows. Section 2 describes the Hopmoc method. Section 3 shows the numerical results. Finally, Sect. 4 addresses the conclusions.

2   Hopmoc Method

The Hopmoc method operates similarly to the Odd-Even Hopscotch method and uses an MMOC-like strategy. One can write a differential operator for the Hopmoc algorithm as

$$L_h \overline{\overline{u}}_{i,j}^n = d \left[\frac{\overline{\overline{u}}_{i-1,j}^n - 2\overline{\overline{u}}_{i,j}^n + \overline{\overline{u}}_{i+1,j}^n}{(\Delta x)^2} + \frac{\overline{\overline{u}}_{i,j-1}^n - 2\overline{\overline{u}}_{i,j}^n + \overline{\overline{u}}_{i,j+1}^n}{(\Delta y)^2} \right]. \tag{1}$$

Using operator (1), the Hopmoc method performs two semi-steps

$$\overline{u}_{i,j}^{n+\frac{1}{2}} = \overline{\overline{u}}_{i,j}^n + \delta t \left(\theta_{i,j}^n L_h \overline{\overline{u}}_{i,j}^n + \theta_{i,j}^{n+\frac{1}{2}} L_h \overline{u}_{i,j}^{n+\frac{1}{2}} \right) \text{ and}$$

$$u_{i,j}^{n+1} = \overline{u}_{i,j}^{n+\frac{1}{2}} + \delta t \left(\theta_{i,j}^{n+\frac{1}{2}} L_h \overline{u}_{i,j}^{n+\frac{1}{2}} + \theta_{i,j}^{n+1} L_h u_{i,j}^{n+1} \right)$$

where

$$\theta_{i,j}^n = \begin{cases} 1, & \text{if } (n+i+j) \text{ is odd, and} \\ 0, & \text{if } (n+i+j) \text{ is even} \end{cases}$$

with $\left(\overline{x}_i^{n+\frac{1}{2}}, \overline{y}_i^{n+\frac{1}{2}} \right) = (x_i - v_1 \delta t, y_j - v_2 \delta t)$ and $\left(\overline{\overline{x}}_i^n, \overline{\overline{y}}_i^n \right) = (x_i - 2v_1 \delta t, y_j - 2v_2 \delta t)$.

Each time step with size Δt in the Hopmoc method encompasses three stages as follows [9].

1. At time step t_n, the Hopmoc method obtains $\overline{\overline{u}}$ for all stencil points $\left(\overline{\overline{x}}_i, \overline{\overline{y}}_j \right)$, for $i = 1, ..., N$ and $j = 1, ..., N$ using an interpolation method.
2. At time semi-step $t_{n+\frac{1}{2}}$, the method computes $\overline{u}_{i,j}^{n+\frac{1}{2}}$ using the explicit (implicit) operator for stencil points for which $n + 1 + i + j$ is odd (even).

3. At time semi-step t_{n+1}, the method computes $u_{i,j}^{n+1}$ using the implicit (explicit) operator for stencil points for which $n + 1 + i + j$ is odd (even).

The method solves neighbor stencil points at the previous time semi-step. Thus, the implicit approach does not demand to solve a linear system in the same way as the Odd-Even method performs. Figure 1 shows two time semi-steps of the Hopmoc algorithm.

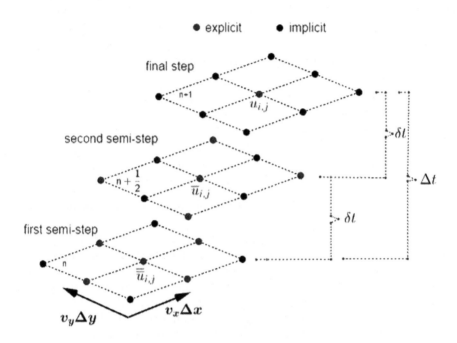

Fig. 1. Two time semi-steps of the Hopmoc method.

3   Computational Analysis

This section shows the results yielded by the Hopmoc method along with a bilinear interpolation [13]. The MacBook Air workstation used in the executions of the simulations featured an Intel® Core™ i5 1.6 GHz processor (Intel; Santa Clara, CA, United States). We implemented the method using the C programming language.

Section 3.1 shows the results yielded by the Hopmoc method when applied to the 2-D advection-diffusion equation and compares the results returned by the Hopmoc method with the Crank-Nicolson and an alternating direction implicit scheme when applied to the same equation. Section 3.2 compares the results delivered by the Hopmoc algorithm with six existing methods when applied to the 2-D Burgers equation.

3.1 Advection-Diffusion Equation

Consider the numerical solution of the 2-D advection-diffusion equation

$$u_t + v_1 u_x + v_2 u_y - \frac{1}{Re}(u_{xx} + u_{yy}) = 0 \tag{2}$$

where $Re = \frac{\rho \cdot \nu \cdot L}{\mu}$ is the Reynolds number, ρ is the density, ν is the characteristic velocity, L is a characteristic linear dimension, and μ is the dynamic viscosity. Eq. (2) is defined in the domain $\Omega = \{(x, y) : 0 \le x \le 1; 0 \le y \le 1\}$ with analytical solution $u(x, y, t) = cos(\pi(x - v_x t))cos(\pi(y - v_y t))e^{-2Re \cdot \pi^2 \cdot t}$ and boundary conditions $u(x, y, t) = cos(\pi(x - v_x t))cos(\pi(y - v_y t))e^{-2Re \cdot \pi^2 \cdot t}$. The accuracy of the approach is measured in terms of errors norms

$$\|\epsilon\|_\infty = \|u_{exact} - u_{computed}\|_\infty = \max_{i,j} \left| u_{i,j}^{exact} - u_{i,j}^{computed} \right|$$

and

$$\|\epsilon\|_2 = \|u_{exact} - u_{computed}\|_2 = \left(\sum_{i=0}^{N} \sum_{j=0}^{N} \left| u_{i,j}^{exact} - u_{i,j}^{computed} \right|^2 \right)^{1/2}$$

where u_{exact} and $u_{computed}$ represent the exact and computed solutions, respectively.

Table 1 shows error norms $\|\epsilon\|_\infty$ and $\|\epsilon\|_2$ when applying the Hopmoc method to Eq. (2). The first part of Table 1 shows the results yielded by the Hopmoc method when varying δt, δx, and δy for the same $\frac{\delta t}{\Delta x \cdot \delta y}$. The second part of Table 1 reveals that the method yields small numerical error when using different Reynolds number regimes. In particular, for a very small Reynolds number regime, the numerical errors are large when compared with the other simulations. This phenomenon indicates that the method presents characteristics to be employed to turbulent flows [14]. Table 2 shows numerical errors when applying the Hopmoc method to Eq. (2) using $t_f = 10^{-4}$, $Re = 5.0$, and $v_x = v_y = 1.0$.

Table 1. Results yielded by the Hopmoc method when applied to Eq. (2) using $\delta t = 10^{-5}$, $t_f = 0.8$, $v_x = v_y = 1.0$, and several Reynolds number regimes (t denotes time and s denotes seconds).

Mesh	Re	$\|\epsilon\|_\infty$	$\|\epsilon\|_2$	t(s)
10 × 10	1.00	2.2104e−02	1.2495e−02	0.4
20 × 20		2.2540e−02	1.2472e−02	0.4
30 × 30		2.2653e−02	1.2453e−02	0.4
40 × 40		2.2703e−02	1.2443e−02	0.6
50 × 50	100.00	1.8454e−02	4.0593e−03	0.7
	10.00	1.8454e−02	4.5660e−03	0.7
	1.00	2.2740e−02	1.2436e−02	0.6
	0.10	3.2200e−02	2.0704e−02	0.6
	0.01	3.5453e−02	2.3400e−02	0.7

Table 3 shows the results yielded by the Hopmoc method in simulations using different values for δt. In simulations using the same Δx and Δy, we conclude that when δt tends to zero, the numerical errors of the method tend to zero. These results show that when respecting conditions of consistency, the Hopmoc is unconditionally stable.

Table 2. Results yielded by the Hopmoc method when applied to Eq. (2) using $Re = 5.0$, $t_f = 10^{-4}$, $v_x = v_y = 1.0$.

$\delta t \times 10^{-5}$	Mesh	$\|\epsilon\|_\infty$	$\|\epsilon\|_2$	t(s)
1.0000	10×10	2.2104e−02	1.2495e−02	0.4
0.2500	20×20	2.2541e−02	1.2473e−02	1.7
0.1111	30×30	2.2655e−02	1.2455e−02	4.3
0.0625	40×40	2.2705e−02	1.2444e−02	10.9
0.0400	50×50	2.2741e−02	1.2437e−02	17.4

Table 3. Results yielded by the Hopmoc method when applied to Eq. (2) using $Re = 1.0$, $t_f = 0.8$, 50×50 grid points, and $v_x = v_y = 1.0$.

$\delta t \times 10^{-5}$	$\|\epsilon\|_\infty$	$\|\epsilon\|_2$	t(s)
100.00	8.2135e+04	1.1126e+04	0.01
10.00	2.2725e−02	1.2419e−02	0.07
1.00	2.2740e−02	1.2436e−02	0.62
0.10	2.2741e−02	1.2437e−02	6.90
0.01	2.2741e−02	1.2437e−02	68.06

3.2   Burgers Equation

The Burgers equation is a simple equation used to comprehend the properties of the Navier-Stokes equations. Furthermore, one obtains the Burgers equation as a result of linking nonlinear wave motion with linear diffusion. As previously mentioned, it is the simplest model for analyzing the combined effect of nonlinear advection and diffusion. As in the Navier-Stokes equations, one can determine a Reynolds number, which indicates the ratio between the advective and the viscous contribution in a flow. The simulation of the flow evolution requires the use of accurate and robust numerical methods [15]. Thus, the Burgers equation arises as a mathematical model from many physical events. Examples of these situations are environmental protection, the flow of a shock wave traveling in a viscous fluid, fluid mechanics, gas dynamic, hydrology, nonlinear acoustics, sedimentation of two kinds of particles in fluid suspensions under the effect of gravity. Other examples are shallow-water waves, acoustic transmission, traffic flow, turbulence, supersonic flows, and wave propagation in a nonlinear thermoelastic media (see [16] and references therein).

This section shows two experiments. The first compares the results yielded by the Hopmoc method with the Crank-Nicolson method and an alternating direction implicit scheme. The second compares the results produced by the Hopmoc algorithm with six other existing methods.

Comparison with Crank-Nicolson Method and an Alternating Direction Implicit Scheme. This section compares the results yielded by the Hopmoc method with Crank-Nicolson (CN) [17] and the alternating direction implicit (ADI) scheme presented by Saqib et al. [18] when applied to the equation $u_t + uu_x + uu_y - \frac{1}{Re}(u_{xx} + u_{yy}) = 0$ where $(x, y, t) \in \Omega \times (0, T]$ with initial conditions $u(x, y, 0) = u_0(x, y)$, $(x, y) \in \Omega$ given by the exact solution $u(x, y, t) = \frac{1}{1 + \frac{(x+y+t)Re}{2}}$ defined in $\Omega = [(x, y) : 0 \leq x, y \leq 1]$. Table 4 (5) shows the results yielded by the Hopmoc method, Crank-Nicolson method, and the alternating direction implicit scheme presented by Saqib et al. [18] in typical grid points when applied to Burgers equation using Reynolds number $Re = 200$, $t_f = 3.0$ $(t_f = 1.0)$, $\delta t = 0.0001$, and 20×20 (30×30) grid points.

Table 4. Results yielded by the Hopmoc method, Crank-Nicolson method, and an alternating direction implicit scheme [18] when applied to Burgers equation using Reynolds number $Re = 200$, $t_f = 3.0$, $\Delta t = 0.0001$, and 20×20 grid points.

(x, y)	Exact	Hopmoc	ADI	CN
(0.20,0.20)	1.0000	1.0000	1.0000	1.0000
(0.05,0.05)	1.0000	1.0000	1.0000	1.0000
(0.20,0.05)	1.0000	1.0000	1.0000	1.0000
(0.15,0.15)	1.0000	1.0000	0.9999	0.9998
(0.30,0.30)	1.0000	1.0000	0.9999	0.9998
(0.30,0.20)	1.0000	1.0000	0.9997	0.9999
(0.70,0.70)	1.0000	1.0000	1.0000	1.0000
(0.70,0.15)	1.0000	1.0000	0.9999	0.9998
(0.80,0.30)	1.0000	1.0000	0.9889	0.9998
(0.90,0.90)	1.0000	1.0000	1.0000	1.0000
(0.25,0.75)	1.0000	0.9988	0.9888	0.9887
(0.75,0.25)	1.0000	1.0000	0.9888	0.9887

Table 5. Results yielded by the Hopmoc method, Crank-Nicolson and an alternating direction implicit scheme [18] when applied to Burgers equation using Reynolds number $Re = 200$, $t_f = 1.0$, $\Delta t = 0.0001$, and 30×30 grid points.

(x,y)	Exact	Hopmoc	ADI	CN
(0.2,0.2)	1.0000	1.0000	1.0000	1.0000
(0.3,0.3)	1.0000	1.0000	0.8515	0.8415
(0.3,0.2)	1.0000	1.0000	0.9000	0.9112
(0.7,0.7)	1.0000	1.0000	0.7515	0.7414
(0.8,0.3)	1.0000	1.0000	0.9876	0.9866
(0.9,0.9)	1.0000	0.9996	0.9996	0.9954

Comparison with Six Other Existing Methods This section illustrates the accuracy of the Hopmoc method when applied to the system of 2-D Burgers equations

$$\begin{cases} u_t + uu_x + vu_y = \frac{1}{Re}\left(u_{xx} + u_{yy}\right) \\ v_t + uv_x + vv_y = \frac{1}{Re}\left(v_{xx} + v_{yy}\right) \end{cases} \tag{3}$$

with initial conditions $u(x,y,0) = f(x,y)$, $v(x,y,0) = g(x,y)$, $(x,y) \in \Omega$, and boundary conditions $u(x,y,t) = \varphi(x,y,t)$, $v(x,y,t) = \psi(x,y,t)$, $x,y \in \partial\Omega, t > 0$ where $\Omega = \{(x,y) : a \le x \le b, a \le y \le b\}$ and $\partial\Omega$ is its boundary, f, g, φ, and ψ are known functions, $u(x,y,t)$ and $v(x,y,t)$ are the velocity componentes to be determined.

We consider the system of 2-D Burgers Eq. (3) with exact solutions $u(x,y,t) = v(x,y,t) = \frac{3}{4} - \dfrac{1}{4\left(1+e^{(-4x+4y-t)\frac{Re}{32}}\right)}$ using $\Delta x = \Delta y = 0.05$ and

$Re = 100$. The initial and boundary conditions are taken from the exact solution. The computational domain is $\Omega = \{(x,y) : 0 \le x, y \le 1\}$. Table 6 compares the numerical solutions yielded by the Hopmoc method with exact solution and six existing methods:

- an exponential finite-difference (Expo-FD) method [19];
- an implicit logarithmic finite-difference method (I-LFDM) [20];
- a modified cubic B-spline differential quadrature (MCBDQ) method [21];
- a splitting method (SM) proposed by Shi et al. [16];
- a Crank-Nicolson/Adams-Bashforth (CNAB) scheme [22]
- the alternating direction implicit (ADI) method proposed by Çelikten and Aksan [23].

Table 6 shows the results yielded by seven methods when applied to the system of Burgers Eq. (3) for $k = 0.0001$ and $t_f = 0.5$ at typical grid points. The table shows that the results yielded by the Hopmoc method are compatible with the six other methods.

Table 6. Comparison of numerical and exact solutions of $u(x,y,t)$ and $v(x,y,t)$ for $Re = 100$, $k = 0.0001$, $\Delta x = \Delta y = 0.05$, and $t_f = 0.5$ at some typical grid points.

	(x,y)	Exact	SM	MCBDQ	Expo-FD	I-LFDM	ADI	CNAB	Hopmoc
$u(x,y,t)$	(0,1,0,1)	0.54332	0.54336	0.54412	0.54300	0.54300	0.54299	0.55979	0.54325
	(0,5,0,1)	0.50035	0.50035	0.50037	0.50034	0.50034	0.50034	0.50160	0.50041
	(0,9,0,1)	0.50000	0.50000	0.50000	0.50000	0.50000	0.50000	0.50001	0.50000
	(0,3,0,3)	0.54332	0.54338	0.54388	0.54270	0.54269	0.54268	0.54844	0.53957
	(0,7,0,3)	0.50035	0.50034	0.50037	0.50032	0.50032	0.50032	0.50155	0.50051
	(0,1,0,5)	0.74221	0.74228	0.74196	0.74215	0.74215	0.74215	0.74510	0.73468
	(0,5,0,5)	0.54332	0.54333	0.54347	0.54252	0.54251	0.54249	0.54733	0.53397
	(0,9,0,5)	0.50035	0.50034	0.50035	0.50030	0.50030	0.50030	0.50155	0.50056
	(0,3,0,7)	0.74221	0.74236	0.74211	0.74212	0.74211	0.74211	0.74324	0.71934
	(0,7,0,7)	0.54332	0.54332	0.54327	0.54247	0.54246	0.54245	0.54733	0.52921
	(0,1,0,9)	0.74995	0.74995	0.74994	0.74994	0.74994	0.74994	0.74996	0.74981
	(0,5,0,9)	0.74221	0.74237	0.74219	0.74210	0.74210	0.74210	0.74330	0.70472
	(0,9,0,9)	0.54332	0.54331	0.54333	0.54229	0.54228	0.54227	0.54616	0.52726
$v(x,y,t)$	(0,1,0,1)	0.95668	0.956650	0.95589	0.95700	0.95700	0.95701	0.95021	0.95675
	(0,5,0,1)	0.99965	0.99965	0.99963	0.99966	0.99966	0.99966	0.99840	0.99959
	(0,9,0,1)	1.00000	1.00000	1.00000	1.00000	1.00000	1.00000	0.99999	1.00000
	(0,3,0,3)	0.95668	0.95662	0.95612	0.95731	0.95731	0.95732	0.95156	0.96043
	(0,7,0,3)	0.99965	0.99966	0.99964	0.99968	0.99968	0.99968	0.99845	0.99949
	(0,1,0,5)	0.75779	0.75772	0.75804	0.75785	0.75785	0.75785	0.75490	0.76532
	(0,5,0,5)	0.95668	0.95667	0.95654	0.95749	0.95749	0.95751	0.95267	0.96603
	(0,9,0,5)	0.99965	0.99966	0.99965	0.99970	0.99970	0.99970	0.99845	0.99944
	(0,3,0,7)	0.75779	0.75764	0.75789	0.75789	0.75789	0.75789	0.75676	0.78066
	(0,7,0,7)	0.95668	0.95668	0.95673	0.95754	0.95754	0.95755	0.95267	0.97079
	(0,1,0,9)	0.75005	0.75005	0.75006	0.75006	0.75006	0.75006	0.75004	0.75019
	(0,5,0,9)	0.75779	0.75763	0.75781	0.75790	0.75790	0.75790	0.75670	0.79528
	(0,9,0,9)	0.95668	0.95669	0.95667	0.95772	0.95772	0.95773	0.95384	0.97274

4 Conclusions

This paper shows the computational results yielded by the Hopmoc method when applied to the two-dimensional advection-diffusion equation. The Hopmoc algorithm yielded better results than did the Crank-Nicolson and the alternating direction implicit scheme presented by Saqib et al. [18] when applied to the advection-diffusion equation. The results yielded by the Hopmoc method when applied to the 2-D Burgers equation are compatible with six other recent studies presented in the literature.

We plan to integrate the Hopmoc method with backward differential formulas in future works. We also intend to study the Hopmoc algorithm applied to the two-dimensional advection-diffusion equation in conjunction with total variation diminishing techniques [24] and flux-limiting procedures to improve its accuracy.

Researchers solve very large-scale problems in the present day with the use of parallel computations. Massive problems today are of the order of billions of degrees of freedom (as examples, see experiments in parallel using structural [25] and computational fluid dynamics problems [26]). We plan to apply the Hopmoc method in parallel implementations using OpenMP, Galois, and Message Passing Interface systems.

References

1. Davis, M.E.: Numerical Methods and Modeling for Chemical Engineers. John Wiley and Sons Inc., Hoboken (2001)
2. Kajishima, T., Taira, K.: Large-eddy simulation. In: Computational Fluid Dynamics, pp. 269–307. Springer, Cham (2017). https://doi.org/10.1007/978-3-319-45304-0_8
3. Duffy, D.J.: Finite Difference Methods in Financial Engineering a Partial Differential Equation Approach. John Wiley and Sons Inc., Hoboken (2006)
4. Ascher, U.M., Ruuth, S.J., Spiteri, R.J.: Implicit-explicit runge-kutta methods for time-dependent partial differential equations. Appl. Numer. Math. **25**(2–3), 151–167 (1997)
5. Pareschi, L., Russo, G.: Implicit-explicit runge-kutta schemes for stiff systems of differential equations. Recent Trends Numer. Anal. **3**, 269–289 (2000)
6. Gourlay, P.: Hopscotch: a fast second order partial differential equation solver. J. Inst. Math. Appl. **6**, 375–390 (1970)
7. Gourlay, A.R.: Some recent methods for the numerical solution of time-dependent partial differential equations. Proc. R. Soc. Math. Phys. Eng. Sci. Ser. A **323**(1553), 219–235 (1971)
8. Gourlay, A.R., Morris, J.L.: Hopscotch difference methods for nonlinear hyperbolic systems. IBM J. Res. Dev. **16**, 349–353 (1972)
9. Oliveira, S., Kischinhevsky, M., Gonzaga de Oliveira, S.L.: Convergence analysis of the Hopmoc method. Int. J. Comput. Math. **86**, 1375–1393 (2009)
10. Douglas, J., Jr. Russell, T.F.: Numerical methods for convection-dominated diffusion problems based on combining the method of characteristics with finite element or finite difference procedures. SIAM J. Numer. Anal. **19**(5), 871–885 (1982)
11. Robaina, D.T.: BDF-Hopmoc: um método implícito de passo múltiplo para a solução de Equações Diferenciais Parciais baseado em atualizações espaciais alternadas ao longo das linhas características. Ph.D thesis, Universidade Federal Fluminense, Niterói, RJ, Brazil, July 2018
12. Robaina, D.T., Gonzaga de Oliveira, S.L., Kischnhevsky, M., Osthoff, C., Sena, A.C.: Numerical simulations of the 1-d modified Burgers equation. In: Winter Simulation Conference (WSC). National Harbor, MD, USA 2019, pp. 3231–3242 (2019)
13. Press, W.H., Flannery, B.P., Teukolsky, S.A., Vetterling, W.T.: Numerical Recipes in C: The Art of Scientific Computing. 2nd edn. Cambridge University Press, Cambridge (1992)
14. Reynolds, O.: An experimental investigation of the circumstances which determine whether the motion of water shall be direct or sinuous, and of the law of resistance in parallel channels. Philos. Trans. R. Soc. **174**, 935–982 (1883)
15. Orlandi, P.: The burgers equation. In: Orlandi, P., ed.: Fluid Flow Phenomena. Fluid Mechanics and Its Applications, vol. 55, pp. 40–50. Springer, Dordrecht (2000). https://doi.org/10.1007/978-94-011-4281-6_4

16. Shi, F., Zheng, H., Cao, Y., Li, J., Zhao, R.: A fast numerical method for solving coupled Burgers' equations. Numer. Methods Partial Differ. Equ. **33**(6), 1823–1838 (2017)
17. Crank, J., Nicolson, P.: A practical method for numerical evaluation of solutions of partial differential equations of the heat conduction type. Math. Proc. Cambridge Philos. Soc. **43**(1), 50–67 (1947)
18. Saqib, M., Hasnain, S., Mashat, D.S.: Computational solutions of two dimensional convection diffusion equation using Crank-Nicolson and time efficient ADI. Am. J. Comput. Math. **7**, 208–227 (2017)
19. Srivastava, V.K., Singh, S., Awasthi, M.K.: Numerical solutions of coupled Burgers' equations by an implicit finite-difference scheme. AIP Adv. **3**(8), 082131 (2013)
20. Srivastava, V.K., Awasthi, M.K., Singh, S.: An implicit logarithm finite difference technique for two dimensional coupled viscous Burgers' equation. AIP Adv. **3**(12), 122105 (2013)
21. Shukla, H.S., Srivastava, M.T.V.K., Kumar, J.: Numerical solution of two dimensional coupled viscous Burgers' equation using the modified cubic B-spline differential quadrature method. AIP Adv. **4**(11), 117134 (2014)
22. Zhang, T., Jin, J., HuangFu, Y.: The Crank-Nicolson/Adams-Bashforth scheme for the Burgers equation with H^2 and H^1 initial data. Appl. Numer. Math. **125**, 103–142 (2018)
23. Çelikten, G., Aksan, E.N.: Alternating direction implicit method for numerical solutions of 2-D Burgers equations. Thermal Sci. **23**(1), S243–S252 (2019)
24. Harten, A.: High resolution schemes for hyperbolic conservation laws. J. Comput. Phys. **49**, 357–393 (1983)
25. Toivanen, J., Avery, P., Farhat, C.: A multilevel FETI-DP method and its performance for problems with billions of degrees of freedom. Int. J. Numer. Method Eng. **116**(10–11), 661–682 (2018)
26. Smith, C.W., Abeysinghe, E., Marru, S., Jansen, K.E.: PHASTA science gateway for high performance computational fluid dynamics. In: PEARC '18 - Proceedings of the Practice and Experience on Advanced Research Computing, Pittsburgh, PA, ACM, vol. 94, July 2018

Optimal ϑ-Methods for Mean-Square Dissipative Stochastic Differential Equations

Raffaele D'Ambrosio and Stefano Di Giovacchino[✉]

Department of Information Engineering and Computer Science and Mathematics,
University of L'Aquila, L'Aquila, Italy
raffaele.dambrosio@univaq.it, stefano.digiovacchino@graduate.univaq.it

Abstract. In this paper, we address our investigation to the numerical integration of nonlinear stochastic differential equations exhibiting a mean-square contractive character along the exact dynamics. We specifically focus on the conservation of this qualitative feature along the discretized dynamics originated by applying stochastic ϑ-methods. Retaining the mean-square contractivity under time discretization is translated into a proper stepsize restriction. Here we analyze the choice of the optimal parameter ϑ making this restriction less demanding and, at the same time, maximizing the stability interval. A numerical evidence is provided to confirm our theoretical results.

Keywords: Stochastic differential equations · Stochastic ϑ-methods · Mean-square dissipativity · Mean-square contractivity

1 Introduction

We consider a system of stochastic differential equations (SDE) of Itô type assuming the following differential form

$$\begin{cases} dX(t) = f(X(t))dt + g(X(t))dW(t), & t \in [0, T], \\ X(0) = X_0, \end{cases} \tag{1}$$

where $f : \mathbb{R}^n \to \mathbb{R}^n$, $g : \mathbb{R}^n \to \mathbb{R}^{n \times m}$ and $W(t)$ is a m-dimensional Wiener process. For theoretical results on the existence and uniqueness of solutions to (1) we refer, for instance, to the classical monograph [25]. Moreover, in the sequel, we assume that the diffusive term in (1) is commutative.

We focus our attention on providing a nonlinear stability analysis of the following stochastic ϑ-methods for (1) that, with reference to the discretized domain

$$\mathcal{I}_{\Delta t} = \{t_n = n\Delta T, \ n = 0, 1, \dots, N, \ N = T/\Delta t\},$$

This work is supported by GNCS-INDAM project and by PRIN2017-MIUR project 2017JYCLSF "Structure preserving approximation of evolutionary problems".

© Springer Nature Switzerland AG 2021
O. Gervasi et al. (Eds.): ICCSA 2021, LNCS 12949, pp. 121–134, 2021.
https://doi.org/10.1007/978-3-030-86653-2_9

assume the following forms:

$$X_{n+1} = X_n + (1 - \vartheta)\Delta t f(X_n) + \vartheta \Delta t f(X_{n+1}) + g(X_n)\Delta W_n, \qquad (2)$$

$$X_{n+1} = X_n + (1 - \vartheta)\Delta t f(X_n) + \vartheta \Delta t f(X_{n+1}) + \sum_{j=1}^{m} g^j(X_n)\Delta W_n^j \qquad (3)$$

$$+ \frac{1}{2}\sum_{j=1}^{m} L^j g^j(X_n)((\Delta W_n^j)^2 - \Delta t) + \frac{1}{2}\sum_{\substack{j_1,j_2=1 \\ j_1 \neq j_2}}^{m} L^{j_1} g^{j_2}(X_n)\Delta W_n^{j_1}\Delta W_n^{j_2},$$

where $\vartheta \in [0,1]$. X_n is the approximate value for $X(t_n)$ and ΔW_n is the discretized Wiener increment, distributed as a gaussian random variable with zero mean and variance Δt. The operator L^j is defined as

$$L^j = \sum_{k=1}^{n} g^{k,j}\frac{\partial}{\partial x^k}, \quad j = 1,...,m,$$

where $g^j(X_n)$ is the j-th column of the matrix $g(X_n)$ and ΔW_n^j the j-th element of vector ΔW_n. We refer to (2) as ϑ-Maruyama methods and to (3) as ϑ-Milstein methods in their componentwise form. We note that, if $m = 1$, (3) reduces to the form

$$X_{n+1} = X_n + (1 - \vartheta)\Delta t f(X_n) + \vartheta \Delta t f(X_{n+1}) + g(X_n)\Delta W_n + \frac{1}{2}g(X_n)g'(X_n)(\Delta W_n^2 - \Delta t).$$

Convergence and linear stability analysis for stochastic ϑ-methods (2)–(3) has been investigated in [3, 20] and reference therein. Successively, a nonlinear stability analysis has been performed in [12], as well as their conservation properties when applied to linear and nonlinear stochastic oscillators [6, 16]. The idea of ϑ-methods has also been extended to stochastic Volterra integral equations, in recent contributions [9, 10].

1.1   Mean-Square Contractivity

Let us focus on the following relevant issue for nonlinear SDEs [22].

Theorem 1. *For a given nonlinear SDE* (1), *let us assume the following properties for the drift f and the diffusion g, by denoting with $|\cdot|$ both the Euclidean norm in $\mathbb{R}^n$ and the trace (or Frobenius) norm in $\mathbb{R}^{n \times m}$:*

(i) $f, g \in \mathcal{C}^1(\mathbb{R}^n)$;
(ii) f satisfies a one-sided Lipschitz condition, i.e. there exists $\mu \in \mathbb{R}$ such that

$$< x - y, f(x) - f(y) > \leq \mu |x - y|^2, \quad \forall x, y \in \mathbb{R}^n; \qquad (4)$$

(iii) g is a globally Lipschitz function, i.e. there exists $L > 0$ such that

$$|g(x) - g(y)|^2 \leq L|x - y|^2 \quad \forall x, y \in \mathbb{R}^n. \qquad (5)$$

Then, any two solutions $X(t)$ and $Y(t)$ of (1), with $\mathbb{E}\,|X_0|^2 < \infty$ and $\mathbb{E}\,|Y_0|^2 < \infty$, satisfy

$$\mathbb{E}\,|X(t) - Y(t)|^2 \leq \mathbb{E}\,|X_0 - Y_0|^2\, e^{\alpha t}, \tag{6}$$

where $\alpha = 2\mu + L$.

We call the inequality (6) *exponential mean-square stability inequality* for (1). If $\alpha < 0$ in (6), then an exponential decay of the mean-square deviation between two solutions of a given SDE (1) occurs. We provide the following definition that appears as the stochastic counterpart of a similar property for deterministic ordinary differentia equations (see, for instance, [18] and references therein).

Definition 1. *A nonlinear SDE (1) whose solutions satisfy the exponential stability inequality (6) with $\alpha < 0$ is said to be exponential mean-square dissipative and to generate exponential mean-square contractive solutions.*

If $g = 0$ in (1), Definition 1 overlaps with the deterministic analog provided by $\mu = 0$ guaranteeing a dissipative behavior of the deterministic solutions to the corresponding ordinary differential problems. Their discretizations have led to the well-known notion of G-stability of numerical methods, introduced by G. Dahlquist in [11]. In the existing literature, there are several papers in which deterministic and stochastic problems with one-sided Lipschitz constants appear. In particular, they find huge application in finance and economics, neural networks, population dynamics, systems synchronization, stochastic differential equations with jumps, dynamical systems, delay ordinary (and stochastic) differential equations (see [4, 8, 24, 26, 28, 30–32, 36] and references therein).

The results presented in [12] shows that it is possible to retain the mean-square contractivity under time discretization through a proper stepsize restriction. A rigorous discussion of the optimal choice of the parameter ϑ, however, is missing in [12] and filling this gap in represents the goal of this contribution. Indeed, we discuss the choice of the optimal parameter ϑ making the aforementioned stepsize restrictions less demanding and, at the same time, maximizing the stability interval. As observed in [12], exponential mean-square contractivity is certainly a relevant property to be inherited also by the discretized problem, since it ensures a long-term damping of the error along the numerical solutions. Specifically, the investigation led to sharp restrictions on the stepsize employed in the numerical discretization to guarantee the conservation of exponential contractive behavior along the numerical dynamics.

It is worth observing that our research is in line with the idea of structure-preserving numerical integration; contributions in this direction and regarding stochastic numerics are given, for instance, in [1, 2, 5, 7, 13, 27, 29, 33, 35] and references therein.

The paper is organized as follows: in Sect. 2 we report results on linear and nonlinear stability of stochastic ϑ-methods (2)–(3), while in Sect. 3 we focus on the analysis of the optimal choice for the parameter ϑ guaranteeing the least demanding stepsize restriction to retain mean-square contractivity along the exact dynamics and maximal linear stability region; some concluding remarks are object of Sect. 4.

2   Stability Analysis of Stochastic ϑ-Methods

In this section, we summarize the main results about linear and nonlinear stability properties of stochastic ϑ-methods (2)–(3), that provide the starting point for our investigation. This section is totally inspired by [12, 19–21].

2.1   Linear Stability

Linear stability analysis relies on considering the following scalar linear test equation

$$\begin{cases} dX(t) = \lambda X(t)dt + \sigma X(t)dW(t), & t \in [0, T], \\ X(0) = X_0, \end{cases} \tag{7}$$

where $\lambda, \sigma \in \mathbb{C}$. We recall the following definition [19–21].

Definition 2. *The solution $X(t)$ to the linear SDE* (7) *is mean-square stable if*

$$\lim_{t \to \infty} \mathbb{E} |X(t)|^2 = 0. \tag{8}$$

It can been shown that, in order to attain (8), the coefficients of (7) have to satisfy

$$\mathrm{Re}(\lambda) + \frac{1}{2}|\sigma|^2 < 0. \tag{9}$$

Then, it looks natural to provide a numerical counterpart of mean-square stability also under time discretization, as follows.

Definition 3. *A numerical solution X_n to linear SDE* (7) *is said to be mean-square stable if*

$$\lim_{n \to \infty} \mathbb{E} |X_n|^2 = 0. \tag{10}$$

It has been proved in [19, 20] that stochastic ϑ-Maruyama methods (2) are mean-square stable if and only if

$$\frac{|1 + (1 - \vartheta)\Delta t\lambda|^2 + \Delta t |\sigma|^2}{|1 - \vartheta\Delta t\lambda|^2} < 1, \tag{11}$$

while the stochastic ϑ-Milstein satisfies (10) if and only if [12]

$$\left| \beta^2 + \frac{\beta\sigma^2\Delta t}{1 - \vartheta\Delta t\lambda} + \frac{\sigma^2\Delta t + \frac{3}{4}\sigma^4\Delta t^2}{(1 - \vartheta\Delta t\lambda)^2} \right| < 1, \tag{12}$$

with

$$\beta = \frac{1 + (1 - \vartheta)\Delta t\lambda - \frac{1}{2}\sigma^2\Delta t}{1 - \vartheta\Delta t\lambda}. \tag{13}$$

2.2 Nonlinear Stability

Let us now recall some of the tools useful to analyze nonlinear stability of stochastic ϑ-methods (2)–(3), i.e., we aim to retain the same behaviour described by the exponential mean-square stability inequality (6). As regards stochastic ϑ-Maruyama methods, the following theorem holds true [12].

Theorem 2. *Under the assumptions (i)–(iii) given in Theorem 1, any two numerical solutions X_n and Y_n, $n \geq 0$, computed by applying the ϑ-Maruyama method (2) to (1) with initial values such that $\mathbb{E}\,|X_0|^2 < \infty$ and $\mathbb{E}\,|Y_0|^2 < \infty$, satisfy the inequality*

$$\mathbb{E}\,|X_n - Y_n|^2 \leq \mathbb{E}\,|X_0 - Y_0|^2\, e^{\nu(\vartheta,\Delta t)t_n}, \tag{14}$$

where

$$\nu(\vartheta, \Delta t) = \frac{1}{\Delta t}\ln \beta(\vartheta, \Delta t) \tag{15}$$

and

$$\beta(\vartheta, \Delta t) = 1 + \frac{\alpha + (1 - \vartheta)^2 M \Delta t}{1 - 2\vartheta\mu\Delta t}\Delta t, \tag{16}$$

with

$$M = \sup_{t\in[0,T]}\,\mathbb{E}|f'(X(t))|^2. \tag{17}$$

We observe that (14) gives the numerical counterpart of the exponential inequality (6) stated for the SDE (1). A negative value of ν gives the exponential mean-square decay of the gap between two numerical solutions computed by (2). Moreover, it has been proved that the numerical exponent ν approaches the exact one α as $\Delta t \to 0$, as stated by the following result [12].

Theorem 3. *Under the same assumptions of Theorem 2, for any fixed value of $\vartheta \in [0,1]$, we have*

$$|\nu(\vartheta, \Delta t) - \alpha| = \mathcal{O}(\Delta t). \tag{18}$$

Analogous results have also been established for stochastic ϑ-Milstein schemes (3) in [12], as provided by the following theorems.

Theorem 4. *Under the assumptions (i)–(iii) given in Theorem 1, any two numerical solutions X_n and Y_n, $n \geq 0$, computed by applying the ϑ-Milstein method (3) to (1) with initial values such that $\mathbb{E}\,|X_0|^2 < \infty$ and $\mathbb{E}\,|Y_0|^2 < \infty$, satisfy the inequality*

$$\mathbb{E}\,|X_n - Y_n|^2 \leq \mathbb{E}\,|X_0 - Y_0|^2\, e^{\varepsilon(\vartheta,\Delta t)t_n}, \tag{19}$$

where

$$\varepsilon(\vartheta, \Delta t) = \frac{1}{\Delta t}\ln \gamma(\vartheta, \Delta t) \tag{20}$$

and

$$\gamma(\vartheta, \Delta t) = \beta(\vartheta, \Delta t) + \frac{3\widetilde{M}\Delta t^2}{4(1 - 2\vartheta\mu\Delta t)}, \tag{21}$$

with $\widetilde{M}$ defined as

$$\widetilde{M} = \sum_{i,j=1}^{m} \sum_{k,l=1}^{n} \widetilde{M}_{i,j}^{k,l},$$ (22)

where

$$\widetilde{M}_{i,j}^{k,l} = \sup_{t \in [0,T]} \frac{\mathbb{E}\left(h_{i,j}^{k,l}(X(t), Y(t))\right)}{\mathbb{E}|X(t) - Y(t)|^2},$$

being

$$h_{i,j}^{k,l}(X(t), Y(t)) = < g^{k,i}(X(t)) \frac{\partial}{\partial x^k} g^j(X(t)) - g^{k,i}(Y(t)) \frac{\partial}{\partial y^k} g^j(Y(t)),$$

$$g^{l,i}(X(t)) \frac{\partial}{\partial x^l} g^j(X(t)) - g^{l,i}(Y(t)) \frac{\partial}{\partial y^l} g^j(Y(t)) >,$$

$i, j = 1, \ldots, m, \ k, l = 1, \ldots, n$.

Theorem 5. *Under the same assumptions of Theorem 4, for any fixed value of $\vartheta \in [0,1]$, we have*

$$|\varepsilon(\vartheta, \Delta t) - \alpha| = \mathcal{O}(\Delta t).$$ (23)

Clearly, inequality (19) reveals a mean-square contractive behavior along the numerical solutions computed by (3), under the condition $\varepsilon < 0$.

Theorems 2 and 4 are the needed ingredient to provide the numerical counterpart of the condition $\alpha < 0$ in (6) under which, according to Definition 1, the nonlinear problem (1) generates mean-square contractive solutions. This issue is described in [12], as follows.

Definition 4. *Consider a nonlinear stochastic differential Eq. (1) satisfying assumptions (i)–(iii) given in Theorem 1 and let X_n and Y_n, $n \geq 0$, be two numerical solutions of (1) computed by the ϑ-methods (2) or (3). Then, the applied method is said to generate mean-square contractive numerical solutions in a region $\mathcal{R} \subseteq \mathbb{R}^+$ if, for a fixed $\vartheta \in [0,1]$,*

$$\nu(\vartheta, \Delta t) < 0, \quad \forall \Delta t \in \mathcal{R}$$

for (2), being $\nu(\vartheta, \Delta t)$ the parameter in (14), or

$$\epsilon(\vartheta, \Delta t) < 0, \quad \forall \Delta t \in \mathcal{R}$$

for (3), where $\epsilon(\vartheta, \Delta t)$ is the parameter in (19).

Definition 5. *A stochastic ϑ-method (2) or (3) is said unconditionally mean-square contractive if, for a given $\vartheta \in [0,1]$, $\mathcal{R} = \mathbb{R}^+$.*

As regards the ϑ-Maruyama method (2), according to Definition 4, mean-square contractive numerical solutions are generated if

$$0 < \beta(\vartheta, \Delta t) < 1,$$

for any $\Delta t \in \mathcal{R}$, i.e.

$$
\mathcal{R} = \begin{cases} \left(0, \dfrac{|\alpha|}{(1-\vartheta)^2 M}\right), & \vartheta < 1, \\ \mathbb{R}^+, & \vartheta = 1. \end{cases} \tag{24}
$$

As a consequence, the ϑ-Maruyama method with $\vartheta = 1$, i.e., the implicit Euler-Maruyama method

$$
X_{n+1} = X_n + \Delta t f(X_{n+1}) + g(X_n) \Delta W_n. \tag{25}
$$

is unconditionally mean-square contractive. In other terms, the stochastic perturbation (25) of the deterministic implicit Euler method preserves its unconditional contractivity property [18].

In analogous way, as regards the ϑ-Milstein method (3), Definition 4 leads to

$$
0 < \gamma(\vartheta, \Delta t) < 1,
$$

for any $\Delta t \in \mathcal{R}$, i.e.

$$
\mathcal{R} = \begin{cases} \left(0, \dfrac{4|\alpha|}{4(1-\vartheta)^2 M + 3\widetilde{M}}\right), & \vartheta < 1, \\ \left(0, \dfrac{4|\alpha|}{3\widetilde{M}}\right), & \vartheta = 1. \end{cases} \tag{26}
$$

Finally, the computation of the regions $\mathcal{R}$ in (24) and (26) require the knowledge of the parameters L, μ, M and $\widetilde{M}$. To make the region $\mathcal{R}$ completely computable, the authors in [12] have adopted the estimation strategy described in [34].

3 Optimal Choice of the Parameter ϑ

We now aim to investigate the choice of the optimal parameter ϑ, for both ϑ-methods (2) and (3), that maximizes the region $\mathcal{R}$ in (24) and (26), as well as the linear stability interval provided by (11) and (12).

To this purpose, let us denote by λ and σ the intercepts of the linearizations of the functions f and g in (1), respectively. In the remainder, we assume λ and σ to be real. In addition, for the nonlinear SDE (1), we assume the conditions of Theorem 1 satisfied, i.e., we deal with exponential mean-square dissipative nonlinear SDEs, according to the Definition 1.

3.1 Analysis of Stochastic ϑ-Maruyama Methdos

We first consider stochastic ϑ-Maruyama methods (2). We note from (11) that, in order to satisfy (10), the following condition must be fulfilled

$$
\Delta t (1 - 2\vartheta) \lambda^2 < -2\left(\lambda + \frac{1}{2}\sigma^2\right). \tag{27}
$$

Moreover, the stability condition for the linear SDE (7) provides that

$$\lambda + \frac{1}{2}\sigma^2 < 0.$$

It is worth separating the analysis of the two cases $\vartheta \in [0, \frac{1}{2})$ and $\vartheta \in [\frac{1}{2}, 1)$. Let us start with the case $\vartheta \in [0, \frac{1}{2})$. Taking into account (11) and (27), for $\vartheta \in [0, \frac{1}{2})$, we define

$$R_{EM} = \begin{bmatrix} -\dfrac{2\lambda + \sigma^2}{(1 - 2\vartheta)\lambda^2} \\[2ex] \dfrac{|\alpha|}{(1 - \vartheta)^2 M} \end{bmatrix}$$

and denote

$$h_{EM}(\vartheta) = ||R_{EM}||^2 = \frac{\left(2\lambda + \sigma^2\right)^2}{(1 - 2\vartheta)^2} + \frac{\alpha^2}{(1 - \vartheta)^4 M^2},$$

whose first derivative is then given by

$$h'_{EM}(\vartheta) = \frac{\left(2\lambda + \sigma^2\right)^2}{(1 - 2\vartheta)^3} + \frac{4\alpha^2}{M^2 (1 - \vartheta)^5}. \tag{28}$$

Hence

$$h_{EM}(\vartheta) > 0, \qquad 0 \leq \vartheta < \frac{1}{2}. \tag{29}$$

Therefore, the function $||R_{EM}||$ is a increasing function of ϑ.

We now analyze the case $\frac{1}{2} \leq \vartheta < 1$. In this case, the linear stability condition (27) is satisfied for any $\Delta t > 0$. This means that the linear stability region is unbounded and the study of the term $||R_{EM}||$ reduces only in the analysis of

$$h_{EM}(\vartheta) = \frac{|\alpha|}{M(1 - \vartheta)^2}, \qquad \frac{1}{2} \leq \vartheta < 1,$$

that is an increasing function of ϑ as well.

Finally, for $\vartheta = 1$, both the linear and the nonlinear stability regions are equal to $\mathbb{R}^+$. Hence, we argue that $\vartheta = 1$ is the optimal choice for ϑ in ϑ-Maruyama methods.

3.2   Analysis of Stochastic ϑ-Milstein Methods

Let us now analyze the family of stochastic ϑ-Milstein methods (3). Taking into account (12) and (13), proper algebraic manipulations suggest that, in order to attain (10), the following condition must hold true

$$\gamma(\vartheta)\Delta t < \left|2\lambda + \sigma^2\right|, \tag{30}$$

where

$$\gamma(\vartheta) = \frac{1}{2}\sigma^4 + \lambda\left(\lambda + 2\vartheta\left(\sigma^2 - \lambda\right) - 2\sigma^2\right). \tag{31}$$

If we define

$$l(\lambda, \sigma) = \frac{\frac{\sigma^4}{2|\lambda|} - \lambda + 2\sigma^2}{2\left(\sigma^2 - \lambda\right)},$$

we have

$$\gamma(\vartheta) > 0$$

if and only if

$$\vartheta < l(\lambda, \sigma).$$

Let us distinguish the following two cases.

Case $l(\lambda, \sigma) < 1$. In this case, for $0 \le \vartheta < l(\lambda, \sigma)$, we define

$$R_{MIL} = \begin{bmatrix} \dfrac{2\left|\lambda + \frac{1}{2}\sigma^2\right|}{\gamma(\vartheta)} \\[2ex] \dfrac{4|\alpha|}{4(1 - \vartheta)^2 M + 3\widetilde{M}} \end{bmatrix}$$

and

$$h_{MIL}(\vartheta) = \|R_{MIL}\|^2 = \frac{4|\lambda + \frac{1}{2}\sigma^2|^2}{\gamma^2(\vartheta)} + \frac{16|\alpha|^2}{(4(1 - \vartheta)^2 M + 3\widetilde{M})^2},$$

whose first derivative is given by

$$h'_{MIL}(\vartheta)\frac{-8|\lambda + \frac{1}{2}\sigma^2|^2\gamma'(\vartheta)}{\gamma^3(\vartheta)} + \frac{32 \cdot 8|\alpha|^2(1 - \vartheta)}{(4(1 - \vartheta)^2 M + 3\widetilde{M})^3}. \tag{32}$$

Since, by (30), $\gamma'(\vartheta) < 0$ for any $0 \le \vartheta < l(\lambda, \sigma)$, we have that R_{MIL} is an increasing function of ϑ, $0 \le \vartheta < l(\sigma, \gamma)$.

Case $l(\lambda, \sigma) \ge 1$. In this case, the linear stability interval is unbounded. Hence, by the second summand in (32), we conclude that R_{MIL} is an increasing function of ϑ, $0 \le \vartheta \le 1$, as well. This shows that also for the stochastic ϑ-Milstein methods, the choice $\vartheta = 1$ is the optimal one.

3.3   A Numerical Evidence: Stochastic Ginzburg-Landau Equation

Ginzburg-Landau equation plays a relevant role in the theory of superconductivity, in particular to describe phase transitions (see [17,23] and references therein). Its stochastic counterpart in presence of multiplicative noise has been provided in [25]. Specifically, we consider the scalar SDE (1) with

$$f(X(t)) = kX(t) - \tau X(t)^3, \qquad g(X(t)) = \rho X(t), \tag{33}$$

being $k, \tau, \rho \in \mathbb{R}$. In our experiments, let us consider $k = -4$, $\tau = \rho = 1$, as in [12,22]. For this problem the constants L and μ are given by $L = 1$ and $\mu = -4$, so $\alpha = -7$. Then, according to Theorem 1, this problem generates mean-square contractive solutions. Moreover, the values of M in (17) and $\widetilde{M}$ in (22) are 16 and 1, respectively. We consider the stochastic ϑ-Maruyama method (2) with $\vartheta = 0.2$. In this case (24) yields

$$\mathcal{R} = (0,\ 0.68).$$

The corresponding estimate on Δt is confirmed in Fig. 1, where the pattern of the mean-square deviation $\mathbb{E}|X_n - Y_n|^2$ in logarithmic scale is depicted for various values of Δt. It is visible that, the more Δt decreases, the more the numerical slope $\nu(0.2, \Delta t)$ in (14) tends to the exact one α in (6). For values of $\Delta t > 0.68$, the mean-square deviation does not exponentially decay.

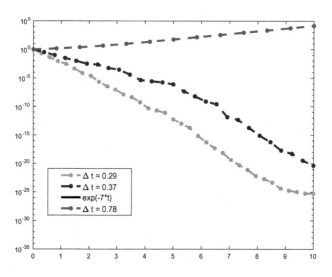

Fig. 1. Mean-square deviations over 2000 paths for the stochastic ϑ-method (2), with $\vartheta = 0.2$, applied to stochastic Ginzburg-Landau equation.

Let us now fix the value of stepsize $\Delta t = 400/2^9$ and perform the simulations over $M = 2000$ paths obtained applying the stochastic ϑ-Maruyama methods (2) and the stochastic ϑ-Milstein methods with $\vartheta = 0, 0.2, 0.5, 0.8, 1$. The numerical evidence provided in Figs. 2 and 3 reveals that, for decreasing values of the parameter ϑ, the mean-square contractivity of the relative stochastic ϑ-Maruyama method is less and less visible, according to the theory here provided.

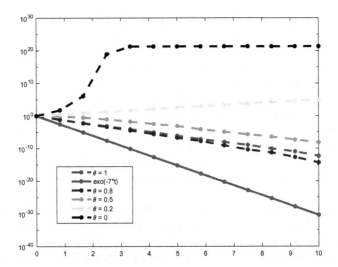

Fig. 2. Mean-square deviations over 2000 paths for stochastic ϑ-Maruyama methods (2), with different values of ϑ and constant stepsize $\Delta t = 400/2^9$, applied to the stochastic Ginzburg-Landau equation.

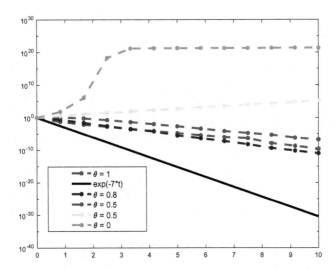

Fig. 3. Mean-square deviations over 2000 paths for stochastic ϑ-Mistein methods (3), with different values of ϑ and constant stepsize $\Delta t = 400/2^9$, applied to stochastic Ginzburg-Landau equation.

4   Conclusions and Future Issues

In this work, we have considered nonlinear stochastic differential equations (1) exhibiting an exponential mean-square dissipative character according to Theorem 1 and Definition 1. We have analyzed the behaviours of stochastic ϑ-methods, i.e., the ϑ-Maruyama (2) and ϑ-Milstein methods (3). As proved in [12], these discretizations are capable of reproducing the mean-square contractivity, according to Definition 4, in accordance to a suitable stepsize restriction. In this paper, we have discussed about to choice of the parameter ϑ in order to maximize both the linear and nonlinear stability intervals. It has been proved that the optimal choice is $\vartheta = 1$ for both families of ϑ-methods, as confirmed by Figs. 2 and 3.

Future developments of this research will be oriented to provide optimal ϑ-methods for SDEs exhibiting a mean-square dissipative character in presence of other types of noise and also in presence of jumps [21,22]. Moreover, the effectiveness of ϑ-methods and their extensions in numerically retaining the characteristic features of nonlinear stochastic oscillators [6,7,14,16,33] deserves further future investigations.

We also highlight that the analysis of nonlinear stability properties of other numerical methods (such as two-step Runge-Kutta methods [15]) also deserves future investigations.

References

1. Buckwar, E., D'Ambrosio, R.: Exponential mean-square stability properties of stochastic linear multistep methods. Adv. Comput. Math. **47**(4), 1–14 (2021)
2. Buckwar, E., Riedler, M.G., Kloeden, P.: The numerical stability of stochastic ordinary differential equations with additive noise. Stoch. Dyn. **11**, 265–281 (2011)
3. Buckwar, E., Sickenberger, T.: A comparative linear mean-square stability analysis of Maruyama- and Milstein-type methods. Math. Comput. Simul. **81**, 1110–1127 (2011)
4. Caraballo, T., Kloeden, P.: The persistence of synchronization under environmental noise. Proc. Roy. Soc. A **46**(2059), 2257–2267 (2005)
5. Chen, C., Cohen, D., D'Ambrosio, R., Lang, A.: Drift-preserving numerical integrators for stochastic Hamiltonian systems. Adv. Comput. Math. **46**, Article Number 27 (2020)
6. Citro, V., D'Ambrosio, R.: Long-term analysis of stochastic ϑ-methods for damped stochastic oscillators. Appl. Numer. Math. **51**, 89–99 (2004)
7. Cohen, D.: On the numerical discretization of stochastic oscillators. Math. Comput. Simul. **82**, 1478–95 (2012)
8. Cont, R., Tankov, P.: Financial Modelling with Jump Processes. Financial Mathematics Series. Champan & All/CRC (2004)
9. Conte, D., D'Ambrosio, R., Paternoster, B.: Improved theta-methods for stochastic Volterra integral equations. Comm. Nonlin. Sci. Numer. Simul. **93**, Article Number 105528 (2021)
10. Conte, D., D'Ambrosio, R., Paternoster, B.: On the stability of theta-methods for stochastic Volterra integral equations. Discr. Cont. Dyn. Syst. B **23**(7), 2695–2708 (2018)

11. Dahlquist, G.: Error analysis for a class of methods for stiff nonlinear initial value problems. Lecture Notes Math. **150**, 18–26 (2020)
12. D'Ambrosio, R., Di Giovacchino, S.: Mean-square contractivity of stochastic ϑ-methods. Commun. Nonlinear Sci. Numer. Simul. **96**, 105671 (2021)
13. D'Ambrosio, R., Di Giovacchino, S.: Nonlinear stability issues for stochastic Runge-Kutta methods. Commun. Nonlinear Sci. Numer. Simul. **94**, 105549 (2021)
14. D'Ambrosio, R., Scalone, C.: Filon quadrature for stochastic oscillators driven by time-varying forces. Appl. Numer. Math. (to appear)
15. D'Ambrosio, R., Scalone, C.: Two-step Runge-Kutta methods for stochastic differential equations. Appl. Math. Comput. **403**, Article Number 125930 (2021)
16. D'Ambrosio, R., Scalone, C.: On the numerical structure preservation of nonlinear damped stochastic oscillators. Numer. Algorithms **86**(3), 933–952 (2020). https://doi.org/10.1007/s11075-020-00918-5
17. Ginzburg, V.L.: On the theory of superconductivity. Il Nuovo Cimento (1955-1965) **2**(6), 1234–1250 (1955). https://doi.org/10.1007/BF02731579
18. Hairer, E., Wanner, G.: Solving Ordinary Differential Equations II. Stiff and Differential-Algebraic Problems. Springer Series in Computational Mathematics, 2nd edn. Springer, Heidelberg (1996). https://doi.org/10.1007/978-3-642-05221-7
19. Higham, D.: An algorithmic introduction to numerical simulation of stochastic differential equations. SIAM Rev. **43**, 525–546 (2001)
20. Higham, D.: Mean-square and asymptotic stability of the stochastic theta method. SIAM J. Numer. Anal. **38**, 753–769 (2000)
21. Higham, D., Kloeden, P.: An Introduction to the Numerical Simulation of Stochastic Differential Equations. SIAM (2021)
22. Higham, D., Kloeden, P.: Numerical methods for nonlinear stochastic differential equations with jumps. Numer. Math. **101**, 101–119 (2005)
23. Hutzenthaler, M., Jentzen, A.: Numerical approximations of stochastic differential equations with non-globally Lipschitz continuous coefficients. Memoirs of the American Mathematical Society, vol. 236, no. 1112 (2015). https://doi.org/10.1090/memo/1112
24. Koleden, P., Lorenz, T.: Mean-square random dynamical systems. J. Differ. Equ. **253**, 1422–1438 (2012)
25. Kloeden, P.E., Platen, E.: Numerical Solution of Stochastic Differential Equations. Stochastic Modelling and Applied Probability, vol. 23. Springer, Berlin (1992). https://doi.org/10.1007/978-3-662-12616-5
26. Liu, X., Duan, J., Liu, J., Kloeden, P.: Synchronization of dissipative dynamical systems driven by non-Gaussian Levy noises. Int. J. Stoch. Anal. 502803 (2010)
27. Ma, Q., Ding, D., Ding, X.: Mean-square dissipativity of several numerical methods for stochastic differential equations with jumps. Appl. Numer. Math. **82**, 44–50 (2014)
28. Majka, M.B.: A note on existence of global solutions and invariant measures for jump SDE with locally one-sided Lipschitz drift. Probab. Math. Stat. **40**, 37–57 (2020)
29. Melbo, A.H.S., Higham, D.J.: Numerical simulation of a linear stochastic oscillator with additive noise. Appl. Numer. Math. **51**, 89–99 (2004)
30. Shen, G., Xiao, R., Yin, X., Zhang, J.: Stabilization for hybrid stochastic systems by aperiodically intermittent control. Nonlinear Anal. Hybrid Syst. **29**, 100990 (2021)
31. Sobczyk, K.: Stochastic Differential Equations with Applications to Physics and Engineering. Mathematics and its Applications, vol. 40. Springer, Dordrecht (1991). https://doi.org/10.1007/978-94-011-3712-6

32. Stuart, A.M., Humphries, A.R.: Dynamical Systems and Numerical Analysis. Part of Cambridge Monographs on Applied and Computational Mathematics. Cambridge University Press, Cambridge (1999)
33. Tocino, A.: On preserving long-time features of a linear stochastic oscillators. BIT Numer. Math. **47**, 189–196 (2007)
34. Wood, G., Zhang, B.: Estimation of the Lipschitz constant of a function. J. Glob. Opt. **8**, 91–103 (1996)
35. Yao, J., Gan, S.: Stability of the drift-implicit and double-implicit Milstein schemes for nonlinear SDEs. Appl. Math. Comput. **339**, 294–301 (2018)
36. Zhao, H., Niu, Y.: Finite-time sliding mode control of switched systems with one-sided Lipschitz nonlinearity. J. Franklin Inst. **357**, 11171–11188 (2020)

Continuous Extension of Euler-Maruyama Method for Stochastic Differential Equations

Dajana Conte[1], Raffaele D'Ambrosio[2], Giuseppe Giordano[1]([✉]), and Beatrice Paternoster[1]

[1] Department of Mathematics, University of Salerno, Fisciano, Italy
{dajconte,gigiordano,beapat}@unisa.it
[2] DISIM, University of L'Aquila, L'Aquila, Italy
raffaele.dambrosio@univaq.it

Abstract. In this work we focus on the development of continuous extension of Euler-Maruyama method, which is used to numerically approximate the solution of Stochastic Differential Equations (SDEs). We aim to provide an approximation of a given SDE in terms of a piecewise polynomial, because, as it is known in the deterministic case, a dense output allows to provide a more efficient error estimate and it is very effective for a variable step-size implementation. Hence, this contribution aims to provide a first building block in such directions, consisting in the development of the scheme.

Keywords: Stochastic Differential Equations · Euler-Maruyama method · Collocation methods · Dense output

1 Introduction

The study of Stochastic Differential Equations (SDEs) has attracted the recent attention of many researchers, due to their applications in many areas such as physics, finance, biology, medicine and chemistry [13,33,34,41,43,47]. The general form of an autonomous SDEs is given by:

$$dX(t) = f(X(t))dt + g(X(t))dW(t), \quad X(0) = X_0, \quad t \in [0, T], \quad (1)$$

where the solution $X(t)$ is a random variable, the coefficient $f(X(t))$ of the deterministic part is the *drift coefficient*, while the coefficient $g(X(t))$ of the stochastic term is denoted as *diffusion coefficient* and $W(t)$ is a multidimensional Wiener process. We remind that a Wiener process $W(t)$ on $[0, T]$ is a real valued continuous-time process with the property that the Wiener increments $W(t) - W(s)$ are independent random variables with

$$W(t) - W(s) \sim \sqrt{t-s} N(0, 1), \quad 0 \leq s < t \leq T,$$

The authors are members of the GNCS group. This work is supported by GNCS-INDAM project and by PRIN2017-MIUR project.

© Springer Nature Switzerland AG 2021
O. Gervasi et al. (Eds.): ICCSA 2021, LNCS 12949, pp. 135–145, 2021.
https://doi.org/10.1007/978-3-030-86653-2_10

where $N(0,1)$ denotes a normally distributed random variable with zero mean and unitary variance.

The integral form of (1) is given by

$$X(t) = X(0) + \int_0^t f(X(s))ds + \int_0^t g(X(s))dW(s), \quad t \in [0, T], \qquad (2)$$

where the first integral is a Riemann integral, while the second one is a stochastic integral, commonly interpreted in either Itô or Stratonovich form. In this paper, we will interpret the second integral of the Eq. (2) as an Itô integral. For a comprehensive treatise of stochastic calculus (also in the direction of numerical integration), as well as on the existence and the uniqueness of solutions to (2), the interested reader can refer, for instance, to [33, 41, 43, 47] and references therein.

Numerical modeling for SDEs is a crucial ingredient to accurately simulate the solutions to nonlinear SDEs, since they cannot be exactly computed in most of the cases. Comprehensive monographs on the topic are, for instance, [33, 41]. Relevant families of stochastic one-step, two-step, Runge-Kutta and two-step Runge-Kutta methods are discussed, for instance, in [20, 32, 33, 39–41, 43, 47–49] and references therein. Moreover, contributions on the preservation of invariance laws along time discretizations are object of [12, 13, 13, 19, 26, 28, 29, 31, 38, 42] and references therein.

With respect to the uniform grid

$$I_h = \{t_n = nh, \ n = 0, 1, \ldots, n, \ Nh = T\}, \quad h > 0, \qquad (3)$$

the well-known Euler-Maruyama method (EM)

$$X_j = X_{j-1} + hf(X_{j-1}) + g(X_{j-1})\Delta W_j, \quad j = 1, 2, \ldots, N, \qquad (4)$$

represents a prototype of the simplest method for SDEs (1). X_j is an approximate solution to $X(t_j)$ and $\Delta W_j = W(t_j) - W(t_{j-1})$ are independent Wiener increments. The present contribution aims to develop a continuous extension to the EM method, useful to provide a building block for a more general theory of continuous methods for SDEs, useful to assess their variable stepsize implementation, particularly effective, for instance, to solve stiff SDEs [1–7, 9–11, 14, 21, 35, 36, 44–46, 50].

The idea we aim to use here is closely related to the analogous scenario of continuous numerical methods for deterministic differential equations at the basis, for instance, of the well-known idea of *deterministic numerical collocation* (refer, for instance, to [16–18, 22–25, 27, 37] and references therein).

Numerical collocation for deterministic problems is a feasible technique to develop dense output methods for functional equations. In other terms, the provided approximant (the *collocation function*) is constructed as linear combination of selected basis functions, spanning a finite dimensional functional space; usually these functions are algebraic polynomials. The collocation function is required to exactly satisfies the given equation in a selected set points of the integration interval, denoted as *collocation points*. It is worth observing that the collocation function can be chosen as a linear combination of ad hoc basis functions, chosen coherently with the qualitative character of the given problem.

Furthermore, the collocation function provides an approximation to the solution of the continuous problem over the entire integration interval. This feature is particularly, for instance, in order to provide high continuous order methods not suffering from order reduction when applied to stiff problems [15,25], as well as to provide efficient and accurate procedure of error estimation useful in variable stepsize implementations [30]. Nonetheless, re-casting discrete numerical methods as collocation methods makes their analysis benefit from their continuous formulation.

Inspired by the idea of collocation for Volterra integral equations [8], we aim to provide here a continuous extension of the EM method (4) applied to the integral form (2) of the SDE (1). The paper is organized as follows: in Sect. 2 we briefly recall the idea of numerical collocation for Volterra Integral Equations; this technique is extended to stochastic differential equations and, for selected values of the collocation parameters, the extension of the EM method (4) is provided. Some concluding remarks are discussed in Sect. 5.

2   Numerical Collocation for Volterra Integral Equations

This section aims to recall the idea of collocation for deterministic Volterra Integral Equations (VIEs)

$$y(t) = g(t) + \int_0^t k(t, \tau, y(\tau)) d\tau, \quad t \in I = [0, T], \tag{5}$$

where $k \in C(D \times R)$, with $D = \{(t, \tau) : 0 \le \tau \le t \le T\}$ and $g \in C(t)$. Let us refer to the set of grid points I_h defined in (3) and let

$$0 \le c_1 < \ldots < c_m \le 1$$

be m collocation parameters. Moreover, let us denote by $t_{nj} = t_n + c_j h$ the so-called collocation points. Collocation methods allow us to approximate the exact solution $y(t)$ of (5) by a piecewise polynomial

$$P(t) \in S_{m-1}^{(-1)}(I_h) = \{v|_{(t_n, t_{n+1})} \in \Pi_{m-1}, n = 0, 1, \ldots, N - 1\},$$

where Π_{m-1} denotes the space of algebraic polynomials of degree not exceeding $m - 1$. The collocation polynomial, restricted to the interval $[t_n, t_{n+1}]$, is of the form

$$P_n(t_n + sh) = \sum_{j=1}^m L_j(s) Y_{nj}, \quad s \in [0, 1], \tag{6}$$

for $n = 0, \ldots, N - 1$, where $L_j(s)$ is the j-th Lagrange fundamental polynomial and $Y_{nj} = P_n(t_{nj})$. The collocation equation is obtained by imposing that the polynomial (6) exactly satisfies the integral equation in the collocation points and requires, at each time step, the solution of a system of m nonlinear equations in the m unknowns Y_{ni}, as follows

$$
\begin{cases}
Y_{ni} = F_{ni} + \Phi_{ni}, \\
y_{n+1} = \displaystyle\sum_{j=1}^{m} L_j(1) Y_{nj},
\end{cases}
$$

where

$$
F_{ni} = g(t_{ni}) + h \sum_{v=0}^{n-1} \int_0^1 k\left(t_{ni}, t_v + sh, P_v(t_v + sh)\right) ds, \tag{7}
$$

$$
\Phi_{ni} = h \int_0^{c_i} k\left(t_{ni}, t_n + sh, P_n(t_n + sh)\right) ds, \tag{8}
$$

for $i = 1, \ldots, m$.

In order to obtain a fully discretized collocation scheme, it is necessary to approximate the integrals appearing in (7) and (8) by suitable quadrature formulae. This procedure leads to the well-known idea of discretized collocation. The discretized collocation polynomial is of the form

$$
\tilde{P}_n(t_n + sh) = \sum_{j=1}^{m} L_j(s) \tilde{Y}_{nj}, \quad s \in [0, 1], \tag{9}
$$

for $n = 0, \ldots, N-1$, where $\tilde{Y}_{nj} = \tilde{P}_n(t_{nj})$. The corresponding discretized collocation method assume the form:

$$
\begin{cases}
\tilde{Y}_{ni} = \tilde{F}_{ni} + \tilde{\Phi}_{ni}, \\
\tilde{y}_{n+1} = \displaystyle\sum_{j=1}^{m} L_j(1) \tilde{Y}_{nj},
\end{cases}
$$

where

$$
\tilde{F}_{ni} = g(t_{ni}) + h \sum_{v=0}^{n-1} \sum_{l=0}^{\mu_1} b_l k\left(t_{ni}, t_v + \xi_l h, \tilde{P}_v(t_v + \xi_l h)\right),
$$

$$
\tilde{\Phi}_{ni} = h \sum_{l=0}^{\mu_2} w_{il} k\left(t_{ni}, t_n + d_{il} h, \tilde{P}_n(t_n + d_{il} h)\right),
$$

for $i = 1, \ldots, m$, where

$$
(\xi_l, b_l)_{l=1}^{\mu_1} \qquad (d_{il}, w_{il})_{l=1}^{\mu_2}
$$

are two quadrature formulas with nodes ξ_l and d_{il} such that

$$
0 \le \xi_1 < \ldots < \xi_{\mu_1} \le 1
$$

and

$$
0 \le d_{i1} < \ldots < d_{i\mu_2} \le 1.
$$

μ_1 and μ_2 are positive integers and b_l and w_{il} are the weights of the formulae.

3   Continuous Extension of the EM Method

Let us now focus our attention to the SDE in integral form (2). We aim to provide an approximation $u(t)$ to the solution $X(t)$ of (2) on the overall integration interval $[0, T]$, such that its restriction $u_n(t)$ to the interval $[t_n, t_{n+1}]$ is a linear polynomial.

To this purpose, let us fix the collocation parameters $c_1, c_2 \in [0, 1]$ and denote by

$$t_{n1} = t_n + c_1 h,$$
$$t_{n2} = t_n + c_2 h$$

the two corresponding collocation points in the interval $[t_n, t_{n+1}]$.

Let us make the following ansatz: in a sufficiently small interval of length h, the solution $X(t)$ to (2) can be approximated by the linear function

$$u_n(t_n + \theta h) = L_1(\theta)U_1^{[n]} + L_2(\theta)U_2^{[n]}, \quad \theta \in [0, 1], \tag{10}$$

where $L_j(\theta)$ are the Lagrange fundamental polynomials with respect the collocation parameters, i.e.,

$$L_1(\theta) = \frac{\theta - c_2}{c_1 - c_2},$$

$$L_2(\theta) = \frac{\theta - c_1}{c_2 - c_1}$$

and $U_i^{[n]} = u_n(t_{ni})$, $i = 1, 2$.

Correspondingly, the following numerical scheme is obtained

$$\begin{cases} U_i^{[n]} = F_n + \Phi_i^{[n]}, & i = 1, 2, \\ X_{n+1} = L_1(1)U_1^{[n]} + L_2(1)U_2^{[n]}, \end{cases} \tag{11}$$

where the lag-term

$$F_n = X(t_0) + \int_0^{t_n} f(u(s))ds + \int_0^{t_n} g(u(s))dW(s), \tag{12}$$

contains all the past history of the dynamics up to the grid point t_n, while the incremental terms $\Phi_1^{[n]}$ and $\Phi_2^{[n]}$ are defined by

$$\Phi_i^{[n]} = \int_{t_n}^{t_{ni}} f(u_n(s))ds + \int_{t_n}^{t_{ni}} g(u_n(s))dW(s).$$

By employing (10) and by a suitable change of variable, the incremental terms assume the form

$$\Phi_1^{[n]} = h \int_0^{c_1} f\left(\sum_{j=1}^{2} L_j(\theta)U_j^{[n]}\right) d\theta + \int_{t_n}^{t_{n1}} g\left(\sum_{j=1}^{2} L_j\left(\frac{s - t_n}{h}\right)U_j^{[n]}\right) dW(s),$$

$$\tag{13}$$

$$\Phi_2^{[n]} = h \int_0^{c_2} f\left(\sum_{j=1}^2 L_j(\theta)U_j^{[n]}\right) d\theta + \int_{t_n}^{t_{n2}} g\left(\sum_{j=1}^2 L_j\left(\frac{s-t_n}{h}\right) U_j^{[n]}\right) dW(s).$$

(14)

We observe that the scheme (11) requires the solution of a system of two nonlinear equations in the unknowns $U_1^{[n]}$ and $U_2^{[n]}$.

By setting $c_2 = 1$, the scheme (11) may be substantively simplified. In fact, in this case, we observe that

$$\Phi_2^{[n]} = \int_{t_n}^{t_{n+1}} f(u_n(s))ds + \int_{t_n}^{t_{n+1}} g(u_n(s))dW(s),$$

(15)

therefore, by (11), (12) and (15),

$$U_n^{[2]} = F_n + \Phi_2^{[n]} = F_{n+1}, \quad n = 0, \dots, N-1.$$

As a consequence, by (10),

$$F_n = U_2^{[n-1]} = u_{n-1}(t_{n-1} + c_2 h) = u_{n-1}(t_n) = X_n,$$

i.e. for each time step the lag term F_n is equal to the numerical solution X_n at the mesh point t_n. Moreover,

$$X_{n+1} = u_n(t_n + h) = u_n(t_n + c_2 h) = U_2^{[n]}$$

and the method (11) assumes the form

$$\begin{cases} U_i^{[n]} = X_n + \Phi_i^{[n]}, & i = 1, 2, \\ X_{n+1} = U_2^{[n]}, \end{cases}$$

(16)

where the expressions of $\Phi_i^{[n]}$ are the same of (13) and (14), for $i = 1, 2$ respectively.

Clearly, the scheme reported in (11) is not a full discretization, unless the involved integrals are replaced by suitable quadrature formulae.

Remaining in the case $c_2 = 1$, we consider the following discretized metod, which is obtained from (16) by approximating the integrals in (13) and (14) by means of the rectangular quadrature rule

$$\begin{cases} \tilde{U}_1^{[n]} = \tilde{X}_n + hc_1 f\left(b_1 \tilde{U}_1^{[n]} + b_2 \tilde{U}_2^{[n]}\right) + g\left(b_1 \tilde{U}_1^{[n]} + b_2 \tilde{U}_2^{[n]}\right) \sqrt{c_1 h} Z_1^{[n]}, \\ \tilde{U}_2^{[n]} = \tilde{X}_n + hf\left(b_1 \tilde{U}_1^{[n]} + b_2 \tilde{U}_2^{[n]}\right) + g\left(b_1 \tilde{U}_1^{[n]} + b_2 \tilde{U}_2^{[n]}\right) \sqrt{h} Z_2^{[n]}, \\ \tilde{X}_{n+1} = \tilde{U}_2^{[n]}, \end{cases}$$

(17)

where $b_1 = L_1(0)$, $b_2 = L_2(0)$ and $Z_1^{[n]}$, $Z_2^{[n]}$ are normal variables of mean 0 and variance 1. The continuous approximant, in the interval $[t_n, t_{n+1}]$, assumes the form

$$\tilde{u}_n(t_n + \theta h) = \frac{1-\theta}{1-c_1} \tilde{U}_1^{[n]} + \frac{\theta - c_1}{1-c_1} \tilde{U}_2^{[n]} \quad \theta \in [0, 1].$$

(18)

The derived family of continuous methods contains the EM method as a particular case when $c_1 = 0$. As a matter of fact, imposing $c_1 = 0$ in (17) and taking into account that $b_1 = 1$ and $b_2 = 0$, the method assumes the form

$$\begin{cases} \tilde{U}_1^{[n]} = \tilde{X}_n, \\ \tilde{U}_2^{[n]} = \tilde{X}_n + hf(\tilde{X}_n) + g(\tilde{X}_n)\sqrt{h}Z_2^{[n]} \end{cases}$$

and the value of $\tilde{X}_{n+1}$ can be evaluated as

$$\tilde{X}_{n+1} = \tilde{X}_n + hf(\tilde{X}_n) + g(\tilde{X}_n)\sqrt{h}Z_2^{[n]}, \tag{19}$$

which corresponds to the EM method (4). In the case of EM method the continuous approximant assumes the form

$$\tilde{u}_n(t_n + \theta h) = (1 - \theta)\tilde{X}_n + \theta\tilde{X}_{n+1} \quad \theta \in [0, 1]. \tag{20}$$

4 Numerical Evidences

In this section we experimentally observe the accuracy of the new class of methods obtained in the previous section, with special emphasis its weak accuracy. We recall that a numerical method has weak order of convergence equal to p if there exists a constant C such that

$$|\mathbb{E}(X_n) - \mathbb{E}(X(t_n))| \leq Ch^p$$

for any fixed $t_n = nh \in I_h$, where $\mathbb{E}(\cdot)$ is the expectation operator.

Let us assume the following SDE as test problem

$$dX(t) = t^2 dt + e^{\frac{d}{2}}\cos(X(t)), \quad X(0) = 0, \quad t \in [0, 1]. \tag{21}$$

The expected value of the exact solution is $\mathbb{E}(X(t)) = \dfrac{t^3}{3}$.

Table 1. Absolute errors in the endpoint $T = 1$ obtained applying (11) to problem (21), for several values of the parameter c_1 .

N	$c_1 = 0$	$c_1 = 1/4$	$c_1 = 1/2$	$c_1 = 3/4$
2^7	$3.0967 \cdot 10^{-3}$	$3.0675 \cdot 10^{-3}$	$3.0095 \cdot 10^{-3}$	$2.8355 \cdot 10^{-3}$
2^6	$6.5909 \cdot 10^{-3}$	$6.5634 \cdot 10^{-3}$	$6.5083 \cdot 10^{-3}$	$6.3435 \cdot 10^{-3}$
2^5	$1.5071 \cdot 10^{-2}$	$1.5044 \cdot 10^{-2}$	$1.5071 \cdot 10^{-2}$	$1.4834 \cdot 10^{-2}$
2^4	$2.8721 \cdot 10^{-2}$	$2.8698 \cdot 10^{-2}$	$2.8653 \cdot 10^{-2}$	$2.8520 \cdot 10^{-2}$

Table 1 shows the absolute errors corresponding to various numerical methods belonging to the family (11), for selected values of c_1 and setting $c_2 = 1$, applied to problem (21). All methods exhibit the same weak order of convergence of EM method (4), which is equal to 1. In other terms, passing to the continuous extension do not deteriorate the order EM.

Let us now set $c_1 = \frac{1}{4}$ and $c_2 = 1$, and apply the corresponding method to the following additional SDE

$$dX(t) = e^{-\frac{t}{2}}dt + e^{X(t)-\frac{t}{2}}dW(t), \quad X(0) = 0, \quad t \in [0,1], \tag{22}$$

whose exact solution has expected value $\mathbb{E}(X(t)) = 2 - 2e^{-\frac{t}{2}}$, and to the problem

$$d(X(t)) = (2X(t) + e^{2t})dt + \mu X(t)dW(t), \quad X(0) = 1, \quad t \in [0,1], \tag{23}$$

whose exact solution has expectation given by $\mathbb{E}(X(t)) = -e^{2t}(1+t)$.

Figure 1 shows the behaviour of the absolute errors for decreasing values of the stepsize. We can observe that the method obtained by imposing $c_1 = 1/4$, applied to (21), (22) and (23), exhibits weak order of convergence equal to 1. So, also for values of c_1 different from 0, we preserve the same weak order of convergence of EM method (4).

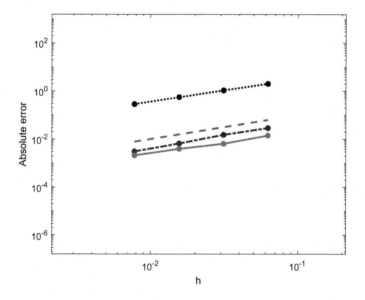

Fig. 1. Absolute weak errors associated to the application of (11) with $c_1 = 1/4$ to problem (21) (dashed-dotted blue line), problem (22) (solid magenta line) and problem (23) (dotted black line). The reference slope of order 1 is depicted by the dashed red line (Color figure online).

5   Concluding Remarks

We have focused on the possibility to continuously extend the EM method (4), by properly revising the idea of numerical collocation method used in the literature for deterministic Volterra integral equations. This continuous extension provides an approximation to the solution in the overall interval of integration.

The present contribution is a first building block for a comprehensive theory of continuous numerical methods for SDEs, useful to assess a variable step-size framework particularly effective, for instance, to solve stiff SDEs [1–7,9–11,21,35,36,44–46,50]. In this direction, further efforts are requested, such as the construction of feasible error estimates by exploiting the continuous approximant. Basis functions other than algebraic polynomials will also be investigated.

References

1. Abdulle, A.: Explicit methods for stiff stochastic differential equations. In: Engquist, B., Runborg, O., Tsai, Y,H. (eds.) Numerical Analysis of Multiscale Computations, LNCSE, vol. 82, pp. 1–22. Springer, Heidelberg (2012). https://doi.org/10.1007/978-3-642-21943-6_1

2. Abdulle, A., Almuslimani, I., Vilmart, G.: Optimal explicit stabilized integrator of weak order 1 for stiff and ergodic stochastic differential equations. SIAM-ASA J. Uncertain. **6**(2), 937–964 (2018)

3. Abdulle, A., Blumenthal, A.: Stabilized multilevel Monte Carlo method for stiff stochastic differential equations. J. Comput. Phys. **251**, 445–460 (2013)

4. Abdulle, A., Cirilli, S.: S-ROCK: Chebyshev methods for stiff stochastic differential equations. SIAM J. Sci. Comput. **30**(2), 997–1014 (2007)

5. Abdulle, A., Cirilli, S.: Stabilized methods for stiff stochastic systems. C. R. Math. **345**(10), 593–598 (2007)

6. Abdulle, A., Vilmart, G., Zygalakis, K.C.: Mean-square A-stable diagonally drift-implicit integrators of weak second order for stiff Itô stochastic differential equations. BIT Numer. Math. **53**(4), 827–840 (2013)

7. Abdulle, A., Vilmart, G., Zygalakis, K.C.: Weak second order explicit stabilized methods for stiff stochastic differential equations. SIAM J. Sci. Comput. **35**(4), A1792–A1814 (2013)

8. Brunner, H.: Collocation Methods for Volterra Integral and Related Functional Equations. Cambridge University Press, Cambridge (2004)

9. Burrage, K., Tian, T.: Stiffly accurate Runge-Kutta methods for stiff stochastic differential equations. Comput. Phys. Commun. **142**(1–3), 186–190 (2001)

10. Burrage, K., Burrage, P.M.: A variable stepsize implementation for stochastic differential equations. SIAM J. Sci. Comput. **24**(3), 848–864 (2003)

11. Burrage, K., Burrage, P.M., Herdiana, R.: Adaptive stepsize based on control theory for stochastic differential equations. J. Comput. Appl. Math. **170**(2–2), 317–336 (2004)

12. Burrage, K., Burrage, P.M.: Low rank Runge-Kutta methods, symplecticity and stochastic Hamiltonian problems with additive noise. J. Comput. Appl. Math. **236**(16), 3920–3930 (2012)

13. Burrage, P.M., Burrage, K.: Structure-preserving Runge-Kutta methods for stochastic Hamiltonian equations with additive noise. Numer. Algorithms **65**(3), 519–532 (2013). https://doi.org/10.1007/s11075-013-9796-6

14. Burrage, K., Tian, T.: The composite Euler method for stiff stochastic differential equations. J. Comput. Appl. Math. **131**(1–2), 407–426 (2001)
15. Butcher, J.C.: Numerical Methods for Ordinary Differential Equations, 3rd edn. Wiley, Chichester (2016)
16. Capobianco, G., Conte, D., Paternoster, B.: Construction and implementation of two-step continuous methods for Volterra Integral Equations. Appl. Numer. Math. **119**, 239–247 (2017)
17. Cardone, A., Conte, D.: Multistep collocation methods for Volterra integro-differential equations. Appl. Math. Comput. **221**, 770–785 (2013)
18. Cardone, A., Conte, D., D'Ambrosio, R., Paternoster, B.: Collocation methods for Volterra integral and integro-differential equations: a review. Axioms **7**(3), 45 (2018)
19. Chen, C., Cohen, D., D'Ambrosio, R., Lang, A.: Drift-preserving numerical integrators for stochastic Hamiltonian systems. Adv. Comput. Math. **46**(2), 1–22 (2020). https://doi.org/10.1007/s10444-020-09771-5
20. Citro, V., D'Ambrosio, R., Di Giovacchino, S.: A-stability preserving perturbation of Runge-Kutta methods for stochastic differential equations. Appl. Math. Lett. **102**, 106098 (2020)
21. Cohen, D., Sigg, M.: Convergence analysis of trigonometric methods for stiff second-order stochastic differential equations. Numer. Math. **121**(1), 1–29 (2012)
22. Conte, D., Del Prete, I.: Fast collocation methods for Volterra integral equations of convolution type. J. Comput. Appl. Math. **196**(2), 652–663 (2006)
23. Conte, D., Paternoster, B.: Multistep collocation methods for Volterra integral equations. Appl. Numer. Math. **59**(8), 1721–1736 (2009)
24. Conte, D., D'Ambrosio, R., Paternoster, B.: Two-step diagonally-implicit collocation based methods for Volterra Integral Equations. Appl. Numer. Math. **62**(10), 1312–1324 (2012)
25. D'Ambrosio, R., Ferro, M., Jackiewicz, Z., Paternoster, B.: Two-step almost collocations methods for ordinary differential equations. Numer. Algorithms **53**(2–3), 195–217 (2010)
26. D'Ambrosio, R., Giordano, G., Paternoster, B., Ventola, A.: Perturbative analysis of stochastic Hamiltonian problems under time discretizations. Appl. Math. Lett. **120**, 107223 (2021)
27. D'Ambrosio, R., Paternoster, B.: Multivalue collocation methods free from order reduction. J. Comput. Appl. Math. **387**, 112515 (2021)
28. D'Ambrosio, R., Di Giovacchino, S.: Mean-square contractivity of stochastic ϑ-methods. Commun. Nonlinear Sci. Numer. Simul. **96**, 105671 (2021)
29. D'Ambrosio, R., Di Giovacchino, S.: Nonlinear stability issues for stochastic Runge-Kutta methods. Commun. Nonlinear Sci. Numer. Simul. **94**, 105549 (2021)
30. D'Ambrosio, R., Jackiewicz, Z.: Construction and implementation of highly stable two-step continuous methods for stiff differential systems. Math. Comput. Simul. **81**(9), 1707–1728 (2011)
31. D'Ambrosio, R., Moccaldi, M., Paternoster, B.: Numerical preservation of long-term dynamics by stochastic two-step methods. Discr. Cont. Dyn. Sys. Ser. B **23**(7), 2763–2773 (2018)
32. D'Ambrosio, R., Scalone, C.: Two-step Runge-Kutta methods for stochastic differential equations. Appl. Math. Comput. **403**, 125930 (2021)
33. Gard, T.C.: Introduction to Stochastic Differential Equations. Marcel Dekker Inc., New York-Basel (1988)
34. Gardiner, C.Q.: Handbook of Stochastic Methods for Physics, Chemistry and the Natural Sciences. Springer Series in Synergetics. Springer, Heidelberg (2009)

35. Haghighi, A., Hosseini, S.M., Rössler, A.: Diagonally drift-implicit Runge-Kutta methods of strong order one for stiff stochastic differential systems. J. Comput. Appl. Math. **293**, 82–93 (2016)
36. Haghighi, A., Rössler, A.: Split-step double balanced approximation methods for stiff stochastic differential equations. Int. J. Comp. Math. **96**(5), 1030–1047 (2019)
37. Hairer, E., Wanner, G.: Solving Ordinary Differential Equations II. Stiff and Differential-Algebraic Problems. Springer, Heidelberg (1996). https://doi.org/10.1007/978-3-642-05221-7
38. Han, M., Ma, Q., Ding, X.: High-order stochastic symplectic partitioned Runge-Kutta methods for stochastic Hamiltonian systems with additive noise. Appl. Math. Comput. **346**, 575–593 (2019)
39. Higham, D.J.: An algorithmic introduction to numerical simulation of stochastic differential equations. SIAM Rev. **43**(3), 525–546 (2001)
40. Higham, D.J.: Mean-square and asymptotic stability of the stochastic theta method. SIAM J. Numer. Anal. **38**(3), 753–769 (2000)
41. Higham, D., Kloeden, P.E.: An Introduction to the Numerical Simulation of Stochastic Differential Equations. SIAM, Philadelphia (2021)
42. Hong, J., Xu, D., Wang, P.: Preservation of quadratic invariants of stochastic differential equations via Runge-Kutta methods. Appl. Numer. Math. **87**, 38–52 (2015)
43. Kloeden, P.E., Platen, E.: The Numerical Solution of Stochastic Differential Equations. Springer, Heidelberg (1992). https://doi.org/10.1007/978-3-662-12616-5
44. Komori, Y., Burrage, K.: A stochastic exponential Euler scheme for simulation of stiff biochemical reaction systems. BIT Numer. Math. **54**(4), 1067–1085 (2014). https://doi.org/10.1007/s10543-014-0485-1
45. Li, T., Abdulle, A., Weinan, E.: Effectiveness of implicit methods for stiff stochastic differential equations. Commun. Comp. Phys. **3**(2), 295–307 (2008)
46. Milstein, G.N., Platen, E., Schurz, H.: Balanced implicit methods for stiff stochastic systems. SIAM J. Numer. Anal. **35**(3), 1010–1019 (1998)
47. Milstein, G.N., Tretyakov, M.V.: Stochastic Numerics for Mathematical Physics. Springer, Heidelberg (2004). https://doi.org/10.1007/978-3-662-10063-9
48. Rössler, A.: Runge-Kutta methods for Itô stochastic differential equations with scalar noise. BIT Numer. Math. **46**(1), 97–110 (2006)
49. Tang, X., Xiao, A.: Efficient weak second-order stochastic Runge-Kutta methods for Itô stochastic differential equations. BIT Numer. Math. **57**, 241–260 (2017)
50. Tian, T., Burrage, K.: Implicit Taylor methods for stiff stochastic differential equations. Appl. Numer. Math. **38**(1–2), 167–185 (2001)

The Mirror Reflection Principle
and Probabilistic Invariants in the Theory
of Multiple Photon Scattering

Oleg I. Smokty[✉]

Federal Research Center – St. Petersburg Institute of Informatics and Automation,
Russian Academy of Sciences, 39, 14th Line, St. Petersburg 198199, Russia
soi@iias.spb.su

Abstract. In the framework of the statistical interpretation of multiple photon scattering as a stationary Markov process, which was given in well-known works by V.V. Sobolev and S. Ueno, the author has formulated the principle of mirror spatial-angular symmetry for total probabilities of photons exiting a homogeneous slab of finite optical thickness $\tau_0 < \infty$. Based on this principle, linear second kind Fredholm integral equations have been obtained, as well as linear singular integral equations, for new objects of the classical radiative transfer theory, namely probabilistic invariants in the case of arbitrary distribution and power of primary energy sources in the medium. Besides, a unified probabilistic function for photons exiting a homogeneous slab was constructed. For finding unique regular solutions of the obtained linear singular integral equations, by analogy with the classical theory, additional integral relations are given, depending on the presence or absence of roots or pseudo-roots in respective characteristic equations. Given the analytic connection between probabilistic and photometric invariants, it has been shown that their calculations by standard computational methods, such as discretization technique or method of successive approximations, lead to substantial savings of computer resources and provide a more convenient form for representing theoretical data and results of numerical modelling of radiation fields of an atmosphere in the visible spectrum range of 0.6–0.8 μm.

Keywords: Mirror reflection principle · Total probability of photon exiting · Probabilistic invariants · Basic boundary-value problem · Phase function · Single scattering albedo · Optical thickness · Mirror symmetrical optical levels · Mirror symmetry vision lines · Unified probabilistic function

1 Basic Probabilistic Functions and Their Main Properties

Let's consider a homogeneous slab of finite optical thickness $\tau_0 < \infty$ (without a reflecting bottom), in which there occur processes of multiple anisotropic photon scattering and absorption with the single-scattering albedo Λ and $(1-\Lambda)$ respectively, and the arbitrary phase function $P(\cos \gamma)$, where γ is a scattering angle. We ignore the effects of radiation polarization and refraction. The distribution and power of primary external

© Springer Nature Switzerland AG 2021
O. Gervasi et al. (Eds.): ICCSA 2021, LNCS 12949, pp. 146–161, 2021.
https://doi.org/10.1007/978-3-030-86653-2_11

and internal energy sources $g(\tau, \eta, \varphi)$ in such a slab are deemed arbitrary and isotropic. Here we focus on the spatial angular structure, nonlocal properties of symmetry and invariance of scalar radiation exiting the medium, as well as on probabilistic interpretation of multiple anisotropic photon scattering in the considered homogeneous slab. We mostly pay attention to one of the problems of the classical radiative transfer theory, in which the medium is illuminated by parallel solar rays falling on its upper $\tau = 0$ or lower $\tau = \tau_0$ border and creating luminosity of a square perpendicular to them, equal to πS. In this case, the basic boundary-value problem of the classical theory of scalar radiative transfer [1, 2] is of a probabilistic nature, meaning that the phase function $P(\cos \gamma)$ which determines the probability of photon scattering at an angle γ (in interaction with the medium's elementary volume dV, and the particles albedo Λ equal to the probability of their "survival" in the initial scattering event), including the basic phenomenological equation of radiative transfer, from the statistical point of view allows to consider the multiple scattering process in terms of a stochastic and uniform Markov chain process with given initial probabilities of distributions of photon absorption and scattering volume coefficients in each current point along their motion path [3–5].

Thus, in the framework of the statistical interpretation of the classical radiative transfer theory, its basic task consists in finding, at each fixed yet arbitrary optical level τ, a spatial-angular distribution of probabilities of photons exiting the medium across the upper ($\tau = 0$) or lower ($\tau = \tau_0$) border at given probabilities that characterize basic optical parameters and initial spatial-angular distribution of probabilities of initial (single) photon scattering and absorption by the medium elementary volume dV. Given the above, to solve the set task, we will use Sobolev' probabilistic method in its original basic statement [6], including generalization to the case of non-isotropic multiple photon scattering [7].

Let the function $p(\tau, \eta', \eta, \varphi' - \varphi, \tau_0)d\varpi$ determine the probability of such an event where a photon moving in the direction $\theta' = \arccos \eta'$ to the outward normal of the given homogeneous slab at the azimuth φ', and then absorbed at the optical depth τ, exits the medium after multiple scatterings through its upper border $\tau = 0$ at the angle $\theta = \arccos \eta$ to the outward normal at the azimuth φ inside the solid angle $d\omega = \sin\theta \, d\theta \, d\varphi$ (Fig. 1).

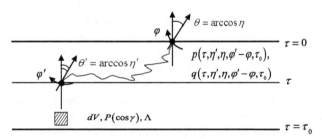

Fig. 1. Probabilities $p(\tau, \eta', \eta, \varphi' - \varphi, \tau_0)$ and $q(\tau, \eta', \eta, \varphi' - \varphi, \tau_0)$ of photons exiting the upper border $\tau = 0$ of the homogeneous slab $[0, \tau_0]$ at their absorption and subsequent radiation at an arbitrary optical thickness τ.

Along with this probabilistic function, we consider the probability that the photon emitted at the optical depth τ in the direction $\theta' = \arccos \eta'$ relative to the outward normal and having the azimuth φ' will exit the medium after multiple scatterings through its upper border $\tau = 0$ at the angle $\theta = \arccos \eta$ to the outward normal at azimuth φ inside the solid angle $d\omega$. Note that according to [1, 6, 7], finding probabilistic functions $p(\tau, \eta', \eta, \varphi' - \varphi, \tau_0)$ and $q(\tau, \eta', \eta, \varphi' - \varphi, \tau_0)$ allows to fully solve the basic boundary-value problem of the scalar radiative transfer theory in the visible range of spectra 0.6–0.8 μm in the case of a homogeneous slab of an arbitrary optical thickness $\tau_0 \leq \infty$ without a reflecting bottom.

2   Probabilistic Invariants in the Theory of Multiple Photon Scattering

Let us turn now to studying the sought-for nonlocal spatial-angular properties of mirror symmetry for the probabilities $p(\tau, \eta', \eta, \varphi' - \varphi, \tau_0)$ and $q(\tau, \eta', \eta, \varphi' - \varphi, \tau_0)$ of photons exiting the homogeneous slab of finite optical thickness $\tau_0 < \infty$, in which the geometric and optical symmetry axes coincide with its middle $\tau_0/2$. In the traditional approach to considering those properties [1, 2] for upgoing $\eta' < 0$ and downgoing $\eta' > 0$ vision lines, usually one arbitrary but fixed optical level τ is selected [6, 7], and then the well-known theorem of optical reciprocity of scattering processes in a given slab [8]. Modification of this standard approach as proposed by the author [9, 10] implies "splitting" of the initially selected arbitrary level τ into two separate optical levels that are mirror symmetrical in relation to the middle of the initial homogeneous slab $\tau_0/2$. In that case, instead of one current optical level τ, the second level $(\tau_0 - \tau)$ is introduced, which is mirror symmetrical to the initial level τ in relation to the level $\tau_0/2$. As demonstrated below, the consideration of two probabilistic functions $p(\tau, \eta', \eta, \varphi' - \varphi, \tau_0)$ and $q(\tau, \eta', \eta, \varphi' - \varphi, \tau_0)$ for two arbitrary upgoing and downgoing mirror symmetrical directions of photon propagation η' and $-\eta'$ leads to the formation, at optical depths τ and $\tau_0 - \tau$ in directions η' and $-\eta'$, of new probabilistic constructions which possess invariance properties at a simultaneous shift of the current optical depths $\tau \leftrightarrow \tau_0 - \tau$ and change of directions $\eta' \leftrightarrow -\eta'$ when vision lines are rotating. This property of spatial-angular mirror symmetry extends and supplements the generally accepted traditional principle of optical reciprocity [11], being a foundation for formulating a more general principle of mirror reflection and spatial-angular mirror symmetry of probabilistic functions $p(\tau, \eta', \eta, \varphi' - \varphi, \tau_0)$ and $q(\tau, \eta', \eta, \varphi' - \varphi, \tau_0)$ in an extended homogeneous slab of finite optical thickness $\tau_0 < \infty$ at any power and arbitrary spatial distribution of primary isotropic energy sources $g(\tau)$ within it.

Following the above-described approach, we consider the behavior of the initial photon (I) absorbed at an arbitrary optical depth τ from the direction of its propagation $\theta' = \arccos \eta'$ at azimuth φ'. At the same arbitrary optical level τ let us introduce the probabilistic function $p_+(\tau_0 - \tau, -\eta', \eta, \varphi' - \varphi, \tau_0)$, which determines the probability of the photon exiting the medium across its lower border $\tau = \tau_0$ in the downgoing direction $\theta = \arccos \eta$ at azimuth $(\varphi' - \varphi)$. Thus, at an arbitrarily selected optical level τ the sum of probabilities $p(\tau, \eta', \eta, \varphi' - \varphi, \tau_0)$ and $q(\tau_0 - \tau, -\eta', \eta, \varphi' - \varphi, \tau_0)$ determines the total probability $p(\tau, \eta', \eta, \varphi' - \varphi, \tau_0)$ of the photon (I) exiting the medium at the angle $\theta = \arccos \eta$ at azimuth $(\varphi' - \varphi)$:

$$p_+(\tau, \eta', \eta, \varphi' - \varphi, \tau_0) = p(\tau, \eta', \eta, \varphi' - \varphi, \tau_0) + p(\tau_0 - \tau, -\eta', \eta, \varphi' - \varphi, \tau_0) \qquad (2.1)$$

Along with the photon (I) absorbed at the level τ from the direction $\theta' = \arccos \eta'$ at azimuth φ', let us consider a "mirror" photon (I_*) absorbed at the mirror symmetrical level $\tau_* = \tau_0 - \tau$ from mirror symmetrical direction $\theta'_* = \pi - \arccos \eta' = \arccos \eta_*$ at azimuth φ' (Fig. 2). Obviously, the function $p(\tau, \eta', \eta, \varphi' - \varphi, \tau_0)$ characterizes the probability of exit from the given homogeneous slab $[0, \tau_0]$ across its upper border $\tau = 0$ from the mirror symmetrical level $\tau_* = \tau_0 - \tau$ in the direction $\theta = \arccos \eta$ at azimuth φ of the "mirror" photon (I_*) moving in the direction $\eta'_* = -\eta'$ at azimuth φ'. The function $p_*[(\tau_0 - \tau)_*, -\eta'_*, \eta, \varphi' - \varphi, \tau_0]$, inversely, determines the probability that the "mirror" photon (I_*) moving in the direction $(-\eta'_*)$ at azimuth φ' will exit the medium across its lower border $\tau = \tau_0$ from the mirror symmetrical level $(\tau_0 - \tau)_* = \tau$.

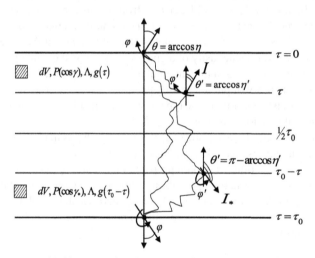

Fig. 2. The geometry of initial (I) and "mirror" (I_*) photons propagation in the homogeneous slab of finite optical thickness $\tau_0 < \infty$.

As follows from Fig. 2, spatial and angular characteristics of the initial (I) and "mirror" (I_*) photons, as well as the propagation directions $(\eta', -\eta'_*)$, and optical levels $\tau, (\tau_0 - \tau)$ and $\tau_*(\tau_0 - \tau)_*$ are mirror symmetrical. In this sense, the photons (I) and (I_*) are photons are "mirror" reflections of each other with respect to the natural level of optical symmetry $\tau_0/2$ of the homogeneous slab under consideration. Therefore, the total probability $p_{\pm,*}$ of the "mirror" photon (I_*) exiting the medium across its upper $\tau = 0$ or lower $\tau = \tau_0$ borders from the mirror symmetrical level $(\tau_0-\tau)$ in the direction (θ, φ) after multiple scatterings is equal to the sum of the following probabilities:

$$\begin{aligned} p_{+,*}(\tau_*, \eta'_*, \eta, \varphi' - \varphi, \tau_0) &= p_*[(\tau_0 - \tau)_*, -\eta'_*, \eta, \varphi' - \varphi, \tau_0] + p_*(\tau_*, \eta'_*, \eta, \varphi' - \varphi, \tau_0) \\ &= p[\tau_0 - \tau, -\eta', \eta, \varphi' - \varphi, \tau_0] + p(\tau, \eta', \eta, \varphi' - \varphi, \tau_0). \end{aligned} \qquad (2.2)$$

The comparison of relations (2.1) and (2.2) leads us to constructing the first probabilistic invariant for the initial photon (I) and its "mirror" reflection (I_*) in the homogeneous slab of finite optical thickness $\tau_0 < \infty$:

$$p_{+,*}\left(\tau_*, \eta'_*, \eta, \varphi' - \varphi, \tau_0\right) = p_+\left(\tau_0 - \tau, -\eta', \eta, \varphi' - \varphi, \tau_0\right) = p_+\left(\tau, \eta', \eta, \varphi' - \varphi, \tau_0\right). \tag{2.3}$$

Thus, the total probabilities of exiting the medium for the initial (I) and "mirror" (I_*) photons from the mirror symmetrical levels τ and τ_* are equal in the mirror directions of their propagation η' и η'_*. Note that relation (2.3) determines the probabilistic equivalence between the mirror optical levels τ and $\tau_0 - \tau$ with respect to the optical symmetry axis of the given homogeneous slab $\tau_0/2$ with simultaneous mirror symmetry of propagation directions η' and $-\eta'$ of the initial photon (I) and its "mirror" analogue (I_*) exiting the medium across any of its outside borders $\tau = 0$ and $\tau = \tau_0$. On the other hand, given the optical homogeneousity of the slab with respect to the phase function $P(cos\gamma)$ and parameter Λ, the relation of probabilistic equivalence (2.3) can be interpreted as a relation of spatial-angular invariance for the total probability of the initial photon (I) exiting the medium from the optical level τ or from its mirror level $\tau_* = \tau_0 - \tau$ in the case of their reciprocal translation $\tau \Leftrightarrow \tau_0 - \tau$ and combined replacement of the photon propagation directions $\eta' \Leftrightarrow -\eta'$. This conclusion is of fundamental nature and allows, in the case of optical homogeneousity of a given slab, to view the noted property for the value $p_+\left(\tau, \eta', \eta, \varphi' - \varphi, \tau_0\right)$ as a fundamental property of spatial-angular invariance [9, 10] for the total probability of photon exiting the initial homogeneous slab $[0, \tau_0]$ for its linear semigroup transformations of the type of spatial shift $\tau \Leftrightarrow \tau_0 - \tau$ and simultaneous spatial rotation of the vision line $\eta' \Leftrightarrow -\eta'$ [12, 13].

Another probabilistic characteristic of scalar radiation fields that allows for its invariance interpretation and respectively symmetrization at arbitrary mirror optical levels τ and $\tau_0 - \tau$ in mirror directions of photon propagation η' and $-\eta'$, is the value of spatial-angular non-symmetry for the total probability of photon existing across the upper $\tau = 0$ or lower $\tau = \tau_0$ borders of the considered homogeneous slab $[0, \tau_0]$. It is obvious that for the initial photon (I) the value of probabilistic non-symmetry can be determined by the following relation:

$$p_-\left(\tau_*, \eta'_*, \eta, \varphi' - \varphi, \tau_0\right) = p\left(\tau_0 - \tau, -\eta', \eta, \varphi' - \varphi, \tau_0\right) - p\left(\tau, \eta', \eta, \varphi' - \varphi, \tau_0\right). \tag{2.4}$$

We have a similar relation for the "mirror" photon (I_*):

$$p_{-,*}\left(\tau_*, \eta'_*, \eta, \varphi' - \varphi, \tau_0\right) = p_*\left(\tau_*, \eta'_*, \eta, \varphi' - \varphi, \tau_0\right) - p\left[(\tau_0 - \tau)_*, -\eta'_*, \eta, \varphi' - \varphi, \tau_0\right] \tag{2.5}$$

From the comparison of (2.4) and (2.5) follows the second probabilistic invariant $p_-\left(\tau, \eta', \eta, \varphi' - \varphi, \tau_0\right)$, which at the mirror symmetrical optical levels τ и $\tau_0 - \tau$ in mirror symmetrical directions of photon propagation η' and $-\eta'$ (2.3) has an additional property of invariance in the case of reciprocal and simultaneous translation of effective variables $\tau \Leftrightarrow \tau_0 - \tau$ and $\eta' \Leftrightarrow -\eta'$ in relation (2.4):

$$p_-\left(\tau, \eta', \eta, \varphi' - \varphi, \tau_0\right) = -p_-\left(\tau_0 - \tau, -\eta', \eta, \varphi' - \varphi, \tau_0\right) \tag{2.6}$$

Furthermore, knowledge of probabilistic invariants $p_+\left(\tau, \eta', \eta, \varphi' - \varphi, \tau_0\right)$ and $p_-\left(\tau, \eta', \eta, \varphi' - \varphi, \tau_0\right)$ allows to uniquely determine values of their constituents according to a simple linear scheme as follows:

$$p(\tau, \eta', \eta, \varphi' - \varphi, \tau_0) = \frac{p_+(\tau, \eta', \eta, \varphi' - \varphi, \tau_0) - p_-(\tau, \eta', \eta, \varphi' - \varphi, \tau_0)}{2}, \quad (2.7)$$

$$p(\tau_0 - \tau, \eta', \eta, \varphi' - \varphi, \tau_0) = \frac{p_+(\tau, \eta', \eta, \varphi' - \varphi, \tau_0) + p_-(\tau, \eta', \eta, \varphi' - \varphi, \tau_0)}{2}. \quad (2.8)$$

Note that the above consideration for $p_\pm(\tau, \eta', \eta, \varphi' - \varphi, \tau_0)$ is also valid for probabilistic invariants $q_\pm(\tau, \eta', \eta, \varphi' - \varphi, \tau_0)$. By reproducing it, we can easily obtain the following relations:

$$q_+(\tau, \eta', \eta, \varphi' - \varphi, \tau_0) = q(\tau, \eta', \eta, \varphi' - \varphi, \tau_0) + q(\tau_0 - \tau, -\eta', \eta, \varphi' - \varphi, \tau_0), \quad (2.9)$$

$$q_-(\tau, \eta', \eta, \varphi' - \varphi, \tau_0) = q(\tau_0 - \tau, -\eta', \eta, \varphi' - \varphi, \tau_0) - q(\tau, \eta', \eta, \varphi' - \varphi, \tau_0), \quad (2.10)$$

$$q_+(\tau, \eta', \eta, \varphi' - \varphi, \tau_0) = q_+(\tau_0 - \tau, -\eta', \eta, \varphi' - \varphi, \tau_0), \quad (2.11)$$

$$q_-(\tau, \eta', \eta, \varphi' - \varphi, \tau_0) = -q_-(\tau_0 - \tau, -\eta', \eta, \varphi' - \varphi, \tau_0). \quad (2.12)$$

It is obvious that all the above remarks regarding the probabilistic invariants $p_\pm(\tau, \eta', \eta, \varphi' - \varphi, \tau_0)$ to the full extent refer to the probabilistic invariants $q_\pm(\tau, \eta', \eta, \varphi' - \varphi, \tau_0)$.

Thus, by using relations (2.3) and (2.4) local properties of optical reciprocity of the phase function $P(\eta', \eta, \varphi' - \varphi)$, which were formulated in [8] for elementary volume dV of a scattering and absorbing medium, and similar properties of angular symmetry of brightness coefficients $\rho(\eta', \eta, \varphi' - \varphi, \tau_0)$ and $\sigma(\eta', \eta, \varphi' - \varphi, \tau_0)$ [14, 15] can be generalized to the whole extended homogeneous slab $[0, \tau_0]$ of finite optical thickness $\tau_0 < \infty$.

3 The Principle of Spatial-Angular Mirror Symmetry for Total Probabilities of Photon Exiting a Homogeneous Slab

The property of equivalence of total probabilities of photon exiting the medium $p_+(\tau, \eta', \eta, \varphi' - \varphi, \tau_0)$ и $q_+(\tau, \eta', \eta, \varphi' - \varphi, \tau_0)$, which are respectively absorbed or radiated at the mirror symmetrical optical levels τ and $\tau_0 - \tau$ in the mirror symmetrical vision lines η' and $-\eta'$, enables us to formulate a general principle of spatial-angular mirror symmetry for their constituents, namely for probabilistic functions $p(\tau, \eta', \eta, \varphi' - \varphi, \tau_0)$ and $q(\tau, \eta', \eta, \varphi' - \varphi, \tau_0)$.

For that purpose, along with the original problem A, let us consider a similar problem A_* which is mirror symmetrical to the original problem A with respect to the location of optical levels τ and $\tau_0 - \tau$ and vision lines η' and $-\eta'$, including the spatial distribution of primary isotropic energy sources $g(\tau)$ (Fig. 3).

By combining the probabilities of the initial (I) and "mirror" (I_*) photon exit p and p_*, respectively, from mirror level (τ, τ_0) and $[(\tau-\tau_0), (\tau - \tau_0)_*]$ across the upper τ

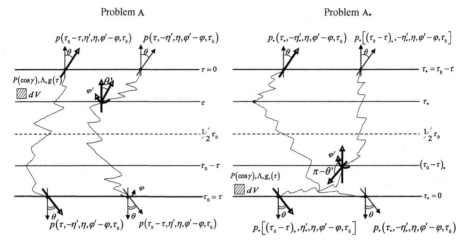

Fig. 3. Total probabilities $p_+(\tau, \eta', \eta, \varphi' - \varphi, \tau_0)$ of photons exiting the homogeneous slab $[0, \tau_0]$ across its upper $\tau = 0$ and lower $\tau = \tau_0$ borders in the initial (A) and symmetrized (A_*) problems.

$= 0$ and lower $\tau = \tau_0$ borders of the given homogeneous slab $[0, \tau_0]$ in the mirror symmetrical vision lines η' и η'_* relative to the origins τ and η' used in the initial non-symmetrized problem A, we can formulate their probabilistic equivalence in the following basic proposition: in a spatial-angular symmetrized problem, the total probabilities $p_+(\tau, \eta', \eta, \varphi' - \varphi, \tau_0)$ of multiply scattered photons exiting the medium across the upper $\tau = 0$ and lower $\tau = \tau_0$ border of a homogeneous slab from its symmetrical levels τ and $\tau_0 - \tau$ are equal in the mirror symmetrical vision lines η' and $-\eta'$ in the case of symmetrized location of primary (isotropic) energy sources with respect to level $\tau_0/2$:

$$p(\tau, -\eta', \eta, \varphi' - \varphi, \tau_0)|_A + p_*(\tau_*, -\eta'_*, \eta, \varphi' - \varphi, \tau_0)|_{A_*}$$
$$= p(\tau_0 - \tau, \eta', \eta, \varphi' - \varphi, \tau_0)|_A + p_*[(\tau_0 - \tau)_*, \eta'_*, \eta, \varphi' - \varphi, \tau_0]|_{A_*} \qquad (3.1)$$

$$p(\tau, \eta', \eta, \varphi' - \varphi, \tau_0)|_A + p_*(\tau_*, \eta'_*, \eta, \varphi' - \varphi, \tau_0)|_{A_*}$$
$$= p(\tau_0 - \tau, -\eta', \eta, \varphi' - \varphi, \tau_0)|_A + p_*[(\tau_0 - \tau)_*, -\eta'_*, \eta, \varphi' - \varphi, \tau_0]|_{A_*} \qquad (3.2)$$

From relations (3.1) and (3.2) it follows that the substitution of the initial problem A by the "mirror" problem A_*, and vice versa, with simultaneous and superposed translation of optical levels $\tau \Leftrightarrow \tau_0 - \tau$, $\tau_* \Leftrightarrow (\tau_0 - \tau)_*$ and vision lines $\eta' \Leftrightarrow -\eta'$, $\eta'_* \Leftrightarrow -\eta_*$ in the case of "mirror" substitution of primary (isotropic) energy sources $g(\tau) \Leftrightarrow g_*(\tau)$ with respect to the middle $\tau_0/2$ of the given homogeneous slab $[0, \tau_0]$, leads to their full probabilistic equivalence. In other words, the total probabilities $p_+(\tau, \eta', \eta, \varphi' - \varphi, \tau_0)$ of photons exiting a homogeneous slab across its upper $\tau = 0$ and lower $\tau = \tau_0$ borders from symmetrical optical levels $(\tau, \tau_0 - \tau)$ in the mirror directions of photon propagation η' and $-\eta'$ in a symmetrized problem, are equal (Fig. 4).

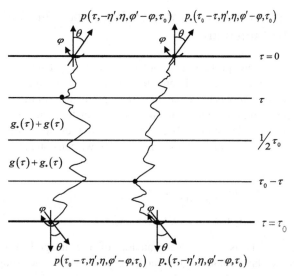

Fig. 4. Combined initial (A) and mirror (A_*) problems to determine the total probability of photon exiting the homogeneous slab [0, τ_0] from its mirror symmetrical optical levels τ и $\tau_0-\tau$ in the symmetrical directions η' and $-\eta'$ of photon propagation

The same conclusion is also true for probabilistic invariants $q_+(\tau, \eta', \eta, \varphi' - \varphi, \tau_0)$. In that case, the following obvious relations:

$$\tau_* = \tau_0 - \tau, \quad (\tau_0 - \tau)_* = \tau, \quad \eta'_* = -\eta', \quad g_*(\tau) = g(\tau_0 - \tau), \tag{3.3}$$

allow   to   directly   and   uniquely   express   probabilistic functions $p_*(\tau_*, \eta'_*, \eta, \varphi' - \varphi, \tau_0)$ and $q_*(\tau_*, \eta'_*, \eta, \varphi' - \varphi, \tau_0)$ in the "mirror" problem A_* through similar probabilistic functions $p(\tau, \eta', \eta, \varphi' - \varphi, \tau_0)$ and $q(\tau, \eta', \eta, \varphi' - \varphi, \tau_0)$ in the initial problem A:

$$p_*(\tau_*, \eta'_*, \eta, \varphi' - \varphi, \tau_0) = p(\tau_0 - \tau, -\eta', \eta, \varphi' - \varphi, \tau_0), \tag{3.4}$$

$$p_*[(\tau_0 - \tau)_*, -\eta'_*, \eta, \varphi' - \varphi, \tau_0] = p(\tau, \eta', \eta, \varphi' - \varphi, \tau_0). \tag{3.5}$$

Thus, the equivalence of photons fully exit from the symmetrical optical levels τ and $\tau - \tau_0$ in the mirror symmetrical directions of their propagation η' and $-\eta'$ in the initial (A) and mirror (A_*) problems helps us to strictly formulate this fundamental property as the following basic proposition: as a result of multiple scattering and absorption of photons in a homogeneous slab of finite optical thickness $\tau_0 < \infty$ the total probabilities of photons exiting across the slab's upper $\tau = 0$ and lower $\tau = \tau_0$ borders from the symmetrical levels τ and $\tau-\tau_0$ in the mirror symmetrical propagation directions η' and $-\eta'$ are equal.

Therefore, it can be stated that the existence of probabilistic invariants $p_\pm(\tau, \eta', \eta, \varphi' - \varphi, \tau_0)$ and $q_\pm(\tau, \eta', \eta, \varphi' - \varphi, \tau_0)$ is resulting from the general statistic interpretation of the multiple photon scattering process in the phenomenological theory of scalar radiative transfer [1, 2, 6, 7].

4 Symmetrization of Main Integral Equations for Basic Probabilistic Functions

The introduction of new probabilistic objects into the classical multiple photon scattering theory, namely the probabilistic invariants $p_\pm(\tau, \eta', \eta, \varphi' - \varphi, \tau_0)$ and $q_\pm(\tau, \eta', \eta, \varphi' - \varphi, \tau_0)$, allows to define them through linear symmetrized integral and integral differential equations which are different from those in [1, 6, 14]. Indeed, by using relations (2.1)–(2.6) and (2.9)–(2.12), after obvious linear transformations of basic integral differential equations considered in [14], we can obtain their symmetrized analogs:

$$\eta' \frac{\partial q_\pm(\tau, \eta', \eta, \varphi' - \varphi, \tau_0)}{\partial \tau} = -q_\pm(\tau, \eta', \eta, \varphi' - \varphi, \tau_0) + p_\pm(\tau, \eta', \eta, \varphi' - \varphi, \tau_0),$$

$$\eta' \in [-1, 1] \cap \eta \in [-1, 1] \cap \varphi \in [0, 2\pi] \cap \tau \in [0, \tau_0]. \tag{4.1}$$

Note that Eq. (4.1) is a complete analog of the integral differential equation that in the scalar radiative transfer theory connects the sought-for radiation intensities $I(\tau, \eta', \eta, \varphi' - \varphi_0, \tau_0)$ and source functions $B(\tau, \eta', \eta, \varphi' - \varphi_0, \tau_0)$ at arbitrary optical depths τ [1, 14]. It is obvious that solutions of Eq. (4.1) formally determine probabilistic invariants $q_\pm(\tau, \eta', \eta, \varphi' - \varphi_0, \tau_0)$ in the case where the values of $p_\pm(\tau, \eta', \eta, \varphi' - \varphi, \tau_0)$ and vice versa. Indeed, taking into account the connection between the probabilistic functions $p(\tau, \eta', \eta, \varphi' - \varphi, \tau_0)$ and $q(\tau, \eta', \eta, \varphi' - \varphi, \tau_0)$ [14], definitions (2.1), (2.4), (2.9), (2.10) of probabilistic variants $p_\pm(\tau, \eta', \eta, \varphi' - \varphi_0, \tau_0)$ and $q_\pm(\tau, \eta', \eta, \varphi' - \varphi_0, \tau_0)$, as well as their invariant properties (2.3), (2.6), (2.11) and (2.12), we can, after simple linear transformation, obtain the following relations:

$$q_\pm(\tau, \eta', \eta, \varphi' - \varphi, \tau_0) = \int_0^\tau p_\pm(\tau', \eta', \eta, \varphi' - \varphi, \tau_0) e^{-\frac{\tau - \tau'}{\eta'}} \frac{d\tau'}{\eta'} \pm \delta(\eta' - \eta)\delta(\varphi' - \varphi)e^{-\frac{\tau}{\eta}},$$

$$\eta' \in [-1, 1] \cap \eta \in [-1, 1] \cap \varphi \in [0, 2\pi] \cap \tau \in [0, \tau_0]; \tag{4.2}$$

$$p_\pm(\tau, \eta', \eta, \varphi' - \varphi, \tau_0) = \frac{\Lambda}{4\pi} \int_0^{2\pi} d\varphi'' \int_{-1}^1 P(\eta'', \eta', \varphi' - \varphi'')q_\pm(\tau, \eta', \eta, d(\varphi'' - \varphi), \tau_0)d\eta'',$$

$$\eta' \in [-1, 1] \cap \eta \in [-1, 1] \cap \varphi \in [0, 2\pi] \cap \tau \in [0, \tau_0]. \tag{4.3}$$

To obtain relations (4.2) and (4.3), we have used the angular symmetry property of phase function $P(-\eta'', -\eta', \varphi - \varphi'') = P(\eta'', \eta', \varphi' - \varphi'')$ [8]. Furthermore, it should be emphasized that the solution of integral differential Eq. (4.1) in view of basic properties (2.1), (2.6), (2.11), and (2.12) for the sought-for probabilistic invariants $p_\pm(\tau, \eta', \eta, \varphi' - \varphi, \tau_0)$ and $q_\pm(\tau, \eta', \eta, \varphi' - \varphi, \tau_0)$ allows to halve the change intervals of effective variable $\tau \in [0, \tau_0]$ and $\eta' \in [-1, 1]$ in comparison to the basic boundary problem of radiative transfer theory [9, 10, 14], and [15], namely: $\tau \in [0, \frac{\tau_0}{2}]$ and $\eta' \in [-1, 1]$, or alternatively $\tau \in [0, \tau_0]$ and $\eta' \in [0, 1]$.

By substituting then (4.2) into (4.3) and vice versa, we can obtain the following linear second kind Fredholm integral equations for the probabilistic invariants $p_\pm(\tau, \eta', \eta, \varphi' - \varphi, \tau_0)$ and $q_\pm(\tau, \eta', \eta, \varphi' - \varphi, \tau_0)$:

$$p_\pm\left(\tau, \eta', \eta, \varphi' - \varphi, \tau_0\right) = \frac{\Lambda}{4\pi}\left[P\left(\eta, -\eta', \varphi' - \varphi\right)e^{-\frac{\tau_0-\tau}{\eta}} \pm P\left(\eta, \eta', \varphi' - \varphi\right)e^{-\frac{\tau}{\eta}}\right]$$

$$+ \frac{\Lambda}{4\pi} \int_0^{2\pi} d\varphi'' \int_0^1 d\eta'' \int_0^\tau [P(\eta', \eta'', \varphi' - \varphi'')p_\pm(\tau', \eta'', \eta, \varphi'' - \varphi, \tau_0)$$

$$+ P(\eta', -\eta'', \varphi' - \varphi'')p_\pm(\tau', -\eta'', \eta, \varphi'' - \varphi, \tau_0)]e^{-\frac{\tau-\tau'}{\eta''}} \frac{d\tau'}{\eta''},$$

$$\eta' \in [-1, 1], \eta \in [0, 1], (\varphi' - \varphi) \in [0, 2\pi] \cap \tau \in [0, \tau_0];$$

(4.4)

$$q_\pm\left(\tau, \eta', \eta, \varphi' - \varphi, \tau_0\right) = \frac{\Lambda}{4\pi} \int_0^{2\pi} d\varphi'' \int_0^1 d\eta' \int_0^\tau [P(\eta', \eta'', \varphi' - \varphi'')q_\pm(\tau', \eta'', \eta, \varphi'' - \varphi, \tau_0)$$

$$+ P(\eta', -\eta'', \varphi' - \varphi'')q_\pm(\tau', -\eta'', \eta, \varphi'' - \varphi, \tau_0)]e^{-\frac{\tau-\tau'}{\eta''}} \frac{d\tau'}{\eta''} \pm \delta(\eta' - \eta)\delta(\varphi' - \varphi)e^{-\frac{\tau}{\eta}},$$

$$\eta' \in [0, 1] \cap \eta \in [0, 1] \cap (\varphi' - \varphi) \in [0, 2\pi] \cap \tau \in [0, \tau_0].$$

(4.5)

Note that integral Eqs. (4.4) and (4.5) have similar kernels and differ only in their absolute terms. Therefore, numerical solutions of these equations can be obtained by using, for example, one-type convergent iterative schemes with zero approximations which are determined by exact formulas of single photon scattering [9, 10, 14, 15].

Thus, to use the probabilistic interpretation of scalar radiative transfer in a homogeneous slab of finite optical thickness $\tau_0 < \infty$ for building basic symmetrized problems of the theory of multiple non-isotropic photon scattering, it is necessary first to carry out the following combined types of spatial-angular symmetrization:

- symmetrization of optical levels that are mirror positioned relative to the middle $\tau_0/2$ of the given homogeneous slab $[0, \tau_0]$,
- symmetrization of the directions of scattered photons propagation η' and $-\eta'$,
- mirror symmetrization of the locations of primary isotropic energy sources $g(\tau)$ relative to the middle $\tau_0/2$ of the given homogeneous slab $[0, \tau_0]$.

The latter type of symmetrization could be ignored since the probabilities of photons exiting the medium, as noted above, do not depend on the location of primary isotropic energy sources.

5 Linear Singular Integral Equations for Probabilistic Invariants of a Homogeneous Slab of Finite Optical Thickness

For implementing the mirror reflection principle and applying it for effectively modelling the probabilistic functions $p(\tau, \eta', \eta, \varphi' - \varphi, \tau_0)$ and $q(\tau, \eta', \eta, \varphi' - \varphi, \tau_0)$, as well as their probabilistic invariants $p_\pm(\tau, \eta', \eta, \varphi' - \varphi, \tau_0)$ and $q_\pm(\tau, \eta', \eta, \varphi' - \varphi, \tau_0)$, our main interest is the class of linear singular equations that were originally introduced into the scalar radiative transfer theory by V.V. Sobolev [1, 15] and T.M. Mallikin [16].

Firstly, in order to reduce the total number of independent angle variables, we can use usual Fourier transformations for phase functions $P(\cos\gamma) \equiv P(\eta', \eta'', \varphi' - \varphi'')$:

$$P(\eta', \eta'', \varphi' - \varphi'') = \wp^0(\eta', \eta'') + 2 \sum_{m=1}^{M_1} \wp^m(\eta', \eta'') \cos m(\varphi' - \varphi''), \qquad (5.1)$$

where azimuthal harmonics $\wp^m(\eta', \eta'')$ are determined as follows:

$$\wp^m(\eta', \eta'') = \frac{1}{2\pi} \int_0^{2\pi} P(\eta', \eta'', \varphi' - \varphi'') \cos m(\varphi' - \varphi'') d\varphi, \quad m = \overline{0, M_1}. \qquad (5.2)$$

Then using transformations similar to (5.1) and (5.2) for probabilistic invariants $p_\pm(\tau, \eta', \eta, \varphi' - \varphi, \tau_0)$ and $q_\pm(\tau, \eta', \eta, \varphi' - \varphi, \tau_0)$, we get

$$p_\pm(\tau, \eta', \eta, \varphi' - \varphi, \tau_0) = \sum_{m=1}^{M_2} p_\pm^m(\tau, \eta', \eta, \tau_0) \cos m(\varphi' - \varphi), \qquad (5.3)$$

$$q_\pm(\tau, \eta', \eta, \varphi' - \varphi, \tau_0) = \sum_{m=1}^{M_3} q_\pm^m(\tau, \eta', \eta, \tau_0) \cos m(\varphi' - \varphi), \qquad (5.4)$$

where azimuthal harmonics $p_\pm^m(\tau, \eta', \eta, \tau_0)$ and $q_\pm^m(\tau, \eta', \eta, \tau_0)$ can be represented as follows:

$$p_\pm^m(\tau, \eta', \eta, \tau_0) = \frac{1}{2\pi} \int_0^{2\pi} p_\pm(\tau, \eta', \eta, \varphi' - \varphi) \cos m(\varphi' - \varphi) d(\varphi' - \varphi), \qquad (5.5)$$

$$q_\pm^m(\tau, \eta', \eta, \tau_0) = \frac{1}{2\pi} \int_0^{2\pi} q_\pm(\tau, \eta', \eta, \varphi' - \varphi) \cos m(\varphi' - \varphi) d(\varphi' - \varphi). \qquad (5.6)$$

Henceforth, when using Fourier series (5.1), (5.3), and (5.4), we will select only one common value of the maximum number of azimuthal harmonics $M = \max\{M_1, M_2, M_3\}$, with required adjustment depending on the relation between the values M_1, M_2, and M_3.

Note that the sought-for linear singular integral equations for azimuthal harmonics of the probabilistic invariants $p_\pm^m(\tau, \eta', \eta, \tau_0)$ and $q_\pm^m(\tau, \eta', \eta, \tau_0)$ follow, in view of basic relations obtained on the basis of [14]:

$$p_\pm^m(\tau, \eta', \varphi - \varphi, \tau_0) = \frac{B_\pm^m(\tau, -\eta', \eta, \varphi' - \varphi, \tau_0)}{\pi S},$$

$$q_\pm^m(\tau, \eta', \eta, \varphi' - \varphi, \tau_0) = \frac{I_\pm^m(\tau, -\eta', \eta, \varphi' - \varphi, \tau_0)}{\pi S}, \qquad (5.7)$$

where the photometric values $B_\pm^m(\tau, \eta', \eta, \tau_0)$ and $I_\pm^m(\tau, \eta', \eta, \tau_0)$ are, respectively, invariants of azimuthal harmonics of source functions $B^m(\tau, \eta', \eta, \tau_0)$ and scalar radiation intensities $I^m(\tau, \eta', \eta, \tau_0)$ [10]:

$$
\begin{aligned}
B_\pm^m(\tau, \eta', \eta, \tau_0) &= B^m(\tau_0 - \tau, -\eta', \eta, \tau_0) \pm B^m(\tau, \eta', \eta, \tau_0), \\
B_\pm^m(\tau, \eta', \eta, \tau_0) &= \pm B_\pm^m(\tau_0 - \tau, -\eta', \eta, \tau_0) \\
I_\pm^m(\tau, \eta', \eta, \tau_0) &= I^m(\tau_0 - \tau, -\eta', \eta, \tau_0) \pm I^m(\tau, \eta', \eta, \tau_0), \\
I_\pm^m(\tau, \eta', \eta, \tau_0) &= \pm I_\pm^m(\tau_0 - \tau, -\eta', \eta, \tau_0)
\end{aligned}
\tag{5.8}
$$

from similar equations for azimuthal harmonics invariants of source functions $B_\pm^m(\tau, \eta', \eta, \tau_0)$ and scalar radiation intensities $I_\pm^m(\tau, \eta', \eta, \tau_0)$, respectively, [9]:

$$
\begin{aligned}
T^m(\eta)p_\pm^m(\tau, \eta', \eta, \tau_0) &= \frac{\Lambda}{4\pi}\left[A^m(\eta, -\eta')e^{-\frac{\tau_0-\tau}{\eta}} \pm A^m(\eta, \eta')e^{-\frac{\tau}{\eta}}\right] \\
\mp \frac{\Lambda}{2}\eta \int_0^1 \frac{p_\pm^m(\tau, \eta', \eta'', \tau_0)A^m(\eta, \eta'')}{\eta - \eta''}d\eta'' &\mp \frac{\Lambda}{2}\eta e^{-\frac{\tau_0}{\eta}}\int_0^1 \frac{p_\pm^m(\tau, \eta', \eta'', \tau_0)A^m(\eta, -\eta'')}{\eta + \eta''}d\eta'',
\end{aligned}
$$
$$
\eta' \in [-1, 1] \cap \eta \in [0, 1] \cap \tau \in [0, \tau_0],
\tag{5.9}
$$

$$
T^m(\eta)q_\pm^m(\tau, \eta', \eta, \tau_0) = \frac{\Lambda}{4\pi}f_\pm^m(\tau, \eta', \eta, \tau_0) \mp \frac{\Lambda}{2}\eta \int_0^1 \frac{q_\pm^m(\tau, \eta', \eta, \tau_0)A^m(\eta, \eta'')}{\eta - \eta''}d\eta'' \mp
$$
$$
\mp \frac{\Lambda}{2}\eta e^{-\frac{\tau_0}{\eta}}\int_0^1 \frac{q_\pm^m(\tau, \eta', \eta, \tau_0)A^m(\eta, -\eta'')}{\eta + \eta''}d\eta'', \quad \eta' \in [-1, 1] \cap \eta \in [0, 1] \cap \tau \in [0, \tau_0].
\tag{5.10}
$$

Functions $f_\pm^m(\tau, \eta', \eta, \tau_0) = f^m(\tau_0 - \tau, -\eta', \eta, \tau_0) \pm f^m(\tau, \eta', \eta, \tau_0)$, entering into Eq. (5.10), are determined on the basis of the following relations:

$$
f^m(\tau, \eta', \eta, \tau_0) = \begin{cases} \frac{\Lambda S}{4}A^m(\eta, \eta')\frac{\eta}{\eta-\eta'}\left(e^{-\frac{\tau}{\eta}} - e^{-\frac{\tau}{\eta'}}\right), \eta' > 0, \\ \frac{\Lambda S}{4}A^m(\eta, \eta')\frac{\eta}{\eta-\eta'}e^{-\frac{\tau}{\eta}}\left[1 - e^{-\left(\frac{1}{\eta}-\frac{1}{-\eta'}\right)(\tau_0-\tau)}\right], \eta' < 0 \end{cases}
\tag{5.11}
$$

Azimuthal harmonics of functions $T^m(\eta)$ are determined by relations [1, 15]:

$$
T^m(\eta) = 1 - \Lambda\eta^2 \int_0^1 \frac{A^m(\eta', \eta')}{\eta^2 - \eta'^2}d\eta', \quad m = \overline{0, M}.
\tag{5.12}
$$

Polynomial kernel functions $A^m(\eta', \eta')$ в (5.11) are found by the formula [15]:

$$
A^m(\eta'', \eta') = P_m^m(\eta')A_m^m(\eta'', \eta') = \sum_{i=m}^N x_i \frac{(i-m)!}{(i+m)!}\mathfrak{R}_i^m(\eta'')P_i^m(\eta'), \quad m = \overline{0, M}, \tag{5.13}
$$

where $P_i^m(\eta')$ are associated Legendre polynomials, and $\mathfrak{R}_i^m T^m(\eta)$ are Sobolev polynomials calculated on the basis of recurrent relations:

$$(i - m + 1)\mathfrak{R}_{i+1,m}^m(\eta'') + (i + m)\mathfrak{R}_{i-1,m}^m(\eta'') = (2i + 1 - \Lambda x_i)\eta''\mathfrak{R}_{i,m}^m(\eta''),$$

$$\mathfrak{R}_m^m(\eta'') = P_m^m(\eta''), \mathfrak{R}_{m+1}^m(\eta'') = (2m + 1 - \Lambda x_m)\eta''P_m^m(\eta''), m = \overline{0, M} \qquad (5.14)$$

The coefficients x_i entering into the auxiliary functions $A^m(\eta, \eta')$ are coefficient of expansion of the given phase function $P(\cos \gamma)$ in terms of Legendre polynomials P_i $(\cos \gamma)$:

$$P(\cos\gamma) = \sum_{i=0}^{N} x_i P_i(\cos\gamma), x_i = \frac{2i + 1}{2} \int_0^\pi P(\cos\gamma)P_i(\cos\gamma)\sin\gamma\, d\gamma, x_i < 2i + 1.$$

$$(5.15)$$

We should, however, pay attention to one important situation connected with possible non-uniqueness of solutions for linear singular integral Eqs. (5.9) and (5.10). As shown in [15, 16] if characteristic equations

$$T^m\left(\frac{1}{k_m}\right) = 0, \quad m = \overline{0, M}, \quad T^m\left(\frac{1}{\tilde{k}_m}\right) = 0, \quad m = \overline{0, M} \qquad (5.16)$$

have neither roots $k_m \subset [0, 1]$ nor pseudo-roots $\tilde{k}_m \subset\,]0, 1[$, then the sought for solutions of Eqs. (5.9) and (5.10) are unique. Otherwise, the solutions of those equations are non-unique, there, according to recommendations in [15], we should assume $\eta = \frac{1}{k_m}$ or $\eta = \frac{1}{\tilde{k}_m}$, considering them together with additional integral relations. These relations can be easily obtained on the basis of defining the probabilistic invariants $p_\pm(\tau, \eta', \eta, \varphi' - \varphi, \tau_0)$ and $q_\pm(\tau, \eta', \eta, \varphi' - \varphi, \tau_0)$ taking into account the respective results in [9, 15], and [10]. For example, for the probabilistic invariants $p_\pm(\tau, \eta', \eta, \varphi' - \varphi, \tau_0)$ in absence of pseudo-roots $\tilde{k}_m$ and existence of only characteristic roots k_m, they appear as follows:

$$\int_0^1 p_\pm(\tau, \eta', \eta'', \tau_0)A^m\left(\frac{1}{k_m}, \eta''\right)\frac{d\eta''}{1 - k_m\eta''}$$

$$+ e^{-k_m\tau_0} \int_0^1 p_\pm(\tau, \eta', \eta'', \tau_0)A^m\left(\frac{1}{k_m}, -\eta''\right)\frac{d\eta''}{1 + k_m\eta''} \qquad (5.17)$$

$$= \frac{1}{2\pi}\left[A^m\left(\frac{1}{k_m}, -\eta'\right)e^{-k_m(\tau_0 - \tau)} \pm A^m\left(\frac{1}{k_m}, \eta'\right)e^{k_m\tau}\right], \quad m = \overline{0, M}$$

6   Unified Probabilistic Function for Photons Exiting a Homogeneous Slab

The methodological importance of probabilistic functions $p^m(\tau, \eta', \eta, \tau_0)$ and $q^m(\tau, \eta', \eta, \tau_0)$, $p_\pm^m(\tau, \eta', \eta, \tau_0)$ and $q_\pm^m(\tau, \eta', \eta, \tau_0)$ considered above lies in the fact that at mirror symmetrical optical levels τ and $\tau_0 - \tau$ of the given homogeneous slab $[0, \tau_0]$ without a reflecting bottom, in the symmetrical directions of photon propagation η' and $-\eta'$, they form, in a way similar to unified photometric functions for intensities $I_\Sigma^m(\tau, \eta', \eta, \tau_0)$ and source functions $B_\Sigma^m(\tau, \eta', \eta, \tau_0)$ [9, 10], new objects of the radiative transfer theory, namely unified probabilistic functions $\mathcal{P}_\Sigma^m(\tau, \eta', \eta, \tau_0)$ and $\mathcal{Q}_\Sigma^m(\tau, \eta', \eta, \tau_0)$ for photons exiting the medium [18, 20]:

$$\mathcal{P}_\Sigma^m(\tau, \eta', \eta, \tau_0) = (\eta' + \eta)p^m(\tau_0 - \tau, -\eta', \eta, \tau_0) + (\eta' - \eta)p^m(\tau, \eta', \eta, \tau_0)$$
$$= \eta' p_+^m(\tau, \eta', \eta, \tau_0) + \eta p_-^m(\tau, \eta', \eta, \tau_0), m = \overline{0, M}, \tag{6.1}$$
$$\mathcal{Q}_\Sigma^m(\tau, \eta', \eta, \tau_0) = (\eta' + \eta)q^m(\tau_0 - \tau, -\eta', \eta, \tau_0) + (\eta' + \eta)q^m(\tau, \eta', \eta, \tau_0)$$
$$= \eta' q_+^m(\tau, \eta', \eta, \tau_0) + \eta q_-^m(\tau, \eta', \eta, \tau_0), m = \overline{0, M}, \tag{6.2}$$

Taking into account relations (5.7) and (5.8), instead of relations (6.1) and (6.2) we get:

$$\mathcal{P}_\Sigma^m(\tau, \eta', \eta, \tau_0) = \frac{1}{\pi S}[(\eta' + \eta)B^m(\tau_0 - \tau, -\eta', \eta, \tau_0) + (\eta' + \eta)B^m(\tau, \eta', \eta, \tau_0)]$$
$$= \frac{1}{\pi S}[\eta' B_+^m(\tau, \eta', \eta, \tau_0) + \eta B_-^m(\tau, \eta', \eta, \tau_0)] = \frac{1}{\pi S}B_\Sigma^m(\tau, \eta', \eta, \tau_0), m = \overline{0, M}, \tag{6.3}$$
$$\mathcal{Q}_\Sigma^m(\tau, \eta', \eta, \tau_0) = \frac{1}{\pi S}[(\eta' + \eta)I^m(\tau_0 - \tau, -\eta', \eta, \tau_0) + (\eta' - \eta)I^m(\tau, \eta', \eta, \tau_0)]$$
$$= \frac{1}{\pi S}[\eta' I_+^m(\tau, \eta', \eta, \tau_0) + \eta I_-^m(\tau, \eta', \eta, \tau_0)] = \frac{1}{\pi S}I_\Sigma^m(\tau, \eta', \eta, \tau_0), m = \overline{0, M}, \tag{6.4}$$

where photometric invariants $B_\pm^m(\tau, \eta', \eta, \tau_0)$ and $I_\pm^m(\tau, \eta', \eta, \tau_0)$ can be determined according to relations (5.8).

Note that knowing non-local properties of spatial-angular symmetry of probabilistic values $p^m(\tau, \eta', \eta, \tau_0)$ and $q^m(\tau, \eta', \eta, \tau_0)$ in case of inversions of effective variables η' and η, similar to photometric values $B^m(\tau, \eta', \eta, \tau_0)$ and $I^m(\tau, \eta', \eta, \tau_0)$, enables us to uniquely define them after simple and obvious linear transformations of the unified probabilistic functions $\mathcal{P}_\Sigma^m(\tau, \eta', \eta, \tau_0)$ and $\mathcal{Q}_\Sigma^m(\tau, \eta', \eta, \tau_0)$ [9, 10].

Thus, strict relations (6.1)–(6.4) form one and common complex of mutually consistent probabilistic and photometric values. These relations enclose the system of unified probabilistic ($\mathcal{P}_\Sigma^m, \mathcal{Q}_\Sigma^m$) and photometric ($B_\Sigma^m, I_\Sigma^m$) functions, including probabilistic ($p_\pm^m$, $q_\pm^m$) and photometric ($B_\pm^m, I_\pm^m$) invariants for finding numerical and analytical solutions of basic boundary-value problems in the classical scalar radiative transfer theory [14, 15] in the case of homogeneous slabs of finite optical thickness $\tau_0 < \infty$; without a reflecting bottom [18, 20]. In that case, the symmetrization of basic structural functions of this theory, such as fundamental Sobolev functions Φ^m, Ambarzumian functions $\varphi^m(\tau, \tau_0)$ and $\psi^m(\tau, \tau_0)$, Chandrasekhar functions X^m and Y^m, conducted in [18, 20], in a certain semantic sense, corresponds to the content of the initial probabilistic interpretation [1, 2] of the phenomenological theory of multiple photon scattering and to the probabilistic meaning of the aforementioned structural functions in the visible spectrum range of 0.6–0.8 μm.

7 Conclusion

From computational viewpoint and taking into account strong evidence [17, 18] and the concept of probabilistic invariants, it follows that the application of numerical methods, for example an angular discretization method, for numerical solutions of linear integral Eqs. (4.4), (4.5), (5.9), (5.10) leads to systems of linear algebraic equations of rank N, i.e. two time lower than for a system of similar integral equations for $p^m(\tau, \eta', \eta, \tau_0)$ and $q^m(\tau, \eta', \eta, \tau_0)$ [14]. Furthermore, from linear second kind Fredholm integral Eqs. (4.4) and (4.5) and linear singular integral Eqs. (5.9) and (5.10) it follows that their regular solutions within the ranges of effective variables $\tau \in [0,\tau_0]$, $\eta' \in [-1,1]$ and $\eta \in [0, 1]$, taking into account basic properties (2.3)–(2.4) and (2.11)–(2.12) of spatial-angular symmetry of probabilistic invariants $p_\pm(\tau, \eta', \eta, \varphi' - \varphi, \tau_0)$ and $q_\pm(\tau, \eta', \eta, \varphi' - \varphi, \tau_0)$, allow to use more narrow ranges of their definition, namely $\tau \in [0, \tau_0/2]$ and $\eta' \in [-1, 1]$, or alternatively $\tau \in [0,\tau_0]$ and $\eta' \in [0,1]$. Understandably, such double decrease in ranges of optical depths τ or angular variables η' will substantially simplify algorithms for numerical solutions of equations like (4.4), (4.5), (5.9), and (5.10), especially in cases of large values $\tau \gg 1$ and $\tau_0 \gg 1$, and also in the case of strongly elongated phase functions $P(\cos \gamma)$ at $m \gg 1$ and $M \gg 1$ [19, 20].

References

1. Sobolev, V.V.: Radiative Transfer in Stellar and Planetary Atmospheres. Gostekhteorizdat, Moscow (1956). (in Russian)
2. Ueno, S.: The Probabilistic Method for Problems of Radiative Transfer. X. Diffuse Reflection and Transmission in a Finite Inhomogeneous Atmosphere. Astrophys. J. **138**(3), 729–745 (1960)
3. Barucha-Reid, A.T.: Elements of the Theory of Markov Processes and their Applications. McGraw-Hill, New York (1960)
4. Kalinkin, A.V.: Markov branching processes with interaction. Russ. Math. Surv. **57**(2), 241–304 (2002) (in Russian)
5. Langville, A.N., Meyer, C.D.: Updating Markov chains with an eye on Google's page rank. SIAM J. Matrix Anal. Appl. **27**(4), 968–987 (2006)
6. Sobolev, V.V.: A new method in the light scattering theory. Astron. J. **28**(5), 355–362 (1951). (in Russian)
7. Minin, I.N.: Radiation diffusion in a slab with anisotropic scattering. Astron. J. **43**(6), 1244–1260 (1966). (in Russian)
8. van de Hulst, H.C.: Light Scattering by Small Particles. Dover Publ., New York (1981)
9. Smokty, O.I.: Radiation Fields Modelling in Problems of Space Spectrophotometry. Nauka, Leningrad (1986). (in Russian)
10. Smokty, O.I.: Development of radiation transfer theory methods on the basis of mirror symmetry principle. In: Current Problems in Atmospheric Radiation (IRS 2000), pp. 341–345. Deepak Publ., Hampton (2001)
11. Ishimaru, A.: Electromagnetic Wave Propagation, Radiation, and Scattering. Prentice Hall Publ, New York (1991)
12. Bhagavantam, S., Venkatarayudu, T.: Theory of Groups and its Application to Physical Problems. Andhra University, Waltair (1951)
13. Gelfand, I.M., Minlos, R.A., Shapiro, Z.Ya.: Representation of Rotation and Lorentz Groups and Their Applications. Dover Publ., New York (2009)
14. Minin, I.N.: Theory of Radiative Transfer in Planetary Atmospheres. Nauka, Moscow (1988).(in Russian)
15. Sobolev, V.V.: Light Scattering in Planetary Atmospheres. Pergamon Press, Oxford (1975)

16. Mullikin, T.M.: Radiative transfer in finite homogeneous atmosphere with anisotropic scattering: I Linear Singular Equations. Astrophys. J. **139**, 379–396 (1964)
17. Smokty, O.I.: Improvements of the methods of radiation fields numerical modeling on the basis of mirror reflection principle. In: Murgante, B., et al. (eds.) ICCSA 2013. LNCS, vol. 7975, pp. 1–16. Springer, Heidelberg (2013). https://doi.org/10.1007/978-3-642-39640-3_1
18. Smokty, O.I.: The Mirror Reflection Principle in the Radiative Transfer Theory. VVM, St. Petersburg (2019). (in Russian)
19. Smokty, O.I.: Analytical approximation for homogeneous slab brightness coefficients in the case of strongly elongate phase functions. In: Radiative Processes in the Atmosphere and Ocean (IRS 2016), pp. 145–149. American Institute of Physics, Ney York (2017)
20. Smokty O.I.: Unified exit function for upgoing and downgoing radiation fields at arbitrary symmetrical levels of a uniform slab. In: Radiative Processes in the Atmosphere and Ocean (IRS 2016). American Institute of Physics, Ney York (2017). https://doi.org/10.1063/1.497 5511

Ensemble Feature Selection Compares to Meta-analysis for Breast Cancer Biomarker Identification from Microarray Data

Bernardo Trevizan[1,2] and Mariana Recamonde-Mendoza[1,2(✉)]

[1] Institute of Informatics, Universidade Federal do Rio Grande do Sul,
Porto Alegre, RS, Brazil
[2] Bioinformatics Core, Hospital de Clínicas de Porto Alegre, Porto Alegre, RS, Brazil
{btrevizan,mrmendoza}@inf.ufrgs.br

Abstract. Identifying stable and precise biomarkers is a key challenge in precision medicine. A promising approach in this direction is exploring omics data, such as transcriptome generated by microarrays, to discover candidate biomarkers. This, however, involves the fundamental issue of finding the most discriminative features in high-dimensional datasets. We proposed a homogeneous ensemble feature selection (EFS) method to extract candidate biomarkers of breast cancer from microarray datasets. Ensemble diversity is introduced by bootstraps and by the integration of seven microarray studies. As a baseline method, we used the random effect model meta-analysis, a state-of-the-art approach in the integrative analysis of microarrays for biomarkers discovery. We compared five feature selection (FS) methods as base selectors and four algorithms as base classifiers. Our results showed that the variance FS method is the most stable among the tested methods regardless of the classifier and that stability is higher within datasets than across datasets, indicating high sample heterogeneity among studies. The predictive performance of the top 20 genes selected with both approaches was evaluated with six independent microarray studies, and in four of these, we observed a superior performance of our EFS approach as compared to meta-analysis. EFS recall was as high as 85%, and the median F1-scores surpassed 80% for most of our experiments. We conclude that homogeneous EFS is a promising methodology for candidate biomarkers identification, demonstrating stability and predictive performance as satisfactory as the statistical reference method.

1 Introduction

Precision medicine aims at tailoring health care testing and medication to each patient according to their individual characteristics. It has become feasible with the growth of digital medical records and high-throughput diagnosis devices [16].

This study was financed in part by the Coordenação de Aperfeiçoamento de Pessoal de Nível Superior - Brasil (CAPES) - Finance Code 001, and Conselho Nacional de Desenvolvimento Científico e Tecnológico (CNPq).

Since the first sequenced human genome, more recent technologies, such as microarray, have helped to understand diseases at the molecular level through the generation of omics data. Microarray technology allows the characterization of the gene expression profile for all genes within an organism, also called transcriptome, which is especially useful when expressions are compared among different groups, such as a disease and a control group.

One of the key applications of transcriptome data for precision medicine is the identification of candidate biomarkers. Biomarkers are measurable parameters that act as indicators of the presence or severity of a disease. The exploration of biomarkers for detecting cancer or selecting the most appropriate therapy for a patient is one of the main branches of precision medicine research [17]. Nonetheless, harnessing the knowledge of high-throughput omics data requires advanced computational methods such as machine learning (ML) algorithms to identify molecular patterns in different patients that may be used to infer hypotheses about diagnostic biomarkers.

As discussed by [15], a fundamental property of biomarkers is stability. Stable biomarkers refer to a set of consistent markers identified from sampling variations. In other words, a biomarker is stable and thus representative of the population if the same can be reproduced in other similar subpopulations. Thus, the integration of multiple transcriptomic datasets has been advocated to improve biomarkers discovery [30]. In addition, biomarkers should have good discriminative power, meaning that based on their application, we can accurately distinguish tumor from non-tumor samples, for instance, if considering diagnostic biomarkers. The identification of stable and precise biomarkers is one of the critical challenges in transcriptome analysis in the scope of precision medicine. One of the state-of-the-art approaches for such investigation is statistical meta-analysis. Several meta-analysis methods exist, and they combine the primary statistics drawn from individual analysis of each experiment, such as p-values or effect sizes (*i.e.*, fold change) [30].

In the field of ML, biomarkers discovery has been frequently approached as a dimensionality reduction problem, for which feature selection (FS) methods may be naturally applied [4]. A plethora of FS methods exist; nonetheless, the high correlation among features, as frequently observed in omics data, leads to instability among results from multiple FS methods or from multiples runs over distinct training samples, reducing the confidence of selected features [19]. One of the proposed solutions to improve the stability of FS on high-dimensional data is ensemble feature selection (EFS). Ensemble learning principles come from the idea of "Wisdom of Crowds," which states that a group of diverse individuals are collectively smarter than any single individual, even expert ones [27]. In ML, ensembles have been explored to produce more robust and accurate learning solutions. In the context of FS, EFS methods use various base selectors and aggregate their results, aiming to obtain a more stable feature subset. Diversity among opinions, an essential property of ensemble learning, is achieved either by using distinct FS methods as base selectors (*i.e.*, function perturbation) or by using sampling variations across base selectors (*i.e.*, data perturbation) [7]. These approaches generate, respectively, heterogeneous and homogeneous ensembles.

In the present work, we aim at exploring homogeneous ensemble FS to identify candidate biomarkers for breast cancer from multiple microarray studies and compare its results to traditional meta-analysis. Our work is motivated by the large volume of transcriptomic data publicly available and the need to jointly analyze different gene expression datasets to derive more stable and precise hypotheses about candidate biomarkers. We believe that homogeneous EFS is as promising as the well-established meta-analysis approaches to extract the most relevant genes for breast cancer diagnosis. Moreover, to the best of our knowledge, there were limited efforts so far towards the comparison of stability and predictive performance among these different analytical approaches.

2   Related Works

The literature regarding EFS has expanded recently, and several previous works have applied homogeneous EFS to high-dimensional data, including microarray. [22] evaluated stability on EFS methods across several domains, including biomedical data, demonstrating that homogeneous EFS approaches lead to a significant gain of stability. An earlier study [23] exploited the performance and stability of homogeneous ensembles applied to high-dimensional omics data. In every case, the EFS approach outperformed other methods. The authors achieved a high accuracy with 3% of selected features and a stability increase with homogeneous EFS compared to other methods. In [1], homogeneous EFS showed the best stability and improvement in performance on identifying biomarkers for cancer diagnosis from microarray datasets. Authors evaluate their approach for four types of cancer (Leukemia, colon, lymphoma, and prostate), using the traditional recursive feature elimination (RFE) as the baseline FS method.

In [5], authors designed an ensemble in which, for each base selector's output, a classifier was trained. The final predictions were combined by majority voting among classifiers. The heterogeneous ensemble approach yielded a stability index of 0.229 for breast cancer data and a predictive error of 28.11%. Using the same ensemble design, [21] achieved an error of 3.09%. However, in the latter, the authors grouped genes using information theory in which a gene is relevant if it is correlated to classes and not to other selected genes. The ensemble selects one gene for each group of highly correlated genes in order to reduce redundancy. The authors also reported an increase in stability compared to other methods. In [3], authors validated the results of genes selected from EFS algorithms with literature meta-analysis results. The authors were able to identify 100 microRNAs (*i.e.,* small non-coding RNAs) that could explain 29 types of cancer, most of which had previous evidence from the literature. Besides, reported accuracies are higher than 90% in all experiments.

In [25], the authors used seven datasets with different FS methods to improve training time and increase accuracy. The best results presented for microarray dataset used a homogeneous ensemble with SVM-RFE and mean aggregation of rankings. According to authors, "[...] *an ensemble approach would seem to be the most reliable approach to feature selection*" [25]. The authors in [6] reviewed the performance of state-of-the-art FS methods across several domains of microarray - breast, prostate, brain, colon, and ovarian cancer. The authors concluded that, in general, performance depends on the FS method, on the classifier, and, mainly, on the problem domain. This emphasizes the importance of EFS methods, which partially alleviates the problem of lack of robustness for single FS methods. We note, however, that none of these works have compared EFS to meta-analysis approaches for dimensionality reduction in microarray data. In the approach presented by [26], authors combined ML with meta-analysis in a sequential manner, in which potential markers identified for mastitis disease in cattle using meta-analysis were further analyzed with decision trees to refine the hypotheses about most relevant features. Their approach was able to identify four biological markers with 83% accuracy in classification; nonetheless, their methodological and biological goals differ from ours. In fact, efforts towards comparing EFS and meta-analysis methodologies for identifying candidate biomarkers are still scarce in the literature.

3   Materials and Methods

3.1   Breast Cancer Microarray Datasets

Microarray datasets containing both tumor and non-tumor (*i.e.,* "normal") samples were collected from the Gene Expression Omnibus (GEO) database. A total of 11 studies were downloaded and pre-processed following the standard bioinformatics pipeline for each manufacturer (Affymetrix or Agilent). Due to limitations in the evaluation process, only datasets containing more than 10 samples in each class were selected as training datasets, defined as $T = \{T_1, T_2, T_3, ..., T_7\}$ (Table 1). The remaining datasets were assign for results' evaluation, defined as $E = \{E_1, E_2, E_3, ..., E_6\}$ (Table 2). To avoid redundancy on gene identification for later comparison of gene sets, all genes were mapped to their intrinsic *Entrez ID* using the R package *biomaRt* [11]. Furthermore, only genes common to all studies were kept for further analyses.

Table 1. Datasets used for training the EFS method.

| | Dataset | Number of samples | | | Tumor/ |
		Tumor	Normal	Total	Total ratio
T_1	GSE38959	30	13	43	0.70
T_2	GSE42568	98	17	115	0.85
T_3	GSE45827	122	36	158	0.77
T_4	GSE53752	46	21	67	0.69
T_5	GSE62944	1119	113	1232	0.91
T_6	GSE70947	148	148	296	0.50
T_7	GSE7904	42	18	60	0.70

Table 2. Datasets used for testing the EFS method.

| | Dataset | Number of samples | | | Tumor/ |
		Tumor	Normal	Total	Total ratio
E_1	GSE10797	27	5	32	0.84
E_2	GSE22820	74	10	84	0.88
E_3	GSE26304	109	6	115	0.95
E_4	GSE57297	25	7	32	0.78
E_5	GSE61304	57	4	61	0.93
E_6	GSE71053	6	12	18	0.33

3.2   Feature Selection Methods

Several FS methods have been proposed to reduce dimensionality. Here, we only considered ranker-based methods, *i.e.,* those that provide as output a ranking of all features ordered by a given score. The completeness in the output of these methods facilitates ranking aggregation among multiple selectors. We focused exclusively on filter methods among ranker-based methods, as they present the best trade-off between predictive and computational performance. This is important when working with high-dimensional data, as well as with homogeneous EFS since it involves multiple applications of the FS methods [22]. In this paper, the following FS methods were explored:

- *Information Gain* (IG) [18] explores the concept of entropy, which quantifies the uncertainty of each feature in the process of decision making. Entropy is used to measure the impurity of information, which in turn allows estimating the relevance of each feature by evaluating the extent to which impurity (and thus uncertainty) decreases when the value of that feature is known.
- Chi squared (ϕ_c) [10] measures the correlation between two random variables A and B based on the χ^2 statistical test. The larger the χ^2, the more important the feature is for the predictive task of interest.
- *Symmetrical Uncertainty* [28] is also based on the information-theoretical concept of entropy. It modifies the IG by introducing a proper normalization factor.
- *Minimum Redundancy Maximum Relevance (mRMR)*, defined by [13], best scores variables with higher correlation to the target value and lower correlation among other variables. For such, it takes the IG between the response variable A given $B \in V$, where V is the set of variables, and subtracts the pairwise IG between other variables and B.
- *Variance* measures the spread of a variable's values from its mean. As a statistical summary function, variance can be expressed as the average squared distances of A from its mean value μ_A.

3.3   Ensemble Feature Selection Design

The idea underlying EFS is to use various base selectors and aggregate their results aiming at more stable and accurate feature subsets. The design of an efficient EFS involves five main decisions: (i) type of base selectors; (ii) number of base selectors; (iii) number and size of different training sets; (iv) aggregation method, and (v) threshold methods [7]. In what follows, we discuss these details regarding the proposed solution.

We implemented a homogeneous EFS in this work, meaning that all base selectors employ the exact same FS method, and diversity among base selectors is introduced at the data level. The FS methods tested and compared in our experiments were listed in Sect. 3.2. In our framework, data diversity is generated in two ways. First, multiple microarray datasets are analyzed in parallel, each of which derives from a distinct study/population of patients with breast cancer (Table 1). Second, we apply data perturbation upon sampling variations for each microarray dataset using the bootstrap method. Given a dataset with n instances, this method randomly selects n instances from the original dataset with replacement to generate a data sample, also called a *bag*. This sampling process is repeated B times, a parameter of our framework, and each bag is analyzed by the chosen FS method to generate a list of features ranked by relevance. We emphasize that only one type of FS method is applied over all bags for a single experiment. We tested a different number of bags (*i.e.*, number of base selectors) per microarray dataset, from 5 to 100, which will be further explained in Sect. 4.

After running all base selectors, multiple rankings are generated and must be aggregated into a consensus ranking. These consensus rankings are referred to as *local rankings* of our framework, as they reflect the consensus among B bags for a given microarray dataset. Many aggregation strategies have been proposed in the literature and were compared by previous works. According to [25], the best aggregation method for microarray datasets is the arithmetic mean, which takes the average of the relevance values yielded by the multiple rankings to reorder features into a consensus ranking. Thus, we adopted mean aggregation to generate a local ranking for each microarray study. Once each study has been individually analyzed, the multiple local rankings (*i.e.*, seven at total) are aggregated into a *global ranking*, which represents our framework's output. We note that both local and global rankings comprise the complete set of features in the datasets. An EFS-based subset of highly discriminative features is obtained by applying a threshold to keep only the top K most relevant features.

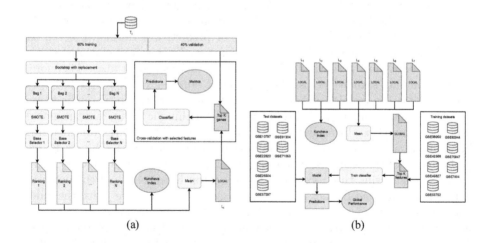

(a)   (b)

Fig. 1. Methodology applied for building and evaluating the proposed EFS approach. a) Homogeneous EFS is applied to each microarray dataset, generating a local ranking. b) The local rankings extracted from the multiple datasets are aggregated into a global ranking.

3.4 Ensemble Feature Selection Evaluation

The main goal in EFS is to improve the stability and predictive performance of the selected feature subset. According to [22], *"stable but not accurate solutions would not be meaningful; on the other hand, accurate but not stable results could have limited utility"*. Thus, our methodology for evaluating FS results adopts both criteria, the stability of the selected features and the predictive performance of the classification models built exclusively with this subset of features. Our methodology to build and evaluate a homogeneous EFS is summarized in Fig. 1. For each training dataset $T_i \in T$, as summarized in Table 1, we created an ensemble following the aforementioned explanation.

To assess the predictive performance of the features subset, we applied a combination of stratified k-fold cross-validation (CV) and holdout to train and test classifiers using the selected top K features. For each iteration of k-fold CV, we split the training data (*i.e.*, training folds) into two parts, one for training (60%) and the other for testing (40%) the classifier, as shown in Fig. 1a. The training part is resampled to perturb data for B bags. B is also the number of base selectors, one for each bag, all running a previously chosen FS method. Since our datasets are characterized by class imbalance, we adopted SMOTE as an over-sampling method to deal with this issue. SMOTE is applied in each bag separately prior to FS analysis, and in the training partition to build the classifiers based on the top K features from the generated local ranking (*i.e.*, mean aggregation of the B rankings). Our EFS approach outputs three objects: (i) local stability for the set of rankings; (ii) final aggregated ranking, also called as local ranking L_i; and (iii) local performance metrics for the top K genes using 5-fold cross-validation.

Applying the basic homogeneous EFS approach to each microarray study only allows the assessment of the stability or predictive performance based on the specific dataset presented to the classifier, *i.e.*, T_i. However, our interest lies in the stability and performance across several microarray studies. Therefore, we also evaluate stability across the set of local rankings L_i for $i = \{1, 2, 3..., 7\}$ generated for the training datasets (Fig. 1b). Moreover, the local rankings are aggregated into a unique global ranking by the arithmetic mean, and the global ranking is finally used to select the EFS-based subset of most relevant features by applying a threshold K (*i.e.*, retaining the top K features). The predictive performance of the proposed homogeneous EFS approach is evaluated by training a classifier using this subset of EFS-based top K features and the union of datasets $T_i \in T$, and further evaluating its generalization power with the independent test datasets $E_i \in E$ (Table 2). Since these datasets contain instances never seen by the classifier, they allow an unbiased estimate of predictive performance. After studies aggregation, the output of our framework is the global stability, global ranking, and global performance metrics.

3.5 Evaluation Metrics

As aforementioned, two evaluation criteria are adopted in our framework: stability and predictive performance. The stability validation step calculates the

Kuncheva Index (Eq. 2) [20], which is the average of pairwise inconsistency indexes between a set of features subsets $\mathcal{S} = \{S_1, S_2, S_3, ..., S_N\}$. An inconsistency index, according to [20], increases proportionally to the intersection's cardinality $r = |A \cap B|$ between two subsets A and B with the same cardinality $|A| = |B| = k$. The maximum value 1 is achieved when $A = B$, and the minimum value is limited to -1. Equation 1 mathematically defines the inconsistency index where $A, B \subset X$ and $|X| = n$. The author also defines a threshold for high and low stability. The stability can be considered high when $\mathcal{I}(\mathcal{S}) \geq 0.5$, and low otherwise.

$$I(A, B) = \frac{rn - k^2}{k(n - k)} \tag{1}$$

$$\mathcal{I}(\mathcal{S}) = \frac{2}{N(N - 1)} \sum_{i=1}^{N-1} \sum_{j=1}^{N} I(S_i, S_j) \tag{2}$$

The predictive performance is estimated by training a classifier solely with the most relevant features provided by the top K features. The top K features are extracted from the local ranking when our goal is to evaluate performance for a single microarray dataset or from the global ranking when we aim to carry out this analysis across multiple datasets. The classifier is evaluated using the independent testing sets to compute the precision, recall, and F1-Score in both cases. *Precision* defines the ratio of true positive (TP) predictions over all positive predictions made by the model, which is the sum of TP and false positives (FP). *Recall*, or sensitivity, defines the ratio of TP predictions over all samples labeled as positive, that is, TP and false negatives (FN). *F1-Score* (*F1*) defines the harmonic mean between precision and recall.

3.6   Classification Algorithms

To evaluate the effectiveness of potential feature sets for classification of breast cancer samples, we estimate the performance using the following supervised learning algorithms as basis for training the classifiers:

- *Decision Tree* (J48) classification algorithm learns decision rules inferred from training data to predict a target variable [8]. J48 iteratively selects the feature that best splits the subset (training data for the first iteration) according to a homogeneity metric, such as information gain.
- *K-Nearest Neighbors* (KNN) is an instance-based algorithm [2] that computes classification through a majority vote among the K nearest neighbors of each new data point. We adopted Euclidean distance as the distance measure.
- *Neural Network* (NNET) uses a multi-layered perceptron algorithm that trains a model using batch gradient descent [14]. Each input neuron represents a feature for a given sample. For training, a weight is assigned for each connection between two neurons. Therefore, a neuron's a_i value is the weighted sum of previous neurons directly connected to a_i. The values are

propagated toward the output layer where the prediction error is calculated. The error is applied on the weights update in the opposite direction using gradient descent and a learning rate α, *i.e.,*, by error *back-propagation*. The model classifies new samples by choosing the classes whose output neuron has the highest probability.

– *Support Vector Machine* (SVM) creates a hyper-plane or a set of hyper-planes that is farthest from the nearest training samples [9]. In this work, a non-linear SVM with a radial kernel is applied. Non-linear SVM applies a kernel function in order to expand the input space into a feature space to make the problem linearly separable.

Experiments were run in R environment, using the *caret* R package to train and evaluate models. Hyperparameter tuning was performed for each algorithm using the package's standard grid set of candidate models. By default, caret automatically chooses the hyperparameters associated with the best performing model.

3.7   Meta-analysis

Meta-analysis represents another viable approach for candidate biomarker identification, being a state-of-the-art method in this type of investigation. In a broad sense, meta-analysis aims at integrating the findings (*i.e.,* primary statistics) from multiple studies to extract more robust hypotheses. The first step of meta-analysis measures the effect size - the strength of a phenomenon - through differential expression analysis (DEA). Here we adopted the LIMMA [24], especially designed for transcriptomics, to extract genes' log fold change (*logFC*) and *p-value* for each study. The *logFC* is simply the difference between the average of expressions of tumor cases and the average of expressions of normal cases. The *p-value* evaluates the significance of a change in expressions between the two groups. In the second step, the meta-analysis method *per se* is applied, which integrates the primary statistics from individual studies to assess the significance of each gene across all populations. In this work, the focus will be on methodologies based on effect size combination due to their performance in meta-analysis for microarray datasets [29]. We used the Random Effect Model (REM) through the *MetaVolcano* R package, which outputs a ranking of genes from the most significant to the least significant to distinguish tumor from normal samples.

As one of ours goals is to compare EFS with meta-analysis methods, we evaluate the results from meta-analysis with a pipeline similar to the one adopted to EFS (Fig. 2). However, instead of the local ranking for a dataset T_i, meta-analysis calculates the effect size for each gene through DEA and the REM generates the global ranking. In this way, we assure a fair comparison between EFS and REM by applying the same data and executing the same evaluation procedure for both methods.

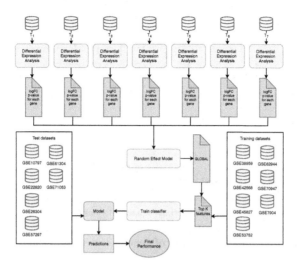

Fig. 2. Evaluation pipeline adopted for the REM meta-analysis method.

4 Results

In what follows we present the results for the homogeneous EFS applied to the selected breast cancer microarray studies in terms of stability and predictive performance. The section ends by discussing the results of the comparative analysis between EFS and the REM meta-analysis. All results are based on the average or values distribution for 5-fold cross-validation.

Our experiments were focused on evaluating every possible combination of the following parameters (and respective values): **Threshold:** 5, 10, 15, 20, 25, 30, 50 75, 100, 150, 200, 250, 500; Number of Bags: 5, 10, 25, 50, 100; **Classifier:** SVM, J48, KNN, NNET; **Base selector:** IG, ϕ_c, SU, mRMR, variance.

4.1 Stability Analysis

High stability for biomarker identification in EFS methods means that the most informative genes for one population are also the most informative for another, *i.e.*, it reflects the reproducibility of the selected genes across the populations. The ideal scenario is stability close to 1 when feature selection methods across different populations choose almost all the same genes. In general, the bigger the subset of selected genes, the higher the stability. Figure 3 shows the expected behavior for global stabilities, with values increasing for larger thresholds. Note that global stabilities, in general, are low, indicating a poor overlap of hypotheses driven from different populations. This may be due to heterogeneity across studies' samples. On the other hand, the stability for the top 5 genes does not follow the trend and is slightly higher than stability achieved until the top 20 genes. This suggests that local ensembles are most agreeing in rankings' top positions.

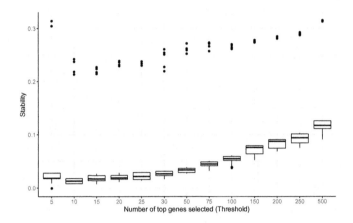

Fig. 3. Stability for different number of genes selected from the global ranking.

To investigate the outliers observed in Fig. 3, we analyzed results separately by base selectors and by the number of bags. Breaking by base selectors (Fig. 4a), it is clear that *variance* presents overall higher stability in comparison to other base selectors, which explains the mentioned outliers. In contrast, the number of bags (Fig. 4b) has an insignificant impact on the overall global stability. Therefore, stability is mainly impacted by the selected genes' subset size and the base selector bias. Since we want to select a minimum set of informative genes - the candidate biomarkers - high stability lies in the choice of the base selector.

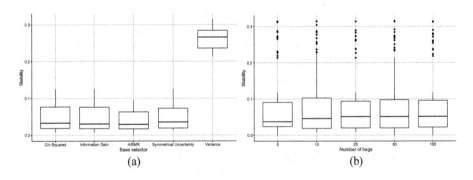

Fig. 4. Analysis of stability for a) each base selector and b) different number of bags.

In GEO, different datasets are deposited from studies within distinct populations. In this work, we will not address the biological implications and challenges of different data sources and sample heterogeneity. However, when stabilities are compared locally within each training dataset (Fig. 5) based on bootstrap samples, the stabilities are much higher. We can see that, once again, variance clearly achieves higher stabilities than other base selectors. Furthermore, the GSE62944 dataset presents higher stability for every base selector in comparison with other datasets. We could hypothesize that a positive correlation exists between the number of samples and stability since this dataset has 1.232 samples while the other datasets have 123 ± 86 samples on average. However, this assumption needs to be validated in further experiments.

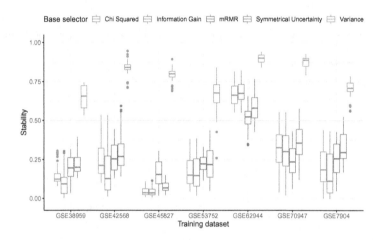

Fig. 5. Local stability for training datasets considering each base selector.

4.2   Predictive Performance Analysis

Classifiers predictive performance can also guide the identification of biomarkers. We observed that the threshold and the base selectors do not impact performance significantly (Fig. 6). We can see a slight increase on performance median until 50 genes are selected. However, in general, the top 5 genes already presented high F1-Scores (Fig. 6a). Among the base selectors, variance presented the most consistent results across multiple evaluations, which corroborates the stability results. Therefore, we will focus our next analyses on results presented by variance as base selector and the top 20 genes as threshold. Although other threshold values could be explored, the top 20 genes seem to show a good compromise between subset size, performance, and stability.

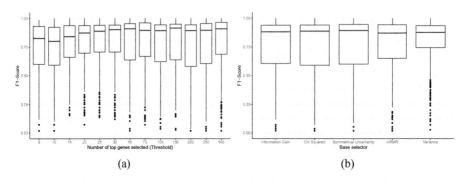

Fig. 6. Analysis of the impact of a) threshold and b) base selector over performance.

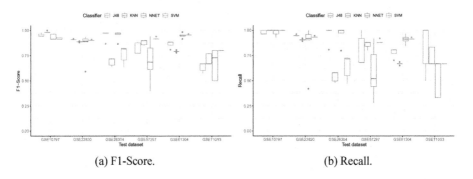

(a) F1-Score.   (b) Recall.

Fig. 7. Predictive performance on test datasets using variance and the top 20 genes.

In general, the EFS method presented satisfactory performance (Fig. 7). In terms of F1-Score (Fig. 7a), we note an insignificant difference between the highest median results within test datasets. We observed more stable results for precision (data not shown) than recall (Fig. 7b), which could be caused by a disproportionately smaller number of negative samples (*i.e.,* normal samples) compared to the number of positive samples, thus making recall more sensitive to prediction variations. Nonetheless, we note that predictive performance mostly varies between different test datasets rather than between different classifiers for the same dataset, due to data heterogeneity.

Table 3 shows the performance summary for F1-Score. There is no unique classifier able to perform satisfactorily for every test dataset. However, the mean absolute error (MAE), which is the mean difference between the highest performance and the other performances, in most cases is less than 0.1. MAE can be seen as a relative measure of how much better the best classifier is among the others. Note that when SVM outperforms other classifiers within test datasets, its MAE is higher if compared to the other cases. In this way, generally, SVM would be the best choice. On the other hand, in terms of recall, KNN seems to be a more robust choice (Table 4).

Table 3. Classifiers mean F1-Score by test dataset for *variance* and the top 20 genes. Bold values indicate the highest performance achieved for a dataset.

Dataset	Classifier				MAE
	SVM	J48	KNN	NNET	
GSE10797	0.92 ± 0.01	0.92 ± 0.03	$\mathbf{0.96 \pm 0.02}$	0.92 ± 0.02	0.04
GSE22820	0.72 ± 0.28	0.87 ± 0.05	$\mathbf{0.88 \pm 0.05}$	0.85 ± 0.16	0.06
GSE26304	0.86 ± 0.12	$\mathbf{0.88 \pm 0.14}$	0.77 ± 0.07	0.84 ± 0.22	0.05
GSE57297	$\mathbf{0.93 \pm 0.04}$	0.79 ± 0.10	0.89 ± 0.03	0.77 ± 0.18	0.11
GSE61304	$\mathbf{0.96 \pm 0.03}$	0.84 ± 0.09	0.84 ± 0.09	0.94 ± 0.04	0.08
GSE71053	$\mathbf{0.76 \pm 0.06}$	0.53 ± 0.14	0.68 ± 0.07	0.62 ± 0.17	0.15

Table 4. Average recall by classifier computed for the test datasets using the *variance* and the top 20 genes. Bold values indicate the highest performance for a dataset.

Dataset	Classifier				MAE
	SVM	J48	KNN	NNET	
GSE10797	$\mathbf{1.00 \pm 0.00}$	0.96 ± 0.04	$\mathbf{1.00 \pm 0.00}$	0.96 ± 0.02	0.03
GSE22820	0.71 ± 0.37	0.87 ± 0.10	$\mathbf{0.89 \pm 0.09}$	0.86 ± 0.23	0.08
GSE26304	0.54 ± 0.43	$\mathbf{0.85 \pm 0.22}$	0.65 ± 0.10	0.81 ± 0.28	0.18
GSE57297	$\mathbf{0.89 \pm 0.09}$	0.78 ± 0.20	$\mathbf{0.89 \pm 0.08}$	0.72 ± 0.25	0.14
GSE61304	$\mathbf{0.92 \pm 0.06}$	0.76 ± 0.16	0.74 ± 0.14	0.90 ± 0.06	0.12
GSE71053	0.62 ± 0.07	0.57 ± 0.29	$\mathbf{0.75 \pm 0.16}$	0.50 ± 0.16	0.19

4.3 EFS Versus Meta-analysis

To validate our EFS framework, we compared it with results generated by the REM meta-analysis approach, considering the top 20 genes in both cases and the variance FS method for EFS. Figure 8 shows the performance a) by classifier across all test datasets and b) by test dataset across all classifiers. Classifiers were able to perform satisfactorily across datasets, with median values surpassing the 0.75 mark in most cases. We note that performance for EFS is more consistent than meta-analysis and, in general, our approach has the interquartile range shifted towards higher values, which indicates better performance. When observing F1-Score per test dataset, our EFS approach outperforms meta-analysis in four out of the six datasets. In particular, for GSE10797, GSE61304, and GSE71053, the improvement brought by EFS over meta-analysis is remarkable. Table 5 summarizes the performance of both methods across all experiments, which corroborates the previous discussion. Note that for GSE613040 and GSE71053, EFS surpasses meta-analysis by 0.331 and 0.139, respectively.

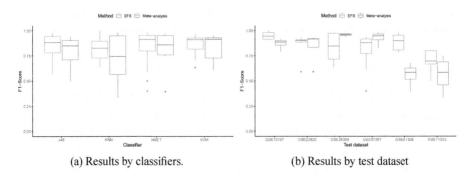

(a) Results by classifiers.   (b) Results by test dataset

Fig. 8. Comparison between the predictive performance of EFS and meta-analysis.

Table 5. Average F1-scores for EFS or meta-analysis based on test datasets, using *variance* and the top 20 genes. Bold values indicate the highest performance for a dataset.

Dataset	Method		MAE
	EFS	Meta-analysis	
GSE10797	**0.94 ± 0.02**	0.87 ± 0.05	0.079
GSE22820	**0.88 ± 0.07**	0.83 ± 0.16	0.048
GSE26304	0.83 ± 0.13	**0.95 ± 0.01**	0.120
GSE57297	0.82 ± 0.13	**0.92 ± 0.06**	0.098
GSE61304	**0.88 ± 0.07**	0.55 ± 0.11	0.331
GSE71053	**0.70 ± 0.10**	0.56 ± 0.18	0.139

When comparing the top 20 ranking generated by EFS and meta-analysis, an interesting finding is that the overlap among them is very low. Only one gene was common to both rankings, namely *S100A7*, which according to literature is highly expressed in breast cancer and may play a role in early tumor progression [12]. The high correlation among genes may justify to some extent the low overlap - a factor that should be further investigated. Among the top 5 genes found by our approach, we identified *TFF1*, *SCGB1D2*, *SCGB2A2*, and *PIP* with previous relation with breast cancer according to Genecards database. A more in-depth analysis of the genes selected by the EFS approach, including the investigation of their biological role, may be useful for better understanding their possible relation with breast cancer.

5   Conclusions

We explored a homogeneous EFS approach for candidate biomarkers identification from microarray data related to breast cancer in this work. A wide range of parameters and configurations were explored. We could note the impact of the FS method chosen as base selectors over performance and the effect of data heterogeneity across distinct, independent studies. By combining several feature rankings extracted from data bootstraps and multiple microarray studies, the classifiers trained with the top EFS-based ranked features distinguished tumor from non-tumor samples with very good and robust performance. A good stability was found for a study-centered analysis, although stability across multiple studies was lower, partially due to high heterogeneity among samples. Interestingly, our approach was shown to outperform in most experiments one of the state-of-the-art approaches for integrative microarray analysis, the REM meta-analysis method. Further analyses of our top-ranked features could introduce new insights into the molecular mechanisms underlying breast cancer.

Provided that feature selection from high dimensional data is a challenging task, mainly when applied in the domain of omics data, we believe that our results are auspicious and encourage further studies in this direction. In particular, for future works, we aim to expand our analysis for other types of cancer, investigate the role of gene expression correlation in the stability of the rankings, perform a further investigation of top-ranked genes, and explore new strategies to improve stability under disease heterogeneity among samples.

Acknowledgments. The authors thank Rodrigo Haas Bueno for his help with microarray data preprocessing.

References

1. Abeel, T., Helleputte, T., Van de Peer, Y., Dupont, P., Saeys, Y.: Robust biomarker identification for cancer diagnosis with ensemble feature selection methods. Bioinformatics **26**(3), 392–398 (2009)
2. Aha, D.W., Kibler, D., Albert, M.K.: Instance-based learning algorithms. Mach. Learn. **6**(1), 37–66 (1991)
3. Alejandro, L.R., Marlet, M.A., Gustavo Ulises, M.R., Alberto, T.: Ensemble feature selection and meta-analysis of cancer miRNA biomarkers. bioRxiv, p. 353201 (2018)
4. Ang, J.C., Mirzal, A., Haron, H., Hamed, H.N.A.: Supervised, unsupervised, and semi-supervised feature selection: a review on gene selection. IEEE/ACM Trans. Comput. Biol. Bioinform. **13**(5), 971–989 (2016)
5. Bolón-Canedo, V., Sánchez-Maroño, N., Alonso-Betanzos, A.: An ensemble of filters and classifiers for microarray data classification. Pattern Recogn. **45**(1), 531–539 (2012)
6. Bolón-Canedo, V., Sánchez-Maroño, N., Alonso-Betanzos, A., Benítez, J., Herrera, F.: A review of microarray datasets and applied feature selection methods. Inf. Sci. **282**, 111–135 (2014)
7. Bolón-Canedo, V., Alonso-Betanzos, A.: Ensembles for feature selection: a review and future trends. Inf. Fusion **52**, 1–12 (2019)
8. Breiman, L., Friedman, J.H., Olshen, R.A., Stone, C.J.: Classification and Regression Trees. Wadsworth and Brooks, Belmont (1984)
9. Cortes, C., Vapnik, V.: Support-vector networks. Mach. Learn. **20**(3), 273–297 (1995)

10. Cramer, H.: Mathematical Methods of Statistics (PMS-9), vol. 9. Princeton University Press, Princeton (1999)
11. Durinck, S., et al.: Biomart and bioconductor: a powerful link between biological databases and microarray data analysis. Bioinformatics **21**, 3439–3440 (2005)
12. Emberley, E.D., Murphy, L.C., Watson, P.H.: S100A7 and the progression of breast cancer. Breast Cancer Res. **6**(4), 1–7 (2004)
13. Peng, H., Long, F., Ding, C.: Feature selection based on mutual information criteria of max-dependency, max-relevance, and min-redundancy. IEEE Trans. Pattern Anal. Mach. Intell. **27**(8), 1226–1238 (2005)
14. Haykin, S.: Neural Networks: A Comprehensive Foundation, 2nd edn. Prentice Hall PTR, Upper Saddle River (1998)
15. He, Z., Yu, W.: Stable feature selection for biomarker discovery. Comput. Biol. Chem. **34**(4), 215–225 (2010)
16. Krieken, J.H.: Precision medicine. J. Hematopathol. **6**(1), 1–1 (2013). https://doi.org/10.1007/s12308-013-0176-x
17. Karley, D., Gupta, D., Tiwari, A.: Biomarker for cancer: a great promise for future. World J. Oncol. **2**(4), 151 (2011)
18. Kent, J.T.: Information gain and a general measure of correlation. Biometrika **70**(1), 163–173 (1983)
19. Khaire, U.M., Dhanalakshmi, R.: Stability of feature selection algorithm: a review. J. King Saud Univ. Comput. Inf. Sci. (2019)
20. Kuncheva, L.: A stability index for feature selection. In: Proceedings of the IASTED International Conference on Artificial Intelligence and Applications, AIA, vol. 2007, pp. 421–427 (2007)
21. Liu, H., Liu, L., Zhang, H.: Ensemble gene selection by grouping for microarray data classification. J. Biomed. Inform. **43**(1), 81–87 (2010)
22. Pes, B.: Ensemble feature selection for high-dimensional data: a stability analysis across multiple domains. Neural Comput. Appl. **32**(10), 5951–5973 (2019). https://doi.org/10.1007/s00521-019-04082-3
23. Pes, B., Dessì, N., Angioni, M.: Exploiting the ensemble paradigm for stable feature selection: a case study on high-dimensional genomic data. Inf. Fusion **35**, 132–147 (2017)
24. Ritchie, M.E., et al.: LIMMA powers differential expression analyses for RNA-sequencing and microarray studies. Nucleic Acids Res. **43**(7), e47 (2015)
25. Seijo-Pardo, B., Porto-Díaz, I., Bolón-Canedo, V., Alonso-Betanzos, A.: Ensemble feature selection: homogeneous and heterogeneous approaches. Knowl.-Based Syst. **118**, 124–139 (2017)
26. Sharifi, S., Pakdel, A., Ebrahimi, M., Reecy, J.M., Fazeli Farsani, S., Ebrahimie, E.: Integration of machine learning and meta-analysis identifies the transcriptomic bio-signature of mastitis disease in cattle. PLoS ONE **13**(2), 1–18 (2018)
27. Surowiecki, J.: The Wisdom of Crowds. Knopf Doubleday Publishing Group, New York (2005)
28. Theil, H.: A note on certainty equivalence in dynamic planning. Econometrica **25**(2), 346–349 (1957)
29. Toro-Domínguez, D., Villatoro-García, J.A., Martorell-Marugán, J., Román-Montoya, Y., Alarcón-Riquelme, M.E., Carmona-Sáez, P.: A survey of gene expression meta-analysis: methods and applications. Brief. Bioinform. **22**(2), 1694–1705 (2021)
30. Walsh, C.J., Hu, P., Batt, J., Santos, C.C.D.: Microarray meta-analysis and cross-platform normalization: integrative genomics for robust biomarker discovery. Microarrays **4**(3), 389–406 (2015)

Structural Characterization of Linear Three-Dimensional Random Chains: Energetic Behaviour and Anisotropy

David R. Avellaneda B.[1], Ramón E. R. González[2(✉)], Paola Ariza-Colpas[3], Roberto Cesar Morales-Ortega[3], and Carlos Andrés Collazos-Morales[4(✉)]

[1] Departamento de Estatística e Informática, Universidade Federal Rural de Pernambuco, Recife, Pernambuco CEP 52171-900, Brazil
[2] Departamento de Física, Universidade Federal Rural de Pernambuco, Recife, Pernambuco CEP 52171-900, Brazil
ramon.ramayo@ufrpe.br
[3] Departamento de Ciencias de la Computación y Electrónica, Universidad de la Costa, Barranquilla, Colombia
[4] Vicerrectoria de Investigaciones, Universidad Manuela Beltrán, Bogotá, Colombia
carlos.collazos@docentes.umb.edu.co

Abstract. In this work, we will make an energetic and structural characterization of three-dimensional linear chains generated from a simple self-avoiding random walk process in a finite time, without boundary conditions, without the need to explore all possible configurations. From the analysis of the energy balance between the terms of interaction and bending (or correlation), it is shown that the chains, during their growth process, initially tend to form clusters, leading to an increase in their interaction and bending energies. Larger chains tend to "escape" from the cluster when they reach a number of "steps" $N >\sim 1040$, resulting in a decrease in their interaction energy, however, maintaining the same behavior as flexion energy or correlation. This behavior of the bending term in the energy allows distinguishing chains with the same interaction energy that present different structures. As a complement to the energy analysis, we carry out a study based on the moments of inertia of the chains and their radius of gyration. The results show that the formation of clusters separated by "tails" leads to a final "prolate" structure for this type of chain, the same structure evident in real polymeric linear chains in a good solvent.

Keywords: Self-avoiding random walk · Linear chains · Interaction energy · Bending energy · Moment of inertia · Radius of gyration · Asphericity · Prolate structure

1 Introduction

The self-excluding random walk (SAW) is a very simple and useful model for understanding some physical phenomena. Under certain conditions, SAWs can model some

© Springer Nature Switzerland AG 2021
O. Gervasi et al. (Eds.): ICCSA 2021, LNCS 12949, pp. 179–191, 2021.
https://doi.org/10.1007/978-3-030-86653-2_13

properties of real linear chains. SAW is not just a useful modeling tool; the field has advanced significantly in the last 50 years, its study has stimulated theoretical developments in areas such as mathematics, physics and computer science and is widely accepted as the main model for polymers diluted [1–4].

We use a model based on a self-excluding walk to generate an ensemble of three-dimensional linear chains. Our model does not consider all possible configurations of the system, avoiding high computational costs. It only considers a number of strings large enough to guarantee acceptable statistical confidence. The cubic network used does not have boundary conditions and the process is carried out in a "finite" time. In this article, we propose and study a conformational energy model based on the flexibility of linear chains that report realistic results and allows you to differentiate chains regardless of their structure. As part of the study, it was examined the balance between the energy of interaction between the "links" in the chain and the bending energy in order to find its relationship with the structure of the chains. This analysis is relevant because it allows differentiating between chains of equal interaction energy but with different structures.

The paper is organized as follows. In the next section, we speak briefly about the proposed algorithm, used to generate the linear chains. We also describe the energy models used, the calculation of the moment of inertia, and the asphericity, tools used for the characterization of the chains. The subsequent section is dedicated to the discussion of the results of our simulations and finally, we expose the conclusions of the work.

2 Methodology

2.1 Numerical Algorithm

The ensemble of linear chains was generated from the sequence of steps showed next, for a d − dimensional network, with free boundaries:

Step 1. Initially, the number of attempts N' is choosen. This variable will define the length of the chain.
Step 2. The origin of the coordinate system is choosen as the the origin of the chain.
Step 3. From the origin, the first step in the generation of the chain is choosen randomly, from the neighbourhood of the origin.
Step 4. The following steps are choosen, randomly, from the rest of the neighbourhood, excluding previously "visited" points. This step is the most important to ensure the SAW.
Step 5. Add the step to the walk (chain).
Step 6. When the number of new possible steps is zero or the number of attemps is reached, the process ends and the chain is saved.

For three-dimensional chains, $N < N'$, this indicated that the "walker" usually gets stuck before the N' attempts.

2.2 Energy of Linear Chains

In our analysis, we use a definition of the interaction energy, which is based on the compactness of the chains [5–7]. We also examine the bending energy as a discrete variable that characterizes the chain flexibility, as well as, its tangential correlations. The total energy of the chain is then, considered as the sum of the bending energy and the interaction energy.

Interaction Energy. Interaction energy accounts for the energy of the chain due to its compactness and it quantifies the short-range interactions (Von Neumann neighbourhood) for non-continuous monomers [5–7]. It reads as,

$$E = \sum_{i<j} e_{ij} \Delta(r_i - r_j). \tag{1}$$

where $\Delta(r_i - r_j) = 1$ if r_i and r_j are attached to the network, but i and j are not adjacent positions along the chain sequence and $\Delta(r_i - r_j) = 0$ for otherwise (see Fig. 1).

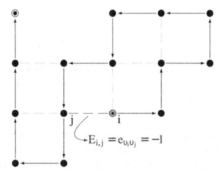

Fig. 1. 14-step 2d-chain that shows the values that the interaction takes Δ(ri - rj), where it adopts the values: 1, for the attachments but not adjacent sites (blue color) and 0 for the otherwise. (Color figure online)

The type of contact between the "links" determines the value of the factor e_{ij}. This variable, in turn, represents the potential energy of interaction between the "links" located in the position r_i and r_j respectively. For attractive "link-link" interaction, $e_{ij} = -1$.

Bending Energy. One of the basic characteristics of all macromolecules is their flexibility [8]. The polymer chains, in the pure state or in dissolution, may adopt different conformations depending on their flexibility. Chains with low flexibility will be more rigid and will tend, in the limit, to behave as a hard stick. On the other hand, when the flexibility is high, the chain may have large changes of direction within a few links. The flexibility of the polymer chain is related to the persistence length l_p. This can be defined as the average value of the maximum linear length of the chain configuration (it is also related to the Kuhn segment as $b = 2l_p$ [9]). For distances greater than l_p, the "memory" associated with the direction, up that point, adopted by the chain is lost. Thus, some energy is always required to fold the chains, which can happen up to at most the l_p. In our model, we use these properties of macromolecules in the definition of our bending energy term.

In the graph of Fig. 2, we show the correlation (in turn related to the flexibility) between u and u', two unit vectors that join three points of the chain (monomers h, i and j), and that are separated by a distance l is given by [10]:

$$\langle u \cdot u' \rangle = \left[\left(1 - (b/l_p)\right)^{1/b} \right]^l \sim \exp(-l/l_p). \tag{2}$$

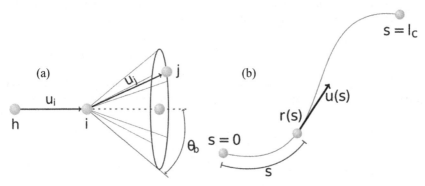

Fig. 2. Schematic representation of a section of the chain that shows high flexibility. Vectors of bond u_i with a fixed link angle between two consecutive monomers, (a). Chain conformation specifying $r(s)$ and the unit vector $u(s)$, (b).

At the limit, the conformation of the chain is a smooth curve as described in Fig. 2b. Using Eq. 2 it can be obtain the correlation function between $u(s)$ and $u(s')$ of two segments of the chain, s and s', as a function of the persistence length l_p. This function shows that the directional correlation of two segments of a macromolecule, decreases exponentially with the growth of the chain length [8, 10, 11]:

$$c(s, s') = \langle u(s) \cdot u(s') \rangle = \exp(-|s - s'|/l_p), \tag{3}$$

We consider our linear chain taking into account interactions with other "links" (monomers) in the same chain and interactions with their environment. These interactions (bending energy) are described by an effective potential that represents the energy cost for their formation. The stability of the chain is determined by two forces, elastic with a negative sign, which leads the chain to a collapse, and another repulsive one with a positive sign, which causes the chain to be stretched. This energy cost is reflected in the chain in the form of free energy, for example, the number of conformations decrease with increasing end-to-end distance, but their free energy increases due to the high correlation that exists in the chain.

In continuous mechanic systems like a rod with stiffness constant k under the action of a force [12, 13], there is a correlated behavior in relation to its structure, described by the Eq. 3. Flexing generates a differentiable curve on the rod, where, at a point $r(s)$ of the curve there is a tangent vector $u(s)$ leading to a behavior similar to that described in Fig. 2b for linear chains. The Hamiltonian describing the internal energy of the rod of length i_c is given by Eq. 4, as follows:

$$\mathcal{H} = \frac{k}{2} \int_0^{l_c} \left(\frac{\partial u(s)}{\partial s} \right)^2 ds. \tag{4}$$

Chains with high flexibility experience changes of direction at a distance of few "links" tending to turn to itself, while low flexibility chains tend to become rigid, because the two-segment correlation function in the chain decreases exponentially with the distance between them as shown in Eq. 3. By virtue of this, we propose a discretization of bending energy in order to be adapted to our chains, in this way, part of the internal energy of the chain which is described in terms of its configuration and which is equivalent to Eq. 4 is given by the following relation:

$$\mathcal{H} \cong H = \frac{k}{b} \sum_{i,j=1}^{N} \varepsilon_{ij} p_i, \tag{5}$$

where the weight function ε_{ij} can take values of (1) or (−1) depending on whether or not the direction of the i-th step changes as compared to the previous step (see Fig. 3) and p_i represents the probability that each step will find any of its accessible microstates, k is a constant of units of energy times distance and finally, our model adopts the $b = 1$ as the length of Kuhn.

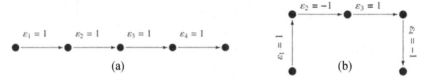

Fig. 3. 4-step 2d-chain that shows the value of the weight function ε_i used to compute the bending energy. Linear chain without deviation with its weight per step equal to 1, (a). Chain with mixed deviations, when direction changes ε_i adopts a weight equal to (−1), as is the case with steps 2 and 4, (b).

The bending energy, in its discrete version H, describes the behavior of the polymer chain from its tangential correlations, for example, for a highly correlated polymer chain (Fig. 3, a) this energy will be purely positive (high free energy) what is expected from the Hamiltonian described by the rod and the correlation given in Eq. 3.

The bending energy can be positive or negative depending on the winding of the chain. In our simulations the calculation of bending energy takes into account the term we call function weight as well as the relative probability of each step of the chain, which has the form of Eq. 6 [14].

$$p_i = \frac{e^{-gn_i}}{\sum_{j=1}^{3d} e^{-gn_i}} = \frac{1}{\sum_{j=1}^{3d} e^{-g(n_j - n_i)}}. \tag{6}$$

In Eq. 6, the sum runs through all possible $3d$ paths from the position occupied by the walker at each instant of time, including the address i, and g is a positive parameter which measures the intensity with which the walk avoids itself. For the sake of simplicity, in this work we implement the limiting case $g = \infty$, which corresponds in the $3d$ case to a discrete domain of probabilities $p_i(g = \infty) = [1/6, 1/5, 1/4, 1/3, 1/2, 1]$.

2.3 Anisotropy in Linear Chains

Inertia, Gyration Tensor and Asphericity. The main moments of the inertia tensor of the chains depend on their structure and they are correlated with their stability. The inertia tensor can be diagonalized, in function of their main moments, how we can see in the Eq. 7, as follows:

$$\hat{I} = \begin{bmatrix} I_x & 0 & 0 \\ 0 & I_y & 0 \\ 0 & 0 & I_z \end{bmatrix}. \tag{7}$$

Each of the main moments of inertia (I_x, I_y, I_z) corresponds to a moment of inertia around one of the main axes, so a relationship between these moments allows us to carry out an analysis on the symmetry or asymmetry of each of the chains. In addition to the inertia tensor, the shape of the chains (like polymeric chains) can be characterized in gyration tensor terms [15, 16]. The gyration tensor is built from the dyadic of the position of the column vector r_α in the center of mass of the system [17],

$$S = \frac{1}{N} \sum_\alpha^N r_\alpha r_\alpha^T = \frac{1}{N} \begin{bmatrix} \sum_\alpha x_\alpha^2 & \sum_\alpha x_\alpha y_\alpha & \sum_\alpha x_\alpha z_\alpha \\ \sum_\alpha y_\alpha x_\alpha & \sum_\alpha y_\alpha^2 & \sum_\alpha y_\alpha z_\alpha \\ \sum_\alpha z_\alpha x_\alpha & \sum_\alpha z_\alpha y_\alpha & \sum_\alpha z_\alpha^2 \end{bmatrix}. \tag{8}$$

The inertia tensor I is directly related with the gyration tensor S,

$$I = \text{Tr}(S)1 - S, \tag{9}$$

were 1 is the unit tensor and $\text{Tr}(S)$ is the gyration tensor trace. From the gyration tensor we can define the asphericity Δ [18] and the nature of asphericity Σ [19], represented by:

$$\Delta = \frac{3}{2} \frac{\text{Tr}(\hat{S}^2)}{(\text{Tr}(\hat{S}))^2}, \quad \Sigma = \frac{4\det(\hat{S})}{(\frac{2}{3}\text{Tr}(\hat{S}^2))^{3/2}}, \tag{10}$$

$\hat{S}$ is the following transformation:

$$\hat{S} = S - \frac{1}{3}\text{Tr}(S)1. \tag{11}$$

For asphericity, we have $0 \le \Delta \le 1$, were $\Delta = 0$ correspond to an object completely symmetric and $\Delta = 1$, correspond to a completely stretched object, similar to a rigid bar. On the other hand, the nature of asphericity is limited by $-1 \le \Sigma \le 1$, were $\Sigma = -1$ represent a oblate object, like a disk, and $\Sigma = 1$, a prolate object [80, 97 Tese]. The combination of asperity and the nature of asphericity allows to identify the approximate shape of the configuration adopted by the three-dimensional chain. For comparative purposes, a transformation of these parameters can be adopted [19, 20] like:

$$\rho = 2\sqrt{\Delta} \in [0, 2], \theta = \frac{1}{3}\arccos(\Sigma) \in [0, \pi/3] \tag{12}$$

Using the parameters ρ and θ related to asphericity and the nature of asphericity we can have a different and more general perspective of the form adopted by the chains.

3 Results and Discussion

3.1 Energy of the Linear Chains

To study the energy of the random chains generated by our simulation, we analyzed the energy given by the interactions between "links" (monomers) in each chain, as well as the energy associated with the bending of the chains. As generated chains are homogeneous (identical links), like "homo-polymeric" chains, they are expected to show a uniform behavior or have a low amount of metastable states [6, 21].

In the contact potential described in Eq. 1, the value of the constant e_{ij} has the information about of interaction energy between the non-continuous and adjacent links. This constant adopts the value of -1, for each contact, thus generating a "folding" force in the chain, known in proteins as a hydrophobic force [7]. The sum of all the interactions in chain defines the energy of the system, called interaction energy E. To take into account the flexibility of the chain and, consequently, its tangent correlations, we proposed adding to the interaction term E, the bending energy H in its discrete version, proposed in Eq. 5.

In the Fig. 4 we can see the characteristic histograms of each energy for three-dimensional chains generated using our algorithm. The shape of the distribution is similar to the results obtained previously for the distribution of number of steps.

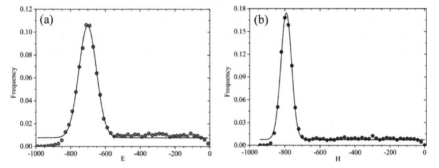

Fig. 4. Interaction energy E (a) and bending energy H (b) histograms, for three-dimensional chains. Gaussian distributions with standard deviations, $\sigma_E \cong 51$ and $\sigma_H \cong 31$, and mean values $\mu_E \cong -702$, and $\mu_H \cong -792$ respectively.

Even when considering both attractive and repulsive behavior of the chains, the resulting chains have negative energy, because in the SAW models for polymers in a good solvent, the attractions prevail over the repulsions [9, 10].

The total energy of the chain is described by the sum of interaction and bending energies, which determine the structural configuration. In Fig. 5a, the total energy distribution is shown. The most probable energy values are between -1650 and -1350 energy units, corresponding to long chains.

The mean values for the interaction energy and bending energy were computed for eleven thousand three-dimensional chains. Although was computed the mean values for the total (interaction plus bending) energy. All the energies were plotted as a function

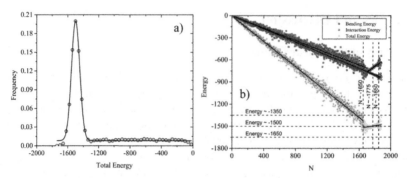

Fig. 5. Total Energy histogram $(E + H)$ for three-dimensional chains with mean with mean $\mu \cong -1499$ and standard deviation $\sigma \cong 53$ (a). Interaction energy E and bending energy H as a function of the number of steps N. The horizontal dotted lines represent the values between which the energies are distributed (see Fig. 4 and Fig. 5a). On the other hand, the vertical ones, represent the values between which distribute all sizes of the three-dimensional chains generated (b).

of N (Fig. 5b). The plots show a linear and uniform behavior for total energy (which is expected [6, 21]). However, when N is large ($\geq\sim 1650$), the behavior of the interaction energy changes from decreasing to increasing, having less negative values, which can be interpreted as representative of more stretched chains, with fewer contacts but with small lengths of persistence still prevailing.

With the objective to better understand the behavior of energy, we have plotted the distribution of the size of chains. The great majority of the three-dimensional chains generated are large ($N > 1700$), how we can see in Fig. 6. These chains, with total energies between -1350 and -1650, belong to the second regime of the graph of the total energy and follow a Gaussian distribution (see Fig. 5). The rest of the chains, with

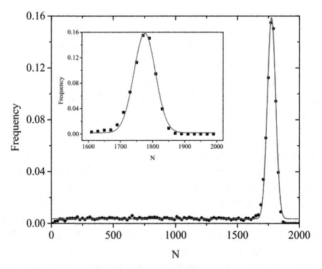

Fig. 6. Normalized histogram of the number of steps N *with* mean $\mu = 1775$ and standard deviation $\sigma = 33$. Behavior obtained for random chains in $3d$, ψ generated from $N' = 2400$ attempts. The inset illustrates the histogram in a wide interval in which a Gausian behavior occurs.

total energy between 0 and -1350, belong to the first regime of the energy graph and follow a uniform distribution.

In the "discrete" bending energy approach, the second regime does not appear (see Fig. 5b). This term has the same linear behavior, regardless of the presence of neighbours interacting with other links (monomers). The folding of the chains in the regime with large N adds negative energy to the system.

In the total energy graph of Fig. 5b, two regimes can also be seen due to the contribution of the interaction term. For chains of size below $N = 1775$ (mean value of N), the energy decreases with N at a rate of 0.86, practically twice the rate of decrease of each term separately, because the two energy contributes in this regime in the same proportion. For larger chains, the energy increases with N and the characteristic rate of this increase is 0.29. Here the different contributions of the two types of energy make the increase less significant than in the case of the interaction energy.

3.2 Inertia and Anisotropy

A typical linear chain with $N = 1772$ steps, generated by the simulation is shown in Fig. 8. This chain forms two clusters separated by a tail. The chain "escapes" from the first cluster and the interaction energy decreases because does not have neighbours that contribute to this energy. The flexibility of the chain is maintained causing the chain to continue folding even in the "bridge" that separates the clusters, and the bending energy remains constant (as we saw earlier). This represents a decrease in the interaction energy, restoring later, during the growth process, the original structure (Fig. 7).

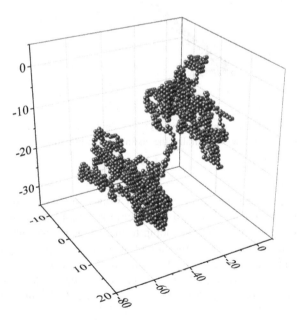

Fig. 7. Typical structure of $N = 1772$ generated by the simulation which shows the formation of two clusters. The spheres represent the links (monomers) of a linear (homo-polymeric) chain.

We plot three different views of main moments of inertia I_x, I_y and I_z and in each plane, a projection of them, see Fig. 8. It is observed that structurally the chains have a preference for adopting a stretched shape along the x axis. Chains like polymeric chains assume, preferably, an ellipsoidal or prolate shape [17, 22–24]. When analyzing our results, from the study of main moments of inertia, we can see that they correspond with this fact.

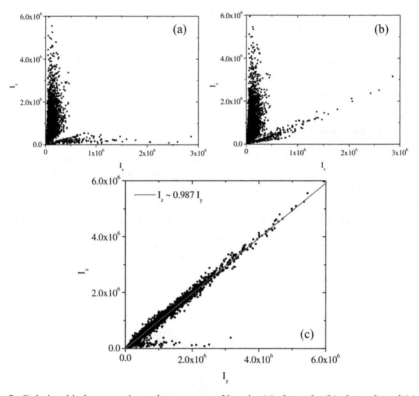

Fig. 8. Relationship between the main moments of inertia, (a): I_z vs. I_x; (b): I_y vs. I_x and (c): I_z vs. I_y. In (c), is illustrated how, structurally, the chains have a preference to adopt a "stretched" shape along the x-axis.

Figure 9 shows the distribution of asphericity values for the three-dimensional generated chains. It's possible to observe a maximum for the asphericity in $\Delta_{max} \approx 0.76$, in accordance with the values corresponding to these type of structures. On the other hand, we see that the occurrences in the case of the nature of asphericity are practically concentrated around the value $\Sigma = 1$, corroborating the same hypothesis. These distributions, as well as the behavior of the main moments of inertia, demonstrated that the three-dimensional chains generated prefer to adopt stretched structures, in general, like real polymeric chains.

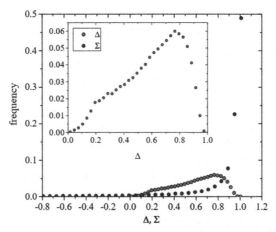

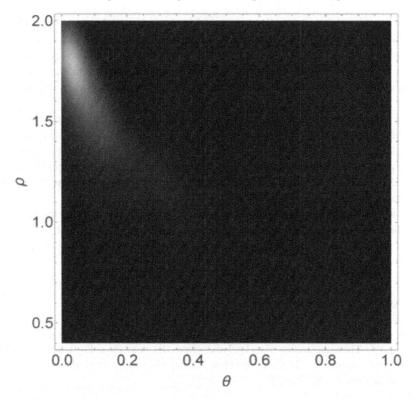

Fig. 9. Distributions of variables: asphericity (Δ), red dots, and nature of asphericity, blue dots, for three-dimensional random linear chains. The behavior of these distributions reveals a preferably prolate structure, from the position of the peaks of the frequencies. (Color figure online)

Fig. 10. Adopted configurations by linear chains. The yellow color represent high density of points, revealing a higher concentration of chains with high values of ρ and low values of θ. This graph shows the most likely prolate configuration that is generated by our model. (Color figure online)

In Fig. 10, we show a different perspective to analyze the form adopted by the three-dimensional linear chains generated. This graph shows a high density of point in the region corresponding to a low values of θ $\left(\theta \approx \frac{\pi}{90}\right)$ and high values of $\rho(\rho \approx 1.78)$, which reflects in the understanding of the preferential behavior of our chains. The region of θ and ρ large is a forbidden region since the eigenvalues of tensor radius of gyration become negative, which is incompatible with real chains. The area below the yellow region (high densities), represents all configurations ranging from a totally prolific geometry at $\theta = 0$ like that of a rigid rod to a totally oblate structure of a rigid ring at $\theta = \frac{\pi}{3}$. From $\frac{\pi}{6}$, the form adopted by the chains is quite noticeable. For $\theta > \frac{\pi}{6}$ the shape is oblate and comparatively prolate for $\theta < \frac{\pi}{6}$. With respect to ρ, the lower this value, the structure becomes more and more spherical, resulting in a spherically symmetrical conformation for $(\pi/6; 0)$. Theoretically, for open polymers flexible, it is known that the adopted form is almost exclusively prolate and rarely spherical, which indicates a peak around $\theta = \frac{\pi}{40}$ and $\rho = 1.55$ [19].

4 Conclusions

Comparing the behavior of the E energy, which characterizes the monomer-monomer interactions of the chains, with the H energy, associated with its flexibility, we found that the chains start with clusters that increase in size, leading to an initial increase in their energy. When the chains reach a large number of steps, the interaction energy revealed that the chain stretches and escapes from the cluster, which results in a loss of interaction energy.

The behavior of the bending energy reveals that, in this escape regime, the chains keep the folding behavior they showed before the escape. This process lasts until the clusters reach a size comparable with the mean radius of gyration. The analysis presented in this work, including both the bending and interaction energies is important because it allows differentiating between chains of interaction energy but with different structures and hence different bending energies.

From analysis of inertia tensor and asphericity, we can conclude that the generated three-dimensional chains present anisotropy, with symmetry around one of the main axes of rotation. As a result, in general, our chains adopt a prolate structure, the same structure evident in real polymeric chains.

References

1. Flory, P.J.: Principles of Polymer Chemistry. Cornell University Press, Ithaca (1953)
2. Flory, P.J.: The configuration of real polymer chains. J. Chem. Phys. **17**(3), 303–310 (1949)
3. Madras, N., Sokal, A.: The pivot algorithm: a highly efficient Monte Carlo method for self-avoiding walk. J. Stat. Phys. **50**(1–2), 109–186 (1988)
4. Slade, G.: Self-avoiding walks. Math. Intell. **16**(1), 29–35 (1994). https://doi.org/10.1007/BF03026612
5. Figueirêdo, P.H., Moret, M.A., Coutinho, S., Nogueira, J.: The role of stochasticity on compactness of the native state of protein peptide backbone. J. Chem. Phys. **133**, 08512 (2010)

6. Boglia, R.A., Tiana, G., Provasi, D.: Simple models of the protein folding and of non-conventional drug design. J. Phys. Condens. Matter **16**(6), 111 (2004)
7. Tang, C.: Simple models of the protein folding problem. Phys. A Stat. Mech. Appl. **288**(1), 31–48 (2000)
8. Grosberg, A.Y., Khokhlov, A.R.: Statistical Physics of Macromolecules. AIP Press, New York (1994)
9. Rubinstein, M., Colby, R.H.: Polymer Physics. Oxford University Press, New York (2003)
10. Teraoka, I.: Polymer Solutions: An Introduction to Physical Properties. Wiley Inter-science, New York (2002)
11. Hsu, H.P., Paul, W., Binder, K.: Standard definitions of persistence length do not describe the local "intrinsic" stiffness of real polymers. Macromolecules **43**(6), 3094–3102 (2010)
12. Landau, L.D., Lifshitz, E.M.: Theory of Elasticity. Elsevier Sciences, New York (1986)
13. Schöbl, S., Sturm, S., Janke, W., Kroy, K.: Persistence-length renormalization of polymers in a crowded environment of hard disks. Phys. Rev. Lett. **113**(23), 238302 (2014)
14. Amit, D.J., Parisi, G., Paliti, L.: Asymptotic behavior of the "true" self-avoiding walk. Phys. Rev. B **27**(3), 1635–1645 (1983)
15. Solc, K.: Shape of random-flight chain. J. Chem. Phys. **55**(1), 335–344 (1971)
16. Rudnick, J., Gaspari, G.: The asphericity of random walks. J. Phys. A: Math. Gen. **30**(11), 3867–3882 (1997)
17. Hadizadeh, S., Linhananta, A., Plotkin, S.S.: Improved measures for the shape of a disordered polymer to test a mean-field theory of collapse. Macromolecules **44**(15), 6182–6197 (2011)
18. Aronovitz, J., Nelson, D.: Universal features of polymer shapes. J. Phys. **47**(9), 1445–1456 (1986)
19. Cannon, J.W., Aronovitz, J.A., Goldbart, P.: Equilibrium distribution of shapes for linear and star macromolecules. J. Phys. I Fr. **1**(5), 629–645 (1991)
20. Alim, K., Frey, E.: Shapes of semi-flexible polymer rings. Phys. Rev. Lett. **99**(19), 198102 (2007)
21. Dokholyan, N.V., Buldyrev, S.V., Stanley, H.E., Shakhnovich, E.I.: Discrete molecular dynamics studies of the folding of a protein-like model. Fold. Des. **3**(6), 577–587 (1998)
22. Theiler, J.: Estimating fractal dimension. J. Opt. Soc. Am. A, OSA **7**(6), 1055–1073 (1990)
23. Blavatska, V., Janke, W.: Shape anisotropy of polymers in disordered environment. J. Chem. Phys. **133**(18), 184903 (2010)
24. Rawdon, E.J., et al.: Effect of knotting on the shape of polymers. Macromolecules **41**(21), 8281–8287 (2008)

Reasonable Non-conventional Generator of Random Linear Chains Based on a Simple Self-avoiding Walking Process: A Statistical and Fractal Analysis

David R. Avellaneda B.[1], Ramón E. R. González[2]($\boxtimes$),
Carlos Andrés Collazos-Morales[3]($\boxtimes$), and Paola Ariza-Colpas[4]

[1] Departamento de Estatística e Informática, Universidade Federal Rural de Pernambuco, Recife, Pernambuco CEP 52171-900, Brazil
[2] Departamento de Física, Universidade Federal Rural de Pernambuco, Recife, Pernambuco CEP 52171-900, Brazil
ramon.ramayo@ufrpe.br
[3] Vicerrectoria de Investigaciones, Universidad Manuela Beltrán, Bogotá, Colombia
carlos.collazos@docentes.umb.edu.co
[4] Departamento de Ciencias de la Computación y Electrónica, Universidad de la Costa, Barranquilla, Colombia

Abstract. Models based on self-excluded walks have been widely used to generate random linear chains. In this work, we present an algorithm capable of generating linear strings in two and three dimensions, in a simple and efficient way. The discrete growth process of the chains takes place in a finite time, in a network without pre-established boundary conditions and without the need to explore the entire configurational space. The computational processing time and the length of the strings depending on the number of trials N'. This number is always less than the real number of steps in the chain, N. From the statistical analysis of the characteristic distances, the radius of gyration (R_g), and the end-to-end distance (R_{ee}), we make a morphological description of the chains and we study the dependence of this quantities on the number of steps, N. The universal critical exponent obtained are in very good agreement with previous values reported in literature. We also study fractal characteristics of the chains using two different methods, Box-Counting Dimension or Capacity Dimension and Correlation Dimension. The studies revealed essential differences between chains of different dimensions, for the two methods used, showing that three-dimensional chains are more correlated than two-dimensional chains.

Keywords: Self-avoiding random walk · Linear chains · Critical exponents · Fractal dimension · Radius of gyration · End-to-end distance

© Springer Nature Switzerland AG 2021
O. Gervasi et al. (Eds.): ICCSA 2021, LNCS 12949, pp. 192–206, 2021.
https://doi.org/10.1007/978-3-030-86653-2_14

1 Introduction

SAWs describes a large spectrum of real systems. One of the first theoretical approaches to this subject is Flory's theory [1]. This theory helped to understand the power laws characteristic of this model and the role of dimensionality, using mean field arguments involving the concept of excluded volume. Other analytical approaches that used rigorous approximate methods such as perturbation theory, self-consistent field theory, and renormalization group, produced estimates with a certain degree of precision for critical exponents and some universal amplitude ratios [2–4].

Numerical methods are an important tool in the study of the properties of long SAW's [5]. These numerical approaches allow us to find the number of possible SAWs of length N. Universal properties are estimated using different techniques such as the ratio method, Pade approximants, or differential approximants, having reported results for square and cubic networks up to $N = 71$ and $N = 21$ steps, respectively [2, 6, 7]. In a SAW of length N, there are $2 \times d$ options for the first step of the walk and at most, $2 \times d - 1$ choices for the rest of the $N - 1$ steps. The number of configurations for as specific value of $N(C_N)$, is in the range: $dN \leq C_N \leq (2d)(2d - 1)^{N-1}$, causing this number to grow with N according to a power law of the type $C_N \sim N^{\gamma-1}$, where γ is an universal critical exponent that depends on network dimension [8–10].

In this work we will not take into account the high number of walks possible for a specific N, but we show that the exponents characteristic of this type of system can be obtained from a subset of all possible configurations. In this work we focus in studying relatively small SAW chain ensembles, which might be relevant in order to reproduce the main characteristics of random linear chains, such as universal critical exponents. It is understood that we are referring to classic SAWs, where the walker tends to take steps at random, avoiding regions of the space already visited by him. These are not self-repelling chains, for which all possible configurations for a given length are considered. In this last type of chain, it is associated with a statistical weight, represented by the Boltzmann factor, with a potential energy proportional to the number of self-intersections [11].

As a complement to the study, we implement a comprehensive analysis of the fractal dimension of the generated chains using two different methods: box-counting dimension and correlation dimension.

The main motivation of this work is to be able to generate sets, statistically representative, of linear chains with known universal characteristics and behaviors, without having to go through the entire configurational space. For a SAW of length N, there are 2d options for the first step of the walk and at least 2d-1 choices for the rest of the N-1 steps, this makes the value of the number of possible configurations (c_N) that a SAW of N steps can adopt, starting from the origin, grow with a power law. A huge effort has been made in the last 6 decades to develop efficient methods for counting SAWs, but knowing the exact values of CN and exponents still represents an open challenge. With our work, we show that only with a subset of all possible configurations we can obtain exponents similar to those reported in the literature.

The paper is organized as follows. In the next section, we briefly review the SAW and its main characteristics, the proposed algorithm, as well as the definitions of the characteristic distances and the methods that will be used in the calculation of the fractal dimension of the chains. The subsequent section is dedicated to the discussion of the results of our simulations and finally, we expose the conclusions of the work.

2 Methodology

The objective of this work is to efficiently generate sets of two- and three-dimensional linear chains, with the typical characteristics of this type of system. Each chain is generated from the path taken by a randomly moving particle within a network without boundary conditions, with the limitation that the particle cannot pass through places previously visited by it.

2.1 Numerical Algorithm

For a d–dimensional network with free boundaries the algorithm to generate a one chain is as follows:

1. Choose the number of attempts N'.
2. Choose the origin of the polymer, which in our case is the origin of the coordinate system.
3. Generate the first step randomly or choose it arbitrarily from a point in the cubic network.
4. Choose the following step randomly from one of the $2 \times d$ possible steps.
5. If the given step leads to self-intersection, go to item 4 and try again with another step. This step is most important to ensure the SAW.
6. If the step leads to an available location, add the step to the walk.
7. If the number of attempts is reached or if the number of possible steps is zero (the walker gets stuck), the simulation is accepted and saved.

Thus, the random chain is formed by N steps generated from N' attempts, and the representative flowchart of this process is shown in the Fig. 1.

After generating the random chain, we store the positions of each of the sites that make up the chain and start computing the characteristic measurements of its configuration. First, we calculate the position r_i of each site to obtain the center of mass r_C, with which we obtain the radius of rotation, R_g. Next, we calculate the end-to-end vector module, R_{ee}, which can be easily derived from the distance of the Nth site to the origin of the chain. All these quantities are properly defined in the next subsection.

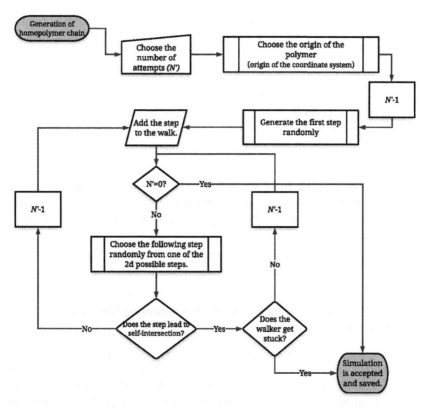

Fig. 1. Flowchart of the computational algorithm used to generate an isolated linear chain in a d-dimensional network.

2.2 Characteristic Distances and Flory Exponent

The proposed algorithm efficiently generates a set of linear chains in uniform $2D$ and $3D$ networks, each of them is formed by N sites in positions $\{r_0, r_1, ..., r_N\}$ in the space of dimension d. The separation distance between the site i and its nearest neighbour is $b = r_i - r_{i-1}$, for $i = 1, 2, ..., N$, which would be equivalent to a Kuhn segment [1, 12]. We describe the chain morphology by exploring the behavior of two characteristic distances, end-to-end distance R_{ee} and the radius of gyration R_g. The end-to-end distance is defined as the mean square variance of the displacement, and the radius of gyration represent the second moment around the center of mass. The expressions of these two quantities are shown below [12]:

$$R_{ee}^2 = \langle (r_N - r_0)^2 \rangle. \tag{1}$$

$$R_g^2 = \frac{1}{N+1} \sum_{i=0}^{N} \langle (r_i - r_C)^2 \rangle, \tag{2}$$

The variables r_N and r_0 are the positions of the end of a chain, and r_C is a centre of mass of the chain. For real chains, a relationship between these distances is [13]:

$$\frac{6R_g{}^2}{R_{ee}{}^2} = 0.952 \tag{3}$$

The dependence of the radius of gyration with N, size of a Kuhn segment b, and exponent γ, for this type of chain, responds to the expression below:

$$R_g = bN^\nu \tag{4}$$

The exponent γ is the Flory exponent [1, 12, 14], which comes from Flory's theory. Flory's exponent does not depend on the type of network, but on the spatial dimensionality d [10]. Generally, Flory's exponent is simply calculated from [14, 15]:

$$\nu = \frac{3}{d+2} \tag{5}$$

2.3 Fractal Dimension of Linear Chains. Box-Counting Dimension and Correlation Dimension

Self-similarity and self-affinity are important characteristics of linear chains. These properties are directly related to their configuration and that are well behaved for a defined variety of scales. Random linear chains in general can behave like random fractals [16–20]. The mean quadratic distance of the end-to-end vector is proportional to the number of connected segments (or Khun lengths), as shown in the following equation:

$$\langle R_{ee}{}^2 \rangle \sim N^{2\nu} \tag{6}$$

For any "r-size" subsection of the chain containing "n-sites", its characteristic size is equal to:

$$\langle r^2 \rangle \sim n^{2\nu} \Rightarrow r \sim bn^\nu \tag{7}$$

The degree of self-similarity for small scales is limited by the Kuhn length b, and the characteristic sizes, (R_g and R_{ee}) of the chain for large scales. Thus, the previous equation is valid for $b < r < R$. The fractal dimension suggests that the number of "segments" within a sphere of radius R is R^{d_F}, and since the behavior of R is known according to Flory's exponent, it is easy define the value of the fractal dimension according to the exponent ν given by [14, 18, 20, 21]:

$$d_F = \nu^{-1}. \tag{8}$$

The capacity dimension (box-counting dimension) is used to calculate the fractal dimension of binary images [16, 18, 19, 22–25]. Since it is a purely geometric method, it was only just used in two-dimensional chains. 2D images was generated, from the walks, considering that each "link" would fill the space between him and his closest

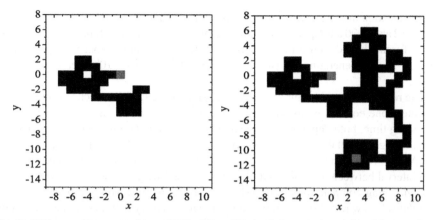

Fig. 2. Filling a path generated by a SAW of $N = 120$ that is born at the origin of the coordinate system. Two "moments" in the evolution of the walk is presented until it complies with the "stop" conditions established in the model.

neighbor, as we see in the Fig. 2. The correlation dimension, introduced in chaos theory [26, 27] is a measure of the dimensionality of the space occupied by an arbitrary set of points. In this case, the points are the positions of the steps of the random walks that generate the linear chains.

The correlation dimension was calculated from the time series of the Euclidean distance between the steps of the walk, both for two-dimensional chains and for three-dimensional chains. Although the strict definition of the d_F fractal dimension requires an "infinite scale reduction", this is not typical of real systems like our linear chains because their size is limited. Thus, $d_{F_{corr}}$ can only be measured within a scale range, which restricts the size of the strings. Because of this, chains with $N < 10$ were not considered in our calculations.

3   Results and Discussion

3.1   Algorithm and Simulations

In the algorithm each chain is generated from an initial number of attempts (N') that remain fixed during a set generation. Then, the random chain is formed by N "steps" generated from these N' attempts, so that, for a squared two-dimensional network, $N' >> N$ and for a cubic three-dimensional network, $N' > N$.

The different behaviors of N in function of N' for the two dimensions of networks respond to the fact that the chains can get trapped before reaching the total number of attempts. In the formation process of two-dimensional chains, there are a few degrees of freedom and the chains can be trapped more easily than in the case of three-dimensional chains.

There is no relationship between the number of attempts and the number of steps in this type of system. It was necessary to define an adequate number of attempts to construct a representative sample of linear chain states, guaranteeing, at the same time, the optimization of the processing time of the computational code. We chose an interval between 2 and 2000 attempts, with a sequence of $N' + 2$ for the two-dimensional case and $N' + 5$ for the three dimensional. Within each interval, 10^4 simulations were performed for each N', thus generating 10^7 two-dimensional chains and 4×10^6 three-dimensional chains.

The results obtained in this system are illustrated in Fig. 3 in which two graphs are presented, one corresponding to the maximum number of steps and the other, to the processing time, both depending on the attempts. For the two-dimensional case (black points), the growth of the maximum number of steps (see Fig. 3a) is approximately linear until $N' = 710$. From this point on, the maximum number of steps generated remains in an interval between $N = 366$ and $N = 690$ with a average value of $N \simeq 468$.

Concerning computational processing time (see Fig. 3b), we see a linear increase at a relatively low rate. For the three-dimensional case (grey diamonds), the maximum number of steps increases linearly (Fig. 3a) as the number of attempts increases. On the other hand, the computational processing time increases according to a power law as we can see in Fig. 3(b). These results were important in choosing a specific number of attempts to generate the chains according to their size, ensuring process efficiency as well as chains of various lengths, both large and small.

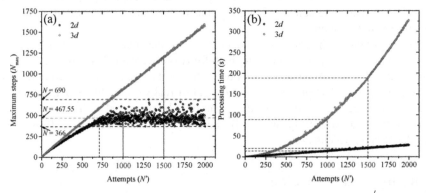

Fig. 3. Relationship between the number of steps N and the number of attempts N' for a sample of 10 thousand simulations in 2D and 3D. Each point in Figure (a) represents the maximum size N_{max} of the series and (b) shows the processing time, in seconds, that it took to generate the 10 thousand simulations for each N'.

We chose $N' = 1000$ attempts for the two-dimensional case, since increasing this value would not benefit the generation of larger chains. For the three-dimensional case, an increase in N' ←benefits the generation of larger chains, but would seriously compromise the processing time, for this reason, we use $N' = 1500$.

3.2 Chain Length, Characteristic Distances and Flory Exponent

The distributions of the continuous variable Nb, that represents the length of a "fully stretched chain" with N "steps", for the generated sets of linear chains, are shown in Fig. 4.

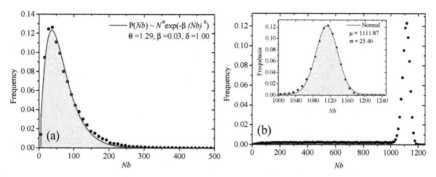

Fig. 4. Normalized histogram of the number of steps N for two-dimensional chains generated from $N' = 1000$ attempts (a), and three-dimensional chains generated from $N' = 1500$ attempts (b). The inset in (b) illustrates the histogram in a wide interval in which a Gaussian behavior occurs, with mean $\mu = 1111.87$ and standard deviation $\sigma = 25.46$.

For the bidimensional case, (Fig. 4a), the resulting curve is adjusted by a Fisher distribution. This curve presents an asymmetry, since the process of formation of two-dimensional chains is characteristic of a low number of degrees of freedom, generating, preferably, chains with small N. Three-dimensional chains (Fig. 4b), on the other hand, are much longer than two-dimensional. This behavior had already been explained previously. The process of forming chains in a three-dimensional network offers more possibilities of "escaping" to a site not yet visited, than in networks of two dimensions, where the probability of finding a "free" site is less, and the process can be interrupted more easily. For this reason, the vast majority are relatively large chains. For three-dimensional chains, the $Nb\psi$ values are uniformly distributed in the case of small chains, while very long chains have a normal distribution. The maximum of the distribution ($N = 1112$ steps for the chains generated from $N' = 1500$), is obtained for large chains (compared to N'). This way, chains with approximately this number of steps appear with greater probability under the conditions established in the simulation.

Figures 5 and 6, show the distribution of the characteristic distances (R_{ee} and R_g) of the chain. The graphs in Fig. 5 represent adjustments using Lhuillier's proposal [28]:

$$P(R) \sim exp\left(-R^{\alpha d} - R^{\delta}\right). \tag{9}$$

The distribution of characteristic distances, in Fig. 5, behaves according to two different exponential laws. The first exponential behavior in Eq. 8, corresponds to small values of characteristic distances, while the second term of this equation represents the behavior for large values. The α and δ exponents are associated with the universal exponent of Flory through the following expressions [29]:

$$\alpha = (vd - 1)^{-1}, \delta = (1 - v)^{-1} \tag{10}$$

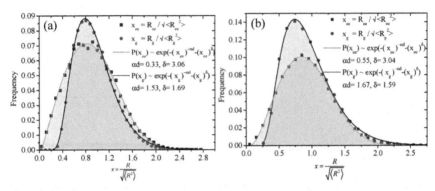

Fig. 5. Normalized histograms of the radius of gyration $R_g \psi$ and end-to-end distance R_{ee}. Behavior obtained for the ensemble of random chains generated from $N' = 1000$ in 2D (a), and $N' = 1500$ in 3Dψ (b). It is observed that a distributions agrees very well with the expression derived by Lhuillier [29, 30], both, the $R_g \psi$ and $R_{ee} \psi$ distribution.

Figure 6 shows the graphs of R_g and R_{ee} distribution, adjusted according to the Fisher-McKenzie-Moore-des-Cloiseaux law [12, 30–32]:

$$P(R) \sim R^\theta \exp\left(-\beta R^\delta\right) \tag{11}$$

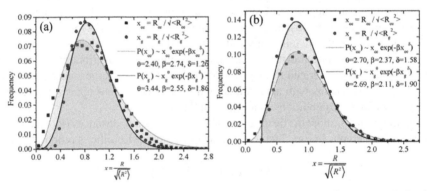

Fig. 6. Normalized histograms of the radius of gyration R_g and end-to-end discance R_{ee}. Behavior obtained for the ensemble of random chains generated from $N' = 1000$ in 2D (a), and $N' = 1500$ in 3D (b). It is observed that a distributions agrees very well with the function proposed by McKenzie and Moore [30] and des Cloizeaux [31, 32], both, the R_g and R_{ee} distribution.

The exponent θ characterizes the shorts-distance intra-chain correlations between two segments of a long chain.

$$\theta = (\gamma - 1)/v \tag{12}$$

The total number of a chain conformations is indirectly determined by the exponent γ. As opposed to ideal chains, where $\gamma = 1$, for real chains, $\gamma > 1$, i.e. there is a reduction of the probabilities in the distribution of R for short chains. The chains in this work were obtained from a random set without fixing the number of steps, unlike the theoretical distribution of Fisher-McKenzie-Moore-des-Cloiseaux [12, 31], although it is a different method to generate the chains, the characteristic distances follow the same distribution.

The value of Flory exponent v, which describes the size of the linear chain, was calculated by computing the mean value of the radius of gyration (see Fig. 7). The values of v obtained from the simulation, incorporated in the graphs, result approximate the expected theoretical value for the Flory exponent that is $v = 0.75$ and $v = 0.60$ for the two and three-dimensional case respectively. The calculation of v of the behavior of R_g results in a value closer to the expected theoretical value than that calculated from the R_{ee}. The Flory exponent was also calculated indirectly from the δ exponent that results from the distribution of characteristic distances, shown in Figs. 5 and 6. The values of α and θ exponents were also obtained from these distributions, in indirect and direct ways, respectively.

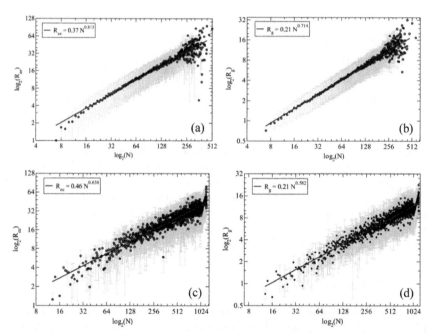

Fig. 7. Log-log scale representation of characteristic distances. Top panels, for 2D-chains: (a), the end-to-end distance (R_{ee}) as a function of the number of steps (N) with its respective value of $v = 0.757$; and (b), the radius of gyration (R_g) as a function of the number of steps (N) with its respective value of $v = 0.743$. Bottom panels, for 3D-chains: (c), the end-toend distance (R_{ee}) as a function of the number of steps (N) with its respective value of $v = 0.638$; and (d), the radius of gyration (R_g) as a function of the number of steps (N) with its respective value of $v = 0.582$.

Table 1 shows the values of exponents reported by our simulations. The above results have shown that although the two behaviors, R_{ee} and R_g, as a function of N fit a power law, the value of the characteristic exponent of R_g is closer to the theoretical value of Flory and to the values reported in [12, 33, 34]. Although, both proposed functions, fit well the distributions (R_{ee} and R_g), the values of the critical exponents obtained for each function, specifically the delta exponent, are different and correspond to different laws. The R_g distribution responds better to the law proposed by Lluillier, and the distribution of R_{ee}, to the Fisher-McKenzie-Moore-des Cloiseaux law.

Table 1. Main exponents calculated directly and indirectly from our simulations. The first and second rows show the exponents calculated from the behavior of the characteristic distances as a function of N. The exponents calculated using the distribution of R_{ee} and R_g appear in the next four rows. The exponents of the third and fourth rows were obtained using the Lhuillier distribution for both distributions. For the calculation of the exponents that appear in the last two rows, the Fisher-McKenzie-Moore-des-Cloiseuax (MMC) distribution was used.

Characteristic distances	$v(2D/3D)$	$\delta(2D/3D)$	$\alpha(2D/3D)$	θ
R_{ee} (vs. N)	0.81/0.64	5.26/2.76	1.61/1.09	–
R_g (vs. N)	0.71/0.58	3.45/2.39	2.38/1.34	–
R_{ee} (Lhuillier)	2.15	3.04	0.18	–
R_g (Lhuillier)	0.93	1.59	0.56	–
R_{ee} (MMC)	0.36	1.58	–	2.70
R_g (MMC)	0.47	1.90	–	2.69

Unlike our model, the previous models reported in the literature, simulate the growth of the chains keeping the N fixed. The parameters, despite being related to a universal growth exponent (Flory exponent), depend strongly of the system; as a result, obtained exponents in our simulations differ from those reported. The most important of these results is the form adopted by the distributions. We studied the distribution for a whole set of fifty thousand two and three-dimensional chains that are distributed as expected, with their own exponents and a positive asymmetry.

3.3 Fractal Dimension

The graphs in Fig. 8 show the results obtained regarding the analysis of the fractal dimension of the two and three-dimensional chains. The histograms reveal that, for the two-dimensional chains (Fig. 8a), the fractal dimension has a normal (Gaussian) behavior for the two method used, box-counting ($d_{F_{Box}}$) and correlation ($d_{F_{Corr}}$). Using the box-counting method (blue dots), the fractal dimension turned out to be less than the spatial dimension d and greater than the correlation dimension (red dots) whose averages were recorded in $\mu_{Box} = 1.698$ and $\mu_{Corr} = 1.189$. In this way, the inequality $d > d_{F_{Box}} > d_{F_{Corr}}$, is valid for our chains, as suggested by fractal theory [22]. Still, for the two-dimensional case, a peak corresponding to $d_{F_{Corr}}$ is closer to the theoretical value of $d_{F2D} = 1.33$. In the tree-dimensional case (Fig. 8b), the distribution of d_{FCorr} shows three peaks, in $d_{F1Corr} \approx 1.45$, $d_{F2Corr} \approx 1.64$ and $d_{F3Corr} \approx 1.86$, with d_{F2Corr} being the most frequent value. In this case, the highest peak in the distribution approach the theoretical value of $d_{F3D} = 1.7$. The quantitative results show that the values obtained from the correlation method are closer to the theoretical ones. On the other hand, the qualitative results reveal more information about the chains. The existence of three peaks reveals that three types of three-dimensional chains are formed during the growth process.

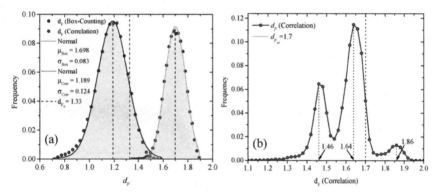

Fig. 8. Frequency distribution of the fractal dimension d_F obtained using the methods of box-counting, d_{FBox} (blue dots), for (a) the two-dimensional case and through the dimension of correlation, d_{FCorr} (red dots), for both (a) two-dimensional and (b) three-dimensional chains. The dashed black lines indicate the maximums obtained in each distribution, in the two-dimensional case it corresponds to the average values of the normal distribution (μ_{Box} and μ_{Corr}). The blue dashed lines represent the theoretical value of d_F which in (a) two-dimensional case is $d_{F2D} = 1.33$ and, (b) three-dimensional is $d_{F3D} = 1.7$. (Color figure online)

Figure 9 show the behavior of fractal dimension as a function of the number of steps, for 2D and 3D chains. Two-dimensional small chains have a high d_{FBox} dimension (Fig. 9a), this is because chains with a small number of steps can fill a large part of the space, which can be interpreted as well-compacted chains. As N increases, it is observed that the dimension d_{FBox} decreases, and for chains greater than $N = 200$, it remains within a region (dashed lines in Fig. 9a) oscillating around $d_{FBox} = 1.587$.

The d_{FCorr} concerns the distribution of "links" within the chain structure. Figure 9, b and c, shows how as N grows, the correlation of the chains increases, but above a certain size ($N = 150$ for two-dimensional chains and $N = 500$ for three-dimensional), d_{FCorr} oscillates around specific average values, $d_{FCorr} = 1.287$ and $d_{FCorr} = 1.590$ for 2D and 3 D case respectively. In a "very" large N region, the correlation between chain "links" decreases uniformly, due to the fact that large chains have a much-stretched structure compared to smaller chains.

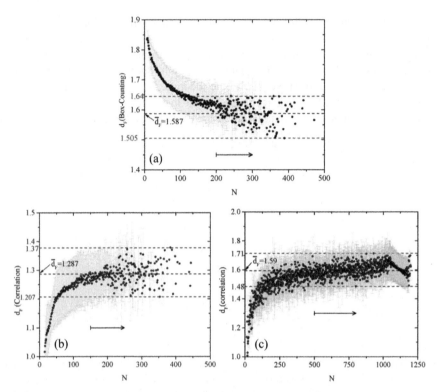

Fig. 9. Fractal dimension via *Box-Counting* ($d_{F_{Box}}$) (a), and via *correlation dimension* ($d_{F_{Corr}}$) (b and c) as a function of the number of steps (N) for two and three-dimensional chains. After reaching a size of $N \approx 200$, the fractal dimension stabilizes, oscillating around the value $d_F = 1.587$, in $2D$ case. In (b), when the strings reach a size of $N \approx 150$ the correlation dimension oscillates around $d_F = 1.287$. In (c), the dimension oscillates around $d_F = 1.59$ when the strings reach a size $N \approx 500$.

4 Conclusions

In this work, we show an algorithm based on natural self-avoiding random walk in square and cubic networks with no boundary was used to generate linear chains. The implementation of this algorithm is simpler and the computational cost is significantly lower than other approaches generally used to generate random linear chains. Although our method does not consider all possible configurations for the N size of the chain the values of the characteristic critical exponent's obtained by us are similar to those reported in the literature.

The study of the number of attempts (N') as a function of N, for two and three dimensions, showed that there are upper limits for N', above which there is no guarantee of better efficiency of the algorithm. This resulted in the optimization of the processing time in the construction of the set of chains. The exponents revealed by the power laws that govern these systems and the distribution of the characteristic magnitudes provides a solid basis for the study of linear chains in square and cubic networks.

The distribution of the characteristic distances for the linear chains obtained from our simulation shows a reasonable correspondence with that reported in the literature. Specifically, we prove that although it is possible to fit the two distributions ($R_g \psi$ and R_{ee}) using both Lhuillier's theory and the law proposed by FisherMcKenzie-Moore-des Cloiseaux, the obtained values of critical exponents such as ψ, ψ, $\psi\psi$ show us that the distribution of $R_g \psi$ responds better to the theory of Lhuillier, while in the case of the distribution of R_{ee}, the best fit is obtained with the Fisher-McKenzie-Moore-des Cloizeaux function. These results are in full agreement with the literature and reinforce the validity of the algorithm used when characterizing this type of system.

Both R_{ee} and R_g resulted in power functions of N and the values of the Flory exponent, in both cases, are quite close to the theoretical value, especially the value corresponding to R_g, showing that this characteristic distance is more appropriate when characterizing structurally this type of chains. Our model does not distinguish between chains of different sizes, showing that structural factors of the chains are invariant of scale. We also show that small chains are more correlated and, consequently, less flexible.

The fractal dimension of the two-dimensional chains is normally distributed, showing that the dimension by Box-counting is greater than the correlation dimension and both, smaller than the spatial dimension, as shown by the theory. In a three-dimensional case, the higher peak in the fractal dimension distribution is very close to the theoretical value. Finally, in very large three-dimensional chains the correlation decreases. This behavior can be associated with a change in the shape of the chains, which can form clusters separated by stretched structures. A more detailed analysis of the results and conclusions can be found in [35].

References

1. Flory, P.J.: Principles of Polymer Chemistry. Cornell University Press, Ithaca (1953)
2. Madras, N., Slade, G.: The Self-Avoiding Walk. Birkhauser, Basel (1953)
3. Yamakawa, H.: Modern Theory of Polymer Solutions. Harper and Row, New York (1971)
4. Wilson, K.G., Kogut, J.: The renormalization group and the expansion. Phys. Rep. **12**(2), 75–199 (1974)
5. Sokal, A.D.: Molecular Dynamics Simulations in Polymer Sciences. Oxford University Press, New York (1995)
6. Guttmann, A.J., Conway, A.R.: Square lattice self-avoiding walks and polygons. Ann. Comb. **5**(3), 319–345 (2001)
7. Jensen, I.: Enumeration of self-avoiding walks on the square lattice. J. Phys. A Math. Gen. **37**(21), 5503–5524 (2004)
8. Li, B., Neal, M, Sokal, A.D.: Critical exponent hyper scaling, and universal amplitude ratios for two and three-dimensional self-avoiding walks. J. Stat. Phys. **80**(3), 661–754 (1995)
9. Hara, T., Slade, G., Sokal, A.D.: New lower bounds on the self-avoiding walk connective constant. J. Stat. Phys. **72**(3), 479–517 (1993)
10. Slade, G.: Self-avoiding walk, spin systems and renormalization. Proc. R. Soc. A **475**(2221), 20180549 (2019)
11. Amit, D.J., Parisi, G., Paliti, L.: Asymptotic behavior of the "true" self-avoiding walk. Phys. Rev. B **27**(3), 1635–1645 (1983)
12. Rubinstein, M., Colby, R.H.: Polymer Physics. Oxford University Press, New York (2003)

13. Teraoka, I.: Polymer Solutions: An Introduction to Physical Properties. Wiley Inter-science, New York (2002)
14. Bhattarcharjee, S.M., Giacometti, A., Maritan, A.: Flory theory for polymers. J. Phys. Condens. Matter **25**, 503101 (2013)
15. Isaacson, J., Lubensky, T.C.: Flory exponent for generalized polymer problems. J. Phys. Lett. **41**(19), 469–471 (1980)
16. Mandelbrot, B.B.: The Fractal Geometry of Nature. W. H. Freeman and company, New York (1982)
17. Banerji, A., Ghosh, I.: Fractal symmetry of proteins interior: what have we learned. Cell. Mol. Life Sci. **68**(16), 2711–2737 (2011)
18. Dewey, T.G.: Fractals in Molecular Biophysics. Oxford University Press, New York (1997)
19. Maritan, A.: Random walk and the ideal chain problem on self-similar structures. Phys. Rev. Lett. **62**(24), 2845–2848 (1989)
20. Kawakatsu, T.: Statistical Physics of Polymers: An Introduction. Springer-Verlag, Heidelberg (2004)
21. Rammal, R., Toulouse, G., Vannimenus, J.: Self-avoiding walks on fractal spaces: exact results and Flory approximation. J. Phys. **45**(3), 389–394 (1984)
22. Takayasu, H.: Fractals in the Physical Sciences. Manchester University Press, New York (1990)
23. Feder, J.: Fractals. Physics of Solids and Liquids. Springer-US, New York (1988)
24. Theiler, J.: Estimating fractal dimension. J. Opt. Soc. Am. A **7**(6), 1055–1073 (1990)
25. Nayfeh, A., Balachandran, B.: Applied Nonlinear Dynamics: Analytical, Computational, and Experimental Methods. Wiley Series in Nonlinear Sciences, Germany (2008)
26. Grassberger, P., Procaccia, I.: Characterization of strange attractors. Phys. Rev. Lett. **50**, 346–349 (1983)
27. Grassberger, P., Procaccia, I.: Measuring the strangeness of strange attractors. Phys. D Nonlin. Phenom. **9**(1), 189–208 (1983)
28. Lhuillier, D.: A simple model for polymeric fractals in a good solvent and an improved version of the Flory approximation. J. Phys. Fr. **49**(5), 705–710 (1988)
29. Victor, J.M., Lhuillier, D.: The gyration radius distribution of two-dimensional polymers chains in a good solvent. J. Chem. Phys. **92**(2), 1362–1364 (1990)
30. McKenzie, D.S., Moore, M.A.: Shape of self-avoiding walk or polymer chain. J. Phys. A Gen. Phys. **4**(5), L82–L85 (1971)
31. des Cloizeaux, J.: Lagrangian theory for self-avoiding random chain. Phys. Rev. A. **10**, 1665 (1974)
32. des Cloizeaux, J., Jannink, G.: Polymers in solution: their modelling and structure. Oxford Science Publications. Clarendon Press, Oxford (1990)
33. Caracciolo, S., Causo, M.S., Pelissetto, A.: End-to-end distribution function for dilute polymers. J. Chem. Phys. **112**(17), 7693–7710 (2000)
34. Vettorel, T., Besold, G., Kremer, K.: Fluctuating soft-sphere approach to coarse-graining of polymer models. Soft Matter **6**, 2282–2292 (2010)
35. Bernal, D.R.: PhD Thesis, http://www.ppgbea.ufrpe.br/sites/www.ppgbea.ufrpe.br/files/doc umentos/tese_david_roberto_bernal.pdf. Accessed 21 June 2021

Digital Analog Converter for the Extraction of Test Signals from Mixed Integrated Circuits

José L. Simancas-García[1], Farid A. Meléndez-Pertuz[1], Ramón E. R. González[2(⊠)], César A. Cárdenas[3], and Carlos Andrés Collazos-Morales[3(⊠)]

[1] Departamento de Ciencias de la Computación y Electrónica, Universidad de Costa, Barranquilla, Colombia
jsimanca3@cuc.edu.co

[2] Departamento de Física, Universidade Federal Rural de Pernambuco, Recife, Pernambuco, Brazil
ramon.ramayo@ufrpe.br

[3] Vicerrectoria de Investigaciones, Universidad Manuela Beltrán, Bogotá, Colombia
carlos.collazos@docentes.umb.edu.co

Abstract. The construction of integrated circuits involves testing the correct operation of its internal blocks. For this, a common practice is the integration of functional blocks to stimulate the internal subsystems and extract the responses to those stimuli. In this article, the design and simulation of a circuit for the extraction of the response signals of the devices under test in analog and mixed-signal integrated circuits is presented. The extraction block is a 2-stage 5-bit segmented A/D converter, operating at a sampling frequency of 10 MHz, implemented in a 0.12 μm technological process, which can be powered with 1.5 Vdc. This proposal offers a reduction in the area consumed, by requiring fewer comparators than other similar solutions found in the literature.

Keywords: Digital analog converter · Signals · Mixed integrated circuits

1 Introduction

The combination of the growing demand for consumer electronics and the constant growth in the packaging density of semiconductor devices, is leading to the integration of more and more functional systems into a single integrated circuit [1]. The result, among other things, is an increased need for the integration of analog and mixed-mode components, e.g. analog-digital, RF-analog-digital, and mechanical-analog-digital, into the same chip or package [17]. Designing such SoC (Systems on Chip) is undoubtedly a challenge, since it links abstraction management at the system level while simultaneously dealing with the physical effects of transistors and the parasitic effects associated with the circuit [2]. Similarly, the next generation of SoC testing represents a real challenge, especially when cost and time to market are usually key requirements. Such mixed-signal integrated circuits contain very complex signal paths and functional specifications, and the test programs developed may not be very viable, as they would be significantly slow

© Springer Nature Switzerland AG 2021
O. Gervasi et al. (Eds.): ICCSA 2021, LNCS 12949, pp. 207–223, 2021.
https://doi.org/10.1007/978-3-030-86653-2_15

in characterizing and debugging the device, which would greatly increase the time to market [2, 28]. The difficulty is accentuated by another aspect of system-level integration, called third- party functional block integration. In order to compete with the complexity of design, the manufacturers of the final system are forced to rely on pre-designed blocks, and carry out the integration of these as part of a larger and more complex system. These functional blocks are obtained from virtual libraries that use software to describe the block as part of the system [3, 7].

In the digital domain, the test mechanisms and techniques are capable of testing most devices in this domain, and the test information can be transported seamlessly throughout the SoC [8–11]. The results can also be extracted in the same way. For this reason, it seems possible to derive a systematic procedure by which the integrator of the final system can access the virtual functional blocks that make up the chip. The problem in the analog domain is how complicated it is to extract test signals over long interconnect lines inside a chip. Rapid degradation of the signals is very likely due to noise and distortion introduced by the parasitic behaviour of some elements within the integrated circuit [4–6].

One of the problems to which an alternative solution is sought is the case of ICs containing analogue functional blocks, and the aim is to test the functioning of these blocks. Test signals from such blocks must travel inside the chip and then be extracted. As is well known, analogue signals are very sensitive to interference, are not immune to noise, and degrade more easily than digital signals [9, 16, 17]. Therefore, the tests on this type of block are not very reliable, since both the input and output signals are disturbed by noise and do not reveal the true behaviour of the block. This problem is compounded by the fact that, at high frequencies, the microband or interconnection lines of each of the elements that make up the system begin to behave like transmission lines, and the following manifestations appear: energy reflection, attenuation, impedance decoupling both at the source and at the load relative to the interconnection line [9, 17]. As a result, manufacturers must perform impedance couplings, source and load balancing, among many other procedures to solve the above mentioned problems. These solutions require design times that manufacturers are not willing to take [2].

Due to the above, one objective of these test techniques is the integration of this block into the chip to be tested, which would pose a problem due to the difficulty associated with the transport of analog signals on an integrated circuit, in particular the fact that these types of signals are very sensitive to noise and the tests could be affected and not show the results that show the real behaviour of the circuit [12–15, 32]. Methods based on the inclusion of data converters, such as D/A and A/D converters, have been presented in the industry as alternatives [4, 5, 27]. The first one for obtaining analog waveforms from digital words, and the second one for extracting in digital format the output signals of the blocks under test, CUT (Core Under Test). For the above, it is clear that the use of D/A converters would be prohibitive, due to the excessive use of area in the silicon tablet, considering that the latter is a critical requirement in the design of VLSI integrated circuits [1]. The aforementioned techniques are static, consisting of predesigned fixed blocks. This has evolved to dynamic solutions based on adaptive [29] and intelligent techniques [30, 31].

The resolution of the waveform digitizer shall be 5 bits and the sampling frequency shall be 10 MHz. The technology for which the circuit proposed here was designed is a 0.12 μm CMOS lithographic process. This process is characterized by having a power level of between 1 V and 1.5 V, and it is assumed that the integrated circuits where this test block is assembled can provide them, since they turn out to be typical of this technology [26]. On the other hand, the dual levels of these voltages are required, these are −1 V and −1.5 V. As the circuit designed in this project is to be used in mixed integrated circuit tests, it is to be assumed that they are also provided by the chip. Other features that characterize this technology are its threshold voltage, V_t, which is around 0.3 V, and an oxide layer thickness of 4 nm.

2   Analogical – Digital Conversion Process

Signal A/D conversion is done in 2 stages, the first is uniform time sampling and the second is amplitude quantization. The samples x[n], evenly spaced time intervals Ts, of the continuous time signal x(t), can be represented by x[n] = x (nTs). In the frequency domain, the sampling process produces periodically repeated versions of the original signal spectrum placed in the integer multiples of the sampling frequency fs = 1/Ts. The above is evidenced in Eq. (1), where Xs (f) represents the spectrum of the sampled signal, and X (f) is the spectrum of the original signal in continuous time [18].

$$Xs(f) = \frac{1}{Ts} \sum_{k=-\infty}^{\infty} X(f - kfs) \tag{1}$$

This process is shown in Fig. 1, for the case where fs = 2W, where W is the highest frequency component of the original signal. In general, the continuous-time signal can be reconstructed from its samples, if repeated versions of the spectrum do not overlap. To achieve the above, the original signal should be band-limited to half the sampling frequency [19].

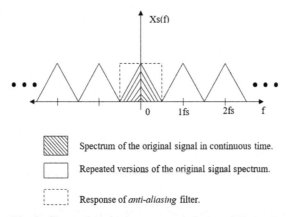

Spectrum of the original signal in continuous time.

Repeated versions of the original signal spectrum.

Response of *anti-aliasing* filter.

Fig. 1. Shape of the power spectrum of a sampled signal.

The overlap between repeated versions of the signal spectrum is known as aliasing, the aliasing does not allow the reconstruction of the signal from its samples. An anti-aliasing filter is regularly used to ensure this requirement [20]. The anti-aliasing filter is an analog low-pass filter that precedes the sampling circuit. The case where the sampling frequency is fs = 2W, is known as Nyquist rate sampling, and is considered critical because that the anti-aliasing filter must have a frequency response with a small transition band [19].

The sampling process is an invertible operation because there is no loss of information. Note that although Fig. 1 shows the sampling process for a baseband signal, Eq. (1) describes the sampled spectrum for any signal spectrum centered on some frequency component other than 0, fc. Assuming the signal has a bandwidth W, the spectrum of the original signal comprises the range [f_c − W/2, f_c + W/2]. To avoid aliasing some authors suggest sampling at rates f_s ≥ 2W [19].

Once the sampling is done, a process called quantization follows. This process involves approximating the values of the samples to a finite set of values, using a close-ness criterion. In this stage the dynamic range of the input signal is divided into equal parts, for uniform quantization. A representative value is assigned to each sector. Samples are assigned one of these values, depending on the sector where it is located. The typical transfer characteristics of A/D Quantizers or Converters are shown in Fig. 2 [19].

Quantization is a non-reversible process. Quantified output amplitudes are usually represented by digital code words with a finite number of bits. For example, for a 1-bit A/D converter as shown in Fig. 2(c), the output values V and −V can be mapped to the digital codes "1" and "0." Another way to visualize this is by using the digital code words instead of the output values on the axis and in Fig. 2. Quantified output values can be considered as the ideal A/D converter output, whose output corresponds to a digit code word [19].

An A/D converter or quantifier with Q output levels is said to have a resolution of N bits if N = Δlog Δ_2 (Q). It should be made clear from Fig. 2 that for an A/D quantifier or converter of Q levels of quantization, only input values separated by at least a distance Δ = 2V/(Q − 1) can be distinguished or designated for different output levels. N bits are required to encode the corresponding Q code words with each output level.

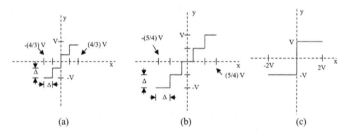

(a)   (b)   (c)

Fig. 2. Transfer characteristics of typical quantifiers.

The transfer curves of the quantifiers in Figs. 2(a) and (b), which are symmetrical, will now be exalted. The half-increase quantifiers do not have an output level of 0 for an input value of 0, generating a DC offset undesirable, as seen in Fig. 2(a). The half-step quantifier, Fig. 2(b), needs an even number of output levels to produce a completely symmetrical transfer curve. This is an advantage because the number of output levels could be a power of 2 and be encoded exactly with $N = log_2 (Q)$ bits. On the other hand, the half-step quantifier needs an odd number of output levels, making Q not a power of two and therefore no efficient coding is performed. For this case, the number of bits needed could be $N = log_2 (Q - 1) + 1$, where $Q - 1$ is taken as a power of 2.

If the number of output levels of the half-step quantifiers are forced to be powers of 2 to use only $Q - 1$ levels, there will be a large amplitude distortion. The distortion, of course, could be negligible when the number of output levels is large.

Most A/D converters provide trade-offs between signal bandwidth, output resolution, and complexity of digital and analog hardware. The qualitative compromises between resolution and band width of some of these techniques are shown in Fig. 3.

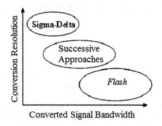

Fig. 3. Commitment resolution vs. bandwidth.

The quantifier used in any A/D converter is a non-linear system, which makes analysis difficult. To make the analysis manageable, the quantifier is usually made a linear system and modeled through a noise source, e[n], added to the signal x[n], to produce the quantified output signal y[n]:

$$y[n] = e[n] + x[n] \tag{2}$$

The block diagram of the model of an A/D converter system displaying the sampling process and the quantifier is shown in Fig. 4. In addition, to simplify the noise analysis of the quantifier.

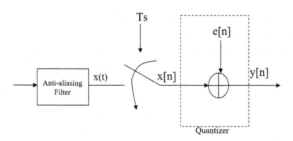

Fig. 4. Block diagram and model of a conventional A/D converter.

3   Convertering Circuits A/D

These are the electronic circuits responsible for carrying out the quantification of the samples obtained from the analogue signal. There are different types of A/D converters, the usefulness of which will depend on the characteristics of the conversion applications [17, 21].

3.1   Flash A/D Converters

The architecture of this converter consists of a string of comparators, which have their inputs connected to a resistance ladder, and their output is fed into a digital encoder circuit, which gives a binary value representative of the highest level of comparison exceeded. It is possible to minimise the quantification error by selecting the values of the resistances at the ends of the chain in an appropriate way [21].

3.2   Successive Approaches

There they are circuits that use as one of their constitutive blocks a D/A converter in a feedback architecture. They deliver an output of n bits in n clock cycles, making successive comparisons of the input with the output of a D/A converter, and dividing the range in half in each cycle, this is achieved by varying from bit to bit, from the most to the least significant [22].

3.3   Converters to Ramp and Double Ramp

The single ramp structure is prone to errors caused by variations in components, particularly the capacitor, and to deviations in the frequency of the oscillating circuit. Therefore, double ramp architectures are much more used, where such errors are compensated by means of a double integration. In the first ramp, the capacitor voltage starts at zero and is charged for a fixed time to a current that depends on the input voltage. Then, in a second phase, it is discharged at constant current for a time which is measured by a meter [23].

4   Proposed Signal Digitalizer

The A/D converter, two-stage ADC, has a conventional Flash A/D converter at its heart. The latter has as constitutive circuits the operational amplifier and the comparator, as well as a convenient logic. A diagram of how the digitizer block is structured can be seen in Fig. 5.

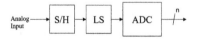

Fig. 5. Block diagram of the signal digitizer

4.1 ADC Block

Alternatives were presented only in the A/D conversion section. Three alternatives were considered, namely. The first was an A/D converter of successive approximations, or SAR, but it was very slow, consuming as many clock pulses as bits of resolution it had. In the case of 5 bits, it would be 5 clock strokes. It was discarded for speed reasons. The second option was a conventional Flash A/D converter, but its area consumption turned out to be excessive and therefore prohibitive for this application. The last consideration, which was the one chosen, is the 2-stage Flash A/D converter. Consumption of small area, similar to that of one of successive approaches, but with the speed of a conventional Flash [24] (Fig. 6).

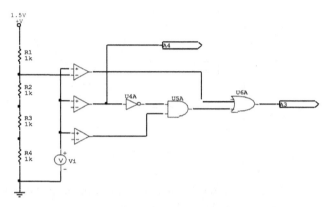

Fig. 6. 2-bit Flash converter.

In this design it has been assumed that the dynamic range of the signal is between 1.5 V and 0 V, in case of a change in the signal range, only the voltage in the resistive network generating the references should be changed. In this project, a standard 1 kΩ resistor type has been used, to facilitate manufacturing in an integrated circuit. The function of this first block is to divide the dynamic range of the signal into four equal sectors, and the 2-bit code on the output will indicate which of the four regions the input sample is in. Table 1 shows this operation.

Table 1. Divide the 2-bit A/D converter into four sectors.

Voltage of analogue in.	A4	A3	Area
0 V < Vin < 0.375 V	0	0	1
0.375 V < Vin < 0.75 V	0	1	2
0.75 V < Vin < 1.125 V	1	0	3
1.125 V < Vin < 1.5 V	1	1	4

This way you get the 2 most significant bits. The comparator used in this section is based on an operational amplifier presented in a previous article [25].

Once the first two bits have been obtained, the sample is taken and pre-processed before obtaining the remaining three bits. The first is to use a D/A converter to transform the two bits already obtained into an analog voltage, a standard configuration for this block called a ladder network is used, and shown in Fig. 7.

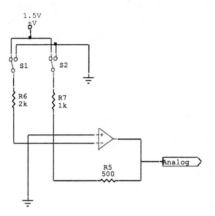

Fig. 7. 2-bit D/A converter.

There is an equation that governs the operation of this circuit, and that was the basis for the elaboration of the design, this is shown below

$$V_o = \frac{-V_{ref} \cdot R5}{R7}(2^1 \cdot B1 + 2^0 \cdot B0) \tag{3}$$

In this equation V_{ref} is the supply voltage, which for this case is 1.5 V, R5 is the feedback resistor, which for this application is 500 Ω, and R7 is the resistor corresponding to the most significant bit, which in the circuit used here has a value of 1 kΩ. The switches S1 and S2 are controlled by the 2 bits already obtained previously, and corresponding to A0 and A1 respectively. The operation of the circuit is shown in Table 2.

Table 2. Input Output ratio of the 2-bit D/A converter.

A1	A0	Output voltage (V)	Area
0	0	0	1
0	1	−0.375	2
1	0	−0.75	3
1	1	−1.125	4

From this table and from Eq. (3) we can deduce the following conclusions: when A0 is active, 0.375 V must be added, and when A1 is active, 0.75 V must be added, therefore it can be concluded that

$$\frac{V_{ref} \cdot R5}{R7} = 0.375 \text{ V} \quad (9.18)$$

If you have to $V_{ref} = 1.5$ V, R7 $= 1$ kΩ, then we get that R5 is 500 Ω, just like we said, getting the circuit in Fig. 7.

The next step is to subtract the analog value obtained previously from the sample value, and to the result of this subtraction apply a conventional 3-bit A/D converter type Flash, and in this way you get the 3 least significant bits, and generally a 5-bit code. Because the analogue levels at the output of the D/A converter are inverted, not a subtractor circuit is used but a summator. The latter process is equivalent to dividing the four zones described above into 8 sub-zones each, and determining the location of the sample related to these sub-zones. The sample summing circuit in Fig. 8.

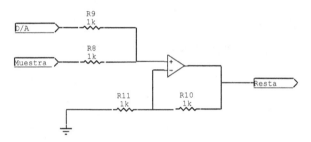

Fig. 8. Remaining circuit.

For this block, as well as for the D/A converter, the operational amplifier designed and presented in a previous article is used [25].

The next step, the one that proceeds to the subtraction, is a 3-bit Flash converter. The converter implemented here is divided into two parts, the first is a comparison circuit, which consists of a series of 7 comparators and a reference generator circuit and can be seen in Fig. 9. The second part is a coding block, which takes the thermometer-type outputs of the comparators and transforms them into a 3-bit binary code, its architecture can be seen in Fig. 10.

4.2 LS Block

It is the first part of the signal digitizer, and is responsible for adding an offset level to the signal coming from the block under test. The aim of this is to ensure that there are no changes in polarity in the signal that is to be digitized, and thus to make the design of the A/D converter more flexible, which will be described later. A summing block was used, which is based on the voltage summing configuration of the operational amplifier. The level slider scheme is shown in Fig. 10. As can be seen in the figure, the added offset voltage is 0.75 V, assuming that the output signal of the block under test does

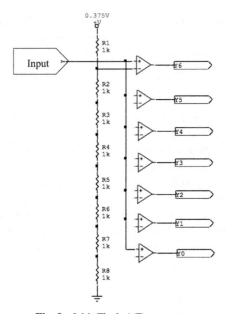

Fig. 9. 3-bit Flash A/D converter.

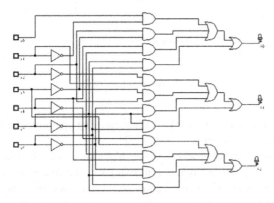

Fig. 10. 3-bit thermometer output encoder.

not exceed 1 V peak to peak. This value may change depending on the application in question. Resistances R3 and R4 must be of the same value for the following ratio to be met:

$$V_o = \left(1 + \frac{R7}{R1}\right)\left(\frac{V_i + 0.75V}{2}\right)$$

Immediately you can see that R7 must be equal to R1, if you want the voltage at the output to be the sum of the input voltages.

4.3 S/H Block

The sam pling frequency was 10 MHz as specified, but for technical reasons it is not possible to sample at this frequency, therefore it was decided to lower the rate to 5 MHz, once seen the results will be concluded because such a change had to be made. This implementation offers a compact and practical solution, especially contributing signifi-cantly to space savings, making it ideal for integrated applications. This circuit has only one drawback, and it is the fact that the value of the sample changes during the whole high level of the clock pulse, a fact that was thought not to represent problems in this project, because as a Flash converter is used, the codes are generated almost instantly, and the code corresponding to the last value that I took the sample during the cycle will be captured. clock. However, the latter is not true, as it takes the A/D converter a while to stabilize the output code. When the simulations and tests were performed, it was found that the use of this type of single circuit was not suitable for the purposes of the work. This could be solved with the inclusion of an isolation circuit, in this case, it was an operational amplifier connected as a buffer or follower. The result is the circuit shown in Fig. 11. These above-mentioned situations will be further clarified in the chapter on tests and results.

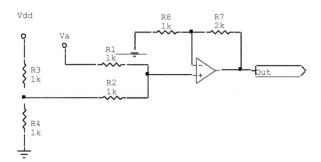

Fig. 11. Level shifter circuit.

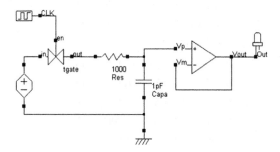

Fig. 12. Sampling and maintenance circuit.

5 Results Overalled

The operation of the signal digitizer is summarized in Table 3, which shows that each range of the input signal corresponds to a binary code. So, what we should expect is that for every sample taken within a given range, we get the corresponding binary code at the encoder output. As can be seen, the static tests of the converter were satisfactory, as all codes were correctly assigned to the corresponding sample. But it should be noted that the simulations showed that it took the converter some time to find the correct code corresponding to the input sample. This time was called the set-up time, and was measured for each test. See Table 4. The ramp sign is shown in Fig. 12. The output codes are shown in Fig. 13. In this last figure it is possible to notice that there is a behavior in time of the countdown type, which corresponds to a ramp type input, which increases its value gradually. The figure shows the codes that are loaded into the log over time.

Table 3. Operation summary of the 5-bit A/D converter.

Voltage ranges	Binary code	Area
0 V–0.047 V	00000	1
0.047 V–0.093 V	00001	2
0.093 V–0.14 V	00010	3
0.14 V–0.187 V	00011	4
0.187 V–0.234 V	00100	5
0.234 V–0.281 V	00101	6
0.281 V–0.328 V	00110	7
0.328 V–0.375 V	00111	8
0.375 V–0.422 V	01000	9
0.422 V–0.469 V	01001	10
0.469 V–0.516 V	01010	11
0.516 V–0.563 V	01011	12
0.563 V–0.61 V	01100	13
0.61 V–0.657 V	01101	14
0.657 V–0.704 V	01110	15
0.704 V–0.751 V	01111	16
0.751 V–0.798 V	10000	17
0.798 V–0.845 V	10001	18
0.845 V–0.892 V	10010	19
0.892 V–0.939 V	10011	20

(continued)

Table 3. (*continued*)

Voltage ranges	Binary code	Area
0.939 V–0.986 V	10100	21
0.986 V–1.033 V	10101	22
1.033 V–1.08 V	10110	23
1.08 V–1.127 V	10111	24
1.127 V–1.174 V	11000	25
1.174 V–1.221 V	11001	26
1.221 V–1.268 V	11010	27
1.268 V–1.315 V	11011	28
1.315 V–1.362 V	11100	29
1.362 V–1.409 V	11101	30
1.409 V–1.456 V	11110	31
1.456 V–1.5 V	11111	32

Table 4. Results of static tests of the A/D converter.

Test sample	Code obtained	Binary code	Status	Area	Error (V)	Error (%)	Time of establishment (ns)
1.48 V	11111	11111	Correct	32	0	0	10
1.43 V	11110	11110	Correct	31	0	0	99
0.54 V	01011	01011	Correct	12	0	0	77.55
0.48 V	01010	01010	Correct	11	0	0	62.97

A final dynamic test was carried out, which consisted of the input of a sinusoidal wave through the sampling circuit. The input wave had 0.5 V peak, an offset level of 0.75, and a frequency of 500 kHz. All sample values were measured at the points where the signal was captured by the converter output log, and it was checked whether the assigned codes were the Corrects. The simulation time used was 5 μs. The results obtained are shown in Figs. 14 and 15, as well as in Table 5 (Fig. 16).

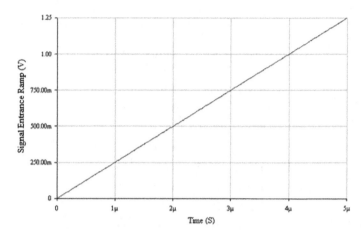

Fig. 13. Signal Entrance Ramp.

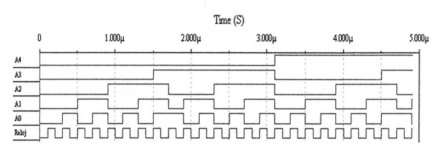

Fig. 14. Output of the A/D converter at a ramp input after being loaded.

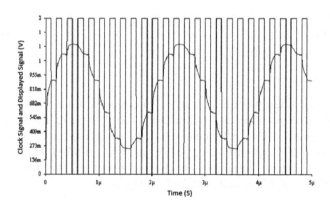

Fig. 15. Output signal from the sampling circuit.

Table 5. Summary of samples taken during the dynamic test of the A/D converter.

Sampling time (ns)	Value of the samples (V)	Code obtained	Binary code	Status	Error (V)	Error (%)
197	0.89	10011	10010	Error	0.046	3.12
1395	0.33	00111	00111	Correct	0	0
1600	0.24	00101	00101	Correct	0	0

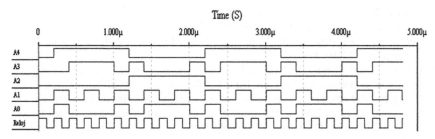

Fig. 16. Waveforms of the output bits of the A/D converter before a sinusoidal input.

Correct performance of the A/D converter is remarkable, and one additional comment can be added, the only sample that gave incorrect results was 0.89 V, but there is a reason for this behavior. If we look at Table 3, we see that the code for zone 19, which is where the 0.89 V sample should be, is 10 010, and the limits of this zone are 0.845 V–0.892 V. Zone 20, which code is 10 011 and is the one assigned to the sample under study, has limits 0.892 V–0.939 V. These errors can occur for several reasons. The first is that the determined limits present some problem and are not in agreement with the reality of the circuit. The other possibility is a problem of numerical approxi mation of the software since the difference separating the sample from one region to another is 0.002, which could result in the error found. If these observations are true, the signal digitizer can be said to have achieved the objective.

6 Conclusions

The signal digitizer was made under the concept of a 2-stage Flash converter. The maximum resolution achieved without significantly increasing the complexity of the hardware was 5 bits. The time of establishment forced a decrease in the sampling frequency, but, even so, the captured signals had an acceptable quality. The use of this architecture led to a noticeable decrease in the use of comparison devices, compared to the conventional Flash conversion technique, as well as an uncomplex operation, which simplifies the design, compared to SAR conversion techniques. This approach is similar to that presented in [32]. In that paper, a signal extraction block is presented using an 8-bit segmented A/D converter with 2 flash-type stages of 4 bits each. Also, it has a 4-bit D/A converter to perform segmentation. The sampling frequency of that converter is 1.7 MHz, unlike the one presented in this paper, which is 10 MHz. The technological process of that approximation is 0.5 μm, which establishes its supply voltage at 4 V, while the one presented here has a 0.12 μm process, which reduced the supply voltage to 1.5 V. This reduced voltage limited the range dynamic of the signal to extract, and allowing to use fewer bits.

References

1. Simancas-García, J.L.: Diagnóstico de Circuitos Integrados Analógicos y de Comunicaciones. INGE@UAN - Tendencias en la Ingeniería **1**(2), 7–19 (2011)
2. Kundert, K., et al.: Design of mixed-signal systems-on-a-chip. IEEE Trans. Comput.-Aided Des. Integr. Circuita Syst. **19**(12), 1561–1571 (2000)
3. Zorian, Y., Marinissen, E., Dey, S.:Testing embedded-core-based sys- tem chips. In: IEEE Computer, pp. 52–60. Junio (1999)
4. Hafed, M., Abaskharoun, N., Roberts, G.: A 4-GHz effective sample rate integrated test core for analog and mixed-signal circuits. IEEE J. Solid-State Circuits **37**(4), 499–514 (2002)
5. Hafed, M., Roberts, G.: A stand-alone integrated excitation/extraction systems for analog BIST application. In: IEEE 2000 Costum Integrated Circuit Conference, p. 4. IEEE (2000)
6. Hafed, M., Roberts, G.: Techniques for high-frequency integrated test and measurement. IEEE Trans. Instrument. Measur. **52**(16), 1780–1786 (2003)
7. Zorian, Y.: System-chips test strategies. In: 35th Design Automation Conference 1998, San Francisco, p. 6. ACM (1998)
8. Albustani, H.: Modelling methods for testability analysis of analog integrated circuits based on pole-zero analysis. Prüfung, 182 p. Dissertation (Ph.D.). Universität Duisburg-Essen. Fakultät für Ingenieurwissenschaften (2004)
9. Dillinger, T.: VLSI Engineering, p. 863. Prentice-Hall, Estados Unidos (1988)
10. Deschamps, J.-P.: Diseño de circuitos integrados de aplicación especifica ASIC, p. 385. Paraninfo, España (1994)
11. Pucknell, D., Eshraghian, K.: Basic VLSI design, 3rd edn., p. 495. Prentice-Hall, Australia (1993)
12. Dufort, B., Roberts, G.: Signal generation using periodic single and multi- bit sigma-delta modulated streams, p. 10 (1997)
13. Dufort, B., Roberts, G.: Optimized periodic sigma-delta bitstreams for analog signal generation, vol. 4, p. 4
14. Haurie, X., Roberts, G.: Arbitrary-precision signal generation for mixed-signal built-in-self-test. IEEE Trans. Circuits Syst.—II Analog Digital Signal Process. **45**(11), 1425–1432 (1998)
15. Hawrysh, E., Roberts, G.: An integration of memory-based analog signal generation into current dft architectures. IEEE Trans. Instrument. Measur. **47**(3), 748–759 (1998)
16. Simancas-García, J.L., Caicedo-Ortiz, J.G.: Modelo computacional de un modulador $\sum$-Δ de 2° orden para la generación de señales de prueba en circuitos integrados analógicos. INGE@UAN - Tendencias en la Ingeniería **5**(9), 43–55 (2014)
17. Rubio, A., et al.: Diseño de circuitos y sistemas integrados, p. 446. Alfaomega, Mexico (2005)
18. Soria Olivas, E., et al.: Tratamiento digital de señales: Problemas y ejercicios resueltos, p. 400. Prentice-Hall, España (2003)
19. Aziz, P., Sorensen, H., Van Der Spiegel, J.: An overview of sigma-delta converters: how a 1-bit ADC achieves more than 16-bit resolution. IEEE Sig. Process. Magazine, 61–84 (1996)
20. Proakis, J., Manolakis, D.: Tratamiento digital de señales: Principios, algoritmos, y aplicaciones, 3rd edn. Prentice-Hall, España (2003)
21. Franco, S.: Design with Operational Amplifiers and Analog Integrated Circuits, 3rd edn., p. 680. MacGraw-Hill, Estados Unidos (2002)

22. Rashid, M.: Circuitos Microelectrónicos: Análisis y Diseño, p. 990. International Thomson, México (1999)
23. Sedra, A., Smith, K.: Circuitos Microelectrónicos, 4th edn., p. 1232. Oxford University, México (1998)
24. Van De Plassche, R., Baltus, M.: An 8-bit 100-MHZ full-Nyquist analog-to-digital converter. IEEE J. Solid-State Circuits **23**(6), 1334–1344 (1988)
25. Simancas-García, J.L.: Diseño de un Amplificador Operacional CMOS de Amplio Ancho de Banda y Alta Ganancia para Aplicaciones de Alta Velocidad. In: IngeCUC, vol. 9, no. 1 (2013)
26. Sicard, E.: Microwind & Dsch User's Manual Version 2, p. 110. National Institute of Applied Sciences, Toulouse, Francia (2002)
27. Pavlidis, A., Louërat, M.-M., Faehn, E., Kumar, A., Stratigopoulos, H.-G.: SymBIST: symmetry-based analog and mixed-signal built-in self-test for functional safety. IEEE Trans. Circuits Syst. I Regul. Pap. **68**(6), 2580–2593 (2021)
28. Thaker, N.B., Ashok, R., Manikandan, S., Nambath, N., Gupta, S.: A cost-effective solution for testing high-performance integrated circuits. IEEE Trans. Compon. Packag. Manuf. Technol. **11**(4), 557–564 (2021)
29. Stratigopoulos, H.-G., Streitwieser, C.: Adaptive test with test escape estimation for mixed-signal ICs. IEEE Trans. Comput.-Aided Des. Integr. Circuits Syst. **37**(10), 2125–2138 (2017)
30. Shi, J., Deng, Y., Wang, Z., He, Q.: A combined method for analog circuit fault diagnosis based on dependence matrices and intelligent classifiers. IEEE Trans. Instrument. Measur. **69**(3), 782–793 (2019)
31. Canelas, A., Póvoa, R., Martins, R., Lourenço, N., Guilherme, J., Carvalho, J.P.: FUZYE: a fuzzy c-means analog IC yield optimization using evolutionary-based algorithms. IEEE Trans. Comput.-Aided Des. Integr. Circuits Syst. **39**(1), 1–13 (2018)
32. Sehgal, A., Liu, F., Ozev, S., Chakrabarty, K.: Test planning for mixed-signal SOCS with wrapped analog cores. In: Design Automation and Test in Europe Conference and Exhibition, Munich, 2005, pp. 50–55. IEEE Computer Society, Munich (2005)

Computational Model for Compressible Two-Phase Flow in Deformed Porous Medium

Evgeniy Romenski[1]([⊠]) [iD], Galina Reshetova[1,2], and Ilya Peshkov[1,3] [iD]

[1] Sobolev Institute of Mathematics SB RAS, Novosibirsk 630090, Russia
evrom@math.nsc.ru
[2] Institute of Computational Mathematics and Mathematical Geophysics SB RAS,
Novosibirsk 630090, Russia
kgv@nmsf.sscc.ru
[3] Department of Civil, Environmental and Mechanical Engineering,
University of Trento, Via Mesiano 77, Trento, Italy
ilya.peshkov@unitn.it

Abstract. A new three-phase model of compressible two-fluid flows in a deformed porous medium is presented. The derivation of the model is based on the application of the Symmetric Hyperbolic Thermodynamically Compatible (SHTC) systems theory to three-phase solid-fluid mixture. The resulting governing equations are hyperbolic and satisfy the laws of irreversible thermodynamics - conservation of energy and growth of entropy. Due to these properties, the formulated model is well suited for the straightforward application of advanced high accuracy numerical methods applicable to the solution of hyperbolic systems, and ensures the reliability of the numerically obtained solutions. On the basis of the formulated nonlinear model, the governing equations for the propagation of small-amplitude waves are obtained, allowing the use of an efficient finite-difference scheme on staggered grids for their numerical solution. Some numerical examples are presented showing the features of wave propagation in a porous medium saturated with a mixture of liquid and gas with their different ratios.

Keywords: Poroelasticity · Three-phase flow · Symmetric hyperbolic thermodynamically compatible model · Wave propagation

The work is supported by Mathematical Center in Akademgorodok, the agreement with Ministry of Science and High Education of the Russian Federation number 075-15-2019-1613. I.P. also acknowledge funding from the Italian Ministry of Education, University and Research (MIUR) under the Departments of Excellence Initiative 2018–2022 attributed to DICAM of the University of Trento (grant L. 232/2016), as well as financial support from the University of Trento under the *Strategic Initiative Modeling and Simulation.* I.P. has further received funding from the University of Trento via the *UniTN Starting Grant initiative.*

© Springer Nature Switzerland AG 2021
O. Gervasi et al. (Eds.): ICCSA 2021, LNCS 12949, pp. 224–236, 2021.
https://doi.org/10.1007/978-3-030-86653-2_16

1   Introduction

Studies on modeling of saturated porous media began in the middle of the last century in the works of Biot [1,2], and then Biot's model of the propagation of elastic waves in porous media and some of its modifications were used in the study of many scientific problems and seismic applications (see, for example [3,4] and references therein). Biot-type models are limited to considering small elastic deformations, while studying modern industrial problems requires new advanced models that take into account finite deformations and complex rheology of saturating fluids.

Multiphase flow approach can be successfully used to design new models of saturated porous media. In papers by Wilmanski [5,6] it is shown that Biot's theory can be interpreted within a two-phase liquid-solid mixture model with some degree of accuracy. In the present paper we formulate a new computational model for compressible two-phase flow in deformed porous medium and its application to simulations of small amplitude wavefields. The derivation of the full nonlinear model generalizes the model of porous medium saturated with single compressible fluid described in [7,8], which is in turn generalizes the so-called unified model of continuum [9,10] and is based on the theory of the Symmetric Hyperbolic Thermodynamically Compatible (SHTC) systems. Two mechanisms of phase interaction are taken into account: phase pressures relaxation to the common value and interfacial friction. The governing equations of the model are hyperbolic, and their solutions satisfy the laws of irreversible thermodynamics.

On the basis of the formulated nonlinear equations under assumption of instantaneous phase pressures relaxation, a model for small amplitude wave propagation in a stationary unstressed state is derived. It turns out that two compression waves are presented in the model, fast and slow, as in the theory of a porous medium saturated with a single fluid. The solution of a series of test problems in which the saturating fluid is a mixture of liquid and gas demonstrates the strong dependence of the waves behavior on the ratio of the volume fractions of fluids.

The presented model and its generalization for the case of arbitrary number of saturating fluids can be used to numerical modeling of compressible fluids flow in deformed porous media at different scales from slow filtration-type to high-rate deformations flows.

2   Three-Phase Thermodynamically Compatible Model for Compressible Two-Phase Flow in Deformed Porous Medium

2.1   Master System for Derivation of the Model

The derivation of the model is based on the synthesis of a unified continuum model [9] and a compressible multiphase flow model [11], as a result of which a three-phase model of a mixture of two fluid phases and a solid phase can be obtained. The presented in this paper model generalizes the two-phase SHTC solid-fluid model for fluid saturated porous medium presented in [7,8] and can be designed for an arbitrary number of fluid mixture constituents in a similar way. Let us consider a three-phase continuum with phase volume fractions $\alpha_1, \alpha_2, \alpha_3$ satisfying the saturation constraint $\alpha_1 + \alpha_2 + \alpha_3 = 1$. Further we assume that phases with number $1, 2$ are fluids and the solid phase is numbered as 3. Thus, the porosity is defined as $\phi = \alpha_1 + \alpha_2 = 1 - \alpha_3$. The generalization of the two-phase solid-fluid flow master system presented in [7] for the three-phase case reads as

$$\frac{\partial \rho v^i}{\partial t} + \frac{\partial (\rho v^i v^k + \rho^2 E_\rho \delta_{ik} + w_n^i E_{w_n^k} + \rho A_{ki} E_{A_{kj}})}{\partial x_k} = 0,$$

$$\frac{\partial \rho}{\partial t} + \frac{\partial \rho v^k}{\partial x_k} = 0,$$

$$\frac{\partial A_{ik}}{\partial t} + \frac{\partial A_{im} v^m}{\partial x_k} + v^j \left(\frac{\partial A_{ik}}{\partial x_j} - \frac{\partial A_{ij}}{\partial x_k} \right) = -\Psi_{ik},$$

$$\frac{\partial \rho c_a}{\partial t} + \frac{\partial (\rho c_a v^k + \rho E_{w_a^k})}{\partial x_k} = 0, \qquad a = 1, 2, \tag{1}$$

$$\frac{\partial w_a^k}{\partial t} + \frac{\partial (w_a^l v^l + E_{c_a})}{\partial x_k} + v^l \left(\frac{\partial w_a^k}{\partial x_l} - \frac{\partial w_a^l}{\partial x_k} \right) = -\Lambda_a^k, \quad a = 1, 2$$

$$\frac{\partial \rho \alpha_a}{\partial t} + \frac{\partial \rho \alpha_a v^k}{\partial x_k} = -\Phi_a, \quad a = 1, 2,$$

$$\frac{\partial \rho s}{\partial t} + \frac{\partial \rho s v^k}{\partial x_k} = Q.$$

Here, $\rho = \alpha_1 \rho_1 + \alpha_2 \rho_2 + \alpha_3 \rho_3$ is the total density of the mixture, ρ_1, ρ_2, ρ_3 are the mass densities of fluids and solid, $c_a = \alpha_a \rho_a / \rho$, $(a = 1, 2)$ are mass fractions of fluids ($c_3 = 1 - c_1 - c_2 = \alpha_3 \rho_3 / \rho$ is the solid phase mass fraction), $v^i = c_1 v_1^i + c_2 v_2^i + c_3 v_3^i$ is the mixture velocity, v_1^i, v_2^i are velocities of fluids and v_3^i is the solid phase velocities, $w_a^k = v_a^i - v_3^i$, $a = 1, 2$ are the relative velocity of motion of fluids phase with respect to solid phase, s is the entropy of the mixture, A_{ik} is the distortion matrix, characterizing the elastic deformation of the mixture. Summation over repeated tensor indices $i, j, k, \ldots = 1, 2, 3$ is implied. No summation over repeated phase indices $a, b, c = 1, 2, 3$ is implied unless it is state otherwise via the summation symbol.

The first and second equations of system (1) are the total momentum and total mass conservation laws for the mixture. The third equation controls the evolution of the mixture distortion matrix (in the case of elastic deformation of a pure solid, this is the inverse deformation gradient). The fourth equation is mass conservation law for fluids. The fifth one is the equation for the velocity of the movement of each fluid relative to the solid phase. The six equation governs the fluid volume fractions and the last one is the mixture entropy balance law. The source term in the latter equation provides the mixture entropy growing as the second law of thermodynamics requires due to the special choice of source terms in (1) for dissipative processes. The dissipative source terms are defined by the thermodynamic forces computed from the given generalized energy E and they are presented below.

To close equations (1) one needs to define a generalized energy E as a function of state variables $\rho, A_{ik}, c_1, c_2, \alpha_1, \alpha_2, w_1^k, w_2^k, s$. Then, thermodynamic forces are computed via derivatives E_ρ, $E_{A_{kj}}$, E_{w_k}, E_{c_1}, E_s as

$$\Phi_a = \rho \sum_{b=1}^{2} \phi_{ab} E_{\alpha_b}, \quad \Lambda_a^k = \sum_{b=1}^{2} \lambda_{ab} E_{w_b^k}, \quad \Psi_{ik} = \frac{1}{\theta} E_{A_{ik}}, \quad a = 1, 2. \quad (2)$$

Here, Φ_a is responsible for the phase pressures relaxation to a common value, Λ_a^k simulates the interfacial friction between phases and Ψ_{ik} is the rate of inelastic deformation of the entire mixture. The entropy production reads as

$$Q = \frac{\rho}{E_s} \sum_{a=1}^{2} \sum_{b=1}^{2} \phi_{ab} E_{\alpha_a} E_{\alpha_b} + \frac{\rho}{E_s} \sum_{a=1}^{2} \sum_{b=1}^{2} \lambda_{ab} E_{w_a^k} E_{w_b^k} + \frac{\rho}{\theta E_s} E_{A_{ik}} E_{A_{ik}} \geq 0, \quad (3)$$

that means that the second law of thermodynamics is satisfied if matrices λ_{ab} and ϕ_{ab} are positive definite.

Pressure p, shear stress tensor σ_{ij} and temperature T are also computed via generalized energy as:

$$p = \rho^2 E_\rho, \quad \sigma_{ij} = -\rho A_{ki} E_{A_{kj}}, \quad T = E_s. \quad (4)$$

The solutions of system (1) also satisfy the first law of thermodynamics – the energy conservation law, which reads as

$$\frac{\partial \rho(E + v^i v^i/2)}{\partial t} + \frac{\partial(\rho v^k(E + v^i v^i/2) + \Pi_k)}{\partial x_k} = 0, \quad (5)$$

where $\Pi_k = v^k p - v^i \sigma_{ik} + \sum_{a=1}^{2} \rho v^l w_a^l E_{w_a^k} + \sum_{a=1}^{2} \rho E_{c_a} E_{w_a^k}$ is the energy flux.

Assume that kinetic coefficients satisfy Onzager's principle, that means that matrices ϕ_{ab}, λ_{ab} are symmetric and can be a functions of state variables. Parameter θ characterizes the rate of inelastic deformation of the mixture and can also be a function of state variables.

The presented equations (1) by construction belong to the class of SHTC systems, that means that they satisfy thermodynamic laws (as we just noted above) and can be transformed into a symmetric form, which is hyperbolic in the sense of Friedrichs, if the generalized energy is a convex function.

For our goals we define the generalized energy E as a sum of the kinematic energy of the relative motion E_1, the energy of volumetric deformation E_2, and the energy of shear deformation E_3:

$$E = E_1(c_1, c_2, w_1, w_2) + E_2(\alpha_1, \alpha_2, c_1, c_2, \rho, s) + E_3(c_1, c_2, \rho, s, A). \quad (6)$$

The kinematic energy of relative motion is defined as

$$E_1 = \frac{1}{2}\sum_{a=1}^{2} c_a w_a^i w_a^i - \frac{1}{2}(\sum_{a=1}^{2} c_a w_a^i)^2. \tag{7}$$

Note that this choice is unique and caused by the fact that with the use of definition of the mixture and relative velocities one can obtain the following identity

$$\rho\left(\frac{v^i v^i}{2} + E_1\right) = \alpha_1 \rho_1 \frac{v_1^i v_1^i}{2} + \alpha_2 \rho_2 \frac{v_2^i v_2^i}{2} + \alpha_3 \rho_3 \frac{v_3^i v_3^i}{2}, \tag{8}$$

and this is exactly the total kinetic energy of the mixture. The energy of volumetric deformation is supposed to be additive:

$$\rho E_2(\alpha_1, \alpha_2, c_1, c_2, \rho, s) = \alpha_1 \rho_1 e_1(\rho_1, s) + \alpha_2 \rho_2 e_2(\rho_2, s) + \alpha_3 \rho_3 e_3(\rho_3, s), \tag{9}$$

with $e_a(\rho_a, s)$, $a = 1, 2, 3$ being the phase internal energies. The latter is equivalent to

$$E_2(\alpha_1, \alpha_2, c_1, c_2, \rho, s) = c_1 e_1\left(\frac{\rho c_1}{\alpha_1}, s\right) + c_2 e_2\left(\frac{\rho c_2}{\alpha_2}, s\right) + c_3 e_3\left(\frac{\rho c_3}{\alpha_3}, s\right). \tag{10}$$

The shear energy E_3 depends on the distortion of the entire mixture and is defined as

$$E_3 = \frac{1}{8} c_{s,M}^2 \left(tr(\mathbf{g}^2) - 3\right), \tag{11}$$

where $c_{s,M}$ is the shear sound velocity of the mixture which is determined below, and $\mathbf{g}$ is the normalized Finger (metric) strain tensor: $\mathbf{g} = \mathbf{G}/(\det \mathbf{G})^{1/3}$, $\mathbf{G} = \mathbf{A}^T \mathbf{A}$.

The shear velocity of the mixture should depend on the phase ratio and we define it by the simple mixture rule

$$c_{s,M}^2 = c_1 c_{s,1}^2 + c_2 c_{s,2}^2 + c_3 c_{s,3}^2, \tag{12}$$

where $c_{s,a}$, $a = 1, 2, 3$ are phase shear sound velocities connected with phase shear moduli μ_a by relation $c_{s,a}^2 = \mu_a/\rho_a$. The fluid shear sound velocities can be finite (non-zero), if we consider the viscous hyperbolic model for fluid as it is done in [7]. In this paper, the influence of the viscosity of saturating fluids on the flow is not taken into account, and we assume that the viscosity affects only the interfacial friction. That is why we take the velocities of shear sound waves of both fluids zero and the mixture shear sound velocity reads as

$$c_{s,M}^2 = c_3 c_{s,3}^2 = (1 - c_1 - c_2) c_{s,3}^2. \tag{13}$$

Using the definition of the generalized energy given above and relationships between mixture and phase parameters, we can compute all thermodynamic functions E_α, E_ρ, $E_{A_{kj}}$, $E_{w_a^k}$, E_{c_a}, $p = \rho^2 E_\rho$, $\sigma_{ij} = -\rho A_{ki} E_{A_{kj}}$, $T = E_s$:

$$E_{\alpha_a} = \frac{p_3 - p_a}{\rho} \ (a = 1, 2), \quad p = \rho^2 E_\rho = \alpha_1 p_1 + \alpha_2 p_2 + \alpha_3 p_3,$$

$$\frac{\partial E}{\partial A} = \frac{c_{s,M}^2}{2} \mathbf{A}^{-T} \left(\mathbf{g}^2 - \frac{tr(\mathbf{g}^2)}{3} \mathbf{I} \right), \quad \sigma_{ij} = -\frac{\rho c_{s,M}^2}{2} \left(g_{ik} g_{kj} - \frac{1}{3} g_{mn} g_{nm} \delta_{ij} \right),$$

$$E_{w_a^i} = c_a w_a^i - c_a (c_1 w_1^i + c_2 w_2^i) = -c_a (v^i - v_a^i), \tag{14}$$

$$E_{c_a} = e_a + \frac{p_a}{\rho_a} - e_3 - \frac{p_3}{\rho_3} - \frac{c_{s,3}^2}{c_{s,M}^2} E_3 + \frac{1}{2} w_a^i (w_a^i - c_1 w_1^i - c_2 w_2^i), \ (a = 1, 2),$$

$$E_s = T = c_1 \frac{\partial e_1}{\partial s} + c_2 \frac{\partial e_2}{\partial s} + c_3 \frac{\partial e_3}{\partial s}.$$

For the closure of the model it is also necessary to define coefficients in the source terms $\Phi_a, \Lambda_a^k, \Psi_{ik}$. The choice of these coefficients may depend on the types of saturating fluids and solid skeleton.

The presented above PDE system along with the closure relations can be used for numerical modeling of compressible two-phase mixture flow in the deformed porous medium in different regimes of flow, from the slow-filtration type flow to high-rate deformation of saturated porous medium. Int the next section we derive the simplified version of the presented model applicable to wavefields simulation.

3 Small Amplitude Wave Propagation in the Porous Medium Saturated by Two-Fluid Mixture

The derivation of the governing PDE system for the small amplitude wave propagation in an immovable medium can be done in a standard linearization procedure, as it is described in [7] for the simpler model of the elastic porous medium saturated with a single fluid. First of all, we assume that the temperature variations in the considered processes are small and, thus, the entropy can be excluded from the set of state variables. This means that the entropy balance law can also be neglected.

Consider an unstressed medium and denote the state variables of this medium by the symbol "0". Assume that the initial value of the fluid volume fractions α_1^0, α_2^0 and solid volume fraction α_3^0 ($\alpha_1^0 + \alpha_2^0 + \alpha_3^0 = 1$) are known. The immovability of the medium means that phase velocities are equal to zero $v_{10}^i = 0, v_{20}^i = 0, v_{30}^i = 0$. In the unstressed state, the pressure and shear stress of the mixture are also equal to zero, that means that $p_1^0 = p_2^0 = p_3^0 = p^0 = 0$, $\sigma_{ik}^0 = 0$. This unstressed state corresponds to reference phase density values $\rho_{10}, \rho_{20}, \rho_{30}$ and to the distortion $A_{ij}^0 = \delta_{ij}$.

We assume that the phase pressure relaxation is instantaneous due to small pore scale. This means that we should replace in the model the evolution equations for the phase volume fractions with the algebraic equations $p_1 = p_2 = p_2 = p$. Then we replace the small variations of distortion by the small deformation tensor and after linearization of stress-strain relations we arrive to Hook's law connecting shear stress and shear strain. Details of this transformation can be found in [7]. As a result, denoting small variations of the velocity of the mixture V^i, phase relative velocities W_1^i, W_2^i, pressure P and shear stress tensor Σ_{ik}, we arrive to the following system for small amplitude wave propagation in a porous medium saturated with two -fluid compressible mixture

$$\rho_0 \frac{\partial V^i}{\partial t} + \frac{\partial P}{\partial x_i} - \frac{\partial \Sigma_{ik}}{\partial x_k} = 0, \tag{15a}$$

$$\frac{\partial W_1^i}{\partial t} + \left(\frac{1}{\rho_1^0} - \frac{1}{\rho_3^0} \right) \frac{\partial P}{\partial x_i} = -\Lambda_1^i = -\lambda_{11} c_1^0 (V^i - V_1^i) - \lambda_{12} c_2^0 (V^i - V_2^i), \tag{15b}$$

$$\frac{\partial W_2^i}{\partial t} + \left(\frac{1}{\rho_2^0} - \frac{1}{\rho_3^0} \right) \frac{\partial P}{\partial x_i} = -\Lambda_2^i = -\lambda_{21} c_1^0 (V^i - V_1^i) - \lambda_{22} c_2^0 (V^i - V_2^i), \tag{15c}$$

$$\frac{\partial P}{\partial t} + K \frac{\partial V^k}{\partial x_k} + K_1' \frac{\partial W_1^k}{\partial x_k} + K_2' \frac{\partial W_2^k}{\partial x_k} = 0, \tag{15d}$$

$$\frac{\partial \Sigma_{ik}}{\partial t} - \mu \left(\left(\frac{\partial V^i}{\partial x_k} + \frac{\partial V^k}{\partial x_i} \right) - \frac{2}{3} \delta_{ik} \left(\frac{\partial V^1}{\partial x_1} + \frac{\partial V^2}{\partial x_2} + \frac{\partial V^3}{\partial x_3} \right) \right) = -\frac{\Sigma_{ik}}{\tau}, \tag{15e}$$

Here, $c_a^0 = \alpha_a^0 \rho_a^0 / \rho_0, (a = 1, 2)$, $\rho_0 = \alpha_1^0 \rho_1^0 + \alpha_2^0 \rho_2^0 + \alpha_3^0 \rho_3^0$, $K = \left(\frac{\alpha_1^0}{K_1} + \frac{\alpha_2^0}{K_2} + \frac{\alpha_3^0}{K_3} \right)^{-1}$ is the bulk modulus of the mixture and $K_a' = (\alpha_a^0 - c_a^0) K$, $a = 1, 2$. The friction source terms Λ_a^i can be rewritten in terms of W_a^i as

$$\Lambda_1^i = \lambda_{11}' W_1^i + \lambda_{12}' W_2^i, \qquad \Lambda_2^i = \lambda_{21}' W_1^i + \lambda_{22}' W_2^i, \tag{16a}$$

$$\begin{pmatrix} \lambda_{11}' & \lambda_{12}' \\ \lambda_{21}' & \lambda_{22}' \end{pmatrix} = \begin{pmatrix} \lambda_{11} & \lambda_{12} \\ \lambda_{21} & \lambda_{22} \end{pmatrix} \begin{pmatrix} -(1 - c_1^0) c_1^0 & c_1^0 c_2^0 \\ c_1^0 c_2^0 & -(1 - c_2^0) c_2^0 \end{pmatrix}. \tag{16b}$$

In the subsequent analysis of wave propagation and numerical examples we assume that both fluids are ideal $c_{s,1}^2 = c_{s,2}^2 = 0$ and solid skeleton is pure elastic. Therefore we take $\tau = \infty$ and $\mu = \rho c_{s,M}^2 = \rho c_3^0 c_{s,3}^2 = \alpha_3^0 \mu_3$, where μ, μ_3 are the shear moduli of the mixture and of the solid phase.

4   Dispersion Relations

In this section we study the dependence of one-dimensional longitudinal waves on their frequency. The dispersion relation $k(\omega)$ for the linear equations (15) can be computed exactly in the same way as in our previous work [7]. Here, k is the complex wavenumber and ω is the real angular frequency. Figure 1 shows phase velocities of the fast $V_{\text{fast}}(\omega)$ and slow $V_{\text{slow}}(\omega)$ compressional waves as a function of angular frequency $\omega = 2\pi f$ (f is the standard frequency) for the porosity $\phi = 0.3 = 1 - \alpha_3^0$ and several values of the volume fraction of gaseous phase: $\alpha_2^0 = 10^{-2}, 10^{-3}, 10^{-4}, 10^{-5}$, and 0. To plot this figure, we set $\lambda_{12}' = \lambda_{21}' = 0$ and took $\lambda_{11}' = 10^3 \sim \eta_{water}^{-1}$ and $\lambda_{22}' = 54 \cdot 10^3 \sim \eta_{air}^{-1}$, where η_{water} and η_{air} are the water and air viscosities. To plot all curves in a single plot, the phase velocities were normalized by the characteristic velocities (eigenvalues of the left hand-side of simplified one-dimensional version of (15)) which are given by

$$C^2_{\text{fast,slow}} = \frac{K + \frac{4}{3}\mu + Z \pm \sqrt{-\frac{16}{3}\mu Z + \left(K + \frac{4}{3}\mu + Z\right)^2}}{2\rho} \tag{17a}$$

$$Z = \rho\left(R_1 K'_1 + R_2 K'_2\right), \qquad R_a = \frac{1}{\rho^0_a} - \frac{1}{\rho^0_3}, \qquad a = 1, 2. \tag{17b}$$

and are equal to $C_{\text{fast}} = 4039$ m/s and $C_{\text{slow}} = 331$ m/s for $\alpha^0_2 = 10^{-2}$, $C_{\text{fast}} = 4037$ m/s and $C_{\text{slow}} = 342.3$ m/s for $\alpha^0_2 = 10^{-3}$, $C_{\text{fast}} = 4096$ m/s and $C_{\text{slow}} = 422.8$ m/s for $\alpha^0_2 = 10^{-4}$, $C_{\text{fast}} = 4301$ m/s and $C_{\text{slow}} = 607.1$ m/s for $\alpha^0_2 = 10^{-5}$, and $C_{\text{fast}} = 4441$ m/s and $C_{\text{slow}} = 692.9$ m/s for $\alpha^0_2 = 0$.

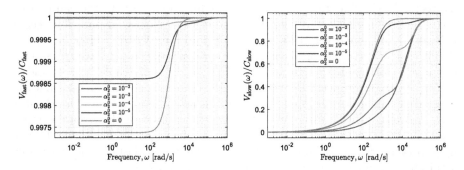

Fig. 1. Normalized phase velocities of the fast and slow compressional waves for various volume fractions of the gaseous phase.

5   Numerical Simulations

In this section we numerically illustrate the main features of small amplitude wavefields described by system (15). For simulations, we use finite difference schemes on staggered grids [12,13]. This approach is most suitable for solving first order symmetric hyperbolic systems in the velocity-stress formulation. By analogy to the Levander approach [8,13], we developed a fourth-order accurate scheme on a staggered grid in space and a second-order accurate scheme in time for modeling the wave propagation in a porous medium saturated with a two-fluid compressible mixture.

We consider a homogeneous porous medium saturated with a mixture of two fluids. Let us call the liquid phase Fluid1, the gas phase Fluid2 and the skeleton Solid with material parameters from Table 1.

The goal of solving the first series of test problems, in some of which a homogeneous medium is defined as one pure phase out of three, is to validate the model. Let us consider four cases: pure Solid ($\alpha^0_1 = \alpha^0_2 = 0, \alpha^0_3 = 1$), pure Fluid1 ($\alpha^0_1 = 1, \alpha^0_2 = \alpha^0_3 = 0$), pure Fluid2 ($\alpha^0_1 = \alpha^0_3 = 0, \alpha^0_2 = 1$), and poroelastic case with Solid and only one Fluid1 ($\alpha^0_1 = 0.3, \alpha^0_2 = 0, \alpha^0_3 = 0.7$).

The computational domain is discretized with $N_x \times N_y$ grid points, $N_x = N_y = 2001$ with spatial step $d_x = d_y = 2.5 \cdot 10^{-5}$ m, which amounts to approximately 13 points per a slow compressional wavelength for a chosen frequency $f_0 = 1$ MHz.

To excite the volumetric-type source we add a source term in the right-hand side of the pressure equation of the system (15). A source function is defined as the product of Dirac's delta function in space and Ricker's wavelet $f(t)$ in time:

$$f(t) = (1 - 2\pi^2 f_0^2 (t - t_0)^2) exp[-\pi^2 f_0^2 (t - t_0)^2], \tag{18}$$

where f_0 is the source central frequency and t_0 is the time wavelet delay. We assumed the frequency $f_0 = 1$ MHz and the source time delay $t_0 = 1/f_0$ s. The source is located in the centre of the computational domain.

Figure 2 shows the wavefield snapshots of the norm of the total mixture velocity vector $\| V \|^2$ at time $t = 4 \cdot 10^{-6}$ s for four cases. As one can expect, the velocities of arising waves coincide with P-wave velocity for pure Solid ($v_p = 6155$ m/s) and sound velocities c_{f1}, c_{f2} for pure Fluid1 ($c_{f1} = 1500$ m/s) and Fluid2 ($c_{f2} = 330 \, m/s$) from Table 1. Only one compressional P-wave arises in the case of pure solid, pure water or pure gas (Fig. 2a–2c). In the case of poroelasticity, we can observe the appearance of fast and slow (Biot's mode) compressional waves (Fig. 2d). Estimated velocities 4440 m/s for the fast wave and 690 m/s for the slow wave are well consistent with the velocity dispersion curves in Fig. 1, obtained with the technique described in [7]. When evaluating the velocities, one should keep in mind Ricker's wavelet delay of $1 \cdot 10^{-6}$ s for chosen poroelastic model with Solid and Fluid1 parameters from Table 1 and porosity $\alpha_1^0 = 0.3$.

Table 1. Physical parameters used for simulation.

State	Property	Parameters	Value	Unit
Fluid1 (water):	Fluid density	$\rho_{f1} = \rho_1^0$	1040	kg/m^3
	Sound velocity	c_{f1}	1500	m/s
	Bulk modulus	$K_{f1} = K_1 = \rho_{f1} c_{f1}^2$	2.34	GPa
Fluid2 (air):	Fluid density	$\rho_{f2} = \rho_2^0$	1.225	kg/m^3
	Sound velocity	c_{f2}	330	m/s
	Bulk modulus	$K_{f2} = K_2 = \rho_{f2} c_{f2}^2$	0.13	GPa
Solid:	Solid density	$\rho_s = \rho_3^0$	2500	kg/m^3
	P-wave velocity	v_p	6155	m/s
	Bulk velocity	c_s	4332	m/s
	Shear velocity	c_{sh}	3787	m/s
	Bulk modulus	$K_s = K_3 = \rho_s c_s^2$	46.91	GPa
	Shear modulus	$\mu_s = \mu_3 = \rho_s c_{sh}^2$	35.85	GPa

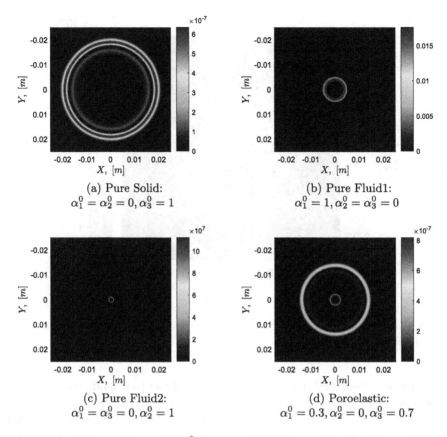

Fig. 2. Wavefield snapshots of $||V||^2$ for pure Solid (a), pure Fluid1 (b), pure Fluid2 (c) and poroelastic (d) media at time $t = 4 \cdot 10^{-6}$ s.

The second test investigates the effect of gas presence on the wave propagation velocity in a three-phase poroelastic medium. To this end, let us vary the value of volume fraction α_2^0 of the air (Fluid2), while the Solid volume fraction being constant $\alpha_3^0 = 0.7$ and porosity is $\alpha_1^0 + \alpha_2^0 = 0.3$. With this consideration, the sample's porosity remains constant, only the ratio of Fluid1 and Fluid2 changes.

Figure 3 shows the wavefield snapshots of the total mixture velocity vector $\| V \|^2$ for poroelastic media with porosity $\alpha_1^0 = 0.299, \alpha_2^0 = 0.001, \alpha_3^0 = 0.7$ at time $t = 4 \cdot 10^{-6}$ s. Comparing phase velocities in the case of two different ratios of liquid and gas volume fractions with $\alpha_1^0 = 0.3, \alpha_2^0 = 0, \alpha_3^0 = 0.7$ and $\alpha_1^0 = 0.299, \alpha_2^0 = 0.001, \alpha_3^0 = 0.7$, one can notice that the velocity of the fast P

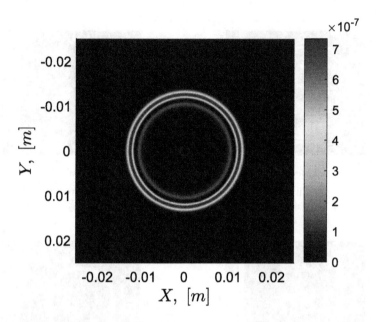

Fig. 3. Wavefield snapshot of $||V||^2$ for poroelastic medium with phase volume fractions $\alpha_1^0 = 0.299, \alpha_2^0 = 0.001, \alpha_3^0 = 0.7$ at time $t = 4 \cdot 10^{-6}$ s.

wave has decreased slightly, while the velocity of the Biot mode has decreased by almost half. That is more, the amplitude of the Biot mode has decreased significantly in comparison with Fig. 2d. In order to trace the dependence on the gas volume fraction α_2^0, we compared snapshot slices of V^1 component passing vertically through $x = 0$ for different parameters $\alpha_2^0 = 0, 0.001, 0.005, 0.01$ and plotted them together in Fig. 4. From a comparison of these cases, the following conclusions can be drawn. With an increase in the parameter α_2^0, the velocity of the fast and slow compressional P waves decreases. Simultaneously, the fast P wave amplitude increases, while the amplitude of Biot's mode decreases very fast with a small increase in α_2^0.

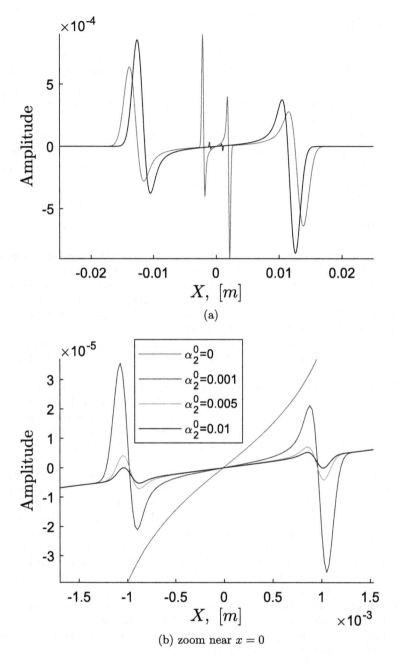

Fig. 4. Snapshot slices of V^1 component at time $t = 4 \cdot 10^{-6}$ s passing through $x = 0$ for different parameters $\alpha_2^0 = 0, 0.001, 0.005, 0.01$.

6 Conclusions

We have presented a new computational model of a deformed porous medium saturated with a compressible mixture of two fluids. The governing equations of the model are hyperbolic and are consistent with two thermodynamic laws: conservation of energy and growth of entropy. A simplified computational model for the propagation of small-amplitude waves is formulated and some test problems are solved numerically using the finite difference method on staggered grids. The applicability of the model is shown for studying the features of wavefields in an elastic porous medium saturated with liquid and gas, including the dependence of wave velocities and amplitudes on their ratio.

References

1. Biot, M.A.: Theory of propagation of elastic waves in fluid-saturated porous solid. I. Low-frequency range. J. Acoust. Soc. Am. **28**(2), 168–178 (1956)
2. Biot, M.A.: Theory of propagation of elastic waves in a fluid-saturated porous solid. II. Higher frequency range. J. Acoust. Soc. Am. **28**(2), 179–191 (1956)
3. Carcione, J.M., Morency, C., Santos, V.: Computational poroelasticity - a review. Geophysics **75**(5), 75A229–75A243 (2010)
4. Pesavento, F., Schrefler, B.A., Sciumè, G.: Multiphase flow in deforming porous media: a review. Arch. Comput. Methods Eng. **24**(2), 423–448 (2016). https://doi.org/10.1007/s11831-016-9171-6
5. Wilmanski, K.: A thermodynamic model of compressible porous materials with the balance equation of porosity. Transp. Porous Media **32**(1), 21–47 (1998)
6. Wilmanski, K.: A few remarks on Biot's model and linear acoustics of poroelastic saturated materials. Soil Dyn. Earthq. Eng. **26**(6–7), 509–536 (2006)
7. Romenski, E., Reshetova, G., Peshkov, I., Dumbser, M.: Modeling wavefields in saturated elastic porous media based on thermodynamically compatible system theory for two-phase solid-fluid mixtures. Comput. Fluids **206**, 104587 (2020)
8. Reshetova, G., Romenski, E.: Diffuse interface approach to modeling wavefields in a saturated porous medium. Appl. Math. Comput. **398**(C), 125978 (2021)
9. Dumbser, M., Peshkov, I., Romenski, E., Zanotti, O.: High order ADER schemes for a unified first order hyperbolic formulation of continuum mechanics: viscous heat-conducting fluids and elastic solids. J. Comput. Phys. **314**, 824–862 (2016)
10. Dumbser, M., Peshkov, I., Romenski, E., Zanotti, O.: High order ADER schemes for a unified first order hyperbolic formulation of Newtonian continuum mechanics coupled with electro-dynamics. J. Comput. Phys. **348**, 298–342 (2017)
11. Romenski, E., Belozerov, A., Peshkov, I.M.: Conservative formulation for compressible multiphase flows. Q. Appl. Math. **74**(1), 113–136 (2016)
12. Virieux, J.: P-SV wave propagation in heterogeneous media: velocity-stress finite-difference method. Geophysics **51**(1), 889–901 (1986)
13. Levander, A.R.: Fourth-order finite-difference P-W seismograms. Geophysics **53**(11), 1425–1436 (1988)

Modeling and Identification of Electromechanical Systems Using Orthonormal Jacobi Functions

Vadim Petrov[✉] [iD]

National Research Technological University "MISiS", Leninsky prospect 4, Moscow 119049, Russian Federation
petrovv@misis.ru

Abstract. Solving parameter identification and structure identification in electromechanical systems is relevant in applications for diagnostics of the technical condition of technical systems, as well as in the development of algorithms for controlling machines and mechanisms. Impulse response functions, correlation and autocorrelation coordinate functions are used as dynamic characteristics. The correspondent models unambiguously give an idea of the mathematical description of a linear or linearized electromechanical system. The methods of spectral description of the dynamic characteristics, based on Fourier transforms, are quite effective in relation to physically realizable systems under normal operating conditions. The paper aims to present new approaches to the development of algorithms for identifying electromechanical systems based on the study of the properties of synthesized transformed orthonormal Jacobi functions and the design of spectral models of the impulse response functions of electromechanical systems. The methods of the theory of spectral and operator transformations, as well as functional analysis, are used in the study. The results demonstrate the success of algorithms for nonparametric identification of linear or linearized electromechanical systems based on spectral models of their impulse response functions in the basis of synthesized generalized orthonormal Jacobi functions. The results of the study can be used in practice for the diagnostics and development of control systems for electric drives of machines and installations with changing parameters or structures.

Keywords: Electromechanical systems · Modeling · Identification · Orthonormal functions · Jacobi functions · Impulse response functions · Fourier transforms · Spectral models · Electric drive

1 Introduction

Solving parameter identification and structure identification in electromechanical systems (EMSs) is relevant in applications for diagnostics of the technical condition of technical systems, as well as the development of algorithms for controlling machines and mechanisms [1–5]. This is particularly important in conditions where the parameters and characteristics of individual elements of the system have variability. As an example of such EMSs, one can consider the electric drive systems of machines and

© Springer Nature Switzerland AG 2021
O. Gervasi et al. (Eds.): ICCSA 2021, LNCS 12949, pp. 237–251, 2021.
https://doi.org/10.1007/978-3-030-86653-2_17

equipment, the quality of management of which determines not only technological processes but also the success of the whole enterprise [1, 6]. Such machines can be found in the mining industry (excavators, conveyors, mine lifting machines, etc.), as well as in other industries (mechanical engineering, transport, etc.) [7–10]. The elements of the EMSs of these machines change their properties under the influence of external factors. Thus, during the operation of long-distance long-line conveyor systems, through which mineral raw materials are transported, the conveyor belt (the elastic element of the EMS that connects its different masses) under different seasonal climatic conditions has different elastic properties, which objectively affects the procedures for the synthesis and implementation of algorithms for controlling the electric drive systems of the entire conveyor.

In most cases, EMSs belong to the category of physically realizable systems, so their dynamic characteristics can be described by classical methods based on systems of differential equations or one of the forms of their representation – operator schemes or operator linear equations [11, 12]. At the same time, in the formulation of the problem of identifying the parameters or structure of a particular EMS in the conditions of its normal functioning, it becomes important to build equivalent models based on the description of its dynamic characteristics. Impulse response functions (IRFs), correlation and autocorrelation coordinate functions are most commonly used as dynamic characteristics that are mathematically unambiguously related to differential equations, Laplace operator equations of the system (response functions), frequency characteristics [1, 13–15]. Their models unambiguously give the idea of the mathematical description of a linear or linearized EMS. The methods of spectral description of the dynamic characteristics, based on Fourier transforms, are quite effective in physically realizable systems under normal operating conditions. Spectral methods using orthonormal or orthogonal functions as a functional basis are of particular interest [1, 15].

If the IRFs of a dynamical system are used as the dynamic characteristic, then its mathematical model can be expressed as [13, 15]:

$$h_\delta(\tau) = \sum_{j=0}^{\infty} \mu_j \Phi_j(\tau)$$

where $h_\delta(\tau)$ are the IRFs; $\Phi_j(\tau)$ is a system of functions, for example, orthogonal or orthonormal ones; μ_j are the coefficients of the Fourier expansion of $h_\delta(\tau)$ in the basis of $\Phi_j(\tau)$ functions. Here, $\Phi_j(\tau)$ has the properties of the orthonormal or orthogonal function on the interval of the argument τ $[0, \infty)$.

The model scheme is shown in Fig. 1.

The use of orthonormalized Chebyshev functions as such a functional basis makes it possible to ensure the uniqueness of the models, their connection with other operator models (for example, Laplace), stability in determining the model parameters, the implementation of computational procedures, etc. [13]. Chebyshev-Laguerre and Chebyshev-Legendre functions, as well as Chebyshev-Hermite ones, are quite reasonably used as orthonormal Chebyshev functions [13, 16–18]. This class of continuous functions is defined on the time interval $(0, \infty)$ that coincides with the interval of determining the IRF, and also allows designing models of stable dynamical systems. However, most of these functions have a limited set of parameters, suitable to implement the optimization

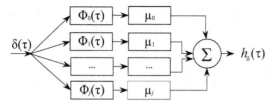

Fig. 1. Model scheme of impulse response functions, where $\delta(\tau)$ is the Dirac function

algorithms for synthesized models. This article is devoted to the study of models of EMSs in the basis of Chebyshev-Legendre functions as a special case of Jacobi functions. The problem of demonstrating new approaches to the design of EMS identification algorithms based on the study of the properties of synthesized transformed orthonormal Jacobi functions and on the construction of spectral models of EMS IRFs is considered in the paper. The methods of the theory of spectral and operator transformations, as well as functional analysis, are used in the paper.

2 Synthesis and Study of Orthonormal Jacobi Functions

Standardized Jacobi polynomials can be expressed as

$$
J_n(x; a, b) = \frac{\Gamma(b+n+1)\cdot\Gamma(a+n+1)}{2^n}
$$
$$
\times \sum_{k=0}^{n} \frac{(x-1)^{n-k}(1+x)^k}{k!\Gamma(n-k+1)\Gamma(a+n-k+1)\Gamma(b+k+1)}, \tag{1}
$$

where $x \in (-1, 1), a > -1, b > -1$; $\Gamma(\cdot)$ is Euler's gamma function [19].

The weight function of the polynomial is defined by the expression

$$
h(x) = (1 - x)^a(1 + x)^b. \tag{2}
$$

The Rodrigue formula for Eq. (1) is represented as

$$
J_n(x; a, b) = \frac{(-1)^n}{2^n n!}(1 - x)^{-a}(1 + x)^{-b}\frac{d^n}{dx^n}\left[(1 - x)^{a+n}(1 + x)^{b+n}\right]. \tag{3}
$$

Using the Rodrigue formula (3), the integration in parts is performed $n-1$ times to obtain the following integral expression of the square of the norm:

$$
\|J_n\|^2 = \frac{(-1)^2}{2^n\cdot\Gamma(n+1)}\int_{-1}^{1}\frac{d^n}{dx^n}\left[(1 + x)^a(1 - x)^b\right]J_n(x; a, b)
$$
$$
= \frac{\Gamma(a+n+1)\Gamma(b+n+1)}{\Gamma(n+1)\cdot(a+b+2n+1)\Gamma(a+b+n+1)}2^{a+b+1}. \tag{4}
$$

In accordance with Eq. (4), the expression for the orthonormal Jacobi polynomial can be written

$$
\hat{J}_n(x; a, b) = \sqrt{\Gamma(n + 1) \cdot (a + b + 2n + 1)\Gamma(a + b + n + 1)\Gamma(b + n + 1) \cdot \Gamma(a + n + 1)}
$$
$$
\times 2^{-\frac{1}{2}(a+b+1)-n}\sum_{k=0}^{n}\frac{(x-1)^{n-k}(1+x)^k}{\Gamma(k+1)\cdot\Gamma(n-k+1)\Gamma(a+n-k+1)\Gamma(b+k+1)}. \tag{5}
$$

In order to obtain functionals that are orthonormal on the interval $(0; \infty)$ and have satisfactory approximation properties, the modification of the standardized Jacobi polynomials was performed.

At the first stage, the substitution of $x = 1 - 2y$ in $J_n(x)$ was realized.

It allows forming the displaced polynomial that has the properties of orthogonality in the interval $(0; 1)$:

$$\dot{J}_n(y; a, b) = (-1)^n \Gamma(a + n + 1)\Gamma(b + n + 1)$$
$$\times \sum_{k=0}^{n} \frac{y^{n-k}(y-1)^k}{\Gamma(k+1)\Gamma(n-k+1)\Gamma(a+n-k+1)\Gamma(b+k+1)}. \tag{6}$$

The binomial formula $(1 - y)^k = \sum_{j=0}^{k} \frac{(-1)^j k! y^j}{j!(k-j)!}$ allows obtaining a different formula for biased orthogonal Jacobi polynomials:

$$\dot{J}_n(y; a, b) = (-1)^n \Gamma(a + n + 1)\Gamma(b + n + 1)$$
$$\times \sum_{k=0}^{n} \frac{1}{\Gamma(k+1)\Gamma(n-k+1)\Gamma(a+n-k+1)\Gamma(b+k+1)} \tag{7}$$
$$\times \sum_{j=0}^{k} \frac{(-1)^j \Gamma(k+1) \cdot y^{j+n-k}}{\Gamma(j+1) \cdot \Gamma(k-j+1)}.$$

The next transformations of the considered polynomials are possible by implementing the substitution of $y = e^{-u \cdot t}$ in Eq. (6) or (7), where u is the scale parameter.

After these transformations, the expression follows:

$$\bar{J}_n(a, b, u, t) = (-1)^n \Gamma(b + n + 1)\Gamma(a + n + 1)e^{-unt}$$
$$\times \sum_{k=0}^{n} \frac{[1-e^{ut}]^k}{\Gamma(k+1)\Gamma(n-k+1)\Gamma(a+n-k+1)\Gamma(b+k+1)}. \tag{8}$$

Equation (8) defines transformed orthogonal Jacobi functions that have orthogonality properties on the interval $(0; \infty)$.

The established functional dependencies allow synthesizing spectral models of dynamic systems, but these models will not be unique. This situation will require additional procedures to check the control algorithms created on the basis of these models and identify EMSs with different properties. For example, for systems with different oscillations. Therefore, it is rational to use orthonormal functions while constructing spectral models.

The norm of the resulting functional can also be determined from Eq. (4).

The necessary orthonormality condition for classical orthogonal complexes is defined as follows [19]:

$$\int_a^b J_n(x)J_m(x)h(x)dx = 0 \text{ for } n \neq m \text{ and } \int_a^b J_n(x)J_m(x)h(x)dx = 1 \text{ for } n = m \tag{9}$$

where a and b are the orthogonality interval, and $h(x)$ is the weight function of the polynomial.

The transformations of the weight function (2) were performed in the same sequence as for the transformation of the classical Jacobi polynomials:

$$H(a, b, t, u) = h(x)_{x=1-2e^{-u \cdot t}} \Big| = 2^{a+b} e^{-uta} \left(1 - e^{-ut}\right)^b. \tag{10}$$

The general formula defining the transformed generalized synthesized orthonormal Jacobi functions (SOJFs) was obtained considering the above conditions of the existence of the orthonormal functional, Eq. (8) for the transformed Jacobi functions, and Eq. (9) for its norm,

$$\hat{J}_n(u, t; a, b) = (-1)^{\frac{2n+b}{2}} \sqrt{u \, \Gamma(a + n + 1)\Gamma(n + 1)\Gamma(a + b + n + 1)(a + b + 2n + 1)}$$
$$\times \sqrt{\Gamma(a + n + 1)} \times e^{-u t \frac{a+b+2n+1}{2}} \cdot \sum_{k=0}^{n} \frac{\left[1 - e^{(ut)}\right]^{\frac{2k+b}{2}}}{\Gamma(k+1)\Gamma(n-k+1)\Gamma(a+n-k+1)\Gamma(b+k+1)}. \tag{11}$$

Using the binomial Newton formula allows obtaining a different expression for SOJF:

$$\hat{J}_n(u, t; a, b) = (-1)^{\frac{2n+b}{2}} \sqrt{u \, \Gamma(a + n + 1)\Gamma(n + 1)\Gamma(a + b + n + 1)(a + b + 2n + 1)}$$
$$\times \sqrt{\Gamma(a + n + 1)} e^{-u t \frac{a+b+2n+1}{2}} \cdot \sum_{k=0}^{n} \frac{1}{\Gamma(k+1)\Gamma(n-k+1)\Gamma(a+n-k+1)\Gamma(b+k+1)}$$
$$\times \sum_{j=0}^{\left(\frac{2k+b}{2}\right)} \frac{(-1)^j \Gamma\left(\frac{2k+b}{2}+1\right) e^{utj}}{\Gamma(j+1)\left[\Gamma\left(\frac{2k+b}{2}\right)-j+1\right]} \tag{12}$$

or

$$\hat{J}_n(u, t; a, b) = (-1)^{\frac{2n+b}{2}} \sqrt{u \, \Gamma(a + n + 1)\Gamma(n + 1)\Gamma(a + b + n + 1)(a + b + 2n + 1)}$$
$$\times \sqrt{\Gamma(a + n + 1)} \sum_{k=0}^{n} \frac{1}{\Gamma(k+1)\Gamma(n-k+1)\Gamma(a+n-k+1)\Gamma(b+k+1)}$$
$$\times \sum_{j=0}^{\left(\frac{2k+b}{2}\right)} \frac{(-1)^j \Gamma\left(\frac{2k+b}{2}+1\right) e^{ut\left(j-\frac{a+b+2n+1}{2}\right)}}{\Gamma(j+1)\,\Gamma\left[\left(\frac{2k+b}{2}\right)-j+1\right]} \tag{13}$$

The graphical interpretation of two orthonormal transformed Jacobi functions for $n = 1$ and $n = 5$ is shown in Fig. 2. The type of the plots indicates that the order of the transformed functions corresponds to the number of transitions of the corresponding function through zero.

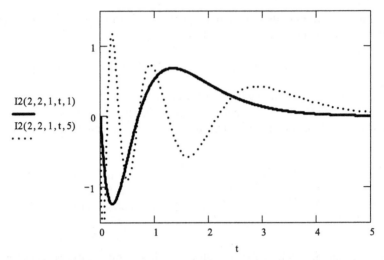

I2(2,2,1,t,1)
———
I2(2,2,1,t,5)
····

Fig. 2. Dependence of transformed Jacobi functions on the parameter t ($J(2, 2, 1, t, 1)$ at $a = 2$, $b = 2$, $u = 1$, and n = 1; $J(2, 2, 1, t, 5)$ at $a = 2$, $b = 2$, $u = 1$, and $n = 5$)

3   Some Properties of Orthonormal Jacobi Functions

Theorem 1. On the recurrent formula of transformed generalized orthonormal Jacobi functions.

For the transformed generalized orthonormal Jacobi functions, the recurrent formula is true

$$2 \cdot \sqrt{\frac{(a+n+1)(a+b+n+1)(n+1)(b+n+1)}{(a+b+2n+3)(a+b+2n+1)(a+b+2n+1)^2}} \hat{J}_{n+1}(u, t; a, b)$$
$$= \left(1 - 2 \cdot e^{-ut} - \frac{b^2-a^2}{(a+b+2n)(a+b+2n+2)}\right) \hat{J}_n(u, t; a, b) \qquad (14)$$
$$- 2 \cdot \sqrt{\frac{n(a+n)(a+b+n)(b+n)}{((a+b+2n-1)(a+b+2n+1)(a+b+2n)^2}} \hat{J}_{n-1}(u, t; a, b).$$

To prove this theorem, the recurrent formula for orthonormal polynomials was used, according to which the condition is fulfilled [19]:

$$\lambda_n \hat{P}_{n+1}(x) = (x - \eta_n)\hat{P}_n(x) - \lambda_{n-1}\hat{P}_{n-1}(x), \qquad (15)$$

where $\lambda_n = \frac{\mu_n}{\mu_{n+1}}$; μ_n is the highest coefficient of the orthonormal polynomial $\hat{P}_n(x)$; η_n is the coefficient determined by the type of the polynomial.

For orthonormal Jacobi polynomials (5), the expansion coefficients $\hat{P}_n(x) = \mu_n x^n + \nu_n x^{n-1} \dots$ can be determined using the following expressions:

$$\mu_n = \frac{\Gamma(a+b+2n+1)}{\sqrt{2^{a+b+2n+1}\Gamma(a+b+n+1)(a+b+2n+1)\Gamma(n+1)\Gamma(a+n+1)\Gamma(b+n+1)}},$$
$$\nu_n = \frac{(a-b)\Gamma(a+b+2n)\sqrt{n(a+b+2n+1)}}{\sqrt{2^{a+b+2n+1}\Gamma(a+b+n+1)\Gamma(n)\Gamma(a+n+1)\Gamma(b+n+1)}}.$$

In the next step, the parameter λ_n in Eq. (15) was determined

$$\lambda_n = \frac{\mu_n}{\mu_{n+1}} = 2\sqrt{\frac{(n+1)(a+n+1)(a+b+n+1)(b+n+1)}{(a+b+2n+2)^2(a+b+2n+3)(a+b+2n+1)}}. \tag{16}$$

Solving the equations composed under the condition that the coefficients are equal for the same powers of the argument, the expression for η_n can be determined as:

$$\lambda_n \nu_{n+1} = \nu_n - \eta_n \mu_n.$$

Solving this equation, the following expression can be obtained:

$$\eta_n = \frac{b^2 - a^2}{(a+b+2n+2)(a+b+2n)}. \tag{17}$$

The expression equivalent to Eq. (14) can be obtained from the three-term recurrent formula taking into account the substitutions carried out earlier when performing transformations on Jacobi polynomials. Thus, the theorem on the recurrent formula of transformed orthonormal Jacobi polynomials can be considered proven.

The recurrent formula (14) can be represented so that it will be possible to calculate the n-th function based on the values of the previous two functions:

$$
\begin{aligned}
2 \cdot &\sqrt{\frac{(a+n+1)(a+b+n+1)(n+1)(b+n+1)}{(a+b+2n+3)(a+b+2n+1)(a+b+2n+1)^2}} \cdot \hat{J}_n(u, t; a, b) \\
&= \left(1 - 2 \cdot e^{-ut} - \frac{b^2-a^2}{(a+b+2n)(a+b+2n+2)}\right) \times \hat{J}_{n-1}(u, t; a, b) \\
&- 2 \cdot \sqrt{\frac{n(a+n)(a+b+n)(b+n)}{(a+b+2n-1)(a+b+2n+1)(a+b+2n)}} \cdot \hat{J}_{n-2}(u, t, a, b).
\end{aligned}
$$

The identified recurrent formula is essential in determining the parameters of spectral models in the conditions of limited information, as well as in the formalization of mathematical operators in the structure of mathematical models.

4 Spectral Models Based on Orthonormal Jacobi Functions

IRF decomposition coefficients are determined using the method of the spectral decomposition of the IRF in the basis of SOJF, Eqs. (11–13), in accordance with the following expression

$$\gamma_i = \int_0^\infty h_\delta(\tau)\hat{J}_i(u, t; a, b)d\tau. \tag{18}$$

The spectral model is defined as

$$h_\delta(\tau) = \sum_{j=0}^\infty \gamma_j \hat{J}_j(u, t; a, b). \tag{19}$$

Equation (11) allows obtaining the general formula for the orthogonal spectral model of the IRF in the basis of orthonormal Jacobi functions:

$$
h_\delta(\tau) = \sum_{j=0}^{\infty} \left\{ \begin{array}{l} \gamma_j(-1)^{\frac{2j+b}{2}}\sqrt{u\,\Gamma(a+j+1)\Gamma(j+1)\Gamma(a+b+j+1)(a+b+2j+1)} \\[4pt] \times \sqrt{\Gamma(a+j+1)}e^{-ut}\frac{a+b+2j+1}{2} \\[4pt] \times \sum_{k=0}^{j} \frac{[1-e^{(ut)}]^{\frac{2k+b}{2}}}{\Gamma(k+1)\Gamma(j-k+1)\Gamma(a+j-k+1)\Gamma(b+k+1)} \end{array} \right\}.
$$

$$(20)$$

The application of the binomial Newton formula allows obtaining the different expression for the IRF model:

$$
h_\delta(\tau) = \sum_{j=0}^{\infty} \left\{ \begin{array}{l} \gamma_j(-1)^{\frac{2j+b}{2}}\sqrt{u\,\Gamma(a+j+1)\Gamma(j+1)\Gamma(a+b+j+1)(a+b+2j+1)} \\[4pt] \times \Gamma(a+j+1)\sum_{k=0}^{j} \frac{1}{\Gamma(k+1)\Gamma(j-k+1)\Gamma(a+j-k+1)\Gamma(b+k+1)} \\[4pt] \times \sum_{i=0}^{\left(\frac{2k+b}{2}\right)} \frac{(-1)^i\Gamma\left(k+\frac{1}{2}b+1\right)e^{-\frac{1}{2}ut(a+b+2n-2i+1)}}{\Gamma(i+1)\Gamma\left(k+\frac{1}{2}b-i+1\right)} \end{array} \right\}.
$$

$$(21)$$

It can be shown that n-th Jacobi function in the space of Laplace transformations can be defined by the following formula:

$$
\begin{aligned} L\left[\hat{J}_n(u,a,b;p)\right] = {}& (-1)^{\frac{2n+b}{2}}\sqrt{u\,\Gamma(a+n+1)\Gamma(n+1)\Gamma(a+b+n+1)} \\ & \times \sqrt{(a+b+2n+1)\Gamma(b+n+1)} \\ & \times \sum_{k=0}^{n}\left\{ \frac{-\Gamma\left(-\frac{p}{u}-\frac{a+b+1}{2}-n\right)\Gamma\left(k+\frac{b}{2}+1\right)}{\Gamma(k+1)\Gamma(n-k+1)\Gamma(a+n-k+1)\Gamma(b+k+1)\Gamma\left(-\frac{p}{u}+k-\frac{a-1}{2}-n\right)} \right\}, \end{aligned}
$$

$$(22)$$

where L $(\cdot)$ is the notation of Laplace transformations.

Some simplifications in order to develop final algorithms were introduced. Let $a \neq 0$, $b = 0$, then on the basis of Eqs. (22), the following operator expression for SOJF was obtained

$$
\begin{aligned} L\left[\hat{J}_n(u,a,b=0;p)\right] = {}& (-1)^{n+1}\sqrt{u\,(a+2n+1)} \\ & \times \frac{\Gamma\left(-\frac{p}{u}-\frac{a-1}{2}-n\right)\Gamma\left(-\frac{p}{u}+\frac{a+1}{2}-n\right)}{u\left(-\frac{p}{u}-\frac{a-1}{2}-n\right)\Gamma\left(-\frac{p}{u}-\frac{a-1}{2}\right)\Gamma\left(-\frac{p}{u}+\frac{a+1}{2}\right)}. \end{aligned}
$$

$$(23)$$

The ratio of operator transforms of the orthonormal functions with the ordinal numbers $j + 1$ and j for SOFJ at $a \neq 0$, $b = 0$ was determined:

$$
\frac{L\left\{\hat{J}_{n+1}(u,a,b=0;p)\right\}}{L\left\{\hat{J}_n(u,a,b=0;p)\right\}} = \sqrt{\frac{a+2n+3}{a+2n+1}}\left(\frac{u^{\frac{a+2n+1}{2}}-p}{u^{\frac{a+2n+3}{2}}+p}\right).
$$

$$(24)$$

The parameters of the chain former of Jacobi functions at the value of the parameter $b = 0$ at each stage were set as:

$$W_{J0}(p) = \frac{L\left\{\hat{J}_1(u, t, a, b = 0; p)\right\}}{L\left\{\hat{J}_0(u, t, a, b = 0; p)\right\}} = \sqrt{\frac{a+3}{a+1}}\left[\frac{\frac{u}{2}(a+1) - p}{\frac{3u}{2}(a+3) + p}\right];$$

$$W_{P1}(p) = \frac{L\left\{\hat{J}_2(u, t, a, b = 0; p)\right\}}{L\left\{\hat{J}_1(u, t, a, b = 0; p)\right\}} = \sqrt{\frac{a+5}{a+3}}\left[\frac{\frac{u}{2}(a+3) - p}{\frac{u}{2}(a+5) + p}\right];$$

$$W_{P2}(p) = \frac{L\left\{\hat{J}_3(u, t, a, b = 0; p)\right\}}{L\left\{\hat{J}_2(u, t, a, b = 0; p)\right\}} = \sqrt{\frac{a+7}{a+5}}\left[\frac{\frac{u}{2}(a+5) - p}{\frac{u}{2}(a+7) + p}\right];$$

$$W_{P3}(p) = \frac{L\left\{\hat{P}_4(u, t)\right\}}{L\left\{\hat{P}_3(u, t)\right\}} = \sqrt{\frac{a+9}{a+7}}\left[\frac{\frac{u}{2}(a+7) - p}{\frac{u}{2}(a+9) + p}\right] \quad \cdots$$

Figure 3 shows the block diagram of the former of orthonormal Jacobi functions for the corresponding parameter values.

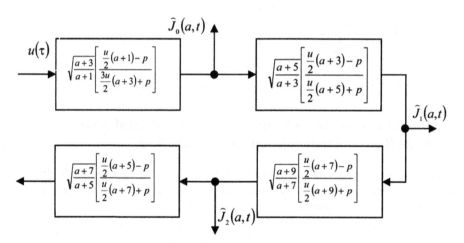

Fig. 3. The block diagram of the former of orthonormal Jacobi functions

If $u(\tau) = \delta(\tau)$, where $\delta(\tau)$ is the Dirac function, then the corresponding Jacobi functions can be observed at the output of each filter cascade.

The formula for the spectral orthogonal model for the IRF, Eq. (20), also allows synthesizing the block diagram of the orthogonal EMS model in the SOJF basis (Fig. 4).

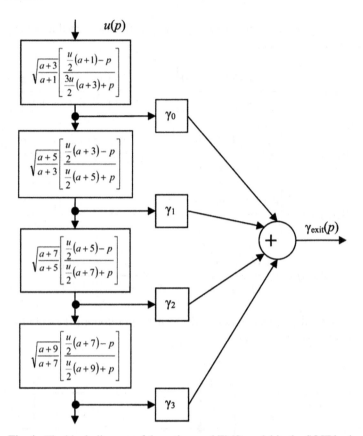

Fig. 4. The block diagram of the orthogonal EMS model in the SOJF basis

5 Nonparametric Identification Based on Spectral Models in the Basis of Orthonormal Jacobi Functions

The relationship between the components of the IRF spectral model in the SOJF basis and the parameters of the transfer function was determined and can be represented as the ratio of two polynomials:

$$W(p) = \frac{A(p)}{G(p)} = \frac{a_0 + a_1 p + a_2 p^2 + \ldots a_n p^n}{g_0 + g_1 p + g_2 p^2 + \ldots g_n p^n}. \tag{25}$$

The identification problem is reduced in this case to determining the coefficients of the polynomials $A(p) = (a_0, a_1, a_2 \ldots a_n)$ and $B(p) = (g_0, g_1, g_2 \ldots g_n)$.

To demonstrate the solution, the nonparametric model based on the first three SOJFs with parameter values $a \neq 0$; $b = 0$ was considered. The general operator expression for the transfer function will have the form:

$$\begin{aligned}
W_1(p) = {} & \gamma_0 2 \frac{\sqrt{u(a+1)}}{2p+u(a+1)} + \gamma_1 2 \frac{[u(a+1)-2p]\sqrt{u(a+3)}}{[2p+u(a+1)][2p+u(a+3)]} \\
& + \gamma_2 2 \frac{\sqrt{u(a+5)}[u(a+1)-2p][u(a+3)-2p]}{[2p+u(a+1)][2p+u(a+3)][2p+u(a+5)]}.
\end{aligned} \tag{26}$$

Equating the coefficients for the same degree of p, the expressions for the coefficients of the polynomials $A(p)$ and $B(p)$ in Eq. (25) were obtained:

$$a_0 = \gamma_0 u^2 \sqrt{u(a+1)}(a+5)(a+3) + \gamma_2 u^2 \sqrt{u(a+5)}(a+3)(a+1)$$
$$+ \gamma_1 \left[u^2 \sqrt{u(a+3)}a(a+6) + \tfrac{5}{2}u^3(a+3) \right],$$
$$a_1 = u^{\frac{3}{2}} \left\{ 4\left[\gamma_0\sqrt{a+1} - \gamma_2\sqrt{a+5}\right] - 8\left[\gamma_2\sqrt{a+5} - 2\gamma_0\sqrt{a+1} + \gamma_1\sqrt{a+3}\right]\right\},$$
$$a_2 = 4\sqrt{u}\left(\sqrt{a+1}\gamma_0 - \sqrt{a+3}\gamma_1 + \sqrt{a+5}\gamma_2\right)$$
$$g_0 = \tfrac{1}{2}u^3(a+1)(a+3)(a+5), \quad g_1 = u^2\left(3a^2 + 18a + 23\right), g_2 = 6u(a+3), \quad g_3 = 4.$$

Similarly, using the above method, the expressions for determining the relationship between the parameters of the approximating transfer functions at different orders of SOJF can be obtained.

6 Synthesis and Study of Spectral Models of Two-Mass EMSs in the Basis of Orthonormal Jacobi Functions

The main aspects of the design of spectral models of two-mass EMSs based on the spectral decomposition of the IRF in the SOJF basis were studied. The case of the open two-mass EMS was considered, the differential equation of which has the form [12]:

$$\begin{cases} J_1 \cdot \frac{d\omega_1}{dt} + c_1 \cdot (\varphi_1 - \varphi_2) = M_d; \\ J_2 \cdot \frac{d\omega_2}{dt} - c_1 \cdot (\varphi_1 - \varphi_2) = -M_c; \\ (u - \omega_1 \cdot k_c) \cdot k_a = M_d + T_e \cdot \frac{dM_d}{dt}, \end{cases} \tag{27}$$

where J_1 and J_2 are the moments of inertia of the first and second EMS masses, respectively; c_1 is the rigidity coefficient of the connection between the first and second EMS masses; ω_1, φ_1, ω_2, and φ_2 are the coordinates (angular velocity and angle of rotation) of the first and second EMS masses, respectively; k_c, k_a and T_e are the design constants of the electric motor device; M_d is the moment of the electric motor device; M_c is the moment of resistance on the executive body; u is the controlling coordinate.

The roots of the characteristic equation have the form:

$$p_{1,2} = -10.36 \pm 31.308 \cdot j,$$

$$p_{1,2} = -7.7 \pm 22.89 \cdot j, \text{ where } j = \sqrt{-1}$$

The coefficients of IRF expansion of the two-mass EMS in the eighth-order SOJF basis, determined with the help of Eq. (18), are presented in Fig. 5 for the fixed value of the scale parameter $u = 7$.

The data presented in Fig. 5 indicate the satisfactory convergence of the spectral model, as well as the effectiveness of the algorithmic support for calculating the model parameters.

To assess the reliability of synthesis of the IRF model, the square of the normalized variance of the IRF spectral model is used, which is determined by the expression [15]:

$$\sigma_n^2(u) = \frac{\int\limits_0^\infty \left[\hat{h}_\delta(\tau) - \sum\limits_{i=0}^n \vartheta_i \hat{H}e_i(u, \tau)\right]^2 d\tau}{\int\limits_0^\infty \hat{h}_\delta^2(\tau)d\tau}. \tag{28}$$

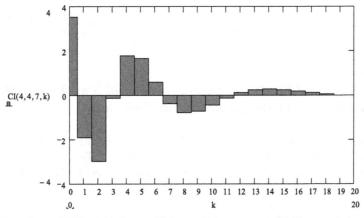

Fig. 5. The values of IRF expansion coefficients of the two-mass EMS control channel (IRF EMS spectral model), established using SOJF ($a = 4, b = 4$) of 8th order at the value of the scale parameter $u = 7$

The best approximation of the spectral model Eq. (19) is achieved by choosing the optimal values of the parameter u. As a criterion, the minimum condition of Eq. (28) is used:

$$\min\left\{\sigma_n^2(u)\right\}.$$

The dependencies of the square of the normalized variance of the spectral IRF model over the control channel of the two-mass EMS on the values of the scale parameters of the transformed orthonormal Jacobi functions for different orders of the spectral model are shown in Fig. 6.

The dependencies of the square of the normalized variance on the scale parameter u indicate the stability of the method of the design of IRF spectral models of EMSs in the basis of orthonormal Jacobi functions presented in this paper.

The proposed methodological approach to the spectral models of electromechanical systems based on orthonormal Jacobi functions has a certain practical significance. The following applications may be noted:

- the possibility of simplifying the linear complex and multi-coupled models of electromechanical multi-mass systems, for example, described by Eq. (27) to the level of the spectral sequential model shown in Fig. 4;

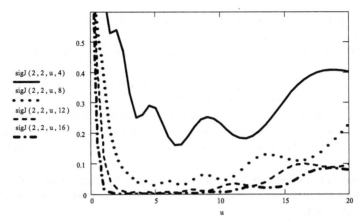

Fig. 6. The dependence of the square of the normalized dispersion of the IRF spectral model of the two-mass EMS control channel on the value of the scale parameter SOJF: ($a = 2$, $b = 0$) sigJ $(0.0, u, 4) - n = 4$; sigJ $(0.0, u, 8) - n = 8$; sigJ $(0.0, u, 12) - n = 12$; sigJ $(0.0, u, 16) - n = 16$, where a, b are the parameters of the spectral model; u is the scale parameter of the optimization procedure; n is the order of the spectral model)

- diagnostics of electric drive systems of machinery (for example, excavators, belt conveyors, etc.) based on the algorithms of parametric identification presented in the paper;
- new opportunities in the interpretation of experimental studies of the dynamic characteristics of electromechanical systems.

The above-mentioned trends will be the subject of further study by the author.

7 Conclusion

The studies presented in this paper demonstrate the success of using the synthesized transformed orthonormal Jacobi functions to solve the problems of constructing IRF spectral models of EMSs.

The parameters of IRF spectral models of EMSs with a high degree of confidence can be used to synthesize algorithms for nonparametric identification of EMSs.

The applicability of the presented Jacobi functions in nonparametric identification algorithms using examples of a two-mass EMS was examined in the paper. This creates the prerequisites for solving the class of problems, for example, the control problems in electric drive systems and the systems of automation of technological processes.

8 Key Focus of Further Study

Having determined the class of EMS mathematical models based on IRF spectral representation in the SOJF basis, it is considered appropriate to solve a number of problems in further study:

- to define parametric identification algorithms based on the data on the IRF EMS spectral model in the SOJF basis;
- to identify the impact of SOJF parameters on the quality of the EMS parametric identification procedure;
- to evaluate the influence of the parameters a and b of SOJF on the optimization procedure of the IRF EMS spectral model;
- to determine the influence of reliability indicators of parametric and nonparametric identification procedures based on IRF EMS spectral models on the quality of control algorithms applied to closed control systems for electric drives of machines and installations;
- to develop algorithmic and methodological support for diagnostics systems of the parameters and structure of electric drive systems of machines and equipment in normal operation.

References

1. Saushev, A., Antonenko, S., Lakhmenev, A., Monahov, A.: Parametric identification of electric drives based on performance limits. In: Murgul, V., Pasetti, M. (eds.) International Scientific Conference Energy Management of Municipal Facilities and Sustainable Energy Technologies EMMFT 2018. EMMFT-2018 2018. Advances in Intelligent Systems and Computing, vol. 982. Springer, Cham. (2020). https://doi.org/10.1007/978-3-030-19756-8_43
2. Brunot, M., Janot, A., Carrillo, F., Garnier, H., Vandanjon, P.O., Gautier, M.: Physical parameter identification of a one-degree-of-freedom electromechanical system operating in closed loop. IFAC-PapersOnLine **48**(28), 823–828 (2015). https://doi.org/10.1016/j.ifacol.2015.12.231
3. Janot, A., Young, P.C., Gautier, M.: Identification and control of electro-mechanical systems using state-dependent parameter estimation. Int. J. Control **90**(4), 643–660 (2017). https://doi.org/10.1080/00207179.2016.1209565
4. Golykov, A.D., Gryzlov, A.A., Bukhanov, S.S.: Parametric identification of mechatronic system with induction and synchronous electric drive. In: Proceedings of the 2017 International Conference "Quality Management, Transport and Information Security, Information Technologies", IT and QM and IS 2017, pp. 319–322 (2017). https://doi.org/10.1109/ITMQIS.2017.8085823
5. Miskin, A.R., Himakuntla, U.M.R., Achary, K.K., Parmar Azan, M., Mungara, H.K., Rao, R.: Simulation diagnostics approach for source identification and quantification in NVH development of electric motors. In: Proceedings of 2020 International Congress on Noise Control Engineering, INTER-NOISE 2020 (2020)
6. Garnier, H., Gilson, M., Young, P.C., Huselstein, E.: An optimal IV technique for identifying continuous-time transfer function model of multiple input systems. Control. Eng. Pract. **15**(4), 471–486 (2007). https://doi.org/10.1016/j.conengprac.2006.09.004
7. Savel'ev, A.N., Kipervasser, M.V., Anikanov, D.S.: Conveyer-belt accidents in mining and metallurgy. Steel Transl. **45**(12), 927–931 (2015). https://doi.org/10.3103/S0967091215120116
8. Sadridinov, A.B.: Analysis of energy performance of heading sets of equipment at a coal mine. Gornye nauki i tekhnologii = Min. Sci. Technol. (Russia) **5**(4), 367–375 (2020). https://doi.org/10.17073/2500-0632-2020-4-367-375

9. Pevzner, L., Dmitrieva, V.: System of automatic load stabilization of mining belt-conveyors. In: Proceedings of the 14th International Symposium on Mine Planning and Equipment Selection, MPES 2005 and the 5th International Conference on Computer Applications in the Minerals Industries, CAMI 2005, pp. 1050–1058 (2005)
10. Melezhik, R.S., Vlasenko, D.A.: Load simulation and substantiation of design values of a pin flexible coupling with a flexible disk-type element. Gornye nauki i tekhnologii = Mining Science and Technology (Russia) 6(2), 133–144 (2021). URL: https://mst.misis.ru/jour/issue/archive
11. Kefal, A., Maruccio, C., Quaranta, G., Oterkus, E.: Modelling and parameter identification of electromechanical systems for energy harvesting and sensing. Mech. Syst. Sig. Process. 121, 890–912 (2019). https://doi.org/10.1016/j.ymssp.2018.10.042
12. Kluchev, V.I.: Theory of Electric Drives. [Teoriya elektroprivoda]. Energoatomizdat, Moscow (1985)
13. Ljung, L.: System Identification: Theory for the User, 2nd edn. Prentice-Hall, Englewood Cliffs (1999)
14. Chen, T., Ljung, L.: Regularized system identification using orthonormal basis functions. In: European Control Conference, ECC 2015, pp. 1291–1296 (2015). https://doi.org/10.1109/ECC.2015.7330716
15. Bessonov, A.A., Zagashvili, Y., Markelov, A.S.: Methods and Means of Identification of Dynamic Objects. Energoatomizdat, Leningrad (1989)
16. Tiels, K., Schoukens, J.: Wiener system identification with generalized orthonormal basis functions. Automatica 50(12), 3147–3154 (2014). https://doi.org/10.1016/j.automatica.2014.10.010
17. Bouzrara, K., Garna, T., Ragot, J., Messaoud, H.: Online identification of the ARX model expansion on Laguerre orthonormal bases with filters on model input and out-put. Int. J. Control 86(3), 369–385 (2013). https://doi.org/10.1080/00207179.2012.732710
18. Karsky, V.: An Improved method for parametrising generalized Laguerre functions to compute the inverse Laplace transform of fractional order transfer functions. In: AIP Conference Proceedings, vol. 2293 (2020). https://doi.org/10.1063/5.0026713
19. Suetin, P.K.: Classical Orthogonal Polynomials. Nauka, Moscow (1979)

Proof of Some Properties of the Cross Product of Three Vectors in $\mathbb{R}^4$ with Mathematica

Judith Keren Jiménez-Vilcherrez[1][iD], Josel Antonio Mechato-Durand[2][iD],
Ricardo Velezmoro-León[3][iD], and Robert Ipanaqué-Chero[3]([✉])[iD]

[1] Universidad Tecnológica del Perú, Av. Vice Cdra 1 – Costado Real Plaza,
Piura, Peru
`C19863@utp.edu.pe`
[2] Universidad Privada Antenor Orrego, Av. Los Tallanes Zona Los Ejidos s/n,
Piura, Peru
`jmechatod1@upao.edu.pe`
[3] Universidad Nacional de Piura, Urb. Miraflores s/n Castilla, Piura, Peru
`{rvelezmorol,ripanaquec}@unp.edu.pe`

Abstract. In this paper, the definition of the cross product of three tangent vectors to $\mathbb{R}^4$ at the same point is stated, based on this definition a lemma is stated in which eight properties of said product are proposed and demonstrated. In addition, four theorems and two corollaries are stated and proved. One of the theorems constitutes an extension of the Jacobi identity. In some of the proofs, programs based on the paradigms: functional, rule-based and list-based from the Wolfram language, incorporated in Mathematica, are used.

Keywords: Cross product in $\mathbb{R}^4$ · Jacobi identity in $\mathbb{R}^4$ · Hyperparalelepiped · Hypervolume · Demonstrations using mathematica

1 Introduction

The cross product of three tangent vectors to $\mathbb{R}^4$ in the same point is formally defined in [6]. From that year to the present it has been used in different research works [1–3].

In this paper, the definition of the cross product of three tangent vectors to $\mathbb{R}^4$ given by Williams and Stein [6] is stated, based on this definition a lemma is stated in which eight properties of said product are proposed and demonstrated. Given five vectors tangent to $\mathbb{R}^4$ at the same point, an extension of the Jacobi identity is established (which is valid for three vectors tangent to $\mathbb{R}^3$ at the same point). Given two vectors tangent to $\mathbb{R}^4$ at the same point, we establish a formula to calculate the area of the parallelogram whose sides are these vectors. Given three vectors tangent to $\mathbb{R}^4$ at the same point, we proof that the volume of the parallelepiped whose edges are these three vectors is equal to the length

© Springer Nature Switzerland AG 2021
O. Gervasi et al. (Eds.): ICCSA 2021, LNCS 12949, pp. 252–260, 2021.
https://doi.org/10.1007/978-3-030-86653-2_18

of the cross product of them. A relation is established between the length of the cross product of three vectors tangent to $\mathbb{R}^4$ in the same point and the areas of the faces of the parallelepiped whose edges are these vectors. Finally, we proof that the absolute value of the quadruple scalar product of four vectors tangent to $\mathbb{R}^4$ at the same point is equal to the hyper-volume of the hyper-parallelepiped whose edges are these four vectors.

The Wolfram language [7], built into Mathematica [8], stands out from traditional computer languages by simultaneously supporting many programming paradigms, such as procedural, functional, rule-based, list-based, pattern-based, and more [4,5]. In this paper, certain demonstrations are carried out with the help of programs implemented based on these first three paradigms; in such a way that these programs are shorter and more efficient.

The structure of this paper is as follows: Sect. 2 introduce the mathematical definition of the Euclidean 4-space, a tangent vector, a vector field, the dot product of two points, the norm of a point, the dot product of two tangent vectors and a frame at a point. Then, Sect. 3 introduces the cross product of three tangent vectors. In this section properties, lemmas, theorems, and corollaries associated with this product are stated and demonstrated. Finally, Sect. 4 closes with the main conclusions of this paper.

2   Mathematical Preliminaries

Definition 1. Euclidean 4-space $\mathbb{R}^4$ *is the set of all ordered quadruples of real numbers. Such a quadruple* $\mathbf{p} = (p_1, p_2, p_3, p_4)$ *is called a* point *of* $\mathbb{R}^4$.

Definition 2. *A* tangent vector $\mathbf{v}_p$ *to* $\mathbb{R}^4$ *consist of two points of* $\mathbb{R}^4$: *its* vector part $\mathbf{v}$ *and its* point of application $\mathbf{p}$.

Definition 3. *Let* $\mathbf{p}$ *be a point of* $\mathbb{R}^4$. *The set* $T_p\left(\mathbb{R}^4\right)$ *consisting of all tangent vectors that have* $\mathbf{p}$ *as point of application is called the* tangent space *of* $\mathbb{R}^4$ *at* $\mathbf{p}$.

Definition 4. *A* vector field V *on* $\mathbb{R}^4$ *is a function that assigns to each point* $\mathbf{p}$ *of* $\mathbb{R}^4$ *a tangent vector* $V(\mathbf{p})$ *to* $\mathbb{R}^4$ *at* $\mathbf{p}$.

Definition 5. *Let* U_1, U_2, U_3 *and* U_4 *be the vector fields on* $\mathbb{R}^4$ *such that*

$$U_1(\mathbf{p}) = (1,0,0,0)_p \qquad U_2(\mathbf{p}) = (0,1,0,0)_p$$
$$U_3(\mathbf{p}) = (0,0,1,0)_p \qquad U_4(\mathbf{p}) = (0,0,0,1)_p$$

for each point $\mathbf{p}$ *of* $\mathbb{R}^4$. *We call* U_1, U_2, U_3, U_4 —*collectively*—*the* natural frame field *on* $\mathbb{R}^4$.

Definition 6. *The dot product of points* $\mathbf{p} = (p_1, p_2, p_3, p_4)$ *and* $\mathbf{q} = (q_1, q_2, q_3, q_4)$ *in* $\mathbb{R}^4$ *is the number*

$$\mathbf{p} \bullet \mathbf{q} = p_1 q_1 + p_2 q_2 + p_3 q_3 + p_4 q_4 \,.$$

Definition 7. *The norm of a point* $\mathbf{p} = (p_1, p_2, p_3, p_4)$ *is the number*

$$\|\mathbf{p}\| = (\mathbf{p} \bullet \mathbf{p})^{1/2} .$$

The norm is thus a real-valued function on $\mathbb{R}^4$; it has fundamental properties $\|\mathbf{p} + \mathbf{q}\| \leq \|\mathbf{p}\| + \|\mathbf{q}\|$ and $\|a\mathbf{p}\| = |a| \|\mathbf{p}\|$, where $|a|$ is the absolute value of the number a.

Definition 8. *The* dot product *of* $\mathbf{v}, \mathbf{w} \in T_p(\mathbb{R}^4)$ *is the number*

$$\mathbf{v}_p \bullet \mathbf{w}_p = \mathbf{v} \bullet \mathbf{w} .$$

A fundamental result of linear algebra is the Schwartz inequality $|\mathbf{v} \bullet \mathbf{w}| \leqq \|\mathbf{v}\| \|\mathbf{w}\|$. This permits us to define the cosine of the angle ϑ between $\mathbf{v}$ and $\mathbf{w}$ by the equation

$$\mathbf{v} \bullet \mathbf{w} = \|\mathbf{v}\| \|\mathbf{w}\| \cos \vartheta ,$$

where $0 \leq \vartheta \leq \pi$.

In particular, if $\vartheta = \pi/2$, then $\mathbf{v} \bullet \mathbf{w} = 0$. Thus we shall define two vectors to be *orthogonal* provided their dot product is zero. A vector of length (norm) 1 is called a *unit vector*.

Definition 9. *A set* $\mathbf{e}_1, \mathbf{e}_2, \mathbf{e}_3, \mathbf{e}_4$ *of four orthogonal unit vectors tangent to* $\mathbb{R}^4$ *at* $\mathbf{p}$ *es called a* frame *at the point* $\mathbf{p}$.

For example, at each point $\mathbf{p}$ of $\mathbb{R}^4$, the vectors $U_1(\mathbf{p}), U_2(\mathbf{p}), U_3(\mathbf{p}), U_4(\mathbf{p})$ of Definition 5 constitute a frame at $\mathbf{p}$.

3   The Cross Product

Definition 10. *If* $\mathbf{v}$, $\mathbf{w}$ *and* $\mathbf{x}$ *are tangent vectors to* $\mathbb{R}^4$ *at the same point* $\mathbf{p}$, *then the* cross product *of* $\mathbf{v}$, $\mathbf{w}$ *and* $\mathbf{x}$ *is the tangent vector*

$$\mathbf{v} \times \mathbf{w} \times \mathbf{x} = - \begin{vmatrix} U_1(\mathbf{p}) & U_2(\mathbf{p}) & U_3(\mathbf{p}) & U_4(\mathbf{p}) \\ v_1 & v_2 & v_3 & v_4 \\ w_1 & w_2 & w_3 & w_4 \\ x_1 & x_2 & x_3 & x_4 \end{vmatrix} .$$

Lemma 1. *If* $\mathbf{v}$, $\mathbf{w}$, $\mathbf{x}$, $\mathbf{y}$ *and* $\mathbf{z}$ *are tangent vectors to* $\mathbb{R}^4$ *at the same point* $\mathbf{p}$, *then*

1. *The cross product* $\mathbf{v} \times \mathbf{w} \times \mathbf{x}$ *is* linear *in* $\mathbf{v}$, *in* $\mathbf{w}$ *and in* $\mathbf{x}$, *and satisfies the alternation rules*

$$\mathbf{v} \times \mathbf{w} \times \mathbf{x} = \mathbf{w} \times \mathbf{x} \times \mathbf{v} = \mathbf{x} \times \mathbf{v} \times \mathbf{w} = -\mathbf{x} \times \mathbf{w} \times \mathbf{v} = -\mathbf{w} \times \mathbf{v} \times \mathbf{x} = -\mathbf{v} \times \mathbf{x} \times \mathbf{w} .$$

2. $\mathbf{v} \times \mathbf{w} \times \mathbf{x} \neq \mathbf{0}$ *if and only if* $\mathbf{v}$, $\mathbf{w}$ *and* $\mathbf{x}$ *are linearly independent,*

3. $\mathbf{v} \times \mathbf{w} \times (\mathbf{x} \times \mathbf{y} \times \mathbf{z}) = \begin{vmatrix} \mathbf{x} & \mathbf{y} & \mathbf{z} \\ \mathbf{x} \bullet \mathbf{w} & \mathbf{y} \bullet \mathbf{w} & \mathbf{z} \bullet \mathbf{w} \\ \mathbf{x} \bullet \mathbf{v} & \mathbf{y} \bullet \mathbf{v} & \mathbf{z} \bullet \mathbf{v} \end{vmatrix}$,

4. $\mathbf{v} \times \mathbf{w} \times (\mathbf{x} \times \mathbf{y} \times \mathbf{z}) \neq \mathbf{v} \times (\mathbf{w} \times \mathbf{x} \times \mathbf{y}) \times \mathbf{z} \neq (\mathbf{v} \times \mathbf{w} \times \mathbf{x}) \times \mathbf{y} \times \mathbf{z}$,

5. $\mathbf{v} \bullet \mathbf{w} \times \mathbf{x} \times \mathbf{y} = \begin{vmatrix} v_1 & v_2 & v_3 & v_4 \\ w_1 & w_2 & w_3 & w_4 \\ x_1 & x_2 & x_3 & x_4 \\ y_1 & y_2 & y_3 & y_4 \end{vmatrix}$,

6. $\mathbf{v} \bullet \mathbf{w} \times \mathbf{x} \times \mathbf{y} \neq 0$ *if and only if* $\mathbf{v}$, $\mathbf{w}$, $\mathbf{x}$ *and* $\mathbf{y}$ *are linearly independent*,

7. *If any two vectors in* $\mathbf{v} \bullet \mathbf{w} \times \mathbf{x} \times \mathbf{y}$ *are reversed, the product changes sign.* *Explicitly*

$$\mathbf{v} \bullet \mathbf{w} \times \mathbf{x} \times \mathbf{y} = \mathbf{v} \bullet \mathbf{x} \times \mathbf{y} \times \mathbf{w} = \mathbf{v} \bullet \mathbf{y} \times \mathbf{w} \times \mathbf{x} = \mathbf{w} \bullet \mathbf{v} \times \mathbf{y} \times \mathbf{x} = \mathbf{w} \bullet \mathbf{x} \times \mathbf{v} \times \mathbf{y} =$$

$$\mathbf{w} \bullet \mathbf{y} \times \mathbf{x} \times \mathbf{v} = \mathbf{x} \bullet \mathbf{v} \times \mathbf{w} \times \mathbf{y} = \mathbf{x} \bullet \mathbf{w} \times \mathbf{y} \times \mathbf{v} = \mathbf{x} \bullet \mathbf{y} \times \mathbf{v} \times \mathbf{w} = \mathbf{y} \bullet \mathbf{v} \times \mathbf{x} \times \mathbf{w} =$$

$$\mathbf{y} \bullet \mathbf{w} \times \mathbf{v} \times \mathbf{x} = \mathbf{y} \bullet \mathbf{x} \times \mathbf{w} \times \mathbf{v} =$$

$$-\mathbf{v} \bullet \mathbf{w} \times \mathbf{y} \times \mathbf{x} = -\mathbf{v} \bullet \mathbf{x} \times \mathbf{w} \times \mathbf{y} = -\mathbf{v} \bullet \mathbf{y} \times \mathbf{x} \times \mathbf{w} = -\mathbf{w} \bullet \mathbf{v} \times \mathbf{x} \times \mathbf{y} = -\mathbf{w} \bullet \mathbf{x} \times \mathbf{y} \times \mathbf{v} =$$

$$-\mathbf{w} \bullet \mathbf{y} \times \mathbf{v} \times \mathbf{x} = -\mathbf{x} \bullet \mathbf{v} \times \mathbf{y} \times \mathbf{w} = -\mathbf{x} \bullet \mathbf{w} \times \mathbf{v} \times \mathbf{y} = -\mathbf{x} \bullet \mathbf{y} \times \mathbf{w} \times \mathbf{v} = -\mathbf{y} \bullet \mathbf{v} \times \mathbf{w} \times \mathbf{x} =$$

$$-\mathbf{y} \bullet \mathbf{w} \times \mathbf{x} \times \mathbf{v} = -\mathbf{y} \bullet \mathbf{x} \times \mathbf{v} \times \mathbf{w},$$

8. $\mathbf{v} \bullet \mathbf{w} \times \mathbf{x} \times \mathbf{y} = -\mathbf{v} \times \mathbf{w} \times \mathbf{x} \bullet \mathbf{y}$.

Proof. The first and second properties follow immediately from Definition 10. The third property is deduced with the assistance of Mathematica. To do this, both terms are calculated and subtracted from the equality. Then the rule-based programming paradigm is used to substitute $\bar{v}$ for Array$[v, 4] = \{v[1], v[2], v[3], v[4]\}$, ..., $\bar{z}$ for Array$[z, 4] = \{z[1], z[2], z[3], z[4]\}$ and the result is the null vector, this validates equality.

$\bar{v} \times \bar{w} \times (\bar{x} \times \bar{y} \times \bar{z}) -$

$$\mathrm{Det}\left[\begin{pmatrix} \bar{x} & \bar{y} & \bar{z} \\ \bar{x}.\bar{w} & \bar{y}.\bar{w} & \bar{z}.\bar{w} \\ \bar{x}.\bar{v} & \bar{y}.\bar{v} & \bar{z}.\bar{v} \end{pmatrix} \right] /. \,\mathrm{OverBar}[a\_] \rightarrow \mathrm{Array}[a, 4] //\mathrm{Simplify}$$

$\{0, 0, 0, 0\}$

The fourth property is deduced from the third property. The sixth and seventh properties are deduced from the fifth property. The last property is deduced from the seventh property and the commutativity of the dot product.

Theorem 1. *The cross product* $\mathbf{v} \times \mathbf{w} \times \mathbf{x}$ *is orthogonal to* $\mathbf{v}$, $\mathbf{w}$ *and* $\mathbf{x}$, *and has length such that*

$$\|\mathbf{v} \times \mathbf{w} \times \mathbf{x}\|^2 = \begin{vmatrix} \mathbf{v} \bullet \mathbf{v} & \mathbf{v} \bullet \mathbf{w} & \mathbf{v} \bullet \mathbf{x} \\ \mathbf{w} \bullet \mathbf{v} & \mathbf{w} \bullet \mathbf{w} & \mathbf{w} \bullet \mathbf{x} \\ \mathbf{x} \bullet \mathbf{v} & \mathbf{x} \bullet \mathbf{w} & \mathbf{x} \bullet \mathbf{x} \end{vmatrix}.$$

Proof. The sixth property of the Lemma 1 shows the perpendicularity of $\mathbf{v} \times \mathbf{w} \times \mathbf{x}$ with $\mathbf{v}$, $\mathbf{w}$ and $\mathbf{x}$. On the other hand, by the definition of norm and the eighth and third properties of Lemma 1, we have to

$$\|\mathbf{v} \times \mathbf{w} \times \mathbf{x}\|^2 = (\mathbf{v} \times \mathbf{w} \times \mathbf{x}) \bullet (\mathbf{v} \times \mathbf{w} \times \mathbf{x}) = -(\mathbf{v} \times \mathbf{w} \times \mathbf{x}) \times \mathbf{v} \times \mathbf{w} \bullet \mathbf{x}$$

$$= -\mathbf{v} \times \mathbf{w} \times (\mathbf{v} \times \mathbf{w} \times \mathbf{x}) \bullet \mathbf{x} = - \begin{vmatrix} \mathbf{v} & \mathbf{w} & \mathbf{x} \\ \mathbf{v} \bullet \mathbf{w} & \mathbf{w} \bullet \mathbf{w} & \mathbf{x} \bullet \mathbf{w} \\ \mathbf{v} \bullet \mathbf{v} & \mathbf{w} \bullet \mathbf{v} & \mathbf{x} \bullet \mathbf{v} \end{vmatrix} \bullet \mathbf{x}$$

$$= \begin{vmatrix} \mathbf{v} \bullet \mathbf{v} & \mathbf{v} \bullet \mathbf{w} & \mathbf{v} \bullet \mathbf{x} \\ \mathbf{w} \bullet \mathbf{v} & \mathbf{w} \bullet \mathbf{w} & \mathbf{w} \bullet \mathbf{x} \\ \mathbf{x} \bullet \mathbf{v} & \mathbf{x} \bullet \mathbf{w} & \mathbf{x} \bullet \mathbf{x} \end{vmatrix}$$

Theorem 2. *Let* $\mathbf{v}$, $\mathbf{w}$, $\mathbf{x}$, $\mathbf{y}$ *and* $\mathbf{z}$ *be tangent vectors to* $\mathbb{R}^4$ *at the same point* $\mathbf{p}$, *then*

$$\mathbf{v} \times \mathbf{w} \times (\mathbf{x} \times \mathbf{y} \times \mathbf{z}) + \mathbf{v} \times \mathbf{x} \times (\mathbf{w} \times \mathbf{y} \times \mathbf{z}) + \mathbf{v} \times \mathbf{y} \times (\mathbf{z} \times \mathbf{x} \times \mathbf{w}) +$$
$$\mathbf{v} \times \mathbf{z} \times (\mathbf{w} \times \mathbf{x} \times \mathbf{y}) + \mathbf{w} \times \mathbf{x} \times (\mathbf{v} \times \mathbf{y} \times \mathbf{z}) + \mathbf{w} \times \mathbf{y} \times (\mathbf{z} \times \mathbf{x} \times \mathbf{v}) +$$
$$\mathbf{w} \times \mathbf{z} \times (\mathbf{v} \times \mathbf{x} \times \mathbf{y}) + \mathbf{x} \times \mathbf{y} \times (\mathbf{z} \times \mathbf{w} \times \mathbf{v}) + \mathbf{x} \times \mathbf{z} \times (\mathbf{v} \times \mathbf{w} \times \mathbf{y}) +$$
$$\mathbf{y} \times \mathbf{z} \times (\mathbf{x} \times \mathbf{w} \times \mathbf{v}) = 0 .$$

Proof. Taking into account that the Jacobi identity exists for three vectors tangent to $\mathbb{R}^3$, we proceeded to develop a Mathematica program in which the possible combinations of the cross product of five vectors tangent to $\mathbb{R}^4$ are considered; in such a way that, an extension of the Jacobi identity was obtained for five vectors tangent to $\mathbb{R}^4$. The program returned twelve possible combinations, of which the third was chosen for representativeness reasons. Obviously, from this choice the other possibilities can be deduced.

```
vecs = {x̄, ȳ, z̄, v̄, w̄} ;

perms = Select[Permutations[vecs, {3}], OrderedQ];

Tu = Tuples[{Identity, Reverse}, {Length[perms]}];

permsTu = MapThread[#1@@{#2}&, {#, perms}]&@Tu;

Table[
    permsk = permsTu[[k]];

    g = Complement[vecs, #]&@permsk;

    f = MapThread[Cross[Sequence@@#1, Cross@@#2]&, {g, permsk}];

    s = Plus@@ (f /. OverBar[a_] → Array[a, 4]) ;

    If[s===Table[0, 4], Plus@@f == 0, Nothing],
    {k, Length[Tu]}][[3]]
```

$$\bar{v}\times\bar{w}\times(\bar{x}\times\bar{y}\times\bar{z})+\bar{v}\times\bar{x}\times(\bar{w}\times\bar{y}\times\bar{z})+\bar{v}\times\bar{y}\times(\bar{z}\times\bar{x}\times\bar{w})+\bar{v}\times\bar{z}\times(\bar{w}\times\bar{x}\times\bar{y})+$$
$$\bar{w}\times\bar{x}\times(\bar{v}\times\bar{y}\times\bar{z})+\bar{w}\times\bar{y}\times(\bar{z}\times\bar{x}\times\bar{v})+\bar{w}\times\bar{z}\times(\bar{v}\times\bar{x}\times\bar{y})+\bar{x}\times\bar{y}\times(\bar{z}\times\bar{w}\times\bar{v})+$$
$$\bar{x}\times\bar{z}\times(\bar{v}\times\bar{w}\times\bar{y})+\bar{y}\times\bar{z}\times(\bar{x}\times\bar{w}\times\bar{v})==0$$

Theorem 3. *If* **v** *and* **w** *are tangent vectors to* $\mathbb{R}^4$ *at the same point* **p***, then the area,* S*, of the parallelogram with sides* **v** *and* **w** *is*

$$S = \sqrt{\left|\begin{matrix} \mathbf{v}\bullet\mathbf{v} & \mathbf{v}\bullet\mathbf{w} \\ \mathbf{w}\bullet\mathbf{v} & \mathbf{w}\bullet\mathbf{w} \end{matrix}\right|}.$$

Proof. Figure 1 shows the parallelogram. If h is the distance from **w** to the segment from **p** to **v** and L is the length of this segment, then the area will be $A = Lh$. The L length can be expressed as

$$L = \sqrt{\mathbf{v}\bullet\mathbf{v}}.$$

The point Q (Fig. 2) is on the segment generated by **v**, i.e.

$$Q = t\mathbf{v}; \quad t \in \mathbb{R}.$$

According to Fig. 2 it is true that

$$(\mathbf{w} - Q)\bullet\mathbf{v} = 0.$$

After solving this equation we obtain

$$t = \frac{\mathbf{v}\bullet\mathbf{w}}{\mathbf{v}\bullet\mathbf{v}}.$$

So that

$$h = \sqrt{(\mathbf{w} - Q)\bullet(\mathbf{w} - Q)}$$
$$= \sqrt{\mathbf{w}\bullet\mathbf{w} + t^2\,\mathbf{v}\bullet\mathbf{v} - 2t\,\mathbf{v}\bullet\mathbf{w}}.$$

Finally, after making the respective substitutions and simplifications with the Mathematica assistance, we obtain

$$S = Lh$$
$$= \sqrt{\left|\begin{matrix} \mathbf{v}\bullet\mathbf{v} & \mathbf{v}\bullet\mathbf{w} \\ \mathbf{w}\bullet\mathbf{v} & \mathbf{w}\bullet\mathbf{w} \end{matrix}\right|}.$$

A more intuitive description of S is

$$S = \sqrt{\left|\begin{matrix} \|\mathbf{v}\|^2 & \|\mathbf{v}\|\,\|\mathbf{w}\|\cos\vartheta \\ \|\mathbf{w}\|\,\|\mathbf{v}\|\cos\vartheta & \|\mathbf{w}\|^2 \end{matrix}\right|}$$
$$= \|\mathbf{v}\|\,\|\mathbf{w}\|\sqrt{\left|\begin{matrix} 1 & \cos\vartheta \\ \cos\vartheta & 1 \end{matrix}\right|}$$
$$= \|\mathbf{w}\|\,\|\mathbf{v}\|\sin\vartheta,$$

where $0 \leqq \vartheta \leqq \pi$ is the smaller of the two angles from **v** to **w** (Fig. 1).

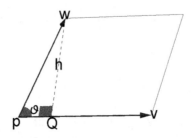

Fig. 1. Parallelogram with edges **v** and **w**.

Theorem 4. *If* **v**, **w** *and* **x** *are tangent vectors to* $\mathbb{R}^4$ *at the same point* **p**, *then* $\|\mathbf{v} \times \mathbf{w} \times \mathbf{x}\|$ *is the volume*, V, *of the parallelepiped with edges* **v**, **w** *and* **x**.

Proof. Figure 2 shows the parallelepiped. If h is the distance from **x** to the parallelogram with sides **v** and **w** and S is the area of this parallelogram, then the volume will be $V = Sh$. The S area can be expressed as

$$S = \sqrt{\begin{vmatrix} \mathbf{v} \bullet \mathbf{v} & \mathbf{v} \bullet \mathbf{w} \\ \mathbf{w} \bullet \mathbf{v} & \mathbf{w} \bullet \mathbf{w} \end{vmatrix}}.$$

The point Q (Fig. 2) is on the plane generated by **v** and **w**, i.e.

$$Q = t\mathbf{v} + s\mathbf{w}; \quad s, t \in \mathbb{R}.$$

According to Fig. 2 it is true that

$$\begin{cases} (\mathbf{x} - Q) \bullet \mathbf{v} = 0, \\ (\mathbf{x} - Q) \bullet \mathbf{w} = 0. \end{cases}$$

After solving this system we obtain

$$t = \frac{\begin{vmatrix} \mathbf{v} \bullet \mathbf{x} & \mathbf{v} \bullet \mathbf{w} \\ \mathbf{w} \bullet \mathbf{x} & \mathbf{w} \bullet \mathbf{w} \end{vmatrix}}{\begin{vmatrix} \mathbf{v} \bullet \mathbf{v} & \mathbf{v} \bullet \mathbf{w} \\ \mathbf{w} \bullet \mathbf{v} & \mathbf{w} \bullet \mathbf{w} \end{vmatrix}}, \quad s = \frac{\begin{vmatrix} \mathbf{v} \bullet \mathbf{v} & \mathbf{v} \bullet \mathbf{x} \\ \mathbf{w} \bullet \mathbf{v} & \mathbf{w} \bullet \mathbf{x} \end{vmatrix}}{\begin{vmatrix} \mathbf{v} \bullet \mathbf{v} & \mathbf{v} \bullet \mathbf{w} \\ \mathbf{w} \bullet \mathbf{v} & \mathbf{w} \bullet \mathbf{w} \end{vmatrix}}.$$

So that

$$h = \sqrt{(\mathbf{x} - Q) \bullet (\mathbf{x} - Q)}$$
$$= \sqrt{\mathbf{x} \bullet \mathbf{x} + t^2\, \mathbf{v} \bullet \mathbf{v} + s^2\, \mathbf{w} \bullet \mathbf{w} - 2t\, \mathbf{v} \bullet \mathbf{x} - 2s\, \mathbf{w} \bullet \mathbf{x} + 2st\, \mathbf{v} \bullet \mathbf{w}}.$$

Finally, after making the respective substitutions and simplifications with the Mathematica assistance, we obtain

$$V = Sh$$

$$= \sqrt{\begin{vmatrix} \mathbf{v} \bullet \mathbf{v} & \mathbf{v} \bullet \mathbf{w} & \mathbf{v} \bullet \mathbf{x} \\ \mathbf{w} \bullet \mathbf{v} & \mathbf{w} \bullet \mathbf{w} & \mathbf{w} \bullet \mathbf{x} \\ \mathbf{x} \bullet \mathbf{v} & \mathbf{x} \bullet \mathbf{w} & \mathbf{x} \bullet \mathbf{x} \end{vmatrix}}$$

$$= \|\mathbf{v} \times \mathbf{w} \times \mathbf{x}\|.$$

A more intuitive description of the length of a cross product is

$$\|\mathbf{v} \times \mathbf{w} \times \mathbf{x}\| = \sqrt{\begin{vmatrix} \|\mathbf{v}\|^2 & \|\mathbf{v}\|\,\|\mathbf{w}\|\cos\vartheta & \|\mathbf{v}\|\,\|\mathbf{x}\|\cos\varphi \\ \|\mathbf{w}\|\,\|\mathbf{v}\|\cos\vartheta & \|\mathbf{w}\|^2 & \|\mathbf{w}\|\,\|\mathbf{x}\|\cos\gamma \\ \|\mathbf{x}\|\,\|\mathbf{v}\|\cos\varphi & \|\mathbf{x}\|\,\|\mathbf{w}\|\cos\gamma & \|\mathbf{x}\|^2 \end{vmatrix}}$$

$$= \|\mathbf{v}\|\,\|\mathbf{w}\|\,\|\mathbf{x}\| \sqrt{\begin{vmatrix} 1 & \cos\vartheta & \cos\varphi \\ \cos\vartheta & 1 & \cos\gamma \\ \cos\varphi & \cos\gamma & 1 \end{vmatrix}}$$

where $0 \leq \vartheta, \varphi, \gamma \leq \pi$ is the smaller of the two angles from $\mathbf{v}$ to $\mathbf{w}$, $\mathbf{w}$ to $\mathbf{x}$ and $\mathbf{v}$ to $\mathbf{x}$, respectively (Fig. 2).

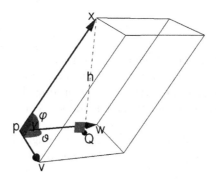

Fig. 2. Parallelepiped with edges $\mathbf{v}$, $\mathbf{w}$ and $\mathbf{x}$.

Corollary 1. *If* $\mathbf{v}_1$, $\mathbf{v}_2$ *and* $\mathbf{v}_3$ *are tangent vectors to* $\mathbb{R}^4$ *at the same point* $\mathbf{p}$ *and* $S_{i,j}$ *is the area of the parallelogram with sides* $\mathbf{v}_i$ *and* $\mathbf{v}_j$, *then*

$$\|\mathbf{v}_1 \times \mathbf{v}_2 \times \mathbf{v}_3\|^2 = S_{23}^2\|\mathbf{v}_1\|^2 + S_{13}^2\|\mathbf{v}_2\|^2 + S_{12}^2\|\mathbf{v}_3\|^2 +$$
$$+ 2\,\mathbf{v}_1 \bullet \mathbf{v}_2\, \mathbf{v}_1 \bullet \mathbf{v}_3\, \mathbf{v}_2 \bullet \mathbf{v}_3 - 2\|\mathbf{v}_1\|^2\,\|\mathbf{v}_2\|^2\,\|\mathbf{v}_3\|^2 .$$

Proof. The proof is immediate from the Theorem 1 and the Theorem 3.

Corollary 2. *If* $\mathbf{v}$, $\mathbf{w}$, $\mathbf{x}$ *and* $\mathbf{y}$ *are tangent vectors to* $\mathbb{R}^4$ *at the same point* $\mathbf{p}$, *then* $|\mathbf{v} \bullet \mathbf{w} \times \mathbf{x} \times \mathbf{y}|$ *is the hypervolume of the hyperparallelepiped with edges* $\mathbf{v}$, $\mathbf{w}$, $\mathbf{x}$ *and* $\mathbf{y}$.

Proof. Figure 3 shows the hyperparallelepiped. By Theorem 4 the volume of the parallelepiped with edges $\mathbf{w}$, $\mathbf{x}$ and $\mathbf{y}$ is $V = \|\mathbf{w} \times \mathbf{x} \times \mathbf{y}\|$. If h is the height, then $h = \|\mathbf{v}\|\,|\cos\theta|$, $0 \leq \theta \leq \pi$ is the smaller of the two angles from $\mathbf{v}$ to $\mathbf{w} \times \mathbf{x} \times \mathbf{y}$. Therefore, the hypervolume of the hyperparalelepiped is

$$W = hV = \|\mathbf{v}\|\,\|\mathbf{w} \times \mathbf{x} \times \mathbf{y}\|\,|\cos\theta| = |\mathbf{v} \bullet \mathbf{w} \times \mathbf{x} \times \mathbf{y}|.$$

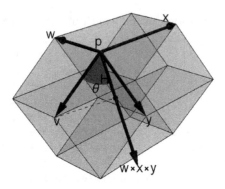

Fig. 3. Hyperparallelepiped with edges **v**, **w**, **x** and **y**.

4   Conclusions

In this paper, the definition of the cross product of three vectors tangent to $\mathbb{R}^4$ at the same point has been stated, based on this definition a lemma has been enunciated in which eight properties of said product have been proposed and demonstrated. Furthermore, four theorems and two corollaries have been stated and proved. One of the theorems constitutes an extension of the Jacobi identity. In some of the proofs, programs based on the paradigms: functional, rule-based and list-based have been used; all of these supported by the Wolfram language, incorporated in Mathematica.

Acknowledgements. The authors would like to thank to the authorities of the Universidad Nacional de Piura for the acquisition of the Mathematica 11.0 license and the reviewers for their valuable comments and suggestions.

References

1. Bayram, E., Kasap, E.: Hypersurface family with a common isogeodesic. Ser. Math. Inform. **24**(2), 5–24 (2014)
2. Düldül, M.: On the intersection curve of three parametric hypersurfaces. Comput. Aided Geom. Des. **27**, 118–127 (2010)
3. Düldül, B.U., Düldül, M.: The extension of Willmore's method into 4-space. Math. Commun. **17**, 423–431 (2012)
4. Gray, J.W.: Mastering Mathematica: Programming Methods and Applications, 2nd edn. Academic Press, Cambridge (1997)
5. Maeder, R.: Programming in Mathematica, 2nd edn. Addison-Wesley, Redwood City (1991)
6. Williams, M.Z., Stein, F.M.: A triple product of vectors in four-space. Math. Mag. **37**(4), 230–235 (1964)
7. Wolfram, S.: An Elementary Introduction to the Wolfram Language, 2nd edn. Wolfram Media, Inc. (2017)
8. Wolfram, S.: The Mathematica Book, 4th edn. Wolfram Media, Champaign; Cambridge University Press, Cambridge (1999)

A Parameterization of the Klein Bottle by Isometric Transformations in $\mathbb{R}^4$ with Mathematica

Ricardo Velezmoro-León[1], Nestor Javier Farias-Morcillo[1],
Robert Ipanaqué-Chero[1(✉)], José Manuel Estela-Vilela[1],
and Judith Keren Jiménez-Vilcherrez[2]

[1] Universidad Nacional de Piura, Urb. Miraflores s/n, Castilla, Piura, Peru
{rvelezmorol,nfariasm,ripanaquec}@unp.edu.pe
[2] Universidad Tecnológica del Perú, Av. Vice Cdra 1-Costado Real Plaza, Piura, Peru
C19863@utp.edu.pe

Abstract. The Klein bottle plays a crucial role in the main modern sciences. This surface was first described in 1882 by the German mathematician Felix Klein. In this paper we describe a technique to obtain the parameterization of the Klein bottle. This technique uses isometric transformations (translations and rotations) and the moving frame associated with the unit circumference lying on the xy-plane. The process we follow is to start with the parametrization of the Euclidean cylinder, then continue with the parameterization of the Möbius strip, after that with the parameterization of the torus of revolution and finally, in a natural way, we describe the aforementioned technique. With the parameterization of the Klien bottle obtained, it is easy to show that it can be obtained by gluing two Möbius strips. Additionally, the parameterizations of the n-twisted and n-turns Klein bottles are obtained. All geometric calculations and geometric interpretations are performed with the Mathematica symbolic calculus system.

Keywords: Parameterization · Klein Bottle · Isometric transformations in $\mathbb{R}^4$ · Moving frame

1 Introduction

The Klein bottle plays a crucial role in the main modern sciences [3,4,12,13,15]. This surface was first described in 1882 by the German mathematician Felix Klein. It may have been originally named the Kleinsche Fläche ("Klein surface") and then misinterpreted as Kleinsche Flasche ("Klein bottle"), which ultimately may have led to the adoption of this term in the German language as well [2]. There are parameterizations in R4 of this surface, the two most popular are: nonintersecting 4D parametrization (modeled from the flat torus) and the so-called 4D Möbius tube [17,18].

© Springer Nature Switzerland AG 2021
O. Gervasi et al. (Eds.): ICCSA 2021, LNCS 12949, pp. 261–272, 2021.
https://doi.org/10.1007/978-3-030-86653-2_19

In this paper we describe a technique to obtain the parameterization of the Klein bottle. This technique uses isometric transformations such as translations and rotations [11]. In addition, it makes use of the moving frame associated with the unit circumference that lies in the xy-plane [7,11]. We have chosen to follow a process that begins with the description of a technique to parameterize the Euclidean cylinder in $\mathbb{R}^3$, continues with the description of the same technique to parameterize the Möbius strip in $\mathbb{R}^3$, continues with the description of another technique to parameterize the torus of revolution in $\mathbb{R}^4$ and ends, inductively, with the description of the technique to parameterize the Klein bottle in $\mathbb{R}^4$. After obtaining the parameterization of the Klein bottle, it became easy to separate it into two parts, both parts being a separate Möbius strip. In addition, the technique was useful to find the parameterization of the n-twisted and n-turns Klein bottles are obtained [5,9]. All geometric calculations and geometric interpretations are performed with the Mathematica v.11.0 symbolic calculus system [19].

The structure of this paper is as follows: Sect. 2 introduces the basic concepts related to surfaces and quotient spaces, as well as the parametrization of the immersion of the Klein bottle in three-dimensional space. Then, Sect. 3 introduces the technique to parameterize the Klein bottle by means of isometric transformations. Here also the parameterizations of the n-twisted and n-fold Klein bottles are obtained. Finally, Sect. 4 closes with the main conclusions of this paper.

2   Basic Concepts

2.1   Surfaces and Quotient Spaces

Definition 1. *A surface or 2-dimensional manifold is a topological space with the same local properties as the familiar plane of Euclidean geometry [10].*

Definition 2. *Let X be a topological space and ρ be an equivalence relation on X. If $p : X \rightarrow X/\rho$, $x \mapsto [x]$ is the natural surjective map, then the collection Ω of all subsets $U \subset X/\rho$ such that $p^{-1}(U)$ is an open set of X, forms the largest topology on X/ρ such that p is continuous. The set X/ρ is called a quotient space of X, with the quotient topology Ω and p is called an identification map [1].*

Example 1. Let $I \times I$ be the unit square.

a. The Euclidean cylinder is the quotient space obtained from unit square $I \times I$ by identifying $(0, t)$ with $(1, t)$, for all $t \in I$ (see Fig. 1).
b. The torus of revolution is the quotient space obtained from unit square $I \times I$ by identifying $(t, 0)$ with $(t, 1)$ and also $(0, t)$ with $(1, t)$, for all $t \in I$ (see Fig. 1).
c. The Möbius strip is the quotient space obtained from unit square $I \times I$ by identifying $(0, t)$ with $(1, 1 - t)$, for all $t \in I$ (see Fig. 1).

d. The Klein bottle is the quotient space obtained from unit square $I \times I$ by identifying $(0, t)$ with $(1, t)$ and also $(t, 0)$ with $(1, 1 - t)$, for all $t \in I$. Since this identification can only be properly represented in $\mathbb{R}^4$, a representation of it is usually made in $\mathbb{R}^3$, which is called immersion of the Klein bottle in $\mathbb{R}^3$ (see Fig. 2).

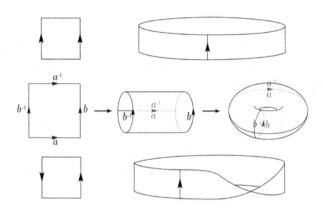

Fig. 1. Cylinder (top), Torus (middle) and Möbius strip (bottom) as the quotient space of unit square.

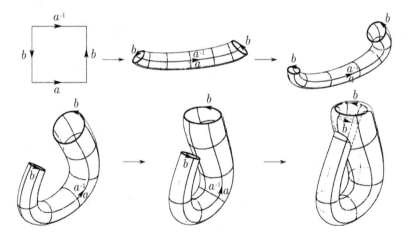

Fig. 2. Klein bottle, immersed in $\mathbb{R}^3$, as the quotient space of unit square.

2.2 Parameterizations of the Immersion of the Klein Bottle in $\mathbb{R}^3$ as a Tubular Surface

Let $\alpha(t) = (\alpha_1(t), \alpha_2(t))$, $t \in [a, b]$, be a curve lying on the xy-plane satisfying $\|\alpha'(t)\| \neq 0$. Let $\mathbf{k} = (0, 0, 1)$ be the z-axis unit vector and $\mathbf{T} = \frac{\alpha'}{\|\alpha'\|}$ be the unit tangent vector field of $\alpha(t)$. Let $\mathbf{N} = \mathbf{k} \times \mathbf{T}$. Then the couple of unit vectors $(\mathbf{N}, \mathbf{k})$ is a moving frame orthogonal to $\alpha'(t)$ and can be used to construct a tube around $\alpha(t)$ as follows:

$$tube(u, v) = \alpha(u) + r(u) \left(\cos v \mathbf{N} + \sin v \mathbf{k} \right), \quad (u, v) \in [a, b] \times [0, 2\pi],$$

where the scalar continuous function $\mathbf{r}(t)$ gives the radius of the tube [8].

M. Trott's [14] proposal is to define:

$$\alpha(t) = \left(\frac{1}{t^4 + 1}, \frac{t^2 + t + 1}{t4 + 1} \right), \quad t \in (-\infty, +\infty),$$

$$r(t) = \frac{84t^4 + 56t^3 + 21t^2 + 21t + 24}{672(1 + t^4)}$$

and the resulting image is shown in Fig. 3.

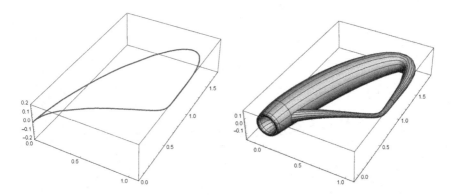

Fig. 3. Central curve α (left) and Klein bottle according to M. Trott's definition (right).

Whereas, G. Franzoni [8] proposes the following definition:

$$\alpha(t) = (a(1 - \cos t), b \sin t(1 - \cos t)), \quad t \in [0, 2\pi],$$

$$r(t) = c - d(t - \pi)\sqrt{t(2\pi - t)}.$$

and the resulting image, with $(a, b, c, d) = \left(20, 8, \frac{11}{2}, \frac{2}{5}\right)$, is shown in Fig. 4.

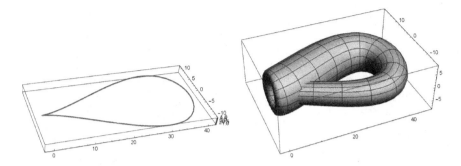

Fig. 4. Central curve α (left) and Klein bottle according to G. Franzoni's definition (right).

3   Parameterizations of Some Surfaces by Isometric Transformations

In this section, two isometric transformations are used: translations and rotations, as well as a moving frame to parameterize the Euclidean cylinder, the Möbius strip, the torus of revolution, and the Klein bottle. The first two in $\mathbb{R}^3$ and the last two in $\mathbb{R}^4$. Although the first three parameterizations are widely known, they are mentioned as a preamble to the fourth parameterization that we propose in this paper. To trace the surfaces construction process in $\mathbb{R}^4$ a trimetric-trimetric model according to [16] is used. This is,

$$O = \{0,0,0\}, \quad \hat{\mathcal{B}} = \left\{ \frac{3}{5\sqrt{3}}(-1,-1,-1), (1,0,0), (0,1,0), (0,0,1) \right\} \quad \text{and}$$

$$\varphi(\mathbf{p}) = \left(p_2 - \frac{3p_1}{5\sqrt{3}}, \ p_3 - \frac{3p_1}{5\sqrt{3}}, \ p_4 - \frac{3p_1}{5\sqrt{3}} \right).$$

3.1   Parameterization of the Euclidian Cylinder in $\mathbb{R}^3$

Let $\beta(t) = (\cos t, \sin t, 0)$, $t \in [0, 2\pi]$, be the unit circumference lying on the xy-plane. Let $\mathbf{k} = (0,0,1)$ be the z-axis unit vector and $\mathbf{T} = \beta'(t) = (-\sin t, \cos t, 0)$ be the unit tangent vector field of $\beta(t)$. Let $\mathbf{N} = \mathbf{k} \times \mathbf{T}$. Let $\delta(s) = \left(0, \frac{s}{2}\right)$, $s \in [-1, 1]$, be a segment lying on the normal plane associated with β; that is, the plane generated by $\mathbf{N}$ and $\mathbf{k}$. Then the couple of unit vectors $(\mathbf{N}, \mathbf{k})$ is a moving frame orthogonal to $\beta'(t)$ and can be used to construct a Euclidean cylinder around $\beta(t)$ as follows:

$$cyl(u,v) = \beta(u) + 0 \cdot \mathbf{N}(u) + \frac{v}{2}\mathbf{k}, \quad (u,v) \in [0, 2\pi] \times [-1, 1].$$

This is

$$cyl(u,v) = \left(\cos(u), \sin(u), \frac{v}{2} \right), \quad (u,v) \in [0, 2\pi] \times [-1, 1] \tag{1}$$

and some resulting images are shown in Fig. 5.

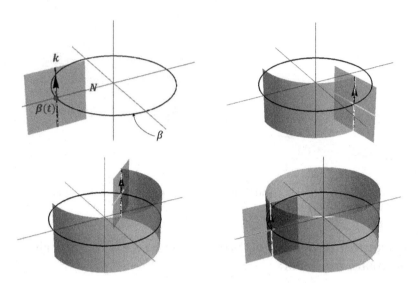

Fig. 5. Circumference β and the moving frame $(\mathbf{N}, \mathbf{k})$ (top-left) and Euclidean cylinder construction process according to Eq. 1 definition (top-right,bottom-left,bottom-right).

3.2 Parameterization of the Möbius strip in $\mathbb{R}^3$

Let $R(\theta) = \begin{pmatrix} \cos\theta & -\sin\theta \\ \sin\theta & \cos\theta \end{pmatrix}$ be the rotation matrix around a point in plane. Let $\beta(t) = (\cos t, \sin t, 0)$, $t \in [0, 2\pi]$, be the unit circumference lying on the xy-plane. Let $\mathbf{k} = (0, 0, 1)$ be the z-axis unit vector and $\mathbf{T} = \beta'(t) = (-\sin t, \cos t, 0)$ be the unit tangent vector field of $\beta(t)$. Let $\mathbf{N} = \mathbf{k} \times \mathbf{T}$. Let $\delta(s) = \left(\frac{s}{2}, 0\right)$, $s \in [-1, 1]$, be a segment lying on the normal plane associated with β; that is, the plane generated by $\mathbf{N}$ and $\mathbf{k}$. Then the couple of unit vectors $(\mathbf{N}, \mathbf{k})$ is a moving frame orthogonal to $\beta'(t)$ and can be used to construct a Möbius strip around $\beta(t)$ as follows:

$$mob(u, v) = \beta(u) + \phi_1(u, v)\mathbf{N}(u) + \phi_2(u, v)\mathbf{k}, \quad (u, v) \in [0, 2\pi] \times [-1, 1],$$

where $\phi(u, v) = R\left(\frac{u}{2}\right) \bullet (\delta(v))^\top$. This is

$$mob(u, v) = \left(\cos(u) - \frac{1}{2}v\cos\left(\frac{u}{2}\right)\cos(u), \sin(u) - \frac{1}{2}v\cos\left(\frac{u}{2}\right)\sin(u), \right.$$
$$\left. \frac{1}{2}v\sin\left(\frac{u}{2}\right) \right), \quad (u, v) \in [0, 2\pi] \times [-1, 1] \quad (2)$$

and some resulting images are shown in Fig. 6.

Also, if we do:

$$\phi(u, v) = R\left(\frac{nu}{2}\right) \bullet (\delta(v))^\top, \quad n = 1, 2, \dots$$

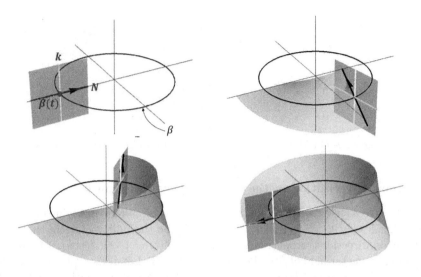

Fig. 6. Circumference β and the moving frame $(\mathbf{N}, \mathbf{k})$ (top-left) and Möbius strip construction process according to Eq. 2 definition (top-right,bottom-left,bottom-right).

and we consider $(u, v) \in [0, 2\pi] \times [0, 2\pi]$, we will obtain n-twisted Möbius strips (see Fig. 7).

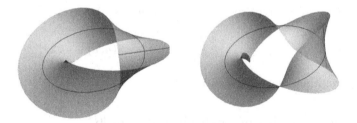

Fig. 7. 2-twisted Möbius strip (left) and 3-twisted Möbius strip (right).

Similarly, if we do:

$$\phi(u, v) = R\left(\frac{u}{2n}\right) \bullet (\delta(v))^\top, \quad n = 1, 2, \ldots$$

and we consider $(u, v) \in [0, 2\pi] \times [0, 2n\pi]$, we will obtain n-fold Möbius strips (see Fig. 8).

Fig. 8. 2-fold Möbius strip (left) and 3-fold Möbius strip (right).

3.3 Parameterization of the Torus of Revolution in $\mathbb{R}^4$

Let $\beta(t) = (\cos t, \sin t, 0, 0)$, $t \in [0, 2\pi]$, be the unit circumference lying on the xy-plane. Let $\mathbf{k}_1 = (0, 0, 1, 0)$ and $\mathbf{k}_2 = (0, 0, 0, 1)$ be the unit vectors in the z-axis and w-axis directions, respectively, and $\mathbf{T} = \beta'(t) = (-\sin t, \cos t, 0, 0)$ be the unit tangent vector field of $\beta(t)$. Let $\mathbf{N} = \mathbf{k}_1 \times \mathbf{k}_2 \times \mathbf{T}$. Let $\delta(s) = \frac{1}{2}(\cos s, \sin s, 0)$, $s \in [0, 2\pi]$, be a segment lying on the normal plane associated with β; that is, the 3D-space generated by $\mathbf{N}$, $\mathbf{k}_1$ and $\mathbf{k}_2$. Then the triad of unit vectors $(\mathbf{N}, \mathbf{k}_1, \mathbf{k}_2)$ is a moving frame orthogonal to $\beta'(t)$ and can be used to construct a torus of revolution around $\beta(t)$ as follows:

$$tor(u, v) = \beta(u) + \frac{1}{2}\cos v \mathbf{N}(u) + \frac{1}{2}\sin u \mathbf{k}_1 + 0 \cdot \mathbf{k}_2, \quad (u, v) \in [0, 2\pi] \times [0, 2\pi].$$

This is

$$tor(u, v) = \left(\frac{1}{2}\cos(u)\cos(v) + \cos(u), \frac{1}{2}\sin(u)\cos(v) + \sin(u), \right.$$

$$\left. \frac{1}{2}\sin(v), 0 \right), \quad (u, v) \in [0, 2\pi] \times [-1, 1] \quad (3)$$

and some resulting image are shown in Fig. 9.

3.4 Parameterization of the Klein Bottle in $\mathbb{R}^4$

Let $R(\theta) = \begin{pmatrix} \cos\theta & 0 & -\sin\theta \\ 0 & 1 & 0 \\ \sin\theta & 0 & \cos\theta \end{pmatrix}$ be the rotation matrix around the y-axis. Let $\beta(t) = (\cos t, \sin t, 0, 0)$, $t \in [0, 2\pi]$, be the unit circumference lying on the xy-plane. Let $\mathbf{k}_1 = (0, 0, 1, 0)$ and $\mathbf{k}_2 = (0, 0, 0, 1)$ be the unit vectors in the z-axis and w-axis directions, respectively, and $\mathbf{T} = \beta'(t) = (-\sin t, \cos t, 0, 0)$ be the unit tangent vector field of $\beta(t)$. Let $\mathbf{N} = \mathbf{k}_1 \times \mathbf{k}_2 \times \mathbf{T}$. Let $\delta(s) = \frac{1}{2}(\cos s, \sin s, 0)$, $s \in [0, 2\pi]$, be a segment lying on the normal plane associated with β; that is, the 3D-space generated by $\mathbf{N}$, $\mathbf{k}_1$ and $\mathbf{k}_2$. Then the triad of unit vectors $(\mathbf{N}, \mathbf{k}_1, \mathbf{k}_2)$ is a moving frame orthogonal to $\beta'(t)$ and can be used to construct a Klein bottle around $\beta(t)$ as follows:

$$kle(u, v) = \beta(u) + \phi_1(u, v)\mathbf{N}(u) + \phi_2(u, v)\mathbf{k}_1 + \phi_3(u, v)\mathbf{k}_2, \quad (u, v) \in [0, 2\pi] \times [0, 2\pi],$$

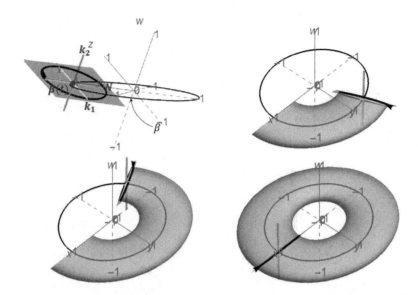

Fig. 9. Circumference β and the moving frame $(\mathbf{N}, \mathbf{k}_1, \mathbf{k}_2)$ (top-left) and torus of revolution construction process according to Eq. 3 definition (top-right, bottom-left, bottom-right). Note that the δ circumference does not leave the plane generated by $\mathbf{N}$ and $\mathbf{k}_1$.

where $\phi(u, v) = R\left(\frac{u}{2}\right) \bullet (\delta(v))^{\top}$. This is

$$kle(u, v) = \left(\frac{1}{2}\cos\left(\frac{u}{2}\right)\cos(u)\cos(v) + \cos(u), \frac{1}{2}\cos\left(\frac{u}{2}\right)\sin(u)\cos(v) + \sin(u),\right.$$

$$\left.\frac{1}{2}\sin(v), -\frac{1}{2}\sin\left(\frac{u}{2}\right)\cos(v)\right), \quad (u, v) \in [0, 2\pi] \times [0, 2\pi] \qquad (4)$$

and some resulting image are shown in Fig. 10.

It is known that the Klein bottle is the result of gluing two Möbius strips together along their boundary circles [6]. If we plot

$$1 + kle(u, v), \quad (u, v) \in [0, 2\pi] \times [0, \pi]$$

and

$$kle(u, v), \quad (u, v) \in [0, 2\pi] \times [\pi, 2\pi]$$

we will obtain two appropriately separated Möbius strips to visualize this result (see Fig. 11).

Also, if we do:

$$\phi(u, v) = R\left(\frac{nu}{2}\right) \bullet (\delta(v))^{\top}, \quad n = 1, 2, \ldots$$

and we consider $(u, v) \in [0, 2\pi] \times [0, 2\pi]$, we will obtain n-twisted Klein bottles (see Fig. 12).

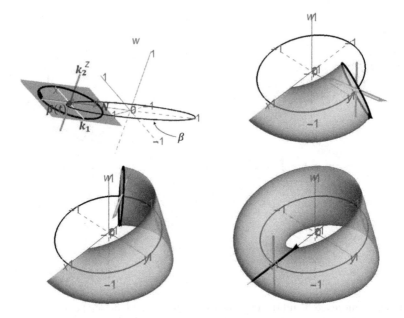

Fig. 10. Circumference β and the moving frame $(\mathbf{N}, \mathbf{k}_1, \mathbf{k}_2)$ (top-left) and torus of revolution construction process according to Eq. 4 definition (top-right,bottom-left,bottom-right). Note that the δ circumference leaves the plane generated by $\mathbf{N}$ and $\mathbf{k}_1$ to effect the rotational motion around the axis generated by $\mathbf{k}_1$.

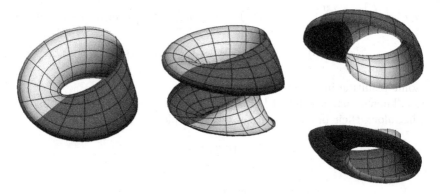

Fig. 11. The Klein bottle as the result of gluing two Möbius strips together along their boundary circles.

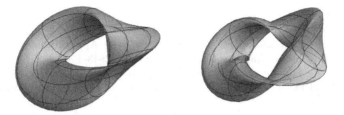

Fig. 12. 2-twisted Klein bottle (left) and 3-twisted Klein bottle (right).

Similarly, if we do:

$$\phi(u,v) = R\left(\frac{u}{2n}\right) \bullet (\delta(v))^\top, \quad n = 1, 2, \ldots$$

and we consider $(u,v) \in [0, 2\pi] \times [0, 2n\pi]$, we will obtain n-fold Klein bottles (see Fig. 13).

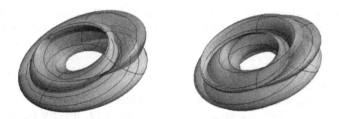

Fig. 13. 2-fold Klein bottle (left) and 3-fold Klein bottle (right).

4   Conclusions

In this paper a technique to obtain the parameterization of the Klein bottle has been described. This technique has used isometric transformations and the moving frame associated with the unit circle lying on the xy-plane. The process followed was to start with the parameterization of the Euclidean cylinder, then continued with the parameterization of the Möbius band, then with the parameterization of the torus of revolution and finally, in a natural way, the aforementioned technique was described. With the parameterization of the Klien bottle obtained, it was easy to show that it can be obtained by gluing two Möbius strips. Additionally, the parameterizations of the n-twisted and n-fold Klein bottles were obtained. All geometric calculations and geometric interpretations were made with the Mathematica symbolic calculation system.

Acknowledgements. The authors would like to thank to the authorities of the Universidad Nacional de Piura for the acquisition of the Mathematica 11.0 license and the reviewers for their valuable comments and suggestions.

References

1. Adhikari, M.R.: Basic Algebraic Topology and its Applications. Springer, New Delhi (2016). https://doi.org/10.1007/978-81-322-2843-1
2. Bonahon, F.: Low-dimensional Geometry: From Euclidean Surfaces to Hyperbolic Knots. Student Mathematical Library, vol. 49. American Mathematical Society (2009)
3. Carlsson, G., et al.: On the local behavior of spaces of natural images. Int. J. Comput. Vision **76**, 1–12 (2008). https://doi.org/10.1007/s11263-007-0056-x
4. Carlsson, G.: Topology and data. Bull. AMS **46**(2), 255–308 (2009). https://doi.org/10.1090/S0273-0979-09-01249-X
5. Cheshkova, M.A.: On a model of the Klein bottle. Math. Mech. **1**(89), 180–184 (2016). https://doi.org/10.14258/izvasu(2016)1-32
6. Carter, M.R.: How Surfaces Intersect in Space: An Introduction to Topology, 2nd edn. World Scientific Publishing Company, Singapore (1995)
7. Clelland, J.N.: From Frenet to Cartan: The Method of Moving Frames. Graduate Studies in Mathematics, vol. 178. American Mathematical Society (2017)
8. Franzoni, G.: The Klein bottle: variations on a theme. Notices AMS **59**(8), 1076–1082 (2012)
9. Hatcher, A.: Algebraic Topology. Cambridge University Press, Cambridge (2002)
10. Massey, W.S.: Algebraic Topology: An Introduction. Springer, New York (1977)
11. O'Neill, B.: Elementary Differential Geometry, Revised 2nd edn. Academic Press, Cambridge (2006)
12. Rapoport D.L.: Klein bottle logophysics: a unified principle for non-linear systems, cosmology, geophysics, biology, biomechanics and perception. J. Phys.: Conf. Ser. **437**(1) (2012). https://doi.org/10.1088/1742-6596/437/1/012024
13. Tanaka, S.: Topology of cortex visual maps. Forma **12**, 101–108 (1997)
14. Trott, M.: Constructing an algebraic Klein bottle. Math. Educ. Res. **8**(1), 24–27 (1999)
15. Swindale, N.V.: Visual cortex: looking into a Klein bottle. Curr. Biol. **6**(7), 776–779 (1996)
16. Velezmoro, R., Ipanaqué, R., Mechato, J.A.: A mathematica package for visualizing objects inmersed in $\mathbb{R}^4$. In: Misra, S., et al. (eds.) ICCSA 2019. LNCS, vol. 11624, pp. 479–493. Springer, Cham (2019). https://doi.org/10.1007/978-3-030-24311-1_35
17. Wikipedia Homepage. https://en.wikipedia.org/wiki/Klein_bottle. Accessed 24 June 2021
18. Wolfram Demonstrations Project Homepage. https://demonstrations.wolfram.com/4DRotationsOfAKleinBottle/. Accessed 24 June 2021
19. Wolfram, S.: The Mathematica Book, 4th edn. Wolfram Media, Champaign; Cambridge University Press, Cambridge (1999)

On the Shooting Method Applied to Richards' Equation with a Forcing Term

Fabio Vito Difonzo[1]($^{(\boxtimes)}$) and Giovanni Girardi[2]

[1] Dipartimento di Matematica, Università degli Studi di Bari Aldo Moro,
Via E. Orabona 4, 70125 Bari, Italy
`fabio.difonzo@uniba.it`
[2] Istituto di Ricerca sulle Acque, Consiglio Nazionale delle Ricerche, Italy,
Via F. De Blasio 5, 70132 Bari, Italy
`giovanni.girardi@ba.irsa.cnr.it`
`https://sites.google.com/site/fabiovdifonzo/home`

Abstract. The problem of modeling water flow in the root zone with plant root absorption is of crucial importance in many environmental and agricultural issues, and is still of interest in the applied mathematics community. In this work we propose a formal justification and a theoretical background of a recently introduced numerical approach, based on the shooting method, for integrating the unsaturated flow equation with a sink term accounting for the root water uptake model. Moreover, we provide various numerical simulations for this method, comparing the results with the numerical solutions obtained by MATLAB `pdepe`.

Keywords: Numerical simulations · Richards' equation · Shooting method

1 Introduction

The problem of water movement in the root zone needs attention in different scientific areas, and with different tools. For instance, this problem is interesting for applied mathematicians, hydrogeologists, agronomists, and numerical analysis scientists. As a matter of fact, mainly when considering the unsaturated condition (which is exactly the case of subsurface and vadose zone applications) the problem becomes challenging from different areas of mathematical research. On the other hand, it is still open the interest in characterizing soils from hydraulic point of view, with laboratory experiments, coupled sometimes with data driven models for approximating hydraulic functions (see for instance [1]).

In the wide range of applications for this physical process, agronomical applications play a primary role, since the understanding of water dynamics into the soil is necessary for efficiently managing any irrigation process (see for instance [2–4]).

The usual tool for modeling the water flow throughout an unsaturated porous medium is the Richards' equation, a parabolic PDE which still deserves a significant research attention (see, for instance, [5–7]). Such a standard infiltration

© Springer Nature Switzerland AG 2021
O. Gervasi et al. (Eds.): ICCSA 2021, LNCS 12949, pp. 273–286, 2021.
https://doi.org/10.1007/978-3-030-86653-2_20

problem is further complicated by possible fractures in the porous media (see [8,9]); or by soil heterogeneities, which can be treated by well-established discontinuous numerical approaches (see for instance [10–12]), giving rise to successful techniques in 1D and 2D spatial domains ([13,14]).

This model could be even endowed with a sink term representing (for instance) the root water uptake (e.g. [15–20]). In this work we provide a theoretical background for a new approach, first proposed in [15]. Such an approach can manage the choice of Gardner's constitutive relations in solving Richards' equation, differently from [21]; can take into account possibly any selection of root uptake functions; and finally, no variable rescaling is necessary, as in [22].

We consider the θ-form of Richards' equation in a one-dimensional spatial domain:

$$\frac{\partial \theta}{\partial t} = \frac{\partial}{\partial z}\left[D(\theta)\frac{\partial \theta}{\partial z}\right] - \frac{dK(\theta)}{d\theta}\frac{\partial \theta}{\partial z} - R(\theta, t, z), \quad t \in [0,T], \ z \in [0,Z], \quad (1)$$

where θ represents the volumetric water content, R is the root-water uptake term, D is the soil water diffusivity and K is the hydraulic conductivity (see [22]); $[0,T]$ is the time domain and $[0,Z]$ is the spatial domain. The soil-water diffusivity is

$$D(\theta) := K(\theta)\frac{dh(\theta)}{d\theta}; \quad (2)$$

as usual h is the suction head in the unsaturated zone, and the plant-root extraction function $-R \in (-R_s, 0]$, with $R_s \geq 0$, depends on $\theta \in [\theta_r, \theta_S]$, where θ_r is the residual water content and θ_s is the water content at saturation, assuming that z is positive downward. As in [22], it is assumed that R has no explicit dependence on time t (differently from [23]), "despite root-water uptake is, in many cases, driven by daily cycles". It is also assumed that $R'(\theta) > 0$ and $R''(\theta) < 0$.

The function R has generally a sigmoid shape, common to many other biological context (see for instance [24]): a classical piecewise linear model for root water uptake function was proposed in [25] and, more recently, a stepwise approach has been introduced in [26], where stationary solutions are studied for Richards' equation with discontinuous root water uptake models. In [27], different piecewise Feddes uptake functions are considered, according to the depth; in [28] different uptake functions are reviewed, whereas in [29] the use of root uptake functions is discussed when in presence of water with high salinity.

2   The *Shooting Method* Approach and Its Theoretical Background

We now endow (1) with initial and Dirichlet boundary conditions, which characterize water contents at the top and bottom of the domain. In particular, let $\theta^0 : [0,Z] \to [0,+\infty)$ be the initial state profile, $\theta_0 : [0,T] \to [0,+\infty]$ and $\theta_Z : [0,T] \to [0,+\infty]$ be sufficiently smooth boundary functions, and let us assume that

$$\theta(0, z) = \theta^0(z), \quad z \in [0, Z], \tag{3a}$$

$$\theta(t, 0) = \theta_0(t), \quad t \in [0, T], \tag{3b}$$

$$\theta(t, Z) = \theta_Z(t), \quad t \in [0, T]. \tag{3c}$$

In order to face this model, we will first resort to a classical TMoL approach and then we will integrate (1) and approximate its solution by means of the *shooting method*, which in light of boundary conditions (3) seems a natural strategy in this context.

2.1   Kirchhoff Transformation on Richards' Equation with Root-Water Uptake

Since, in practice, numerical methods for directly solving (1)–(3) with respect to θ could be highly unstable because of the nonlinear nature of Richards' equation (see [30]), it is reasonable to transform the θ-based form by a suitable change of variable, to obtain a new equation easier to handle using classical numerical schemes (see also the recent papers [31,32]). Resorting to matric flux potential allows to express (1) in a handier way, which comes from the classical Kirchhoff transformation used in heat equation; more precisely, we introduce the variable

$$\mu(\theta(t, z)) := \int_{\theta_r}^{\theta(t,z)} D(\tau) \, \mathrm{d}\tau, \tag{4}$$

which we plug into the boundary value problem (1)-(3).
Considering that

$$\frac{\mathrm{d}\mu}{\mathrm{d}\theta} = D(\theta)$$

and replacing this in (1) it follows that

$$F(\mu)\frac{\partial \mu}{\partial t} = L\mu - R(\mu), \tag{5}$$

where

$$L := \frac{\partial^2}{\partial z^2} - G(\cdot)\frac{\partial}{\partial z} \tag{6}$$

is the *Kirchhoff operator*,

$$F(\mu) := \frac{1}{D(\theta)}, \quad G(\mu) := \frac{1}{D(\theta)}\frac{\mathrm{d}K}{\mathrm{d}\theta},$$

and $R(\mu)$ is identified with $R(\theta(\mu))$. Finally, we also get

$$\mu(0, z) = \mu^0(z), \quad z \in [0, Z], \tag{7a}$$

$$\mu(t, 0) = \mu_0(t), \quad t \in [0, T], \tag{7b}$$

$$\mu(t, Z) = \mu_Z(t), \quad t \in [0, T]. \tag{7c}$$

Remark 1. If Gardner soil model is considered, then

$$K(h) = K_s e^{\alpha h},$$

where $\alpha \in \mathbb{R}$, $\alpha \neq 0$. Let us stress that $\frac{dK}{dh} = \alpha K(h)$. Now, from (2), it follows that $D(\theta) = \frac{1}{\alpha}\frac{dK}{d\theta}$, and so

$$G(\mu) = \alpha.$$

Thus, within the Gardner model, the Kirchhoff operator $L := \frac{\partial^2}{\partial z^2} - \alpha\frac{\partial}{\partial z}$ is a *linear* elliptic differential operator.

2.2 Solving Richards' Equation Applying Shooting Method

For applying TMoL to (5)–(7), the time derivative operator in the left-hand side in (1) is approximated through a first order finite difference, choosing a time-step Δt and a time mesh $t_n = n\Delta T$, $n = 0, \ldots, N$, for $[0, T]$ such that $T = N\Delta t$. Let $\mu^0(z)$ $z \in [0, Z]$ be a known guess of the solution to (5) at time $t = 0$, we focus on the following second order ordinary differential equation in space

$$F(\mu^n)\frac{\mu^n - \mu^{n-1}}{\Delta t} = L\mu^n - R(\mu^n), \quad n = 1, \ldots, N \tag{8}$$

for $z \in [0, Z]$, with boundary conditions

$$\mu^n(0) = \mu_0(t^n), \quad \mu^n(Z) = \mu_Z(t^n), \quad n = 1, \ldots, N. \tag{9}$$

Rearranging (8) and exploiting (6), we obtain

$$\frac{d^2\mu^n(z)}{dz^2} = \frac{F(\mu^n(z))(\mu^n(z) - \mu^{n-1}(z))}{\Delta t} + R_n + \alpha\nu^n, \tag{10}$$

where $\mu^n(0)$ and $\mu^n(Z)$ are given.

Next we will prove that, under the given assumptions on the functions R and F, solution to the above problem (8) and (9) exists and is unique.

For sake of clarity, we state the argument in a more general setting. Let us assume we have a nonlinear second-order boundary value problem

$$y''(x) = f(x, y, y'), \quad a \leq x \leq b, \quad y(a) = \alpha, \quad y(b) = \beta. \tag{11}$$

It is just an observation that (10) is of the form described by (11).
We have the following result.

Theorem 1 ([33]). *Assume that $f(x, y, z)$ is continuous on the region $R = \{(x, y, z) : a \leq x \leq b, y, z \in \mathbb{R}\}$ and that $\frac{\partial f}{\partial y} = f_y(x, y, z)$ and $\frac{\partial f}{\partial z} = f_z(x, y, z)$ are continuous on R; further, let $\alpha, \beta \in \mathbb{R}$. If there exists a constant $M > 0$ for which f_y and f_z satisfy*

$$f_y(x, y, z) > 0, \quad for\ all\ (x, y, z) \in R, \tag{12a}$$

$$|f_z(x, y, z)| \leq M, \quad for\ all\ (x, y, z) \in R, \tag{12b}$$

then the boundary value problem

$$y'' = f(x, y, y'), \quad y(a) = \alpha, \quad y(b) = \beta$$

has a unique solution $y = y(x)$ *for* $a \leq x \leq b$.

In order to numerically solve (11) we consider, for each $k \in \mathbb{N}$, the following initial value problem

$$y''(x) = f(x, y, y'), \quad a \leq x \leq b, \quad y(a) = \alpha, \quad y'(a) = \alpha'_k, \tag{13}$$

where $\alpha'_k \in \mathbb{R}$. The idea is to select the sequence of initial value problems, depending on $\{\alpha'_k\}_{k \in \mathbb{N}}$, such that

$$\lim_{k \to \infty} y(b, \alpha'_k) = y(b) = \beta, \tag{14}$$

where $y(x, \alpha'_k)$ denotes the solution to (13), for a given $k \in \mathbb{N}$, and $y(x)$ is the solution to (11). We select α'_0 for starting the limit process (14) solving

$$y''(x) = f(x, y, y'), \quad a \leq x \leq b, \quad y(a) = \alpha, \quad y'(a) = \alpha'_0.$$

Then, if $y(b, \alpha'_0)$ is not sufficiently close to β, we keep solving (13) with different values α'_1, α'_2 and so on until $y(b, \alpha'_k)$ is sufficiently close to β. Determining the sought parameter α'_k requires solving a zero-finding problem of the form

$$y(b, \alpha') - \beta = 0$$

with respect to α'. Under the assumptions of Theorem 1, last problem could be solved by using Newton's method, considering the sequence

$$\alpha'_k = \alpha'_{k-1} - \frac{y(b, \alpha'_{k-1}) - \beta}{\eta(b, \alpha'_{k-1})}, \tag{15}$$

where $\eta(x, \alpha')$ solves the following initial value problem

$$\eta''(x, \alpha') = \frac{\partial f}{\partial y}(x, y, y')\eta(x, \alpha') + \frac{\partial f}{\partial y'}(x, y, y')\eta'(x, \alpha'), \quad \eta(a, \alpha') = 0, \quad \eta'(a, \alpha') = 1,$$

for $a \leq x \leq b$.

In order to leverage (15) for solving (8)–(9), we need that

$$f(z, \mu^n, \nu^n) := \frac{F(\mu^n(z))(\mu^n(z) - \mu^{n-1}(z))}{\Delta t} + R_n + \alpha \nu^n$$

fulfills assumptions of Theorem 1. Easy computations provide

$$\frac{\partial f}{\partial \mu^n} = \frac{F'(\mu^n)(\mu^n - \mu^{n-1}) + F(\mu^n)}{\Delta t} + R'(\mu^n), \tag{16}$$

$$\frac{\partial f}{\partial \nu^n} = \alpha. \tag{17}$$

From Eqs. (16) and (17) it follows that (12b) is always satisfied; whereas, in order for (12a) to hold, it suffices to prove that $\mu F(\mu)$ is increasing, since in common scenarios root-water uptake term R is usually increasing as a function of μ, so that $R'(\mu) > 0$. This is a consequence of the following.

Proposition 1. *Let $I \subseteq \mathbb{R}$, $f : I \to \mathbb{R}$ be differentiable over I and let us assume that, for all $x \in I$, $f(x) > 0$, $f'(x) < 0$ and $f''(x) > 0$. Then the function $x \mapsto xf(x)$ is strictly increasing for all $x \in I$.*

Proof. Let $x_0, x_1 \in I$ and $x_0 < x_1$. By contradiction, assume $x_0 f(x_0) \geq x_1 f(x_1)$. Thus

$$x_0 f(x_0) \geq (x_1 - x_0)f(x_1) + x_0 f(x_1).$$

Rearranging term yields

$$-x_0(f(x_1) - f(x_0)) \geq (x_1 - x_0)f(x_1),$$

and using the fact that

$$f(x_1) - f(x_0) \geq f'(x_0)(x_1 - x_0)$$

it follows that

$$-x_0 f'(x_0) \geq f(x_1),$$

since $x_1 - x_0 > 0$. But $f(x_1) > 0$, giving a contradiction.

We observe now that, in Gardner model, F satisfies conditions of Proposition 1; moreover, since $\mu^{n-1}(z)$ is always positive – on the account of (4), then $F(\mu^n)(\mu^n - \mu^{n-1})$ is increasing as a function of μ^n, since so is $F(\mu^n)\mu^n$. Hence, Theorem 1 guarantees existence and uniqueness of solution to (10).

The idea now is to express the boundary value problem (8) and (9) as a first order differential equation endowed with suitable initial conditions, so that this new differential problem yields the same solution as the first one: this is the core idea of *shooting method* for solving boundary value problems.

Similarly to [15], we notice that Eq. (8) can be written as a first order system of differential equations of the form

$$\frac{d\mu^n}{dz} = \nu^n, \tag{18a}$$

$$\frac{d\nu^n}{dz} = F(\mu^n)\frac{\mu^n - \mu^{n-1}}{\Delta t} + G(\mu^n)\nu^n + R(\mu^n), \tag{18b}$$

which we endow with the following *initial conditions*

$$\mu^n(0) = \mu_0(t^n), \quad \nu^n(0) = \hat{\nu}, \tag{19}$$

$\hat{\nu}$ being computed by standard numerical methods (see, for instance, [34]) in such a way that second condition in (9) is met, that is

$$\mu^n(Z) = \mu_Z(t_n). \tag{20}$$

3 Numerical Simulations

Example 1. From [35] a loamy sand is considered with the following hydraulic parameters:

$$\theta_r = 0.057, \quad \theta_S = 0.41, \quad K_S = 350.2\,\text{cm/d}. \tag{21}$$

For this example, we consider a simulation time of $T = 3$ h and a spatial domain of $Z = 20$ cm.

In order to apply our numerical methods to (8) for this specific soil, it is advisable to first scale quantities and operators so to simplify the equations, as in [22]; in particular, we set

$$\Theta := \frac{\theta - \theta_r}{\theta_S - \theta_r}. \tag{22}$$

For exemplification purposes, Dirichlet boundary conditions will only be considered in the numerical computations displayed in Fig. 1.

Moreover, we will restrict ourselves to the setting of non-classical symmetry classification: that is, soil-water diffusivity and root-water uptake satisfy the condition

$$- R(\mu) = A\frac{\mu}{D(\mu)} + \kappa\mu. \tag{23}$$

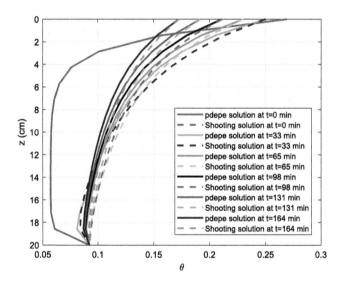

Fig. 1. Numerical simulations for the loamy sand with parameters (21), obtained by shooting method, compared with the solution arising from MATLAB **pdepe** tool.

Specifically, as in [22], we will consider the functions[1]:

$$D(\mu) := ce^{\mu},$$
$$R(\mu) := \mu(1 - e^{-\mu}),$$

with $c = 1 - e^{-1}$ and $k = -1$.

Further, dealing with such re-scaled parameters, for all $t \geq 0$ we consider a time-varying top boundary condition at $z = 0$ given by

$$\Theta(t, 0) = \frac{2T - t}{2T} 0.6, \tag{24}$$

meaning that $\theta(0, 0) = 0.2688$; and a constant bottom condition at $z = Z$ given by

$$\Theta(t, Z) = 0.1, \tag{25}$$

implying that $\theta(0, Z) = 0.0923$.

Let us point out that, after determining the normalized solution $\Theta(t, z)$ (either by shooting or pdepe method) we go back to original water content by computing

$$\theta(t, z) = (1 - \Theta(t, z))\theta_r + \Theta(t, z)\theta_S.$$

Within this specific setting, we use as step-sizes $\Delta t = 8.8974$ min, and $\Delta z \approx 1.3$ cm when comparing pdepe and the shooting method: we point out that the shooting method runs using a temporal step-size of one order less than pdepe, while retaining similar performances (see Figs. 1 and 2).

In Fig. 1 we build numerical solutions of (18) with parameters as in (21) and initial condition as suggested in [22] and according to the aforementioned hydraulic functions and parameters, for comparing shooting solution to pdepe solution. The initial condition at time $t = 0$ is a linear combination of exponential functions of the form

$$\Theta^0(z) := c_1 e^{\lambda_1 z} + c_2 e^{\lambda_2 z}, \tag{26}$$

such that $\Theta^0(0) = 0.6$, $\Theta^0(Z) = 0.1$. Thus, after straightforward computations, we get

$$c_1 = \frac{0.1e^{-\lambda_2 Z} - 0.6}{e^{\lambda_2 Z} - e^{\lambda_1 Z}},$$

$$c_2 = \frac{0.1 - 0.6e^{-\lambda_1 Z}}{e^{\lambda_2 Z} - e^{\lambda_1 Z}},$$

with $\lambda_{1,2}$ chosen arbitrarily but with different sign, so to make (26) retain a physical meaning.

Figure 2 summarizes the behavior of different methods in this example.

---

[1] We highlight that the uptake function R should be used as reported here, instead as the one originally proposed in [22], in order for (23) to hold; the function D stays the same.

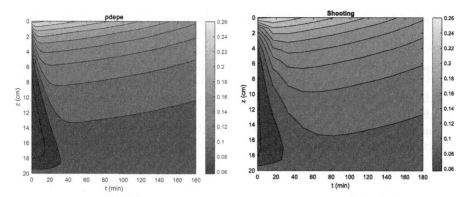

(a) Pdepe solution: Example 1.

(b) Shooting method solution: Example 1.

Fig. 2. Dynamics of water content over time for the aforementioned problem, referring to a loamy sand in 1, with an uptake function as in [22]. The figures refer, respectively, to the MATLAB **pdepe** solution and the shooting method, described in Sect. 2.

Example 2. Here, we consider a silty loam Swift current wheat site. We fix $Z = 70$ cm and $T = 3$ days, while the following hydraulic parameters are as follows:

$$\theta_r = 0, \quad \theta_S = 0.48, \quad K_S = 100.1 \, \text{cm/d}$$

We select boundary conditions as

$$\theta(t, 0) = 0.3610, \; \theta(t, Z) = 0.2051, \quad t \in [0, T],$$

while initial condition is chosen to be

$$\theta(0, z) = \theta(0, 0) + \frac{\theta(0, Z) - \theta(0, 0)}{Z} z.$$

In this case example, we consider a different root uptake model that in Example 1, which would not require any re-scaling in order to be fruitfully handled. More specifically, we adopt a Feddes-type stress function with the following form:

$$R = R_{\max} \alpha (h), \tag{27}$$

where $R_{\max}$ is the so-called *maximum uptake*, representing the potential root water uptake (see [21]), and the function $\alpha(h)$ is defined, for $h_4 < h_3 < h_2 < h_1 < 0$, as

$$\alpha(h) := \begin{cases} 0, & \text{if } h_1 \leq h \leq 0 \text{ or } h \leq h_4, \\ \frac{h - h_1}{h_2 - h_1}, & \text{if } h_2 < h < h_1, \\ 1, & \text{if } h_3 \leq h \leq h_2, \\ \frac{h - h_4}{h_3 - h_4}, & \text{if } h_4 < h < h_3, \end{cases} \tag{28}$$

and represents the stress response function.

Such a function is piecewise linear on the interval $[h_4, 0]$, so that $\alpha(h)$ is just continuous with finitely many jumps in its derivative. Let us stress that results in Sect. 2 still applies, and that - from a numerical point of view - a wise selection of the discretized grid allows to overlook such points, where the numerical integration may become stiff: this example shows, actually, that the shooting method is also robust with respect to possibly mildly nonsmooth uptake models.

Figure 3 depicts the solutions computed through MATLAB `pdepe` and the Shooting method (18) for this specific soil setting, showing their similar behaviors.

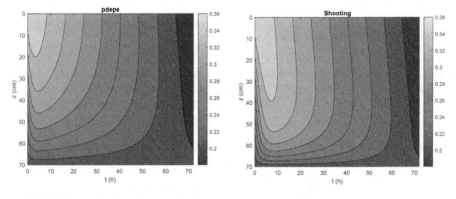

(a) Pdepe solution: Example 2.   (b) Shooting method solution: Example 2.

Fig. 3. Dynamics of water content over time for the Example in 2. The soil is a silty loam, and the uptake function is of Feddes type, cropped with wheat. The figures refer, respectively, to the MATLAB `pdepe` solution and the shooting method, described in Sect. 2.

Example 3. Let us consider the same soil parameters as in Example 2, with again $T = 3$ days and $Z = 70$ cm. Same initial and boundary conditions as in Example 2 are selected.

For this example a Li uptake function (see [36]) is now modeling $\alpha(h)$ in (27). More specifically, we select

$$\alpha(h) := \begin{cases} \dfrac{1}{1+e^{12.25\left(0.504 - \frac{\theta(h_0) - \theta(h)}{\theta(h_0) - \theta(h_{fc})}\right)}}, & h_{fc} < h < 0, \\[2ex] \min\left(1, \dfrac{(H(h)\tau(h))^{0.5}}{T_p/T_m}\right), & h_{pwp} \le h \le h_{fc}, \\[2ex] 0, & h < h_{pwp}, \end{cases}$$

being

$$H(h) := \frac{h - h_{pwp}}{h_{fc} - h_{pwp}}, \quad \tau(h) := \frac{\theta(h) - \theta(h_{pwp})}{\theta(h_{fc}) - \theta(h_{pwp})}, \quad h_{pwp} < h_{fc} < 0.$$

Again, such an uptake model $\alpha(h)$ is only continuous over the interval $[h_{pwp}, 0]$, but nonetheless shooting method performs well and provides convergence, as expected from Sect. 2.

Here, as in [36], the values of field capacity and permanent wilting point are, respectively, $h_{fc} = -3$ m, $h_{pwp} = -160$ m, the maximum potential transpiration rate $T_m = 7.06 \cdot 10^{-8}$ m/s and the potential transpiration rate $T_p = 7.407 \cdot 10^{-8}$ m/s; finally, $\theta(h)$ is defined, as in [32],

$$\theta(h) = \begin{cases} \theta_r + (\theta_S - \theta_r)e^{\lambda h}, & \text{for } h \leq 0, \\ 0, & \text{for } h > 0. \end{cases} \tag{29}$$

A comparison between pdepe and the shooting method is depicted in Fig. 4. As it can be seen, the shooting method keeps retaining a solution shape similar to pdepe but, as shown in [15], it can manage more general and specific uptake models and boundary conditions than pdepe.

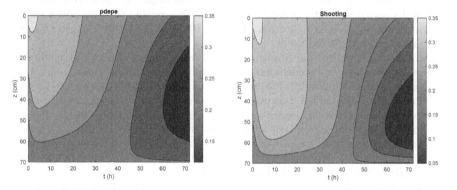

(a) Pdepe solution: Example 3.   (b) Shooting method solution: Example 3.

Fig. 4. Water movement over time for the problem in Example 3, still regarding a silty loam soil. The figures depict, respectively, MATLAB pdepe and the shooting method solutions.

4 Conclusions and Future Works

In this paper we propose the theoretical background for the shooting method applied to Richards' equation with a sink term as proposed in [15]. We show convergence results and exemplify on several soils with different uptake models. A new approach, recently proposed in [37], rewrites (1) as an integral equation and solves it by standard quadrature schemes. We anticipate to tackle such an integral equation in presence of layered soils, and integrate it using a peridynamics formulation, leveraging techniques developed in [38–40]). Modeling memory terms of root water uptake (e.g. [41–43]) can also be faced in future mathematical works, maybe also considering variations due to heat transfer [44].

Acknowledgments. The first author has been funded by *REFIN* Project, grant number 812E4967. The second author acknowledges the partial support of the project RIUBSAL (grant number 030_DIR_2020_00178), funded by Regione Puglia through the P.S.R. Puglia 2014/2020 - Misura 16 – Cooperazione - Sottomisura 16.2 call.

References

1. Maggi, S.: Estimating water retention characteristic parameters using differential evolution. Comput. Geotech. **86**, 163–172 (2017). https://doi.org/10.1016/j.compgeo.2016.12.025
2. Friedman, S.P., Communar, G., Gamliel, A.: DIDAS - user-friendly software package for assisting drip irrigation design and scheduling. Comput. Electron. Agricult. **120**, 36–52 (2016). https://doi.org/10.1016/j.compag.2015.11.007
3. Brunetti, G., Šimůnek, J., Bautista, E.: A hybrid finite volume-finite element model for the numerical analysis of furrow irrigation and fertigation. Comput. Electron. Agricult. **150**, 312–327 (2018). https://doi.org/10.1016/j.compag.2018.05.013
4. Arbat, G., Puig-Bargués, J., Duran-Ros, M., Barragán, J., de Cartagena, F.R.: Drip-Irriwater: computer software to simulate soil wetting patterns under surface drip irrigation. Comput. Electron. Agricult. **98**, 183–192 (2013). https://doi.org/10.1016/j.compag.2013.08.009
5. Berardi, M., Difonzo, F., Notarnicola, F., Vurro, M.: A transversal method of lines for the numerical modeling of vertical infiltration into the vadose zone. Appl. Numer. Math. **135**, 264–275 (2019). https://doi.org/10.1016/j.apnum.2018.08.013
6. Berardi, M., Difonzo, F.V.: Strong solutions for Richards' equation with Cauchy conditions and constant pressure gradient. Environ. Fluid Mech. **20**(1), 165–174 (2019). https://doi.org/10.1007/s10652-019-09705-w
7. Casulli, V.: A coupled surface-subsurface model for hydrostatic flows under saturated and variably saturated conditions. Int. J. Numer. Meth. Fluids **85**(8), 449–464 (2017). https://doi.org/10.1002/fld.4389
8. Nordbotten, J.M., Boon, W.M., Fumagalli, A., Keilegavlen, E.: Unified approach to discretization of flow in fractured porous media. Comput. Geosci. **23**(2), 225–237 (2018). https://doi.org/10.1007/s10596-018-9778-9
9. Berre, I., et al.: Verification benchmarks for single-phase flow in three-dimensional fractured porous media. Adv. Water Resour. **147**, 103759 (2021). https://doi.org/10.1016/j.advwatres.2020.103759
10. Berardi, M.: Rosenbrock-type methods applied to discontinuous differential systems. Math. Comput. Simul. **95**, 229–243 (2014). https://doi.org/10.1016/j.matcom.2013.05.006
11. Lopez, L., Maset, S.: Time-transformations for the event location in discontinuous ODEs. Math. Comp. **87**(3), 2321–2341 (2018). https://doi.org/10.1090/mcom/3305
12. Del Buono, N., Elia, C., Lopez, L.: On the equivalence between the sigmoidal approach and Utkin's approach for piecewise-linear models of gene regulatory. SIAM J. Appl. Dyn. Syst. **13**(3), 1270–1292 (2014). https://doi.org/10.1137/130950483
13. Berardi, M., Difonzo, F., Vurro, M., Lopez, L.: The 1D Richards' equation in two layered soils: a Filippov approach to treat discontinuities. Adv. Water Resour. **115**, 264–272 (2018). https://doi.org/10.1016/j.advwatres.2017.09.027
14. Berardi, M., Difonzo, F., Lopez, L.: A mixed MoL-TMoL for the numerical solution of the 2D Richards' equation in layered soils. Comput. Math. Appl. **79**(7), 1990–2001 (2019). https://doi.org/10.1016/j.camwa.2019.07.026

15. Difonzo, F.V., Masciopinto, C., Vurro, M., Berardi, M.: *Shooting* the numerical solution of moisture flow equation with root water uptake models: a Python tool. Water Resour. Manage. **35**(8), 2553–2567 (2021). https://doi.org/10.1007/s11269-021-02850-2

16. Camporese, M., Daly, E., Paniconi, C.: Catchment-scale Richards equation-based modeling of evapotranspiration via boundary condition switching and root water uptake schemes. Water Resour. Res. **51**(7), 5756–5771 (2015). https://doi.org/10.1002/2015WR017139

17. Manoli, G., Bonetti, S., Scudiero, E., Morari, F., Putti, M., Teatini, P.: Modeling soil-plant dynamics: assessing simulation accuracy by comparison with spatially distributed crop yield measurements. Vadose Zone J. **14**(2) (2015), vzj2015-05. https://doi.org/10.2136/vzj2015.05.0069

18. Roose, T., Fowler, A.: A model for water uptake by plant roots. J. Theor. Biol. **228**(2), 155–171 (2004). https://doi.org/10.1016/j.jtbi.2003.12.012

19. Couvreur, V., Vanderborght, J., Javaux, M.: A simple three-dimensional macroscopic root water uptake model based on the hydraulic architecture approach. Hydrol. Earth Syst. Sci. **16**(8), 2957–2971 (2012). https://doi.org/10.5194/hess-16-2957-2012

20. Albrieu, J.L.B., Reginato, J.C., Tarzia, D.A.: Modeling water uptake by a root system growing in a fixed soil volume. Appl. Math. Model. **39**(12), 3434–3447 (2015). https://doi.org/10.1016/j.apm.2014.11.042

21. Simunek, J., Hopmans, J.W.: Modeling compensated root water and nutrient uptake. Ecol. Model. **220**(4), 505–521 (2009). https://doi.org/10.1016/j.ecolmodel.2008.11.004

22. Broadbridge, P., Daly, E., Goard, J.: Exact solutions of the Richards equation with nonlinear plant-root extraction. Water Resour. Res. **53**(11), 9679–9691 (2017). https://doi.org/10.1002/2017WR021097

23. Mathur, S., Rao, S.: Modeling water uptake by plant roots. J. Irrig. Drain. Eng. **125**(3), 159–165 (1999). https://doi.org/10.1061/(ASCE)0733-9437(1999)125:3(159)

24. D'Abbicco, M., Buono, N.D., Gena, P., Berardi, M., Calamita, G., Lopez, L.: A model for the hepatic glucose metabolism based on hill and step functions. J. Comput. Appl. Math. **292**, 746–759 (2016). https://doi.org/10.1016/j.cam.2015.01.036

25. Feddes, R.A., Kowalik, P., Kolinska-Malinka, K., Zaradny, H.: Simulation of field water uptake by plants using a soil water dependent root extraction function. J. Hydrol. **31**(1), 13–26 (1976). https://doi.org/10.1016/0022-1694(76)90017-2

26. Berardi, M., Girardi, G., D'Abbico, M., Vurro, M.: Optimizing water consumption in Richards' equation framework with step-wise root water uptake: a simplified model (2021, submitted)

27. Jarvis, N.: A simple empirical model of root water uptake. J. Hydrol. **107**(1), 57–72 (1989). https://doi.org/10.1016/0022-1694(89)90050-4

28. Babazadeh, H., Tabrizi, M., Homaee, M.: Assessing and modifying macroscopic root water extraction basil (ocimum basilicum) models under simultaneous water and salinity stresses. Soil Sci. Soc. Am. J. **81**, 10–19 (2017). https://doi.org/10.2136/sssaj2016.07.0217

29. Coppola, A., Chaali, N., Dragonetti, G., Lamaddalena, N., Comegna, A.: Root uptake under non-uniform root-zone salinity. Ecohydrology **8**(7), 1363–1379 (2015). https://doi.org/10.1002/eco.1594

30. Bergamaschi, L., Putti, M.: Mixed finite elements and Newton-type lineariza-tions for the solution of Richards' equation. Int. J. Numer. Meth. Eng. **45**, 1025–1046 (1999). https://doi.org/10.1002/(SICI)1097-0207(19990720)45:8⟨1025::AID-NME615⟩3.0.CO;2-G

31. Li, N., Yue, X., Ren, L.: Numerical homogenization of the Richards equation for unsaturated water flow through heterogeneous soils. Water Resour. Res. **52**(11), 8500–8525 (2016). https://doi.org/10.1002/2015WR018508

32. Suk, H., Park, E.: Numerical solution of the Kirchhoff-transformed Richards equa-tion for simulating variably saturated flow in heterogeneous layered porous media. J. Hydrol. **579**, 124213 (2019). https://doi.org/10.1016/j.jhydrol.2019.124213

33. Mathews, J.H., Fink, K.K.: Numerical Methods Using MATLAB, 4th edn. Pearson, New York (2004)

34. D'Ambrosio, R., Jackiewicz, Z.: Construction and implementation of highly stable two-step continuous methods for stiff differential systems. Math. Comput. Simul. **81**(9), 1707–1728 (2011). https://doi.org/10.1016/j.matcom.2011.01.005

35. Carsel, R.F., Parrish, R.S.: Developing joint probability distributions of soil water retention characteristics. Water Resour. Res. **24**(5), 755–769 (1988). https://doi.org/10.1029/WR024i005p00755

36. Li, K.Y., De Jong, R., Coe, M.T., Ramankutty, N.: Root-water-uptake based upon a new water stress reduction and an asymptotic root distribution function. Earth Interact. **10**(14), 1–22 (2006)

37. Berardi, M., Difonzo, F.: A quadrature-based scheme for numerical solutions to Kirchhoff transformed Richards' equation (2021, submitted)

38. Lopez, L., Pellegrino, S.F.: A spectral method with volume penalization for a nonlinear peridynamic model. Int. J. Numer. Meth. Eng. **122**(3), 707–725 (2020). https://doi.org/10.1002/nme.6555

39. Lopez, L., Pellegrino, S.F.: A space-time discretization of a nonlinear peridynamic model on a 2D lamina (2021)

40. Lopez, L., Pellegrino, S.F., Coclite, G., Maddalena, F., Fanizzi, A.: Numerical methods for the nonlocal wave equation of the peridynamics. Appl. Numer. Math. **155**, 119–139 (2018). https://doi.org/10.1016/j.apnum.2018.11007

41. Carminati, A.: A model of root water uptake coupled with rhizosphere dynamics. Vadose Zone J. **11**, vzj2011-0106 (2012). https://doi.org/10.2136/vzj2011.0106

42. Kroener, E.: How mucilage affects soil hydraulic dynamics. J. Plant Nutr. Soil Sci. **184**, 20–24 (2021). https://doi.org/10.1002/jpln.202000545

43. Wu, X., Zuo, Q., Shi, J., Wang, L., Xue, X., Ben-Gal, A.: Introducing water stress hysteresis to the Feddes empirical macroscopic root water uptake model. Agricul. Water Manage. **240**, 106293 (2020). https://doi.org/10.1016/j.agwat.2020.106293

44. Tubini, N., Gruber, S., Rigon, R.: A method for solving heat transfer with phase change in ice or soil that allows for large time steps while guaranteeing energy conservation. Cryosphere Discuss. **2020**, 1–41 (2020). https://doi.org/10.5194/tc-2020-293

Preliminary Analysis of Interleaving PN-Sequences

Sara D. Cardell[1]([✉]), Amparo Fúster-Sabater[2], and Verónica Requena[3]

[1] Centro de Matemática, Computação e Cognição, UFABC, Santo André-SP, Brazil
s.cardell@ufabc.edu.br
[2] Instituto de Tecnologías Físicas y de la Información (CSIC), Madrid, Spain
amparo@iec.csic.es
[3] Departamento de Matemáticas, Universidad de Alicante, Alicante, Spain
vrequena@ua.es

Abstract. Some pseudorandom sequences with good crytographic features can be obtained from the interleaving of other families of sequences with unsuitable properties. PN-sequences obtained from maximum-length Linear Feedback Shift Registers exhibit good statistical aspects, such as balancedness, large period, adequate distribution of 0s and 1s and excellent autocorrelation, although their linearity makes them vulnerable against cryptographic attacks. In this work, we present a preliminary analysis on the random features of the interleaving of shifted versions of a PN-sequence. The application of statistical and graphic tests and their corresponding results complete the work.

Keywords: PN-sequence · t-interleaved sequence · Statistical tests · Randomness

1 Introduction

The interleaving of sequences has been a topic quite studied in the literature [12,25,26]. Interesting algebraic and geometric results can be obtained by interleaving term by term integer sequences [17]. In [36], the authors obtained new families of almost balanced binary sequences with an optimal auto-correlation value. As an extension of this work, in [19] the author investigated at what extent interleaved binary sequences with period $4p$ obtained with the technique given in [36] from Legendre and Hall sequences have high linear complexity. Furthermore, in [27] the authors explored the linear complexity of three families of binary sequences built from interleaving Legendre sequences, generalized GMW sequences and twin-prime sequences (see also [38]). The interleaving of pseudorandom sequences has been also used for signal synchronization (see [6]). Here, the author constructed a binary timing sequence produced by interleaving shorter pseudorandom sequences and discovered the parameter values giving the minimum acquisition time. This paper also shows that there exists an optimum number of component sequences for a proper interleaving.

© Springer Nature Switzerland AG 2021
O. Gervasi et al. (Eds.): ICCSA 2021, LNCS 12949, pp. 287–297, 2021.
https://doi.org/10.1007/978-3-030-86653-2_21

The output sequences of a maximum-length LFSR (Linear Feedback Shift Register) are called PN-sequences (or m-sequences) [24]. Some keystream generators combine these PN-sequences using nonlinear functions in order to increase their linear complexity and destroy their linearity producing keystream sequences for cryptographic purposes. A very popular group of keystream generators is the family of decimation-based sequence generators. Inside this class, the shrinking generator was the first one to be introduced [16]. This generator is very easy to implement and its formation rule is based on the irregular decimation of one PN-sequence using the position of the ones of another PN-sequence. In [8], the authors proved that the sequence produced by the shrinking generator (the shrunken sequence) can be also generated interleaving shifted versions of a PN-sequence. The vulnerability of this generator consists in the fact that the shifts are known (see [8]), i.e., a shrunken sequence cannot be generated from random shifted versions of a given PN-sequence. Some authors used this fact to perform cryptanalytic attacks in order to break this generator (see for instance [7–11]). A natural way to get over this weakness is to modify the shifts or to interleave PN-sequences produced by different primitive polynomials.

In [12], the authors introduced the concept of t-interleaved sequence defined as the sequence resultant of interleaving t sequences. They focused on the study of the characteristics of the sequences obtained interleaving shifted versions of a single PN-sequence with different shifts. In this work, we present a preliminary study on the randomness of these sequences. Furthermore, we perform an important battery of tests such as FIPS $140 - 2$, created by the National Institute of Standard and Technology [37], Maurer's Universal Statistical Test [31], and Lempel-Ziv Complexity Test [18] as well as some graphic tests effective to investigate the behaviour of these sequences. The sequences that pass all the tests above and have pseudorandom behaviour are potential candidates to be used as keystream sequences in stream cipher applications [4,33]. Recall that in stream ciphers the encryption is made XOR-ing the plaintext and the keystream. The decryption is also made XOR-ing the ciphertext and the keystream. This keystream must satisfy certain randomness characteristics, since the security of the cipher depends completely on such a sequence. For example, the shrinking generator is part of the internal structure of different stream ciphers as the EP0619659A2 [15], an European patent application; or the Decim^{v_2}, a hardware oriented stream cipher submitted to the ECRYPT Stream CipherProject (eSTREAM) [1], among other applications [5,14]. In addition, the shrunken sequence has been considered as a particular solution of a kind of linear difference equations. In [13,22] the authors analyzed cryptographic parameters of this decimated sequence in terms of solutions to linear equations. This sequence has been also studied as the output sequence of linear elementary cellular automata (CA) [8]. In [8] the authors used these CA and their properties to recover the complete shrunken sequence. Therefore, it would be interesting to study these applications using the t-interleaving sequences instead of the shrunken sequence and compare the results.

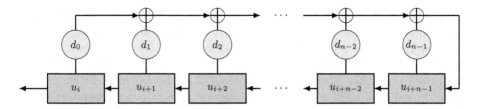

Fig. 1. An LFSR of length n

This paper is organized as follows: In Sect. 2, we introduce some basic notions and generalities about PN-sequences and the interleaving of such sequences. In Sect. 3, we perform a randomness analysis of the interleaving of shifted versions of a PN-sequence. Tables, figures and numerical results of the applied tests are also provided. At last, in Sect. 4, we conclude our research with some finale statements and some directions for the future.

2   Fundamentals and Basic Concepts

Let $\mathbb{F}_2 = \{0, 1\}$ be the binary field, also known as the Galois field of two elements. Let $\{u_i\} = \{u_0, u_1, u_2, u_3, \ldots\}$ be a binary sequence, that is, each term of the sequence satisfies $u_i \in \mathbb{F}_2$, $i \geq 0$. We say that $\{u_i\}$ is a periodic sequence if there exists a positive integer τ such that $u_{i+\tau} = u_i$, for all $i \geq 0$. This number τ is known as the period of the sequence.

Let n be a positive integer and $d_0, d_1, \ldots, d_{n-1}$ some elements of $\mathbb{F}_2$. A sequence $\{u_i\}$ is called a binary n-th order linear recurring sequence if it satisfies

$$u_{i+n} = d_{n-1}u_{i+n-1} + d_{n-2}u_{i+n-2} + \ldots + d_1u_{i+1} + d_0u_i, \qquad i \geq 0 \quad (1)$$

The relation in (1) is known as an n-th order linear recurrence relationship (l.r.r). The polynomial of degree n given by

$$p(x) = x^n + d_{n-1}x^{n-1} + d_{n-2}x^{n-2} + \ldots + d_1x + d_0 \in \mathbb{F}_2[x],$$

is called the characteristic polynomial of the l.r.r. (and the characteristic polynomial of $\{u_i\}$).

The generation of linear recurring sequences can be carried out through Linear Feedback Shift Registers (LFSRs) [24]. An LFSR of length n is a device composed of n interconnected stages. The initial state (stage contents at round zero) is the seed, and since the register operates in a deterministic form, the resultant sequence is completely determined by the initial state. The input of each round is a bit resultant from applying a linear transformation function to a previous state (see Fig. 1). If the characteristic polynomial $p(x)$ of the corresponding l.r.r. is primitive, then the LFSR is said to be a maximum-length. Furthermore, the resultant sequence $\{u_i\}$ obtained from a maximum-length LFSR of n stages is called PN-sequence (or m-sequence) and its period is $\tau = 2^n - 1$ (2^{n-1} ones and $2^{n-1} - 1$ zeros) [24].

The linear complexity of a sequence, denoted by LC, can be defined as the lowest order l.r.r. that produces the sequence (in other words, the length of the shortest LFSR that produces it). The Berlekamp-Massey algorithm can determine the characteristic polynomial and the LC of the sequence given $2 \cdot LC$ bits of the sequence [29]. Therefore, it is obvious that the LC must be as large as possible and that LFSRs should never operate as keystream generators (since their LC is quite low). Their linearity has to disappear; that is, their LC has to grow in order to consider their output sequences for cryptographic goals.

The following definition introduces the concept of t-interleaved sequence, the main notion of this paper.

Definition 1. *The sequence $\{v_j\}$ is a t-**interleaved sequence** if it is obtained interleaving t sequences $\{a_i^{(1)}\}$, $\{a_i^{(2)}\}$, ..., $\{a_i^{(t)}\}$, all of them with the same period τ. As a consequence, the sequence $\{v_j\}$ has the following form:*

$$\{v_j\} = \left\{ a_0^{(1)}, a_0^{(2)}, \ldots, a_0^{(t)}, a_1^{(1)}, a_1^{(2)}, \ldots, a_1^{(t)}, \ldots, a_{\tau-1}^{(1)}, a_{\tau-1}^{(2)}, \ldots, a_{\tau-1}^{(t)} \right\}.$$

In this work, we consider that each one of the sequences $\{a_i^{(k)}\}$, $k = 1, \ldots, t$, is a shifted version a PN-sequence $\{a_i\}$. It is worth noticing that if the characteristic polynomial of $\{a_i\}$ is primitive with degree n, then the resultant t-interleaved sequence is nearly balanced with $t \cdot 2^{n-1}$ ones.

In [12], the authors studied the period and LC of the interleaving of PN-sequences. They mostly focused on the interleaving of 2^r (a power of two) PN-sequences produced by the same primitive polynomial, that is, shifted versions of one single PN-sequence. In Sect. 3, we perform a deeper analysis on the random characteristics of these new interleaving sequences, maintaining the focus on the interleaving of shifted versions of a PN-sequence.

3 Randomness of the t-interleaved Sequences

Currently, an important and widely studied topic is to design non-linear lightweight cryptography generators that produce sequences with good distribution and statistical properties [8,20]. The study of randomness and unpredictability of the sequences obtained by random number generators is essential to assess the quality of them and their use in many applications as cryptography, numerical analysis, information transmission, or game theory, among others [1,21,23]. For that, we need some tools to evaluate whether a generator is random or not.

There are different statistical batteries and graphic techniques to carry out this purpose. In this work, we present some results obtained through some of the main statistical tests, as FIPS 140-2, Maurer's Universal Statistical Test, and Lempel-Ziv Complexity Test.

Moreover, we show some graphic tests, as chaos game or return map, which help us to visualize the sequences generated and determine if they can be considered random. It is worth highlighting that although the sequences pass all these tests, it does not mean we have a perfect random number generator, but the confidence in the generator increases. Moreover, we should expect some sequences to appear nonrandom and fail at least some of the tests. However, if many sequences fail the tests, we should be suspicious that our generator is not good.

In our study, we have analyzed hundreds of t-interleaved sequences for different values of t and generated from characteristic polynomials of degree L, with $15 \leq L \leq 21$. In this section, we show the media of the values obtained for t-interleaved sequences obtained with various characteristic polynomials of degree 21.

Firstly, we expose results obtained for our t-interleaved sequences using the statistical tests mentioned above.

FIPS 140-2
The FIPS $140 - 2$ is composed of four statistical tests which evaluate whether a given sequence of numbers is random. For this goal, we need to assess binary streams of 20000 bits. Any failure in some of these tests means that the sequence must be rejected. Next, we give a little introduction about these tests and the criteria used to determine whether a sequence pass each of the tests.

1. *Monobit test*: Count the number of 0 and 1 in the bitstream. If we denote by X the number of 1s, then the test is passed if $X \in (9725, 10275)$.
2. *Poker test*: Divide the sequence into blocks of 4 bits. Count and save the number of coincidences of the 16 possible decimal values for the 4 bit blocks. Let $f(i)$ be the number of times that each one of these values appear, with $i \in [0, 15]$. Compute

$$X = \frac{16}{5000} \sum_{i=1}^{16} f(i)^2 - 5000.$$

If $X \in (1.03, 57.4)$, then the test is passed.
3. *Runs test*: A run is a maximal succession of consecutive identical bits (all ones or all zeros) within the bitstream. Count and store the frequencies of runs of all lengths (of lengths 1 through 6) in the sequence. For the intentions of this test, runs with length >6 are considered in the range of length 6. If the runs occurred are each within the corresponding interval specified in Table 1, then the test is passed.
4. *Long runs test*: Long runs are runs of length ≥ 26 (either zeros or ones).

We show the results of the Runs test by a graphic representation. In Fig. 2, we determine that the test is passed if the runs (runs of zeros, represented by a red line, and runs of ones, represented by a blue line) that occur are each within the green lines which represent the specific intervals determined in Table 1.

Table 1. Required intervals for the different lengths of runs in the Runs Test

Length of the run	Required interval
1	2,315–2,685
2	1,114–1,386
3	527–723
4	240–384
5	103–209
6+	103–209

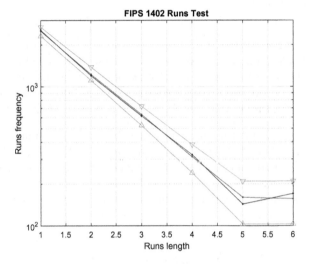

Fig. 2. Runs test for a 4-interleaved sequence of degree $L = 16$. (Color figure online)

Maurer's Universal Test
This test is one of the 15 tests from the NIST statistical suite [3] for cryptographic use. It locates the most general class of statistical faults. The randomness of a binary sequence is checked by the entropy of the sequence, which is calculated from the repetition intervals of blocks of length l in the sequence. The sequence can be considered random, if the p-valued obtained is ≥ 0.01.

Lempel-Ziv Compression Test
This is a compression algorithm which replace strings that repeat in the sequence with references to locations where they appeared first or with the length of repeated pattern. The aim of this test is to determine how far the sequence evaluated can be compressed. For this, the number of distinct patterns in the sample stream will be analyzed. It is considered to be random if it can not be compressed. If the p-value is greater or equal to 0.01, then Lempel-Ziv Test is passed.

All the analyzed t-interleaved sequences have successfully passed each one of the previous tests. Table 2 presents a small sample of the results obtained in Monobit test, Poker test, Maurer's Universal test, and Lempel-Ziv test, given different t-interleaved sequences obtained by characteristic polynomials of degree 21. All these values are the average of the results obtained for any sample of t-interleaved sequences studied.

Table 2. Statistical test results for t-interleaved sequences obtained by characteristic polynomials of degree 21.

t-interleaved	Monobit test	Poker test	Maurer's test	Lempel-Ziv test
3	10008	20.6272	0.5778	0.9719
4	10001	16.2688	0.5715	0.6703
5	10018	16.9600	0.0594	0.0931
6	9838	10.7584	0.2842	0.5584
7	9945	17.4656	0.1482	0.6703
8	9934	15.4048	0.1422	0.6703

Next, we present two powerful graphic tools, the return application and chaos game, which allow us to analyze the behaviour of our t-interleaved sequences from a visual representation.

Return map
The return map visually measures the entropy of the sequence. This method is an important cryptanalysis tool for chaotic cryptosystems introduced in [35] to destroy diverse schemes based on the Lorenz system.

A interesting way to explore a random number generator is to create a visualisation of the numbers that it produces. This type of methods should not be considered as a formal analysis of the study of pseudorandom sequences, but it can get a rough impression of the functioning of these. Obtaining a cloud with patterns or fractals means that the sequence is non-random.

Figure 3 exhibits the return application of the 5-interleaved sequence generated by $p(x) = x^{16} + x^{14} + x^{12} + x + 1$. We observe that the graph is a messy cloud without patterns which means that our sequences are random.

However, Figs. 4a and 4b are return applications of two 8-interleaved sequences generated with a characteristic polynomial of degree 16 and 17, respectively; we can observe the lack of randomness of these sequences since that present clear patterns. At first glance, it seems that these imperfect sequences only appear in some particular cases where t is a power of 2.

Chaos game
The chaos game is a most popular algorithm for approximating IFS fractals [2, 28, 34]. This algorithm allows to carry out the conversion of a given sequence into a $2D$ representation of that. As return application, this method is used

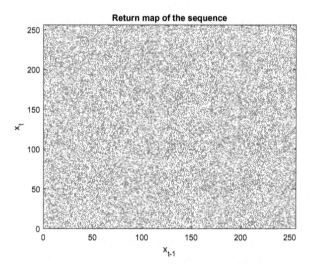

Fig. 3. Return map of a 5-interleaved sequence of degree 16.

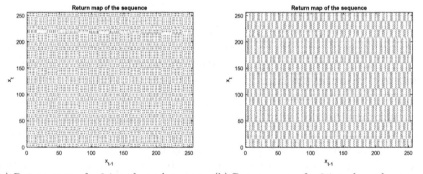

(a) Return map of a 8-interleaved sequence of degree 17. (b) Return map of a 8-interleaved sequence of degree 16.

Fig. 4. Return maps of imperfect 8-interleaved sequences.

to analyze random number generators using visual representations of the data generated [30, 32].

In Fig. 5a, we show the chaos game representation of the 5-interleaved sequence given previously. In this case, there is no any shape or figure, it is a disordered cloud of points which implies good randomness. However, in Fig. 5b, we obtain the chaos game for an 8-interleaved sequence generated with the characteristic polynomial of degree 16, $q(x) = x^{16} + x^{13} + x^{11} + x^{10} + x^9 + x^6 + x^3 + x^2 + 1$. In fact, we can observe a fractal structure repeated in this representation. It means that our sequence cannot been considered random. These chaos game representations with such a well-defined patterns have only occurred on rare occasions, in a very insignificant percentage of t-interleaved sequences. However, as in the return map case, these imperfect sequences have the peculiarity that the value of t is a power of 2.

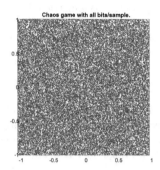
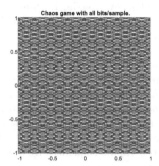

(a) Chaos game of a 5-interleaved sequence of degree 16.

(b) Chaos game of a 8-interleaved sequence of degree 16.

Fig. 5. Chaos game of t-interleaved sequences.

4   Conclusions

In this paper, we perform a preliminary analysis on the randomness of the t-interleaved sequences. We use statistical and graphic tools to analyze the random behaviour of these sequences. All our sequences have passed the statistical tests successfully. Therefore, these sequences are suitable as keystream sequences in stream cipher applications. Using two important graphic representations in the analysis of pseudorandom number generators, we have observed that for some particular cases, as for instance when t is a power of 2, we can obtain sequences that cannot be considered random. As a future work, we would like to analyze with more detail these cases where our sequences fail in order to determine the best values of t for which we can generate t-interleaved sequences with good cryptographic properties. We would also like to study the interleaving of PN-sequences generated from different primitive polynomials since that it could contribute to obtain better cryptographic results.

Acknowledgments. This work was in part supported by Comunidad de Madrid (Spain) under project CYNAMON (P2018/TCS-4566), co-funded by FSE and European Union FEDER funds. The third author is partially supported by Spanish grant VIGROB-287 of the Universitat d'Alacant.

References

1. Wu, H.: The stream cipher HC-128. In: Robshaw, M., Billet, O. (eds.) New Stream Cipher Designs. LNCS, vol. 4986, pp. 39–47. Springer, Heidelberg (2008). https:// doi.org/10.1007/978-3-540-68351-3_4

2. Barnsley, M.: Fractals Everywhere. Academic Press, Cambridge (1988)
3. Bassham, L., et al.: A statistical test suite for random and pseudorandom number generators for cryptographic applications (16 September 2010). https://tsapps.nist.gov/publication/get_pdf.cfm?pub_id=906762
4. Biryukov, A., Perrin, L.: State of the art in lightweight symmetric cryptography. IACR Cryptol. ePrint Arch. **2017**, 511 (2017)
5. Bishoi, S.K., Senapati, K., Shankar, B.: Shrinking generators based on σ-LFSRs. Discret. Appl. Math. **285**, 493–500 (2020). https://www.sciencedirect.com/science/article/pii/S0166218X20303346
6. Bluestein, L.I.: Interleaving of pseudorandom sequences for synchronization. IEEE Trans. Aerosp. Electron. Syst. AES **4**(4), 551–556 (1968)
7. Caballero-Gil, P., Fúster-Sabater, A., Pazo-Robles, M.E.: New attack strategy for the shrinking generator. J. Res. Pract. Inf. Technol. **41**(2), 171–180 (2009)
8. Cardell, S.D., Aranha, D.F., Fúster-Sabater, A.: Recovering decimation-based cryptographic sequences by means of linear CAs. Logic J. IGPL **28**(4), 430–448 (2020)
9. Cardell, S.D., Fúster-Sabater, A.: Cryptanalysing the shrinking generator. Procedia Comput. Sci. **51**, 2893–2897 (2015)
10. Cardell, S.D., Fúster-Sabater, A.: Performance of the cryptanalysis over the shrinking generator. In: Herrero, Á., Baruque, B., Sedano, J., Quintián, H., Corchado, E. (eds.) International Joint Conference. CISIS 2015. Advances in Intelligent Systems and Computing, vol. 369. Springer, Cham (2015). https://doi.org/10.1007/978-3-319-19713-5_10
11. Cardell, S.D., Fúster-Sabater, A., Ranea, A.: Linearity in decimation-based generators: an improved cryptanalysis on the shrinking generator. Open Math. **16**(1), 646–655 (2018)
12. Cardell, S.D., Fúster-Sabater, A., Requena, V.: Interleaving shifted versions of a PN-sequence. Mathematics **9**(687), 1–23 (2021)
13. Cardell, S.D., Fúster-Sabater, A.: Linear models for high-complexity sequences. In: Gervasi, O., et al. (eds.) ICCSA 2017, Part I. LNCS, vol. 10404, pp. 314–324. Springer, Cham (2017). https://doi.org/10.1007/978-3-319-62392-4_23
14. Díaz Cardell, S., Fúster-Sabater, A.: Cryptography with Shrinking Generators. SM, Springer, Cham (2019). https://doi.org/10.1007/978-3-030-12850-0
15. Coppersmith, D., Herzberg, A., Krawczyk, H.M., Kutten, S., Mansour, Y.: A shrinking generator for cryptosystems (1987). https://patents.google.com/patent/EP0619659A2/en
16. Coppersmith, D., Krawczyk, H., Mansour, Y.: The shrinking generator. In: Stinson, D.R. (ed.) CRYPTO 1993. LNCS, vol. 773, pp. 22–39. Springer, Heidelberg (1994). https://doi.org/10.1007/3-540-48329-2_3
17. Crilly, T.: Interleaving integer sequences. Math. Gaz. **91**(520), 27–33 (2007)
18. Doğanaksoy, A., Göloğlu, F.: On Lempel-Ziv complexity of sequences. In: Gong, G., Helleseth, T., Song, H.-Y., Yang, K. (eds.) SETA 2006. LNCS, vol. 4086, pp. 180–189. Springer, Heidelberg (2006). https://doi.org/10.1007/11863854_15
19. Edemskiy, V.: On the linear complexity of interleaved binary sequences of period 4p obtained from hall sequences or Legendre and hall sequences. Electron. Lett. **50**(8), 604–605 (2014)
20. Eisenbarth, T., Kumar, S., Paar, C., Poschmann, A., Uhsadel, L.: A survey of lightweight-cryptography implementations. IEEE Des. Test Comput. **24**(6), 522–533 (2007)

21. Fluhrer, S.R., McGrew, D.A.: Statistical analysis of the alleged RC4 keystream generator. In: Goos, G., Hartmanis, J., van Leeuwen, J., Schneier, B. (eds.) FSE 2000. LNCS, vol. 1978, pp. 19–30. Springer, Heidelberg (2001). https://doi.org/10.1007/3-540-44706-7_2

22. Fúster-Sabater, A.: Generation of cryptographic sequences by means of difference equations. Appl. Math. Inf. Sci. **8**, 475–484 (2014)

23. Gennaro, R.: Randomness in cryptography. IEEE Secur. Priv. **4**(02), 64–67 (2006)

24. Golomb, S.W.: Shift Register-Sequences. Aegean Park Press, Laguna Hill (1982)

25. Gong, G.: Theory and applications of q-ary interleaved sequences. IEEE Trans. Inf. Theory **41**(2), 400–411 (1995)

26. Jiang, S., Dai, Z., Gong, G.: On interleaved sequences over finite fields. Discret. Math. **252**, 161–178 (2002)

27. Li, N., Tang, X.: On the linear complexity of binary sequences of period $4n$ with optimal autocorrelation value/magnitude. IEEE Trans. Inf. Theory **57**(11), 7597–7604 (2011)

28. Martyn, T.: The chaos game revisited: yet another, but a trivial proof of the algorithm's correctness. Appl. Math. Lett. **25**(2), 206–208 (2012). https://www.sciencedirect.com/science/article/pii/S0893965911003922

29. Massey, J.: Shift-register synthesis and BCH decoding. IEEE Trans. Inf. Theory **15**(1), 122–127 (1969)

30. Mata-Toledo, R.A., Willis, M.A.: Visualization of random sequences using the chaos game algorithm. J. Syst. Softw. **39**(1), 3–6 (1997). https://www.sciencedirect.com/science/article/pii/S0164121296001586

31. Maurer, U.M.: A universal statistical test for random bit generators. J. Cryptol. **5**(2), 89–105 (1992). https://doi.org/10.1007/BF00193563

32. Orúe, A.B., Fúster-Sabater, A., Fernández, V., Montoya, F., Hernández, L., Martín, A.: Herramientas gráficas de la criptografía caótica para el análisis de la calidad de secuencias pseudoaleatorias, p. 180–185. Actas de la XIV Reunión Española sobre Criptología y Seguridad de la Información, RECSI XIV, Menorca, Illes Balears, Spain (October 2016)

33. Orúe, A.B., Hernández, L., Martín, A., Montoya, F.: A lightweight pseudorandom number generator for securing the Internet of Things. IEEE Access **5**, 27800–27806 (2017)

34. Peitgen, H.O., Jurgens, H., Saupe, D.: Chaos and Fractals: New Frontiers of Science. Springer, Heidelberg (2004). https://doi.org/10.1007/b97624

35. Pérez, G., Cerdeira, H.A.: Extracting messages masked by chaos. Phys. Rev. Lett. **74**, 1970–1973 (1995). https://link.aps.org/doi/10.1103/PhysRevLett.74.1970

36. Tang, X., Ding, C.: New classes of balanced quaternary and almost balanced binary sequences with optimal autocorrelation value. IEEE Trans. Inf. Theory **56**(12), 6398–6405 (2010)

37. U.S. Department of Commerce: FIPS 186, Digital signature standard. Federal Information Processing Standards Publication 186, N.I.S.T., National Technical Information Service, Springfield, Virginia (1994)

38. Xiong, H., Qu, L., Li, C., Fu, S.: Linear complexity of binary sequences with interleaved structure. IET Commun. **7**(15), 1688–1696 (2013)

Numerical Prediction Model
of Runway-Taxiway Junctions for Optimizing
the Runway Evacuation Time

Misagh Ketabdari[1]([:envelope:]) [:id:], Ignacio Puebla Millán[2] [:id:], Maurizio Crispino[1] [:id:],
Emanuele Toraldo[1] [:id:], and Mariano Pernetti[3] [:id:]

[1] Transportation Infrastructures and Geosciences Section, Department of Civil and
Environmental Engineering, Politecnico di Milano, 20133 Milan, Italy
{misagh.ketabdari,Maurizio.crispino,emanuele.toraldo}@polimi.it
[2] Department of Bioengineering and Aerospace Engineering, University Carlos III de Madrid
UC3M, Madrid, Spain
[3] Department of Engineering, Università della Campania "Luigi Vanvitelli", 81031 Aversa, Italy
mariano.pernetti@unicampania.it

Abstract. It is vital to ensure efficient and fast connection networks that guarantee
the development of society in the global atmosphere. Air transport and airport
operations play fundamental roles that help to achieve the needs of globalized
society, constituting a pillar for providing essential services to a great number
of passengers and goods around the world. In the coming years, air industry is
expected to expand due to the imminent increase of passengers. This growth
in air passenger journeys demands several measures to be applied in airports
worldwide. Air traffic must comply with great levels of safety and efficiency
according to existing standards to guarantee optimal airport operations. Moreover,
airport infrastructures, such as runways and taxiways, should be continuously
improved to minimize the possible costs and probability of associated risks, and to
maximize their capacities by re-evaluating their designs to cope with the demands
derived from the growth in traffic forecasted for the future.

In this regard, a prediction model to simulate the behavior of aircraft in landing was developed, allowing to predict the optimum locations of runway-taxiway
junctions, enhancing the efficiency of the runway, and minimizing the runway
evacuation time. This model is based on tire-fluid-pavement interactions that govern the dynamic behavior of the aircraft during landing in wet and dry pavement
conditions. The results provide the accurate landing distance required for operating aircraft inside the airport, which can be used to design enhanced capacity
runways, while guaranteeing the safety of operations by minimizing the related
accidents probabilities.

Keywords: Runway-taxiway junction optimization · Runway capacity · Airport
design

O. Gervasi et al. (Eds.): ICCSA 2021, LNCS 12949, pp. 298–308, 2021.
https://doi.org/10.1007/978-3-030-86653-2_22

1 Introduction

Due to the continuous international growth of society, air traffic is expected to experience a significant increase by the end of the next twenty years, in terms of passengers and goods [1]. However, this growth needs to be followed by the improvements in airport infrastructures and enhancement of their related services, in order to satisfy the expected market demands. Thus, it is necessary to continuously monitor the current needs of the airports and to predict the possible future requirements.

Safety and efficiency are main objectives in air transport operations that are required to be guaranteed by airport infrastructures such as runways and taxiways. These infrastructures should be continuously improved in order to minimize the related operations costs and the risks derived from their use. Different scientific studies and standards have been published worldwide trying to investigate possible solutions to these issues [2], evaluating the safe operations on runways and the different possibilities to manage the volume of passengers at airports worldwide. As a result, the efficiency of airports infrastructures can be maximized by re-evaluating their designs and characteristics and providing alternative design solutions in order to cope with the demands derived from the growth in traffic forecasted for the future.

In this context, runway and taxiway designs and their junctions are playing a vital role in determining the runway capacity, evacuation time, fuel consumption, and associated costs of related operations. Therefore, it is essential to develop a prediction model that offers optimized solutions of the taxiway junctions' locations according to the operating fleet.

Furthermore, landings are complex operations that nowadays contribute to one of the major phases of flights in which accidents occur with higher probabilities [3, 4]. Therefore, this study focused on the aircraft behavior during landings in order to provide a deeper understanding of the underlying physical principles which set the basis for the runway and taxiway designs.

2 Background Review and Scope

The available methods to predict the behavior of aircraft in landing are mainly based on empirical methods [5–7] and [8], which were trying to predict pavement friction based on models derived from experimental data [9, 10] and [11]. These approaches generally include the effects of pavement contamination (e.g. presence of water-film on pavement) in their computations. The main issue in the development of these methods is the lack of reference data regarding operating aircraft to be adopted in the simulations of these models. The results obtained in these experimental studies are tightly bounded to a strict range of values in final approach speeds, tire pressures and water-film thicknesses, which lead to major difficulties when the same approaches are required to be extrapolated to other areas of interest with values that fall out of range of those adopted by the original studies. Thus, the aim of this study is to extend those limitations, allowing to reach a deeper understanding and evaluating the influences of a wider range of involved parameters [12, 13].

Among analytical models available in literature, the one developed by Ong and Fwa [5, 6] was selected as a reference for this study. This model proposes a three-dimensional approach based on Finite Element Analysis (FEA) using computational simulation, which integrate three sub-models of fluid, tire and pavement in order to analyze the relative influencing parameters that govern the dynamics of the operations. Consequently, it is possible to determine precisely the optimal distances of runway-taxiway junctions from the runway's threshold, according to numerous impact parameters such as airport climate pattern, operating aircraft categories, infrastructure type and capacity, route connections, and operating costs.

3 Methodology

Determining the most accurate model among the existing ones in the literature, which can portray the dynamic behavior of aircraft in landing, was one of the main challenges for this study. Most of the current available studies focused on experimentally derived empirical models due to simpler parameters to be considered in their methodologies. However, in this study a computational finite element model centered on a theoretical basis was developed, in order to simulate the interactions between aircraft tire, fluid, and different types of pavements.

The selected theoretical approach is based on the research conducted by Horne et al., published by NASA [9, 10]. In this context, the dynamic influences of the fluid forces acting on the main gear tire throughout the whole landing process can be evaluated.

In this regard, the crucial forces, which should be implemented in fluid equations, are the fluid drag and uplift forces, whose values vary with the change in aircraft velocity. The fluid drag force acting on the main gear tire in presence of various water-film thicknesses on the runway surface can be expressed for a single aircraft wheel as:

$$F_{drag} = \frac{1}{2}C_{Dh}\rho_w t_w f(w)V^2 \tag{1}$$

Where, F_{drag} stands for the fluid drag/retardation force; C_{Dh} represents the hydrodynamic drag coefficient; ρ_w is the water density; t_w is the water-film thickness; V stands for the aircraft speed; and $f(w)$ is a function which stands for the chord length of the tire cross section at the water surface, which has been presented by Horne et al. [9] as:

$$f(w) = 2w\left[\frac{\delta + t_w}{w} - \left(\frac{\delta + t_w}{w}\right)^2\right]^{\frac{1}{2}} \tag{2}$$

Where, w stands for the tire width; and δ represents the vertical tire deflection. As a result, it can be deduced that the fluid drag force varies linearly respect to the fluid density and thickness, the tire width, and the landing velocity squared. It must be highlighted that in order to avoid higher complexity, this study considered the evaluation of the drag force for a single wheel or front mounted tandem wheel ignoring the effects of other possible landing gear components in the surroundings. Additionally, the possible effects induced by hydroplaning in the drag force are not considered for simplification purposes.

The fluid uplift force also generates a significant effect over the main gear tire, which can be the cause of the phenomena called hydroplaning in its higher values. This force, which also needs to be modeled accurately, can be expressed for a single aircraft wheel as:

$$F_{uplift} = \frac{1}{2}C_{Lh}\rho_w AV^2 \tag{3}$$

Where, F_{uplift} stands for the fluid uplift force; C_{Lh} represents the hydrodynamic uplift coefficient; ρ_w is the water density; A is the tire-ground gross contact area; and V is the aircraft landing velocity. However, Eq. 3 only portrays the effect of the hydrodynamic pressure on the tire footprint area. According to [11], the front region of the water-film layer impacting the tire, known as the water wedge, needs to be considered as the hydrodynamic pressure in this region contributes also to the total vertical force. This way, this region is defined by the water wedge length l until the water reaches the tire footprint. If the effects of water spray are neglected, the effect of the hydrodynamic pressure in this region can be expressed through the Eq. 4.

$$F_{wedge} = wl\rho_w V^2 \tag{4}$$

Where F_{wedge} stands for the fluid uplift force in the wedge region; w represents the tire width, l is the length of the wedge; and V is the aircraft velocity. The length of wedge can be calculated by modeling the tire profile, as presented in Fig. 1.

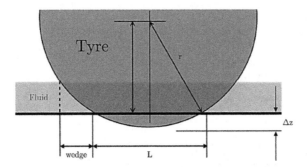

Fig. 1. Tire contact simplified model [14]

Finally, the overall vertical force acting on main gear tire can be computed by combining both contributions represented by Eqs. 3 and 4. This way, when the F_{uplift} equals the aircraft wheel load, it causes a complete loss of braking potential and consequently aircraft sliding with the wheel locked up on the existing water-film on the runway surface (i.e. hydroplaning phenomenon). Therefore, fluid uplift force has a great impact on the friction between tire and pavement in presence of water-film on the surface, and consequently on the final landing required distance. Both drag and uplift fluid forces have been used inside the developed model in order to predict the optimal locations of runway-taxiway junctions, as illustrated in Fig. 2.

After determining accurate models to evaluate the related fluid forces acting on the tire and pavement surface, a MATLAB®-based numerical model has been developed in order to simulate the dynamic behavior of aircraft in landing based on the aerodynamics

```
%% FLUID FORCES ON LANDING GEAR

pr = 0.25.*p_max; % rated tire inflation pressure [Pa]

delta = Fz./(2.4.*(p+0.08.*pr).*(w.*d).^(1/2)) + w.*C_z; % vertical tire
f_w   = 2.*w.*(((delta+dL)./w)-((delta+dL)./w).^2).^(1/2);

F_drag = 1/2.*CD_h.*rho_water.*dL.*f_w.*V.^2; % Fluid drag force [N]

 F_uplift = rho_water.*V.^2.*(A_G.*CL_h.*0.5+ x_under.*w);
```

Fig. 2. A part of developed numerical script to compute the fluid drag and uplift forces

of the aircraft components, their structural mechanics, and the fluid dynamics based on the guidelines established previously by Ong et al. [5, 6], Pasindu et al. [15] and Ketabdari et al. [16, 17], as illustrated in Fig. 3.

```
while V(i)> exit_velocity

% Obtain the SNv for each timestep
[SNv] = computation_SNv_NASA_distribution(aircraft,V(i),dL,L_average);

% dynamic friction coefficient
mu = SNv./100;

% deceleration rate
a = mu.*g + (0.5.*rho_air.*V(i).^2.*A.*(CD-mu.*CL))/(W);
V_next = V(i) - a.*dt;

delta_v = V(i) - V_next;
delta_x = delta_v.*dt;

V_vector(i,counter) = V(i);

V(i) = V_next;

% distance travelled on one interation
x(i,counter) = x0 + V(i).*dt;
t(counter) = t(end) + dt;
x0 = x(i,counter);


counter = counter + 1;
end
end
```

Fig. 3. A part of developed numerical script to compute the aircraft braking distance in landing

4   Validation

In order to evaluate the accuracy of the developed model results for computing the dynamic skid resistance in wet pavement, and to evaluate the model's reliability for aircraft tires, a preliminary validation study has been performed. In this procedure, the skid resistance values obtained by the developed computation model were compared to the experimental data extracted from Horne et al. study [18] to check the applicability of the selected approaches. The conditions set for this validation process are gathered in Table 1.

Table 1. Boundary conditions set for validation process of skid resistance model

Tire dimensions	Tire type	Wheel load	Water-film thickness	Inflation pressure
32 × 8.8 inches	VII	9979 kg	3.8 mm	1999.5 kPa

Two types of pavements were considered in the computations of this study; rigid pavement runway (i.e. concrete) with tested sliding speed range from 10 to 40 m/s; and flexible pavement runway (i.e. asphalt) with tested sliding speeds range from 10 to 50 m/s. The extracted experimental Skid Number values (SNv) and those obtained by the developed model of this study are presented in Table 2.

Table 2. Skid number values computed versus those collected from literature

V (m/s)	SNv (literature)		SNv computed (model)		Percentage error (%)	
	Concrete	Asphalt	Concrete	Asphalt	Concrete	Asphalt
10	23	39	21	36.5	−8.7	−6.4
20	20	34	19.5	33.5	−2.5	−1.4
30	16.5	29	17	30	+3.0	+3.4
40	13	24	14	24.5	+7.7	+2.1
50	–	19	–	17.5	–	−7.9

As it can be interpreted from Table 2, the SN values obtained for asphalt and concrete pavements from the developed computation models present sufficient similarity to those extracted from the literature [18], which can guarantee a significant accuracy of the developed fluid-tire-pavement interaction model. The absolute percentage errors obtained for the predictions were resulted below ±9% with low numerical differences. Therefore, it can be stated that the chosen methodology constitutes a suitable approach for the simulation of the behavior of aircraft tires over the wet runway pavement.

5 Traffic Data and Aircraft Classification

Once the applicability of the methodology has been validated, it is possible to find a solution for the taxiway design. Nevertheless, in order to provide an efficient solution, a database of operating aircraft should be provided to be adopted by the model.

These data can consist of commercial airplanes and general aviation. The character-istics of the aircraft belonging to these two categories generally differ from each other in term of dimension, weight, fuel consumption, and operation costs, however, both cat-egories are assumed to operate on the same runway. Therefore, same simulation steps and circumstances should be applied on both aircraft categories.

According to the simulation model described in previous sections, the main tech-nical specifications of aircraft that are required in the proposed prediction model are Maximum Landing Weight (MLW), landing gear arrangement, tire inflation pressure, tire dimensions, aerodynamic lift and drag coefficients, final approach speed (V), and wing area (S). These data are generally available through the Airport Planning Manu-als (APMs) or reliable aircraft database sources such as BADA-Eurocontrol [19], and Aircraft Performance Database (APD) [20].

The database used for the simulations is composed by 95 aircraft, classified into different groups based on their final approach speeds, according to the criteria established by ICAO [21]. In Table 3 a partial list of these aircraft is presented.

Table 3. A partial preview of provided aircraft database to be adopted by the proposed model

Operating aircraft	Aircraft category	MLW (kg)	V (m/s)	S (m^2)
Airbus A319	C	62500	64,82	122,60
Airbus A320	C	66000	69,96	122,60
Embraer 175	C	34000	63,79	72,72
Embraer 190	C	43000	63,79	92,53
Boeing 737-800 Winglets	D	65317	73,05	124,65
Airbus A320 Sharklets	C	66000	69,96	122,60
Airbus A321	D	75500	72,02	122,60
Boeing 737-700 Winglets	C	58060	66,88	124,65
McDonnell Douglas MD-82	C	58967	68,93	112,32
Airbus A318	C	57500	62,25	122,60
Boeing 737-400	C	56245	71,51	91,09
Boeing 737-300	C	51710	68,42	91,09
Boeing 737-800	D	65317	73,05	124,65
McDonnell Douglas MD-83	D	63276	74,08	112,32
De Havilland DHC-8 (8-400)	B	28009	62,63	63,10
ATR 72	B	22350	58,65	61,00
Embraer Brasilia E120	A	11700	58,13	39,40

According to ICAO [21], aircraft should be classified respect to their final app-roach/threshold speeds. This way, four official groups can be formed: group A comprises those aircraft with final approach speeds lower than 91 kts (46 m/s); group B comprises those whose speeds are higher than 91 kts and lower or equal to 121 kts (61 m/s); group C comprises the models with speeds ranging from 122 kts to 140 kts (72 m/s); and group D includes those models with final approach speeds higher than 141 kts up to 165 kts (85 m/s). As a result, four groups are established, as presented in Fig. 4.

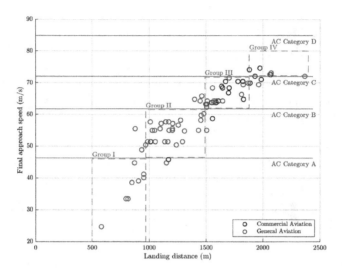

Fig. 4. Aircraft categorisation based on final approach speeds

6   Results and Discussion

The inputs required to be adopted in the simulation process as the boundary conditions of this study are air density (ρ: 1.225 kg/m^3), water density at 25 °C (ρ_w: 997.1 kg/m^3), landing speed reduction (k_s: 2 m/s), water-film thickness (t_w: 3 mm), dry friction coefficient (μ: 0.5), flap induced drag variation (Δf: 0.12), and runway exit speed (V_{exit}: 30 kts).

Based on the selected boundary conditions and aircraft categorization approach, the proposed model can simulate the aircraft behavior in landing operations for wet and dry pavement scenarios. The results will be used in numerical prediction of the locations of taxiway exits with optimal operations capacity for the airport design.

The behavior of aircraft in landing can be predicted by calculating the aircraft Landing Distance Required (LDR). Once the LDRs of selected operating aircraft are computed based on the finite element model explained in methodology chapter, an individual Optimal Turn-off Point (OTP) for each of these aircraft will be determined, which are needed in order to determine the closest location of runway exit junction.

As a result of simulations of landing aircraft behavior for selected operating aircraft (Table 3) through proposed model, the predicted locations of runway-taxiway junctions for wet and dry pavement scenarios are presented in Figs. 5 and 6. It is assumed that the landing direction is to be only from left toward right designation. According to ICAO

[21], a minimum separation of 450 m should be guaranteed between parallel runway exits.

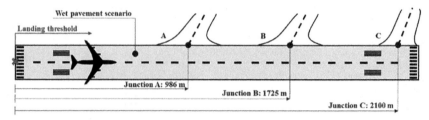

Fig. 5. Predicted optimal locations of three runway-taxiway junctions for designing a runway in wet pavement scenario (not to scale)

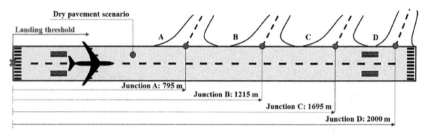

Fig. 6. Predicted optimal locations of four runway-taxiway junctions for designing a runway in dry pavement scenario (not to scale)

In dry pavement scenario, shorter landing distances resulted in an accumulation of OTPs over areas closer to the runway landing threshold respect to the wet pavement scenario.

Aircraft mean rolling distances can be calculated in order to evaluate the effectivity of both proposed solutions. These distances are defined as the travelled distance from each individual OTP to the actual runway exit. Lower rolling distances will logically lead to lower evacuation times. Therefore, thanks to OTP calculations, it is possible to maximize the runway capacity and performance level, which would increase the number of operations. These mean rolling distances related to the predicted locations of proposed model (proposed configurations) should be compared with common runway taxiway configurations for selected operating aircraft, in order to evaluate the solutions efficiency. Therefore, current runway-taxiway configuration of an Italian international airport was selected for comparative purposes, and the results are presented in Table 4.

As it can be interpreted from the obtained results, new runway-taxiway configurations for both wet and dry surfaces offer improvements respect to the common runway configurations in term of reduction in mean rolling distances and consequently reduction in total runway evacuation time. A reduction of 27% (95 m) for wet pavement condition and a reduction of 42% (149 m) for dry pavement condition have been registered in mean rolling distances for landing operations.

Table 4. Calculated mean rolling distances for wet and dry pavement scenarios

Results	Mean rolling distance		Distance variation		Percentage variation	
	Wet	Dry	Wet	Dry	Wet	Dry
Proposed configurations	275 m	207 m	−95 m	−149 m	−27%	−42%
Existing configurations	370 m	356 m				

7 Conclusion

The accurate estimation of the aircraft landing distances constitutes a complex process in which diverse parameters of different nature play significant roles. The main scope of this study was to develop a precise numerical prediction model to simulate the interactions between tire-fluid-pavement during landing phase of flight and finally offers a possibility to extrapolate this methodology and apply it on different airports with various boundary conditions.

This study is a complementary work to the previous attempts by the authors [15, 16], which had to face different criticalities encountered during the model development stages, such as lack of precise reference models and complete experimental databases. In this study, the prediction model, which is based on the analysis of pavement friction and fluid properties, has been developed by adopting the equations offered by Horne et al. [9, 10] and Sinnamon et al. [11], providing an alternative way to the existing models proposed by Ong et al. [5, 6], Pasindu et al. [15], and finally validated according to existing experimental data from available literature [18].

The developed model can simulate the aircraft dynamic behavior in landing operations for wet and dry pavement conditions. As a result of the simulation, LDR for aircraft in dry pavement condition demonstrate noticeable shorter values respect to the values for wet pavement, which clarify the conspicuous impact of contaminants over the runway surface.

Eventually, the proposed methodology successfully proved to be a valid approach in order to calculate the accurate aircraft braking distance in landing phase of flight, which can be adopted for commercial airports with various types of operating aircraft to perform similar studies, trying to optimize the operation capacity of their infrastructures (e.g. runways and taxiways). Therefore, it will constitute one of the possible ways to improve the safety and efficiency of the operations, and to achieve noticeable reductions in terms of overall airport operations costs and infrastructures evacuation times, while ensuring and matching the needs of growth in air passenger journeys and goods freight demands.

Acknowledgements. There is no conflict of interest to disclose. This research did not receive any specific grant from funding agencies in the public, commercial, or not-for-profit sectors.

References

1. IATA Homepage. https://www.iata.org/en/publications/store/20-year-passenger-forecast. Accessed 03 Feb 2021
2. ICAO Homepage. https://www.icao.int/safety/RunwaySafety/Documents/Forms/AllItems. Accessed 07 Dec 2020.
3. Airplanes, Boeing Commercial: Statistical summary of commercial jet airplane accidents, worldwide operations 1959–2019 (2020)
4. Van Es, G.W.H.: A study of runway excursions from a European perspective. Eurocontrol, Tech. Rep. NLR-CR-2010-259 (2010)
5. Ong, G.P., Fwa, T.F.: Wet-pavement hydroplaning risk and skid resistance: modeling. J. Transp. Eng. **133**(10), 590–598 (2007)
6. Ong, G.P., Fwa, T.F.: Prediction of wet-pavement skid resistance and hydroplaning potential. Transp. Res. Rec. **2005**(1), 160–171 (2007)
7. Pasindu, H.R., Fwa, T.F.: Improving wet-weather runway performance using trapezoidal grooving design. Transp. Develop. Econ. **1**(1), 1–10 (2015)
8. Fwa, T.F., Chan, W.T., Lim, C.T.: Decision framework for pavement friction management of airport runways. J. Transp. Eng. **123**(6), 429–435 (1997)
9. Horne, W.B., Joyner, U.T., Leland, T.J.: Studies of the Retardation Force Developed on an Aircraft Tire Rolling in Slush Or Water, vol. 552. National Aeronautics and Space Administration (1960)
10. Horne, W.B., Dreher, R.C.: Phenomena of Pneumatic Tire Hydroplaning, vol. 2056. National Aeronautics and Space Administration (1963)
11. Sinnamon, J.F.: Hydroplaning and Tread Pattern Hydrodynamics. Highway Safety Research Institute University of Michigan (1974)
12. Ketabdari, M., Giustozzi, F., Crispino, M.: Sensitivity analysis of influencing factors in probabilistic risk assessment for airports. Saf. Sci. **107**, 173–187 (2018)
13. Ketabdari, M., Toraldo, E., Crispino, M.: Numerical risk analyses of the impact of meteorological conditions on probability of airport runway excursion accidents. In: Gervasi, O., et al. (eds.) ICCSA 2020. LNCS, vol. 12249, pp. 177–190. Springer, Cham (2020). https://doi.org/10.1007/978-3-030-58799-4_13
14. Blundell, M., Harty, D.: Multibody Systems Approach to Vehicle Dynamics. Elsevier, Amsterdam (2004)
15. Pasindu, H.R., Fwa, T.F., Ong, G.P.: Computation of aircraft braking distances. Transp. Res. Record: J. Transp. Res. Board **2214**(1), 126–135 (2011). https://doi.org/10.3141/2214-16
16. Ketabdari, M., Crispino, M., Giustozzi, F.: Probability contour map of landing overrun based on aircraft braking distance computation. In: World Conference on Pavement and Asset Management. WCPAM 2017, vol. 1, pp. 731–740. CRC Press, Boca Raton (2019)
17. Ketabdari, M., Toraldo, E., Crispino, M., Lunkar, V.: Evaluating the interaction between engineered materials and aircraft tyres as arresting systems in landing overrun events. Case Stud. Construct. Mater. **13**, e00446 (2020)
18. Horne, W.B., J., Y.T., Taylor, G.R.: Recent Research to Improve Tire Traction on Water, Slush and Ice. Publication N66–83984. NASA Langley Research Center, Hampton (1965)
19. Nuic, A.: User Manual for the Base of Aircraft Data (BADA). Eurocontrol Experimental Centre, Cedex, France, Revision, 3 (2011)
20. Eurocontrol Homepage: Aircraft performance database. https://contentzone.eurocontrol.int/aircraftperformance. Accessed 17 June 2020
21. International Civil Aviation Administration, ICAO: Manual, A. D. Part 2 Taxiways, Apron and Holding Bays. DOC, 9157 (2005)

Jacobian-Dependent Two-Stage Peer Method for Ordinary Differential Equations

Dajana Conte⬭, Giovanni Pagano$^{(\boxtimes)}$⬭, and Beatrice Paternoster⬭

Department of Mathematics, University of Salerno, Via Giovanni Paolo II,
132 - 84084 Fisciano, SA, Italy
{dajconte,gpagano,beapat}@unisa.it,
https://www.dipmat.unisa.it

Abstract. In this paper we derive new explicit two-stage peer methods for the numerical solution of ordinary differential equations by using the technique introduced in [2] for Runge-Kutta methods. This technique allows to re-determine the order conditions of classical methods, obtaining new coefficients values. The coefficients of new methods are no longer constant, but depend on the Jacobian function of the ordinary differential equation. The new methods preserve the order of classical peer methods, and are more accurate and with better stability properties. Numerical tests highlight the advantage of new methods especially for stiff problems.

Keywords: Peer methods · Jacobian-dependent coefficients

1 Introduction

Peer methods are two-step s-stage methods for the numerical solution of Ordinary Differential Equations (ODEs)

$$y'(t) = f\big(t, y(t)\big). \tag{1}$$

After the work of R. Weiner et al. [1], much more research has been conducted on peer methods, since based on the choice of their coefficients you can obtain explicit parallelizable methods [7–10], simply explicit methods [11–16] or implicit methods [17–21]. It is also possible to use particular techniques, such as the Exponential Fitting (EF) [6], obtaining peer methods adapted to the problem [22–24], as they follow the apriori known trend of the real solution.

In this paper we focus on solving the problem (1), and we derive the coefficients of the peer methods in a different way, using the approach that was applied on explicit two- and three-stage Runge-Kutta methods in [2]. Subsequently, the same technique was also applied by other authors on explicit four-stage Runge-Kutta methods [4]. In all these cases, for particular choices of the coefficients,

The authors Conte, Pagano and Paternoster are members of the GNCS group. This work is supported by GNCS-INDAM project and by PRIN2017-MIUR project.

O. Gervasi et al. (Eds.): ICCSA 2021, LNCS 12949, pp. 309–324, 2021.
https://doi.org/10.1007/978-3-030-86653-2_23

A-stable versions of the methods are obtained, with accuracy order that increases by one compared to the standard case. In the paper [3], specific numerical tests have been carried out in order to confirm these theoretical observations.

We consider explicit peer methods of the form

$$Y_{n,i} = \sum_{j=1}^{s} b_{ij} Y_{n-1,j} + h \sum_{j=1}^{s} a_{ij} f(t_{n-1,j}, Y_{n-1,j}) + h \sum_{j=1}^{i-1} r_{ij} f(t_{n,j}, Y_{n,j}),$$

$$i = 1, ..., s,$$ (2)

where the stages are $Y_{n,i} \approx y(t_{n,i})$, with $t_{n,i} = t_n + h c_i$, and the coefficients matrices are

$$A = (a_{ij})_{i,j=1}^{s}, \qquad B = (b_{ij})_{i,j=1}^{s}, \qquad R = (r_{ij})_{i,j=1}^{s},$$ (3)

with R lower triangular matrix.

The coefficients are obtained by modifying the classical order conditions for peer methods, according to the approach proposed in [2], and, as a consequence, they depend on the Jacobian of the function f. This new Jacobian-dependent methods have better stability properties than the classical ones.

The organization of this work is the following: in Sect. 2 we show the classic form of the explicit two-stage peer methods; in Sect. 3 we derive the Jacobian-dependent coefficients of the new two-stage peer methods by imposing different order conditions than the standard case; in Sect. 4 we analyze the linear stability properties of the classic and Jacobian-dependent methods, obtaining for the latter a particular version characterized by a bigger absolute stability region; in Sect. 5 we show numerical tests in order to confirm our theoretical observations; in Sect. 6 we discuss the results and the future research.

2   Classic Peer Methods

Given a discretization $\{t_n, \, n = 1, ..., N\}$ of the time interval $[t_0, T]$ associated to the problem (1), classic explicit two-stage ($s = 2$) peer methods assume the form

$$Y_{n,1} = b_{11} Y_{n-1,1} + b_{12} Y_{n-1,2} + h a_{11} f(t_{n-1,1}, Y_{n-1,1}) + h a_{12} f(t_{n-1,2}, Y_{n-1,2}),$$
$$Y_{n,2} = b_{21} Y_{n-1,1} + b_{22} Y_{n-1,2} + h a_{21} f(t_{n-1,1}, Y_{n-1,1}) + h a_{22} f(t_{n-1,2}, Y_{n-1,2})$$
$$+ h r_{21} f(t_{n,1}, Y_{n,1}),$$ (4)

where $c_1 \in [0, 1)$ and $c_2 = 1$ (it is convention to use $c_s = 1$ for s-stage peer methods). Remembering that $Y_{n,i} \approx y(t_{n,i})$, $t_{n,i} = t_n + h c_i$, the solution at the generic grid point $t_n + h$ is determined by $Y_{n,2}$, in each time step from t_n to $t_{n+1} = t_n + h$.

The classical order conditions of explicit peer methods are obtained by annihilating the necessary number of residuals, defined as

$$
h\,\Delta_i := y(t_{n,i}) - \sum_{j=1}^{s} b_{ij}\,y(t_{n-1,j}) - h\sum_{j=1}^{s} a_{ij}\,y'(t_{n-1,j}) + \sum_{j=1}^{i-1} r_{ij}y'(t_{n,j}),
\tag{5}
$$

$$
i = 1, ..., s.
$$

In fact, the following definition applies:

Definition 1. *The peer method*

$$
Y_{n,i} = \sum_{j=1}^{s} b_{ij}\,Y_{n-1,j} + h\sum_{j=1}^{s} a_{ij}\,f(t_{n-1,j}, Y_{n-1,j}) + h\sum_{j=1}^{i-1} r_{ij}\,f(t_{n,j}, Y_{n,j}),
\tag{6}
$$

$$
i = 1, ..., s,
$$

is consistent of order p if

$$
\Delta_i = O(h^p), \qquad i = 1, ..., s.
\tag{7}
$$

To familiarize with the technique we will apply in the next section, in the current section we collect the coefficients of classic explicit two-stage peer methods using it already.

2.1 Two-Stage Classic Version

In this paragraph we impose differently the same order conditions as obtained in [1] in the general case of s stages. In order to derive them, we define the Local Truncation Errors (LTEs) related to the stages as in (9), replacing $Y_1(t)$ and $Y_2(t)$ with the continuous functions defined in (10).

The method we analyze in this work is (4), and to have visibility of its free coefficients, we consider them in the following matrices:

$$
A = \begin{pmatrix} a_{11} & a_{12} \\ a_{21} & a_{22} \end{pmatrix}, \
B = \begin{pmatrix} b_{11} & b_{12} \\ b_{21} & b_{22} \end{pmatrix}, \
R = \begin{pmatrix} 0 & 0 \\ r_{21} & 0 \end{pmatrix}, \
c = \begin{pmatrix} c_1 & 1 \end{pmatrix}.
\tag{8}
$$

There are ten free coefficients, and we're going to determine some of them by requiring that the accuracy order of the method be equal to two. After that, in Sect. 4, we'll assign the coefficients left free with the aim of achieving optimal linear stability properties.

Remembering that the stages $Y_{n,1}$ and $Y_{n,2}$ determine the numerical solution at the mesh points $t_n + h\,c_1$ and $t_n + h$ ($= t_{n+1}$), respectively, we define the linear operators

$$
\underline{L_1}\big(y(t)\big) = y(t + h\,c_1) - Y_1(t), \qquad \underline{L_2}\big(y(t)\big) = y(t + h) - Y_2(t),
\tag{9}
$$

that are functions by which you can measure the error of (4). In fact, with $Y_1(t)$ and $Y_2(t)$ we indicate the continuous expressions of the discrete stages $Y_{n,1}$ and $Y_{n,2}$, respectively:

$$
\begin{aligned}
Y_1(t) &= b_{11}\, y\big(t + h(c_1 - 1)\big) + b_{12}\, y(t) + h\, a_{11}\, y'\big(t + h(c_1 - 1)\big) + h\, a_{12}\, y'(t), \\
Y_2(t) &= b_{21}\, y\big(t + h(c_1 - 1)\big) + b_{22}\, y(t) + h\, a_{21}\, y'\big(t + h(c_1 - 1)\big) + h\, a_{22}\, y'(t) \\
&\quad + h\, r_{21} y'(t + h\, c_1).
\end{aligned}
\tag{10}
$$

We evaluate $\underline{L_1}\big(y(t)\big)$ and $\underline{L_2}\big(y(t)\big)$ for $y(t) = t^k$, $k = 0,1,2,...$, but, as explained in [2,5,6], for linear operators only the moments (i.e. the expressions of $L_i(t^k)$ corresponding to $t = 0$) are of concern. The notation we use to indicate the moments of order k associated with operator $\underline{L_i}$ is $L_{i,k} := \underline{L_i}(t^k)$. Linear operators can be written in a form similar to Taylor series expansion, the terms of which are their respective moments, so the following property holds:

$$
\underline{L_i}\big(y(t)\big) = \sum_{k=0}^{\infty} \frac{1}{k!}\, L_{i,k}\, y^{(k)}(t).
\tag{11}
$$

These operators represent the LTEs committed, i.e. a measure to determine how much the solution of the differential Eq. (1) fails to solve the difference Eq. (4).

The complete expression of $\underline{L_1}(t^k)$ is, combining (9) and (10),

$$
\begin{aligned}
\underline{L_1}(t^k) &= (t + h\, c_1)^k - b_{11}\big(t + h(c_1 - 1)\big)^k - b_{12}\, t^k - h\, k\, a_{11}\big(t + h(c_1 - 1)\big)^{k-1} \\
&\quad - h\, k\, a_{12} t^{k-1},
\end{aligned}
\tag{12}
$$

and the moments $L_{1,0}$, $L_{1,1}$, $L_{1,2}$ and $L_{1,3}$ are

$$
\begin{aligned}
L_{1,0} &= 1 - b_{11} - b_{12}, \\
L_{1,1} &= h\, c_1 - h\, b_{11}(c_1 - 1) - h\, a_{11} - h\, a_{12}, \\
L_{1,2} &= (h\, c_1)^2 - b_{11}\big(h(c_1 - 1)\big)^2 - 2\, h\, a_{11}\big(h(c_1 - 1)\big), \\
L_{1,3} &= (h\, c_1)^3 - b_{11}\big(h(c_1 - 1)\big)^3 - 3\, h\, a_{11}\big(h(c_1 - 1)\big)^2.
\end{aligned}
\tag{13}
$$

Cancelling the first three moments leads to the following three equations system, which the coefficients of (4) must satisfy so that the accuracy order of the first stage $Y_{n,1}$ is equal to two:

$$
\begin{cases}
1 - b_{11} - b_{12} = 0, \\
c_1 - b_{11}(c_1 - 1) - a_{11} - a_{12} = 0, \\
c_1^2 - b_{11}(c_1 - 1)^2 - 2\, a_{11}(c_1 - 1) = 0.
\end{cases}
\tag{14}
$$

In fact, from (9) and applying (11) to $\underline{L_1}(t^k)$, leads to

$$
y(t + h\, c_1) = Y_1(t) + \frac{h^3}{3!}\big(c_1^3 - b_{11}(c_1 - 1)^3 - 3\, a_{11}(c_1 - 1)^2\big) y'''(t) + O(h^4). \tag{15}
$$

We indicate the LTE on $Y_{n,1}$ with $err_1(t)$, and the Principal term of the Local Truncation Error (PLTE) on $Y_{n,1}$ with $t_{err_1}(t)$:

$$err_1(t) = t_{err_1}(t) + O(h^4), \text{ with}$$

$$t_{err_1}(t) = \frac{h^3}{3!}\left(c_1^3 - b_{11}(c_1 - 1)^3 - 3\,a_{11}(c_1 - 1)^2\right)y'''(t). \tag{16}$$

By doing the same on the second stage $Y_{n,2}$ which, like the first stage $Y_{n,1}$, we calculate with order of accuracy equal to two, we obtain:

$$\underline{L_2}(t^k) = (t + h)^k - b_{21}\left(t + h(c_1 - 1)\right)^k - b_{22}t^k - h\,k\,a_{21}\left(t + h(c_1 - 1)\right)^{k-1}$$
$$- h\,k\,a_{22}\,t^{k-1} - h\,k\,r_{21}\left(t + c_1\,h\right)^{k-1}, \tag{17}$$

$$\begin{aligned}
L_{2,0} &= 1 - b_{21} - b_{22},\\
L_{2,1} &= h - h\,b_{21}(c_1 - 1) - h\,a_{21} - h\,a_{22} - h\,r_{21},\\
L_{2,2} &= h^2 - b_{21}(h(c_1 - 1))^2 - 2\,h\,a_{21}(h(c_1 - 1)) - 2\,h^2 r_{21}\,c_1,\\
L_{2,3} &= h^3 - b_{21}(h(c_1 - 1))^3 - 3\,h\,a_{21}(h(c_1 - 1))^2 - 3\,h\,r_{21}\,(h\,c_1)^2,
\end{aligned} \tag{18}$$

$$\begin{cases}
1 - b_{21} - b_{22} = 0,\\
1 - b_{21}(c_1 - 1) - a_{21} - a_{22} - r_{21} = 0,\\
1 - b_{21}(c_1 - 1)^2 - 2\,a_{21}(c_1 - 1) - 2\,r_{21}\,c_1 = 0,
\end{cases} \tag{19}$$

$$y(t + h) = Y_2(t) + \frac{h^3}{3!}\left(1 - b_{21}(c_1 - 1)^3 - 3\,a_{21}(c_1 - 1)^2 - 3\,r_{21}\,c_1^2\right)y'''(t) + O(h^4). \tag{20}$$

In conclusion, requiring that the global accuracy order of the peer method (4) be equal to two, means deriving the coefficients by solving the systems (14) and (19). Invoking (8), we observe that four free coefficients remain to be assigned later.

The extension of this procedure in the case of s stages is possible and allows to get the coefficients of classical s-stage peer methods with accuracy order $p = s$. It is customary to assign fixed values to the coefficients of B (respecting $B1 = 1$, with $1 = (1, ..., 1)^T$), R and c, obtaining $(a_{ij})_{i,j=1}^s$ as a function of them [1]:

$$A = (C\,V_0\,D^{-1} - R\,V_0)V_1^{-1} - B(C - I)V_1\,D^{-1}\,V_1^{-1},$$

$$V_0 = (c_i^{j-1})_{i,j=1}^s, \quad V_1 = \left((c_i - 1)^{j-1}\right)_{i,j=1}^s, \quad I = I_s \text{ (Identity matrix of order s)},$$

$$D = diag(1, ..., s), \quad C = diag(c_i). \tag{21}$$

3   New Jacobian-Dependent Peer Methods

The new Jacobian-dependent methods are obtained by defining differently the functions $Y_1(t)$ and $Y_2(t)$ in (10), and therefore the operators (9) with whom we calculate the LTEs from which the stages are affected. In doing so, it will change the definition of $\underline{L_2}(y(t))$, but not that of $\underline{L_1}(y(t))$.

In fact, we are assuming that the following localizing assumption applies only at the previous grid points $\{t_0, ..., t_{n-1}\}$, but not at t_n:

$$Y_{n-1,j} = y(t_{n-1} + h\,c_j), \; \forall j = 1, ..., s. \tag{22}$$

Therefore, we're going to consider the LTEs committed in the calculation of the previous stages $Y_{n,j}$, $j = 1, ..., i-1$ in determining $Y_{n,j}$, $j = i, ..., s$. This change leads to the dependence of the coefficients on the Jacobian function of the problem f.

3.1   Two-Stage Jacobian-Dependent Version

As mentioned before, the application of the new hypothesis (22) on peer methods (4) doesn't produce any changes in $Y_{n,1}$, as it depends exclusively on $Y_{n-1,j}$, $j = 1, 2$. The only variation concerns $Y_{n,2}$, because it depends on $Y_{n,1}$, that is affected by the LTE (16). Therefore, the expression of $L_1(t^k)$ remains the same (12) as in the classic case, as well as the moments (13), the order conditions (14), and the LTE (16).

However, now we need to recalculate the order conditions of the second stage $Y_{n,2}$, bearing in mind, this time, the error made in the calculation of the first stage $Y_{n,1}$ (16). The new definition of $Y_2(t)$ is

$$\begin{aligned} Y_2(t) =\, & b_{21}\, y\big(t + h(c_1 - 1)\big) + b_{22}\, y(t) + h\, a_{21}\, y'\big(t + h(c_1 - 1)\big) + h\, a_{22}\, y'(t) \\ & + h\, r_{21} f\big(t + h\,c_1, Y_1(t)\big), \end{aligned} \tag{23}$$

where, unlike the definition of $Y_2(t)$ in (10), there is $f\big(t + h\,c_1, Y_1(t)\big)$ instead of $y'(t + h\,c_1)$.

In fact, it now applies that

$$y'(t + h\,c_1) = f\big(t + h\,c_1, y(t + h\,c_1)\big) = f\big(t + h\,c_1, Y_1(t) + err_1(t)\big). \tag{24}$$

Remembering that the Taylor series expansion of a generic function $g(x)$ at x_0 is

$$g(x) = g(x_0) + g'(x_0)\,(x - x_0) + ... + \frac{g^n(x_0)}{n!}(x - x_0)^n + O\big((x - x_0)^n\big), \tag{25}$$

we get the Taylor series expansion of f in (24) at $Y_1(t)$ as follows:

$$\begin{aligned} f\big(t + h\,c_1, Y_1(t) + err_1(t)\big) &= f\big(t + h\,c_1, Y_1(t)\big) + j_1(t)\, err_1(t) + O\big(err_1(t)^2\big) \\ &= f\big(t + h\,c_1, Y_1(t)\big) + j_1(t)\, t_{err_1}(t) + O(h^4), \end{aligned}$$

with $j_1(t) = f_y(t + h\,c_1, y)_{|y = Y_1(t)}$ (Jacobian function).

$$\tag{26}$$

By combining (24) and (26), we finally get

$$f(t + h\,c_1, Y_1(t)) = y'(t + h\,c_1) - j_1(t)\,t_{err_1}(t) + O(h^4). \tag{27}$$

The replacement of the expression just found (27) in $Y_2(t)$ (23) leads to the new shape of $\underline{L_2}(t^k)$:

$$\begin{aligned}
\underline{L_2}(t^k) = {} & (t + h)^k - b_{21}\big(t + h(c_1 - 1)\big)^k - b_{22}t^k - h\,k\,a_{21}\big(t + h(c_1 - 1)\big)^{k-1} \\
& - h\,k\,a_{22}t^{k-1} - h\,r_{21}\Big(k(t + h\,c_1)^{k-1} - j_1(t)\Big(\frac{h^3}{3!}\big(c_1^3 - b_{11}(c_1 - 1)^3 \\
& - 3\,a_{11}(c_1 - 1)^2 k(k - 1)(k - 2)t^{k-3}\big)\Big)\Big).
\end{aligned} \tag{28}$$

The new moments $L_{2,0}$, $L_{2,1}$, $L_{2,2}$ and $L_{2,3}$ are

$$\begin{aligned}
L_{2,0} = {} & 1 - b_{21} - b_{22}, \\
L_{2,1} = {} & h - h\,b_{21}(c_1 - 1) - h\,a_{21} - h\,a_{22} - h\,r_{21}, \\
L_{2,2} = {} & h^2 - b_{21}\big(h(c_1 - 1)\big)^2 - 2\,h\,a_{21}\big(h(c_1 - 1)\big) - 2\,h^2 r_{21}\,c_1, \\
L_{2,3} = {} & h^3 - b_{21}\big(h(c_1 - 1)\big)^3 - 3\,h\,a_{21}\big(h(c_1 - 1)\big)^2 - 3\,h\,r_{21}\,(h\,c_1)^2 \\
& + m_1(t)\,r_{21}\big(h^3(c_1^3 - b_{11}(c_1 - 1)^3 - 3\,a_{11}(c_1 - 1)^2)\big),
\end{aligned} \tag{29}$$

where $m_1(t) = h j_1(t)$. This time we cancel all the four moments, thus obtaining the second stage $Y_{n,2}$ with accuracy order equal to three. Numerical experiments will show that, despite this, the global accuracy order of the new peer methods remains two:

$$\begin{cases}
1 - b_{21} - b_{22} = 0, \\
1 - b_{21}(c_1 - 1) - a_{21} - a_{22} - r_{21} = 0, \\
1 - b_{21}(c_1 - 1)^2 - 2a_{21}(c_1 - 1) - 2r_{21}c_1 = 0, \\
1 - b_{21}(c_1 - 1)^3 - 3a_{21}(c_1 - 1)^2 - 3r_{21}c_1^2 + m_1(t)r_{21}\big(c_1^3 \\
\quad - b_{11}(c_1 - 1)^3 - 3a_{11}(c_1 - 1)^2\big) = 0.
\end{cases} \tag{30}$$

The resolution of the first three equations in (30) with (14) leads exactly to the same order conditions as the standard methods. Calculating instead the coefficients by solving the systems (30) and (14) in full, we get the new Jacobian-dependent peer methods. Jacobian dependency is evidenced by the presence of $m_1(t) = h\,j_1(t)$ in the last equation of (30).

Always keeping in mind (8), as there are seven independent equations to be solved now, the number of free coefficients for new peer methods is three and no longer four. By fixing b_{11}, b_{21} and c_1, the coefficients of new peer methods are shown in (31). We observe that the coefficients a_{21}, a_{22} and r_{21} of new methods depend on $m_1(t)$, i.e. the Jacobian function of the problem at time t. This leads to two important considerations:

– a_{21}, a_{22} and r_{21} must be updated at each time-step;

- in the multi-dimensional case, a_{21}, a_{22} and r_{21} become matrices with the same size as $m_1(t)$.

$$a_{11} = \left(-(b_{11}(-1+c_1)^2) + c_1^2 \right) / \left(2(-1+c_1) \right),$$

$$a_{12} = \left(-(b_{11}(-1+c_1)^2) + (-2+c_1)c_1 \right) / \left(2(-1+c_1) \right),$$

$$\begin{aligned}
a_{21} = &\left(b_{11}(-1+b_{21})m_1(t) + (-1+b_{11}+7b_{21}-10b_{11}b_{21})c_1^3 m_1(t) \right.\\
&- (-1+b_{11})b_{21}c_1^5 m_1(t) + b_{21}c_1^4(2+5(-1+b_{11})m_1(t))\\
&+ c_1(4+3b_{11}m_1(t)+b_{21}(4-5b_{11}m_1(t))) + c_1^2(-6 \\
&\left. + 3m_1(t) - 3b_{11}m_1(t) + b_{21}(-6-3\,m_1(t)+10b_{11}m_1(t))) \right)\\
&/\left(2(-1+c_1)(-(b_{11}m_1(t)) - 3(-1+b_{11})c_1^2 m_1(t) + (-1 \right.\\
&\left. + b_{11})c_1^3 m_1(t) + 3c_1(-2+b_{11}m_1(t))) \right),
\end{aligned}$$

$$\begin{aligned}
a_{22} = &\left(-10 + 3b_{11}m_1(t) - 9(-1+b_{11})c_1^3 m_1(t) + 2(-1+b_{11})c_1^4 m_1(t) \right.\\
&+ c_1(24 - 11b_{11}m_1(t)) + 3c_1^2(-4-3m_1(t)+5b_{11}m_1(t))\\
&- b_{21}(-1+c_1)^2(-2 - b_{11}m_1(t) - 3(-1+b_{11})c_1^2 m_1(t)\\
&\left. + (-1+b_{11})c_1^3 m_1(t) + c_1(-4+3b_{11}m_1(t))) \right) / \left(2(-1 \right.\\
&+ c_1)(-(b_{11}m_1(t)) - 3(-1+b_{11})c_1^2 m_1(t) + (-1\\
&\left. + b_{11})c_1^3 m_1(t) + 3c_1(-2+b_{11}m_1(t))) \right),
\end{aligned}$$

$$\begin{aligned}
r_{21} = &\left(-5 - b_{21}(-1+c_1)^3 + 3c_1 \right) / \left(-(b_{11}m_1(t)) - 3(-1+b_{11})c_1^2 m_1(t) \right.\\
&\left. + (-1+b_{11})c_1^3 m_1(t) + 3c_1(-2+b_{11}m_1(t)) \right).
\end{aligned}$$

(31)

4   Linear Stability Analysis

The family of explicit s-stage peer methods can be written in the compact form

$$Y_n = B Y_{n-1} + h A F(Y_{n-1}) + h R F(Y_n),$$
$$\text{where } Y_n = (Y_{n,i})_{i=1}^s, F(Y_n) = \left(f(Y_{n,i}) \right)_{i=1}^s.$$

(32)

In order to perform linear stability analysis of classic peer methods and new Jacobian-dependent peer methods, applying (32) to the test equation $y' = \lambda y$, $Re(\lambda) < 0$, results in

$$Y_n = (I - zR)^{-1}(B + zA)Y_{n-1} =: M(z)Y_{n-1}, \text{ with } z = h\lambda.$$

(33)

Therefore, $M(z)$ is the stability matrix of the method, i.e. the numerical method (32) is absolutely stable if $\rho(M(z)) < 1$, where ρ is the spectral radius of $M(z)$. The short analysis just shown is detailed in [1].

4.1   Absolute Stability Regions of the New Two-Stage Methods

We have used *Mathematica* to fix the coefficients left free after the imposition of the order conditions, with the aim of maximizing the size of the absolute stability region of classic and new two-stage peer methods.

Referring to new two-stage Jacobian-dependent peer methods, we need to fix b_{11}, b_{21} and c_1, then finding the other coefficients of the method by exploiting (31). The exploration range for these parameters is:

- $c_1 \in [0, 1)$, because the intermediate stages determine the numerical solution within the subinterval $[t_n, t_{n+1}]$,
- b_{11} and $b_{21} \in [-2, 2]$, that is usually the range of values in which the coefficients of the matrix B are considered for peer methods.

In doing so, we get the best result with

$$b_{11} = -0.59, \qquad b_{21} = -1, \qquad c_1 = 0.3. \tag{34}$$

Trying to do the same thing for classic peer methods as well, using the same intervals for parameter exploration (this time obviously including r_{21} as well, which is a free parameter for the classic methods), we get the largest real axis of the corresponding absolute stability region by using

$$b_{11} = -0.52, \qquad b_{21} = -1.3, \qquad c_1 = 0.3, \qquad r_{21} = 0.8. \tag{35}$$

Finally, in order to compare the new peer methods with the classical ones, we consider two-stage classic peer method with the same coefficients as the peer Jacobian-dependent (34), fixing the best possible value for r_{21}:

$$b_{11} = -0.59, \qquad b_{21} = -1, \qquad c_1 = 0.3, \qquad r_{21} = 1.17. \tag{36}$$

Figure (1) shows the absolute stability regions of Jacobian-dependent and classic methods, using as parameters those just reported.

We note that, although we have fixed the coefficients of the classic peer method (35) in such a way as to maximize the amplitude of the real axis in its stability region, the stability region of the new Jacobian-dependent peer method contains it. In addition, the absolute stability region of the classic method has a rather strange shape, narrowing in some places to just the real axis. The stability region of the classic version with non-optimal coefficients (36) is smaller than that of the other two methods.

5   Numerical Tests

In this section, we conduct numerical tests on two scalar ODEs, by solving them with Jacobian-dependent and classic peer methods, using as coefficient values (34) (for Jacobian-dependent version), (35), and (36) (for classic versions), in order to verify the theoretical properties we derived in the previous sections. Since the exact solution of the following problems is known, we evaluate the

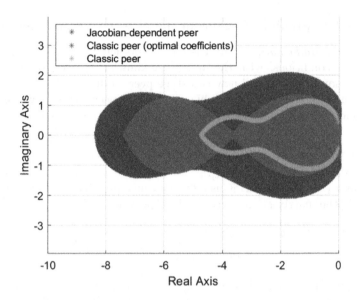

Fig. 1. Absolute stability regions of the Jacobian-dependent peer method with optimal coefficients (34), the classic peer method with optimal coefficients (35) and the classic peer method set the coefficients of the Jacobian-dependent method (36), respectively.

absolute error at the last grid point, and the order estimate of the methods using the formula

$$p(h) = \frac{cd(h) - cd(2\,h)}{log_{10}2}, \tag{37}$$

where $cd(h)$ is the achieved number of correct digits at the endpoint T of the integration interval $[t_0, T]$ with step-size equal to h. For more information on the quantities taken into account for numerical tests consult [3].

5.1 Prothero-Robinson Equation

Let's solve the Prothero–Robinson scalar equation

$$\begin{cases} y'(t) = \lambda\big(y(t) - sin(t)\big) + cos(t), \\ y(0) = 0, \end{cases} \tag{38}$$

in the interval $[0, \pi/2]$, for different values of λ. The exact solution of the problem (38) is $y(t) = sin(t)$. This equation is widely used to test numerical methods as it becomes more and more stiff as $|\lambda|$ increases.

The results shown in Tables (1), (2), (3), (4), (5) and (6) allow for the following important observations. The Jacobian-dependent method is much more accurate than the classic method with optimal coefficients (35) (there is a difference of two orders of magnitude between their respective absolute errors) and slightly more accurate than the classic method with the same coefficients (36).

As the stiffness of the problem increases, it is necessary to use the Jacobian-dependent method, which has better stability properties than the other two, especially the classic peer with the same coefficients. Finally, the global accuracy order estimation of the three methods tends to two, so the classic methods and the new method don't suffer from order reduction on stiff problems.

Figure (2) confirms the greater accuracy of the Jacobian-dependent method than the other two, for $\lambda = -50$ and $\lambda = -100$. In Fig. (3), we represent the trend of the three numerical solutions and the exact one, fixing the problem (i.e. $\lambda = -10^3$) and choosing two different values for h. For the first value of h, numerical solutions calculated with the two classical methods explode, and only the Jacobian-dependent method provides a good result. For a smaller value of h, even the classic method with coefficients (35) provides an acceptable solution. Finally, looking at Fig. (4), we can appreciate the fact that when the problem is very stiff (i.e. $\lambda = -10^4$), the Jacobian-dependent method becomes more convenient than the others.

Table 1. Absolute errors at the endpoint T on Prothero-Robinson problem (38), in correspondence of several values of the number $N + 1$ of mesh-points, with $\lambda = -50$.

N	Jacobian-dependent (34)	Classic (optimal coefficients) (35)	Classic (36)
2^8	1.3525e−09	5.3149e−08	8.8738e−09
2^9	3.0154e−10	9.6927e−09	2.0059e−09
2^{10}	6.8472e−11	1.9597e−09	4.6980e−10
2^{11}	1.6040e−11	4.3112e−10	1.1317e−10
2^{12}	3.8507e−12	1.0031e−10	2.7743e−11

Table 2. Estimated order $p(h)$ on Prothero-Robinson problem, with $\lambda = -50$.

N	Jacobian-dependent (34)	Classic (optimal coefficients) (35)	Classic (36)
2^9	2.1653	2.4551	2.1453
2^{10}	2.1387	2.3063	2.0942
2^{11}	2.0939	2.1845	2.0536
2^{12}	2.0585	2.1037	2.0282

5.2   Non-linear Scalar Equation

Let's solve the non-linear scalar equation

$$\begin{cases} y'(t) = -y(y+1), \\ y(0) = 2, \end{cases} \tag{39}$$

in the interval $[0, 5]$. The exact solution of the problem (39) is $y(t) = \dfrac{(2/3)e^{-t}}{1 - (2/3)e^{-t}}$.

Table 3. Absolute errors at the endpoint T on Prothero-Robinson problem (38), in correspondence of several values of the number $N + 1$ of mesh-points, with $\lambda = -10^2$.

N	Jacobian-dependent (34)	Classic (optimal coefficients) (35)	Classic (36)
2^8	3.7746e−10	2.0039e−08	2.4883e−09
2^9	8.4564e−11	3.3246e−09	5.5482e−10
2^{10}	1.8851e−11	6.0605e−10	1.2541e−10
2^{11}	4.2772e−12	1.2251e−10	2.9374e−11
2^{12}	9.9920e−13	2.6911e−11	7.0786e−12

Table 4. Estimated order $p(h)$ on Prothero-Robinson problem, with $\lambda = -10^2$.

N	Jacobian-dependent (34)	Classic (optimal coefficients) (35)	Classic (36)
2^9	2.1582	2.5915	2.1651
2^{10}	2.1654	2.4557	2.1453
2^{11}	2.1399	2.3066	2.0941
2^{12}	2.0978	2.1866	2.0530

Table 5. Absolute errors at the endpoint T on Prothero-Robinson problem (38), in correspondence of several values of the number $N + 1$ of mesh-points, with $\lambda = -10^3$.

N	Jacobian-dependent (34)	Classic (optimal coefficients) (35)	Classic (36)
2^8	1.3301e−10	6.6674e−10	−
2^9	3.4133e−12	1.1099e−10	1.5820e−12
2^{10}	3.2963e−13	2.2591e−11	1.3811e−12
2^{11}	6.0840e−14	3.6213e−12	3.9857e−13
2^{12}	1.2546e−14	5.8420e−13	9.0372e−14

Table 6. Estimated order $p(h)$ on Prothero-Robinson problem, with $\lambda = -10^3$.

N	Jacobian-dependent (34)	Classic (optimal coefficients) (35)	Classic (36)
2^9	5.2843	2.5867	−
2^{10}	3.3723	2.2966	0.1959
2^{11}	2.4377	2.6412	1.7929
2^{12}	2.2779	2.6320	2.1409

The results (Tables (7) and (8)) related to the non-linear problem (39) confirm the observations and comments made previously, although in this case the stability advantage of Jacobian-dependent methods is not observed as the equation is non-stiff. The greater accuracy of new methods is confirmed.

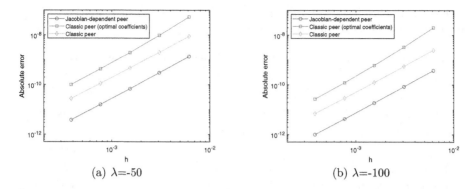

Fig. 2. Absolute errors as a function of h, for different values of λ.

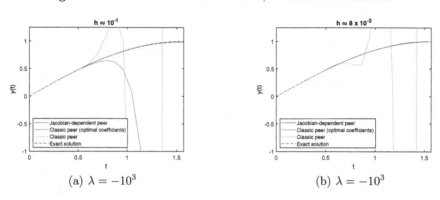

Fig. 3. Solution of the problem (38) for different values of h, with $\lambda = -10^3$.

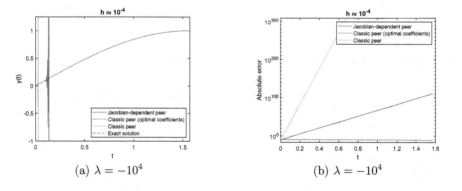

Fig. 4. Solution and absolute errors related to the problem (38), with $h \approx 10^{-4}$ and $\lambda = -10^4$.

Table 7. Absolute errors at the endpoint T on non-linear scalar problem (39), in correspondence of several values of the step-size h.

N	Jacobian-dependent (34)	Classic (optimal coefficients) (35)	Classic (36)
1/2	1.3608e−02	1.7557e−01	1.2034e−02
1/4	1.1327e−03	4.4501e−02	2.7823e−03
1/8	1.7397e−04	5.6674e−03	6.3510e−04
1/16	3.0933e−05	8.3379e−04	1.4977e−04
1/32	6.2091e−06	1.5820e−04	3.6144e−05
1/64	1.3654e−06	3.4293e−05	8.8605e−06
1/128	3.1827e−07	7.9767e−06	2.1923e−06
1/256	7.6699e−08	1.9235e−06	5.4518e−07
1/512	1.8817e−08	4.7231e−07	1.3593e−07

Table 8. Estimated order $p(h)$ on non-linear scalar problem.

N	Jacobian-dependent (34)	Classic (optimal coefficients) (35)	Classic (36)
1/2	–	–	–
1/4	3.5866	1.9802	2.1128
1/8	2.7029	2.9731	2.1312
1/16	2.4916	2.7649	2.0842
1/32	2.3167	2.3980	2.0509
1/64	2.1850	2.2057	2.0283
1/128	2.1010	2.1041	2.0149
1/256	2.0530	2.0520	2.0077
1/512	2.0271	2.0260	2.0039

6   Conclusions and Future Research

In this paper, we have derived new Jacobian-dependent peer methods with better stability properties than the classic ones. In addition, although we have focused on this new methods, we would like to stress the fact that we have also determined the coefficients of classic explicit peer methods (35) in order to maximize the relative absolute stability region.

Updating the coefficients that depend on the Jacobian function at each step comes at a higher cost. However, this additional cost is acceptable, given the benefits obtained both in terms of stability and accuracy.

The next research will focus on further improving the stability properties of Jacobian-dependent peer methods by investigating the possibility of deriving A-stable methods. Finally, we'll adapt the methods obtained in this paper to the multi-dimensional case, transforming the coefficients that depend on the Jacobian function into matrices.

References

1. Weiner, R., Biermann, K., Schmitt, B., Podhaisky, H.: Explicit two-step peer methods. Comput. Math. Appl. **55**, 609–619 (2008). https://doi.org/10.1016/j.camwa.2007.04.026
2. Ixaru, L.: Runge-Kutta methods with equation dependent coefficients. Comput. Phys. Commun. **183**, 63–69 (2012). https://doi.org/10.1016/j.cpc.2011.08.017
3. Conte, D., D'Ambrosio, R., Pagano, G., Paternoster, B.: Jacobian-dependent vs Jacobian-free discretizations for nonlinear differential problems. Comput. Appl. Math. **39**(3), 1–12 (2020). https://doi.org/10.1007/s40314-020-01200-z
4. Fang, Y., Yang, Y., You, X., Wang, B.: A new family of A-stable Runge-Kutta methods with equation-dependent coefficients for stiff problems. Numer. Algorithms **81**(4), 1235–1251 (2018). https://doi.org/10.1007/s11075-018-0619-7
5. Ixaru, L.: Operations on oscillatory functions. Comput. Phys. Commun. **105**, 1–19 (1997). https://doi.org/10.1016/S0010-4655(97)00067-2
6. Ixaru, L., Berghe, G.: Exponential Fitting (2004). https://doi.org/10.1007/978-1-4020-2100-8
7. Kulikov, G., Weiner, R.: Doubly quasi-consistent parallel explicit peer methods with built-in global error estimation. J. Comput. Appl. Math. **233**, 2351–2364 (2010). https://doi.org/10.1016/j.cam.2009.10.020
8. Schmitt, B., Wiener, R.: Parallel start for explicit parallel two-step peer methods. Numer. Algorithms **53**, 363–381 (2010). https://doi.org/10.1007/s11075-009-9267-2
9. Schmitt, B., Weiner, R., Jebens, S.: Parameter optimization for explicit parallel peer two-step methods. Appl. Numer. Math. **59**, 769–782 (2009). https://doi.org/10.1016/j.apnum.2008.03.013
10. Weiner, R., Kulikov, G.Y., Podhaisky, H.: Variable-stepsize doubly quasi-consistent parallel explicit peer methods with global error control. J. Comput. Appl. Math. **62**, 2351–2364 (2012). https://doi.org/10.1016/j.apnum.2012.06.018
11. Horváth, Z., Podhaisky, H., Weiner, R.: Strong stability preserving explicit peer methods. J. Comput. Appl. Math. **296**, 776–788 (2015). https://doi.org/10.1016/j.cam.2015.11.005
12. Jebens, S., Weiner, R., Podhaisky, H., Schmitt, B.: Explicit multi-step peer methods for special second-order differential equations. Appl. Math. Comput. **202**, 803–813 (2008). https://doi.org/10.1016/j.amc.2008.03.025
13. Klinge, M., Weiner, R.: Strong stability preserving explicit peer methods for discontinuous Galerkin discretizations. J. Sci. Comput. **75**(2), 1057–1078 (2017). https://doi.org/10.1007/s10915-017-0573-x
14. Klinge, M., Weiner, R., Podhaisky, H.: Optimally zero stable explicit peer methods with variable nodes. BIT Numer. Math. **58**(2), 331–345 (2017). https://doi.org/10.1007/s10543-017-0691-8
15. Montijano, J.I., Rández, L., Van Daele, M., Calvo, M.: Functionally fitted explicit two step peer methods. J. Sci. Comput. **64**(3), 938–958 (2014). https://doi.org/10.1007/s10915-014-9951-9
16. Weiner, R., Schmitt, B., Podhaisky, H., Jebens, S.: Superconvergent explicit two-step peer methods. J. Comput. Appl. Math. **223**, 753–764 (2009). https://doi.org/10.1016/j.cam.2008.02.014
17. Jebens, S., Knoth, O., Weiner, R.: Linearly implicit peer methods for the compressible Euler equations. J. Comput. Phys. **230**, 4955–4974 (2011). https://doi.org/10.1016/j.jcp.2011.03.015

18. Kulikov, G.Y., Weiner, R.: Doubly quasi-consistent fixed-stepsize numerical integration of stiff ordinary differential equations with implicit two-step peer methods. J. Comput. Appl. Math. **340**, 256–275 (2018). https://doi.org/10.1016/j.cam.2018.02.037

19. Lang, J., Hundsdorfer, W.: Extrapolation-based implicit-explicit peer methods with optimised stability regions. J. Comput. Phys. **337**, 203–215 (2016). https://doi.org/10.1016/j.jcp.2017.02.034

20. Schneider, M., Lang, J., Hundsdorfer, W.: Extrapolation-based super-convergent implicit-explicit peer methods with A-stable implicit part. J. Comput. Phys. **367**, 121–133 (2017). https://doi.org/10.1016/j.jcp.2018.04.006

21. Schneider, M., Lang, J., Weiner, R.: Super-convergent implicit-explicit Peer methods with variable step sizes. J. Comput. Appl. Math. **387**, 112501 (2019). https://doi.org/10.1016/j.cam.2019.112501

22. Conte, D., D'Ambrosio, R., Moccaldi, M., Paternoster, B.: Adapted explicit two-step peer methods. J. Numer. Math. **27**, 69–83 (2018). https://doi.org/10.1515/jnma-2017-0102

23. Conte, D., Mohammadi, F., Moradi, L., Paternoster, B.: Exponentially fitted two-step peer methods for oscillatory problems. Comput. Appl. Math. **39**(3), 1–19 (2020). https://doi.org/10.1007/s40314-020-01202-x

24. Conte, D., Paternoster, B., Moradi, L., Mohammadi, F.: Construction of exponentially fitted explicit peer methods. Int. J. Circuits **13**, 501–506 (2019)

HTTP-DTNSec: An HTTP-Based Security Extension for Delay/Disruption Tolerant Networking

Lucas William Paz Pinto[1], Bruno L. Dalmazo[2]([✉]) [iD], André Riker[3], and Jéferson Campos Nobre[4] [iD]

[1] University of Vale do Rio dos Sinos, São Leopoldo, Brazil
`lwppinto@unisinos.br`
[2] Federal University of Rio Grande, Rio Grande, Brazil
`dalmazo@furg.br`
[3] Federal University of Pará, Belém, Brazil
`afr@ufpa.br`
[4] Federal University of Rio Grande do Sul, Porto Alegre, Brazil
`jcnobre@inf.ufrgs.br`

Abstract. Communication networks operating in challenging environments can be grouped by the concept of Delay/Disruption-Tolerant Networking (DTN). Different protocols can be used in DTN, such as the Bundle Protocol (BP) and the HyperText Transfer Protocol - DTN (HTTP-DTN). In this context, security properties are of fundamental importance in DTN like in regular networks. However, the challenges in DTN hamper the use of traditional security mechanisms. Although BP has been extended to include such mechanisms, there is still no analogous extension for HTTP-DTN. In this paper, we propose the HTTP-DTNSec, a security extension for HTTP-DTN. This extension improves the confidentiality and integrity of HTTP-DTN as well as updates the base protocol for HTTP/2. The proposed extension was implemented as a proof of concept and it was used to perform experiments in a simulated environment. These experiments show that HTTP-DTNSec performed the transfer of packages (i.e., a group of related objects) in a safe manner and with an increase in performance concerning HTTP-DTN. Finally, we provide some concluding remarks and future directions.

Keywords: DTN · HTTP-DTN · Network security

1 Introduction

It is important for many applications to support communication in challenging environments such as InterPlanetary Networking, Battlefield Networking, Rural communications, and Underwater Communications. The Delay/Disruption-Tolerant Networking (DTN) emerged to covers all these environments [8,20], since they present difficulties for the operation of a traditional network, such

© Springer Nature Switzerland AG 2021
O. Gervasi et al. (Eds.): ICCSA 2021, LNCS 12949, pp. 325–340, 2021.
https://doi.org/10.1007/978-3-030-86653-2_24

as the Internet. The conditions found in DTN environments cause some problems in computer networks, such as long periods of disconnections that hinder end-to-end communication.

In the context of the IETF, the Bundle Protocol (BP) [5, 17] has been developed in order to enable the transmission of messages between DTN nodes, through a logic of storage and sending of packets. This allows the nodes to store all the packets they receive locally, and send them when a connection is available with another node. However, some authors point out problems in the BP, for example, problems of synchronization and complexity [22]. Thus, alternative approaches to BP were created in order to avoid such difficulties. One of these alternatives is Hyper Text Transfer Protocol - Delay-Tolerant Networking (HTTP-DTN) [23].

Security properties are critical in DTN as in traditional networks [7]. However, there are additional difficulties in implementing delay or disconnected-tolerant security mechanisms [13, 16]. In this context, an extension of BP has been developed, named *Bundle Protocol Security Specification* (BPSec) [4], to support end-to-end integrity and confidentiality. However, this extension only applies to BP, which means that other DTN protocols, such as HTTP-DTN, do not support these important security aspects.

The present paper proposes *HTTP-DTNSec*, a security extension proposal for HTTP-DTN. Such a proposal incorporates the security properties found in BPSec, maintaining the desirable characteristics of HTTP-DTN. In addition, HTTP-DTNSec updates the HTTP version used in HTTP-DTN (i.e., from HTTP/1.1 to HTTP/2). Such an update allows a performance increase in the exchange of messages. The results show that the extension is feasible, incorporating new features as well as maintaining adequate performance.

This article is structured as follows. Section 2 presents the theoretical background on DTN and security challenges that exist in this type of network. Section 3 describes the HTTP-DTNSec specification. Section 4 contains the experiments carried out as well as an analysis of the obtained results. Section 5 presents the related work. Finally, Sect. 6 presents the conclusions and future work.

2   Background

The DTN architecture proposed by the IETF [20] describes the Bundle Layer, which interfaces with the lower layers using Convergence Layer Adapters (CLAs) to carry out the transport of information. The end-to-end transport semantics are redefined and are now confined within each DTN hop. Thus, CLA protocols for different transport protocols have been developed, (e.g., UDP [18] and TCP [19]). This makes it possible to use the most suitable transport protocol for each hop.

BP defines the concept of blocks, which are a set of information referring to data and metadata that make up a bundle. The IETF defines three blocks [17]: the Primary Block, the first block in a bundle which contains information regarding source and destination addressing; the *Payload* Block, which carries the data that the application wants to send via the BP; and the Extension Block, which has a variety of data that can provide additional information needed to process a bundle along the network.

This section is organized as follows. Initially, aspects related to security in DTN will be presented. Next, HTTP-DTN will be described in more detail.

2.1 DTN Security

There are several challenges in implementing security mechanisms in DTN. Some of these challenges occur due to the computational processing of such mechanisms [12,21]. For example, the use of digital certificates is difficult since a trusted third party for verifying the authenticity of certificates may not be available at the time of processing such verification.

Managing cryptographic keys presents challenges in DTN [3]. This is due to the fact that the acceptance of a session key, or even the exchange of public keys between the origin and the end of an end-to-end communication may not be possible, due, for example, to the lack of connection when sending the message [6]. Thus, a mechanism for the distribution of keys must be defined considering the communication problems in DTN. However, some insecure strategies can be adopted due to these problems. An example of such strategies is the use of keys shared with the nodes, which would reduce the complexity of managing them. This sharing is not adequate, since if the compromise occurs on any node, the entire DTN would be compromised and the key assignment process would need to be carried out again [14].

In the context of the IETF, security policies are applied through additional blocks to the bundle, which add end-to-end integrity and confidentiality services to BP [4]. Additional BPSec Blocks are: Block Integrity Block (BIB), used to guarantee the authenticity and integrity of the package block that the BIB sender wishes to ensure, and the Block Confidentiality Block (BCB), used to indicate that a part of the packet has been encrypted at the node originating from the BCB. BPSec is applied by definition only to nodes that accept its implementation, these nodes are known as Security Aware Nodes. There may be nodes in a DTN that do not implement BSP and that can communicate with all DTN nodes.

The authentication of the information in the BIB can be verified by any node that is in the path of the BIB sender and the recipient of the bundle. When a node other than the one originating from the bundle adds a safety block, it will be called the Security Origin Node of that block. The Security Origin Node can define a destination for the block you just added, which will not necessarily be the final destination of the bundle. This feature allows security policies to be applied on specific parts of the network, preventing unnecessary security processing at nodes that are not at risk.

2.2   HTTP-DTN

HTTP-DTN is a protocol developed as an alternative to BP in DTN. After several failures were pointed out in BP [22,23] proposed the use of HTTP as a protocol for exchanging messages in DTN. Such proposal is accomplished through some small changes of headings and addition of a store and forward logic. HTTP in this case is not used as on the Internet where it works end-to-end, instead the HTTP-DTN architecture is hop-by-hop (i.e., HTTP transfers between neighbouring nodes). HTTP requires only a reliable form of delivery, regardless of how the transport protocol does it, thus allowing it to operate over a wide variety of transport protocols [24].

HTTP-DTN is used as a Peer-to-Peer (P2P) protocol in the sense that multiple files can be transferred in two directions simultaneously between two communicating nodes. Thus, the DTN nodes do not have a client and server relationship as there is in the use of HTTP on the conventional Internet. For example, if a package sent from one node to another needs to pass through 4 other nodes on the way, there will be 5 separate HTTP transactions, one between each pair of nodes. Thus, the control of the delivery of the message passes from the source node to the destination node, for each hop between nodes. When two nodes contact each other, a request for files to be copied, saved, or sent can be made between nodes using the HTTP PUT and HTTP GET commands. A node also always has the option to reject a package by checking for error codes.

The addition of headers to HTTP is necessary for the operation of HTTP-DTN, they are:

- *Content-Destination*: specifies the final destination of the package being sent. This header can be used for routing applications that use HTTP-DTN, so that those applications decide which network paths to use.
- *Content-Source*: contains information about the origin identifier of the package. The existence of this header is also mandatory on packets, as well as the ability to process it on each node.
- *Content-Length*: loads the size in bytes of the information being carried in the package.
- *Content-ID*: allows files to be identified only within the context of the Source Information Header.
- (*Content-MD5*: adds the ability to verify end-to-end integrity. This header must be generated at the origin node that first sent the packet, and can be checked at any node before the destination to identify errors or corrupted data in the packet. If there is a problem with the package, the node can then either discard it or request the shipment of the damaged parts in the package.
- *Content-Host*: enables the identification of the node that sent the information and how to find it in the DTN. This header can be left blank in the package if you only need to request files available from a neighboring node. When a node sends a packet on the network, the information contained in the Source Information Header must be the same as that existing in the Host Header field.

Many other HTTP headers are directly useful for HTTP-DTN. For example, *ETag* which provides an ID for unique copies of files on the network, and can be used to provide global identifiers (*GUID*) for each version of a file.

2.3   HTTP/2

The specifications described by [2], define that HTTP/2 requests are made in frame format. There are several types of framework defined in the protocol, each with a different function. However, they all share a series of headers that are: size (*Length*) of the *payload* frame (24 bits) represented in bytes; type (*Type*) (8 bits) that determines the format and semantics of the frame; *flags* (8 bits) is reserved for Boolean *flags* specific to the frame type; R (1 bit) is not changed at any time and must exist when sending a request and be ignored when handling the frame; Stream identifier (*Stream Identifier*) (31 bits) represents an identification for the *stream* to which the frame in question belongs; the structure and content of the *payload* of the *frame* is entirely dependent on the value defined in the Type field. In Fig. 1 you can see the structure of the headers.

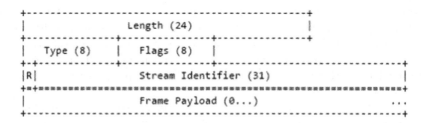

Fig. 1. Headers shared between all frames. Adapted from [2]

2.4   Headers-Type Frames in Requests

During an encounter with another security-aware node, there are a number of predefined requests that will occur. Such requests are intended to ensure that packages in transit on the network reach their final destinations.

A GET-type request is a HEADERS-type frame. It has the objective of obtaining from the paired node, any package whose recipient is the node that made the request. The headers sent in the *Header Block Fragment* field of the frame are:

- *method*: GET string.
- *Content-Destination*: identification of the destination node.

A PUT request is used to inform the paired node that a packet is being sent. Unlike a GET, this request consists of two frames, one of the HEADERS type and the other of the DATA type. In the HEADERS table, a series of headers is informed, containing values referring to the packet being sent. The DATA table contains the content to be sent. The headers sent in the *Header Block Fragment* field of the frame are:

- *method*: PUT string.
- *Date*: date of the package generation.
- *Host*: identification of the custody node of the package.
- *Content-Source*: identification of the source node.
- *Content-Destination*: identification the destination node.
- *Content-Integrity*: hash of the package content.
- *Content-Length*: size of the package content.

For each request made in a meeting, there are 3 types of possible responses. All are HEADERS type frames, followed by 0 or more DATA frames. Only one header of status type is sent in each HEADERS frame. The answers are: confirmation of the existence of packages (status header with a value of 200), used to answer a GET request, indicating the presence of packages for the node that made the request; denial response (status header with a value of 404), used as a response to a GET request, responding to the requesting node that there are no packages for it; and receipt confirmation (status header with value 202), used to respond to PUT requests and signalling that the package was successfully received and processed.

All frames are transported through a stream. A stream is by definition a sequence of frames, independent and bidirectional, exchanged between a client and a server during an HTTP/2 connection. The characteristics of a stream are: a single HTTP/2 connection can contain multiple open streams; streams can be established unilaterally by one of the points, or shared by the client and server; streams can be terminated by both the client and the server; streams are processed in the order they are received; streams are identified by an integer. This value is assigned to the stream by the point that started the request. Client-initiated streams will have odd values, and even streams when they are initiated by the server.

3   HTTP-DTNSec: Security and Performance Enhancements for the HTTP-DTN

This section aims to present the proposed solution in the context of the present work. Such solution is the extension of HTTP-DTN to provide support for security properties. HTTP-DTNSec is a protocol that allows packets to be exchanged securely on HTTP-based DTN, both hop-by-hop and end-to-end.

To ensure integrity and security in the transport of packets, HTTP-DTNSec implements an encryption solution in its solution when sending messages. Such a solution enables authentication, confidentiality and integrity of messages through its structure. Each package sent from one node to another goes through the encryption process, ensuring a level of security hop-by-hop. Besides, with the addition of extra headers in HTTP-DTNSec, it is also possible to guarantee encryption in end-to-end messages.

HTTP-DTNSec is based on HTTP/2 because it allows some performance improvements, since it support multiple requests and responses on the same connection. With HTTP/1.1, it is possible to have a client send more than one request, it is necessary that for each of them there is a different connection [2]. With HTTP/1, there are the occurrence of repeated and verbose requests, which cause unnecessary network traffic and also excessive use of the TCP congestion window.

Another important aspect for HTTP-DTNSec is that it can use an efficient coding of the HTTP header fields. So, it is possible for HTTP-DTNSec to add priority to some requests, leaving more important packages to be processed first.

The rest of this section is organized as follows. Firstly, it presents the hop-by-hop and end-to-end encryption aspects. Next, it describes how two nodes negotiate to establish a communication.

3.1   Hop-by-Hop and End-to-End Encryption

HTTP-DTNSec was developed to ensure hop-by-hop and end-to-end security in DTN. In order to achieve this goal, all packets sent from one node to another are encrypted and processed. In addition to ensuring confidentiality, HTTP-DTNSec also ensures the integrity and authenticity of a message. The message in this way also has its integrity checked at the destination, and in the event of an inconsistency a new request to obtain the package again is made.

The encryption process is performed when the message is first sent and the decryption occurs when it reaches its destination. This process is addressed in the HTTP-DTNSec protocol using the Content-Destination header from the HEADERS table. The session key is then also encrypted, but using the public key of the destination. The destination uses its private key to decrypt the session key. With the session key, the destination can decrypt the message and the signature. The destination generates a hash of the message, and with the public key of sender verifies in the signature that the hash matches the sent one.

Figure 2 demonstrates a process of sending a package with end-to-end security. First, Node A encrypts the message using public key C shared between all nodes. Then, it sends the message to Node B and it verifies that its identifier does not correspond to the final destination of the package. Node B stores the package with the encrypted message and waits for an opportunity to send it to Node C. When a connection is established with Node C, Node B sends the package. Finally, Node C verifies the final destination of the package and with the private key C obtains the decrypted message.

HTTP-DTNSec does not perform cryptographic key management. However, for the use of an asymmetric encryption process, it is essential that the source and destination nodes have their respective public keys. The way in which this

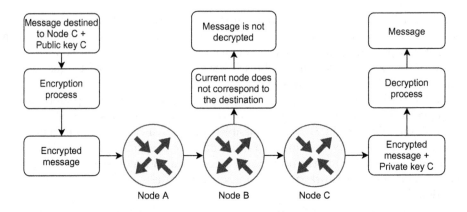

Fig. 2. End-to-end secure flow in HTTP-DTN

exchange will take place is outside the scope of HTTP-DTNSec (some alternatives are presented in [14]). Using end-to-end encryption, it is not possible to verify the integrity of a hop-by-hop package, as the hash of the message because it is encrypted can only be verified at the destination. In this way, if there was any corruption of the package during the end-to-end shipment, it would only be detected at the final destination.

The procedure to obtain the value to guarantee the integrity, occurs after the encryption of the package. In order to be able to verify the hop-by-hop integrity, an extra field is used in the PUT request HEADER frame, called Content-Integrity. First, a hash of the encrypted content is generated and the result is inserted in the Content-Integrity field. The node that receives the PUT request, must then generate a hash of the encrypted content of the received DATA frame. Finally, compare the value you obtained with the value sent in the Content-Integrity field. In case of inconsistent values, the node will respond with a Negative Response. In Fig. 3, it is possible to observe the sequence diagram of the message exchange.

HTTP-DTNSec, by definition, must be executed by all nodes that are part of the DTN. If there is a node that does not support it, it cannot be considered as a source or destination of packets. Besides, HTTP-DTNSec does not define a particular solution for encryption, so different solutions can be implemented according to the limitations found in a specific DTN.

3.2 Negotiation for Communication

A meeting between two nodes using HTTP-DTNSec can be divided into two distinct moments, the negotiation and the exchange of messages. The negotiation moment uses the HTTP/2 connection opening to initiate communication between nodes. HTTP-DTNSec does not change the way messages are exchanged at this time, it only defines default configuration values for the connection.

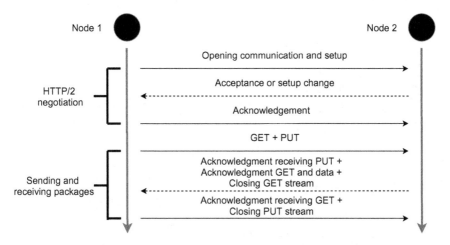

Fig. 3. Sequence diagram of message exchange

The node that initiates the negotiation will be considered as an HTTP/2 client, while the other paired node will be the server. The client node opens the connection by sending a PREAMBLE frame together with 2 SETTINGS frames. The first SETTINGS frame contains the configuration values for the connection, while the second is an empty SETTINGS frame that signals the end of the message. At this moment the communication is made through a single stream, this value is 0. The server node when receiving the message, using the same stream confirms the settings defined by the client and responds with 2 SETTINGS frames. The first frame can be used to modify some value of the connection configuration, while the second is an empty frame SETTINGS to signal the end of the message. Finally, the client node responds with an empty SETTINGS frame, signalling that the connection is now defined and is ready to send HTTP-DTNSec requests.

After the negotiation, when the client node has a package to be sent ahead, it will open a new stream with a value equals 1 and execute the logic of Fig. 4 (a). First, the node obtains all packages saved locally and for each one, it performs the encryption process and creation of the HTTP/2 frames used by HTTP-DTNSec. When all frames are created, the node then sends all packages to another paired node. The node waits for a receipt response for each sent package, deleting those that were successfully received locally, and keeping those that received problems.

The server node, upon receiving the PUT request, will check for integrity of each package received, in the case of a corrupted package, the server will return an error message. Then, the node will then check if it is the final destination of the message, if not, it will just store the package, in the other case it will decrypt the contents of the packages. Finally, it returns a receipt message to the paired node. At the end of the request, the client node will close the stream of value 1.

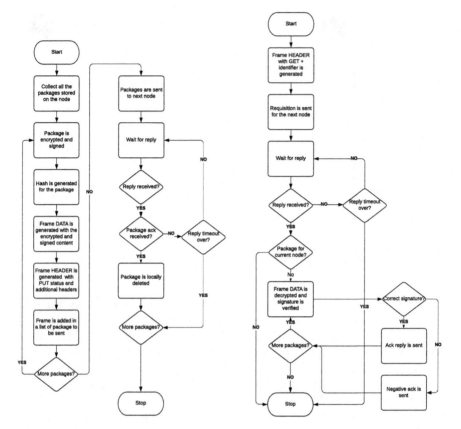

a) Execution flow diagram for PUT on HTTP-DTN for sending packages

b) Execution flow diagram for GET on HTTP for sending packages

Fig. 4. Execution flow diagrams

At the same time the PUT request is made, the client node will also carry out the GET request. So, the node opens a new stream of value 3 and executes the logic of Fig. 4. First the node sends a request to the paired node, asking if there are packages that it is the destination. When there is no response or there are no packages for the node, the stream is closed. In case of packages, the node decrypts the package's content and sends a response confirming receipt to the paired node.

The server node when receiving the GET request will later check for the existence of packages whose destination is from the requesting node, otherwise, it sends a denial message. When packages exist, the node encrypts the content of the package and creates the HTTP/2 frames to perform the upload. Finally, the server waits for a receipt response and in case of confirmation it will delete the packages received correctly. At the end of the request, the server node will close the stream of value 3.

4 Evaluation

This section presents the implementation and gives further details on the evaluation settings used to evaluate HTTP-DTNSec. In addition, we present the preformed experiments and the obtained results. Finally, we discuss the obtained results.

4.1 Implementation and Environment Settings

HTTP-DTNSec was based on an existing implementation of HTTP-DTN [15]. Our codes were developed in Python due to the main libraries and the framework used to produce itself were elaborated on in this language. The following resources were used for its development:

- Twisted[1]: development framework for networking applications.
- Hyper-H2[2]: library for the development of HTTP/2 applications.
- GnuPG[3]: library that implements the OpenPGP format (encryption and integrity).
- SQLite[4]: simple database employed for package storage.

The proposed scenario simulates an environment where two nodes meet and exchange messages by sending and receiving packages. These received the identification of Node A and Node B, in both the Windows operating system was used and the HTTP-DTNSec protocol was installed. Node B is a virtual machine, whose host is Node A. Both were configured with fixed IPs in an isolated network, where only the two machines exist. Two text files of different sizes were created in node B, so that they could be sent to node A. The file sizes are respectively File1 with 24 bytes, File2 with 1216 bytes. Finally, to verify the packets exchanged between nodes, Wireshark packet capture software was used. Figure 5 represents the environment defined for the execution of experiments.

---

[1] Twisted - https://twistedmatrix.com.

[2] Hyper-H2 - https://python-hyper.org/projects/hyper-h2/.

[3] GnuPG - https://www.gnupg.org/.

[4] SQLite https://www.sqlite.org.

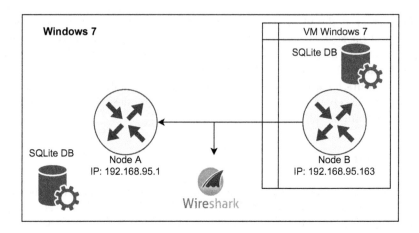

Fig. 5. Environment Settings for the HTTP-DTNSec evaluation

Fig. 6. HTTP-DTNSec send process

4.2   Experiments

Three meetings between the nodes were simulated in the proposed environment. At each meeting, one of the existing files in Node B was sent to Node A. In order to verify the changes to the HTTP-DTN protocol suggested in the 3 section, each meeting was also executed using it. The following captures made by Wireshark, correspond to the sending of the two files of different sizes. In the captures, only the DATA frames are shown using HTTP-DTNSec and beside the content of the Data field. Figure 6 represents the sending of File 2.

Finally, the following captures represent the number of messages exchanged by the nodes during a meeting, using HTTP-DTNSec and HTTP-DTN. Figure 7 shows the sending of a package using HTTP-DTN, while Fig. 8 shows the exchange of HTTP-DTNSec messages.

4.3   Discussion

When observing the packet captures of the meetings that used HTTP-DTNSec, it was possible to verify that the protocol is applying the encryption proposed in the 3 section when sending the frames that contains the packages. In addition, when analyzing the size of the packages, it was noticed that the OpenPGP encryption is actually using the compression described by [11]. This compression is causing small files, when going through the encryption process, to have the size of their package increased, but large files are reduced in size. In Fig. 9, a comparison on the content sent size considering HTTP-DTNSec and HTTP-DTN.

Through the captures that demonstrate the amount of messages exchanged between nodes using HTTP-DTNSec and HTTP-DTN, it can be seen that the HTTP-DTN protocol during a meeting, only sends one message while HTTP-DTNSec sends four. However, it is important to understand that HTTP-DTNSec performs PUT and GET requests during a single connection, while HTTP-DTN would need to establish different connections for each of the requests.

5   Related Work

Several feasibility studies about using security properties in DTN point out the network communication overheads as a significant barrier. In light of this, this section presents the most prominent research initiatives related to DTN security.

Duarte *et al.* [9] present a solution for the autonomous management of computer networks. The authors also propose applying the solution in DTN environments using a proof of concept of HTTP-DTN to carry out the storage and sending messages between the network nodes. The results achieved by the authors show the operation of the HTTP-DTN protocol and that its use on the network did not increase significantly the traffic data flow. In addition, the open-source implementation of HTTP-DTN made by the authors served as a basis for the practical development of the present work.

Asokan *et al.* [1] discuss the application of integrity assurance in DTN bundles. This paper presents the problem of verifying integrity in fragmented bundles (e.g., a large bundle is divided into several smaller bundles). However, the hash function that is applied to guarantee integrity is used over the entire bundle, thus, in the case of a fragment loss, the receiving node is unable to obtain the same hash value. The authors suggest that the hash function be applied to each bundle fragment individually. Therefore, the integrity check can be performed even in the case of lost bundles and the receiving node can request selective retransmission. The authors state that their proposal has the disadvantage of increasing the network traffic and the cpu load of nodes.

```
15:33:30 192.168.95.163        192.168.95.1            HTTP         1514 Continuation
```

Fig. 7. Message exchange using HTTP-DTN

```
15:30:16 192.168.95.163        192.168.95.1        HTTP2        117 Magic, SETTINGS
15:30:16 192.168.95.1          192.168.95.163      HTTP2        102 SETTINGS, SETTINGS
15:30:16 192.168.95.163        192.168.95.1        HTTP2        721 SETTINGS, HEADERS, HEADERS, DATA, DATA
15:30:16 192.168.95.1          192.168.95.163      HTTP2        90 HEADERS, HEADERS
```

Fig. 8. Message exchange using HTTP-DTNSec

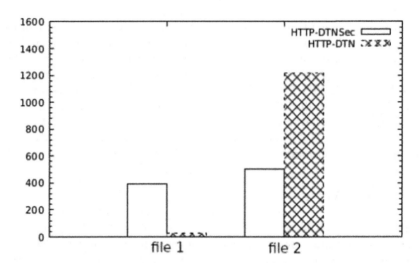

Fig. 9. Comparison on the content sent size considering HTTP-DTNSec and HTTP-DTN

Farrell and Cahill [10] analyze several issues related to DTN security. More specifically, they discuss some security decisions regarding the BP and BPSec. Also, the authors point out the reason for not using widespread security mechanisms, such as TLS, is often due to network limitations. Such limitations also hamper managing DTN keys, in which several messages are necessary to carry out the exchange of keys among nodes. The use of digital certificates may also not be feasible in many places, as there may be unavailability of connections when checking the validity with a trusted third party.

6   Final Remarks

Several security challenges can be highlighted in DTNs. Although security mechanisms are important in all communication networks, vulnerabilities are even more critical in some DTN environments, such as InterPlanetary Networking and Battlefield Networking. In the context of the IETF, the Bundle Protocol (BP) has been extended to enable safe messages exchange in challenging environments through the BPSec. In order to deal with the complexity of BP, alternatives with HTTP-DTN were developed. However, the security properties found at BPSec have not been extended to such alternatives.

In this paper, we propose the HTTP-DTNSec, a security extension for HTTP-DTN. Our extension aims to improve the confidentiality and integrity of HTTP-DTN, allowing it to have a security level comparable to BP. Besides, HTTP-DTNSec updates the version of HTTP used in HTTP-DTN. The performed evaluation shows that HTTP-DTNSec performed the transfer of packages in a safe manner and with an increase in performance concerning HTTP-DTN. The main reason behind such increase is due to HTTP-DTNSec being able to send more than one request per connection through the use of HTTP/2.

Besides the encouraging results found in the experiments carried out, some future works can also be proposed. First, it is necessary to deploy and analyze HTTP-DTNSec on an operational DTN scenario. Secondly, key management mechanisms should be considered in HTTP-DTNSec, similarly to what occurs in BPSec. Finally, it would be interesting to investigate the use of HTTP/3 as a transport for HTTP semantics since the use of QUIC as a transport protocol can lead to different impacts on HTTP-DTNSec.

References

1. Asokan, N., Kostiainen, K., Ginzboorg, P., Ott, J., Luo, C.: Towards securing disruption-tolerant networking. Nokia Research Center, Tech. rep. NRC-TR-2007-007 (2007)
2. Belshe, M., Peon, R., Thomson, M.: Hypertext Transfer Protocol Version 2 (HTTP/2). RFC 7540 (May 2015). https://doi.org/10.17487/RFC7540, https://rfc-editor.org/rfc/rfc7540.txt
3. Bhutta, M.N.M., Cruickshank, H., Nadeem, A.: A framework for key management architecture for DTN (KMAD): requirements and design. In: 2019 International Conference on Advances in the Emerging Computing Technologies (AECT), pp. 1–4. IEEE (2020)
4. Birrane, E.J., McKeever, K.: Bundle Protocol Security Specification. Internet-Draft draft-ietf-dtn-bpsec-27, Internet Engineering Task Force (February 2021). https://datatracker.ietf.org/doc/html/draft-ietf-dtn-bpsec-27, work in Progress
5. Burleigh, S., Fall, K., Birrane, E.J.: Bundle Protocol Version 7. Internet-Draft draft-ietf-dtn-bpbis-31, Internet Engineering Task Force (January 2021). https://datatracker.ietf.org/doc/html/draft-ietf-dtn-bpbis-31, work in Progress
6. Dalmazo, B.L., Vilela, J.P., Curado, M.: Online traffic prediction in the cloud. Int. J. Netw. Manag. **26**(4), 269–285 (2016)
7. Dalmazo, B.L., Vilela, J.P., Curado, M.: Triple-similarity mechanism for alarm management in the cloud. Comput. Secur. **78**, 33–42 (2018)
8. Ranjan Das, S., Sinha, K., Mukherjee, N., Sinha, B.P.: Delay and disruption tolerant networks: a brief survey. In: Mishra, D., Buyya, R., Mohapatra, P., Patnaik, S. (eds.) Intelligent and Cloud Computing. SIST, vol. 194, pp. 297–305. Springer, Singapore (2021). https://doi.org/10.1007/978-981-15-5971-6_32
9. Duarte, P.A.P.R., Nobre, J.C., Granville, L.Z., Tarouco, L.M.R.: A p2p-based self-healing service for network maintenance. In: 12th IFIP/IEEE International Symposium on Integrated Network Management (IM 2011) and Workshops, pp. 313–320. IEEE (2011)
10. Farrell, S., Cahill, V.: Security considerations in space and delay tolerant networks. In: 2nd IEEE International Conference on Space Mission Challenges for Information Technology (SMC-IT 2006), pp. 8-pp. IEEE, Pasadena (2006)
11. Finney, H., Donnerhacke, L., Callas, J., Thayer, R.L., Shaw, D.: OpenPGP Message Format. RFC 4880 (November 2007). https://doi.org/10.17487/RFC4880, https://rfc-editor.org/rfc/rfc4880.txt
12. Ivancic, W.D.: Security analysis of DTN architecture and bundle protocol specification for space-based networks. In: 2010 IEEE Aerospace Conference, pp. 1–12. IEEE (2010)

13. Kulkarni, L., Mukhopadhyay, D., Bakal, J.: Analyzing security schemes in delay tolerant networks. In: Satapathy, S., Bhateja, V., Joshi, A. (eds.) Proceedings of the International Conference on Data Engineering and Communication Technology. Advances in Intelligent Systems and Computing, vol. 468. Springer, Singapore (2017). https://doi.org/10.1007/978-981-10-1675-2_60

14. Menesidou, S.A., Katos, V., Kambourakis, G.: Cryptographic key management in delay tolerant networks: a survey. Future Internet **9**(3), 26 (2017)

15. Nobre, J.C., Duarte, P.A.P.R., Granville, L.Z., Tarouco, L.M.R.: Self-* properties and p2p technology on disruption-tolerant management. In: 2013 IEEE Symposium on Computers and Communications (ISCC), pp. 000676–000681 (July 2013). https://doi.org/10.1109/ISCC.2013.6755026

16. Paul, A.B., Biswas, S., Nandi, S., Chakraborty, S.: MATEM: a unified framework based on trust and MCDM for assuring security, reliability and QoS in DTN routing. J. Netw. Comput. Appl. **104**, 1–20 (2018)

17. Scott, K., Burleigh, S.C.: Bundle Protocol Specification. RFC 5050 (November 2007). https://doi.org/10.17487/RFC5050, https://rfc-editor.org/rfc/rfc5050.txt

18. Sipos, B.: Delay-Tolerant Networking UDP Convergence Layer Protocol. Internet-Draft draft-sipos-dtn-udpcl-01, Internet Engineering Task Force (March 2021). https://datatracker.ietf.org/doc/html/draft-sipos-dtn-udpcl-01, work in Progress

19. Sipos, B., Demmer, M., Ott, J., Perreault, S.: Delay-Tolerant Networking TCP Convergence Layer Protocol Version 4. Internet-Draft draft-sipos-dtn-tcpclv4-02, Internet Engineering Task Force (July 2016). https://datatracker.ietf.org/doc/html/draft-sipos-dtn-tcpclv4-02, work in Progress

20. Torgerson, L., et al.: Delay-Tolerant Networking Architecture. RFC 4838 (April 2007). https://doi.org/10.17487/RFC4838, https://rfc-editor.org/rfc/rfc4838.txt

21. Urunov, K., Vaqqasov, S., Namgung, J.I., Park, S.H.: Security issues for DTN mechanism of UIoT. In: The 18th IEEE International Symposium on Consumer Electronics (ISCE 2014), pp. 1–3. IEEE (2014)

22. Wood, L., Eddy, W.M., Holliday, P.: A bundle of problems. In: 2009 IEEE Aerospace conference, pp. 1–17. IEEE (2009)

23. Wood, L., Holliday, P.: Using HTTP for delivery in Delay/Disruption-Tolerant Networks. Internet-Draft draft-wood-dtnrg-http-dtn-delivery-09, Internet Engineering Task Force (June 2014). https://datatracker.ietf.org/doc/html/draft-wood-dtnrg-http-dtn-delivery-09, work in Progress

24. Wood, L., Holliday, P., Floreani, D., Psaras, I., England, G.: Moving data in DTNS with http and mime making use of http for delay-and disruption-tolerant networks with convergence layers (2009)

Multivariate Conditional Transformation Models. Application to Thyroid-Related Hormones

Carla Díaz-Louzao[1]([✉])(ID), Óscar Lado-Baleato[2](ID), Francisco Gude[3](ID), and Carmen Cadarso-Suárez[2](ID)

[1] Department of Psychiatry, Radiology, Public Health, Nursing, and Medicine, Universidade de Santiago de Compostela, Santiago de Compostela, Spain
carla.diaz.louzao@usc.es
[2] Department of Statistics, Mathematical Analysis, and Optimization, Universidade de Santiago de Compostela, Santiago de Compostela, Spain
{oscarlado.baleato,carmen.cadarso}@usc.es
[3] Clinical Epidemiology Unit, Complexo Hospitalario Universitario de Santiago de Compostela, Santiago de Compostela, Spain
francisco.gude.sampedro@sergas.es

Abstract. Multivariate Conditional Transformation Models (MCTMs) were recently proposed as a new multivariate regression technique. These models characterize jointly the covariates effects on the marginal distributions of the responses and their correlations without requiring parametric assumptions. Flexibility, in both the responses and covariates effects are achieved using Bernstein basis polynomials. In this paper we compare MCTMs estimations with the well established Copula Generalized Additive Models (CGAMLSS). MCTMs conditional correlation estimations outperform the CGAMLSS ones, showing lower estimation error, and variability. Finally, MCTMs were applied to the joint modelling of three thyroid hormones concentrations – Thyroid Stimulating Hormone (TSH), triiodothyronine (T3), and thyroxine (T4) – conditionally on age. Our results show how the marginal distribution and correlations of the hormones concentrations are influenced by the age of the patients.

Keywords: Multivariate regression · Bernstein basis · Most likely transformations · Thyroid hormones

Supported by Grant from the Program of Aid to the Predoctoral Stage (ED481A-2018/154) of the Galician Regional Authority (Consellería de Cultura, Educación e Ordenación Universitaria, Xunta de Galicia) and European Social Fund 2014/2020. Developed under the project MTM2017-83513-R and co-financed by the Ministry of Economy and Competitiveness (SPAIN) and by the European Regional Development Fund (ERDF). Also supported by the project ED431C 2020/20, financed by the Competitive Research Unit Consolidation 2020 Programme of the Galician Regional Authority (Xunta de Galicia).

© Springer Nature Switzerland AG 2021
O. Gervasi et al. (Eds.): ICCSA 2021, LNCS 12949, pp. 341–351, 2021.
https://doi.org/10.1007/978-3-030-86653-2_25

1   Introduction

The thyroid gland is a small gland located in front of the neck, that performs a vitally important function: the control of the metabolism. This is fulfilled by taking the iodine found in many foods and converting it into two thyroid hormones: triiodothyronine (T3), and thyroxine (T4). The normal thyroid gland produces about 20% T3, and about 80% T4 and every cell in the body depends upon thyroid hormones for regulation of their metabolism. Only a small fraction of the thyroid hormones is free to enter tissues and has a biologic effect.

The thyroid gland is under the control of the pituitary gland. This small gland located at the base of the brain, produces Thyroid Stimulating Hormone (TSH) when the T3 and T4 levels drop too low, in order to stimulate the thyroid to produce more hormones (see Fig. 1).

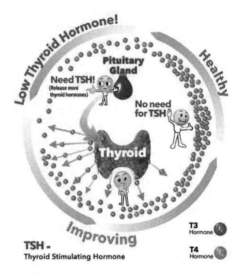

Fig. 1. Diagram of the functioning of thyroid hormones (Source: [13]).

But this mechanism does not always work well, and when that happens, it can lead to profound adverse effects if undiagnosed or untreated. Thyroid dysfunction is a common disease. For instance, the estimated prevalence of hyperthyroidism ranges from 0.2% to 1.3% in parts of the world with enough iodine [12], being larger in those places with a lack of iodine.

Primary thyroid failure brings with it a number of different symptoms, raging from fatigue, sensitivity to cold or poor memory and concentration, to dyspnea, weight gain or even depression. As this gland is in charge of the regularization of the metabolism, its malfunction is harmful to the whole body.

The relationship between TSH and T4 was broadly studied since it seems that this tandem has the greatest diagnostic value for primary thyroid failure [9].

Many studies suggested an inverse, linear relationship between log-TSH and T4, being this taken as a generally accepted principle of thyroid physiology thanks to Spencer et al. (1990) [11]. It was not until 2010 that studies began to emerge refuting the linearity of the relationship [4,5].

It is obvious that the simultaneous study of the thyroid-related hormone may improve the diagnosis of primary thyroid failure. In addition, it is also possible that the relationship between these hormones varies with age and should be included as a covariate. To this aim, Multivariate Conditional Transformation Models (MCTM, [8]) were considered in this paper. These models characterize the covariates effects not only on the correlations (relationship) between the outcomes, but also on the Cumulative Distribution Functions (CDF) of each response. It is also possible to obtain percentile curves from these models for each biomarker, facilitating interpretability.

The remainder of the paper is structured as follows: in Sect. 2, the structure of MCTM is briefly explained; a simulation study is described in Sect. 3, comparing MCTM performance in the bivariate case with conditional Copula Generalized Additive Models, for Location, Scale, and Shape (CGAMLSS, [10]); Sect. 4 presents the results of the clinical study; and, finally, the paper ends with some conclusions (Sect. 5).

2 Multivariate Conditional Transformation Models

Multivariate Transformation Models were developed for a J-dimensional, absolutely continuous response vector $\mathbf{Y} = (Y_1, \ldots, Y_J)^\top \in \mathbb{R}^J$. In these models, an unknown, bijective, strictly monotonically increasing transformation function $h : \mathbb{R}^J \to \mathbb{R}^J$ maps the outcome $\mathbf{Y}$ to a set of J independent and identically distributed, absolutely continuous random variables $Z_j \sim \mathbb{P}_Z, j \in \{1, \ldots, J\}$, where $\mathbb{P}_Z$ is a pre-defined distribution (usually, $Z_j \sim N(0, 1)$). That is:

$$h(\mathbf{Y}) = (h_1(\mathbf{Y}), \ldots, h_J(\mathbf{Y}))^\top = (Z_1, \ldots, Z_J)^\top = \mathbf{Z} \in \mathbb{R}^J. \qquad (1)$$

MCTM are obtained by extending the transformation function (1) to include a vector of potential covariates $\mathbf{X}$, resulting in:

$$h(\boldsymbol{y}|\boldsymbol{x}) = (h_1(\boldsymbol{y}|\boldsymbol{x}), \ldots, h_J(\boldsymbol{y}|\boldsymbol{x}))^\top. \qquad (2)$$

A triangular structure is imposed on the transformation function h in order to simplify calculations, without imposing any limitation on the estimation of the responses. This means that the jth component of the transformation function depends only on the first j elements of $\boldsymbol{y}$. Therefore, for each $j \in \{1, \ldots, J\}$, the components of (2) can be written as:

$$h_j(\boldsymbol{y}|\boldsymbol{x}) = h_j((y_1, \ldots, y_J)|\boldsymbol{x}) = h_j((y_1, \ldots, y_j)|\boldsymbol{x}). \qquad (3)$$

Finally, the h_j transformation functions in (3) are set to be linear combinations of marginal transformation functions $\tilde{h}_j : \mathbb{R} \to \mathbb{R}$:

$$h_j(\boldsymbol{y}|\boldsymbol{x}) = \sum_{j=1}^{j-1} \lambda_{jj}(\boldsymbol{x})\tilde{h}_j(y_j|\boldsymbol{x}) + \tilde{h}_j(y_j|\boldsymbol{x}), \qquad (4)$$

where $\lambda_{jj}(x)$ and $\tilde{h}_j(y_j|x)$ are expressed in terms of basis function expansions. Thus:

$$\tilde{h}_j(y_j|x) = a_j(y_j)^\top \vartheta_{j,1} - b(x)^\top \vartheta_{j,2}, \tag{5}$$

and:

$$\lambda_{jj}(x) = x^\top \gamma_{jj}, \quad 1 \leq j < j \leq J. \tag{6}$$

Here, a_j and b are polynomial Bernstein basis functions for the response and the covariates respectively, and $(\vartheta_{j,1}, \vartheta_{j,2}, \gamma_{jj})$ are the parametric coefficients of the model. Their estimation and inference are based on the model log-likelihood. In Fig. 2, a representation of the Bernstein basis of four different orders is shown.

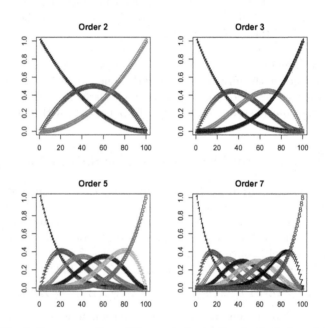

Fig. 2. Representation of Bernstein basis of order 2, 3, 5, and 7.

Summing up the model's specifications, the MCTM are characterised by a set of marginal conditional transformations $\tilde{h}_j(y_j|x)$, $j \in \{1, \dots, J\}$ (computed as in (5)), each of which applying to only one component of $\mathbf{Y}$ and the covariates ($\mathbf{X}$), and by a lower triangular ($J \times J$) matrix of transformation coefficients $\Lambda(x)$:

$$\Lambda(x) = \begin{pmatrix} 1 & & & & 0 \\ \lambda_{21}(x) & 1 & & & \\ \lambda_{31}(x) & \lambda_{32}(x) & 1 & & \\ \vdots & \vdots & & \ddots & \\ \lambda_{J1}(x) & \lambda_{J2}(x) & \dots & \lambda_{J,J-1}(x) & 1 \end{pmatrix},$$

with $\lambda_{jj}(\boldsymbol{x})$, $1 \leq j < j \leq J$, defined as in (6). Under the standard normal reference distribution $\mathbb{P}_Z = N(0,1)$, the coefficients of $\boldsymbol{\Lambda}(\boldsymbol{x})$ characterise the dependence structure of the responses. This is so because $\tilde{\boldsymbol{Z}}(\boldsymbol{x}) = (\tilde{Z}_1(\boldsymbol{x}), \ldots, \tilde{Z}_J(\boldsymbol{x}))^{\top}$, defined by the random variables $\tilde{Z}_j(\boldsymbol{x}) = \tilde{h}_j(Y_j \mid \boldsymbol{x})$, follows a multivariate Gaussian distribution $\tilde{\boldsymbol{Z}}(\boldsymbol{x}) \sim N(\boldsymbol{0}_J, \boldsymbol{\Sigma}(\boldsymbol{x}))$, with $\boldsymbol{\Sigma}(\boldsymbol{x}) = \boldsymbol{\Lambda}(\boldsymbol{x})^{-1}(\boldsymbol{\Lambda}(\boldsymbol{x})^{-1})^{\top}$. Inference on the responses correlations is done by using parametric bootstrap (see [8] for more details).

In MCTM, the conditional Cumulative Distribution Function (CDF) for each outcome, $F_Y(Y_j \mid \boldsymbol{x})$, is given by:

$$\mathbb{P}(Y_j \leq y_j \mid \mathbf{X} = \boldsymbol{x}) = \Phi_{0,\sigma_j^2}\left(\tilde{h}_j(y_j \mid \boldsymbol{x})\right) = F_Z\left(\boldsymbol{a}_j(y_j)^{\top}\boldsymbol{\vartheta}_{j,1} - \boldsymbol{b}_j(\boldsymbol{x})^{\top}\boldsymbol{\vartheta}_{j,2}\right).$$

3   Simulation Study for the Bivariate Case

In Sect. 5 of their paper, Klein et al. (2019) [8] evaluated the performance of MCTM with a simulation of a bivariate response depending on one covariate, comparing the results of MCTM models with two other approaches: Bayesian structured additive distributional regression models, and vector generalised additive models (VGAM). In this Section, we are going to replicate that simulation study, considering now as a competitor the conditional Copula Generalized Additive Models, for Location, Scale, and Shape (CGAMLSS, [10]).

CGAMLSS are bivariate copula models with parametric marginal distributions, one-parameter copulas, and semi-parametric specification for the predictors of all parameters of both the marginal and the copula jointly, using a penalised likelihood framework. Unlike MCTM, this approach is only limited to the bivariate case for continuous responses. Even so, we do not see that this is a good enough reason not to compare them when we have two possibly related responses.

3.1   Simulation Design

In the simulation set-up, $R = 100$ data sets of size $n = 1000$ were simulated and one single covariate $x \sim U[-0.9, 0.9]$ was considered. The two continuous outcomes were considered following a Dagum distribution. The bivariate response dependence structure depends on x, being $\lambda(x) = x^2$ (to see all of the details, go to [8], Subsect. 5.1). With this construction, the first margin is independent of the covariate x, but the scale parameter of the second margin is not (not so the shape parameters a and p, which remain constant).

For the MCTM, we applied Bernstein polynomials of order 8 and 3 in the responses and the covariate respectively.

For the CGAMLSS, we used the true specification of the model, i.e. a Gaussian Copula and Dagum marginals, with the correlation parameter and the b_2 parameter of the Dagum distribution for y_2 dependent on x, while all of the parameters of the first marginal, as well as the shape parameters a_2 and p_2 of the second one did not depend on any covariate. For this purpose, both the predictor for b_2 and the correlation parameter ρ of the copula were specified using cubic B-splines with 20 inner knots, as it was done for the other approaches compared in the original paper. Likewise, the other parameters of the model were estimated as constants.

3.2   Evaluation of the Performance

The performance of the methods was evaluated by estimating the lower element of $\boldsymbol{\Lambda}^{(r)}$, $\hat{\lambda}^{(r)}(x)$, from every data replicate $r \in \{1, \ldots, R\}$, on a grid of length $G = 100$ within the range of x, and comparing it to the real one, $\lambda(x)$. On the one hand, we considered the estimated curves as they were; on the other hand, we computed the root mean squared error $\mathrm{RMSE}\left(\lambda, \hat{\lambda}^{(r)}\right) = \sqrt{\frac{1}{G}\sum_{g=1}^{G}\left(\lambda(x_g) - \hat{\lambda}(x_g)\right)^2}$ and the standard deviation $\mathrm{SD}\left(\hat{\lambda}^{(r)}\right)$ of each sample as a numeric measure of discordance.

3.3   Results of the Simulation Study

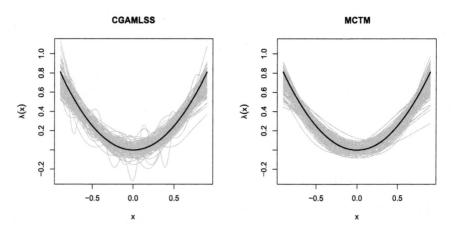

Fig. 3. Function estimates $\hat{\lambda}(x)$ for the effect $\lambda(x) = x^2$ on the correlation parameter for CGAMLSS (left) and MCTM (right). The black line represents the true function and the grey lines are the estimations of the R = 100 simulation samples.

Figure 3 shows the estimations for $\lambda(x) = x^2$ of the 100 simulated data sets for CGAMLSS (left panel) and MCTM (right panel). Even though both models seemed to capture the general form properly, MCTM led to better smoothed curves. Although CGAMLSS was fitted using the actual model specification, in terms of RMSE and SD both approaches were very similar (Fig. 4). This is highly remarkable because MCTM was fitted without specifying parametric marginal distributions, a requirement that is restrictive, and an obstacle in practice. Besides this, we have found a big difference in the execution times, since, on average, the adjustment of MCTM

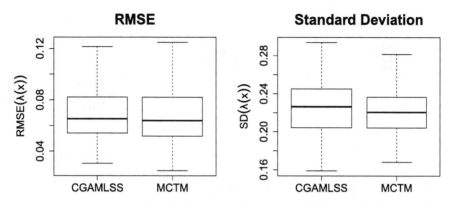

Fig. 4. Left: RMSE $\left(\lambda, \hat{\lambda}^{(r)}\right)$ for CGAMLSS and MCTM. Right: SD $\left(\lambda, \hat{\lambda}^{(r)}\right)$ for CGAMLSS and MCTM.

was less than a second (0.480 s), while a CGAMLSS model took almost 5 s (4.807 s). Therefore, a MCTM approach seemed to be more advisable than a CGAMLSS approach.

4 Joint Modelling for Thyroid-Related Hormones

As we stated in Sect. 1, a good control of the activity of the thyroid gland is vital for the overall functioning of our bodies. In order to do so in the best way, it is really important to know the relationship between the hormones involved (TSH, T3 and T4)as thoroughly as possible, adjusting for age. In this Section, we used the statistical methodology of MCTM for that purpose.

The data we have available comes from A-Estrada Glycation and Inflammation Study (AEGIS), which is a cross-sectional population-based study being performed in the municipality of A Estrada (NW, Spain). Its aim is to investigate the association between glycation, inflammation, lifestyles, and their association with common diseases and to study discordances between markers for glycaemia. In this database, we have a total of 1516 subjects who agreed to participate in the study, and the data collection and recruiting phase were completed in March 2015. After excluding extreme values and the treated ones due to high variability of the values, a total of 1282 healthy subjects remains.

We studied the relationship between TSH (mIU/L), free T3 (pmol/L), and free T4 (ng/dL). We considered the logarithm of the TSH (log-TSH) because of previous studies [4,5,11]. In Fig. 5, the two-by-two scatter plots of the variables are shown.

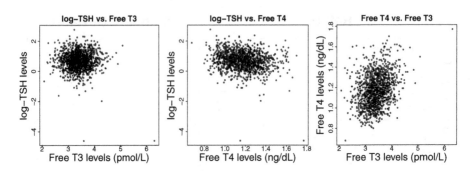

Fig. 5. Two-by-two scatter plot of the thyroid-related hormones. Left: log-TSH vs. Free T3. Centre: log-TSH vs. Free T4. Right: Free T4 vs. Free T3.

4.1 Trivariate Conditional Transformation Models for Log-TSH, Free T3, and Free T4 Hormones

We applied MCTM to study the relationship between log-TSH, free T3, and free T4, adjusting for the age of the patient to take into account a possible variation of these hormones and their relationship with age.

Model Specification. Based on a $\mathbb{P}_Z = N(0,1)$ distribution, our marginal conditional distribution is parametrised as:

$$\tilde{h}_j(y_j|\text{Age}) = \boldsymbol{a}_j(y_j)^\top \boldsymbol{\vartheta}_{j,1} - \boldsymbol{b}(\text{Age})\boldsymbol{\vartheta}_{j,2}, \; j \in \{\text{log-TSH, Free T3, Free T4}\}, \quad (7)$$

where basis functions $\boldsymbol{a}$ are Bernstein polynomials of order eight and $\boldsymbol{b}$ is a Bernstein polynomial of order three. The coefficients of $\boldsymbol{\Lambda}$ were parametrised as:

$$\lambda_{jj}(\text{Age}) = \boldsymbol{b}(\text{Age})\boldsymbol{\gamma}_{jj}, \quad \jmath < j \in \{\text{log-TSH, Free T3, Free T4}\}, \quad (8)$$

Results for Marginal Distributions. The estimated marginal conditional CDFs $F_j(y_j|\text{age})$, and densities $f_j(y_j|\text{age})$, are shown in Fig. 6. The different colours represent the ages of the individuals. We can see that the shapes differ for the three hormones, and that hormone levels decrease as age increases for log-TSH and Free T3, specially for the latter. For Free T4, we can see that higher values are obtained in middle-aged individuals. Another way of seeing this is as depicted in Fig. 7, with the scatter plot of the hormones as a function of age, and the percentile lines shown in red. Here it can be clearly seen that, although log-TSH levels appear to decrease with age, this change is very small, and the log-TSH level be considered invariant with respect to age.

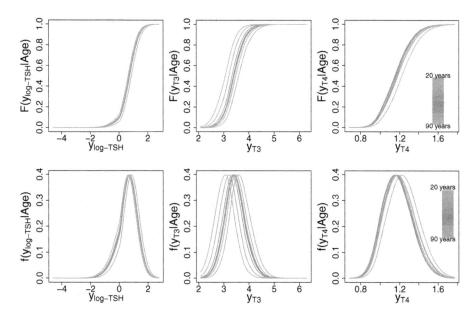

Fig. 6. Estimated marginal conditional CDFs $F_j(y_j|\text{age})$, and conditional densitiy functions $f_j(y_j|\text{age})$, for $j = $ log-TSH (left), $j = $ Free T3 (centre), and $j = $ Free T4 (right), from age = 20 (brown) to age = 90 (purple). (Color figure online)

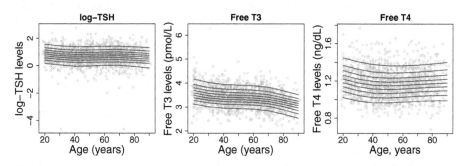

Fig. 7. Scatter plot of thyroid-related hormones vs. age. In red, percentile curves are shown. Left: log-TSH levels. Centre: Free T3 levels. Right: Free T4 levels. (Color figure online)

Results for Dependence Structure. In Fig. 8 it is shown that the two-by-two conditional Spearman's rho $\rho^S(\text{Age})$ for the three hormones, is statistically significant in all three cases. The grey area represents the confidence interval estimates at 95% from 1000 parametrically drawn bootstrap as explained in Subsect. 3.4 in [8]. We can see that the relationship between log-TSH and Free T3 is negative, and statistically significant for individuals between 30 and 50 years old. Correlation between log-TSH and Free T4 is negative (as expected). Finally, correlation between Free T3 and Free T4 is positive and decreases with age.

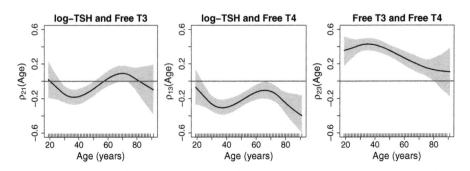

Fig. 8. Two-by-two conditional Spearman's rho $\rho^S(\text{age})$ for the thyroid-related hormones (left: log-TSH and Free T3; centre: log-TSH and Free T4; right: Free T3 and Free T4). Grey area represents the 95% bootstrap confidence interval, while black line represents the bootstrapped point estimate. Red line indicates the 0. (Color figure online)

5   Conclusions

In this work we demonstrate the usefulness of MCTMs in the biomedical setting. MCTMs allow for a joint estimation of the covariates effects on the distribution of each response, and on their correlations. MCTMs performance was evaluated in comparison with the frequentist conditional copula regression models. Finally, MCTMs allowed us to model jointly, for the first time, the age effect on the concentrations of the three thyroid hormones, offering a better understanding of these endocrinological measurements.

Thyroid dysfunction diagnosis have been the object of research in several works. However, given the state of art of regression analysis, these studies have been conducted considering thyroid hormone concentrations as mutually independent variables. For instance, Jonklaas et al. (2019) [7] focused on the clinical variables effect over the thyroid hormones distribution, with independence of any pathological status. Hoermann et al. (2016) [6] discussed the convenience of estimating a trivariate reference region for TSH, T3, and T4. Indeed, they proved that univariate reference curves for each hormone cause thyroid dysfunction over-diagnosis. Therefore, trivariate regression modelling might be advantageous for thyroid dysfunction diagnosis. In this context, the proposal of trivariate reference regions – i.e., regions characterizing the 95% of healthy patients' results - are desirable in the clinical setting. In future work, we will investigate the MCTMs extension for conditional trivariate reference region estimation.

MCTMs represent a leap forward in the multivariate regression analysis application to real clinical problems. To date, multivariate regression usage in the clinical setting is anecdotal [1–3]. Indeed, currently, when several correlated measurements are available for the same patient, they are treated as mutually independent variables, or a pre-specified transformation is applied to them resulting in a univariate problem. This common strategy may depreciate useful information arising from the continuous measurements joint distribution. Hence,

MCTMs may enhance the application of multivariate regression for diseases whose diagnosis is based on several continuous biomarkers, or clinical assessments that are based on correlated measurements.

References

1. Duarte, E., et al.: Applying spatial copula additive regression to breast cancer screening data. In: Gervasi, O., et al. (eds.) ICCSA 2017. LNCS, vol. 10405, pp. 586–599. Springer, Cham (2017). https://doi.org/10.1007/978-3-319-62395-5_40

2. Espasandín-Domínguez, J.: Assessing the relationship between markers of glycemic control through flexible copula regression models. Stat. Med. **38**(27), 5161–5181 (2019). https://doi.org/10.1002/sim.8358

3. Espasandín-Domínguez, J., et al.: Bivariate copula additive models for location, scale and shape with applications in biomedicine. In: Gil, E., Gil, E., Gil, J., Gil, M.Á. (eds.) The Mathematics of the Uncertain. SSDC, vol. 142, pp. 135–146. Springer, Cham (2018). https://doi.org/10.1007/978-3-319-73848-2_13

4. Hadlow, N.C., Rothacker, K.M., Wardrop, R., Brown, S.J., Lim, E.M., Walsh, J.P.: The relationship between TSH and free T4 in a large population is complex and nonlinear and differs by age and sex. J. Clin. Endocrinol. Metab. **98**(7), 2936–2943 (2013). https://doi.org/10.1210/jc.2012-4223

5. Hoermann, R., Eckl, W., Hoermann, C., Larisch, R.: Complex relationship between free thyroxine and TSH in the regulation of thyroid function. Eur. J. Endocrinol. **162**(6), 1123–1129 (2010). https://doi.org/10.1530/EJE-10-0106

6. Hoermann, R., Larisch, R., Dietrich, J.W., Midgley, J.E.: Derivation of a multivariate reference range for pituitary thyrotropin and thyroid hormones: diagnostic efficiency compared with conventional single-reference method. Eur. J. Endocrinol. **174**(6), 735–743 (2016). https://doi.org/10.1530/EJE-16-0031

7. Jonklaas, J., Razvi, S.: Reference intervals in the diagnosis of thyroid dysfunction: treating patients not numbers. Lancet Diab. Endocrinol. **7**(6), 473–483 (2019). https://doi.org/10.1016/S2213-8587(18)30371-1

8. Klein, N., Hothorn, T., Barbanti, L., Kneib, T.: Multivariate conditional transformation models. Scand. J. Stat. (2019). https://doi.org/10.1111/sjos.12501

9. Kumar, M.S., Safa, A.M., Deodhar, S.D., Schumacher, O.P.: The relationship of thyroid-stimulating hormone (TSH), thyroxine (T4), and triiodothyronine (T3) in primary thyroid failure. Am. J. Clin. Pathol. **68**(6), 747–751 (1977). https://doi.org/10.1093/ajcp/68.6.747

10. Marra, G., Radice, R.: Bivariate copula additive models for location, scale and shape. Comput. Stat. Data Anal. **112**, 99–113 (2017). https://doi.org/10.1016/j.csda.2017.03.004

11. Spencer, C., LoPresti, J., Patel, A., Guttler, R., Eigen, A., Shen, D., Gray, D., Nicoloff, J.: Applications of a new chemiluminometric thyrotropin assay to subnormal measurement. J. Clin. Endocrinol. Metab. **70**(2), 453–460 (1990). https://doi.org/10.1210/jcem-70-2-453

12. Taylor, P.N., et al.: Global epidemiology of hyperthyroidism and hypothyroidism. Nat. Rev. Endocrinol. **14**(5), 301 (2018). https://doi.org/10.1038/nrendo.2018.18

13. UCLA Endocrine Center (UCLA Health): What are normal thyroid hormone levels? (nd). https://www.uclahealth.org/endocrine-center/normal-thyroid-hormone-levels#:~:text=TSH%20normal%20values%20are%200.5%20to%205.0%20mIU%2FL. Accessed 15 June 2021

A Self-interpolation Method for Digital Terrain Model Generation

Leonardo Ramos Emmendorfer[5]($\boxtimes$)(ID), Isadora Bicho Emmendorfer[1],
Luis Pedro Melo de Almeida[2,4](ID), Deivid Cristian Leal Alves[3](ID),
and Jorge Arigony Neto[2](ID)

[1] Oceanology, Graduate Program, Federal University of Rio Grande, Rio Grande,
RS, Brazil
[2] Instituto de Oceanografia, Federal University of Rio Grande, Av. Itália, KM 8,
Campus Carreiros, Rio Grande, RS 96203-270, Brazil
[3] Departament of Geoprocessing, Federal Institute of Rio Grande do Sul,
Rio Grande, RS, Brazil
[4] ATLANTIC, Edifício LACS, Estrada da Malveira da Serra 920,
2750-834 Cascais, Portugal
[5] Centro de Ciências Computacionais, Federal University of Rio Grande, Av. Itália,
KM 8, Campus Carreiros, Rio Grande, RS 96203-270, Brazil

Abstract. This work presents an iterative method for obtaining a digital terrain model (DTM) from a digital surface model (DSM) given as input. The novel approach is compared to a state-of-the-art method from the literature using three case studies that represent diverse situations and landscapes including a coastal region composed of dunes, a mountain region, and also an urban area. The proposed method was revealed to be a promising alternative in terms of a better root-mean-square error. Input surface artifacts are successfully removed with the adoption of the proposed method.

Keywords: Digital terrain model · Digital surface model · Interpolation

1 Introduction

Topographic data are widely adopted as supporting information for research and civil applications, both in the context of environmental management, as well as in landscape and urban planning. Advances in remote sensing techniques and computational methods in the recent decades led to the widespread adoption of elevation data in several domains such as hydrological modeling, forest modeling, among others [11].

A Digital Elevation Model (DEM) is a digital representation of the Earth's surface, usually as a digital image that represents elevation values. The type of sensor used to collect elevation data determines the quality of the final result [12]. Data collected with surveying equipment, such as GPS, theodolites, and total

© Springer Nature Switzerland AG 2021
O. Gervasi et al. (Eds.): ICCSA 2021, LNCS 12949, pp. 352–363, 2021.
https://doi.org/10.1007/978-3-030-86653-2_26

stations, collect topographic elevations of the terrain precisely. However, the cost of the final product and time spent to obtain the information might be higher, therefore the number of data points collected might be smaller. Those factors usually restraint the adoption of this type of equipment for large-scale data collection [12]. In that sense, remote sensing techniques such as LiDAR (Light Detection and Ranging), orbital sensors, UAVs (Unmanned Aerial Vehicles), and radar type sensors such as SAR (Synthetic Aperture Radar) and InSAR (Interferometric Synthetic Aperture Radar) are some of the alternatives available for the collection of topographic data. Those approaches, however, collect land surface information instead of terrain information purely [12].

A grid DEM represents a continuous regular surface interpolated through the elevation points (discrete points). Generally, this grid is a set of data that builds up an image, where each pixel value corresponds to an elevation. A triangular irregular network (TIN) represents an irregular surface composed of non-overlapping triangles that connect at irregularly spaced measurement points, these triangles are represented as vector data structures.

When a DEM includes structures above the ground, such as trees, houses, and other buildings, it should be referred to as a Digital Surface Model (DSM) which represents the "first" measured surface. A DSM, therefore, models the reflective response of the terrain and also all objects and surfaces above it.

Digital Terrain Models (DTMs), on the other side, digitally represent the terrain's ground surface without any objects or constructions. This differentiation between ground and non-ground elevation data is of wide interest due to the high applicability of terrain data for environmental and civil construction purposes. DTMs are often adopted in topographic analysis, flood models, urban planning, among other applications [9]. A DSM, on the other hand, is useful when the aim is to represent and analyze surfaces above the ground, such as a 3D map of a city, for instance [2].

A DTM can be obtained as a refined version of a DSM, resulting from specific processing and/or filtering which aim to represent more accurately bare earth without the artifacts or objects on the surface [14]. Several DTM-from-DSM filtering algorithms have been proposed, which are built upon diverse numerical and analytical approaches. The choice of a method depends, among other factors, on the type of data structure considered. Typically, for point cloud data, filtering is used to make this separation between terrain and surface [10].

Under existing DTM-from-DSM methods the user must set a small to moderate number of parameters, which are at least three, typically. This work proposes a simple DTM-from-DSM filtering algorithm that relies on the iterative interpolation of the elevation values in the DTM. The number of parameters of the novel method is smaller when compared to other approaches, which might place the novel method as an easier-to-use alternative for practitioners from diverse areas. We show the applicability of the novel approach by comparing it to a state-of-the method from the literature for three test sites.

The paper is organized as follows. Related works are briefly reviewed in Sect. 2. The proposed approach is presented in Sect. 3. The evaluation methodology is proposed in Sect. 4. The resulting model is evaluated and compared to a state-of-the art method in Sect. 5. Section 6 concludes the paper.

2   Related Work

Several filtering methods for obtaining the DTM from grid-type DSMs. In [6] a DTM-from-DSM method is proposed, which is especially applicable in steep, forested areas where other filtering algorithms typically have problems when distinguishing between ground and off-ground elements. Some methods rely on the extraction of morphological characteristics of the terrain [5,8,13]. In [7] a method is proposed which relies on extracting the most contrasted connected components from LiDAR data to remove non-ground objects. Recently in [1] a method based on the interpolation matrix of the weighted thin-plate spline is proposed.

The Terra (Terrain Extraction from elevation Rasters through Repetitive Anisotropic filtering) filtering method [9] adopts the terrain aspect to guide the direction of anisotropic filtering, maximizing the preservation of the terrain edges. A smoothing operation is locally directed along the terrain slope. Initially, the slope direction (aspect) at a node $Z_{i,j}$ is computed from a set of nodes inside a squared window centered at (i, j), which contains $\eta \times \eta$ nearby nodes. Later at each step of the iterative process, the smoothing itself occurs considering only the upstream grid nodes within a smaller $\lambda \times \lambda$ filter window centered at (i, j). The process is repeated M times, where M is a parameter. Higher values for M lead to smoother surfaces. The iterative update for each node at each step m is performed as:

$$Z_{i,j}^m = min(Z_{i,j}^{m-1}, mean(\{u|u \in U((i,j), \eta, \lambda)\})) \tag{1}$$

where the window U contains the upstream grid nodes as mentioned above, determined for the neighborhood of each (i, j), iteratively at step. Other kernel averaging functions could be used, as mentioned in the paper[1].

3   Proposed Self-interpolation Method

A simple iterative interpolation approach is proposed for the generation of a DTM from a DSM given (denoted as S). Elevation values at coordinates (i, j) in S are successively replaced by the result of the interpolation computed from the elevation values nearby. The iterative process starts at step $m = 0$ with the initial version of the DTM set as $T^{m=0} = S$, which corresponds to the DSM given. Successive refined versions of the DTM are represented as $T^0, T^1, \cdots T^M$, where M represents the number of iterative steps. Theoretically, all the remaining nodes apart from (i, j) itself could be adopted for the computation of the interpolation

---

[1] The R source code of Terra is available at: https://www.umr-lisah.fr/?q=fr/scriptsr/ terra-script-r.

at (i, j). Therefore, the updated elevation value for each node $t_{i,j}^m \in T^m$ at each step m could be computed as the minimum between the previous value $t_{i,j}^{m-1}$ and its interpolated value as:

$$t_{i,j}^m = min(t_{i,j}^{m-1}, \mathcal{M}((i,j), T^{m-1} - \{t_{i,j}^{m-1}\})) \tag{2}$$

where $\mathcal{M}(p, V)$ represents the interpolation computed from input nodes in $V = T^{m-1} - \{t_{i,j}^{m-1}\}$ for the output coordinates given by $p = (i, j)$. However, majorly for the purpose of computational efficiency, only a subset of the nearby nodes will be adopted for the computation of the interpolation $\mathcal{M}$ at each node (i, j). A squared filter window $W((i, j), \lambda)$, centered at (i, j), containing $(\lambda \times \lambda) - 1$ nodes, is defined as the neighborhood of a node at coordinates (i, j). The filter window W is therefore defined as:

$$W((i,j), \lambda, m) = \left\{ t_{a,b}^{m-1} \text{for } a \in \left[i - \left\lfloor \frac{\lambda}{2} \right\rfloor, i + \left\lfloor \frac{\lambda}{2} \right\rfloor \right], b \in \left[j - \left\lfloor \frac{\lambda}{2} \right\rfloor, j + \left\lfloor \frac{\lambda}{2} \right\rfloor \right] \right\}$$
$$- \{t_{i,j}^{m-1}\}$$

which leads to expression for the iterative process of the proposed method as:

$$t_{i,j}^m = min(T_{i,j}^{m-1}, \mathcal{M}((i,j), W((i,j), \lambda, m))) \tag{3}$$

Algorithm 1 illustrates the novel approach[2].

Algorithm 1: Self-Interpolation Method (SIM)

Input: A DSM matrix S
Input: The number of iterations M
Input: The window size λ
Output: The DTM Z
1 $T^0 \leftarrow S$;
2 **for** $m = 1, 2, \cdots, M$ **do**
3   **for** *all* $t_{i,j}^m \in T^m$, (i, j) = *the coordinates of all nodes in S* **do**
4          $W' \leftarrow \{$all $t_{a,b}^{m-1}$ in a $(\lambda \times \lambda)$ filter window centered at $(i, j)\}$
5          $W'' \leftarrow W' - \{t_{i,j}^{m-1}\}$
6          $t_{i,j}^m \leftarrow min(t_{i,j}^{m-1}, \mathcal{M}((i,j), W''))$;
7 $Z \leftarrow T^M$
   return Z

The method $\mathcal{M}$ adopted in the step 4 of Algorithm 1 could be any point interpolation of choice. Here in this study the Inverse Distance Weighting (IDW) method was used, which is widely adopted in the spatial context and provides

---

[2] The R source code of SIM is available at https://github.com/emmendorfer/sim.

a computationally efficient approach [4]. The IDW method computes the interpolation $\hat{y}_{new}$ for a given location new from a set of n given values $y_{l_1}, y_{l_2}, \cdots, y_{l_n}$ at locations $l_1, l_2, \cdots, l_n$ as:

$$\hat{y}_{new} = \sum_{k=1}^{n} w_{k,new} \times y_{l_k} \tag{4}$$

where the weights $w_{k,new}$ for input data point at l_k are given as:

$$w_{k,new} = \frac{D_{l_k,new}^{-\alpha}}{\sum_{k=1}^{n} D_{l_k,new}^{-\alpha}} \tag{5}$$

where $D_{l_k,new}$ is the Euclidean distance between nodes at locations l_k and new. In this work, $\alpha = 2$ is assumed, which is the most commonly adopted value.

4   Methodology

Two test sites were considered for the evaluation of the method, which provided three distinct application scenarios due to multiple topographic situations covered. The first corresponds to a low-altitude UAV photogrammetric survey using a DJI Phantom 4 PRO drone[3]. The aerial survey, which occurred during the years 2018 and 2019, covers a 424ha urbanized coastal area in southern Brazil with an overlap of 80/60, frontal and side percentage respectively. The photogrammetric bundle consisted of 3096 photos properly calibrated and adjusted through 68 control points acquired by GNSS-RTK [7]. The 1 m *Dunes* DSM corresponds to a 100 × 100 m squared region centered at 32° 11'10"S 52°09'13"W, which is composed by sand dunes covered with very low vegetation.

Also from the same site, an urban area was considered. It corresponds to a 100 × 100m squared region centered at 32°11'20"S 52°09'49"W which is denoted here as the *UrbanRG* DSM, also with 1 m resolution.

Finally, a 5 m resolution DSM derived from interferometric synthetic aperture radar (IFSAR)[4] was also included. This 2017 data corresponds to an area in Alaska, USA. A 500 × 500 m area centered at 61°15'00"N 157°16'39"W was denoted here as *Alaska*. Elevations in this area vary from 115 m to 130 m.

A quantitative evaluation was performed for the *Dunes* and *Alaska* DSMs, as follows. Since both cases correspond to low vegetation areas, we will consider both original DSMs as being the reference DTMs for the evaluation of the DSM-to DTM methods considered. From each DTM two noisy DSMs are obtained by performing the increment in the elevation values for a portion of the nodes. Two alternative percentages were adopted: 25% and 50%. This allows one to verify

---

[3] All ellipsoidal height coordinates obtained by the survey were adjusted to orthometric heights using the geoid heights provided by the MAPGEO2015 software, which were made available by the Instituto Brasileiro de Geografia e Estatística (IBGE).

[4] Available at: https://www.sciencebase.gov/catalog/item/5cc3ccd0e4b09b8c0b760 969.

the robustness of the filter on varying levels of artifacts inserted. For each node that is randomly selected to be altered, a random increment uniformly sampled from the interval $[0\,m, 5\,m]$ is added to the corresponding elevation value.

Two DTMs are computed for each noisy DSM, from both Terra and SIM methods. Finally, the Root Mean Squared Error (RMSE) between the reconstructed DTM and the original reference DTM is computed using all the elevation values available. This type of evaluation differs from what is adopted in other works where the evaluation is based on "reference DTMs" which are DSMs that were processed using some reference method.

5   Results

Initially, we evaluate the effect of the parameters of the algorithms on the quality of the resulting DTMs. Figure 1 shows the RMSE computed from the resulting DTMs from both Terra and SIM as a function of the parameters M and λ, for varying proportions of nodes altered (25% and 50%), considering both *Dunes* and *Alaska* areas. The Terra algorithm, besides M and λ has also the parameter η to be set. Since η should be greater than λ, we adopted an ad-hoc rule where $\eta = \lambda + 2$, in this first set of quantitative evaluations. Odd values were adopted for both window sizes, which guarantees that the center of the windows is exactly at (i, j).

In general, the RMSE response surfaces obtained from SIM attained lower values when compared to the corresponding RMSE surfaces obtained from Terra. Table 1 shows the optimal RMSE computed for all cases, along with the corresponding parameter values.

Table 1. Optimal RMSE values from Terra and SIM applied to *Dunes* and *Alaska* areas. The best result for each case is shown in bold.

Area	% of randomized elevations	Method	Best parameter values		Optimal RMSE
			λ	M	
Dunes	25%	Terra	5	3	0.5897
		SIM	3	3	**0.1097**
Dunes	50%	Terra	5	3	0.9537
		SIM	3	6	**0.1451**
Alaska	25%	Terra	3	3	0.6533
		SIM	3	2	**0.1994**
Alaska	50%	Terra	3	7	0.9652
		SIM	3	5	**0.2684**

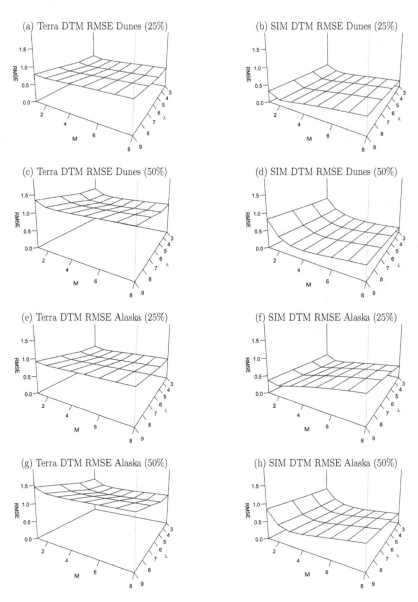

Fig. 1. RMSE computed from the result of Terra and SIM algorithms, as a function of the parameters M and λ, for varying proportions of nodes altered (25% and 50%), considering both the *Dunes* and *Alaska* areas.

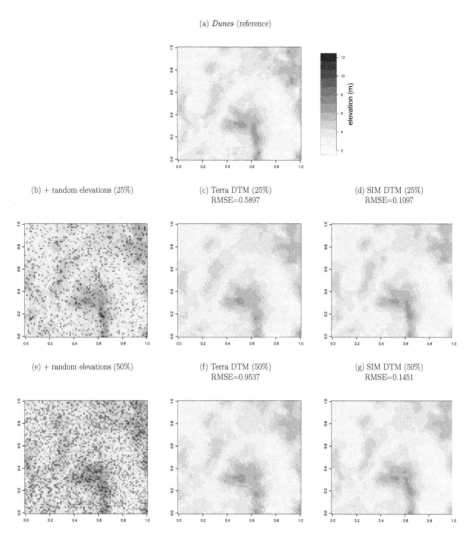

(a) *Dunes* (reference)

(b) + random elevations (25%)

(c) Terra DTM (25%)
RMSE=0.5897

(d) SIM DTM (25%)
RMSE=0.1097

(e) + random elevations (50%)

(f) Terra DTM (50%)
RMSE=0.9537

(g) SIM DTM (50%)
RMSE=0.1451

Fig. 2. Optimal results from Terra and SIM when applied to the *Dunes* area.

Figures 2 and 3 show the results obtained from both methods Terra and SIM using the optimal parameters shown in Fig. 1, applied to noisy DSMs. Figure 2 shows the results obtained for the *Dunes* region. The RMSE obtained from SIM was smaller than RMSE from TERRA. One can notice that the smoothing effect performed by the SIM method (Figs. 2d and 2g) preserves some dune features, while the TERRA method was not as conservative, resulting in a roughly "spotted" aspect, which makes certain dune features less evident in the reconstructed maps (Figs. 2c and 2f).

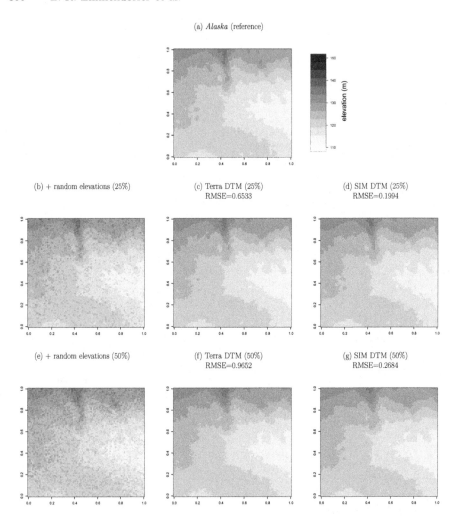

Fig. 3. Optimal results from Terra and SIM when applied to the *Alaska* area.

Figure 3 shows the reconstruction results obtained for the *Alaska* region. The RMSE values obtained from SIM were lower when compared to Terra, as in the previous case. The reconstructions obtained using the SIM filtering method (Fig. 3d and 3g) better preserve morphological characteristics of the original data when compared to the reconstructions obtained from Terra (Fig. 3c and 3f). The borders of an image are not affected by the Terra method, therefore border pixels were excluded from all images shown in the figures.

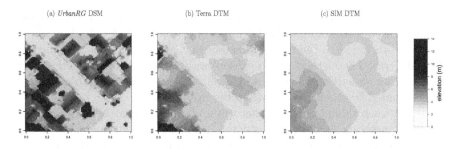

Fig. 4. Results from Terra and SIM when applied to the *UrbanRG* area.

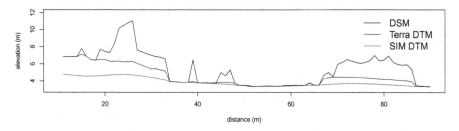

Fig. 5. A transect for *UrbanRG* area, considering the original DSM and resulting DTMs shown in Fig. 4.

To illustrate the adoption of the novel method in another context, Fig. 4 shows the DTM resulting from both Terra and SIM when applied to the *UrbanRG* area, which corresponds to an urban area. Parameters adopted for Terra in this case were $\lambda = 7$, $\eta = 10$. SIM was executed using $\lambda = 5$. For both methods, $M = 70$ was used. Other parameter settings for both methods led to worse results. Figure 5 shows transects for the same area, considering the original DSM and resulting DTMs. One can notice that SIM was able to achieve a further realistic result in this case when compared to Terra.

6   Conclusion

The choice of an appropriate method for generating a DTM given a DSM depends on the type of terrain, the type of objects to be removed, the amount of computational resources to be afforded, among other factors. Each method has its advantages and drawbacks: a method that fits well with some type of situation can be unsuited for another. This also motivates the improvement of existing methods and the search for alternatives. This paper described a novel DTM-from-DSM method which is very simple but rather effective when compared to a state-of-the-art reference approach. The method is evaluated on three scenarios that represent diverse situations that can be found in real applications. Optimal RMSE from the novel method was better when compared to the reference Terra method for all test cases considered, after performing an exhaustive search

on the parameter space for both methods. Results from Terra, however, could
be improved since the number of parameters of the method is higher and the
search was restricted to the two parameters that are common to both methods
SIM and Terra. The method proposed here was revealed as an attractive initial
alternative for practitioners since its parameter setting is easier to accomplish
when compared to other methods from the literature.

Further work should explore other types of applications. Also, other DTM-
from-DSM methods should be considered in further evaluations. Besides that,
other alternatives could be considered for the interpolation method adopted in
SIM besides just the IDW approach used here. For instance, variations of IDW
itself such as the IDWR method recently proposed [3,4] could be used.

References

1. Chen, C., Li, Y.: A fast global interpolation method for digital terrain model generation from large lidar-derived data. Remote Sens. **11**(11), 1324 (2019)
2. Croneborg, L., Saito, K., Matera, M., McKeown, D., van Aardt, J.: A guidance note on how digital elevation models are created and used-includes key definitions, sample terms of reference and how best to plan a DEM-mission. International Bank for Reconstruction and Development, vol. 104, New York (2015)
3. Emmendorfer, L.R., Dimuro, G.P.: A novel formulation for inverse distance weighting from weighted linear regression. In: Krzhizhanovskaya, V.V., et al. (eds.) ICCS 2020. LNCS, vol. 12138, pp. 576–589. Springer, Cham (2020). https://doi.org/10.1007/978-3-030-50417-5_43
4. Emmendorfer, L.R., Dimuro, G.P.: A point interpolation algorithm resulting from weighted linear regression. J. Comput. Sci. **50**, 101304 (2021)
5. Geiß, C., Wurm, M., Breunig, M., Felbier, A., Taubenböck, H.: Normalization of TanDEM-X DSM data in urban environments with morphological filters. IEEE Trans. Geosci. Remote Sens. **53**(8), 4348–4362 (2015)
6. Kobler, A., Pfeifer, N., Ogrinc, P., Todorovski, L., Oštir, K., Džeroski, S.: Repetitive interpolation: a robust algorithm for DTM generation from aerial laser scanner data in forested terrain. Remote Sens. Environ. **108**(1), 9–23 (2007)
7. Mongus, D., Žalik, B.: Computationally efficient method for the generation of a digital terrain model from airborne lidar data using connected operators. IEEE J. Sel. Topics Appl. Earth Observ. Remote Sens. **7**(1), 340–351 (2013)
8. Perko, R., Raggam, H., Gutjahr, K., Schardt, M.: Advanced DTM generation from very high resolution satellite stereo images. ISPRS Ann. Photogramm. Remote Sens. Spatial Inf. Sci. **2**, 165–172 (2015)
9. Pijl, A., Bailly, J.S., Feurer, D., El Maaoui, M.A., Boussema, M.R., Tarolli, P.: Terra: terrain extraction from elevation rasters through repetitive anisotropic filtering. Int. J. Appl. Earth Observ. Geoinf. **84**, 101977 (2020)
10. Serifoglu Yilmaz, C., Gungor, O.: Comparison of the performances of ground filtering algorithms and DTM generation from a UAV-based point cloud. Geocarto Int. **33**(5), 522–537 (2018)
11. Tarolli, P.: High-resolution topography for understanding earth surface processes: opportunities and challenges. Geomorphology **216**, 295–312 (2014)
12. Wilson, J.P.: Digital terrain modeling. Geomorphology **137**(1), 107–121 (2012)

13. Zhang, Y., Zhang, Y., Yunjun, Z., Zhao, Z.: A two-step semiglobal filtering approach to extract DTM from middle resolution DSM. IEEE Geosci. Remote Sens. Lett. **14**(9), 1599–1603 (2017)
14. Zhang, Y., Zhang, Y., Zhang, Y., Li, X.: Automatic extraction of DTM from low resolution DSM by two-steps semi-global filtering. ISPRS Ann. Photogramm. Remote Sens. Spatial Inf. Sci. **3**(3), 249–255 (2016)

Energy Performance of a Service Building: Comparison Between EnergyPlus and TRACE700

José Brito[1] ⓘ, João Silva[1,2](✉) ⓘ, José Teixeira[1] ⓘ, and Senhorinha Teixeira[2] ⓘ

[1] Mechanical Engineering and Resource Sustainability Centre (MEtRICs),
University of Minho, Guimarães, Portugal
js@dem.uminho.pt
[2] ALGORITMI Centre, University of Minho, Guimarães, Portugal
st@dps.uminho.pt

Abstract. The amount of energy consumed by a building can be estimated by performing dynamic simulations. In this study, two building simulation energy software, EnergyPlus, and TRACE700 were used to assess the energy performance of an existing service building. The building, placed in a specific zone of Portugal, has thirty people and a floor area of 2,000 m^2. It consumes mainly electricity, natural gas, and solar energy. The dynamic simulation started with the weather file upload, and then the construction, illumination, interior equipment, and HVAC systems were defined. The results were compared with the actual energy consumption values, and the deviation was 2% in the case of EnergyPlus and 0.5% in the case of TRACE700.

Keywords: Energy performance · Services building · Simulation

1 Introduction

Nowadays, the energy analysis market is huge due to the European Union (EU) legislation to improve energy efficiency. The energy efficiency in buildings will be, for sure, a significant step forward in the next few years. Estimating the energy consumption in existing buildings will be part of the process [1].

The future EU plans predict that the existing buildings must be transformed into nearly zero-energy buildings. In other words, the EU, with the directives implemented, intend to turn these buildings into energy efficient infrastructures that consume a reduced amount of energy [2].

The application of accurate simulation methods [3] will be crucial in achieving highly energy-efficient buildings and the EU goals. The dynamic simulation of buildings consists of predictions based on computer programs of the amount of energy that a building consumes during a period [4]. Computer simulation software allows the analysis of the energy efficiency of buildings and the estimation of energy consumption necessary to provide the comfort needs of its occupants [5]. Usually, all the available tools work by the same principles and require the same information sequentially. Firstly, the user inserts

© Springer Nature Switzerland AG 2021
O. Gervasi et al. (Eds.): ICCSA 2021, LNCS 12949, pp. 364–375, 2021.
https://doi.org/10.1007/978-3-030-86653-2_27

the structural construction parameters, which allows the software to calculate the thermal transmittance coefficient, obtaining, this way, the heating flow exchanged between the inside and outside of the building [6]. Once this flow is calculated, the programs calculate the amount of energy consumed to maintain a comfortable temperature inside of the building [7–10]. In this way, the software computes the energy consumption by all active systems [8–14]. In order to perform the calculations, the user must provide specific data such as the climatic file that contains information about the air temperature, humidity, radiation of the region, construction, occupation, illumination, equipment, and HVAC (Heat Ventilation and Air Conditioning) systems [10, 11]. Then, the software will use a simulation management tool that works as an interface of data exchange, aggregating two modules responsible for the simulation: the heat and mass balance simulation module and the building's system simulation module [15]. The first one performs the thermal balance simulation based on the phenomenon of heat transfer. The second one is responsible for the building's systems simulation; in other words, it does the simulation of HVAC and DHW (Domestic Hot Water) systems. Finally, the output data are analyzed [12, 13]. This process plays an essential role in determining the optimal design variables since the building's response is highly sensitive to the input data provided for the computer simulations. In the building energy sector, whole-building energy simulation programs that mainly consider key performance indicators such as energy demand and costs are the significant tools [17]. De Boeck et al. [18] provided a literature review on improving the energy performance of residential buildings. The author identified the most widely used tools in the literature, and EnergyPlus is one of them. More recently, Hashempour et al. [19] presented a complete, precise, up-to-date literature review on energy performance optimization of existing buildings. The author analyzed the different existing building simulation energy tools such as DOE-2, EnergyPlus, TRNSYS, eQUEST, and IDA ICE. EnergyPlus is the software that presents the highest percentage of the use amongst the different energy dynamic simulation tools and building modeling and/or visualization platforms, respectively. Crawley reported the same conclusion et al. [20] and Sousa [21]. The former compared the top20 simulation tools for building energy performance simulation. The last author reviewed five different simulation tools and concluded that although TRNSYS is the most complete tool, EnergyPlus, and IDA ICE are more adequate to import or export files from AutoCAD Software tools.

There are some research papers in the literature using building energy simulation tools for the analysis of different problems. Shrivastava et al. [22] presented a comparative study of popular simulation tools and their architecture from the perspective of TRNSYS. The author found that this software provided a good agreement within error between 5 and 10%. Furthermore, the author recommended the simultaneous analysis of the same system on different programs to avoid bias results. Sadeghifam et al. [23] examined the energy saving in building elements and how its coupling with effective air quality factors can contribute towards an ultimate energy efficient design. These works were based on a typical house in Malaysia, and the building was modeled using Revit software and then imported to EnergyPlus software to evaluate the best option in terms of energy savings.

Overall, despite different building energy simulation tools, choosing the appropriate tool depends on the final user's objective and perspective. In this way, this paper aims to study the energy performance in an existing building in Portugal and perform the same analysis using two different programs to avoid biased results as suggested by Srivastava et al. [22]. This is an important work to provide answers to particular questions regarding the energy performance of an existing building. Further, the influence of some key variables can be analyzed quickly and effectively. This is an added value since the owners of the service building will see the building energy demand and what are its cost and energy performance rating.

2   Case Study

This chapter describes the building selected as a test case, its location, construction characteristics, occupation, illumination, equipment, and HVAC systems. Then, the energy modeling methodology used in EnergyPlus and TRACE700 is presented, highlighting, and explaining the details of the two models.

2.1   Building Description

The building is a nursing home for seniors located in Penafiel, Oporto district, Portugal. The building, made up of floors, is an old construction, which was never retrofitted with any improvement. The first floor, referred to as Floor -1, includes bedrooms, bathrooms, corridors, kitchen, canteen, church, fridge room, and laundry.

The second floor, referred to as Floor 0, comprises bedrooms, a corridor, bathroom, waiting room, and an administrative office. The third floor, Floor 1, also includes bedrooms, hall, bathroom, meeting room, gym, and infirmary.

Each floor has an average height of 3 m. Regarding the building's orientation, the main facade is oriented to the southeast, as shown in Fig. 1. The arrow points to the north.

Main front

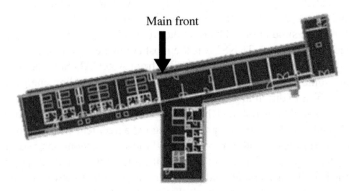

Fig. 1. Illustration of the building orientation plan.

HVAC and DHW Systems. The building has only one heating system and no cooling systems. It consists of one natural gas boiler that heats the water up to 70 °C. Two electrical pumps pump the hot water to the radiators and fan coils. As far as the domestic hot water system is concerned, this consists of one boiler (the same as the heating system) that heats the water up to 50 °C, subsequently pumped by two electrical pumps to 2 storage tanks, with a capacity is 300 L each. Besides, there is a supplementary system to the boiler system previously described, consisting of 14 solar collectors that heat water at 50 °C, subsequently pumped by one electrical pump to a 600 L storage (Table 1).

Table 1. Characteristics of the building's HVAC and DHW system.

Equipment	Power (kW)
Boiler	203.6
Pump HVAC	0.33
Pump HVAC	0.65
Pump DHW	0.093
Pump DHW	0.093
Pump DHW	0.165

2.2 Software Implementation

EnergyPlus and TRACE700 are energy simulation software that allows the user to simulate the building energy consumption under different thermomechanical conditions. By introducing the constructive building characteristics, they can calculate the building thermal transmission coefficient and, in this way, calculate the heat losses and gains of the building with the environment. Knowing the occupation thermal comfort needs and the thermal load, the software calculates the energy required for operation.

There are a few differences between these two tools. EnergyPlus allows the user to know the thermal load generated by illumination, occupation, and equipment in non-thermal zones. On the other hand, TRACE700 does not calculate the solar contribution on domestic hot water simulation and does not allow a 3D modeling of the buildings.

The next subsections will present the construction of each energy model on this different software.

3D Model. In the EnergyPlus software, the 3D building modeling was developed with the Sketchup graphical interface. Sketchup has an OpenStudio Plug-in, which allows the user to draw the building and, at the same time, to build the energy model. OpenStudio is a data insertion interface that enables the data to insertion to construct the energy model.

The 3D drawing on Sketchup started with the design of the floor, followed by the spaces boundary conditions, the existing solar collectors, and, finally, the building orientation was defined. Figure 2 a) shows the final building 3D modeling, where the cover is brown, the external walls are yellow, and the windows are represented in blue. Figure 2 b) shows the compartment's boundary conditions, wherein green represents the inner walls and floors, grey is used for the floor adjacent to the ground, and blue is for the outer walls and roof.

The TRACE700 software does not allow direct 3D modeling.

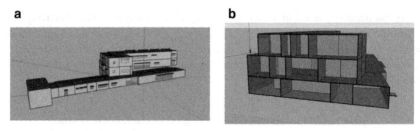

Fig. 2. 3D external view of the building: a) EnergyPlus final model and b) compartments boundary conditions. (Color figure online)

Input Data. *Climatic Zone.* In terms of data insertion, the climatic file is the first step of data insertion into the software when the user wants to build energy models. In the EnergyPlus, the climatic file upload is carried out on OpenStudio, which is the EnergyPlus input data interface. In the TRACE700, the upload is made directly on TRACE700. In these two tools, the climatic file insertion is very similar.

Firstly, the climatic EPW file is downloaded from an app called CLIMAS-SCE. This app allows the user to obtain the EPW file by inserting the city, its region, and its altitude. From the EnergyPlus website, it is downloaded the DDY. In the specific case of the DDY file, the Oporto DDY was selected as the nearest location to the actual site because it is the only city from the north of Portugal that had a DDY file.

The main difference between the EPW and DDY files is that the first contains information concerning the outer temperature, relative humidity, solar radiation, opaque cloudiness. This information is used by the software to simulate the building's thermal behavior for 8670 h, but this file does not possess any information about the critical days of the year in terms of external air temperature. For this reason, the software needs the DDY file, which provides this information. Figure 3 shows the EPW file down-load on CLIMAS-SCE.

Thermal Zones. In this subchapter, the variables that characterize the use of the different thermal zones were defined: the occupation, its metabolic rate, the use of equipment, and lighting. In both softwares, the definition of these variables is very similar. It starts with the introduction of the values related to occupation, lighting, and equipment. This introduction is made for every thermal zone, but in the case of EnergyPlus, it is possible to introduce the values of these variables in non-thermal zones. This introduction is followed by the definition of the lighting, occupation, and electrical equipment working schedules. The service building management provided these data.

Construction Details. In both tools, the materials present in the construction were created because these software are much more adapted to the American specifics than European. For the EnergyPlus, firstly, the material layers were formed. In this stage, the variables required for each material from EnergyPlus are the thickness of each material layer, its conductivity, and its thermal resistance. Then the layers are grouped into walls, here identified as the external walls, floors, cover, internal walls, etc. Finally, on the Sketchup platform, using the OpenStudio Plug-in, these constructions were associated with the 3D model, previously created.

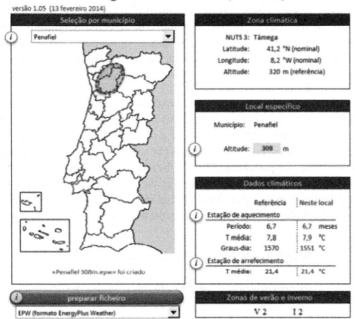

Fig. 3. Climas-SCE Software.

On Trace700, the creation of the material layers is not needed. The wall creation is immediately done, and only its thermal transmittance coefficient and its area is required from the software to create the walls. Then the constructions were associated with each thermal zone. Figure 4 shows the construction definition on TRACE700.

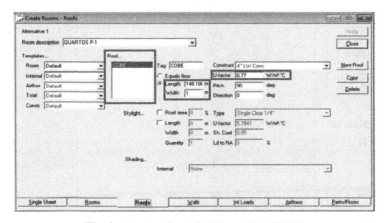

Fig. 4. Construction definition on TRACE700.

Loads. The loads correspond to the internal gains existing in the thermal zones of the building. These gains are associated with the number of occupants, their expected activity, the power of the equipment, and lighting. The selected building possesses both natural gas and electric powered devices. In both tools, the definition of these variables that contribute to the internal gains is identical. This definition is made for every thermal zone. Table 2 shows the load definition in the energy simulation software.

HVAC and DHW Systems. Regarding the HVAC and DHW definition, this task is quite different in both software. On EnergyPlus, this definition starts with the building of an HVAC circuit. Here, firstly, the boiler is added to the circuit, then a pump (the HVAC circuit possesses 2 pumps, but on EnergyPlus has selected only an equivalent pump that corresponds to these two pumps), and finally the radiators and fan coils. When the pump is selected, the option "Auto sized" concerning the pumped water flow and the electric power of the pump must be determined. In this way, the EnergyPlus will calculate the electric power and water flow needed to satisfy the building heating needs. The definition

Table 2. Loads defined in the energy simulation software.

Zone	Number of persons	Light (W)	Electric equipment (W)	Natural gas equipment (W)
Bathroom	0	406	–	–
Hall	6	545	–	–
Bedrooms	12	774	–	–
Canteen	20	384	–	–
Kitchen	4	144	3,600	43,000
Fridge room	0	72	37,200	21,000
Laundry	2	72	1,470	–
Church	8	406	–	–
Hall	0	168	–	–
Secretary	2	108	–	–
Manager's office	1	72	–	–
Waiting room	2	60	–	–
Bathroom (Floor 0)	0	46	–	–
Hall (Floor 0)	0	112	–	–
Bedrooms (Floor 0)	9	1,129	–	–
Hall (Floor 1)	0	316	–	–
Nursery	2	60	–	–
Gym	3	108	–	–
Meeting room	4	60	–	–
Doctor's office	2	78	–	–
Bedrooms (Floor 1)	9	1,129	–	–
Bathroom (Floor 1)	0	23	–	–

of an HVAC system on TRACE700 is similar to the EnergyPlus, but here all the HVAC equipment is defined and its characteristics. In the end, the setup temperatures for each month are defined too in each software.

Regarding the DHW system definition, on EnergyPlus is quite similar to the HVAC system definition. Here the circuit is created and then added all the equivalent collectors, pumps, storages (taken as an equivalent of all collectors, pumps, and storages). TRACE700 does not simulate the solar contribution on the DHW system. Having the system of solar collectors, a great contribution to the DHW system, another software (SCE.ER) was used to simulate the DHW system instead of TRACE700. The results of SCE.ER was added to the TRACE700 results.

3 Results and Discussion

In this section, the simulation results on EnergyPlus and TRACE700 are presented, discussed, and compared also with real energy consumption results. Regarding the results of the simulations on both software, the EnergyPlus presented a total energy consumption in a year of 1,435.84 GJ and TRACE700, 1,457.72 GJ. The deviation to the real consumption values is respectively 2 and 0.5% for these softwares. The results of both tools were very identical which shows that either software can be used to simulate this type of buildings.

An analysis was made to compare both energy models in terms of natural gas and electricity consumption in the building. Figure 5 shows the monthly variation of the natural gas consumption in both models and the real natural gas consumption. Figure 6 shows the consumption of electricity in each month, comparing both models and the real values recorded.

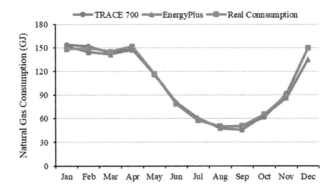

Fig. 5. Comparison of natural gas consumption in each month for TRACE700 and EnergyPlus.

The analysis of Fig. 5 shows that either simulation provided identical values, which demonstrates that the use of each tool is reliable. Looking at the values and the curve's path, it is perceptive that the natural gas consumption is higher during the months when the heating needs are higher, so then the boiler has a far higher load during these months.

Observing the data in Fig. 6, the results of each software in terms of electricity consumption are also very similar.

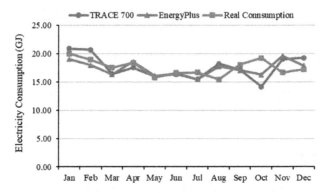

Fig. 6. Comparison of electricity consumption in each month between TRACE700 and Energy-Plus.

The differences observed are explained by the "Auto-sized" option for the pump's power, on EnergyPlus. During the heating months, TRACE700 electricity consumption results are higher than with the EnergyPlus. During these months, the pumps present on the TRACE700 model have higher power than those present on the EnergyPlus model.

When the heating system is not working at its highest power, the electricity consumption on EnergyPlus is higher or equal to those with the TRACE700 because the TRACE700 model did not simulate the DHW system.

Figure 7 shows the results of the natural gas consumption and compares these in both models. Analyzing this figure indicates that the kitchen and laundry equipment consumption is constant during the year in both models.

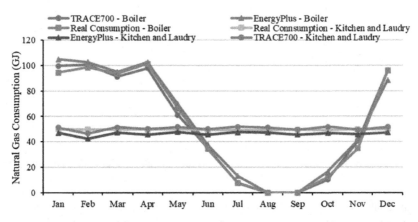

Fig. 7. Results of the natural gas consumption in EnergyPlus and TRACE700 for different components.

Figures 8 and 9 show the results of electricity consumption, respectively, in EnergyPlus and TRACE700. In both models, analyzing these two graphics shows that the interior equipment is the major source of electricity consumption. The Fan coils and illumination consumption is constant all the year, and the pumps consume low power during the months when the heating system is off. In the TRACE700 this consumption

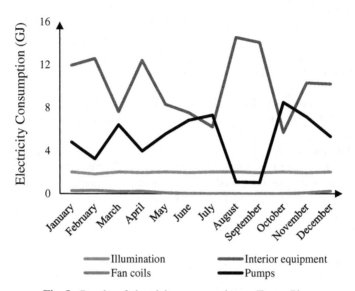

Fig. 8. Results of electricity consumption on EnergyPlus.

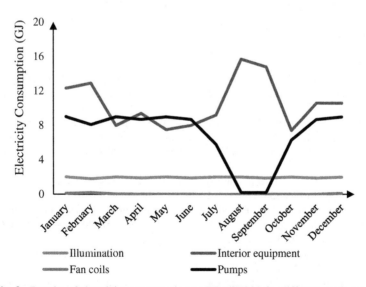

Fig. 9. Results of electricity consumption on TRACE700 for different equipment.

is zero in August and September because the heating system is off and, in this model, the DHW system was not simulated.

4 Conclusions

After analyzing the results, it is concluded that each software has similar results when simulating the energy consumption in this kind of buildings. The EnergyPlus and TRACE700 models obtained, respectively, 1,435.84 GJ and 1,457.72 GJ. Compared with the real energy consumption values, TRACE700 and EnergyPlus models results had a deviation of 0.5% and 2%, which supports the conclusion of being reliable and suitable to reality.

However, the most significant differences during the model construction are related to the 3D modeling, the use of different interfaces in EnergyPlus, the simulation of the DHW system.

Regarding the negative aspects of these tools, they are much more adapted to the American market than to the European. It can be observed in the libraries where most of the materials and equipment needed are not available. Another negative aspect is the fact that these tools do not calculate thermal bridges.

In conclusion, a brief overview is that both software is a great tool to predict the energy consumption of a services building.

Acknowledgments. The authors would like to express their gratitude for the support given by FCT within the R&D Units Project Scope UIDB/00319/2020 (ALGORITMI) and R&D Units Project Scope UIDP/04077/2020 (MEtRICs).

References

1. Energia. Comissão Europeia
2. Economidou, M., Todeschi, V., Bertoldi, P., et al.: Review of 50 years of EU energy efficiency policies for buildings. Energy Build. **225**, 110322 (2020). https://doi.org/10.1016/j.enbuild.2020.110322
3. Sultanguzin, I.A., Kruglikov, D.A., Yatsyuk, T.V., Kalyakin, I.D., Yavorovsky, Y., Govorin, A.V.: Using of BIM, BEM and CFD technologies for design and construction of energy-efficient houses. E3S Web Conf. **124**, 03014 (2019). https://doi.org/10.1051/e3sconf/201912403014
4. Gao, H., Koch, C., Wu, Y.: Building information modelling based building energy modelling: a review. Appl. Energy **238**, 320–343 (2019). https://doi.org/10.1016/j.apenergy.2019.01.032
5. Amani, N., Soroush, A.A.R.: Effective energy consumption parameters in residential buildings using Building Information Modeling. Glob. J. Environ. Sci. Manage. **6**, 467–480 (2020). https://doi.org/10.22034/gjesm.2020.04.04
6. Neymark, J., Judkoff, R.: International energy agency building energy simulation test and diagnostic method. Natl. Renew. Energy Lab. (2008)
7. Guzmán Garcia, E., Zhu, Z.: Interoperability from building design to building energy modeling. J. Build. Eng. **1**, 33–41 (2015). https://doi.org/10.1016/J.JOBE.2015.03.001

8. Esteves, D., Silva, J., Rodrigues, N., Martins, L., Teixeira, J., Teixeira, S.: Simulation of PMV and PPD thermal comfort using energyplus. In: Misra, S., et al. (eds.) ICCSA 2019. LNCS, vol. 11624, pp. 52–65. Springer, Cham (2019). https://doi.org/10.1007/978-3-030-24311-1_4

9. Noversa, R., Silva, J., Rodrigues, N., Martins, L., Teixeira, J., Teixeira, S.: Thermal simulation of a supermarket cold zone with integrated assessment of human thermal comfort. In: Gervasi, O., et al. (eds.) ICCSA 2020. LNCS, vol. 12254, pp. 214–227. Springer, Cham (2020). https://doi.org/10.1007/978-3-030-58817-5_17

10. Silva, J., et al.: Energy performance of a service building: comparison between energyplus and revit. In: Gervasi, O., et al. (eds.) ICCSA 2020. LNCS, vol. 12254, pp. 201–213. Springer, Cham (2020). https://doi.org/10.1007/978-3-030-58817-5_16

11. Lanzisera, S., Dawson-Haggerty, S., Cheung, H.Y.I., et al.: Methods for detailed energy data collection of miscellaneous and electronic loads in a commercial office building. Build. Environ. **65**, 170–177 (2013)

12. Palmero-marrero, A.I., Gomes, F., Sousa, J., Oliveira, A.C.: Energetic analysis of a thermal building using geothermal and solar energy sources. Energy Rep. **6**, 201–206 (2020)

13. Gao, Y., Li, S., Xingang, F., Dong, W., Bing, L., Li, Z.: Energy management and demand response with intelligent learning for multi-thermal-zone buildings. Energy **210**, 118411 (2020)

14. Queiroz, N., Westphal, F.S., Ruttkay Pereira, F.O.: A performance-based design validation study on EnergyPlus for daylighting analysis. Build. Environ. **183**, 107088 (2020)

15. Al-janabi, A., Kavgic, M., Mohammadzadeh, A., Azzouz, A.: Comparison of EnergyPlus and IES to model a complex university building using three scenarios: free-floating, ideal air load system, and detailed. J. Build. Eng. **22**, 262–280 (2019)

16. Chen, Y., Deng, Z., Hong, T.: Automatic and rapid calibration of urban building energy models by learning from energy performance database. Appl. Energy **277**, 115584 (2020)

17. Stevanović, S.: Optimization of passive solar design strategies: a review. Renew. Sustain. Energy Rev. **25**, 177–196 (2013). https://doi.org/10.1016/j.rser.2013.04.028

18. De Boeck, L., Verbeke, S., Audenaert, A., De Mesmaeker, L.: Improving the energy performance of residential buildings: a literature review. Renew. Sustain. Energy Rev. **52**, 960–975 (2015). https://doi.org/10.1016/J.RSER.2015.07.037

19. Hashempour, N., Taherkhani, R., Mahdikhani, M.: Energy performance optimization of existing buildings: a literature review. Sustain Cities Soc. **54**, 101967 (2020). https://doi.org/10.1016/j.scs.2019.101967

20. Crawley, D.B., Hand, J.W., Kummert, M., Griffith, B.T.: Contrasting the capabilities of building energy performance simulation programs. Build. Environ. **43**, 661–673 (2008). https://doi.org/10.1016/j.buildenv.2006.10.027

21. Sousa, J.: Energy simulation software for buildings: review and comparison. In: International Workshop on Information Technology for Energy Applicatons-IT4Energy, Lisabon, p. 12 (2012)

22. Shrivastava, R.L., Kumar, V., Untawale, S.P.: Modeling and simulation of solar water heater: a TRNSYS perspective. Renew. Sustain. Energy Rev. **67**, 126–143 (2017). https://doi.org/10.1016/j.rser.2016.09.005

23. Sadeghifam, A.N., Zahraee, S.M., Meynagh, M.M., Kiani, I.: Combined use of design of experiment and dynamic building simulation in assessment of energy efficiency in tropical residential buildings. Energy Build. **86**, 525–533 (2015). https://doi.org/10.1016/j.enbuild.2014.10.052

CFD Prediction of Shock Wave Impacting a Cylindrical Water Column

Viola Rossano[ID] and Giuliano De Stefano[(✉)][ID]

Engineering Department, University of Campania Luigi Vanvitelli,
81031 Aversa, Italy
giuliano.destefano@unicampania.it

Abstract. Computational fluid dynamics (CFD) analysis is carried out to evaluate the early stages of the aerobreakup of a cylindrical water column due to the impact of a travelling plane shock wave. The mean flow in a shock tube is simulated by adopting the compressible unsteady Reynolds-averaged Navier-Stokes modelling approach, where the governing equations are solved by means of a finite volume-based numerical technique. The volume of fluid method is employed to track the transient interface between air and water on the fixed numerical mesh. The present computational modelling approach for industrial gas dynamics applications is verified to have significant practical potential by making a comparison with reference experiments and numerical simulations.

Keywords: Aerobreakup · Computational fluid dynamics · Industrial gas dynamics · Shock tube

1 Introduction

The aerodynamic breakup of liquid droplets impacted by shock waves has a number of important industrial applications including, for instance, supersonic combustion airbreathing jet engines (scramjets), where the behavior of fuel droplets in supersonic airflow plays a crucial role [1,2]. Experimentally, this physical phenomenon can be studied by using a shock tube device, wherein a planar shock wave is produced that travels towards the downstream test section, with uniform gaseous flow conditions being established, e.g. [3,4]. Computationally, the droplet breakup can be approximated in two spatial dimensions by considering a water cylinder with circular section in the high-speed flow behind a travelling shock wave. The latter can be obtained by numerically reproducing the shock tube flow conditions that are given and known.

The main goal of the present work is the computational evaluation of the aerodynamic breakup of a cylindrical water column. The study is focused on the initial stages in the interaction process between the travelling plane shock wave and the water cylinder at relatively high Mach number. In fact, the gas flow Mach number that is $\mathrm{Ma} = 1.73$ corresponds to one of the cases recently studied in [5]. The numerical experiments are conducted by employing one of the CFD solvers

© Springer Nature Switzerland AG 2021
O. Gervasi et al. (Eds.): ICCSA 2021, LNCS 12949, pp. 376–386, 2021.
https://doi.org/10.1007/978-3-030-86653-2_28

that are commonly and successfully used for building virtual wind tunnels in the industrial gas dynamics research, e.g. [6,7]. The compressible unsteady Reynolds-averaged Navier-Stokes (URANS) equations [8], supplied with a suitable closure model, are solved by means of a finite volume-based numerical technique. The transient interface between air and water is approximated on the fixed numerical mesh by using the volume of fluid (VOF) method [9]. The overall computational modelling procedure is validated against reference experimental and numerical data corresponding to the same flow configuration.

In the following, after introducing the main aspects of the present CFD modelling procedure, the results of the numerical simulation are presented and discussed. Finally, some concluding remarks are provided.

2   CFD Model

2.1   Flow Geometry

The simplified two-dimensional geometry of the shock tube device is reproduced by means of a rectangular domain. The coordinates system (x, y), with the x-axis coinciding with the shock tube axis, is chosen so that $-8 < x/D_0 < 15$ and $-10 < y/D_0 < 10$, where D_0 represents the initial diameter of the cylindrical water column. The origin of the system corresponds to the initial position of the leading edge of the column, which is immersed in supersonic air crossflow. The flow configuration is sketched in Fig. 1, where V_1 represents the constant velocity of the travelling planar shock wave.

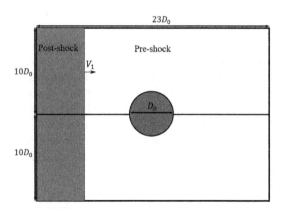

Fig. 1. Sketch of physical domain, with travelling air shock wave.

The shock tube flow conditions are simulated by initially imposing two very different pressure levels into the two different tube sections, separated by a virtual cross diaphragm, which is located upstream of the column at $x/D_0 = -2$. Initially, both tube sections contain still air, which is treated as ideal gas, at the same temperature and very different densities. Given and known the pressure level, say p_1, in the driven section (right side), as well as the compression ratio across the shock, p_2/p_1 (associated to the prescribed Mach number), the pressure level p_4 to be imposed in the driver section (left side), is analytically determined by exploiting the theoretical shock tube relations [10]. As is demonstrated in the next section, the present CFD model is able to reproduce a plane shock wave travelling at the prescribed velocity that is $V_1 = 594$ m/s. For brevity, the flow parameters corresponding to the simulated shock tube device are reported in Table 1.

Table 1. Shock tube parameters.

Parameter	Value
Driven section pressure	$p_1 = 101.3$ kPa
Driven section density	$\rho_1 = 1.204$ kg/m^3
Shock compression ratio	$p_2/p_1 = 3.32$
Driver section pressure	$p_4 = 1.50$ MPa
Driver section density	$\rho_4 = 17.8$ kg/m^3
Temperature	$T_1 = T_4 = 293.15$ K
Travelling shock velocity	$V_1 = 594$ m/s

It is worth noting that the actual computational domain corresponds to half the physical domain sketched in Fig. 1, with a symmetry condition at the centerline ($y = 0$) being imposed. In fact, previous experiments conducted at comparable characteristic flow numbers demonstrated that flow symmetry is maintained in the early stages of the aerobreakup process [11]. Also, wall boundary conditions are set at the other three edges of the computational domain.

For the current flow configuration, the Reynolds and Weber numbers are defined based on the reference length D_0 and the air conditions behind the shock, specifically,

$$\text{Re} = \frac{\rho_2 V_2 D_0}{\mu_2}, \qquad \text{We} = \frac{\rho_2 V_2^2 D_0}{\sigma}.$$

Given the values for the flow parameters that are reported in Table 2, where the surface tension σ at the interface between air and water is assumed constant, the above characteristic numbers result in being $\text{Re} = 2.38 \times 10^5$ and $\text{We} = 1.93 \times 10^4$. Therefore, based on the present high Weber number, in the early stages of the aerobreakup, capillary effects can be neglected with respect to inertial ones. In fact, according to the recent reclassification of the droplet breakup regimes

Table 2. Water column and post-shock air flow parameters.

Parameter	Value
Initial column diameter	$D_0 = 4.80 \times 10^{-3}$ m
Liquid water density	$\rho_w = 1000$ kg/m^3
Surface tension	$\sigma = 7.286 \times 10^{-2}$ N/m
Air density	$\rho_2 = 2.706$ kg/m^3
Air pressure	$p_2 = 0.336$ M Pa
Air viscosity	$\mu_2 = 1.80 \times 10^{-5}$ Pa s
Air velocity	$V_2 = 329$ m/s
Mach number	Ma$_2 = 0.789$

proposed in [12], the current flow configuration belongs to the shear-induced entrainment regime, where the gas goes around the liquid mass producing a surface layer peeling-and-ejection action [13]. Finally, differently from what done in [5], the viscous effects are not ignored in the present study.

2.2 VOF Method

To approximate the transient interface between the gas (air) and the liquid (water), while tracking it on the fixed numerical mesh, a classical methodology that is based on the concept of fractional volume of fluid is used. This technique was shown to be more flexible and efficient for treating the complex interface between two immiscible fluids, when compared to other methods [9]. Following the VOF approach, the same governing equations are solved for the two different fluid phases, while assuming that they share the velocity, pressure and temperature fields. Practically, the balance equations are solved for an effective fluid, whose averaged properties are evaluated according to the two volume fractions. For instance, the averaged density is defined as

$$\rho = \alpha \rho_w + (1 - \alpha)\rho_a,$$

where ρ_a stands for the air density, and α represents the volume fraction of water. The latter variable is calculated throughout the computational domain by solving the associated continuity equation.

2.3 CFD Solver

The shock tube flow, along with the aerodynamic breakup of the water column, is numerically predicted by solving the URANS equations, which describe the unsteady mean turbulent compressible flow. The governing equations are supplied with the k–ω SST two-equation turbulence model [14]. The resolved compressible URANS equations, which are not reported here for brevity, can be found, for instance, in [8].

The current numerical simulations are performed by means of the industrial solver ANSYS Fluent 19, which has been successfully employed by the authors in previous industrial CFD studies using Reynolds-averaging based models, e.g. [15–17]. The pressure-based solver utilizes the Finite Volume (FV) method to approximate the mean flow solution, with the conservation principles being applied over each control volume, e.g. [18]. The results presented in the following have been obtained using a mesh consisting of about 2.8×10^5 FV cells. The mesh has been suitably refined in the space region close to the position of the water column, including the near wake, where the numerical resolution corresponds to about 50 cells per cylinder diameter. The constant time-step size $\Delta t = 1$ µs has been used for the explicit transient calculation.

3   Results

The shock tube flow is simulated starting from the initial conditions discussed in the previous section. Specifically, a planar shock wave develops and travels towards the driven section, where the water column is located, while a set of expansion waves propagate in the opposite direction, towards the driver section. The robustness of the CFD model and the accuracy of the time-dependent solution have been assessed through comparison with the analytical ideal shock-tube theory [10]. For instance, the instantaneous pressure profile along the tube at the moment when the incident shock impacts on the column, say $t = t_0$, is reported in Fig. 2. Note that, at the contact surface, which travels in the same direction of the shock, though at lower velocity, it holds $p_3 = p_2$, whereas $\rho_3 > \rho_2$.

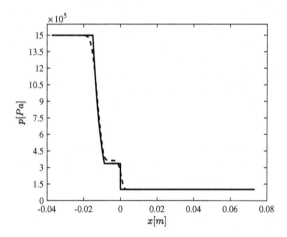

Fig. 2. Pressure profile along the shock tube at the impact time instant: present CFD solution (dashed line) compared to theoretical profile (solid line).

As a consequence of the impact of the air shock wave, the deformation and the breakup of the water column are observed. In the following, as is usually done in droplet aerobreakup studies, the independent time variable is non-dimensionalized as follows

$$t^* = \frac{V_2}{D_0} \sqrt{\frac{\rho_2}{\rho_w}} \, (t - t_0),$$

where ρ_2 and V_2 represent post-shock air flow conditions, corresponding to the Mach number $Ma_2 = 0.789$ in the shock-moving reference frame.

In the early stages of the interaction process, the water cylinder is flattened in the streamwise direction, which corresponds to a decreasing centerline width, together with an increasing cross-stream length. The initial deformation of the water column is illustrated in Fig. 3, where the contour maps of density (left side) and pressure (right side) are reported at different time instants between $t^* = 0.008$ (top) and 0.72 (bottom). By inspection of this figure, in the very early stage, the air flow field resembles the flow past a circular cylinder. Then, the streamwise flattening of the liquid column is observed, with the successive formation of peripheral tips. Note that, due to the finite numerical resolution, the interface between the two different fluids appears to be diffuse. In Fig. 4, the density and pressure contours are shown at later time instants, namely, between $t^* = 0.96$ (top) and 1.4 (bottom). Apparently, the liquid sheet becomes thinner and thinner until, finally, the rupture indicates the breakup initiation. Then, the water body is continuously eroded at the periphery, with the appearance of microdroplets stripped from the parent column. During the successive stage, the water column would disintegrate into fragments distributed widely in the flow field. These pictures confirm the qualitative features of the so-called shear stripping breakup mechanism, as is expected at current high Weber number [13].

From a quantitative point of view, the present CFD solution is validated against the corresponding numerical solution of Meng and Colonius [5], as well as the experimental findings of Igra and Takayama [11]. Here, the deformed column shape is evaluated as that corresponding to the isoline $\alpha = 0.5$ for the volume fraction of water. The time-dependent centerline width of the deforming water column, say w, is shown in Fig. 5 (left side). Apparently, a good agreement is observed with respect to both numerical and experimental reference data. Note that the two different lines that are reported for the reference numerical solution correspond to $\alpha = 0.25$ (dashed line) and 0.99 (dashed-dotted line). When examining the time-dependent cross-stream length of the column, say d, which is drawn in the same Fig. 5 (right side), the current CFD result, while matching the reference numerical data, appears to be overestimated with respect to experiments. Similarly, a good agreement with reference data is found for the time-dependent section area of the coherent water body, which is shown in Fig. 6 (left side), where A_0 stands for the initial circular section area. When examining the time-dependent position of the water column leading-edge, say Δx, which is reported in the same Fig. 6 (right side), again, the CFD result appears to be overestimated with respect to the experimental findings, while agreeing with the reference numerical data. For a more detailed discussion of the performance of the proposed computational model, the reader is referred to [19], where a more comprehensive study has been completed.

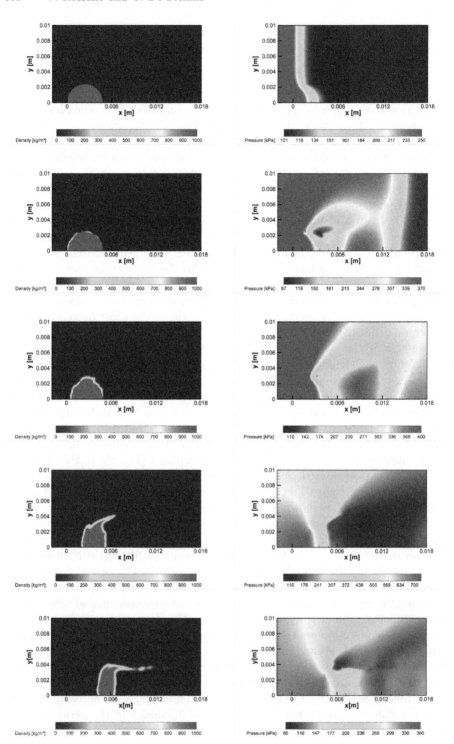

Fig. 3. Contour maps of density (left) and pressure (right) at different time instants corresponding to $t^* = 0.008$, 0.09, 0.24, 0.48, and 0.72 (from top to bottom).

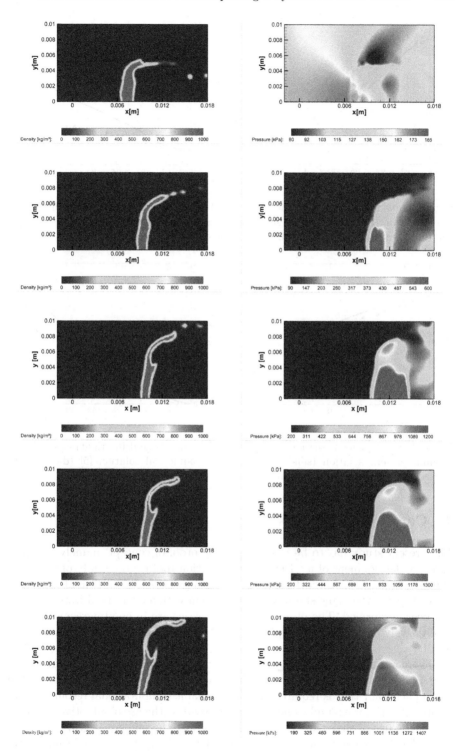

Fig. 4. Contour maps of density (left) and pressure (right) at different time instants corresponding to $t^* = 0.96$, 1.2, 1.32, 1.35, and 1.4 (from top to bottom).

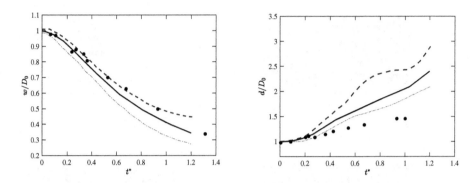

Fig. 5. Time-dependent centerline width (left) and cross-stream length (right) of the water column: present solution (solid line), reference numerical solution [5] (dashed and dashed-dotted lines), and experimental data [11] (symbols).

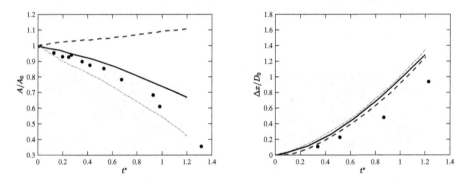

Fig. 6. Time-dependent section area (left) and leading-edge drift (right) of the water column: present solution (solid line), reference numerical solution [5] (dashed and dashed-dotted lines), and experimental data [11] (symbols).

4   Concluding Remarks

The present study has to be intended as the proof-of-concept, namely, the preliminary development of a CFD based prediction tool for the study of the aerodynamic breakup of water droplets in supersonic gas flows. CFD analysis of the early stages of the plane shock wave interaction with a cylindrical water column has been performed. The observed characteristics are in good agreement, both qualitatively and quantitatively, with the reference experimental and numerical results.

There remains the possibility of developing more sophisticated computational models for this particular industrial gas dynamics application, depending on the level of accuracy that is required and the available computational resources. Following [20, 21], for example, one possibility could be the use of adaptive methods that make use of the wavelet transform to adjust the numerical grid resolution to the local flow conditions [22–24]. Wavelet-based adaptive URANS modelling procedures that have been recently developed [25–27] could be explored in this context. Moreover, in order to simulate the appearance of microdroplets, the use of VOF-Lagrangian hybrid methods [28] could be considered, in the framework of either the present industrial CFD solver or the wavelet-based approach [29].

References

1. Liu, N., Wang, Z., Mingbo, S., Wang, H., Wang, B.: Numerical simulation of liquid droplet breakup in supersonic flows. Acta Astronaut. **145**, 116–130 (2018)
2. Villermaux, E.: Fragmentation. Annu. Rev. Fluid Mech. **39**, 419–446 (2017)
3. Wang, Z., Hopfes, T., Giglmaier, M., Adams, N.A.: Effect of Mach number on droplet aerobreakup in shear stripping regime. Exp. Fluids **61**, 193 (2020)
4. Poplavski, S., Minakov, A., Shebeleva, A., Boiko, V.: On the interaction of water droplet with a shock wave: experiment and numerical simulation. Int. J. Multiph. Flow **127**, 103273 (2020)
5. Meng, J.C., Colonius, T.: Numerical simulations of the early stages of high-speed droplet breakup. Shock Waves **25**, 399–414 (2015)
6. Sembian, S., Liverts, M., Tillmark, N., Apazidis, N.: Plane shock wave interaction with a cylindrical water column. Phys. Fluids **28**, 056102 (2016)
7. Rapagnani, D., Buompane, R., Di Leva, A., et al.: A supersonic jet target for the cross section measurement of the 12C(α, γ)16O reaction with the recoil mass separator ERNA. Nucl. Instrum. Meth. Phys. Res. B **407**, 217–221 (2017)
8. Wilcox, D.C.: Turbulence Modeling for CFD, 3rd edn. DCW Industries Inc, La Canada (2006)
9. Hirt, C.W., Nichols, B.D.: Volume of fluid (VOF) method for the dynamics of free boundaries. J. Comput. Phys. **39**, 201–225 (1981)
10. Sod, G.A.: A survey of several finite difference methods for systems of nonlinear hyperbolic conservation laws. J. Comput. Phys. **27**, 1–31 (1978)
11. Igra, D., Takayama, K.: A study of shock wave loading on a cylindrical water column. Technical report, vol. 13, pp. 19–36. Institute of Fluid Science, Tohoku University (2001)
12. Theofanous, T.G., Li, J.G.: On the physics of aerobreakup. Phys. Fluids **20**, 052103 (2008)
13. Theofanous, T.G.: Aerobreakup of Newtonian and viscoelastic liquids. Annu. Rev. Fluid Mech. **43**, 661–690 (2011)
14. Menter, F.R.: Two-equation eddy-viscosity turbulence models for engineering applications. AIAA J. **32**, 1598–1605 (1994)
15. Reina, G.P., De Stefano, G.: Computational evaluation of wind loads on sun-tracking ground-mounted photovoltaic panel arrays. J. Wind Eng. Ind. Aerodyn. **170**, 283–293 (2017)
16. De Stefano, G., Natale, N., Reina, G.P., Piccolo, A.: Computational evaluation of aerodynamic loading on retractable landing-gears. Aerospace **7**, 68 (2020)
17. Natale, N., Salomone, T., De Stefano, G., Piccolo, A.: Computational evaluation of control surfaces aerodynamics for a mid-range commercial aircraft. Aerospace **7**, 139 (2020)
18. Iannelli, P., Denaro, F.M., De Stefano, G.: A deconvolution-based fourth-order finite volume method for incompressible flows on non-uniform grids. Int. J. Numer. Methods Fluids **43**, 431–462 (2003)

19. Rossano, V., De Stefano, G.: Computational evaluation of shock wave interaction with a cylindrical water column. Appl. Sci. **11**, 4934 (2021)
20. Regele, J.D., Vasilyev, O.V.: An adaptive wavelet-collocation method for shock computations. Int. J. Comput. Fluid Dyn. **23**, 503–518 (2009)
21. Hosseinzadeh-Nik, Z., Aslani, M., Owkes, M., Regele, J.D.: Numerical simulation of a shock wave impacting a droplet using the adaptive wavelet-collocation method. In: Proceedings of the ILASS-Americas 28th Annual Conference on Liquid Atomization and Spray Systems, Dearborn, MI, USA, May 2016
22. De Stefano, G., Vasilyev, O.V.: Wavelet-based adaptive large-eddy simulation with explicit filtering. J. Comput. Phys. **238**, 240–254 (2013)
23. De Stefano, G., Brown-Dymkoski, E., Vasilyev, O.V.: Wavelet-based adaptive large-eddy simulation of supersonic channel flow. J. Fluid Mech. **901**, A13 (2020)
24. De Stefano, G., Vasilyev, O.V.: Hierarchical adaptive eddy-capturing approach for modeling and simulation of turbulent flows. Fluids **6**, 83 (2021)
25. De Stefano, G., Vasilyev, O.V., Brown-Dymkoski, E.: Wavelet-based adaptive unsteady Reynolds-averaged turbulence modelling of external flows. J. Fluid Mech. **837**, 765–787 (2018)
26. Ge, X., Vasilyev, O.V., De Stefano, G., Hussaini, M.Y.: Wavelet-based adaptive unsteady Reynolds-averaged Navier-Stokes computations of wall-bounded internal and external compressible turbulent flows. In: Proceedings of the 2018 AIAA Aerospace Sciences Meeting, Kissimmee, FL, USA, January 2018
27. Ge, X., Vasilyev, O.V., De Stefano, G., Hussaini, M.Y.: Wavelet-based adaptive unsteady Reynolds-averaged Navier-Stokes simulations of wall-bounded compressible turbulent flows. AIAA J. **58**, 1529–1549 (2020)
28. Shen, B., Ye, Q., Tiedje, O., Domnick, J.: Simulation of the primary breakup of non-Newtonian liquids at a high-speed rotary bell atomizer for spray painting processes using a VOF-Lagrangian hybrid model. In: Proceedings of the 29th European Conference on Liquid Atomization and Spray Systems, Paris, France, September 2019
29. Nejadmalayeri, A., Vezolainen, A., De Stefano, G., Vasilyev, O.V.: Fully adaptive turbulence simulations based on Lagrangian Spatio-temporally varying wavelet thresholding. J. Fluid Mech. **749**, 794–817 (2014)

A MATLAB Implementation of Spline Collocation Methods for Fractional Differential Equations

Angelamaria Cardone$^{(\boxtimes)}$ ⓘ, Dajana Conte ⓘ, and Beatrice Paternoster ⓘ

Dipartimento di Matematica, Università di Salerno, Fisciano, Italy
{ancardone,dajconte,beapat}@unisa.it

Abstract. The present paper illustrates a MATLAB program for the solution of fractional differential equations. It is based on a spline collocation method on a graded mesh, introduced by Pedas and Tamme in [J. Comput. Appl. Math. **255**, 216–230 (2014)]. This is the first program proposed to implement spline collocation methods for fractional differential equations, and it is one of the few algorithms available in the literature for these functional equations. An explicit formulation of the method is derived, and the computational kernel is a nonlinear system to be solved at each time step. Such system involves some fractional integrals, whose analytical expression is given; their computation requires the knowledge of the coefficients of some polynomials and the evaluation of some special functions. The method is written in a compact matrix form, to improve the efficiency of the MATLAB implementation. The overall algorithm is outlined and then the attention is focused on some routines, which are given. In particular, some MATLAB native routines are used to evaluate special functions and to compute the coefficients of some polynomials. The complete list of the input and output parameters is available. Finally, an example of usage of the MATLAB program on a test problem is provided and some numerical experiments are shown.

Keywords: Fractional differential equations · Spline collocation · Algorithm · Implementation · MATLAB

1 Introduction

The aim of the present paper is to describe a MATLAB implementation of a spline collocation method for the solution of a nonlinear fractional differential equation (FDE):

$$\begin{cases} D^\alpha y(t) = f(t, y(t)), \, 0 \le t \le b, \\ y^{(i)}(0) = \gamma_i, \qquad i = 0, \dots, n-1, \end{cases} \tag{1}$$

Authors are members of the INdAM Research group GNCS and are supported by GNCS-INDAM project. A. Cardone and Dajana Conte are supported by PRIN2017-MIUR project.

© Springer Nature Switzerland AG 2021
O. Gervasi et al. (Eds.): ICCSA 2021, LNCS 12949, pp. 387–401, 2021.
https://doi.org/10.1007/978-3-030-86653-2_29

where $n - 1 < \alpha < n$, $n \in \mathbb{N}$; $b, \gamma_i \in \mathbb{R}$; $b > 0$. We assume that f is a continuous real-valued function defined in $[0, b]$ and that the fractional derivative is meant in the Caputo sense: [16, 21, 32]:

$$D^\alpha y(t) = \frac{1}{\Gamma(n - \alpha)} \int_0^t \frac{y^{(n)}(s)}{(t - s)^{\alpha + 1 - n}} ds.$$

Under suitable hypotheses, there exists a unique continuous solution of the initial value problem (IVP) (1). However the solution is generally not smooth around the initial point 0 [16, 21, 29, 32].

In the last twenty years an increasing number of fractional differential models have been proposed in various fields, e.g. anomalous diffusion [24], anomalous transport models [3], viscoelastic materials [36], option pricing models [34]. Therefore, great attention was paid also to the numerical solution of FDEs and many methods have been proposed so far, e.g. Adomian decomposition methods [14], Petrov-Galerkin methods [25], product integration methods [20], fractional linear multistep methods [23], time-splitting schemes [6], collocation methods [1, 11, 22, 29, 30], spectral methods [4, 37]. In particular, here we focus our attention on spline collocation methods, which is a very popular class of methods for ODEs and Volterra integral equations [2, 7, 8, 13, 15, 26]. In the application to FDEs, spline collocation methods are highly accurate and have also good stability properties, as proved in [9–12, 27–30].

On the other side, the development of mathematical software for FDEs is at an initial stage. An updated overview is given in [18] and some other references are [19, 31, 33, 35]. In this scenario, our aim is to offer a new software tool based on a spline collocation method proposed in [29] and also studied in [9]. This software was applied in [9, 11, 12] for numerical experiments, but was nowhere illustrated so far.

The paper is organized as follows. Section 2 illustrates the spline collocation method introduced in [29] and gives a compact matrix form of it, suitable for implementation. We illustrate the mathematical implementation in Sect. 3, i.e. the main program, the auxiliary routines, the special MATLAB functions used in the codes. Input and output parameters are described is in Sect. 4. Section 5 contains an example of use of the MATLAB program and some numerical experiments. Some conclusion are drawn in Sect. 6.

2   The Spline Collocation Method

The solution of (1) can be written as

$$y = J^\alpha z + Q, \tag{2}$$

where $z = D^\alpha y$ is the solution of the nonlinear equation

$$z = f(t, J^\alpha z + Q). \tag{3}$$

In the previous relations:

$$(J^\alpha z)(t) = \frac{1}{\Gamma(\alpha)} \int_0^t (t-s)^{\alpha-1} z(s)\, ds, \quad t > 0, \tag{4}$$

$$Q(t) = \sum_{i=0}^{\lceil \alpha \rceil - 1} \frac{\gamma_i}{i!} t^i, \tag{5}$$

with $\lceil \alpha \rceil$ equal to the smallest integer not less than α.

In [29], authors solve problem (3) by a one-step collocation method and then compute the solution of (1) by (2). In particular, the approximate collocation solution $v(t) \in S_k^{(-1)}(I_N)$, with

$$S_k^{(-1)}(I_N) = \left\{ v : v|_{\sigma_j} \in \pi_k,\, j = 1, \dots, N \right\}, \tag{6}$$

where π_k is the space of algebraic polynomials of degree not exceeding k.

Given a mesh of points $0 = t_0 < t_1 < \cdots < t_N = b$ with $h_j = t_j - t_{j-1}$ and a set of collocation abscissae $0 \le \eta_1 < \dots < \eta_m \le 1$, the collocation solution $v(t)$ is represented as follows:

$$v(t) = \sum_{l=1}^{N} \sum_{k=1}^{m} z_{lk} L_{lk}(t), \quad t \in [0, b], \tag{7}$$

where $z_{jk} = v(t_{jk})$ and

$$L_{lk}(t) = \begin{cases} k-\text{th Lagrange fund. pol. wrt to } \{t_{li}\}_{i=1}^m, & t \in [t_{l-1}, t_l] \\ \\ 0 & \text{otherwise} \end{cases}$$

$v(t)$ verifies equation (3) at the collocation points $t_{jk} = t_{j-1} + h_j \eta_k,\, j = 1, \dots, N, k = 1, \dots, m$, i.e.:

$$z_{jk} = f\left(t_{jk}, (J^\alpha v)(t_{jk}) + Q(t_{jk})\right), \quad k = 1, \dots, m, \tag{8}$$

$j = 1, \dots, N$. By definition of $L_{lk}(t)$, the system (8) may be recast as:

$$z_{jk} = f\left(t_{jk}, \sum_{\mu=1}^{m} (J^\alpha L_{j\mu})(t_{jk}) z_{j\mu} + \sum_{l=1}^{j-1} \sum_{\mu=1}^{m} (J^\alpha L_{l\mu})(t_{jk}) z_{l\mu} + Q(t_{jk}) \right), \tag{9}$$

$k = 1, \dots, m$.

To ensure the fastest error decay, a suitable graded mesh must be adopted, i.e.

$$t_j = b \left(\frac{j}{N} \right)^r, \quad j = 0, \dots, N. \tag{10}$$

The best value of the grading exponent $r \ge 1$ depends on the smoothness of the problem data (compare [12, 29]).

In [9] it is provided an explicit expression for the fractional integrals appearing in (9). Namely:

$$(J^\alpha L_{j\mu})(t_{jk}) = h_j^\alpha \sum_{\nu=0}^{m-1} a_\nu^{(\mu)} \eta_k^{\nu+\alpha} \frac{\Gamma(1+\nu)}{\Gamma(1+\nu+\alpha)}, \tag{11}$$

$$(J^\alpha L_{l\mu})(t_{jk}) = \frac{1}{\Gamma(\alpha)} h_l^\alpha \sum_{\nu=0}^{m-1} a_\nu^{(\mu)}$$

$$\left(\frac{t_{j-1} + \eta_k h_j - t_{l-1}}{h_l} \right)^{\nu+\alpha} \mathrm{B}\left(\frac{h_l}{t_{j-1} + \eta_k h_j - t_{l-1}}; 1+\nu, \alpha \right). \tag{12}$$

Function $B(x; b, c)$ is the incomplete beta function:

$$\mathrm{B}(x; a, b) := \int_0^x \sigma^{a-1} (1-\sigma)^{b-1} \, d\sigma,$$

while $\{a_\nu^{(\mu)}\}_{\nu=0}^{m-1}$ are the coefficients of the following Lagrange polynomial:

$$\prod_{\substack{i=1 \\ i\neq\mu}}^{m} \frac{\tau - \eta_i}{\eta_\mu - \eta_i} = \sum_{\nu=0}^{m-1} a_\nu^{(\mu)} \tau^\nu, \qquad \mu = 1, \ldots, m. \tag{13}$$

We observe that the coefficients of a polynomial $p(x) = q(x)r(x)$ are found by convolving the vectors $\mathbf{q}$ and $\mathbf{r}$ of the coefficients of polynomials $q(x)$ and $r(x)$, respectively. Therefore, coefficients $a_\nu^{(\mu)}$ can be computed by iteratively computing the convolution products of vectors $\left[\dfrac{1}{\eta_\mu - \eta_i}, \dfrac{-\eta_i}{\eta_\mu - \eta_i} \right]$, $i = 1, \ldots, m$, $i \neq \mu$ (cfr. (13)). This technique is applied in the routine `matrix Lagrange.m` (see Fig. 2).

The matrix form of system (9) is the following

$$\mathbf{z}_j = f\left(\mathbf{t}_j, h_j^\alpha \mathbf{A} \mathbf{z}_j + \sum_{l=1}^{j-1} h_l^\alpha \mathbf{E}^{[l,j]} \mathbf{z}_l + \mathbf{q}_j \right), \tag{14}$$

where

$$\mathbf{z}_j = [z_{j1}, \ldots, z_{jm}]^T, \quad \mathbf{t}_j = [t_{j1}, \ldots, t_{jm}]^T, \quad \mathbf{q}_j = [Q(t_{j1}), \ldots, Q(t_{jm})]^T, \tag{15}$$

$$(f(\mathbf{t}_j, \mathbf{x}))_k = f(t_{jk}, x_k), k = 1, \ldots, m, \mathbf{x} \in \mathbb{R}^m,$$

$$A_{k\mu} = \sum_{\nu=0}^{m-1} a_\nu^{(\mu)} \eta_k^{\nu+\alpha} \frac{\Gamma(1+\nu)}{\Gamma(1+\nu+\alpha)}, \tag{16}$$

$$E_{k\mu}^{[l,j]} = \frac{1}{\Gamma(\alpha)} \sum_{\nu=0}^{m-1} a_\nu^{(\mu)} \left(\frac{t_{j-1} + \eta_k h_j - t_{l-1}}{h_l} \right)^{\nu+\alpha} \mathrm{B}\left(\frac{h_l}{t_{j-1} + \eta_k h_j - t_{l-1}}; 1+\nu, \alpha \right), \tag{17}$$

$$k, \mu = 1, \ldots, m.$$

Once the approximate solution $v(t)$ of problem (3) has been computed, the numerical solution $y_N \approx y$ of problem (1) is found via (2), i.e.:

$$y_N = J^\alpha v + Q.$$

Therefore

$$y_N(t) = \sum_{\mu=1}^{m} (J^\alpha L_{j\mu})(t) z_{j\mu} + \sum_{l=1}^{j-1} \sum_{\mu=1}^{m} (J^\alpha L_{l\mu})(t) z_{l\mu} + Q(t), \quad t \in [t_{j-1}, t_j]. \quad (18)$$

By setting $\sigma = (t - t_j)/h_j$, the fractional integrals in (18) may be computed similarly as (11) and (12). Then, we obtain:

$$y_N(t) = h_j^\alpha \mathbf{b}^T \mathbf{z}_j + \sum_{l=1}^{j-1} h_l^\alpha (\mathbf{b}^{[l,j]})^T \mathbf{z}_l, \quad t \in [t_{j-1}, t_j], \quad (19)$$

where m-dimensional vectors $\mathbf{b}$ and $\mathbf{g}^{[l,j]}$ are defined as follows:

$$\mathbf{b}_\mu = \sum_{\nu=0}^{m-1} a_\nu^{(\mu)} \sigma^{\nu+\alpha} \frac{\Gamma(1+\nu)}{\Gamma(1+\nu+\alpha)}, \quad (20)$$

$$\mathbf{g}_\mu^{[l,j]} = \frac{1}{\Gamma(\alpha)} \sum_{\nu=0}^{m-1} a_\nu^{(\mu)} \left(\frac{t_{j-1} + \sigma h_j - t_{l-1}}{h_l} \right)^{\nu+\alpha} B \left(\frac{h_l}{t_{j-1} + \sigma h_j - t_{l-1}}; 1+\nu, \alpha \right), \quad (21)$$

$\mu = 1, \ldots, m.$

3   The MATLAB Implementation

The main program fcoll.m is written in Fig. 1, auxiliary functions are shown in Figs. 2, 3, 4, 5, 6 and 7, auxiliary MATLAB built in functions are listed in Table 1. Input and output arguments are described in Sect. 4.

Lines 2 and 3 of the main program fcoll.m construct the vector t containing the points of the graded mesh (10) and the vector h of the stepsizes, by using the MATLAB function diff. The line 4 makes use of the auxiliary function matrix_Lagrange.m which computes the matrix $Alagr$, whose components contain the coefficients of Lagrange polynomial (13).

$$Alagr(\mu, \nu) = a_{\nu-1}^{(\mu)}.$$

At line 5, function matrix_A.m constructs matrix $\mathbf{A}$ (defined in (16)). Vector $\mathbf{b}$ defined in (20), is computed in correspondence of $\sigma = 1$, by function vector_b.m, at line 6.

The for loop at lines 11–20 constructs and solves the nonlinear system (14); evaluates the approximate solution at the mesh points $y_N(t_0) = \gamma_0$, $y_N(t_1)$, ..., $y_N(t_N)$, by formula (19). The nonlinear system is computed by function F.m (see Fig. 7), whose input argument Bj (equal to the summation in (14) is constructed by function lag.m at line 13, and input argument Qj (array equal

```
1 function [t,y]=fcoll(f,b,gam,alpha,eta,r,N)
2     t=b*([0:N]'/N).^r;
3     h=diff(t);
4     Alagr=matrix_Lagrange(eta);
5     A=matrix_A(alpha,eta,Alagr);
6     bvect=vector_b(alpha,Alagr);
7     options=optimset('TolFun',1e-14,'TolX',1e-14,'Display','off');
8     m=length(eta);
9     Z=zeros(m,N); y=zeros(N+1,1); y(1)=gam(1);
10    iniz=feval(f,t(1)+eta*h(1),gam(1)*ones(m,1))
11    for j=1:N
12        tj=t(j)+eta*h(j);
13        Bj=lag(Z,Alagr,t,h,eta,alpha,j);
14        Qj=Q(tj,alpha,gam);
15        Z(:,j)=fsolve(@system_F,iniz,options,f,tj,A,Bj,Qj,h,j,alpha);
16        Wj=lag_y(Z,Alagr,t,h,alpha,j);
17        Qjp1=Q(t(j+1),alpha,gam);
18        y(j+1)=h(j)^(alpha)*(bvect')*Z(:,j)+Wj+Qjp1;
19        iniz=Z(:,j);
20    end
21 end
```

Fig. 1. Main program `fcoll.m`

to the vector $\mathbf{q}_j$ defined in (15)) is constructed at line 14 by function `Q.m`. The nonlinear system (14) is solved by MATLAB inbuilt function `fsolve`, with high accuracy requested (10^{-14}) and initial value equal to the approximate solution in the previous time step. Finally, the approximate solution is derived by (19), where the summation is computed by function `lag_y.m`.

In all the functions, the formulation of the method in terms of products of vectors and matrices was adopted, to obtain a gain in efficiency of the overall algorithm.

4   Input and Output Parameters

The MATLAB code fcoll.m has the input arguments `f,b,gam,alpha,eta,r,N` and the output arguments `t,y`. The type and the meaning of each of them are illustrated in Subsects. 4.1 and 4.2.

4.1   Input Parameters

1. **f** - function handle or string containing name of m-file
   f must return the value of the function $f(t, y)$ at a given point (t, y).

```
1 function [Alagr]=matrix_Lagrange(eta)
2    m=length(eta);
3    mu=1;
4    a=[1/(eta(mu)-eta(2)),-eta(2)/(eta(mu)-eta(2))];
5    for i=3:m
6        b=[1/(eta(mu)-eta(i)),-eta(i)/(eta(mu)-eta(i))];
7        a=conv(a,b);
8    end
9    Alagr(mu,:)=a;
10   for mu=2:m
11       a=[1/(eta(mu)-eta(1)),-eta(1)/(eta(mu)-eta(1))];
12       for i=[2:mu-1,mu+1:m]
13           b=[1/(eta(mu)-eta(i)),-eta(i)/(eta(mu)-eta(i))];
14           a=conv(a,b);
15       end
16       Alagr(mu,:)=a;
17   end
18   Alagr=fliplr(Alagr);
19 end
```

Fig. 2. Function `matrix_Lagrange.m`

```
1 function[A]=matrix_A(alpha,eta,Alagr)
2    m=length(eta);
3    interv=[0:m-1];
4    w=gamma(interv+1)./gamma(interv+1+alpha);
5    A=((eta.^((alpha+interv))).*w)*Alagr';
6 end
```

Fig. 3. Function `matrix_A.m`

```
1 function[b]=vector_b(alpha,Alagr)
2    m=size(Alagr,1);
3    interv=[0:m-1]';
4    w=gamma(interv+1)./gamma(interv+1+alpha);
5    b=Alagr*((1.^((alpha+interv))).*w);
6 end
```

Fig. 4. Function `vector_b.m`

```
1   function Bj=lag(Z,Alagr,t,h,eta,alpha,j)
2     m=length(eta);
3     Bj=zeros(m,1);
4     interv=[0:m-1];
5     w=gamma(interv+1)./gamma(interv+1+alpha);
6     for lambda=1:j-1
7         d=(t(j)-t(lambda)+eta*h(j))/h(lambda);
8         for k=1:m
9             W(k,:)=betainc(1/d(k),1+interv,alpha);
10        end
11        E=(((d.^((alpha+interv))).*w).*W)*Alagr';
12        Bj=Bj+(h(lambda)^alpha)*E*Z(:,lambda);
13    end
14 end
```

Fig. 5. Function lag.m

```
1 function z=Q(t,alpha,gam)
2   interv=0:ceil(alpha)-1;
3   z=sum(((t.^interv)./factorial(interv)).*(gam(interv+1).'),2);
4 end
```

Fig. 6. Function Q.m

```
1   function F=system_F(x,f,tj,A,Bj,Qj,h,j,alpha)
2     F=x-feval(f,tj,h(j)^(alpha)*A*x+Bj+Qj);
3   end
```

Fig. 7. Function system_F.m

```
1   function Wj=lag_y(Z,Alagr,t,h,alpha,j)
2     Wj=0;
3     m=size(Alagr,1);
4     interv=[0:m-1]';
5     w=gamma(interv+1)./gamma(interv+1+alpha);
6     for lambda=1:j-1
7         d=(t(j+1)-t(lambda))/h(lambda);
8         W=betainc(1/d,1+interv,alpha);
9         g=Alagr*(((d.^((alpha+interv))).*w).*W);
19        Wj=Wj+(h(lambda)^alpha)*g'*Z(:,lambda);
11    end
12 end
```

Fig. 8. Function lag_y.m

Table 1. Auxiliary MATLAB functions adopted in the algorithm in alphabetical order

Function	Task
betainc	The incomplete beta function
ceil	Function $\lceil \cdot \rceil$
conv	Convolution of two vectors
diff	Differences between adjacent elements of array
gamma	The gamma function
factorial	The factorial
fliplr	Flips array left to right
fsolve	Solves a nonlinear system

```
[result] = f(t,y)
```

Input Parameters

- **t** - double scalar
  The current value of the independent variable t.
- **y** - double scalar
  The current value of the independent variable y.

Output Parameters

- **result** - double scalar
  The value of $f(t,y)$.

2. **b** - double scalar
   b, the end point of the integration interval $[0, b]$.
3. **gam** - double array
   The vector of the initial values $[\gamma_0, \ldots, \gamma_{n-1}]^T$ of the IVP (1).
   Constraint: The length n of the array **gam** should be equal to $\lceil \alpha \rceil$.
4. **alpha** - double scalar
   α, the fractional index.
5. **eta(m)** - double array
   eta is equal to the vector $[\eta_1, \ldots, \eta_m]^T$ of the collocation parameters.
   Constraint: $0 \leq$ **eta(1)** $\leq \cdots \leq$ **eta(m)**.
6. **r** - double scalar
   r, the grading exponent.
7. **N** - double scalar
   N, the number of mesh points.

4.2 Output Parameters

1. **t(N+1)** - double array
   t is equal to the graded mesh $[t_0, \ldots, t_N]^T$, defined in (10).

2. **Y(N+1)** - double array
   $Y(j)$ is the approximate value of the solution $y(t_j)$.

5   Numerical Examples

For the users' convenience, we report an example of usage of our program, on the Test problem 1 (see Figs. 9 and 10). Then, we compare the exact and the numerical solution, compute the absolute error for Test problems 1, 2 and 3 (see Figs. 11, 12 and 13). In all cases, we applied the spline collocation method (14) (19) with $m = 2$, collocation parameters $\eta = \left[\frac{3 \pm \sqrt{3}}{6}\right]$, the grading exponent r suitably chosen in order to obtain superconvergence (compare [9,12,29]). The number of the mesh points is set to $N = b \cdot 2^5$.

Test Problem 1 [29]

$$D^{1/2}y(t) = y^2(t) + \frac{1}{\Gamma(1.5)}t^{0.5} - t^2, \ t \in [0,1] \quad y(0) = 0,$$

The exact solution is $y(t) = t$. The value of the grading exponent is $r = 2.5$.

Test Problem 2 [17]

$$D^\alpha y(t) = \frac{40320}{\Gamma(9-\alpha)}t^{8-\alpha} - 3\frac{\Gamma(5+\frac{\alpha}{2})}{\Gamma(5-\frac{\alpha}{2})}t^{4-\frac{\alpha}{2}} + \left(\frac{3}{2}t^{\frac{\alpha}{2}} - t^4\right)^3 \quad t \in [0,1],$$

$$+\frac{9}{4}\Gamma(\alpha+1) - (y(t))^{\frac{3}{2}},$$

$$y(0) = 0,$$

```
1 f=@(t,y) y.^2+1/gamma(1.5)*t.^0.5-t.^2;
2 b=10;
3 gam=[0];
4 alpha=1/2;
5 eta=[(3-sqrt(3))/6; 1-((3-sqrt(3))/6)];
6 r=2.6;
7 N=b/2^-5;
8 [t,y]=fcoll(f,b,gam,alpha,eta,r,N);
9 err=abs(y(end)-b)
```

Fig. 9. Script: test_example.m

```
>> test_example

err =

    1.2416e-05
```

Fig. 10. Output: test_example.m

$\alpha = 1/2$. Exact solution $y = t^8 - 3t^{4+\alpha/2} + \frac{9}{4}t^\alpha$. The method is applied with $r = 1.8$

Test Problem 3 [5]

$$D^\alpha y(t) = \lambda y + \rho y(1 - y^2) + g(t), \quad t \in (0, 8], \qquad y(0) = 2,$$

with $\alpha = 0.15$, $\lambda = -3$, $\rho = 0.8$. $g(t)$ is set in such a way that the solution is

$$y(t) = y_0 + \sum_{k=1}^{6} t^{\sigma_k}, \text{ with } \sigma_k = k\alpha, \ k = 1, \ldots, 5, \text{ and } \sigma_6 = 2 + \alpha. \text{ The chosen}$$

value of the grading exponent is $r = 7.17$.

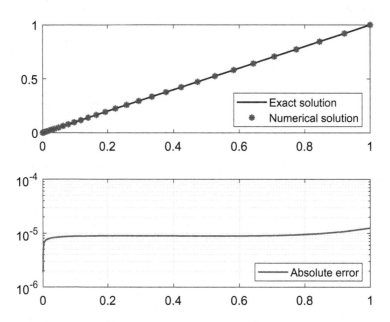

Fig. 11. Test example 1. Numerical and exact solution (on the top), absolute error (at the bottom).

6 Conclusions

We proposed a MATLAB algorithm to solve FDEs by spline collocation methods. We obtained an explicit formulation of the method, and explained how to compute some fractional integrals with high accuracy. To exploit MATLAB efficiency with matrix manipulation, we derived a matrix formulation of the method. We listed all the codes of the program, the input and output parameters. An example of use was included and some numerical examples were provided.

A further development of this work may regard the implementation of two-step collocation methods [10, 11], which will require a suitable starting procedure, too.

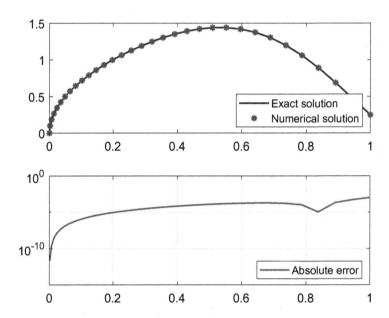

Fig. 12. Test example 2. Numerical and exact solution (on the top), absolute error (at the bottom).

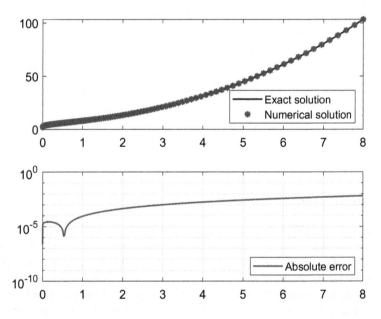

Fig. 13. Test example 3. Numerical and exact solution (on the top), absolute error (at the bottom).

References

1. Blank, L.: Numerical treatment of differential equations of fractional order. Technical report, University of Manchester, Department of Mathematics (1996). Numerical Analysis Report
2. Brunner, H.: Collocation methods for Volterra integral and related functional differential equations. Cambridge Monographs on Applied and Computational Mathematics, vol. 15. Cambridge University Press, Cambridge (2004). https://doi.org/10.1017/CBO9780511543234
3. Bueno-Orovio, A., Kay, D., Grau, V., Rodriguez, B., Burrage, K.: Fractional diffusion models of cardiac electrical propagation: role of structural heterogeneity in dispersion of repolarization. J. R. Soc. Interface **11**(97), 20140352 (2014)
4. Burrage, K., Cardone, A., D'Ambrosio, R., Paternoster, B.: Numerical solution of time fractional diffusion systems. Appl. Numer. Math. **116**, 82–94 (2017). https://doi.org/10.1016/j.apnum.2017.02.004
5. Cao, W., Zeng, F., Zhang, Z., Karniadakis, G.E.: Implicit-explicit difference schemes for nonlinear fractional differential equations with nonsmooth solutions. SIAM J. Sci. Comput. **38**(5), A3070–A3093 (2016). https://doi.org/10.1137/16M1070323
6. Cao, W., Zhang, Z., Karniadakis, G.E.: Time-splitting schemes for fractional differential equations I: smooth solutions. SIAM J. Sci. Comput. **37**(4), A1752–A1776 (2015). https://doi.org/10.1137/140996495
7. Cardone, A., Conte, D., Paternoster, B.: A family of multistep collocation methods for Volterra Integro-differential equations. AIP Conf. Proc. **1168**, 358–361 (2009). https://doi.org/10.1063/1.3241469
8. Cardone, A., Conte, D.: Multistep collocation methods for Volterra Integro-differential equations. Appl. Math. Comput. **221**, 770–785 (2013). https://doi.org/10.1016/j.amc.2013.07.012
9. Cardone, A., Conte, D.: Stability analysis of spline collocation methods for fractional differential equations. Math. Comput. Simulat. **178**, 501–514 (2020). https://doi.org/10.1016/j.matcom.2020.07.004
10. Cardone, A., Conte, D., Paternoster, B.: Stability analysis of two-step spline collocation methods for fractional differential equations, submitted
11. Cardone, A., Conte, D., Paternoster, B.: Two-step collocation methods for fractional differential equations. Discrete Contin. Dyn. Syst. Ser. B **23**(7), 2709–2725 (2018). https://doi.org/10.3934/dcdsb.2018088
12. Cardone, A., Conte, D., Paternoster, B.: Numerical treatment of fractional differential models. In: Abdel Wahab, M. (ed.) FFW 2020 2020. LNME, pp. 289–302. Springer, Singapore (2021). https://doi.org/10.1007/978-981-15-9893-7_21
13. Conte, D., Paternoster, B.: Multistep collocation methods for Volterra Integral equations. Appl. Numer. Math. **59**(8), 1721–1736 (2009). https://doi.org/10.1016/j.apnum.2009.01.001
14. Daftardar-Gejji, V., Jafari, H.: Adomian decomposition: a tool for solving a system of fractional differential equations. J. Math. Anal. Appl. **301**(2), 508–518 (2005). https://doi.org/10.1016/j.jmaa.2004.07.039
15. D'Ambrosio, R., Paternoster, B.: Multivalue collocation methods free from order reduction. J. Comput. Appl. Math. **387**, 112515 (2021). https://doi.org/10.1016/j.cam.2019.112515
16. Diethelm, K.: The Analysis of Fractional Differential Equations. Lecture Notes in Mathematics, vol. 2004. Springer, Berlin (2010)

17. Diethelm, K., Ford, N.J., Freed, A.D.: Detailed error analysis for a fractional Adams method. Numer. Algorithms **36**(1), 31–52 (2004). https://doi.org/10.1023/B:NUMA.0000027736.85078.be
18. Garrappa, R.: Numerical solution of fractional differential equations: a survey and a software tutorial. Mathematics **6**(2), 16 (2018). https://doi.org/10.3390/math6020016
19. Garrappa, R.: Short Tutorial: Solving Fractional Differential Equations by Matlab Codes. Department of Mathematics University of Bari, Italy (2014)
20. Garrappa, R., Popolizio, M.: On accurate product integration rules for linear fractional differential equations. J. Comput. Appl. Math. **235**(5), 1085–1097 (2011). https://doi.org/10.1016/j.cam.2010.07.008
21. Kilbas, A.A., Srivastava, H.M., Trujillo, J.J.: Theory and Applications of Fractional Differential Equations, North-Holland Mathematics Studies, vol. 204. Elsevier Science B.V, Amsterdam (2006)
22. Li, X.: Numerical solution of fractional differential equations using cubic B-spline wavelet collocation method. Commun. Nonlinear Sci. Numer. Simul. **17**(10), 3934–3946 (2012). https://doi.org/10.1016/j.cnsns.2012.02.009
23. Lubich, C.: Fractional linear multistep methods for Abel-Volterra integral equations of the second kind. Math. Comp. **45**(172), 463–469 (1985). https://doi.org/10.2307/2008136
24. Metzler, R., Klafter, J.: The random walk's guide to anomalous diffusion: a fractional dynamics approach. Phys. Rep. **339**(1), 77 (2000). https://doi.org/10.1016/S0370-1573(00)00070-3
25. Mustapha, K., Abdallah, B., Furati, K.: A discontinuous Petrov-Galerkin method for time-fractional diffusion equations. SIAM J. Numer. Anal. **52**(5), 2512–2529 (2014). https://doi.org/10.1137/140952107
26. Paternoster, B.: Phase-fitted collocation-based Runge-Kutta-Nystrom method. Appl. Numer. Math. **35**(4), 339–355 (2000). https://doi.org/10.1016/S0168-9274(99)00143-9
27. Pedas, A., Tamme, E.: On the convergence of spline collocation methods for solving fractional differential equations. J. Comput. Appl. Math. **235**(12), 3502–3514 (2011). https://doi.org/10.1016/j.cam.2010.10.054
28. Pedas, A., Tamme, E.: Spline collocation methods for linear multi-term fractional differential equations. J. Comput. Appl. Math. **236**(2), 167–176 (2011). https://doi.org/10.1016/j.cam.2011.06.015
29. Pedas, A., Tamme, E.: Numerical solution of nonlinear fractional differential equations by spline collocation methods. J. Comput. Appl. Math. **255**, 216–230 (2014). https://doi.org/10.1016/j.cam.2013.04.049
30. Pedas, A., Tamme, E.: Spline collocation for nonlinear fractional boundary value problems. Appl. Math. Comput. **244**, 502–513 (2014). https://doi.org/10.1016/j.amc.2014.07.016
31. Petrás, I.: Fractional derivatives, fractional integrals, and fractional differential equations in MATLAB. In: Assi, A.H. (ed.) Engineering Education and Research Using MATLAB, Chapter 10. IntechOpen, Rijeka (2011). https://doi.org/10.5772/19412
32. Podlubny, I.: Fractional differential equations, Mathematics in Science and Engineering, vol. 198. Academic Press Inc, San Diego, CA (1999). An introduction to fractional derivatives, fractional differential equations, to methods of their solution and some of their applications

33. Sowa, M., Kawala-Janik, A., Bauer, W.: Fractional differential equation solvers in Octave/MATLAB. In: 2018 23rd International Conference on Methods & Models in Automation & Robotics (MMAR), pp. 628–633. IEEE (2018)
34. Wang, W., Chen, X., Ding, D., Lei, S.L.: Circulant preconditioning technique for barrier options pricing under fractional diffusion models. Int. J. Comput. Math. **92**(12), 2596–2614 (2015). https://doi.org/10.1080/00207160.2015.1077948
35. Wei, S., Chen, W.: A MATLAB toolbox for fractional relaxation-oscillation equations. arXiv preprint arXiv:1302.3384 (2013)
36. Yang, W., Chen, Z.: Fractional single-phase lag heat conduction and transient thermal fracture in cracked viscoelastic materials. Acta Mechanica **230**(10), 3723–3740 (2019). https://doi.org/10.1007/s00707-019-02474-z
37. Zayernouri, M., Karniadakis, G.E.: Exponentially accurate spectral and spectral element methods for fractional ODEs. J. Comput. Phys. **257**(part A), 460–480 (2014). https://doi.org/10.1016/j.jcp.2013.09.039

The k-Colorable Unit Disk Cover Problem

Monith S. Reyunuru[1], Kriti Jethlia[2], and Manjanna Basappa[2]($\boxtimes$) (iD)

[1] Amazon Development Center, Hyderabad, India
f20160006@hyderabad.bits-pilani.ac.in
[2] CSIS Department, Birla Institute of Technology and Science Pilani,
Hyderabad Campus, Hyderabad, India
{f20180223,manjanna}@hyderabad.bits-pilani.ac.in

Abstract. We consider the k-Colorable Discrete Unit Disk Cover (k-$CDUDC$) problem as follows. Given a parameter k, a set P of n points, and a set D of m unit disks, both sets lying in the plane, the objective is to compute a set $D' \subseteq D$ such that every point in P is covered by at least one disk in D' and there exists a function $\chi : D' \to C$ that assigns colors to disks in D' such that for any d and d' in D' if $d \cap d' \neq \emptyset$, then $\chi(d) \neq \chi(d')$, where C denotes a set containing k distinct colors.

For the k-$CDUDC$ problem, our proposed algorithms approximate the number of colors used in the coloring if there exists a k-colorable cover. We first propose a 4-approximation algorithm in $O(m^{7k}n \log k)$ time for this problem, where k is a positive integer. The previous best known result for the problem when $k = 3$ is due to the recent work of Biedl et al. [Computational Geometry: Theory & Applications, 2021], who proposed a 2-approximation algorithm in $O(m^{25}n)$ time. For $k = 3$, our algorithm runs in $O(m^{21}n)$ time, faster than the previous best algorithm, but gives a 4-approximate result. We then generalize the above approach to yield a family of ρ-approximation algorithms in $O(m^{\alpha k}n \log k)$ time, where $(\rho, \alpha) \in \{(4, 7), (6, 5), (7, 5), (9, 4)\}$. We also extend our algorithm to solve the k-Colorable Line Segment Disk Cover (k-$CLSDC$) and k-Colorable Rectangular Region Cover (k-$CRRC$) problems, in which instead of the set P of n points, we are given a set S of n line segments, and a rectangular region $\mathcal{R}$, respectively.

Keywords: Colorable unit disk cover · Approximation algorithm · Grid-partitioning

1 Introduction

Our motivation for studying the problem arises from practical applications in the frequency/channel assignment problem in wireless/cellular networks. In ad-hoc mobile networks, each host(station/tower) is equipped with a Radio-Frequency (RF) transceiver to provide reliable transmission inside a circular range, represented by a disk, within some distance. Each wireless client is equipped with corresponding receivers. The clients themselves are represented by a set of points

© Springer Nature Switzerland AG 2021
O. Gervasi et al. (Eds.): ICCSA 2021, LNCS 12949, pp. 402–417, 2021.
https://doi.org/10.1007/978-3-030-86653-2_30

P in a plane. The disks representing the range (which is presumably the same for all stations) of each potential host is represented by the set D. In the spirit of reducing interference in broadcast and other energy-saving measures, we aim to limit or reduce the number of different frequencies(channels) assigned to each, represented by coloring. Typically, (Wi-Fi) networks are built with 3 independent channels [4], hence the motivation for a study on the 3-$CDUDC$ problem. In the same spirit, we generalize the 3-$CDUDC$ to the k-$CDUDC$ problem, where $k > 0$ is an integer. We further generalize the problem by considering line segments and a continuous rectangular region as representing potential wireless clients (resp. the k-$CLSDC$ and k-$CRRC$ problems), instead of points.

1.1   Related Work

The 3-$CDUDC$ problem, to the best of our knowledge, was first studied by Biedl et al., [3]. They gave a 2-approximation algorithm in $O(nm^{25})$ time for the 3-$CDUDC$ problem. Their approach first partitions the plane into horizontal strips, solves the problem for every strip optimally, then returns the union of solutions of all strips. To solve the problem for any strip they show that at most a constant number of disks of an optimal solution intersect any vertical line. Based on this, they define a directed acyclic graph such that there exists a path from source to a destination corresponding to this optimal solution. In this paper, we attempt to improve upon this impractical $O(nm^{25})$ running time. Our approach, however, focuses on the specific geometric properties that arise from the dual conditionals of the problem statement. Although both of the approaches, initially, begin by dividing the plane, we recognize a unique bound that exists in our need to bound the colorability and provide a novel solution in the same regard.

A notion of *conflict-free coloring (CF-coloring)* was introduced by Even et al. [8]. and Smorodinsky [13]. In the *CF-coloring* problem we are given a set of points (representing client locations) and a set of base stations, the objective is to assign colors (representing frequencies) to the base stations such that any client lying within the range of at least one base station is covered by the base station whose color is different from the colors of the other base stations covering the client, and the number of colors used should be as minimum as possible. Here, the range of base stations is modeled as regions e.g., disks or other geometric objects. Even et al., [8] proved that $O(\log n)$ colors are always sufficient to *CF-color* a set of disks in the plane, and in the worst case, $\Omega(\log n)$ colors are required. Note that this *CF-coloring* of disks is different from our notion of k-colorable disk cover of points. In the former overlapping disks may be given the same color if they don't share a client, whereas in the k-$CDUDC$ overlapping disks must be colored with distinct colors. A generalization of *CF-coloring* is called a *k-fault-tolerant CF-coloring*. Cheilaris et al. [7] presented a polynomial-time $(5 - \frac{2}{k})$-approximation algorithm for the *k-fault-tolerant CF-coloring* in 1-dimensional space. Horev et al. [11] proved that $O(k \log n)$ colors are sufficient for any set of n disks in the plane. For *dynamic CF-coloring* and results on *CF-coloring* of other geometric objects, we refer to [5] and references therein.

A related problem [2] to the k-$CDUDC$ problem in the literature is the *Discrete Unit Disk Cover* ($DUDC$) problem. In the $DUDC$ problem, we are given a set P of n points and a set D of m unit disks, our goal is to select as the smallest number of disks from D as possible such that the union of these selected disks covers all points in P. As in the k-$CDUDC$, here also, the sets P and D can be considered as representing a set of wireless clients and a set of base stations or towers, respectively. The $DUDC$ problem is NP-hard and is a very well studied one. There is a polynomial time approximation scheme (PTAS) with impractical running time for this problem [12]. The current best approximation algorithm with reasonable running time is $(9 + \epsilon)$ for any $\epsilon > 0$ [2]. However, a series of approximation algorithms have been proposed for this problem by various authors over the past two decades, and a complete survey on this can be found in [10]. When a line segment is used to represent a potential wireless client, the $DUDC$ problem becomes a *Line Segment Disk Cover* ($LSDC$) problem. In a similar line, there is another variant of the $DUDC$ problem, a *Rectangular Region Cover (RRC)* problem, in which all the continuous set of points lying in a rectangular region represent wireless clients. All the available results for the $DUDC$ problem also extend to the $LSDC$ and RRC problems [1], with slightly different running time. We also extend our results for the k-$CDUDC$ problem to solve the colorable variants of the $LSDC$ and RRC problems, namely, the k-$CLSDC$ and k-$CRRC$ problems.

2   The k-CDUDC Problem

In this section we consider the following problem.

– k-*Colorable Discrete Unit Disk Cover* (k-$CDUDC$): Given a set P of n points, and a set D of m unit disks (of radius=1), both lying in the plane, and a parameter k, the objective is to compute a set $D' \subseteq D$ that covers all points in P such that the set D' can be partitioned into $\{D'_1, D'_2, \ldots, D'_k\}$, where for each $a \in \{1, 2, \ldots, k\}$ the disks in D'_a are pairwise disjoint, i.e., the disks in D' can be colored with at most k colors such that the overlapping disks receive distinct colors and every point in P is covered by a disk in D'.

As it was pointed out in [3] that there is a related problem, namely, *Unit Disk Chromatic Number (UDCN)* problem, that aims to color all nodes in a given unit disk graph with at most k colors. The $UDCN$ problem is NP-hard for any $k \geq 3$ [6]. Similar to Biedl et al. [3], we can center a set D of m unit disks in the plane such that there are at least $k + 1$ pairwise non-disjoint disks that have a common intersection region and a unit disk graph $G_D = (V_D, E_D)$ induced by D is connected. Let us then place a set P of n points in this intersection region. Now observe that the set P has a cover which is at most k-colorable, whereas the graph G_D is at least $(k+1)$-colorable. Hence, the k-$CDUDC$ problem is different from the $UDCN$ problem. Biedl et al. [3] showed that the *3-CDUDC* problem is NP-hard by carefully incorporating a set P of n points in the NP-hard proof of the $UDCN$ problem with $k = 3$ in [6]. This directly implies that the k-$CDUDC$

is NP-hard since the k-$CDUDC$ is a generalization of 3-$CDUDC$. It is also easy
to see that the k-$CDUDC$ problem belongs to the class NP, as follows: Here, the
certificate for any Yes instance of k-$CDUDC$ is a set of k distinct colors identified
by non-negative integers $1, 2, \ldots, k$, and a mapping $\chi : D' \to \{1, 2, \ldots, k\}$, where
$D' \subseteq D$. A polynomial time verifier checks if every point in P is covered by a
disk in D' and for every pair of disks $d, d' \in D'$ if $d \cap d' \neq \emptyset$, whether it is the
case that $\chi(d) \neq \chi(d')$.

2.1   A 4-Approximate Algorithm

Here, our algorithm is based on partitioning the plane containing points into a
grid and then determining bound on the number of unit disks that can participate
in any k-colorable covering of points lying within any square of the grid. We first
define a grid of width two units that partitions the plane into squared regions.
Each of these squared regions is a grid cell with a size 2×2. For simplicity assume
no point of P lies on the boundary of these grid cells. Let us associate a unique
ID id_C to each grid cell C as follows; let $p = (x_p, y_p)$ be a point in C and τ be
the grid width, then $id_C = (\lfloor \frac{x_p}{\tau} \rfloor, \lfloor \frac{y_p}{\tau} \rfloor)$, (see Fig. 1). Note that each grid cell has
a unique ID associated with it but multiple points can be associated with the
same ID (if they lie within the corresponding grid cell). Let id_{C_1} and id_{C_2} be any
two arbitrary grid cells with base points (x_1, y_1) and (x_2, y_2) respectively. We
define the greater than operator for an ID as follows: $(id_{C_1} > id_{C_2}) \iff ((x_1 >
x_2) \wedge (y_1 > y_2)) \vee ((x_1 = x_2) \wedge (y_1 > y_2)) \vee ((x_1 > x_2) \wedge (y_1 = y_2))$. Note that
our definition of id_C implies that the grid cells are indexed from bottom-left to
top-right, this operator simply indicates the order of iteration that is followed by
our algorithm. We move left to right row-wise. A pre-defined order is essential
to our handing-over logic.

C_1 $[0, 2]$	C_2 $[1, 2]$	C_1 $[2, 2]$
C_3 $[0, 1]$	C_4 $[1, 1]$	C_3 $[2, 1]$
C_1 $[0, 0]$	C_2 $[1, 0]$	C_1 $[2, 0]$

Fig. 1. Assignment of unique ID's id_C and color sets for a grid $\mathcal{G}$ with $\tau = 2$

Given a grid cell C', the following lemma provides a bound on the cardinality of the k-colorable unit disks covering all points lying within the grid cell C'. Let $D_{C'} \subseteq D$ be the set of k-colorable unit disks covering all points of P lying within the grid cell C'. The proof of the lemma is based on the observation that determining this bound is the same as determining a maximum number of disjoint unit disks that could potentially intersect C'.

Observation 1. *If C' is a grid cell of size $\tau \times \tau$, then the maximum number of pairwise disjoint unit disks that could potentially intersect C' is at most $2\tau + 2 + (\frac{\tau}{2})^2$ if τ is even, and is atmost $4 \times \lceil \frac{\tau}{2} \rceil + 4 + (\frac{\tau}{2})^2$ if τ is odd.*

Proof. We will provide an upper bound to the number of pairwise disjoint unit disks that can cover a square C' of side length τ. Let us prove this by considering the two cases: τ being even and odd.

Since we aim at bringing an upper bound to the number of disks that have a common intersection point with C', we divide the region of C' into two parts; the inner part of the square and the outer edge on which these common intersection points can lie. To maximize the number of disks, it is intuitive to keep them as far as possible to increase the spacing between disks and thereby trying to increase the number of disks.

C' **with Even Side Length:** When τ is a multiple of 2, it is quite intutive that a symmetric pattern is likely to give the best results. So we attempt two types of symmetric pattern.

Case 1: Considering the square C' to be symmetric along the vertical axis, we arrange the disks in two possible cases: either a disk is arranged with edge of C' as tangent such that the center of C' lies vertically above/below the disk (see Fig. 2a), or the vertical partition is tangent to some of the disks (see Fig. 2b). In the first case the maximum number of disks along horizontal part of outer edge would be $2 \times \frac{\tau}{2}$, since the diameter of disk is 2. In the second case it can be shown that the maximum number of disks along the horizontal part would be $2 \times (\frac{\tau}{2} + 1)$.

However, as we see for the case $\tau = 2$ (shown in Fig. 2a and Fig. 2b), both the cases break down to the same case since if for a given pair of parallel edges of C' if one of the above case is true, then for the other pair the other case stands true. By induction we can prove that this stands true for all the even values of $\tau > 2$. Through this we get an upper limit in the number of disks along the edges. For the inner area of C' the maximum number of disjoint disks are $(\frac{\tau}{2})^2$. So the maximum number of disjoint disks are $(\tau/2)^2 + 2 \times \tau/2 + 2 \times (\tau/2 + 1)$, which is $2\tau + 2 + (\tau/2)^2$.

Case 2: Considering C' to be diagonally symmetric, here we consider one of its diagonals as the symmetry line and place the disk such that their center is on the diagonal and intersects the square at its corner. As the disk covers a part of the edge of the square, say δ where $\delta < 1$ (i.e., radius of disk) and $\delta > 0$, apart from the case where the disks are arranged at the diagonals again (see Fig. 3a), the maximum number of disks on an edge is still equivalent to $\frac{\tau}{2}$. The number of

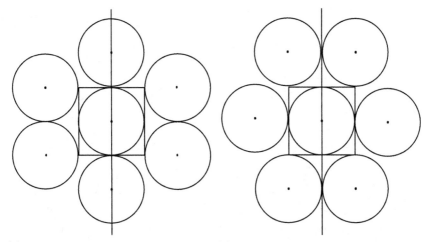

(a) Center of disk along the vertical axis (b) Vertical line dividing the square as tangent

Fig. 2. Possible symmetric arrangement along vertical axis for even values of τ.

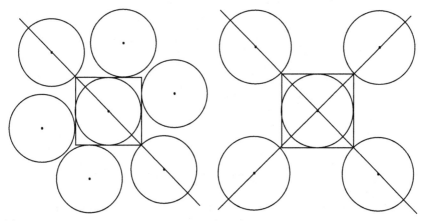

(a) Disks arranged along one of the di-(b) Disks arranged are along both the agonal diagonals

Fig. 3. Possible symmetric arrangement along diagonal for even values of τ.

disks within the interiors of $\mathcal{C}'$ as stated above in the previous case will remain same. Therefore, the total number of maximum possible pairwise disjoint disks are $2 + 4 \times (\frac{\tau}{2}) + (\frac{\tau}{2})^2$, which is again equivalent to $2\tau + 2 + (\frac{\tau}{2})^2$ (see Fig. 3a, Fig. 3b, Fig. 4a, Fig. 4b, and Fig. 4c for illustration of the case $\tau = 2$).

$\mathcal{C}'$ **with Odd Side Length:** When τ is not a multiple of 2, again it is quite intuitive that a symmetric pattern is likely to give the best results. So we attempt two types of symmetric pattern.

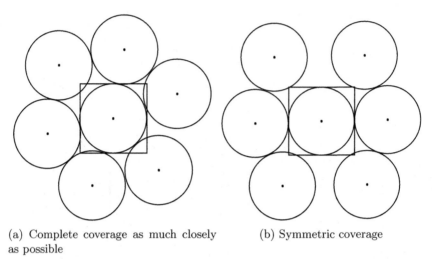

(a) Complete coverage as much closely as possible

(b) Symmetric coverage

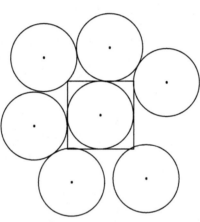

(c) Asymmetric coverage

Fig. 4. Various possibilities by trial and error for even values of τ.

Case 1: Like the case for even values of τ we consider the symmetric distribution along the horizontal and vertical axes. Again we have two possibilities either the center of disk lying along the axes or symmetric about the axes for both the pairs of edges. It can be shown that for both the pair of edges we would have only one amongst the two configurations at a time for getting the minimum number of disks. In the first case, $\lceil \frac{\tau}{2} \rceil$ disks can completely be accommodated on one edge and one disk as a common disk between two adjacent edges. So, there will be $4 \times \lceil \frac{\tau}{2} \rceil + 4$ disks in the exterior part for this case. The number of disks in the interior part as in the even case would be $(\frac{\tau}{2})^2$. So, the total becomes $4 \times \lceil \frac{\tau}{2} \rceil + 4 + (\frac{\tau}{2})^2$ (see Fig. 5a and Fig. 5b).

Case 2: When diagonally symmetric, the case is quite similar to the previous case with 4 disks at the corners and $\lceil \frac{\tau}{2} \rceil$ among the four edges of $\mathcal{C}'$. The total again is calculating to $4 \times \lceil \frac{\tau}{2} \rceil + 4 + (\frac{\tau}{2})^2$. $\qquad\qquad\Box$

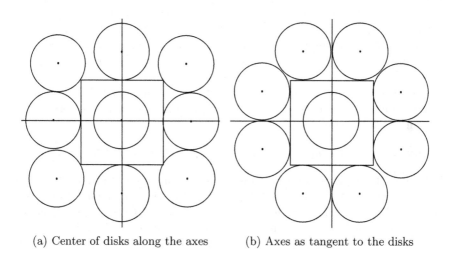

(a) Center of disks along the axes   (b) Axes as tangent to the disks

Fig. 5. Possible symmetric arrangement for odd values of τ.

Lemma 1. *If $\mathcal{C}'$ is a grid cell of size 2×2 and $D_{\mathcal{C}'} \subseteq D$ is a k-colorable solution for $P \cap \mathcal{C}'$, then $|D_{\mathcal{C}'}| \leq 7k$.*

Proof. From Observation 1 the cardinality of any set $S_{\mathcal{C}'}$ of pairwise disjoint unit disks intersecting with a grid cell $\mathcal{C}'$ is at most 7 as $\tau = 2$. Therefore, $|S_{\mathcal{C}'}| \leq 7$. Now consider another $k-1$ sets $S_{\mathcal{C}'}^1, S_{\mathcal{C}'}^2, \ldots, S_{\mathcal{C}'}^{k-1}$, each of which can either be a replica of the same collection of disks in $S_{\mathcal{C}'}$ or a rotation or transformation of $S_{\mathcal{C}'}$ such that disks within each set remain pairwise disjoint and intersect $\mathcal{C}'$. Hence, any k-colorable solution $D_{\mathcal{C}'} \subseteq S_{\mathcal{C}'} \cup S_{\mathcal{C}'}^1 \cup S_{\mathcal{C}'}^2 \cup \ldots \cup S_{\mathcal{C}'}^{k-1}$. Thus, the lemma follows. $\qquad\qquad\Box$

The outline of our algorithm (Algorithm 1) for computing a cover $D' \subseteq D$ of the points P is as follows. We first partition the rectangular region containing the objects in D and P into individual grid cells of size $\tau \times \tau$. By utilizing the bound obtained in Observation 1 (for e.g., for $\tau = 2$ the actual bound is in Lemma 1) we compute a k-colorable cover of the points lying in each grid cell, in an exhaustive manner. To ensure that there is no conflict in the overall aggregate solution, we use a handing-over logic. Only disks of any particular grid cell cover centered within the same grid cell are colored with the associated color set. If a disk is required to be a part of this grid cell cover, but is centered in another grid cell, it is handed-over to that grid cell. Based on the grid width τ and the diameter of the disk, we then define a coloring scheme χ that assigns a color to each disk in the union D' of all the individual grid cell covers computed.

Finally, we return the pair (D', χ). Since the diameter of the disks is fixed to be two units, the approximation factor of the algorithm is implied by the choice of the value τ. If the value of τ is 2, then a unit disk can participate in the k-colorable covers of points lying in four adjacent grid cells. Hence, we prove that Algorithm 1 is a 4-approximate algorithm (see Theorem 1). Later, we show that by varying the grid width τ, which results in a unit disk participating in more than four individual grid cell covers, we can obtain a family of algorithms with approximation factors corresponding to the choice of the value of τ (see Subsect. 2.2).

We now define any coloring function that assigns colors to disks to be conflict-free if for any pair of non-disjoint disks (i.e., overlapping disks) the colors assigned to them are different.

Lemma 2. *The coloring χ defined by Algorithm 1 is conflict-free.*

Proof. For the sake of contradiction, let us assume that there are two disks $d, d' \in D'$ such that $d \cap d' \neq \emptyset$, and $\chi(d) = \chi(d')$, where D' along with χ is the output of Algorithm 1. Since $d \cap d' \neq \emptyset$, the distance between the centers of d and d' is at most 2. Let the centers of d and d' be lying in the grid cells C and C', respectively. Observe that C and C' are either linearly or diagonally adjacent. If d and d' are chosen to cover points lying only in the respective grid cells, then d and d' are assigned colors from different color sets because the row and column numbers mod 2 in their ID's are not the same for both (see for-loop at Line 17) (contradicting that $\chi(d) = \chi(d')$). Therefore, the only possibility for color-conflict to arise between d and d' is that when both d and d' are centered in the same grid cell C, where d covers a point lying in the cell above C and d' covers a point lying in the cell below C and each disk is initially chosen by the respective grid cell by means of the algorithm (Note that a similar case can be studied for horizontally and diagonally opposite grid cells). As per our color scheme (Line 16–28), these grid cells are the nearest to have the same color set (say C_1) associated with them (see Fig. 1). Step 11 in the algorithm solves the conflict that arises in this case as follows. By means of the grid cell ID condition, disk d' is handed over to grid cell C as its ID is greater and will receive the color set associated with that cell (C_3 in this case, see Fig. 1). Disk d however will not be handed over, but retains a color from the color set C_1 (contradicting that $\chi(d) = \chi(d')$). Thus, the lemma follows.   □

Theorem 1. *Algorithm 1 is a 4-approximation algorithm that runs in $O(m^{7k} n \log k)$ time for the k-CDUDC problem.*

Proof. In line 2 of Algorithm 1, the rectangular region $\mathcal{R}$ is partitioned into $O(n)$ grid cells. For each grid cell C, let n_C denote the number of points of P lying in C, i.e., $n_C = |P \cap C|$. For each grid cell C we then enumerate all $O(m^7)$ subsets of D such that each such subset contains at most 7 pairwise disjoint disks intersecting with the cell C. In each iteration of the for-loop at line 3, among $\binom{cm^7}{k}$ subsets of at most $7k$ disks for some constant c, we compute a set D_C that covers all n_C

Algorithm 1. K_Colorable_Cover(P, D, k)

Input: A set P of n points, a set D of m unit disks in the plane, and an integer $k(> 0)$ such that $P \subset \cup_{d \in D} d$ and D can provide a k-colorable cover of the points in P.

Output: A k-colorable set $D' \subseteq D$ that covers all the points in $\mathcal{P}$ and a color mapping $\chi : D' \to \kappa$, where κ denotes the color set of distinct colors, and $|\kappa| \leq 4k$.

1: Let the points in P and disks in D be lying entirely within the first quadrant of the coordinate system, and $\mathcal{R}$ be an axis-aligned rectangular region containing P and D, whose left and bottom boundary lines coincide with the y- and x-axes of the coordinate system, respectively.

2: Define a grid $\mathcal{G}$ that partitions $\mathcal{R}$ such that each grid cell is of size 2×2 and for each point $p = (x_p, y_p)$ lying in such a cell $\mathcal{C}$, let the unique id associated with $\mathcal{C}$ be $id_{\mathcal{C}} = [\lfloor \frac{x_p}{2} \rfloor, \lfloor \frac{y_p}{2} \rfloor]$. For each such cell $\mathcal{C}$, we also define a handover set $H_{\mathcal{C}} \leftarrow \emptyset$.

    /* the grid cells in the following loop are considered in row-wise order from bottom-left to top-right, as defined in Subsection 2.1 */

3: **for** each grid cell $\mathcal{C}$ if $P \cap \mathcal{C} \neq \emptyset$ **do**

4:     **if** $H_{\mathcal{C}} = \emptyset$ **then**

5:         Let $D'' = \{d \in D \mid d \cap \mathcal{C} \neq \emptyset\}$

6:         Generate all subsets $D_1, D_2, \ldots, D_{O(m^7)} \subseteq D''$, each containing at most 7 pairwise disjoint disks, and among these, choose k subsets $S_{\mathcal{C}}^1, S_{\mathcal{C}}^2, \ldots, S_{\mathcal{C}}^k$, whose union covers all the points in $P \cap \mathcal{C}$.

7:     **else**

8:         Let $D'' = \{d \in D \mid d \cap \mathcal{C} \neq \emptyset, d \notin H_{\mathcal{C}}\}$.

9:         Generate all subsets $D_1, D_2, \ldots, D_{O(m^7)} \subseteq D''$, each containing at most 7 pairwise disjoint disks, and among these, choose k subsets $S_{\mathcal{C}}^1, S_{\mathcal{C}}^2, \ldots, S_{\mathcal{C}}^k$, whose union covers all the points in $P \cap \mathcal{C}$ but also contains all disks $d \in H_{\mathcal{C}}$.

    /* $|S_{\mathcal{C}}^1 \cup S_{\mathcal{C}}^2 \cup \ldots S_{\mathcal{C}}^k \cup H_{\mathcal{C}}| \leq 7k$ due to Lemma 1 */

10:     **end if**

11:     If any disk d in any subset $S_{\mathcal{C}}^i$ (for $i = 1, \ldots, k$) is centered in another grid cell $\mathcal{C}'$ whose ID $id_{\mathcal{C}'} > id_{\mathcal{C}}$, we remove that disk from $S_{\mathcal{C}}^i$ and add it to the handover set of that cell $H_{\mathcal{C}'}$.

12:     $D_{\mathcal{C}} \leftarrow S_{\mathcal{C}}^1 \cup S_{\mathcal{C}}^2 \cup \ldots \cup S_{\mathcal{C}}^k$

13:     For every point $p \in P$ that is covered by a disk $d \in D_{\mathcal{C}}$ we remove it from P.

14: **end for**

15: $D' \leftarrow \bigcup\limits_{\mathcal{C}, P \cap \mathcal{C} \neq \emptyset} D_{\mathcal{C}}$

16: Let C_1, C_2, C_3, C_4 be four disjoint color sets, each containing k distinct colours.

17: **for** every grid cell $\mathcal{C}$ with $id_c = [i, j]$ **do**

18:     **if** $((i \bmod 2 = 0) \wedge (j \bmod 2 = 0))$ **then**

19:         Assign C_1 to $\mathcal{C}$.

20:     **else if** $((i \bmod 2 = 0) \wedge (j \bmod 2 \neq 0))$ **then**

21:         Assign C_2 to $\mathcal{C}$.

22:     **else if** $((i \bmod 2 \neq 0) \wedge (j \bmod 2 = 0))$ **then**

23:         Assign C_3 to $\mathcal{C}$.

24:     **else if** $((i \bmod 2 \neq 0) \wedge (j \bmod 2 \neq 0))$ **then**

25:         Assign C_4 to $\mathcal{C}$.

26:     **end if**

27: **end for**

28: For every grid cell $\mathcal{C}$ and its assigned color set C_i we allot one color each to the subsets $S_{\mathcal{C}}^1, S_{\mathcal{C}}^2, \ldots S_{\mathcal{C}}^k$. Every disk centered within that subset will now be colored with the corresponding color from the color set.

29: For any disk $d \in D'$ let $\chi(d)$ represents the color assigned to the disk d in the above coloring assignment process.

30: **return** (D', χ)

---

points lying in $\mathcal{C}$. In order to do this, we first compute the voronoi diagram $VOD_{\mathcal{C}}$ on the center points of these $7k$ disks. We then do point location queries for each of these $n_{\mathcal{C}}$ points on $VOD_{\mathcal{C}}$ to determine the closest center point. We will then test whether the corresponding disk centered at the closest center point covers this point. This step will take $O(k \log k + n_{\mathcal{C}} \log k)$ time. Therefore, we invest $O(m^{7k}(k \log k + n_{\mathcal{C}} \log k))$ time to compute k-colorable cover of the points lying in the cell $\mathcal{C}$. The total time the for-loop takes over all the nonempty grid cells is $O(m^{7k}k \log k) + \sum_{\mathcal{C}} O(m^{7k}n_{\mathcal{C}} \log k) = O(m^{7k}(n + k) \log k)$. The remaining steps of the algorithm (including the preprocessing in steps 1 and 2) will take no more than $O(mn)$ time. Under the assumption that n is much larger than the number k of colors, the time complexity of the algorithm in total is $O(m^{7k}n \log k)$. In order to assign the colors at a later step of the algorithm, with each grid cell $\mathcal{C}$ we also associate and store the corresponding cover $D_{\mathcal{C}}$ computed in the for-loop at line 3. Thus, the additional space the algorithm requires is $O(nk)$.

The handover behaviour at line 11 of the algorithm and Lemma 2 ensure that any color-conflict is resolved in an elegant way. Since we use four disjoint sets of k distinct colors and for each grid cell we compute k-colorable unit disk cover (from Lemma 1), the approximation factor of the algorithm is 4. Thus, the theorem follows. $\qquad\square$

The following corollary says that Theorem 1 yields a faster algorithm for 3-$CDUDC$ than that of Biedl et al. [3], but at the cost of increase in approximation factor.

Corollary 1. *There exists a 4-approximate algorithm that runs in $O(m^{21}n)$ time for the 3-CDUDC problem.*

Proof. Follows from Theorem 1 by applying $k = 3$. $\qquad\square$

2.2   Generalization

In this subsection, we generalize the results from the preceding subsection to a general case observations for possible values of the width τ of each cell in the grid partitioning approach. We begin by attempting to generalize the potential coloring schemes for each grid cell $\tau \times \tau$ similar to the observations presented in Lemma 1. We first define a parameter ρ that represents the factor indicating the number of additional color sets needed to satisfy a union of independent solution sets. As a result, we obtain a family of approximation algorithms for the k-$CDUDC$ problem, depending upon a different possible values for ρ and τ.

Lemma 3. *The union of all independently optimal k-colorable solution sets for points lying in each grid cell $\mathcal{C}$ of size $\tau \times \tau$ is ρk colourable, where*

$$\rho = \begin{cases} 4 & \text{if } \tau \geq 2 \\ 6 & \text{if } \frac{8}{5} \leq \tau < 2 \\ 7 & \text{if } \sqrt{2} \leq \tau < \frac{8}{5} \\ 9 & \text{if } 1 \leq \tau < \sqrt{2} \end{cases}$$

Proof. We prove this by verifying the number of grid cells a unit disk can maximally intersect while considering each case. This determines the number of disjoint color sets, each consisting of at most k distinct colors. Clearly, this number is the same as ρ. Note that we are interested in finding the upper bound of such intersections.

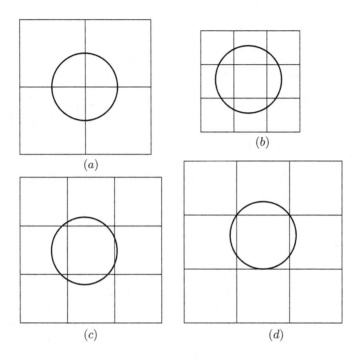

Fig. 6. Proof of Lemma 3

$\tau \geq 2$: Since the diameter of each disk is 2 and the width of grid cell is also $(\geq)2$, certainly, no disk can span more than 2 linearly adjacent disks (see Fig. 6(a)). Thus, maximal intersection count is achieved by placing the disk in any of the grid intersection corners. Here the disk will certainly intersect 4 grid cells regardless of the width.

$\frac{8}{5} \leq \tau < 2$: If the width of the grid cell is less than 2, surely, a unit disk can span 3 linearly adjacent grid cells. However, if the width is $\frac{8}{5}$, the disk cannot span more than 2 diagonally adjacent grid cells (see Fig. 6(d)). Hence, if the width is greater than or equal to $\frac{8}{5}$, then no matter where the disk is centered (in Fig. 6(d)), it can not intersect more than 6 grid cells simultaneously.

$\sqrt{2} \leq \tau < \frac{8}{5}$: If the width of the grid cell is $\sqrt{2}$, surely, a unit disk can span 3 linearly adjacent grid cells and the middle cell from the next adjacent column or row of the linearly adjacent grid cells (see Fig. 6(c)). Thus, we use the position shown in Fig. 6(c) to indicate maximal count possible.

$1 \leq \tau < \sqrt{2}$: For width ≥ 1, a unit disk could potentially intersect 3 linearly and diagonally adjacent grid cells (see Fig. 6(b)).

We do not consider the grid partitioning with grid cells of width $\tau < 1$ as it is inconsequential to our study and provides no substantial results (as the colorability increases substantially with no real improvement to the number of intersecting disks). □

In the same spirit, we provide a generalization of the observation presented in Lemma 3 for the maximum number of pairwise disjoint unit disks that can intersect a square of size $\tau \times \tau$. We define a parameter α that is the count of mutually non-intersecting unit disks that can intersect a grid cell of width τ.

Lemma 4. *The bound on the cardinality of a set S_C^i of pairwise disjoint unit disks that can intersect a grid cell C of size $\tau \times \tau$ is α, for an integer $1 \leq i \leq k$, where*

$$
\alpha = \begin{cases}
4 & \text{if } \tau = 1 \\
5 & \text{if } \frac{8}{5} \geq \tau \geq \sqrt{2} \\
7 & \text{if } \tau = 2 \\
10 & \text{if } \tau = 3 \\
14 & \text{if } \tau = 4 \\
17 & \text{if } \tau = 5
\end{cases}
$$

Proof. Consider a grid cell C of size 1×1. Imagine placing four unit disks, each centered farthest apart from one another, but outside the cell C, and touching one of the four corners of the cell C. Since the cell size is 1×1, we can not place any more disk that intersects the cell, but at the same time disjoint from each of these four disks. Hence, the bound $\alpha = 4$ for the case of width $\tau = 1$. Proof for each of the other cases can be done similar to the proofs provided for the cases $\tau = 1$ given above, and $\tau = 2$ given in Lemma 1 (see, for example, the cases $\tau = 1$ and $\tau = 8/5$ being illustrated in Fig. 7). We do not consider grid cells of width $\tau > 5$ as it provides no improvement to the running time while the approximation factor stays the same after $\tau = 2$ (as seen in Lemma 3). On the other hand, $\tau < 1$ is also not considered because the approximation factor ρ becomes arbitrarily very high for $\tau < 1$. One can observe here that the upper bound for $|S_C^i|$ obtained in Observation 1 is tight in case of τ being even, whereas in case of odd, it is a loosely bound. □

Theorem 2. *There exists a ρ-approximation algorithm to solve the k-CDUDC problem, that has a running time of $O(m^{\alpha k} n \log k)$ for a given grid width $1 \leq \tau \leq 5$, where*

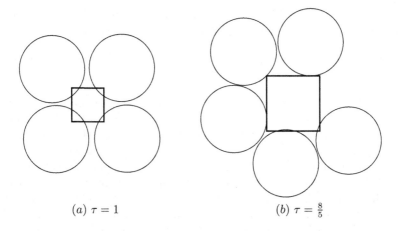

(a) $\tau = 1$ (b) $\tau = \frac{8}{5}$

Fig. 7. Proof of Lemma 4

$$\rho = \begin{cases} 4 & \text{if } \tau \geq 2 \\ 6 & \text{if } \frac{8}{5} \leq \tau < 2 \\ 7 & \text{if } \sqrt{2} \leq \tau < \frac{8}{5} \\ 9 & \text{if } 1 \leq \tau < \sqrt{2} \end{cases}, \qquad \alpha = \begin{cases} 4 & \text{if } \tau = 1 \\ 5 & \text{if } \frac{8}{5} \geq \tau \geq \sqrt{2} \\ 7 & \text{if } \tau = 2 \\ 10 & \text{if } \tau = 3 \\ 14 & \text{if } \tau = 4 \\ 17 & \text{if } \tau = 5 \end{cases}$$

Proof. The input of Algorithm 1, in addition to a set P of n points, a set D of m unit disks, and an integer $k(> 0)$, also consists of a grid partitioning parameter τ. From Lemmata 3 and 4, it is clear that any reasonable value for the parameter τ will imply the values of ρ and α. Hence, we have a ρ-approximation algorithm in $O(m^{\alpha k} n \log k)$ time. □

3   Line Segment and Rectangular Region Cover

In the same spirit as the k-*CDUDC* problem is considered due to its practical application in frequency/channel assignment in wireless networks, we also define two problems, that generalize the locations of potential wireless clients from discrete set of points to line segments and from discrete set of points to a continuous rectangular region, namely, the k-*Colorable Line Segment Disk Cover (k-CLSDC)* and k-*Colorable Rectangular Region Cover (k-CRRC)* problems, respectively.

We begin our approach using a fundamental combinatorial result involving unit disks that helps us to transform the above problems into our original k-*CDUDC* problem. Given a set D of m unit disks in the plane, a sector is the smallest region bordered by the boundary lines of disks and is covered by the same set of disks in D. Thus, the arrangement of all disks of D subdivides the plane into many sectors. It is not hard to show that the worst-case complexity of the arrangement of any set of m unit disks is quadratic, as stated below.

Observation 2. (Funke et al. [9]). *The number of sectors created by intersection of m unit disks in D is $O(m^2)$.*

To develop approximation algorithms for the k-*CLSDC* problem, we transform every instance of k-*CLSDC* problem into an instance of k-*CDUDC* problem as follows. In an instance of k-*CLSDC* problem, we have a set D of m unit disks covering a finite union of n line segments of arbitrary length with arbitrary orientation, and an integer k, The objective, here, is to compute a k-colorable cover of all the line segments. We split each of these line segments into slices such that each such slice lies within some sector. Now, for each subset of slices lying within a single sector, we add one point into the same sector and remove all the slices. This collection of points is referred to as P'. Hence, from Observation 2 we have that $|P'| = O(m^2)$. This can be taken as an instance of the k-*CDUDC* problem, where P' is taken as the input set of points P.

Similarly, we can do a similar transformation for the k-*CRRC* problem. Here, we have a set D of m unit disks covering a continuous rectangular region $\mathcal{R}$. Our objective is to compute a k-colorable set of units disks such that $\mathcal{R}$ is covered by the union of these disks. As above, we split $\mathcal{R}$ into $O(m^2)$ sectors, as induced by the union of disks in D, and add one point into each sector.

For the k-*CLSDC* problem, the construction of the set P' of points can be done as follows. We first preprocess the given set D of m unit disks into any reasonable data structure, e.g., a doubly connected edge list $(DCEL)$, in $O(m^2)$ time [14]. We can build the Voronoi diagram VOR_D on the center points of disks in D in $O(m \log m)$ time. We store cross pointers between Voronoi cells (or disks) in VOR_D and the faces (or sectors) of $DCEL$ that are contributed by the corresponding disks. For each of the given n line segments, we do point location query on VOR_D for the left endpoint of the line segment to determine the disk in which it lies. We then follow the cross pointer to access the sector that contains it. Subsequently, we traverse the adjacent sectors of this sector in $DCEL$. As we do, we add points into those sectors (also, into P') that covers a portion of the line segment and mark the corresponding faces (or sectors) as processed in $DCEL$. This step will take $O(n \log m + m^2)$ time. In the case of the k-*CRRC* problem, the construction of P' takes $O(m^2)$ time as we have to test whether each sector is intersected by $\mathcal{R}$.

Therefore, we have the following results for the k-*CLSDC* and k-*CRRC* problems.

Theorem 3. *We have a ρ-approximation algorithm to solve the k-CLSDC problem, that has a runing time of $O(n \log m + m^{\alpha k+2} \log k + m^{\alpha k} k \log k)$ for a given grid width $1 \le \tau \le 5$, where ρ and α are defined as in Theorem 2.*

Proof. Follows from Theorem 2 by applying $|P| = |P'| = O(m^2)$, where $O(n \log m)$ is due to the preprocessing of the input before we run the algorithm of Theorem 2. ☐

Theorem 4. *We have a ρ-approximation algorithm to solve the k-CRRC problem, that has a runing time of $O(m^{\alpha k+2} \log k + m^{\alpha k} k \log k)$ for a given grid width $1 \le \tau \le 5$, where ρ and α are defined as in Theorem 2.*

Proof. Follows from Theorem 2 by applying $|P| = |P'| = O(m^2)$. □

4   Conclusion

In this paper, we have proposed constant-factor approximation algorithms for computing k-colorable unit disk covering of points, line segments, and a rectangular region. In the future work, we wish to improve the running time of the local algorithm for a grid cell instead of brute-forcing for optimal cover of points lying in the grid cell. This will improve the overall running time of the approximation algorithms.

References

1. Basappa, M.: Line segment disk cover. In: Panda, B.S., Goswami, P.P. (eds.) CALDAM 2018. LNCS, vol. 10743, pp. 81–92. Springer, Cham (2018). https://doi.org/10.1007/978-3-319-74180-2_7
2. Basappa, M., Acharyya, R., Das, G.K.: Unit disk cover problem in 2D. J. Discrete Algorithms **33**, 193–201 (2015)
3. Biedl, T., Biniaz, A., Lubiw, A.: Minimum ply covering of points with disks and squares. Comput. Geom. **94**, 101712 (2021)
4. Brass, A., Hurtado, F., Lafreniere, B.J., Lubiw, A.: A lower bound on the area of a 3-coloured disk packing. Int. J. Comput. Geom. Appl. **20**(3), 341–360 (2010)
5. de Berg, M., Markovic, A.: Dynamic conflict-free colorings in the plane. Comput. Geom. **78**, 61–73 (2019)
6. Clark, B.N., Colbourn, C.J., Johnson, D.S.: Unit disk graphs. Discrete Math. **86**(1–3), 165–177 (1990)
7. Cheilaris, P., Gargano, L., Rescigno, A., Smorodinsky, S.: Strong conflict-free coloring for intervals. Algorithmica **70**(4), 732–749 (2014)
8. Even, G., Lotker, Z., Ron, D., Smorodinsky, S.: Conflict-free colorings of simple geometric regions with applications to frequency assignment in cellular networks. SIAM J. Comp. **33**(1), 94–136 (2003)
9. Funke, S., Kesselman, A., Kuhn, F., Lotker, Z., Segal, M.: Improved approximation algorithms for connected sensor cover. Wirel. Netw. **13**(2), 153–164 (2007)
10. Fraser, R., López-Ortiz, A.: The within-strip discrete unit disk cover problem. Theor. Comput. Sci. **674**, 99–115 (2017)
11. Horev, E., Krakovski, R., Smorodinsky, S.: Conflict-free coloring made stronger. In: Kaplan, H. (ed.) SWAT 2010. LNCS, vol. 6139, pp. 105–117. Springer, Heidelberg (2010). https://doi.org/10.1007/978-3-642-13731-0_11
12. Mustafa, N.H., Ray, S.: Improved results on geometric hitting set problems. Discrete Comput. Geom. **44**(4), 883–895 (2010)
13. Smorodinsky, S.: Combinatorial problems in computational geometry. Ph.D. thesis. Tel-Aviv University (2003)
14. de Berg, M., Cheong, O., van Kreveld, M., Overmars, M.: Computational Geometry: Algorithms and Applications, 3rd edn. Springer, Heidelberg (2008). https://doi.org/10.1007/978-3-540-77974-2

An Improved Accuracy Split-Step Padé Parabolic Equation for Tropospheric Radio-Wave Propagation

Mikhail S. Lytaev$^{(\boxtimes)}$ (iD)

St. Petersburg Federal Research Center of the Russian Academy of Sciences,
14-th Linia, V.I., No. 39, Saint Petersburg 199178, Russia

Abstract. This paper is devoted to modeling the tropospheric electromagnetic waves propagation over irregular terrain by the higher-order finite-difference methods for the parabolic equation (PE). The proposed approach is based on the Padé rational approximations of the propagation operator, which is applied simultaneously along with longitudinal and transversal coordinates. At the same time, it is still possible to model the inhomogeneous tropospheric refractive index. Discrete dispersion analysis of the proposed scheme is carried out. A comparison with the other finite-difference methods for solving the parabolic equation and the split-step Fourier (SSF) method is given. It is shown that the proposed method allows using a more sparse computational grid than the existing finite-difference methods. This in turn results in more fast computations.

Keywords: Parabolic wave equation · Finite difference methods · Tropospheric propagation · Electromagnetic propagation · Numerical dispersion

1 Introduction

When solving several problems of wireless communications, navigation and radar, it is essential to predict the characteristics of radio-wave propagation in the constantly varying troposphere. The propagation of radio-waves is significantly affected by spatial variations in the tropospheric refractive index, irregular terrain and its dielectric properties, rough sea surface [9], vegetation [4,24], and urban development [16]. Often, the solution to this problem is an integral part of complex real-time software systems [3]. Given the above challenges, developing both appropriate mathematical models and efficient numerical methods remains actual.

The most widely used deterministic approach to solving this class of problems is the parabolic equation (PE) method [13,14]. Modern formulation of the PE method in terms of the one-way Helmholtz equation is directly derived from Maxwell's equations, which makes it rather reliable and verified. The PE method can take into account all the above-mentioned features of tropospheric propagation [9,10,23,28,29], but its effective usage requires fast and reliable numerical methods.

© Springer Nature Switzerland AG 2021
O. Gervasi et al. (Eds.): ICCSA 2021, LNCS 12949, pp. 418–433, 2021.
https://doi.org/10.1007/978-3-030-86653-2_31

The fastest and most widely used numerical method for solving the PE in tropospheric radio-wave propagation problems is the split-step Fourier (SSF) method [14,22]. Its main drawbacks include instability and errors when modeling the upper and lower boundary conditions [11,19]. Finite-difference methods for solving PE [1,5,19], in contrast, provide efficient boundary conditions modeling. However, the existing finite-difference methods for PE are more time-consuming. When performing numerical radio coverage predictions over rugged terrain or in urban canyons, it is necessary to account for large propagation angles, which requires a dense computational grid and a large approximation order [17]. This, in turn, leads to an increase in computational costs. A number of important results concerning the numerical solution of PE by the finite-difference methods were obtained in computational hydroacoustics studies [6]. The use of the principle of universality of mathematical models [25] makes it possible to partially reuse the corresponding numerical methods in the problem under consideration.

This article continues a series of works [17–19] devoted to improving the performance of finite-difference schemes for the PE. Previously, a higher-order approximation was applied along the longitudinal coordinate, while the diffraction part of the propagation operator was approximated by the 2nd or 4th order scheme. In this paper, it is shown that under certain conditions it is possible to simultaneously apply the higher-order Padé approximation to the exponential propagation operator and its diffraction part. At the same time, the new scheme differs only in coefficients, which means that it inherits all the properties of existing finite-difference schemes.

The paper is organized as follows. The next section presents a solution to the one-way Helmholtz equation using the split-step Padé method. In Sect. 3, we introduce a joint Padé approximation along with the longitudinal and transversal coordinates. Section 4 provides a comparative analysis of the numerical schemes under various propagation conditions.

2   Split-Step Padé PE

Following the generally accepted methodology of splitting the wave field into forward and backward propagating waves [8], the equation for waves propagating along the positive x direction can be written as follows

$$\frac{\partial u}{\partial x} = ik(\sqrt{1+L} - 1)u, \tag{1}$$

where

$$Lu = \frac{1}{k^2}\frac{\partial^2 u}{\partial z^2} + hu,$$

$$h(x,z) = m^2(x,z) - 1,$$

$k = 2\pi/\lambda$ is the wavenumber, λ is the wavelength, $m(x,z)$ is the modified refractivity of the troposphere [14].

Function $u(x, z)$ is subject to the impedance boundary condition at the lower boundary

$$\left(q_1 u + q_2 \frac{\partial u}{\partial z}\right)\Bigg|_{z=h(x)} = 0,$$

where $h(x)$ is the terrain profile, complex coefficients q_1 and q_2 are determined from the dielectric properties of the ground [14].

The wave propagation process is generated by the Dirichlet initial condition of the form

$$u(0, z) = u_0(z), \tag{2}$$

with known function $\psi_0(z)$, which corresponds to the antenna pattern.

Step-by-step solution of Eq. (1) can be formally written using the pseudo-differential [26] propagation operator as follows

$$u(x + \Delta x, z) = P(L)\, u(x, z), \tag{3}$$

$$P(L) = \exp\left(ik\Delta x \left(\sqrt{1 + L} - 1\right)\right).$$

The following notation is further used for discrete versions of functions

$$u_j^n = u(n\Delta x, j\Delta z),$$

where Δx and Δz are the longitudinal and transversal grid steps respectively.

Considering propagation operator (3) as a function of operator L, we can apply a rational Padé approximation of the order $[m/n]$ [2,5] as follows

$$P(L)\, u = \frac{1 + \sum_{l=1}^{m} \tilde{a}_l L^l}{1 + \sum_{l=1}^{n} \tilde{b}_l L^l} u + O(L^{n+m})u. \tag{4}$$

Next, we will represent the rational approximation in the form of a product

$$\frac{1 + \sum_{l=1}^{m} \tilde{a}_l L^l}{1 + \sum_{l=1}^{n} \prod \tilde{b}_l L^l} = \prod_{l=1}^{p} \frac{1 + a_l L}{1 + b_l L},$$

where $p = \max(m, n)$. The pseudo-differential square root operator, which is responsible for the diffraction, and the operator exponent, which is responsible for integration along the longitudinal coordinate x, are simultaneously approximated. Any desired propagation angle can be achieved by selecting a sufficient Padé approximation order [17].

The action of propagation operator (3) is approximately reduced to the sequential application of the rational operator of the form

$$v = \frac{1 + a_l L}{1 + b_l L} u,$$

which in turn is equivalent to solving the following one-dimensional differential equation

$$(1 + bL)\, v = (1 + aL)\, u, \tag{5}$$

where $u(z)$ is known field, obtained on a previous step, $v(z)$ is the desired field. Replace operator L is Eq. (5) with its finite-difference analog [18]

$$\hat{L}u_j = \frac{1}{k^2 \Delta z^2} \frac{\delta^2}{1 + \alpha \delta^2} + h_j u_j,$$

where $\alpha = 0$ for the 2nd order approximation and $\alpha = 1/12$ for the 4th order approximation, second difference operator δ^2 is defined as follows

$$\delta^2 u = u(z - \Delta z) - 2u(z) + u(z + \Delta z) = u_{j-1} - 2u_j + u_{j+1}.$$

Then we obtain the following finite-difference equation

$$\left(\tau \left(1 + \alpha \delta^2 \right) + b_l \delta^2 + b_l \tau h_j \right) v_j + b_l \tau \alpha \delta^2 \left(h_j v_j \right)$$
$$= \left(\tau \left(1 + \alpha \delta^2 \right) + a_l \delta^2 + a_l \tau h_j \right) u_j + a_l \tau \alpha \delta^2 \left(h_j u_j \right),$$

where $\tau = (k \Delta z)^2$.

The above equation, along with discrete boundary conditions [18,19], is a tridiagonal system of linear algebraic equations and is solved in linear time. The result is a numerical scheme that has an arbitrary order of accuracy when integrated along the longitudinal coordinate x and up to 4th order accuracy in z.

Note that case $p = 1$ in approximation (4) is equivalent to the well known Crank-Nikolson scheme for PE [10]. At the same time, the Padé approximation scheme is significantly superior to the Crank-Nikolson method both in terms of computational speed and smoothness of the obtained solutions [14,19].

3   Modified Padé Approximation

A certain disadvantage of the previously obtained scheme is the approximation of transversal differentiation operator L. While an arbitrary order of accuracy is achievable for the approximation of the exponential operator, the order of approximation does not exceed 4 for operator L. It is reasonable to assume that increasing the approximation order of operator L can further improve the performance of the scheme. The obvious solution to this problem is the use of multi-point schemes for the approximation of the differential operator. However, it will also increase the complexity of the scheme, since Eq. (5) will no longer be reduced to a tridiagonal matrix. Instead, in this paper, we propose to use Padé approximation (4) simultaneously by the longitudinal and transversal coordinates.

Following [7,12], the differentiation operator can be expressed in terms of the second difference operator as follows

$$\frac{\partial^2 u}{\partial z^2} = g(\delta^2)u,$$

$$g(\xi) = -\frac{1}{\Delta z^2} \ln^2 \left(1 + \frac{\xi}{2} + \sqrt{\left(1 + \frac{\xi}{2}\right)^2 - 1}\right).$$

For the case of a homogeneous medium $(m(x, z) \equiv 1)$ we can write down the following approximation of the propagation operator with respect to operator δ^2

$$P\left(\frac{1}{k^2}\frac{\partial^2}{\partial z^2}\right) u = P\left(\frac{1}{k^2}g\left(\delta^2\right)\right) u \approx \prod_{l=1}^{p} \frac{1 + a_l'\delta^2}{1 + b_l'\delta^2} u. \tag{6}$$

There is no explicit approximation of operator L in this expression. Thus, the longitudinal and transversal coordinate discretization is performed using single high-order Padé approximation. The disadvantage of this approach, which was previously considered for solving the computational hydroacoustics problems [5], is its limitation to the case of a homogeneous medium.

Let us now return to the case of an inhomogeneous troposphere. Function g is presented according to the Taylor series as follows

$$g(\xi) = \frac{1}{\Delta z^2}\left[\xi - \frac{1}{12}\xi^2 + \frac{1}{90}\xi^3 - \cdots\right].$$

Put $\delta^2 + \tau h$ as an argument of function g to the above decomposition

$$\frac{1}{k^2}g(\delta^2 + \tau h)u = \frac{1}{\tau}\left[\delta^2 + \tau h - \frac{1}{12}\left(\delta^2 + \tau h\right)^2 + \frac{1}{90}\left(\delta^2 + \tau h\right)^3 - \cdots\right]u$$

$$= \frac{1}{\tau}\left[\delta^2 u + \tau h u - \frac{1}{12}\left(\delta^4 u + \ldots\right) + \frac{1}{90}\left(\delta^6 u + \ldots\right) - \cdots\right].$$

In tropospheric propagation problems, $h(x, z)$ is a slowly varying function both in range and height, and its value rarely exceeds 0.0005 [14]. This makes it possible to neglect the terms containing h (denoted as $\ldots$ in the above expansion) and obtain the following approximate relation

$$\frac{1}{k^2}g(\delta^2 + \tau h)u \approx \frac{1}{k^2}g(\delta^2)u + hu = \frac{1}{k^2}\frac{\partial^2 u}{\partial z^2} + hu.$$

Substituting this approximation into the propagation operator, we obtain

$$P(L) = P\left(\frac{1}{k^2}\frac{\partial^2}{\partial z^2} + h\right)u \approx P\left(\frac{1}{k^2}g(\delta^2 + \tau h)\right)u.$$

Next, using Padé approximation (6) with new coefficients a_l' and b_l', we can write the following approximation of the propagation operator, that takes into account the inhomogeneous refractive index

$$P\left(\frac{1}{k^2}g(\xi)\right)\bigg|_{\xi=\delta^2+\tau h} u \approx \prod_{l=1}^{p} \frac{1 + a_l'\xi}{1 + b_l'\xi}\bigg|_{\xi=\delta^2+\tau h} u. \tag{7}$$

As a result, we come to a numerical scheme that is equivalent to the 2nd order one ($\alpha = 0$) except for coefficients a'_l and b'_l.

Note that the computational complexity of the proposed scheme is equal to the complexity of the 2nd order scheme. The difference is only in the Padé approximation coefficients. Approximation (7) simultaneously includes the differentiation operator, the square root operator, and the exponential operator. Thus, single approximation is responsible simultaneously for three parameters of the numerical scheme: grid steps Δx, Δz and the maximum propagation angle.

The idea of joint use of the Padé approximation for sampling along the longitudinal and transversal coordinates was expressed in the works [5, 7]. However, previously it was limited to the case of a homogeneous medium. In this paper, we show that when the refractive index is sufficiently small, it is possible to generalize this idea to the case of an inhomogeneous medium without additional computational costs. A small refractive index just arises in the problems of tropospheric propagation. Of course, such a scheme is not applicable in the case of strong spatial variations of the refractive index. Latter occurs, for example, when modeling the water-ground boundary in hydroacoustics.

It should be noted the smallness of the refractive index is also significantly used in the derivation of the widely used SSF method, so this assumption is not a shortcoming compared to the SSF method. At the same time, the proposed scheme, as well as other finite-difference schemes, makes it possible to effectively model both the upper transparent boundary [19] and the lower impedance boundary [14].

It is recommended to use Padé approximations of the order $[p-1/p]$ or $[p-2/p]$ [14, 19].

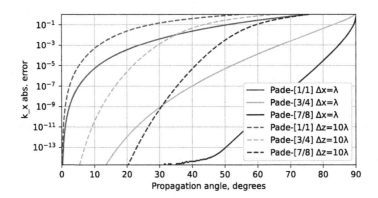

Fig. 1. Dependence of discrete dispersion relation error R on propagation angle θ. $\Delta z \to 0$.

Table 1. Optimal values of Δz for various PE numerical methods (larger is better). $\Delta x = 1\lambda$, rational approximation order is $[7/8]$, acceptable error $tol = 10^{-3}$.

$\theta_{max},°$	$\Delta z, \lambda$ (2nd order)	$\Delta z, \lambda$ (4th order)	$\Delta z, \lambda$ (joined)	$\Delta z, \lambda$ (SSF)
10	0.8	1.5	2.8	2.88
20	0.2	0.5	1.3	1.46
30	0.09	0.2	0.9	1.0
45	0.01	0.1	0.6	0.70
60	0.01	0.1	0.4	0.58
70	0.01	0.01	0.3	0.53

4   Numerical Results and Discussion

The first two subsections demonstrate the advantage of the proposed approach in terms of the maximum propagation angle and density of the computational grid. Then we demonstrate on a concrete example the consistency of the made assumption about the sufficient smallness of the refractive index. In the last example, we show the application of the proposed method in a realistic propagation scenario.

4.1   Error Analysis

In this section, we provide a discrete dispersion analysis [17,28] of the proposed scheme. The discrete horizontal wavenumber for the class of numerical schemes under consideration is written as follows [17]

$$k_x^d(\Delta x, \Delta z, a_1 \ldots a_p, b_1 \ldots b_p, \theta) = k + \frac{\ln \prod_{l=1}^{p} t_l}{i \Delta x},$$

$$t_l = \frac{1 - \frac{4a_l}{\tau} \left(\sin^2 \left(\frac{k \Delta z \sin \theta}{2} \right) + 4\alpha \sin^4 \left(\frac{k \Delta z \sin \theta}{2} \right) \right)}{1 - \frac{4b_l}{\tau} \left(\sin^2 \left(\frac{k \Delta z \sin \theta}{2} \right) + 4\alpha \sin^4 \left(\frac{k \Delta z \sin \theta}{2} \right) \right)},$$

where θ is the angle between the direction of the wave and the positive x direction.

The horizontal wavenumber for original one-way Helmholtz equation (1) is written as follows

$$k_x = \begin{cases} \sqrt{k^2 - k_z^2}, & |k_z| \leq k, \\ i\sqrt{k_z^2 - k^2}, & |k_z| > k, \end{cases}$$

where $k_z = k \sin \theta$ is the vertical wavenumber.

Comparing the discrete and original dispersion relation

$$R\left(\Delta x, \Delta z, a_1 \ldots a_p, b_1 \ldots b_p, \theta\right)$$
$$= |k_x^d \left(\Delta x, \Delta z, a_1 \ldots a_p, b_1 \ldots b_p, \theta\right) - k_x \left(k_z = k \sin \theta\right)|$$

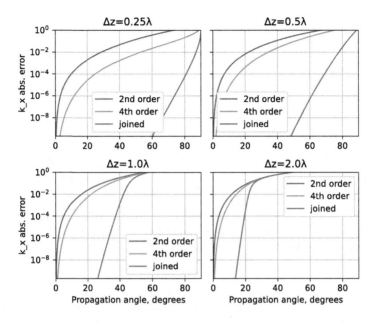

Fig. 2. Dependence of discrete dispersion relation error R on propagation angle θ. $\Delta x = 1\lambda$, Rational approximation order is equal to [7/8].

makes it possible to estimate the accuracy of a numerical scheme depending on its parameters [17]. Figure 1 demonstrates the dependence of discrete dispersion relation error R on propagation angle θ for the following three approximation orders: [1/1], [3/4], [7/8] and two values of longitudinal grid step Δx: 1.0λ and 10.0λ. In this example we set $\Delta z \to 0$, so the discretization by z is not considered. It is clearly seen that the accuracy monotonically decreases with increasing propagation angle. Increasing the order of rational approximation and decreasing grid step Δx gives a more accurate scheme. Thus, there is a direct dependency between the accuracy of the numerical scheme and the propagation angle.

Next, we consider the effect of the approximation of operator L on the accuracy of the numerical scheme. Figure 2 demonstrates the dependence of R on the propagation angle for the four different values of transversal grid step Δz: 0.25λ, 0.5λ, 1.0λ, 2.0λ and three variants of approximation by z: 2nd order, 4th order and the proposed joined approximation. Longitudinal grid step $\Delta x = 1.0\lambda$, Padé approximation order is [7/8]. It can be seen that, as before, the accuracy monotonically decreases with increasing propagation angle. It is also clearly observable how increasing the order of approximation by z increases the accuracy of the numerical scheme. At the same time, it should be noted that all three considered variants of approximation by z are computationally equivalent since they differ only in the coefficients of the scheme.

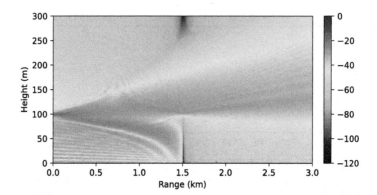

Fig. 3. Knife-edge diffraction. Two-dimensional distribution of the electromagnetic field ($20 \log |u(x,z)|$), obtained by the integral equation method [27]. Operational frequency is equal to 600 MHz.

Next, we will consider in more detail how the accuracy of the approximation by z affects the required density of the computational grid and, accordingly, the computational time. To determine the optimal density of the computational grid, we will solve the following maximization problem [17]

$$\Delta z \to \max$$

under condition

$$\max_{\theta \in [0,\theta_{max}]} R\left(\Delta x, \Delta z, a_1 \ldots a_p, b_1 \ldots b_p, \theta\right) < tol,$$

where θ_{max} is the maximum required propagation angle, *tol* is the acceptable error at each step.

For the SSF method, it is somewhat easier to determine the optimal Δz by using the Nyquist-Shannon sampling theorem

$$\Delta z \le \frac{\lambda}{2 \sin \theta_{max}}.$$

Table 1 demonstrates the optimal values of Δz for several propagation angles and numerical methods. The acceptable error *tol* was taken equal to 10^{-3}. It can be seen that the proposed scheme allows the use of a much more sparse computational grid than other finite difference methods, and is almost equal to the SSF method by this criterion. The advantages of the proposed method are particularly visible on large propagation angles.

4.2   Knife-Edge Diffraction

In this subsection, we consider the classical knife-edge diffraction problem [21,27]. Here we also consider the backscattering from the obstacle. Modeling of

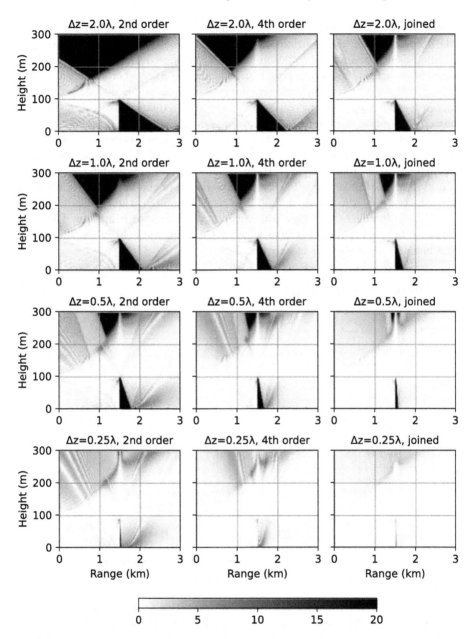

Fig. 4. Knife-edge diffraction. Two-dimensional distribution of the error between reference solution u_e and finite-difference solution u_{fd} $(|20\log|u_e(x,z)| - 20\log|u_{fd}(x,z)||)$. In all examples, the order of the Padé approximation is equal to $[7/8]$, longitudinal grid step $\Delta x = \lambda$.

diffraction on such obstacles at large propagation angles is essential when calculating the field in dense urban area. As a reference, we will use the solution obtained using the integral equation for the knife-edge diffraction [27]. The horizontally polarized monochromatic source is located at the height of 100 m over the perfectly conducted surface. The operation frequency is 600 MHz ($\lambda = 0.5$ m). Knife-edge with a height of 100 m is located at a distance of 1.5 km from the source of radiation. The reference solution is depicted in Fig. 3.

Figure 4 shows the spatial distribution of the error between the reference solution and the various configurations of the finite difference method. Padé approximation of order [7/8] with three various approximations by z and various values of Δz are considered. The backscattering from the knife-edge is modeled using the two-way PE approach [20]. It is clearly seen how the maximum propagation angle increases with increasing accuracy of the approximation. The need to handle large propagation angles when calculating the field in the diffraction zone behind the obstacle is clearly visible.

4.3   Propagation in a Duct

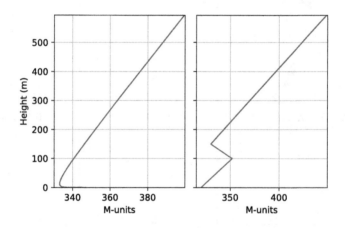

Fig. 5. M-profile of the evaporation duct (left) and elevated duct (right).

In this example, we model the radio-wave propagation in a tropospheric waveguide. The aim of this example is to demonstrate that the assumptions made in the previous section about the sufficient smallness of the tropospheric refractive index do not really affect the accuracy of the proposed method.

Modified refractivity m is determined by the refractive index of the troposphere n and the radius of the Earth R as follows

$$m = n + 1/R.$$

Function m makes it possible to take into account the over-the-horizon diffraction along with tropospheric refraction [14]. We further express the tropospheric refraction profile by the M-profile $M = 10^6(m - 1)$. Figure 5 shows the tropospheric refractive index profiles for the two widely occurring waveguides: evaporated duct and elevated duct [14]. Figure 6 demonstrates the results of modeling in the mentioned profiles by the proposed method. The radiation frequency in this example is equal to 10 GHz ($\lambda = 0.03$ m). We will compare the results obtained in this example with the SSF method implemented in the PETOOL program [22]. Figure 7 shows the spatial distribution of the error between the proposed method and the SSF method. It can be seen that the results of both methods actually coincide. Thus, the proposed method can properly handle the inhomogeneities of the troposphere. It is also clear that the proposed method can properly handle the discrete transparent boundary condition [19], posed on the upper boundary of the computational domain.

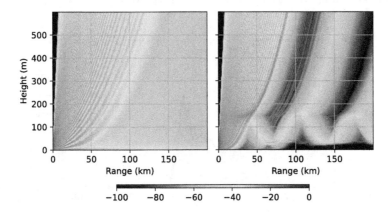

Fig. 6. Propagation in the evaporation duct (left) and elevated duct (right). Two-dimensional distribution of the electromagnetic field ($20 \log |u(x, z)|$), obtained by the proposed method. In both examples, Padé approximation order is $[7/8]$, $\Delta x = 1000\lambda$, $\Delta z = 0.5\lambda$. Operational frequency is equal to 10 GHz.

4.4   Propagation over Irregular Terrain

In the last example, we consider the propagation of radio-waves over irregular terrain in the elevated duct from the previous example. The radiation frequency in this example is equal to 1.5 GHz ($\lambda = 0.2$ m). Figure 8 shows the two-dimensional distribution of the electromagnetic field obtained by the proposed method with the following parameters: Padé approximation order in equal to $[7/8]$, $\Delta x = 100\lambda$, $\Delta z = 3\lambda$. The inhomogeneities of the landscape are approximated by a piecewise constant function. Figure 9 and 10 show a comparison with

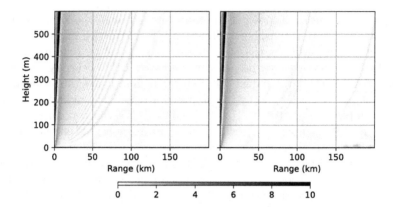

Fig. 7. Propagation in the evaporation duct (left) and elevated duct (right). Two-dimensional distribution of the error between reference solution u_{ssf}, obtained by the SSF method, and solution u_{fd}, obtained by the proposed method ($|20\log|u_{ssf}(x,z)| - 20\log|u_{fd}(x,z)||$).

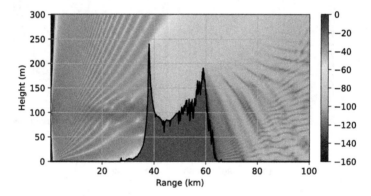

Fig. 8. Propagation in the elevated duct over irregular terrain. Two-dimensional distribution of the electromagnetic field ($20\log|u(x,z)|$), obtained by the proposed method. Operational frequency is equal to 1.5 GHz.

the 2nd order approximation. It is clearly seen that for the same computational parameters, the 2nd order scheme yields a significant underestimation of the field in the diffraction zone, as well as a noisy solution in several areas. To obtain a correct solution by the 2nd order scheme, it is necessary to use six times more dense grid by z.

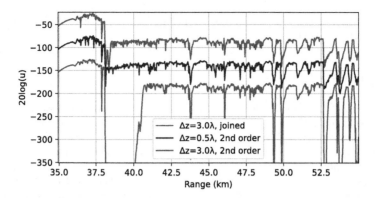

Fig. 9. Propagation in the elevated duct over irregular terrain. Distribution of the electromagnetic field ($20 \log |u(x,z)|$) at the height of 5 m over the surface. The results are shifted by 50 dB for convenience.

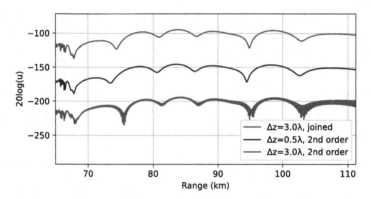

Fig. 10. Propagation in the elevated duct over irregular terrain. Distribution of the electromagnetic field ($20 \log |u(x,z)|$) at the height of 5 m over the surface. The results are shifted by 50 dB for convenience.

5   Conclusion

The proposed method gives an opportunity to significantly improve the performance of finite-difference methods for solving PE in the inhomogeneous troposphere. At the same time, no significant changes to the existing software implementations are required, since the new scheme differs only in coefficients of the rational approximation. The new scheme retains all the advantages of the finite-difference schemes, in particular, the correct boundary conditions modeling. At the same time, the joint Padé approximation along the longitudinal and transversal coordinates made it possible to reduce the density of the computational grid and, accordingly, the computational time.

The proposed method is implemented as a Python 3 library and is freely available [15].

Acknowledgements. This study was supported by the Russian Science Foundation grant No. 21-71-00039.

References

1. Apaydin, G., Ozgun, O., Kuzuoglu, M., Sevgi, L.: A novel two-way finite-element parabolic equation groundwave propagation tool: tests with canonical structures and calibration. IEEE Trans. Geosci. Remote Sens. **49**(8), 2887–2899 (2011)
2. Baker, G.A., Graves-Morris, P.: Pade Approximants, vol. 59. Cambridge University Press, Cambridge (1996)
3. Brookner, E., Cornely, P.R., Lok, Y.F.: AREPS and TEMPER-getting familiar with these powerful propagation software tools. In: IEEE Radar Conference, pp. 1034–1043. IEEE (2007)
4. Cama-Pinto, D., et al.: Empirical model of radio wave propagation in the presence of vegetation inside greenhouses using regularized regressions. Sensors **20**(22), 6621 (2020)
5. Collins, M.D.: A split-step pade solution for the parabolic equation method. J. Acoust. Soc. Am. **93**(4), 1736–1742 (1993)
6. Collins, M.D., Siegmann, W.L.: Parabolic Wave Equations with Applications. Springer, New York (2019). https://doi.org/10.1007/978-1-4939-9934-7
7. Ehrhardt, M., Zisowsky, A.: Discrete non-local boundary conditions for split-step padé approximations of the one-way Helmholtz equation. J. Comput. Appl. Math. **200**(2), 471–490 (2007)
8. Fishman, L., McCoy, J.J.: Derivation and application of extended parabolic wave theories. I. The factorized Helmholtz equation. J. Math. Phys. **25**(2), 285–296 (1984)
9. Guo, Q., Long, Y.: Two-way parabolic equation method for radio propagation over rough sea surface. IEEE Trans. Antennas Propag. **68**, 4839–4847 (2020)
10. Guo, Q., Zhou, C., Long, Y.: Greene approximation wide-angle parabolic equation for radio propagation. IEEE Trans. Antennas Propag. **65**(11), 6048–6056 (2017)
11. Kuttler, J.R., Janaswamy, R.: Improved Fourier transform methods for solving the parabolic wave equation. Radio Sci. **37**(2), 1–11 (2002)
12. Lee, D., Schultz, M.H.: Numerical Ocean Acoustic Propagation in Three Dimensions. World Scientific, Singapore (1995)
13. Leontovich, M.A., Fock, V.A.: Solution of the problem of propagation of electromagnetic waves along the earth's surface by the method of parabolic equation. J. Phys. USSR **10**(1), 13–23 (1946)
14. Levy, M.F.: Parabolic Equation Methods for Electromagnetic Wave Propagation. The Institution of Electrical Engineers, UK (2000)
15. Lytaev, M.S.: Python wave prorogation library (2020). https://github.com/mikelytaev/wave-propagation
16. Lytaev, M., Borisov, E., Vladyko, A.: V2I propagation loss predictions in simplified urban environment: a two-way parabolic equation approach. Electronics **9**(12), 2011 (2020)
17. Lytaev, M.S.: Automated selection of the computational parameters for the higher-order parabolic equation numerical methods. In: Gervasi, O., et al. (eds.) ICCSA 2020. LNCS, vol. 12249, pp. 296–311. Springer, Cham (2020). https://doi.org/10.1007/978-3-030-58799-4_22

18. Lytaev, M.S.: Numerov-pade scheme for the one-way Helmholtz equation in tropospheric radio-wave propagation. IEEE Antennas Wirel. Propag. Lett. **19**(12), 2167–2171 (2020)
19. Lytaev, M.S.: Nonlocal boundary conditions for split-step padé approximations of the Helmholtz equation with modified refractive index. IEEE Antennas Wirel. Propag. Lett. **17**(8), 1561–1565 (2018)
20. Mills, M.J., Collins, M.D., Lingevitch, J.F.: Two-way parabolic equation techniques for diffraction and scattering problems. Wave Motion **31**(2), 173–180 (2000)
21. Nguyen, V.D., Phan, H., Mansour, A., Coatanhay, A., Marsault, T.: On the proof of recursive Vogler algorithm for multiple knife-edge diffraction. IEEE Trans. Antennas Propag. **69**(6), 3617–3622 (2020)
22. Ozgun, O., et al.: PETOOL v2.0: parabolic equation toolbox with evaporation duct models and real environment data. Comput. Phys. Commun. **256**, 107454 (2020)
23. Permyakov, V.A., Mikhailov, M.S., Malevich, E.S.: Analysis of propagation of electromagnetic waves in difficult conditions by the parabolic equation method. IEEE Trans. Antennas Propag. **67**(4), 2167–2175 (2019)
24. Ramos, G.L., Pereira, P.T., Leonor, N., Caldeirinha, F.R.: Analysis of radiowave propagation in forest media using the parabolic equation. In: 2020 14th European Conference on Antennas and Propagation (EuCAP). IEEE (2020)
25. Samarskii, A.A., Mikhailov, A.P.: Principles of Mathematical Modelling: Ideas, Methods, Examples. Taylor and Francis, Routledge (2002)
26. Taylor, M.: Pseudodifferential Operators and Nonlinear PDE, vol. 100. Springer, Cham (2012). https://doi.org/10.1007/978-1-4612-0431-2
27. Vavilov, S.A., Lytaev, M.S.: Modeling equation for multiple knife-edge diffraction. IEEE Trans. Antennas Propag. **68**(5), 3869–3877 (2020)
28. Wang, D.D., Pu, Y.R., Xi, X.L., Zhou, L.L.: An analysis of narrow-angle and wide-angle shift-map parabolic equations. IEEE Trans. Antennas Propag. **68**(5), 3911–3918 (2020)
29. Zhang, P., Bai, L., Wu, Z., Guo, L.: Applying the parabolic equation to tropospheric groundwave propagation: a review of recent achievements and significant milestones. IEEE Antennas Propag. Mag **58**(3), 31–44 (2016)

Numerical Algorithm of Seismic Wave Propagation and Seismic Attenuation Estimation in Anisotropic Fractured Porous Fluid-Saturated Media

Mikhail Novikov[(✉)], Vadim Lisitsa[iD], Tatyana Khachkova, Galina Reshetova, and Dmitry Vishnevsky

Sobolev Institute of Mathematics SB RAS,
4 Koptug Avenue, Novosibirsk 630090, Russia
novikovma@ipgg.sbras.ru

Abstract. We present a numerical algorithm of seismic wave propagation in anisotropic fractured fluid-saturated porous media and estimation of seismic attenuation. The algorithm is based on numerical solution of anisotropic Biot equations of poroelasticity. We use finite-difference approximation of Biot equations on the staggered grid. We perform a set of numerical experiments of wave propagation in fractured media. Fractures in the media are connected and filled with anisotropic material providing wave induced fluid flow within connected fractures. Numerical estimations of inverse quality factor demonstrate the effect of fracture-filling material anisotropy on seismic wave attenuation.

Keywords: Biot model · Poroelasticity · Anisotropy · Wave propagation · Seismic attenuation · Inverse quality factor · Finite differences

1 Introduction

On of the most popular research topics in modern geophysics is estimation of reservoir transport properties and fluid mobility. Accurate information about these properties is essential for fractured reservoirs exploration, in particular, CO_2 sequestration [6,15], geothermal energy exploration [12,20]. One of the attributes that can potentially be used as an indicator of transport properties in fractured formations is seismic attenuation. Seismic attenuation in fractured fluid-saturated rock is provided by many essentially different mechanisms. In particular, when a seismic wave propagates through such media, it generates pressure gradients and so cause so-called wave-induced fluid flow (WIFF). Usually these fluid flows are divided in two types. First type is fracture-to-background

The work is supported by Mathematical Center in Akademgorodok, the agreement with Ministry of Science and High Education of the Russian Federation number 075-15-2019-1613.

WIFF (FB-WIFF). FB-WIFF is defined mostly by high physical properties contrast between background and fracture-filling material [8,17,19]. Most intense FB-WIFF develops at low frequencies. Conversely, in presence of connected fractures second type of WIFF, fluid flow between fractures (fracture-to-fracture WIFF, FF-WIFF) occurs. In contrast to FB-WIFF, flow of the second type has its intensity maximum at high frequencies. Most important factors affecting FF-WIFF are physical properties of fracture-filling material [5,8,19]. Both types of WIFF result in frequency-dependent seismic wave attenuation. So adequate seismic attenuation estimation can become the base for suggestions about transport properties of the reservoir.

Unfortunately, theoretical studies of WIFF involve relatively simple fracture systems. In particular, studied fracture connectivity is often limited by pairwise intersections of fractures of two orientations [8]. Even if fracture systems are more complex [7] long-distance percolation is not guaranteed. Transport properties of reservoirs depend significantly on the presence of long length chains formed by connected fractures, and it is necessary to study dependencies between fracture connectivity and seismic attenuation. So proper modeling of wave propagation in fractured models with given percolation length is needed to study FF-WIFF and resulting attenuation. Another aspect that must also be taken into consideration is anisotropy. Fracture-filling material should correspond to fracture orientation to provide more realistic fractured media model and obtain more precise correlations between fracture system structure and attenuation.

In this paper, we present a numerical algorithm of seismic wave propagation in anisotropic fractured fluid-saturated porous media and estimation of seismic attenuation. We use presented algorithm to study fracture connectivity effect on seismic attenuation.

2   Anisotropic Biot Equations

Biot theory of poroelasticity is still the most common approach to model seismic waves propagation in porous fluid-saturated media up to the present day in rock physics [2,3,13]. This theory involve two-phase continuous media, where both solid and fluid introduced. Moreover, it involves physical parameters defining the microstructure of the rock and so its transport properties (porosity, tortuosity, permeability) and these parameters can adequately estimated with laboratory experiments for real rocks. In particular, we apply dynamic Biot equations for 2D orthotropic media (in O_{xz} plane) [4]:

$$\rho_f \frac{T_x}{\phi} \frac{\partial q_x}{\partial t} + \frac{\eta}{k_x} q_x = -\frac{\partial p}{\partial x} - \rho_f \frac{\partial v_x}{\partial t} \tag{1}$$

$$\rho_f \frac{T_z}{\phi} \frac{\partial q_z}{\partial t} + \frac{\eta}{k_z} q_z = -\frac{\partial p}{\partial z} - \rho_f \frac{\partial v_z}{\partial t} \tag{2}$$

$$\rho \frac{\partial v_x}{\partial t} = \frac{\partial \sigma_{xx}}{\partial x} + \frac{\partial \sigma_{xz}}{\partial z} - \rho_f \frac{\partial q_x}{\partial t} \tag{3}$$

$$\rho \frac{\partial v_z}{\partial t} = \frac{\partial \sigma_{xz}}{\partial x} + \frac{\partial \sigma_{zz}}{\partial z} - \rho_f \frac{\partial q_z}{\partial t} \tag{4}$$

$$\frac{\partial \sigma_{xx}}{\partial t} = c_{11} \frac{\partial v_x}{\partial x} + c_{13} \frac{\partial v_z}{\partial z} + M\alpha_1 \left(\frac{\partial q_x}{\partial x} + \frac{\partial q_z}{\partial z} \right) \tag{5}$$

$$\frac{\partial \sigma_{zz}}{\partial t} = c_{13} \frac{\partial v_x}{\partial x} + c_{33} \frac{\partial v_z}{\partial z} + M\alpha_3 \left(\frac{\partial q_x}{\partial x} + \frac{\partial q_z}{\partial z} \right) \tag{6}$$

$$\frac{\partial \sigma_{xz}}{\partial t} = c_{55} \left(\frac{\partial v_z}{\partial x} + \frac{\partial v_x}{\partial z} \right) \tag{7}$$

$$-\frac{\partial p}{\partial t} = M\alpha_1 \frac{\partial v_x}{\partial x} + M\alpha_3 \frac{\partial v_z}{\partial z} + M \left(\frac{\partial q_x}{\partial x} + \frac{\partial q_z}{\partial z} \right) \tag{8}$$

Biot model describes solid particles velocity (v_x v_z) and relative fluid velocity (q_x q_z), as well as stress tensor components σ_{xx}, σ_{zz}, σ_{xz}, and fluid pressure p. Here ρ_f is the fluid density, η is fluid viscosity, ρ is the density of saturated material. Parameters T_x, T_z and k_x, k_z represent anisotropic tortuosity and permeability components for x-axis and z-axis directions, respectively. Equations also include undrained material modulus tensor components c_{ij}, fluid-solid coupling modulus M (also known as fluid storage coefficient) and Biot effective stress components α_x, α_z (Biot-Willis constants).

3   Numerical Scheme

To perform numerical experiments we present explicit second order finite-difference scheme approximating equations (1–8) on the staggered grid [23], where different fields components are defined in different nodes (Fig. 1). Let i, j, n be integer and correspond to grid node indices in x- and z-direction and number of time step, respectively. Then normal stresses σ_{xx} and σ_{zz}, are stored in nodes $(i\Delta x, j\Delta z, (n+1/2)\Delta t)$. Fluid and solid velocities x-components v_x, q_x are stored in nodes $((i+1/2)\Delta x, j\Delta z, n\Delta t)$, and z-components are defined in nodes $(i\Delta x, (j+1/2)\Delta z, n\Delta t)$. Shear stress σ_{zz} is stored in nodes $((i+1/2)\Delta x, (j+1/2)j\Delta z, (n+1/2)\Delta t)$. Note that physical parameters of the media are stored in nodes $(i\Delta x, j\Delta z)$ and need to be properly averaged in following equations.

Therefore we approximate equations (1–8) by the system of finite-difference equations

$$D_t[q_x]_{i+\frac{1}{2}j}^{n-\frac{1}{2}} = (C_1)_{i+1/2j} \left(D_x[\sigma_{xx}]_{i+\frac{1}{2}j}^{n-\frac{1}{2}} + D_z[\sigma_{xz}]_{i+\frac{1}{2}j}^{n-\frac{1}{2}} \right)$$
$$+ (C_2)_{i+1/2j} D_x[p]_{i+\frac{1}{2}j}^{n-\frac{1}{2}} + A_x \left[\frac{\eta}{k_x} \right]_{i+1/2j}^{n-1/2} C_2 A_t[q_x]_{i+1/2j}^{n-1/2}, \tag{9}$$

$$D_t[q_z]_{ij+\frac{1}{2}}^{n-\frac{1}{2}} = (C_3)_{i+j1/2} \left(D_x[\sigma_{xz}]_{ij+\frac{1}{2}}^{n-\frac{1}{2}} + D_z[\sigma_{zz}]_{ij+\frac{1}{2}}^{n-\frac{1}{2}} \right)$$
$$+ (C_4)_{i+j1/2} D_z[p]_{ij+\frac{1}{2}}^{n-\frac{1}{2}} + A_z \left[\frac{\eta}{k_z} \right]_{ij+\frac{1}{2}}^{n-\frac{1}{2}} C_4 A_t[q_z]_{ij+\frac{1}{2}}^{n-\frac{1}{2}}, \tag{10}$$

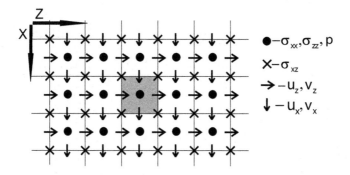

Fig. 1. Schematic illustration of staggered grid.

$$D_t[v_x]_{i+\frac{1}{2}j}^{n-\frac{1}{2}} = (C_5)_{i+1/2j} \left(D_x[\sigma_{xx}]_{i+\frac{1}{2}j}^{n-\frac{1}{2}} + D_z[\sigma_{xz}]_{i+\frac{1}{2}j}^{n-\frac{1}{2}} \right)$$
$$- (C_1)_{i+1/2j} D_x[p]_{i+\frac{1}{2}j}^{n-\frac{1}{2}} - A_x \left[\frac{\eta}{k_x} \right]_{i+1/2j}^{n-1/2} C_1 A_t[q_x]_{i+1/2j}^{n-1/2}, \tag{11}$$

$$D_t[v_z]_{ij+\frac{1}{2}}^{n-\frac{1}{2}} = (C_6)_{i+j1/2} \left(D_x[\sigma_{xz}]_{ij+\frac{1}{2}}^{n-\frac{1}{2}} + D_z[\sigma_{zz}]_{ij+\frac{1}{2}}^{n-\frac{1}{2}} \right)$$
$$- (C_3)_{i+j1/2} D_z[p]_{ij+\frac{1}{2}}^{n-\frac{1}{2}} - A_z \left[\frac{\eta}{k_z} \right]_{ij+\frac{1}{2}}^{n-\frac{1}{2}} C_3 A_t[q_z]_{ij+\frac{1}{2}}^{n-\frac{1}{2}}, \tag{12}$$

$$D_t[\sigma_{xx}]_{ij}^n = (c_{11}^u)_{ij} D_x[v_x]_{ij}^n + (c_{13}^u)_{ij} D_z[v_z]_{ij}^n$$
$$+ (\alpha_1)_{ij} M_{ij} (D_x[q_x]_{ij}^n + D_z[q_z]_{ij}^n), \tag{13}$$

$$D_t[\sigma_{zz}]_{ij}^n = (c_{13}^u)_{ij} D_x[v_x]_{ij}^n + (C_{33}^u)_{ij} D_z[v_z]_{ij}^n$$
$$+ (\alpha_3)_{ij} M_{ij} (D_x[q_x]_{ij}^n + D_z[q_z]_{ij}^n), \tag{14}$$

$$D_t[\sigma_{xz}]_{i+\frac{1}{2}j+\frac{1}{2}}^n = (\tilde{c}_{55}^u)_{i+1/2j+1/2}(D_z[v_x]_{i+\frac{1}{2}j+\frac{1}{2}}^n + D_x[v_z]_{i+\frac{1}{2}j+\frac{1}{2}}^n), \tag{15}$$

$$D_t[p]_{ij}^n = -\alpha_{ij} M_{ij} (D_x[v_x]_{ij}^n + D_z[v_z]_{ij}^n) + M_{ij} (D_x[q_x]_{ij}^n + D_z[q_z]_{ij}^n), \tag{16}$$

where D_x, D_z and D_t - second order central difference operators, and operators A_x and A_z represent spatial averaging in x- and z-axis, respectively. Operator A_t is time averaging. For discontinuous coefficients we apply the volume balance technique [21] to obtain the second order of convergence of numerical solution in models with sharp interfaces aligned to the grid lines [9,16]. So use following notations for coefficients:

$$(C_1^{-1})_{i+1/2j} = A_x[\rho_f]_{i+1/2j} - \frac{A_x[\rho]_{i+1/2j}A_x[T_x]_{i+1/2j}}{A_x[\phi]_{i+1/2j}},$$

$$(C_2^{-1})_{i+1/2j} = \frac{(A_x[\rho_f]_{i+1/2j})^2}{A_x[\rho]_{i+1/2j}} - \frac{A_x[\rho_f]_{i+1/2j}A_x[T_x]_{i+1/2j}}{A_x[\phi]_{i+1/2j}},$$

$$(C_3^{-1})_{ij+1/2} = A_z[\rho_f]_{ij+1/2} - \frac{A_z[\rho]_{ij+1/2}A_z[T_z]_{ij+1/2}}{A_z[\phi]_{ij+1/2}},$$

$$(C_4^{-1})_{ij+1/2} = \frac{(A_z[\rho_f]_{ij+1/2})^2}{A_z[\rho]_{ij+1/2}} - \frac{A_z[\rho_f]_{ij+1/2}A_z[T_z]_{ij+1/2}}{A_z[\phi]_{ij+1/2}}, \qquad (17)$$

$$(C_5^{-1})_{i+1/2j} = A_x[\rho]_{i+1/2j} - \frac{A_x[\rho_f]_{i+1/2j}A_x[\phi]_{i+1/2j}}{A_x[T_x]_{i+1/2j}},$$

$$(C_6^{-1})_{ij+1/2} = A_z[\rho]_{ij+1/2} - \frac{A_z[\rho_f]_{ij+1/2}A_z[\phi]_{ij+1/2}}{A_z[T_z]_{ij+1/2}}.$$

Finally, to average undrained stiffness tensor component c_{55} in nodes $((i + 1/2)\Delta x, (j + 1/2)\Delta z)$ we apply the formula of the harmonic mean:

$$(\tilde{c}_{55}^u)_{i+1/2j+1/2}^{-1} = \frac{1}{4}\left(\frac{1}{(c_{55}^u)_{ij}} + \frac{1}{(c_{55}^u)_{i+1j}} + \frac{1}{(c_{55}^u)_{ij+1}} + \frac{1}{(c_{55}^u)_{i+1j+1}}\right).$$

4   Problem Set Up

We apply numerical modeling of plane P-wave propagation within fractured porous fluid-saturated media. As a computational domain we consider rectangular area of poroelastic medium (Fig. 2). This medium consists of homogeneous background containing inhomogeneous layer, where heterogeneities are presented by the fractures filled with more permeable highly porous material to provide the flow of fluid. The signal is propagated in z-direction through fractured layer of interest and also through two receiver lines located at both sides of it. Thus, we record the wave before and after its interaction with the fractured layer.

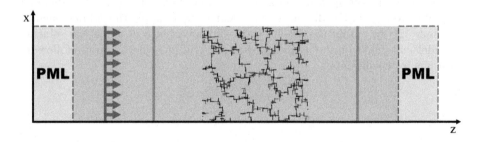

Fig. 2. A sketch of the computational domain and the acquisition system. Red line correspond to the source positions and wave propagation direction, two green lines represent the receiver positions (Color figure online).

The size of computational domain in x-direction is 1 m, and z-direction size L_z vary depending on the signal wavelength. At the boundaries $x = 0$ m and $x = 1$ m we impose periodic conditions, and perfectly matched layers [1,10,11] are set at two other boundaries $z = 0$ m and $z = L_z$ m. Grid step in both directions is $\Delta x = \Delta z = 2$ mm.

5   Numerical Estimation of Seismic Attenuation

As we want to observe correlation between fracture connectivity and seismic attenuation, proper estimation should be calculated for parameter characterizing this attenuation. Standard parameter demonstrating wave attenuation is inversed quality factor, and it can be calculated by many techniques [14]. One approach to calculate the inverse quality factor is using complex frequency-dependent phase velocity of seismic wave $c(\omega)$:

$$Q^{-1} = \frac{\Im(c(\omega))}{\Re(c(\omega))}. \tag{18}$$

Having recorded signals from numerical modeling of wave propagation in media of interest, it is convenient to apply spectral ratio method to estimate total attenuation.

To estimate inverse quality factor we consider a plane wave propagating in z-direction through the fractured domain

$$v(t, z) = v_0 e^{i\omega(t-z/c(\omega))} = v_0 e^{i\omega(t-zs(\omega))}, \tag{19}$$

where $c(\omega)$ is the complex frequency-dependent effective phase velocity of the medium, and $s(\omega) = 1/c(\omega)$ is the slowness. During wave propagation, wave amplitude v_0 is changing due to attenuation within the fractured media. Assume that first receiver line is located in $z = z_1$. Then the average (in x-direction) of the recorded signal in first receiver line is given by

$$\langle v(t, x, z_1)\rangle_x = v_1 = v_0 e^{i\omega(t-z_1 s(\omega))}. \tag{20}$$

In our numerical experiments, domain between two receiver lines contains three layers - two homogeneous layers on either side of third, fractured layer. So, first, it is necessary to take into account transmission losses in two homogeneous layers. Let us denote size in z-direction for left homogeneous layer, fractured layer, and right homogeneous layer by l_1, L_f and l_2, respectively, and also denote the slowness within background media by $s_b(\omega)$. Then we can represent averaged signal recorded in second receiver line

$$\langle v(t, x, z_2)\rangle_x = v_2 = v_0 e^{i\omega(t-z_1 s(\omega))} e^{-i\omega(l_1+l_2)s_b(\omega)} e^{-i\omega L_f s_f(\omega)}, \tag{21}$$

where $s_f(\omega)$ is effective slowness in fractured layer.

Let us denote factor representing wave propagation in homogeneous background by v_b, so

$$\langle v(t, x, z_2)\rangle_x = v_2 = v_1 v_b e^{-i\omega L_f \Re(s_f(\omega))} e^{\omega L_f \Im(s_f(\omega))}. \tag{22}$$

Hence we obtain the imaginary part of effective slowness in the layer of interest as follows:

$$\Im(s_f(\omega)) = \frac{1}{\omega L_f} ln \left| \frac{v_2}{v_1 v_b} \right|. \tag{23}$$

However, the real part of $s(\omega)$ is many-valued, and to determine the phase uniquely, we use the approximation of phase velocity in fractured layer. We pick the travel times t_1 and t_2 at the maximum amplitude of the two signals for the two receiver positions, and, since we know the background velocity, we can approximate phase velocity within the fractured layer by

$$V_P^0 = \frac{L_f}{t_2 - t_1 - t_b}, \tag{24}$$

where $t_b = (l_1 + l_2)/\Re(v_b)$ is the time of wave propagation in two homogeneous layers between the receiver lines. Therefore, the phase $\Re(s_f)$ is chosen to satisfy the condition

$$\left| \frac{\Re(s_f)}{\omega L_f} - \frac{\omega L_f}{V_P^0} \right| < \pi. \tag{25}$$

However, some error in attenuation estimation can be also associated with the fact that the fractured layer is finite. Moreover, the set up of experiment causes reflections on interfaces between homogeneous and fractured layer. To avoid these effects, consider two numerical experiments, where fractured layer lengths L_f and L'_f are different. Also assume that considered error is presented by the factor $e^{-i\phi(\omega)}$ (regardless of layer length) in the signal expression, so we have

$$v_2 = v_1 v_b e^{-i\omega(L_f \Re(s_f(\omega)) + \Re(\phi)/\omega)} e^{\omega(L_f \Im(s_f(\omega)) + \Im(\phi)/\omega)}, \tag{26}$$

$$v'_2 = v_1 v_b e^{-i\omega(L'_f \Re(s_f(\omega)) + \Re(\phi)/\omega)} e^{\omega(L'_f \Im(s_f(\omega)) + \Im(\phi)/\omega)}$$

for the averaged signals recorded in second receiver line for first and second experiment, correspondingly. Set these new signal representations equal to ones analogous to (21)

$$v_2 = v_0 e^{i\omega(t - z_1 s(\omega))} e^{-i\omega(l_1 + l_2) s_b(\omega)} e^{-i\omega L_f \hat{s}_f(\omega)}, \tag{27}$$

$$v'_2 = v_0 e^{i\omega(t - z_1 s(\omega))} e^{-i\omega(l_1 + l_2) s_b(\omega)} e^{-i\omega L'_f \hat{s}'_f(\omega)}, \tag{28}$$

where $\hat{s}_f$ and $\hat{s}'_f$ are two known different slowness estimations for two experiments. As a result, we obtain a system of two equations for our new estimation for imaginary part of the slowness

$$\Im(s_f(\omega)) + \frac{\Im(\phi(\omega))}{\omega L_f} = \Im(\hat{s}_f), \tag{29}$$

$$\Im(s_f(\omega)) + \frac{\Im(\phi(\omega))}{\omega L'_f} = \Im(\hat{s}_f'), \tag{30}$$

and its real part

$$\Re(s_f(\omega)) + \frac{\Re(\phi(\omega))}{\omega L_f} = \Re(\hat{s}_f), \tag{31}$$

$$\Re(s_f(\omega)) + \frac{\Re(\phi(\omega))}{\omega L'_f} = \Re(\hat{s}_f'). \tag{32}$$

Obviously, in such case we can obtain exact solution for $s_f(\omega)$, and then estimate inverse quality factor using formula (18). However, in case of three and more signals propagated in fractured layers of different length, one may apply, for example, least squares method to retrieve complex slowness.

6   Numerical Experiments

First we perform numerical experiments to verify the proposed attenuation estimation method. We model the propagation of seismic wave through homogeneous anisotropic layer located in more stiff and less porous and permeable background. For the fluid we set density $\rho_f = 1090$ kg/m^3, and fluid viscosity $\eta = 0.001$ Pa·s. Other physical properties of both materials are given in Table 1. We perform two sets experiments for two different lengths of anisotropic layer of interest $L = d$ and $L = 0.5d$, where d is approximately four wavelengths in the background.

Resulting numerical estimations of inverse quality factor and phase velocity are presented in Fig. 3. We can clearly observe that lesser length of the layer provide bigger error in comparison with both bigger length and analytical solution, which is obtained from dispersion relation of Biot model (1–8). However, applying our method described in Subsect. 5 results in more precise estimations, what is observed in better agreement with analytical estimations, especially for inverse quality factor. Note that although our numerical Q^{-1} estimation relative error (in comparison with analytical one) for worst case ($L = d$) at some frequencies almost reach 1, we observe good agreement of all velocity estimations with relative error less than 10^{-3}.

Next we perform a set of numerical experiments to study the anisotropy effect on seismic attenuation. We apply discrete fracture networks formed by two sets of microscale fractures parallel to x-axis and z-axis. Fractures are represented by rectangulars with length 30 mm and width 4 mm. Moreover, fracture systems are generated by simulated annealing approach [22, 24] to provide given distance, that can be traveled within continuous chain of fractures throughout the fractured model. In other words, we consider models with different connectivity degree. Fracture network generation process is described in detail in [18], and examples of models of different connectivity degrees are demonstrated in Fig. 4. We consider homogeneous isotropic extremely low permeable media as the background (physical properties are provided in the first column in Table 2),

Table 1. Material properties.

Parameter	Background	Homogeneous layer
Drained rock		
Porosity ϕ	0.1	0.3
x-axis Permeability k_x, m^2	10^{-13}	5^{-10}
z-axis Permeability k_z, m^2	10^{-13}	$2 \cdot 10^{-13}$
x-axis Tortuosity T_x	1.83	2.2
z-axis Tortuosity T_z	1.83	3.29
x-axis Biot and Willis constant α_1	0.2962	0.6759
z-axis Biot and Willis constant α_3	0.2962	0.9991
Undrained rock		
Density ρ, kg/m^3	2494	2182
Undrained modulus c_{11}, GPa	69.1	36.59
Undrained modulus c_{33}, GPa	69.1	6.885
Undrained modulus c_{13}, GPa	7.159	4.629
Undrained modulus c_{55}, GPa	31	0.02
Fluid storage coefficient M, GPa	20.102	6.8304

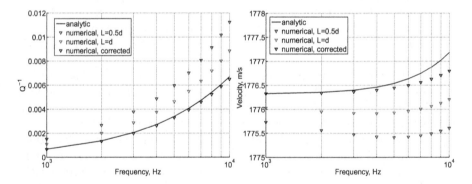

Fig. 3. Inverse quality factor (left) and phase velocity (right) for homogeneous anisotropic layer.

and fractures are filled with anisotropic material. Moreover, to intensify fluid flow within fractures along fractures of both orientation, we set proper physical parameters for x-axis oriented fractures, z-axis oriented fractures and its intersections. In particular, we set bigger permeability, smaller tortuosity and reduce stiffness of the material in fracture direction. Physical properties of fracture-filling materials for x-axis oriented fractures, z-axis oriented fractures and intersections of fractures of two different orientations are given in second, third, and fourth columns in Table 2, respectively. Fluid density is $1000\,\text{kg/m}^3$, and fluid viscosity is 0.001 Pa·s. Numerical experiments of wave propagation are performed for fractured models of six different degrees of connectivity (10 realizations for each case) and for 1–10 kHz initial signal central frequency range.

Table 2. Fractured media materials

Parameter	BG	FX	FZ	FXZ	ISO1	ISO2
Drained rock						
Porosity ϕ	0.1	0.225	0.225	0.225	0.2250	0.1425
x-axis Permeability k_x, m^2	10^{-15}	$1.414 \cdot 10^{-10}$	$5.3 \cdot 10^{-13}$	$1.414 \cdot 10^{-10}$	$1.414 \cdot 10^{-10}$	$5.3 \cdot 10^{-13}$
z-axis Permeability k_z, m^2	10^{-15}	$5.3 \cdot 10^{-13}$	$1.414 \cdot 10^{-10}$	$1.414 \cdot 10^{-10}$	$1.414 \cdot 10^{-10}$	$5.3 \cdot 10^{-13}$
x-axis Tortuosity T_x	1.83	1.17	1.83	1.17	1.17	1.83
z-axis Tortuosity T_z	1.83	1.83	1.17	1.17	1.17	1.83
x-axis Biot and Willis constant α_1	0.2962	0.6896	0.6418	0.6915	0.7346	0.7346
z-axis Biot and Willis constant α_3	0.2962	0.6418	0.6896	0.6915	0.7346	0.7346
Undrained rock						
Density ρ, kg/m^3	2494	1870	2458	2318	2318	2318
Undrained modulus c_{11}, GPa	69.097	38.958	46.432	38.958	38.958	46.432
Undrained modulus c_{33}, GPa	69.097	46.432	38.958	38.958	38.958	46.432
Undrained modulus c_{13}, GPa	7.159	19.811	19.811	19.811	16.337	23.81
Undrained modulus c_{55}, GPa	30.969	11.311	11.311	11.311	11.311	11.311
Fluid storage coefficient M, GPa	20.1	9.4287	9.4287	9.4026	9.33	9.488

Resulting fluid pressure field snapshots for six fractured media models with different connectivity degree are demonstrated in Fig. 5. As background permeability is very low, fluid flow in fractures is induced by pressure contrasts at the background/fracture interfaces. So, we observe intensive pressure gradients behind propagated wave only within the connected fractures. Moreover, pressure gradients increase more intensively when fracture connectivity is relatively low (snapshots A, B and C in Fig. 5). Difference in pressure gradients in fracture systems with high connectivity are almost unobservable.

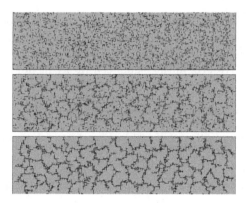

Fig. 4. Models of fractured media with different percolation length (increasing from top to bottom).

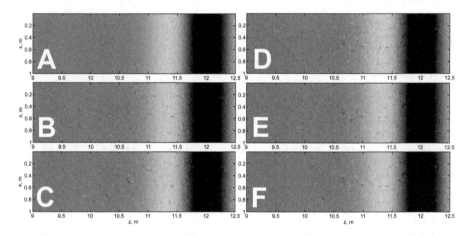

Fig. 5. Fluid pressure in fractured media with different percolation (percolation length increases from A to F). Signal central frequency is 3 kHz.

To study anisotropy effect on seismic attenuation we also perform a set of numerical experiments considering isotropic fracture-filling material. All fractured media models are the same as described above except physical properties for material in fractures. We consider two models of material. First model is soft and have high permeability (similar to properties along the fractures in previous experiments) in both axis directions (fifth column in Table 2). In contrast, second model represent stiff and low permeable isotropic material with lower porosity (sixth column in Table 2).

To compare two isotropic fracture-filling materials we demonstrate fluid pressure field snapshots in Fig. 6 for almost not connected fractures and higher connectivity considered. In soft material case pressure gradients within fractures can be easily observed. However, stiff material provide almost no pressure gradients. Additionally, we again observe higher pressure gradients for high fracture connectivity, at least for soft material in fractures.

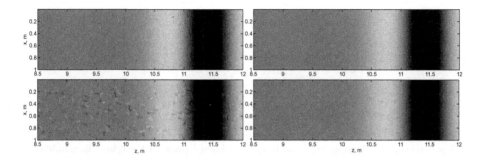

Fig. 6. Fluid pressure in isotropic fractured media. Top images correspond to the lowest fracture connectivity, bottom images represent results for the highest fracture connectivity; row correspond to fracture-filling material: left - soft, right - stiff. Signal central frequency is 3 kHz.

Finally, we estimate wave attenuation for all three fracture-filling materials and all six degrees of connectivity. Obtained frequency-dependent inverse quality factor estimates are presented in Fig. 7. First, we observe the increase of attenuation with fracture connectivity increase. Also attenuation predictably increases with frequency increase in general, and attenuation peak for all cases at the frequency near to 5 kHz can be caused by fractured model periodicity. Additionally, we see that most significant increase of attenuation occurs from almost non-intersecting fractures to first two considered connectivity stages. Furthermore, the highest values of Q^{-1} are obtained in case of isotropic soft material for almost all frequencies and connectivity degrees. Estimations for isotropic stiff fracture-filling material are generally smaller than other estimations, except the case of non-intersecting fractures. Such effect is probably observed because with negligible wave induced fluid flow (both between fractures and background and within fractures) scattering is the only significant attenuation mechanism, which is independent of permeability. As expected, attenuation in the case of anisotropy almost always stays in range limited by attenuation for isotropic stiff material from below and by attenuation for isotropic soft material from above. However, attenuation observably differs between anisotropic and isotropic fracture filler. Thus it is important to take into account the anisotropy of material providing fluid transport in fractured porous fluid-saturated media.

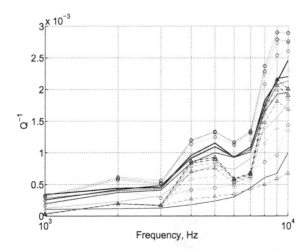

Fig. 7. Inverse quality factor for fractured media. Line style corresponds to fracture-filling material: thick solid lines - anisotropic, dashed lines with circles - isotropic soft, dash-and-dot lines with triangles - isotropic stiff. Colors correspond to fracture connectivity degrees in increasing order: pink, yellow, red, green, blue, black (Color figure online)

7   Conclusion

In our work we present a numerical algorithm of seismic wave propagation in anisotropic fractured porous fluid-saturated media and estimate wave attenuation in such media. The algorithm is based on numerical solution of Biot equations of poroelasticity using staggered-grid finite-difference scheme. Seismic attenuation estimation is provided by inverse quality factor calculation based on spectral ratio approach. Presented algorithm is used to study the effect of fracture-filling material anisotropy and fracture connectivity on seismic wave attenuation. We numerically model wave propagation through fractured media with complex two-scale structure in frequency range from 1 to 10 kHz. Numerical experiments include different degrees of fracture connectivity as well as fracture-filling material properties. We consider anisotropic material in fractures providing better transport properties along the fractures than transversely to fractures. Results show that both fracture connectivity and anisotropy significantly affects seismic wave attenuation.

References

1. Berenger, J.P.: A perfectly matched layer for the absorption of electromagnetic waves. J. Comput. Phys. **114**(2), 185–200 (1994)
2. Biot, M.A.: Theory of propagation of elastic waves in fluid-saturated porous solid. I. Low-frequency range. J. Acoust. Soc. Am. **28**, 168–178 (1956)

3. Carcione, J.M., Morency, C., Santos, J.E.: Computational poroelasticity – a review. Geophysics **75**(5), 75A229–75A243 (2010)
4. Cheng, A.H.D.: Material coefficients of anisotropic poroelasticity. Int. J. Rock Mech. Min. Sci. **342**, 199–205 (1997)
5. Guo, J., Rubino, J.G., Glubokovskikh, S., Gurevich, B.: Effects of fracture intersections on seismic dispersion: theoretical predictions versus numerical simulations. Geophys. Prospect. **65**(5), 1264–1276 (2017)
6. Huang, F., et al.: The first post-injection seismic monitor survey at the Ketzin pilot CO_2 storage site: results from time-lapse analysis. Geophysical Prospect. **66**(1), 62–84 (2018)
7. Hunziker, J., et al.: Seismic attenuation and stiffness modulus dispersion in porous rocks containing stochastic fracture networks. J. Geophys. Res. Solid Earth **123**(1), 125–143 (2018)
8. Kong, L., Gurevich, B., Zhang, Y., Wang, Y.: Effect of fracture fill on frequency-dependent anisotropy of fractured porous rocks. Geophys. Prospect. **65**(6), 1649–1661 (2017)
9. Lisitsa, V., Podgornova, O., Tcheverda, V.: On the interface error analysis for finite difference wave simulation. Comput. Geosci. **14**(4), 769–778 (2010)
10. Lisitsa, V.: Optimal discretization of PML for elasticity problems. Electron. Trans. Numer. Anal. **30**, 258–277 (2008)
11. Martin, R., Komatitsch, D., Ezziani, A.: An unsplit convolutional perfectly matched layer improved at grazing incidence for seismic wave propagation in poroelastic media. Geophysics **73**(4), T51–T61 (2008)
12. Marty, N.C.M., Hamm, V., Castillo, C., Thiery, D., Kervevan, C.: Modelling water-rock interactions due to long-term cooled-brine reinjection in the dogger carbonate aquifer (paris basin) based on in-situ geothermal well data. Geothermics **88**, 101899 (2020)
13. Masson, Y.J., Pride, S.R.: Finite-difference modeling of Biot's poroelastic equations across all frequencies. Geophysics **75**(2), N33–N41 (2010)
14. Mavko, G., Mukerj, T., Dvorkin, J.: The Rock Physics Handbook, 2nd edn. Cambridge University Press, Cambridge (1999)
15. Menke, H.P., Reynolds, C.A., Andrew, M.G., Pereira Nunes, J.P., Bijeljic, B., Blunt, M.J.: 4D multi-scale imaging of reactive flow in carbonates: assessing the impact of heterogeneity on dissolution regimes using streamlines at multiple length scales. Chem. Geol. **481**, 23–37 (2018)
16. Moczo, P., Kristek, J., Galis, M.: The Finite-Difference Modelling of Earthquake Motion: Waves and Ruptures. Cambridge University Press, Cambridge (2014)
17. Muller, T.M., Gurevich, B., Lebedev, M.: Seismic wave attenuation and dispersion resulting from wave-induced flow in porous rocks – a review. Geophysics **75**(5), 75A147–75A164 (2010)
18. Novikov, M.A., Lisitsa, V.V., Bazaikin, Y.V.: Wave propagation in fractured porous media with different percolation length of fracture systems. Lobachevskii J. Math. **41**(8), 1533–1544 (2020)
19. Rubino, J.G., Muller, T.M., Guarracino, L., Milani, M., Holliger, K.: Seismoacoustic signatures of fracture connectivity. J. Geophys. Res. Solid Earth **119**(3), 2252–2271 (2014)
20. Salaun, N., et al.: High-resolution 3D seismic imaging and refined velocity model building improve the image of a deep geothermal reservoir in the Upper Rhine Graben. Lead. Edge **39**(12), 857–863 (2020)
21. Samarskii, A.A.: The Theory of Difference Schemes. (Pure and Applied Mathematics), vol. 240. CRC Press, Bosa Roca (2001)

22. Tran, N.: Simulated annealing technique in discrete fracture network inversion: optimizing the optimization. Comput. Geosci. **11**, 249–260 (2007)
23. Virieux, J.: P-SV wave propagation in heterogeneous media: velocity-stress finite-difference method. Geophysics **51**(4), 889–901 (1986)
24. Xu, C., Dowd, P.: A new computer code for discrete fracture network modelling. Comput. Geosci. **36**(3), 292–301 (2010)

Record-to-Record Travel Algorithm for Biomolecules Structure Prediction

Ioan Sima$^{(\boxtimes)}$ ⓘ and Daniela-Maria Cristea ⓘ

Babes-Bolyai University, 400084 Cluj-Napoca, Romania
{sima.ioan,danielacristea}@cs.ubbcluj.ro,
http://www.cs.ubbcluj.ro/

Abstract. The biomolecules structure prediction problem (BSP) - especially the protein structure prediction (PSP) and the nucleic acids structure prediction - was introduced in the computational biology field approximately 45–50 years ago. The PSP on hydrophobic-polar lattice model (HP model) is a combinatorial optimisation problem, and consists in aims to minimize an arbitrary energy function associated with every native structure.

To solve the PSP problem, many metaheuristic methods were applied. Although the record-to-record travel algorithm (RRT) has proven useful in solving combinatorial optimisation problems, it has not been applied so far to solve the PSP problem.

In this paper, a mathematical modeling for PSP on the 2D HP rectangular lattice is developed and an adapted record-to-record travel algorithm (aRRT) is applied to address the combinatorial optimisation problem. For candidate solutions perturbation, a rotation and a diagonal move mutation operators were used.

A benchmark data set is used to test the RRT algorithm. The results obtained show that the algorithm is competitive when compared to the best published results. The main advantage of the RRT algorithm is that it is time-efficient, and requires small computational resources to obtain the same results as swarm intelligence algorithms.

Keywords: Record-to-record travel algorithm · Local search · Metaheuristic · Biomolecules · Protein structure prediction · HP lattice model

1 Introduction

Biomolecules are small or large molecules, being essential to biological processes present in living organisms. In chemical language, the large molecules are called macromolecules. All macromolecules are composed of tens to hundreds of thousands, even millions, of smaller subunits called monomers [8].

Macromolecules have several levels of structural organization. Biomacromolecular compounds such as proteins or nucleic acids (DNA and RNA) have four structure levels: primary, secondary, tertiary and quaternary structure,

© Springer Nature Switzerland AG 2021
O. Gervasi et al. (Eds.): ICCSA 2021, LNCS 12949, pp. 449–464, 2021.
https://doi.org/10.1007/978-3-030-86653-2_33

respectively. In particular, all proteins reach the tertiary structure, but not all get to the quaternary structure. In the last two structure levels (tertiary or quaternary) the proteins and nucleic acids fold up to the functional state, called native structure (or folded conformation) [5].

Biomolecules structure prediction (BSP) consists in predicting the native structure based on the information stored in the primary structure. Myriads of tertiary conformations can be generated from only one primary structure. Only one conformation from this huge set of conformations is native conformation (i.e. it has the correct biological function). Even with today's most powerful computers, it is unfeasible in terms of time consumption to go through and check all the tertiary structures, even for very short macromolecules of dozens of monomers.

To address this problem by computational methods, a multitude of models have been proposed over time. One of the simplest models is hydrophobic-polar (HP), suitable for protein structure prediction problem (PSP), was presented in 1985 by Dill and developed in the next years [4,11]. It uses discretized spaces to represent monomers positions on the lattice. PSP on lattice model is a problem where a set of beans corresponding to the monomers of a protein must be arranged on a certain type of lattice. Even if this model is very simple, the exhaustive search in the combinatorial space is not feasible because PSP on HP model is proved to be NP-complete [2,3].

Consequently, a series of non-deterministic search techniques and metaheuristics have been applied to solving PSP so far. There were applied several classes of such techniques: evolutionary, metaheuristics, machine and deep learning, Monte-Carlo simulation and so on.

Among the first authors to address this problem were Unger and Moult, who applied genetic algorithm (GA) to the HP model [24]. GA-based researches continue to this day in a variety of variants. A combination of GA with Particle Swarm Optimization (PSO) was presented by Zhou [26], and Huang applied GA based on Optimal Secondary Structures (GAOSS) [10]. Recently, another combination of GA with Great Deluge Algorithm (GDA) proposed by Turabieh [23], and Evolutionary Monte Carlo (EMC) was applied on 2D HP by Liang and Wong [12]. Metaheuristic techniques used include Ant Colony Optimization Algorithm (ACO) [20], (PSO) [16], Ions Motion Optimization (IMO) algorithm combined with the Greedy Algorithm (IMOG) [25], etc.

Record-to-Record Travel algorithm (RRT) was introduced by Dueck in 1993 [6] as a variant of threshold accepting (TA) local search (LS) method [7]. It was applied with good results to many non-polynomial problems such as traveling salesman, knapsack, linear programming, quadratic programming, minimization of spin-glass Hamiltonian's, rough set theory [13,14], vehicle routing problem [19], etc.

This work applies an RRT algorithm to solve the PSP problem on the HP model. To our knowledge, although RRT is a fairly old algorithm, it has not yet been applied to this problem.

This paper is organized as follows. After the introduction, the optimisation problem formulation is given in Sect. 2. The third section explains the adapted Record-to-Record Travel algorithm. Section 4 contains the computational results and discussion, and the final section summarizes conclusions and further work.

2   Optimisation Problem

2.1   Biochemical Background

Biomolecules (or biological molecules) are molecules present in living organisms, frequently produced by them; they are essential to biological processes. Famous biomolecules include small molecules like amino acids (AA's), lipids, carbohydrates, nucleotides, antioxidants, polyols, vitamins, etc. and large macromolecules such as polynucleotides (DNA, RNA), proteins or polysaccharides [8].

All macromolecules are composed of tens to hundreds of thousands, even millions, of smaller subunits called monomers. For proteins, the monomers are the amino acids (20 types), for carbohydrates, the monomers are the monosaccharides, and nucleic acids (DNA, RNA) are build of the nucleotides. The proteins are organized in the form of complex structures that can be layered on four levels, called: primary, secondary, tertiary and quaternary structure. Protein primary structure is represented by the linear sequence (protein chain) of twenty types of amino acids. Secondary structures are represented by small local folding motifs, and tertiary structures of macromolecules are the three-dimensional shape of polymers [5]. This structure level is very important because proteins have biological function only in this form. The quaternary structure is a less common level, in which interactions between multiple 3D tertiary units occur.

The role of nucleic acids is to store genetic information, while proteins have multiple biological functions. Proteins are the molecules that provide life (the living). All other biomolecules participate in the proper functioning of the proteins. In conclusion, proteins are the most important constituents of biological life, hence the importance of knowing the native structure (folded) and the possibility of predicting it from the sequence of AA's (primary structure).

Biomolecules structure prediction consists of predicting the native structure based on the information stored in the primary structure, i.e. folded protein structure predicted from its linear sequence of amino acids.

For proteins, to one amino acids sequence (primary structure) there are many corresponding tertiary (or quaternary) structures. The problem is that the number of possible tertiary structures increases exponentially related to the length of the AA's sequence. Only one structure (called the native conformation) of this set of tertiary structures is physiological, aka is a "good" conformation. All other tertiary structures are pathological forms ("bad" conformations), causing various diseases [9]. The study of the protein folding phenomenon began in the '60s. Protein folding is the physical process by which a protein acquires its native conformation. As a result, researches on protein structure prediction are emerging in the next decade. The energy landscape theory of protein folding assumes that a protein's native state corresponds to its free energy minimum [1]. Figure 1

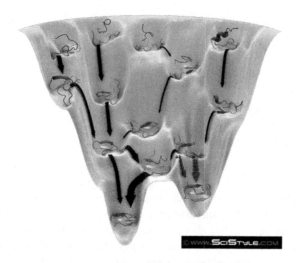

Fig. 1. An artistic representation of the funnel folding. The image was created by Thomas Splettstoesser [22]

shows the artistic image of the folding funnel, where many conformations fall down in local energy minimums. "Good" native conformation is found in global energy minimum.

Over time, starting with the '70s, to simplify and solve the PSP problem, various types of models have been proposed: all-atoms models, bean models, full energy models, statistical backbone potential models, lattice models, etc. [17].

2.2   The Hydrophobic-Polar Model

One of the simplest and well-known models for protein structure prediction is hydrophobic polar lattice model [11]. PSP on HP lattice is a discrete combinatorial optimisation problem. This model is a standard in testing the efficiency of the folding algorithms.

It introduces two main simplifications: i) the continuous Euclidean space is discretized, and every AA occupies one node of a specific regular lattice; ii) the 20 types of amino acids residues are reduced to two types of letters, based on their hydrophobicity:

1) "H": (Gly - G, Ala - A, Pro - P, Val - V, Leu - L, Ile - I, Met - M, Phe - F, Tyr - Y, Trp - W) and
2) "P": (Ser - S, Thr - T, Cys - C, Asn - N, Gln - Q, Lys - K, His - H, Arg - R, Asp - D, Glu - E) [18].

Note that, AA's can be denoted with 1-letter or 3-letter symbols. The "H" letter represents hydrophobic AA, and the "P" letter represents polar AA.

A *lattice*, Λ, in $\mathbb{R}^n$, is a set of n-dimensional vectors that form an additive group for any two points $\mathbf{u}, \mathbf{v} \in \Lambda$, i.e. it holds $(\mathbf{u} \pm \mathbf{v}) \in \Lambda$ [15].

A typical lattice Λ in $\mathbb{R}^n$ has the form:

$$\Lambda = \left\{ \sum_{i=1}^{n} b_i \mathbf{v}_i \; \middle| \; b_i \in \mathbb{Z} \right\} \tag{1}$$

where $\{\mathbf{v}_1, \mathbf{v}_2, ..., \mathbf{v}_n\}$ is a basis for vector space $\mathbb{R}^n$.

Common lattices that have been studied in the context of lattice protein models are: 2D rectangular (in particular, 2D square), 2D triangular, 3D cubic and 3D face-centered-cubic (FCC) lattice.

At the primary structure level, a protein is represented by the ordered sequence Pr, of length n, where $Pr = (a_1, ..., a_n)$, and a_i is the ith AA in the sequence with $a_i \in \{A, R, N, D, C, E, Q, G, H, I, L, K, M, F, P, S, T, W, Y, V\}$.

In the HP model, the lattice coordinates define possible placements for AA's, and the primary structure of the protein will be represented by a string called HP, where $HP = (h_1, h_2, ..., h_n)$, and $h_i \in \{H, P\}$. The HP string is a constant because the order of the AA's in the primary structure of the protein sequence remains unchanged throughout the algorithm run. It can be put on the lattice, forming self-avoided walks (SAWs). A SAW is a sequence of moves on a lattice that does not visit the same point more than once. There are many ways to arrange an HP string to the lattice. Every path is encoded by a sequence of directions on a lattice, which is a string of length n or $n-1$. This string, which corresponds to a protein tertiary structure (rarely, to the quaternary structure), can be encoded in many ways: relative directions (SRL string; Straight, Right and Left), absolute directions ($RULD$ string; Right, Up, Left and Down), binary representations, etc. All paths on the lattice must comply with the SAW rule.

Two adjacent letters of HP string are *sequence neighbours (or connected neighbours)*. Two non-adjacent letters of HP string, but which are neighbours after arranging on the lattice, are called *topological neighbours*.

In biological proteins, the folding process is strongly influenced by the hydrophobicity, i.e. they fold in the form of almost spherical globules with the hydrophobic amino acid residues in the protein kernel, and the polar ones, on the protein surface. To simulate this phenomenon, a distance-based arbitrary energy function is introduced. There are two types of letters in the HP string, thus, there are 2^2 types of contacts between topological neighbours. For every kind of contact, a value of free energy is introduced, as follows: e(H,H) = -1; e(H,P) = 0; e(H,P) = 0; e(P,P) = 0. Thus, H-H contacts are encouraged, and the others are neither encouraged nor penalized. A H-H contact (h_i, h_j) between the i-th and j-th AA from the original sequence can be formed iff $|i-j| > 2$.

The total free-energy, $E(c)$, of a walk (or conformation) c is computed as the number of H-H contacts, taken with the minus sign:

$$E(c) = - \sum_{i,j=1, |i-j|>2}^{n} e(a_i, a_j) \tag{2}$$

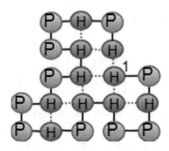

Fig. 2. Optimal conformation with the hydrophobic kernel (red) and polar surface (blue). Legend: *Hp* string (Input): HPHPPHHPHPPHPHHPPHPH. *RULD* string (Output): CRDDLULDLLURURULURRD. E(c) = −9 (number of dot lines). H - hydrophobic amino acid. P - polar amino acid. E(c) - energy of conformation (Color figure online)

In the optimisation approach, the conformation with the smallest energy has the greatest chance of forming a hydrophobic kernel (see Fig. 2); the search of native conformation from tertiary structures set is transformed into a search for the optimal walk on lattice (the SAW with the minimum total free-energy). In the HP model, the finding of native conformation in both 2D and 3D is NP-complete [2,3].

Table 1 summarizes the simplifications brought by the HP model and the correspondences to biological proteins.

Table 1. HP model summary

Biological proteins	Protein HP data structures
Primary structure	*HP string*
Tertiary structure (conformation)	*RULD string* - for absolute directions encoding
	SLR string - for relative directions encoding
Physical free-energy	Conventional free-energy
n - number of AA's	n - number of letter in HP string
Number of biological conformations $\geq 9^{n-1}$	Number of walks on 2D HP lattice: 4^{n-1} - absolute encoding 3^{n-1} - relative encoding
	Number of SAW: unknown

2.3   Mathematical Model for Protein Structure Prediction

In this work, the 2D HP square lattice, absolute encoding of directions and space coordinates of lattice positions have been used to denote the protein conformations (viz. SAWs).

The 2D square lattice is:

$$\Lambda = \{b_1 \mathbf{v}_1 + b_2 \mathbf{v}_2\} \tag{3}$$

where $b_1, b_2 \in \mathbb{Z}$, $\{\mathbf{v}_1, \mathbf{v}_2\}$ is a basis for vector space $\mathbb{R}^2$, and the vectors $\mathbf{v}_1$ and $\mathbf{v}_2$ meet the conditions: 1) $\mathbf{v}_1 \perp \mathbf{v}_2$, and 2) $\|\mathbf{v}_1\| = \|\mathbf{v}_2\|$. Thus, the primitive cell of a 2D HP lattice is a square whose sides, by convention, have length 1 (see Fig. 3).

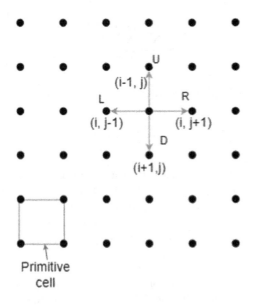

Fig. 3. The 2D square lattice

The main role of the lattice is to divide the continuous space in a discrete space. Thereby, amino acids spatial position is restricted to precise points in space.

For an *HP* sequence with n AA's, the *RULD* is a string of length n, where $RULD = (d_1, d_2, ..., d_k, ..., d_n) \in \Lambda$, and $d_k \in \{C, R, U, L, D\}$. It represents a path on the 2D square lattice which must comply with the SAW restriction: in space coordinates, each *RULD* letter is associated with a distinct 2D vector ($\forall k \neq l : \mathbf{v}_k \neq \mathbf{v}_l$). In 2D HP lattice, each node (point or vector) from a given position *(i, j)* has four neighbours, as you can see in Fig. 3. Hence, it follows that each AA can have a maximum of 4 AA's neighbours (extreme AA's in the *HP* sequence have one sequence neighbour and can have maximum three topological neighbours, and the others have two sequence neighbours and can have maximum two topological neighbours).

Based on the minimum energy principle and the abstraction presented above, the following optimisation model (OM) for PSP problem on 2D square lattice can be established:

$$min \quad E(c) \tag{4}$$

subject to:

$$x^1_{n,n} = 1 \tag{5}$$

$$x^2_{n,n+1} = 1 \tag{6}$$

$$\sum_{k=1}^{n} \sum_{i=1}^{2n-1} \sum_{j=1}^{2n-1} x^k_{i,j} = n \tag{7}$$

$$0 \le \sum_{k=1}^{n} x^k_{i,j} \le 1, \quad \forall i, j \in \{1, 2, ..., 2n-1\} \tag{8}$$

$$x^{k+1}_{i,j} \le x^k_{i,j-1} + x^k_{i,j+1} + x^k_{i-1,j} + x^k_{i+1,j} \tag{9}$$

where:

$E(c)$ is the free energy of $RULD$ string c (protein conformation), $x^k_{i,j}$ is a three-dimensional variable where i and j are 2D vector coordinates of the lattice, and k is an kth AA in the HP sequence, such that:

$$x^k_{i,j} = \begin{cases} 1, & if \quad x^k = (i,j) \\ 0, & else \end{cases} \tag{10}$$

Here, "0" means that at position (i, j) there is no AA, and "1" means that at position (i, j) there is only one AA. Equation (5) fixes the first AA in the HP string at the position (n, n) on the lattice, and (6) fixes the second AA at the position $(n, n+1)$ on the lattice (right direction). This second constraint reduces the combinatorial space four times. Equations (7), (8) and (9) constrain that a walk in the grid occupies exactly n nodes, each lattice node cannot contain more than one AA and the two sequence neighbours AA's occupy the adjacent nodes in the lattice, respectively.

3   Record-to-Record Travel Algorithm

3.1   The Basic Algorithm

Record-to-Record Travel algorithm (RRT) as well as Great Deluge Algorithm (GDA), introduced by Dueck in 1993 [6], belong to the Threshold Accepting algorithm (TA) [7]. All these methods (RRT, TA, GDA) are variants of Local Search (LS) metaheuristics. They are suitable for discrete optimisation; they are essentially *one-parameter* algorithms, meaning that it is necessary to tune only a single parameter, called *Deviation*.

The RRT and GDA algorithms are different from their predecessors, like Hill-Climbing or Simulated Annealing, in the acceptance way of a candidate solution from a neighbourhood. The RRT algorithm is very similar to the GDA algorithm, which is described by his author, Dueck, as follows: "Imagine, the GDA is to find the maximum point on a certain surface, for instance, the highest point in a fictitious empty country. Then, we let it rain without end in this country. The algorithm walks around in this country, but it never makes a step beyond the ever-increasing water level. And it rains and rains.... Our idea is that in the end the GDA gets wet feet when it has reached one of the very highest points in the country, so that is has found a point close to the optimum."

RRT differs from GDA by two points: i) RRT water level increase is given by the quality of the best solution found until that iteration; ii) a variable *Record* is updated only if the quality of the best solution is greater than its value.

To avoid the local minimum, RRT algorithm allows a poorer solution than the current solution ("submerged" solutions up to the value of the *Deviation* variable are accepted).

3.2 The Adapted Record-to-Record Travel Algorithm

In this paper, the RRT algorithm was adapted for PSP problem-solving. The description of adapted RRT includes three stages: representation and generation of the random initial conformation, evaluation of the conformations, and mutation.

Representing and Generating of the Random Initial Conformation. In this work, absolute directions encoding was used. The important data structures are: "*HP*" string, "*RULD*" string and "Chromosome" class, which encapsulates the two types of strings - a borrowed notion from the GA language. The *HP* string is the RRT input parameter, and the *RULD* string is the output one; both have the same length.

The generation of the first random solution (*RULD* string) starting from the *HP* string, is done through a stochastic backtracking variant in the same way as in our previous work [21]. For *HP* string (e.g. a chain of 6 AA's is: HHPHHP) it is built one *RULD* string (it can be CRRUUR). The first AA, "C", is placed in the center of the lattice, and the other AA's follow the next directions: right, right, up, up, and right. The first and the second letter ("CR") are always fixed. The next letter is chosen randomly from the {R, U, L, D} alphabet. Then the SAW condition is checked. If the point on the lattice is unoccupied, the next direction is generated and so on. Otherwise, the other letters of the alphabet {R, U, L, D} are chosen. If no direction passes the SAW rule, then it returns to the previous letter and the algorithm is resumed.

Evaluation of the Conformations. The energy of newly generated conformation is computed by:

$$E(c) = f(HPstring, RLUDstring) \tag{11}$$

Function f doesn't have an analytical formula. The algorithm counts the
number of H-H contacts (topological neighbours) of the walk on the lattice.
It depends on all the arguments shown in Eq. (11). In Fig. 2 is an example
of optimum conformation well-arranged on the 2D HP lattice of an input HP
strings with n = 20 AA's, the output $RULD$ string, and the minimum energy.

Mutation. In the adapted RTT algorithm is implemented a k-point rotation
and a diagonal move mutation operator. A conformation (or chromosome), $\mathbf{C} =$
$\{d_1, d_2, ..., d_n\}$, where $d_i \in \{C, R, U, L, D\}$ is mutated to a new chromosome $\mathbf{C}$'.
The position g ($3 \leq g \leq$ n), called mutation point, is random chosen in every
iteration, and the letter at position g is replaced by one letter sampled uniformly
from the $\{R, U, L, D\}$ set. The number of mutation points in an iteration is
equal with the value of k ($1 \leq k \leq n - 2$). The letters at positions 1 and 2 (d_1
and d_2) remain unaltered by mutation because the first AA, C, is fixed in the
lattice center, and the second AA is always R (positioned to the right relative to
the first) because at the second position the other three directions (Up, Left or
Down) give the same conformations, rotated by 90°, 180° and 270°, respectively.
When $k = d$, the mutation produces a completely new random conformation,
this way helping to explore the combinatorial space.

If the new letter, from position g, produces a feasible conformation, then the
second part of the chain is rotated with 90°, 180° or 270°, respectively. The
rotation occurs for k positions in each iteration.

The mutation operation (which is a small disturbance) can produce a conformation of quality close to the original conformation or one of very different
quality, either extremely weak or optimal one. This is a way to see the chaotic
behavior of the energy function on the HP model (Eq. 2). I.e., for sequence 1
from the data set, a mutation that substitutes U → R at the 12th position of
a weak conformation (E = −4), CRDDLULDLLUULULDLUUR, produces an
optimum conformation (E = −9), CRDDLULDLLURURULURRD (see Fig. 4).

In the second phase, the RRT algorithm searches the existence of one of the
16 $RULD$ substrings that designate the square corner: UUR; UUL; DDR; DDL;
RRU; RRD; LLU; LLD; URU; ULU; DRD; DLD; RUR; RDR; LUL; LDL. For
the last substring found a diagonal move is executed, as shown in Fig. 5. If the
new conformation is feasible, it is kept.

Finally, the mutation operation is:

$$\mathbf{C}(t+1) = Mutation(\mathbf{C}(t), g) \tag{12}$$

If the mutation operation fails (i.e. the resulting conformation is not feasible),
a new letter is chosen from the directions not yet explored for the rotation
mutation. If no variant is feasible, it proceeds to the next iteration and the old
conformation is kept.

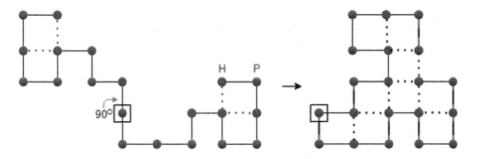

Fig. 4. The rotation mutation

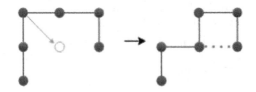

Fig. 5. The diagonal mutation

The Algorithm. After generating the first solution, its energy is computed, which is a negative value. The *Record* parameter gets this value. The *Deviation* parameter is set to a negative value close to zero. Afterward, for each iteration, a stochastic small perturbation is applied to the old configuration. If the energy of the new solution is better than the energy of the old solution, then the *Record* parameter is updated to the new energy. If the new energy is better than old energy or fall between old *Deviation* and *Record* parameters, then the new solution is accepted, even if it is worse solution. This step ensures the exploration of the solution space.

Because PSP on HP lattice is a minimization problem, the adapted RRT algorithm is mirrored, unlike the basic algorithm. Pseudo-code of the adapted RRT algorithm can be seen in Algorithm 1.

RRT algorithm works as follows: let **C** be the current solution with a *Record* equal to its free-energy and **C'** as an alternative solution derived from the first one through a small variation, like a mutation. The *Deviation* is set to a negative value close to zero. If the new free energy of **C'** is smaller than (*Record* - *Deviation*), then **C'** is the new solution in the iteration.

The main advantages of this algorithm are: it is time-efficient, requires small computational resources to obtain the same results as the other algorithms, and it has only two tuning parameters: *Deviation* and *MaxIter* (the number of iteration). Besides that, the RRT algorithm is easy to implement.

Algorithm 1: RRT algorithm

    Input   : $MaxIter$, HP sequence
    Output: **C*** (optimum RULD string)

1 $t \longleftarrow 0$;
2 Generate a random initial conformation, $\mathbf{C}(t)$ (a RULD string);
3 Update $\mathbf{C*} \longleftarrow \mathbf{C}(t)$;
4 Compute $Record \longleftarrow$ free-energy of the $\mathbf{C}(t)$;
5 Set an allowed $Deviation \leq 0$;
6 **while** $(t < MaxIter)$ **do**
       /* Mutation by eq (12)   */
7   **if** $Exist(\mathbf{C} \leftarrow Mutation(\mathbf{C}(t)))$ **then**
8   Compute $E \longleftarrow$ free-energy of new $\mathbf{C}$;
9   **if** $(E < (Record - Deviation))$ **then**
10   Update $\mathbf{C}(t+1) \longleftarrow \mathbf{C}$;
11   **if** $(E < Record)$ **then**
12            $Record \longleftarrow E$;
13   Update $\mathbf{C*} \longleftarrow \mathbf{C}$;
14          $t \longleftarrow t + 1$;

15 return $\mathbf{C*}$;

---

4   Computational Results and Discussion

So far, to our knowledge, the RRT algorithm was not applied to the PSP problem. We implemented the RRT algorithm in Python 3.7. and we ran the experiments on the next hardware configuration: Intel Core i5, 1.8 GHz CPU, 8 GB RAM under macOS Mojave (version 10.14) operating system.

The experiments were repeated 50 times independently for each HP sequence from the benchmark data set, and, for every experiment, were run 100,000 iterations. The $Deviation$ parameter value is $Round(0.4 * MinEnergy)$ for every conformation. The MinEnergy is taken from literature.

The algorithm implementation will be publicly available at the https://github.com/simaioan/RRT_PSP

4.1   Benchmark Proteins Used in RRT Evaluation

The current RRT algorithm was applied for nine benchmark sequences taken from Unger [24] and Huang [10]. These benchmarks, shown in Table 2, are used to test RRT algorithm. The table contains information about sequence number, protein length (denotes the number of amino acids residues), the sequence of amino acids (H and P) and the best known free-energy (minimum or optimum).

4.2   Comparison to Other Methods

The RRT results are compared with the other four best optimal solutions obtained by other authors. Table 3 contains GA results taken from Unger and

Table 2. Benchmark data set

No seq	No of AA	Sequence	Optimal energy
1	20	HPHP PHHP HPPH PHHP PHPH	−9
2	24	HHPP HPPH PPHP PHPP HPPH PPHH	−9
3	25	PPHP PHHP PPPH HPPP PHHP PPPHH	−8
4	36	PPPH HPPH HPPP PPHH HHHH HPPH HPPP PHHP PHPP	−14
5	48	PPHP PHHP PHHP PPPP HHHH HHHH HHPP PPPP HHPP HHPP HPPH HHHH	−23
6	50	HHPH PHPH PHHH HPHP PPHP PPHP PPPH PPPH PPPH PHHH HPHP HPHPHH	−21
7	60	PPHH HPHH HHHH HHPP PHHH HHHH HHHP HPPP HHHH HHHH HHHH PPPP HHHH HHPH HPHP	−36
8	64	HHHH HHHH HHHH PHPH PPHH PPHH PPHP PHHP PHHP PHPP HHPP HHPP HPHP HHHH HHHH HHHH	−42
9	85	HHHH PPPP HHHH HHHH HHHH PPPP PPHH HHHH HHHH HHPP PHHH HHHH HHHH HPPP HHHH HHHH HHHH PPPH PPHH PPHH PPHPH	−52

Note: The 9th sequence (85 AA's) is taken from Huang [10].
The other sequences and optimal energy are taken from Unger [24].

Moult [24]; Genetic Algorithm based on Optimal Secondary Structures (GAOSS) taken from Huang C. et al. [10]; Evolutionary Monte Carlo (EMC) taken from Liang and Wong [12]; and Genetic Algorithm hybridized with Great Deluge Algorithm (GAGDA) which was taken from [23].

For relative short sequences (n ≤ 50), RRT produces the best results (finds the conformation with the minimum free-energy known). For n > 50, RRT did not reach the optimal solution in the performed experiments. These results show that RRT requires improvements to achieve better results.

Table 3. Comparison of aRRT with other algorithms

No seq	Length	GA	GAOSS	EMC	GAGDA	aRRT
1	20	−9	−9	−9	−9	**−9**
2	24	−9	−9	−9	−9	**−9**
3	25	−8	−8	−8	−8	**−8**
4	36	−14	−14	−14	−14	**−14**
5	48	−22	−23	−23	−23	**−23**
6	50	−21	−21	−21	−21	**−21**
7	60	−34	−36	−35	−33	−31
8	64	−37	−42	−39	−42	−35
9	84	−	−52	−52	−52	−43

5   Conclusion and Further Work

The BSP is a combinatorial search problem. The PSP on HP lattice is a type of BSP that focuses on finding the most packaged protein conformations with the hydrophobic kernel.

A mathematical model was developed for PSP on the 2D HP lattice and a variant of local search, called Record-to-Record Travel algorithm was proposed and applied to predict the optimum protein conformations for nine benchmark sequences on a 2D HP square lattice, a well-known ab-initio model. The stochastic small perturbation of the old conformation was performed using rotation and diagonal move mutation operators.

Experimental results show that RRT algorithm finds the best solutions for relatively short sequences. For longer sequences, the results are promising. The interest is in the building of a more efficient mutation operator like crankshaft, rigid rotations, bead flip and so on. Further, experiments with other *Deviation* parameter values can be conducted to find better-energy.

Future researches will focus on the development of a population-based RRT algorithm (or GA hybridized with RRT) that can be applied to other lattice types: 2D triangular, 3D cubic, FCC, etc. The application of classic RRT and population-based RRT to other classes of biomolecules, like nucleic acids (DNA and RNA) is another proposed desideratum.

Acknowledgment. The authors would like to thank Prof. Univ. Dr. Bazil Parv for his professional guidance and valuable support.

References

1. Anfinsen, C.B.: Principles that govern the folding of protein chains. Science **181**(4096), 223–230 (1973)

2. Berger, B., Leighton, T.: Protein folding in the hydrophobic-hydrophilic (HP) model is np-complete. J. Comput. Biol. **5**(1), 27–40 (1998). https://doi.org/10. 1089/cmb.1998.5.27, https://doi.org/10.1089/cmb.1998.5.27, pMID: 9541869

3. Crescenzi, P., Goldman, D., Papadimitriou, C., Piccolboni, A., Yannakakis, M.: On the complexity of protein folding. J. Comput. Biol. **5**(3), 423–465 (1998). https:// doi.org/10.1089/cmb.1998.5.423, pMID: 9773342

4. Dill, K.A.: Theory for the folding and stability of globular proteins. Biochemistry **24**, 1501 (1985)

5. Dinu, V., Trutia, E., Popa-Cristea, E., Popescu, A.: Biochimie medicală - mic tratat. Ed. Medicală, Bucureşti (2006)

6. Dueck, G.: New optimization heuristics: the great deluge algorithm and the record-to-record travel. J. Comput. Phys. **104**(1), 86–92 (1993). https://doi. org/10.1006/jcph.1993.1010, http://www.sciencedirect.com/science/article/pii/ S0021999183710107

7. Dueck, G., Scheuer, T.: Threshold accepting: a general purpose optimization algorithm appearing superior to simulated annealing. J. Comput. Phys. **90**(1), 161–175 (1990). https://doi.org/10.1016/0021-9991(90)90201-B, http:// www.sciencedirect.com/science/article/pii/002199919090201B

8. Garett, R.H., Grisham, C.M.: Biochemistry. Emily Barrosse, John J. Vondeling, 2nd edn. (1999)

9. Harrison, P.M., Chan, H.S., Prusiner, S.B., Cohen, F.E.: Thermodynamics of model prions and its implications for the problem of prion protein folding 11edited by a. r. fersht. J. Mol. Biol. **286**(2), 593–606 (1999). https://doi.org/10.1006/jmbi.1998. 2497, http://www.sciencedirect.com/science/article/pii/S0022283698924974

10. Huang, C., Yang, X., He, Z.: Protein folding simulations of 2D HP model by the genetic algorithm based on optimal secondary structures. Comput. Biol. Chem. **34**(3), 137–142 (2010). https://doi.org/10.1016/j.compbiolchem.2010.04. 002, http://www.sciencedirect.com/science/article/pii/S147692711000040X

11. Lau, K.F., Dill, K.A.: A lattice statistical mechanics model of the conformational and sequence spaces of proteins. Macromolecules **22**(10), 3986–3997 (1989). https://doi.org/10.1021/ma00200a030

12. Liang, F., Hung Wong, W.: Evolutionary Monte Carlo for protein folding simulations. J. Chem. Phys. **115**, 3374–3380 (2001). https://doi.org/10.1063/1.1387478

13. Mafarja, M., Abdullah, S.: A fuzzy record-to-record travel algorithm for solving rough set attribute reduction. Int. J. Syst. Sci. **46**, 1–10 (2015). https://doi.org/ 10.1080/00207721.2013.791000

14. Mafarja, M.M., Abdullah, S.: Record-to-record travel algorithm for attribute reduction in rough set theory. J. Theor. Appl. Inf. Technol. **49**, 507–513 (2013)

15. Mann, M., Backofen, R.: Exact methods for lattice protein models. Bio-Algorithms Med-Syst. **10**(4), 213–225 (2014). https://doi.org/10.1515/bams-2014-0014, https://www.degruyter.com/view/journals/bams/10/4/article-p213.xml

16. Mansour, N., Kanj, F., Khachfe, H.: Particle swarm optimization approach for protein structure prediction in the 3D HP model. Interdiscip. Sci.: Comput. Life Sci. **4**(3), 190–200 (2012). https://doi.org/10.1007/s12539-012-0131-z

17. Neumaier, A.: Molecular modeling of proteins and mathematical prediction of protein structure. SIAM Rev. **39**(3), 407–460 (1997). https://doi.org/10.1137/ S0036144594278060

18. Rashid, M.A., Khatib, F., Hoque, M.T., Sattar, A.: An enhanced genetic algorithm for ab initio protein structure prediction. IEEE Trans. Evol. Comput. **20**(4), 627–644 (2016)

19. Schittekat, P.: Deconstructing record-to-record travel for the capacitated vehicle routing problem. Oper. Res. Manag. Sci. Lett. **1**, 17–27 (2018)
20. Shmygelska, A., Hoos, H.H.: An ant colony optimisation algorithm for the 2D and 3D hydrophobic polar protein folding problem. BMC Bioinform. **6**(1), 30 (2005). https://doi.org/10.1186/1471-2105-6-30
21. Sima, I., Parv, B.: Protein folding simulation using combinatorial whale optimization algorithm. In: 2019 21st International Symposium on Symbolic and Numeric Algorithms for Scientific Computing (SYNASC), pp. 159–166 (2019)
22. Splettstoesser, T.: "© [Thomas Splettstoesser]/ Adobe Stock" Behance. Protein Folding Funnel. https://www.behance.net/gallery/10952399/Protein-Folding-Funnel. Accessed 02 Mar 2021
23. Turabieh, H.: A hybrid genetic algorithm for 2D protein folding simulations. Int. J. Comput. Appl. **139**, 38–43 (2016). https://doi.org/10.5120/ijca2016909127
24. Unger, R., Moult, J.: Genetic algorithms for protein folding simulations. J. Mol. Biol. **231**(1), 75–81 (1993). https://doi.org/10.1006/jmbi.1993.1258, http://www.sciencedirect.com/science/article/pii/S0022283683712581
25. Yang, C.H., Wu, K.C., Lin, Y.S., Chuang, L.Y., Chang, H.W.: Protein folding prediction in the HP model using ions motion optimization with a greedy algorithm. BioData Min. **11**(1), 17 (2018). https://doi.org/10.1186/s13040-018-0176-6
26. Zhou, C., Hou, C., Zhang, Q., Wei, X.: Enhanced hybrid search algorithm for protein structure prediction using the 3D-HP lattice model. J. Mol. Model. **19**(9), 3883–3891 (2013)

Another Dubious Way to Compute the Exponential of a Matrix

Jerzy Respondek[(⊠)] [iD]

Faculty of Automatic Control, Electronics and Computer Science, Department of Applied Informatics, Silesian University of Technology, ul Akademicka 16, 44-100 Gliwice, Poland
jerzy.respondek@polsl.pl

Abstract. In this paper we analyzed recent works on inverting Vandermonde matrix, both classical and generalized, which were unknown during the publication of Moler's and Van Loan's paper 'Nineteen Dubious Ways to Compute the Exponential of a Matrix'. Upon that analysis we proposed the Vandermonde method as the fourth candidate for calculating exponent of generic matrices. On this basis we also proposed the Vandermonde based method to compute the exponential of certain class of special matrices, i.e. the companion matrices.

Keywords: Matrix exponential · Confluent Vandermonde matrix · Vandermonde matrix · Structured matrix · Special matrix · Companion matrix

1 Introduction

In the paper [1] Moler and Van Loan have concluded that the exponential of a matrix could be computed in many ways, but none are completely satisfactory; though they noticed that some of the methods are preferable to others. In this comments in general we agree with this opinion, but give arguments that the set of the preferable methods should be extended.

Particularly, we proposed the algorithms for inverting the generalized Vandermonde matrix as a main tool to find the exponent for the companion matrices, a kind of special (also called structured) form matrices.

The paper is organized as follows: in Sect. 2 we presented the range of applications of the Vandermonde matrices, in Sect. 3 we gave its definition followed by the efficient algorithm to find the inversion, in Sect. 4 we proposed a method for computing the exponential of a companion matrix, in Sect. 5 we illustrated the proposed method by the example of a decent, 10×10 dimensionality, in Sect. 6 we discussed the known Vandermonde method for generic matrices, with special focus to a degree, in which the new algorithms for the Vandermonde matrices directly improved the quality of that

This work was supported by Statutory Research funds of Department of Applied Informatics, Silesian University of Technology, Gliwice, Poland (02/100/BK_21/0008).

© Springer Nature Switzerland AG 2021
O. Gervasi et al. (Eds.): ICCSA 2021, LNCS 12949, pp. 465–478, 2021.
https://doi.org/10.1007/978-3-030-86653-2_34

exponent algorithm. In Sect. 7 we discussed some other methods for generic matrices. In Sect. 8 by the example we showed how to apply the algorithms for inverting the Vandermonde matrices to obtain speed up the matrix multiplications, and in Sect. 9 we gave some final conclusions and proposed a potential ways of further progress in the area.

2   Importance of the Vandermonde Matrices

The confluent Vandermonde matrix is a generalization of classical Vandermonde matrix, allowing multiple nodes, thus the formulas and algorithms designed for them also works in the classical case. We formally defined it in Sect. 3. They arise in a broad range of both theoretical and practical problems. Below we surveyed the problems which make it necessary to inverse the confluent Vandermonde matrices.

- Control problems: investigating of the so-called controllability [2] of the higher order systems leads to the problem of inverting the classic Vandermonde matrix [3] (in case of distinct zeros of the system characteristic polynomial) and the confluent Vandermonde matrix [4] (for systems with multiply characteristic polynomial zeros). As the examples of the higher order models of the physical objects may be mentioned the Timoshenko's elastic beam equation [5] (4th order) and the Korteweg-de Vries equation of waves on shallow water surfaces [6] (3rd, 5th and 7th order).
- Interpolation: apart from the ordinary polynomial interpolation with single nodes (which leads to classical, well-known Newton interpolation) we consider the Hermite interpolation, allowing the multiple interpolation nodes. That problem leads to the system of linear equations, with the confluent Vandermonde matrix ([7] pp. 363–373, [8]).
- Information coding: the confluent Vandermonde matrix is used in coding and decoding the information in the Hermitian's code [9].
- Cryptography: decoding the Reed-Solomon codes [10].
- Optimization of the non-homogeneous differential equations [11].
- So-called fast matrix multiplication: transformation of so-called arbitrary precision approximating-algorithms to exactly-computing algorithms is presented in [12]. An example of approximating algorithms is given in [13] and [14]. More on this we gave in Sect. 8.
- The classic Vandermonde matrix still remains a subject of research e.g. [15, 16].
- Other branches of Vandermonde generalization also are investigated e.g. [17].

3   Theoretical Background

First let us explain what the confluent Vandermonde matrix V is. Let $\lambda_1, \lambda_2, \ldots, \lambda_r$ be given pair wise distinct zeros of the characteristic polynomial $p(s) = (s - \lambda_1)^{n_1} \ldots (s - \lambda_r)^{n_r}$ with $n_1 + \ldots + n_r = n$. The confluent Vandermonde matrix V

related to the zeros of $p(s)$ is defined to be the $n \times n$ matrix $V = [V_1 \; V_2 \; \ldots \; V_r]$, where the block matrix $V_k = V(\lambda_k, n_k)$ is of order $n \times n_k$ having elements [18]:

$$
[V(\lambda_k, n_k)]_{ij} = \begin{cases} \binom{i-1}{j-1} \lambda_k^{i-j}, & \text{for } i \geq j \\ 0, & \text{otherwise} \end{cases}
$$

for $k = 1, 2, \ldots, r$; $i = 1, 2, \ldots, n$ and $j = 1, 2, \ldots, n_k$. For $n_1 = \ldots = n_r = 1$ we get the classical Vandermonde matrix.

The articles [4, 18] present the following theorem for inverting the confluent Vandermonde matrices:

3.1 Theorem – Inverting Confluent Vandermonde Matrices

The inverse of the confluent Vandermonde matrix V has the form $V^{-1} = \begin{bmatrix} W_1^T \; W_2^T \; \cdots \; W_r^T \end{bmatrix}^T$. The column vectors h_{ki} of the block matrix $W_k = [h_{kn}, \ldots, h_{k1}]$ in the inverse confluent Vandermonde matrix V^{-1} may be recursively computed by the following scheme:

$$
\begin{cases} h_{k1} = \begin{bmatrix} K_{k,1} \cdots K_{k,n_k} \end{bmatrix}^T \\ h_{k2} = J_k(\lambda_k, n_k)h_{k1} + a_1 h_{k1} \\ \quad \vdots \\ h_{kn} = J_k(\lambda_k, n_k)h_{k(n-1)} + a_{n-1} h_{k1} \end{cases} , \quad k = 1, 2, \ldots, r \qquad (1)
$$

where λ_k is the eigenvalue, a_k are the coefficients of the characteristic polynomial $p(s)$, $J_k(\lambda_k, n_k)$ is the elementary Jordan block (2):

$$
J_k(\lambda_k, n_k) = \begin{bmatrix} \lambda_k & 1 & & \cdots & 0 \\ & \lambda_k & 1 & & \vdots \\ & & \ddots & \ddots & \\ & & & \lambda_k & 1 \\ 0 & & & & \lambda_k \end{bmatrix}_{n_k \times n_k} , \quad k = 1, 2, \ldots, r \qquad (2)
$$

and $K_{k,j}$ are the auxiliary coefficients that can be calculated recursively in a numerical way by the following algorithm [19]:

3.2 Algorithm – Fast Inversion of Confluent Vandermonde Matrices

The $K_{k,j}$, $j = n_k - 1, \ldots, 1$ (4) coefficients may be recursively computed by the following recursive scheme (3):

$$\begin{cases} L_{ki}^{(q+1)}(\lambda_k) = q!(\lambda_k - \lambda_i)^q \cdot K_{k,n_k-q} - q \cdot L_{ki}^{(q)}(\lambda_k), \quad i = 1, .., k-1, k+1, .., r \\ K_{k,n_k-q-1} = -\frac{1}{(q+1)!} \sum_{i=1,\ i \neq k}^{r} n_i \frac{L_{ki}^{(q+1)}(\lambda_k)}{(\lambda_k - \lambda_i)^{q+1}} \end{cases}$$

(3)

for $q = 0, 1, .., n_k - 2$ and $k = 1, 2, \ldots, r$. The K_{k,n_k} coefficients may be computed directly from the formula (4):

$$K_{k,n_k}(s)\big|_{s=\lambda_k} = \left[\frac{1}{p(s)} (s - \lambda_k)^{n_k} \right]_{s=\lambda_k}, \quad k = 1, 2, \ldots, r$$

(4)

4 Computing the Exponential of a Companion Matrix

The proposed method bases on the rule, that the Jordan canonical form similarity matrix of the companion[1] matrix is a confluent Vandermonde matrix ([20] pp. 86–95). Let the companion matrix is defined by the characteristic polynomial $p(s) = (s - \lambda_1)^{n_1} \ldots (s - \lambda_r)^{n_r}$. Then its matrix form is as follows (5):

$$C = \begin{bmatrix} 1 & 0 & 0 & 0 \\ \vdots & \ddots & \vdots & \vdots & \vdots \\ 0 & \cdots & 1 & 0 & 0 \\ 0 & \cdots & 0 & 1 & 0 \\ 0 & \cdots & 0 & 0 & 1 \\ -a_0 & \cdots & -a_{n-3} & -a_{n-2} & -a_{n-1} \end{bmatrix}$$

(5)

To calculate the coefficients $a_0, \ldots, a_{n-1}$ at first we need to rewrite the characteristic polynomial $p(s)$ in a "flattened" form. That step can be formalized in the way proposed in [21] p. 278. Namely, we define the series of n numbers $s_1, .., s_n$ as follows:

$$\begin{cases} s_1 = \ldots = s_{n_1} & := \lambda_1 \\ s_{n_1+1} = \ldots = s_{n_1+n_2} & := \lambda_2 \\ \quad \vdots \\ s_{n_1+\ldots+n_{r-1}+1} = \ldots = s_n := \lambda_r \end{cases}$$

(6)

Thus for the characteristic polynomial the following equality (7) holds the true:

$$p(s) = (s - \lambda_1)^{n_1} (s - \lambda_2)^{n_2} \ldots (s - \lambda_r)^{n_r} = (s - s_1)(s - s_2) \cdot \ldots \cdot (s - s_n), \ n_1 + \ldots + n_r = n$$

(7)

---

[1] In mathematical literature in some languages instead is used the term 'Frobenius'. But in the English Literature the class of Frobenius matrices is more general, encompassing the companion matrices as its special case.

After the expansion of the polynomial $p(s)$ and sorting it with respect to decreasing powers of the s variable, we can receive (8):

$$p(s) = 1 \cdot s^n - w_1^{(n)}(s_1, \ldots, s_n)s^{n-1} + w_2^{(n)}(s_1, \ldots, s_n)s^{n-2} + \ldots + (-1)^n w_n^{(n)}(s_1, \ldots, s_n) \qquad (8)$$

where $w_j^{(n)}(s_1, \ldots, s_n)$ are the so-called elementary symmetric[2] functions, defined by the formula (9) ([22] pp. 25):

$$\begin{cases} w_1^{(n)}(s_1, \ldots, s_n) = s_1 + s_2 + \ldots + s_n \\ w_2^{(n)}(s_1, \ldots, s_n) = s_1 s_2 + s_1 s_3 + \ldots + s_1 s_n + \ldots + s_{n-1} s_n \\ w_3^{(n)}(s_1, \ldots, s_n) = s_1 s_2 s_3 + s_1 s_2 s_4 + \ldots + s_{n-2} s_{n-1} s_n \\ \cdots\cdots\cdots\cdots\cdots\cdots\cdots\cdots\cdots\cdots\cdots\cdots\cdots\cdots\cdots\cdots \\ w_n^{(n)}(s_1, \ldots, s_n) = s_1 s_2 \cdot \ldots \cdot s_n \end{cases} \qquad (9)$$

And the searched coefficients $a_0, \ldots, a_{n-1}$ in the companion matrix (5) we can express by choosing the sign of the proper elementary symmetric function. Observe (8) and (9):

$$a_i = (-1)^{n-i} w_{n-i}^{(n)}(s_1, \ldots, s_n), \qquad i = 0, 1, \ldots, n - 1 \qquad (10)$$

Algorithm for calculating the elementary symmetric functions (9) can be found e.g. in [17]. Now the companion matrix can be rewritten in the – convenient - Jordan canonical form (11):

$$C = T \cdot J \cdot T^{-1} \qquad (11)$$

The core of the proposed method is fact, that the similarity matrix T after the Jordan decomposition of the companion matrix is just the Vandermonde matrix ($T = V$, classical Vandermonde for single nodes and confluent, for multiple nodes, respectively), combined with the latest algorithms for their inverting. We perform the inversion of the Vandermonde matrix V by the algorithm 3.2.

Exponent of a matrix in a Jordan canonical form is easy to compute. A general function f of the matrix in a Jordan canonical form is expressed by the Eq. (12):

$$f(C) = V \cdot f(J) \cdot V^{-1} \qquad (12)$$

Thus the searched exponent expresses the formula (13):

$$e^C = V \cdot e^J \cdot V^{-1} \qquad (13)$$

Considering that the differential of the exponential function is equal to itself, matrix equality for the function $f(\psi) = e^\psi$ (here ψ stands for both a scalar or a matrix) is:

$$f(J_k(\lambda_k, n_k)) = \begin{bmatrix} \frac{f(\lambda_k)}{0!} & \frac{f'(\lambda_k)}{1!} & \cdots & \frac{f^{(n_k)}(\lambda_k)}{n_k!} \\ 0 & \frac{f(\lambda_k)}{0!} & \cdots & \frac{f^{(n_k-1)}(\lambda_k)}{(n_k-1)!} \\ \vdots & \vdots & \ddots & \vdots \\ 0 & 0 & \cdots & \frac{f(\lambda_k)}{0!} \end{bmatrix} = \begin{bmatrix} e^{\lambda_k} & e^{\lambda_k} & \cdots & \frac{e^{\lambda_k}}{n_k!} \\ 0 & e^{\lambda_k} & \cdots & \frac{e^{\lambda_k}}{(n_k-1)!} \\ \vdots & \vdots & \ddots & \vdots \\ 0 & 0 & \cdots & e^{\lambda_k} \end{bmatrix} \qquad (14)$$

---

[2] Any function we call symmetric, if and only if after any arbitrary permutation of its independent variables we receive the same polynomial ([23] pp. 77–84).

where $J_k(\lambda_k, n_k)$ is the elementary Jordan block of the form (2).

To sum up, finding the exponent of a companion matrix by the proposed in this article method requires the following steps, with their corresponding time complexities:

- Building the confluent Vandermonde matrix V and the Jordan block matrix J - time complexity of those both operations are $O(n^2)$.
- Inverting the confluent Vandermonde matrix V - algorithm 3.2 assures, that this step is of an $O(n^2)$ time class, regardless of the matrix parameters and particularly their multiplicities.
- Calculating the exponent of the Jordan block – also quadratic time (formula (14)).
- Applying the formula $e^C = V \cdot e^J \cdot V^{-1}$ (13) requires two matrix multiplications, which by standard algorithm can be performed in $O(n^3)$ time. We also have faster algorithms, and in Sect. 8 we discussed in details that one of them, which just involves Vandermonde matrices.
- Additionally, if we want to have the companion matrix in the explicit form, we need to calculate the elementary symmetric functions (9). The time complexity of this step is – at worse - $O(n^2)$ (17).

Thus the overall time complexity of the proposed method to calculate the exponent of a companion matrix is determined by the applied matrix multiplication algorithm, being - at worst – of the cubic time.

5   Example – Calculating the Exponent of a Companion Matrix

Let us consider companion matrix defined by the following characteristic polynomial:

$$p(s) = (s + 0.5)^1 (s + 3.0)^2 (s + 2.0)^3 (s + 1.0)^4 \tag{15}$$

Its eigenvalues are summarized in the Table 1:

Table 1. Eigenvalues of the companion matrix with their respective multiplicity.

i	1	2	3	4
λ_i	−0.5	−3.0	−2.0	−1.0
n_i	1	2	3	4

To obtain the explicit form of the companion matrix, defined by the characteristic polynomial (15), we have to obtain the $s_1, .., s_n$ roots, in compliance with the scheme (6). Observe:

$$\begin{cases} s_1 := & \lambda_1 = -0.5 \\ s_2 = s_3 := & \lambda_2 = -3.0 \\ s_4 = s_4 = s_6 := & \lambda_3 = -2.0 \\ s_7 = s_8 = s_9 = s_{10} := \lambda_4 = -1.0 \end{cases} \tag{16}$$

Execution of the algorithm for calculating the elementary symmetric functions (17) for the s_i nodes with values (16), together with formula (10), leads to the following form of the companion matrix, the exponent to be calculated of (17):

$$
C =
\begin{bmatrix}
0 & 1 & 0 & 0 & 0 & 0 & 0 & 0 & 0 & 0 \\
0 & 0 & 1 & 0 & 0 & 0 & 0 & 0 & 0 & 0 \\
0 & 0 & 0 & 1 & 0 & 0 & 0 & 0 & 0 & 0 \\
0 & 0 & 0 & 0 & 1 & 0 & 0 & 0 & 0 & 0 \\
0 & 0 & 0 & 0 & 0 & 1 & 0 & 0 & 0 & 0 \\
0 & 0 & 0 & 0 & 0 & 0 & 1 & 0 & 0 & 0 \\
0 & 0 & 0 & 0 & 0 & 0 & 0 & 1 & 0 & 0 \\
0 & 0 & 0 & 0 & 0 & 0 & 0 & 0 & 1 & 0 \\
0 & 0 & 0 & 0 & 0 & 0 & 0 & 0 & 0 & 1 \\
36.0 & 294.0 & 1039.0 & 2098.5 & 26807.0 & 22801.5 & 13002.0 & 493.5 & 119.0 & 16.5
\end{bmatrix}
$$

(17)

The generalized, confluent Vandermonde matrix, corresponding to the eigenvalues given in Table 1, have the form (18):

$$
V =
\begin{bmatrix}
1.00000 & 1.00 & 0.00 & 1.00 & 0.00 & 0.00 & 1.00 & 0.00 & 0.00 & 0.00 \\
-0.5000 & -3.00 & 1.00 & -2.00 & 1.00 & 0.00 & -1.00 & 1.00 & 0.00 & 0.00 \\
0.25000 & 9.00 & -6.00 & 4.00 & -4.00 & 1.00 & 1.00 & -2.00 & 1.00 & 0.00 \\
-0.1250 & -27.00 & 27.00 & -8.00 & 12.00 & -6.00 & -1.00 & 3.00 & -3.000 & 1.00 \\
0.06250 & 81.00 & -108.00 & 16.00 & -32.00 & 24.00 & 1.00 & -4.00 & 6.00 & -4.00 \\
-0.0312 & -243.00 & 405.00 & -32.000 & 80.00 & -80.00 & -1.00 & 5.00 & -10.00 & 10.00 \\
0.01562 & 729.00 & -1458.00 & 64.00 & -192.00 & 240.00 & 1.00 & -6.00 & 15.00 & -20.00 \\
-0.0078 & -2187.00 & 5103.00 & -128.00 & 448.00 & -672.00 & -100 & 7.00 & -21.00 & 35.00 \\
0.00390 & 6561.00 & -17496.0 & 256.00 & -1024.00 & 1792.00 & 1.00 & -8.00 & 28.00 & -56.00 \\
-0.0019 & -19683.00 & 59049.0 & -512.00 & 2304.00 & -4608.00 & -100 & 9.00 & -36.00 & 84.00
\end{bmatrix}
$$

(18)

and the Jordan matrix built by the eigenvalues given in Table 1, have the form of a block diagonal matrix (19):

$$
J = diag[J_1(\lambda_1, n_1), \ J_2(\lambda_2, n_2), \ J_3(\lambda_3, n_3), \ J_4(\lambda_4, n_4)]
$$

(19)

To invert the Vandermonde matrix (18) we apply the algorithm 3.2, and directly we obtain (20):

$$
V^{-1} =
\begin{bmatrix}
54.61 & 336.78 & 902.63 & 1378.23 & 1319.82 & 821.47 & 332.23 & 84.195 & 12.136 & 0.758 \\
1.72 & 13.440 & 44.90 & 84.42 & 98.58 & 74.20 & 36.005 & 10.860 & 1.8475 & 0.135 \\
0.300 & 2.3500 & 7.87 & 14.86 & 17.43 & 13.20 & 6.4500 & 1.9625 & 0.3375 & 0.025 \\
-100.3 & -758.22 & -2433.04 & -4362.15 & -4820.22 & -3405.55 & -1539.11 & -429.55 & -67.29 & -4.518 \\
-38.00 & -288.33 & -929.55 & -1675.22 & -1861.33 & -1322.33 & -600.66 & -168.33 & -26.44 & -1.77 \\
-12.00 & -92.00 & -300.33 & -549.33 & -621.00 & -450.0 & -209.00 & -60.00 & -9.66 & -0.66 \\
45.00 & 408.00 & 1485.50 & 2899.50 & 3401.81 & 2509.87 & 1170.86 & 334.50 & 53.312 & 3.625 \\
-85.50 & -612.75 & -1872.88 & -3222.06 & -3439.06 & -2358.75 & -1039.00 & -283.56 & -43.56 & -2.875 \\
18.00 & 147.00 & 501.5 & 938.25 & 1064.00 & 761.50 & 345.50 & 96.25 & 15.00 & 1.00 \\
-18.00 & -129.00 & -390.5 & -658.75 & -684.75 & -456.00 & -195.00 & -51.75 & -7.75 & -0.50
\end{bmatrix}
$$

(20)

Finally, the searched exponent for the companion matrix (17) have the following form, by the matrix formulas (13) and (14):

$$
e^C = \begin{bmatrix}
116.33 & 712.2 & 1906.64 & 2910.64 & 2787.18 & 1734.76 & 701.59 & 177.8 & 25.62 & 1.60 \\
-57.66 & -354.6 & -952.07 & -1454.74 & -1393.4 & -867.33 & -350.78 & -88.89 & -12.8143 & -0.80 \\
28.83 & 177.79 & 477.53 & 728.61 & 697.27 & 433.84 & 175.43 & 44.45 & 6.40 & 0.40 \\
-14.41 & -88.90 & -238.29 & -362.86 & -347.46 & -216.41 & -87.57 & -22.20 & -3.20 & -0.20 \\
7.20 & 44.41 & 118.99 & 181.61 & 174.80 & 109.06 & 44.11 & 11.17 & 1.60 & 0.10 \\
-3.61 & -22.35 & -60.05 & -91.99 & -88.54 & -54.58 & -21.84 & -5.4996 & -0.79 & -0.04 \\
1.77 & 10.88 & 28.88 & 43.44 & 40.53 & 23.98 & 9.63 & 2.49 & 0.36 & 0.02 \\
-0.85 & -5.17 & -13.68 & -20.72 & -20.07 & -13.40 & -6.79 & -2.035 & -0.31 & -0.02 \\
0.73 & 5.17 & 16.10 & 29.30 & 34.31 & 26.65 & 13.26 & 3.30 & 0.40 & 0.02 \\
-0.69 & -4.96 & -14.99 & -24.61 & -22.83 & -9.95 & 1.39 & 3.68 & 1.00 & 0.08
\end{bmatrix}
$$

6 Discussion of the Vandermonde Method for Generic Matrices

As the main drawback of the Vandermonde method 11 Moler and Van Loan [1] points out weak time complexity, claimed to be $O(n^4)$, resulting from the necessity of calculating the matrix series A^{k-1}.

With available now algorithms enabling fast matrix multiplication that statement is not precise enough to compare the matrix exponential calculation ways. Nowadays the Strassen's algorithm is well known since decades with $n \cdot O(n^{2.807}) = O(n^{3.807})$. Thus we can decrease the time complexity to $O(n^{3.807})$ effectiveness. In general, with the help of $O(n^k)$ matrix multiplication we can improve the method 11 from [1] to the $O(n^{k+1})$ time efficiency algorithm. Since 1978 that exponential has been decreased a number of times, from 2.78041 in 1978 (Pan [24]), 2.7799 in 1979 (Bini et al. [13]), 2.376 in 1990 (Coppersmith, Winograd [25]) to 2.373 in 2014 (Vassilevska [26]).

Thus, for this time, involving the fastest matrix multiplication algorithm to the method 11 gives the $O(n^{3.373})$ algorithm, being much closer to $O(n^3)$ than to – the considered in [1] - $O(n^4)$.

Here one can complain that those algorithms are useful only in case of large matrices, but that is just the case indicated by 1, i.e. the efficiency gains importance for large n.

Another issue related to the time complexity of the method 11 which is analyzed by the Moler's and Van Loan's work [1], is the necessity to invert the Vandermonde matrix. The efficiency of the used algorithm to invert the Vandermonde matrices is no less important than the efficiency of involved matrix multiplication algorithms, because in case of applying the generic matrix inversion algorithms we have, besides calculating the A^{k-1} series, the $O(n^3)$ operation to be performed by the method 11, which is a not negligible time.

Moler and Van Loan as the example of implementing method 11 proposes expressions developed by Vidyasagar [27]. The selected work of a matrix exponential copes with distinct eigenvalues. The topic in the article [28] allowed multiple eigenvalues. At the time the Moler and Van Loan work [1] appeared, there were available, now a bit obsolete, references [29] and [30] for inverting the Vandermonde matrix. They contain a generic symbolic methodology, which cannot be assessed in the view of numerical computing.

Another fast but dubious way to invert a Vandermonde matrix is to apply the theorem given by Cormen et al. [31], showing how to invert a matrix in a time no worse than the applied matrix multiplication algorithm. Thus applying e.g. the matrix multiplication way proposed in [26] we directly gain the method to obtain the v_{jk} entries in $O(n^{2.373})$ time.

In practice, such a way of proceeding is worthless, because Vandermonde matrices are known for their ill-conditioning, caused by their very structure. That is one of the reasons why there is a necessity to develop specialized algorithms for the structured matrices, besides usually also with better efficiency[3].

In the opposite to - though indirectly - referenced works [29] and [30], nowadays we dispose by fast and reliable algorithms designed specially to invert Vandermonde matrices, both in the single eigenvalues case as well as the multiple ones. In the next paragraphs we selected, to our best knowledge, the fastest together with a decent stability - importantly - applicable to a general form of Vandermonde matrix, with allowed single or multiple each of the eigenvalues, with independent, different multiplicity for each of the eigenvalues[4].

For the single eigenvalues case the $O(n^2)$ algorithms can be found in [15] and [16]. Like one could expect, a more delicate is the problem for the multiple eigenvalues case. Articles [4, 18, 19] together give the universal $O(n^2)$ algorithm, covering every input parameters allowed by the mathematical definition of the confluent Vandermonde matrix.

Another approach is used in the work Zhong, Zhaoyong [32] which achieves $O(n^2)$ time only in a special case of small eigenvalues multiplicities, but in return copes with more general form of the type of structured Vandermonde matrix in question.

When it goes to the space requirements, storage of the method 11 is indeed n^3 even if the spanning matrices $A_0, \ldots, A_{n-1}$ are not saved, because to avoid it still it's necessary to save the consecutive A^{k-1} powers. But a short lookout on operating memory market shows that n^3 space complexity is not an issue now.

7 Discussion Other Methods for Generic Matrices

7.1 The Langrange Interpolation by the Eigenvectors Method

Moler and Van Loan [1] in section 14 propose another approach to utilize the Langrange interpolation by the eigenvectors. They conclude that the $O(n^4)$ work involved in the computations of the consecutive A^{k-1} powers is unnecessary.

---

[3] The problem is neatly defined in the very title of the classical in the structured matrices field monograph [35].

[4] The trait of generality is - surprisingly - not a standard for the algorithms available in the literature. For example the classical in the associated problem of solving the Vandermonde linear systems article [33] only sketches algorithm for a very peculiar version of the confluence, i.e. with allowed multiplicity equal to only of the first eigenvalue, with all the rest single. The same work [33] suggests that the general case cannot be easily treated, stating in the second paragraph of the page 900: "(can be treated easily)... *with only the two endpoints of confluency greater than one, or that with all points of the same order of confluency*.". Significantly, all of the four Pascal-like codes in the appendix of [33] copes only with a classical Vandermonde linear systems, with single eigenvalues (pp. 901–902).Obviously algorithms with such an artificial restrictions are worthless in the view of computing the exponential of a matrix.

In general that observation is proper, but it says nothing about the numerical price of that gain. The numerical stability of inverting a matrix consisting of eigenvectors, which has no special structure, in fact is a lottery, since we cannot apply any specialized algorithm. Also due to lack of structure of the matrix to the inverted, its $O(n^3)$ cost is inevitable, unless we use even more dubious theorem on inverting the matrix by multiplication from Cormen et al. [31].

Thus the eigenvectors method in our opinion is more dubious than the Vandermonde approach, also in the view of the new works on inverting that kind of matrix, that appeared after first publishing of [1].

7.2 The Schur Decomposition Method

Beside scaling and squaring and ODE-based methods, Moler and Van Loan [1] as the one of the best method sees the Schur decomposition with eigenvalue clustering.

Unfortunately also the Schur decomposition suffers from numerical stability. It can be only partially mitigated by the block clustering, merging into a separate blocks nearly confluent eigenvalues. Higham in [34] pp. 228–229 proposes to set the blocking parameter $\delta = 0.1$ and notices, that the optimal choice of δ is problem-dependent. What's worse instability can be present for all choices of δ and all orders of blocking.

Moreover, the blocking operation itself is costly. When computing the separation on $m \times m$ blocks by exact methods it costs $O(m^4)$ (!), whereas by approximate methods it is still $O(m^3)$ [34] p. 226.

In the Vandermonde method the numerical stability lets know only during inverting this kind of matrix. But this matrix is known for its very ill-conditioning and thanks to it each algorithm must be followed by the numerical stability analysis, or at least by thorough experimental tests in this matter. That is necessary since for any choice of points its grow rate is at least exponential, and for the harmonic points it grows faster than factorial. That is worse than Pascal and Hilbert matrices, also known for their ill-conditioning ([35] p. 428).

Moreover, for the Vandermonde matrix with single nodes the well-known algorithm by Bjorck And Pereyra [33] achieves (very) high accuracy, despite 'embedded' into this matrix ill-conditioning ([35] pp. 434–436, [36]). The fair price we pay for it is to calculate the A^{k-1} series.

In the view of the above analysis in our opinion the Vandermonde method is worthy to be consider method to calculate the exponent of a generic matrix.

8   Vandermonde Matrices in Fast Matrix Multiplication

Both the groups of methods to compute the exponential of a matrix, i.e. for the generic and structured matrices, requires to multiply matrices. An the effectiveness of that operation decides on the overall efficiency of finding the exponential. The topic of fast matrix multiplication have already a broad literature; survey of most known algorithms we gave in the beginning of Sect. 6.

In certain class of the fast matrix multiplication (FMM) algorithms a Vandermonde matrix plays important role, namely to transform so-called arbitrary precision FMM algorithm (abbreviated as APA) to its exact-precision equivalent (abbreviated as EC) we need to invert the Vandermonde matrix. The first APA algorithm in the literature was given in [13], for the base size matrices 12×12. In this section as the example of APA algorithm we present an algorithm for 3×3 base dimension, discovered in work [14] p. 438.

To calculate the matrix product $D = AB$ of two matrices, at first we calculate the following 21 auxiliary – so-called – aggregated products, for $i, j \in \{1, 2, 3\}$:

$$
\begin{cases}
u_{i,i} = (a_{i,1} + \varepsilon^2 a_{i,2})(\varepsilon^2 b_{1,i} + b_{2,i}) \\
v_{i,i} = (a_{i,1} + \varepsilon^2 a_{i,3}) b_{3,i} \\
w_i = a_{i,1}(b_{2,i} + b_{3,i}) \\
u_{i,j} = (a_{i,1} + \varepsilon^2 a_{j,2})(b_{2,i} - \varepsilon b_{1,j}), \quad (i \neq j) \\
v_{i,j} = (a_{i,1} + \varepsilon^2 a_{j,3})(b_{3,i} + \varepsilon b_{1,j}), \quad (i \neq j)
\end{cases}
\tag{21}
$$

The product is called aggregated, because in a single multiplications the APA algorithm groups a series of the A, B matrix factors entries. The crucial issue in the APA algorithm is to select small ε, but simultaneously the ε must not be equal to zero.

We the use of the products (21), and only with that products, we can express the desired product $D = AB$ as follows:

$$
\begin{cases}
d'_{i,i} = \frac{1}{\varepsilon^2}(u_{i,i} + v_{i,i} - w_i) \\
d'_{j,i} = \frac{1}{\varepsilon^2}(u_{i,j} + v_{i,j} - w_i) + \frac{1}{\varepsilon}(v_{j,i} - v_{j,j}), \quad (i \neq j)
\end{cases}
\tag{22}
$$

Poking (21) into (22), in matrix notation, gives:

$D=AB+$

$$
+ \begin{bmatrix}
0 & a_{12}b_{11} - a_{13}b_{11} + a_{13}b_{31} - a_{23}b_{31} & a_{12}b_{11} - a_{13}b_{11} + a_{13}b_{31} - a_{33}b_{31} \\
a_{22}b_{12} - a_{23}b_{12} - a_{13}b_{32} + a_{23}b_{32} & 0 & a_{22}b_{12} - a_{23}b_{12} + a_{23}b_{32} - a_{33}b_{32} \\
a_{32}b_{13} - a_{33}b_{13} - a_{13}b_{33} + a_{33}b_{33} & a_{32}b_{13} - a_{33}b_{13} - a_{23}b_{33} + a_{33}b_{33} & 0
\end{bmatrix} \varepsilon
$$

$$
+ \begin{bmatrix}
-a_{12}b_{11} & -a_{23}b_{12} & -a_{33}b_{13} \\
-a_{13}b_{11} & -a_{22}b_{12} & -a_{33}b_{13} \\
-a_{13}b_{11} & -a_{23}b_{12} & -a_{32}b_{13}
\end{bmatrix} \varepsilon^2 = AB + R_1 \varepsilon + R_2 \varepsilon^2
$$

$$\tag{23}$$

Thus if $\varepsilon \to 0$ and $\varepsilon \neq 0$, then D converges to exact value AB but never is exactly equal. There is a question whether it is possible to obtain the exact value. The general, method how to transform any APA matrix multiplication algorithm to its exact version is presented in [12] p. 92.

The highest degree of the ε parameter in (23) is equal to $d = 2$. By the method presented in [12] to construct an exact algorithm we select $d + 1 = 3$ pairwise distinct parameters ε, named ε_1, ε_2, ε_3, and make a linear combination of (23) with coefficients α_1, α_2, α_3:

$$
\begin{aligned}
D &= \alpha_1(AB + R_1\varepsilon_1 + R_2\varepsilon_1^2) + \alpha_2(AB + R_1\varepsilon_2 + R_2\varepsilon_2^2) + \alpha_3(AB + R_1\varepsilon_3 + R_2\varepsilon_3^2) \\
&= (\alpha_1 + \alpha_2 + \alpha_3)AB + (\alpha_1\varepsilon_1 + \alpha_2\varepsilon_2 + \alpha_3\varepsilon_3)R_1 + (\alpha_1\varepsilon_1^2 + \alpha_2\varepsilon_2^2 + \alpha_3\varepsilon_3^2)R_2
\end{aligned}
\tag{24}
$$

We can now observe, that if $\alpha_1 + \alpha_2 + \alpha_3 = 1$ and $\alpha_1\varepsilon_1^k + \alpha_2\varepsilon_2^k + \alpha_3\varepsilon_3^k = 0$ for $1 \le k \le d + 1$, we obtain an exact result for the desired product.

That condition can be rewritten in the following matrix form:

$$\begin{bmatrix} 1 & 1 & 1 \\ \varepsilon_1 & \varepsilon_2 & \varepsilon_3 \\ \varepsilon_1^2 & \varepsilon_2^2 & \varepsilon_3^2 \end{bmatrix} \begin{bmatrix} \alpha_1 \\ \alpha_2 \\ \alpha_3 \end{bmatrix} = \begin{bmatrix} 1 \\ 0 \\ 0 \end{bmatrix} \tag{25}$$

We can notice, that the system matrix in (25) is just the classical Vandermonde matrix. Article [37] proved, that if an APA algorithm is efficient on the level $O(n^k)$, the corresponding EC algorithm, constructed by the method proposed by [12], have efficiency $O(n^k \log n)$. That means that the cost of that is an additional logarithmic factor, what is not a big price.

Particularly, for the proposed in current section APA algorithm, requiring 21 multiplications to find the product of 3×3 dimension[5] matrices, we have the efficiency:

$$O(n^{\log_3 21} \cdot \log n) \approx O(n^{2.77} \cdot \log n) \tag{26}$$

instead of "standard" $O(n^3)$. And in the same level the algorithms to compute the exponential of a matrix are accelerated, both for the generic as well as structured matrices.

9 Summary

In this article we showed how to apply the Vandermonde matrices to the aim of finding an exponent of a matrix, both generic as well as certain class of special matrices, i.e. the companion matrices. We use the algorithms for inverting two classes of Vandermonde matrices:

- The classical Vandermonde, in the case that all of the eigenvalues are distinct
- The so-called confluent Vandermonde, in the case of multiple eigenvalues.

The new algorithms to invert both types of matrices assures to return the result after quadratic time in every case. Thus the meaning of the methods to find the matrix exponents, which bases on Vandermonde special matrices, raised. To this fact is devoted this article.

Interestingly, we can use the Vandermonde matrices in two different stages of the algorithm:

- To calculate the sole exponent.
- To speed up matrix multiplication, necessary in the main exponent algorithm.

As the further directions of research we can see:

- Involving the vector hardware units, which are embedded into contemporary CPUs.
- Designing algorithms for the GPU units to find the exponent.

---

[5] Higher dimensions 'can be multiplied by the 3×3 algorithm by recursion.

References

1. Moler, C., Van Loan, C.: Nineteen dubious ways to compute the exponential of a matrix, twenty-five years later. SIAM Rev. **45**(1), 3–49 (2003)
2. Klamka, J.: Controllability of Dynamical Systems. Kluwer Academic Publishers, Dordrecht (1991)
3. Respondek, J.: Approximate controllability of infinite dimensional systems of the n-th order. Int. J. Appl. Math. Comput. Sci. **18**(2), 199–212 (2008)
4. Respondek, J.: On the confluent Vandermonde matrix calculation algorithm. Appl. Math. Lett. **24**, 103–106 (2011)
5. Timoshenko, S.: Vibration Problems in Engineering, 3rd edn. D. Van Nostrand Company, London (1955)
6. El-Sayed, S.M., Kaya, D.: An application of the ADM to seven order Sawada-Kotara equations. Appl. Math. Comput. **157**, 93–101 (2004)
7. Kincaid, D.R., Cheney, E.W.: Numerical Analysis: Mathematics of Scientific Computing, 3rd edn. Brooks Cole, California (2001)
8. Spitzbart, A.: A generalization of Hermite's interpolation formula. Am. Math. Mon. **67**(1), 42–46 (1960)
9. Lee, K., O'Sullivan, M.E.: Algebraic soft-decision decoding of Hermitian codes. IEEE Trans. Inf. Theory **56**(6), 2587–2600 (2010)
10. Reed, I.S., Solomon, G.: Polynomial codes over certain finite fields. SIAM J. Soc. Ind. Appl. Math. **8**(2), 300–304 (1960)
11. Gorecki, H: On switching instants in minimum time control problem. One dimensional case n-tuple eigenvalue. Bull. de L'Acad. Pol. Des. Sci. **16**, 23–30 (1968)
12. Bini, D.: Relations between exact and approximate bilinear algorithms. Applications. Calcolo **17**, 87–97 (1980)
13. Bini, D., Capovani, M., Lotti, G., Romani, F.: $O(n^2.7799)$ complexity for approximate matrix multiplication. Inform. Process. Lett. **8**(5), 234–235 (1979)
14. Schönhage, A.: Partial and total matrix multiplication. SIAM J. Comput. **10**(3), 434–455 (1981)
15. Eisinberg, A., Fedele, G.: On the inversion of the Vandermonde matrix. Appl. Math. Comput. **174**, 1384–1397 (2006)
16. Yan, S., Yang, A.: Explicit algorithm to the inverse of Vandermonde matrix. In: International Conference on Test and Measurements, pp. 176–179 (2009)
17. El-Mikkawy, M.E.A.: Inversion of a generalized Vandermonde matrix. Int. J. Comput. Math. **80**, 759–765 (2003)
18. Hou, S., Hou, E.: Recursive computation of inverses of confluent Vandermonde matrices. Electron. J. Math. Technol. **1**, 12–26 (2007)
19. Respondek, J.: Numerical recipes for the high efficient inverse of the confluent Vandermonde matrices. Appl. Math. Comput. **218**, 2044–2054 (2011)
20. Gorecki, H.: Optimization of the Dynamical Systems. PWN, Warsaw (1993)
21. Hou, S., Hou, E.S.: A recursive algorithm for triangular factorization of inverse of confluent Vandermonde matrices. In: AIP Conference Proceedings, vol. 1089, no. 277, pp. 277–288 (2009)
22. Barbeau, E.J.: Polynomials, Problem Books in Mathematics. Springer, Berlin (1989). https://www.springer.com/gp/book/9780387406275. ISBN 978-0-387-40627-5. Series ISSN 0941-3502
23. Prasolov, V.: Polynomials, Algorithms and Computation in Mathematics. Springer, Berlin (2004). https://doi.org/10.1007/978-3-642-03980-5

24. Pan, V.Ya: Strassen's algorithm is not optimal - trilinear technique of aggregating uniting and cancelling for constructing fast algorithms for matrix operations. In: Proceedings of 19th Annual Symposium on Foundations of Computer Science, Ann Arbor, Mich (1978)
25. Coppersmith, D., Winograd, S.: Matrix multiplication via arithmetic progressions. J. Symb. Comput. **9**(3), 251–280 (1990)
26. Williams, V.V.: Multiplying matrices in $O(n^{2.373})$ time. Tech. rep. Stanford University (2014)
27. Vidyasagar, M.: A novel method of evaluating e^{At} in closed form. IEEE Trans. Automatic Contr. AC **15**, 600–601 (1970)
28. Luther, U., Rost, K.: Matrix exponentials and inversion of confluent Vandermonde matrices. Electron. Trans. Numer. Anal. **18**, 91–100 (2004)
29. Tou, J.T.: Determination of the inverse Vandermonde matrix. IEEE Trans. Autom. Contr. AC **9**, 314 (1964)
30. Wu, S.H.: On the inverse of Vandermonde matrix. IEEE Trans. Autom. Contr. AC **11**, 769 (1966)
31. Cormen, T.H., Leiserson, Ch.E., Rivest, R.L., Stein, C.: Introduction to Algorithms, 3rd edn. The MIT Press (2009)
32. Zhong, X., Zhaoyong, Y.: A fast algorithm for inversion of confluent Vandermonde-like matrices involving polynomials that satisfy a three-term recurrence relation. SIAM J. Matrix Anal. Appl. **19**(3), 797–806 (1998)
33. Bjorck, A., Pereyra, V.: Solution of Vandermonde systems of equations. Math. Comp. **24**(112), 893–903 (1970)
34. Higham, N.J.: Functions of Matrices. Theory and Computation, SIAM, Philadelphia (2008)
35. Higham, N.J.: Accuracy and Stability of Numerical Algorithms. SIAM, Philadelphia (1996)
36. Higham, N.J.: Error analysis of the Bjorck-Pereyra algorithms for solving Vandermonde systems. Numer. Math. **50**(5), 613–632 (1987)
37. Romani, F.: Complexity measures for matrix multiplication algorithms. Calcolo **17**, 77–86 (1980)

Application of Mathematical Models to Estimate the Water Level Stability in a Two-Tank System in Series with Service Demand Variations

Nilthon Arce[1], José Piedra[1], Eder Escobar[2], Edwar Lujan[3], Luis Aguilar[2], and Flabio Gutierrez[2(✉)]

[1] Universidad Nacional de Jaen, Jr. Cuzco N° 250, Pueblo Libre, Jaén, Cajamarca, Peru
{nilthon_arce,jpiedrat}@unj.edu.pe
[2] Universidad Nacional de Piura, Urb. Miraflores s/n, Castilla, Piura, Peru
{eescobarg,laguilari,flabio}@unp.edu.pe
[3] Universidad Católica de Trujillo, Panamericana Norte Km. 555, Moche, Trujillo, La Libertad, Peru
elujan@uct.edu.pe

Abstract. In this research work, a two-tank system in series was developed. In this system, water flows from the first tank to the second. Opening variations were present in the inlet valve, while the opening flow was fixed in the outlet valve. Using differential equations, the Jacobian matrix and eigenvalues, it was possible to analyze the stability of the system and obtain the equilibrium points. To estimate the adequate water levels, a linear and nonlinear model were implemented. Both models were able to estimate in real time the tanks water levels, identifying the system equilibrium points, this ensured the service demand. A simulation process was applied to evaluate the model's behavior over time. This simulation was performed using the Simulink tool of Matlab. Both linear and nonlinear models obtained similar results. Results analysis suggested the use of the linear model to implement the control system, since its transfer functions compositions are less complex and more efficient to deploy.

Keywords: Couple tank system · Equilibrium points · Stability · Control systems

1 Introduction

An interconnected tank system is composed by various inlets and outlets, where a required liquid level must be controlled at all time. These systems are used in a wide range of industries such as: food and beverage processing, power plants, pharmaceuticals, water treatment, cement and concrete processing, petrochemical, paper mills, among others.

© Springer Nature Switzerland AG 2021
O. Gervasi et al. (Eds.): ICCSA 2021, LNCS 12949, pp. 479–492, 2021.
https://doi.org/10.1007/978-3-030-86653-2_35

Tank systems are also used to experiment with: new controllers used to determine the tank levels [1–3], tank shape design [4–6]; systems robustness against failures or parameter alterations [7–9]; fuzzy controllers which simulate tanks handling by humans or under unknow models [10–12]; teaching tools for control systems [13, 14].

During modelling the physical system through differential equations, it is necessary to perform a stability analysis, which is plotted on the phase plane, where regions or stability points in the system are shown alongside with the direction from which stability is achieved. In the case of a tanks system, it is very important to determine the moment in which the system is stabilized by changes in the opening valves, which can be triggered by system request or failures present in the supply liquid.

In this research work, a two tanks system in series is studied. Water levels in the tanks are controlled by two valves (inlet and outlet), where the flow opening constitutes the system input variables. Openings states are represented accordingly with zero and one when totally close or open. Water levels in tanks represent the system output variables, meaning a solution must be obtained for a real problem represented by multiple inputs and outputs (MIMO).

Regarding the two-tank system, the Lyapunov method is used in [15] to perform the system stability analysis, however model simulation was not implemented. In [16] authors shown in an interactive fashion the water level control in two tanks. Model linearization and simulation were performed using the Taylor series and LabVIEW software respectively. Results were only shown for liquid control for a single variation in the inlet valve. In [17] a system of two nonlinear differential equations, which were solved through a numeric method were obtained. However, equilibrium points in the system nor the detailed data results were found.

In this research work, the system is composed by two tanks in series. These tanks are operated by valves, where water flows from the first tank to second. Opening variations are perform in the inlet valve, while maintaining a fix opening in the outlet one. A pair of two mathematical models are applied to estimate the water level. These are nonlinear and linear models, where the former is represented by ordinal differential equations, whereas the second by transfer functions. In order to ensure the demand of service, a stability analysis is performed over the hyperbolic points of the system. Finally, a simulation using the Simulink tool of Matlab is performed to observe the model's behaviors according with time.

2 Preliminaries

2.1 Single Tank Model

Using the Bernoulli equation [18], the outlet fluid velocity value of a tank is given by:

$$v = 2gh \tag{1}$$

Where h: water level height, g: gravity acceleration.

Considering that the inlet flow Q_e is proportional to the inlet valve opening V_e, the following is obtained:

$$Q_e = k_1 a_1 \tag{2}$$

Where k_1 is constant, a_1 represents the opening in the inlet valve V_e.

To compute the outlet flow Q_s a linear approximation between Q_s and the opening in the outlet valve V_s is perform, obtaining:

$$Q_s = k_2 a_2 \sqrt{2gh} = k_2 \sqrt{2g} a_2 \sqrt{h} = k a_2 \sqrt{h} \tag{3}$$

Where k is constant, a_2 represents the opening in the outlet valve V_s.

The outlet flow is given by the product between the fluid velocity and transversal area through for which circulates, such that:

$$Q = Av = A\sqrt{2gh} = A\sqrt{2g}\sqrt{h} = k\sqrt{h} \tag{4}$$

The accumulated liquid inside is the result of the difference between the inlet and outlet flow of the tank. Then, the following equation must be satisfied:

$$A\frac{dh}{dt} = Q_e - Q_s \tag{5}$$

This implies that water level variations in the tank base area over time are defined as the difference between inlet and outlet flow Q_e and Q_s respectively.

3 Problem Formulation

The introduction section described a wide range of problems involving tank systems. As observed, for such systems, in order to achieve stability and maintain the liquid service demand, it is necessary to perform opening variations in the inlet valve, while maintaining a fix opening in the outlet valve.

Figure 1, describes a two-tank system in series which are operated by valves. The opening percentages in the inlet and outlet valves V_e and V_s are controlled by α and β respectively. Liquid levels in tanks 1 and 2 are represented by H and h respectively. Also, tanks 1 and 2 base areas are defined by A_1 and A_2.

In the setting described before, a stable equilibrium point must be obtained. This must be achieved with opening variations in the inlet valve V_e while a fix opening is maintained in the outlet valve V_s. Finding the stable equilibrium points will allow us to stabilize the system. Then the adequate liquid levels H and h of tanks 1 and 2 will be known.

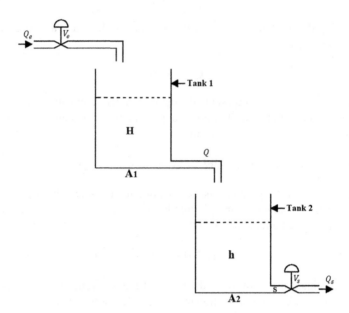

Fig. 1. A two-tank series system operated by valves.

4   Nonlinear Model

Applying Eqs. (2), (3), (4) y (5) in the system described in Fig. 1, a differential equation for the nonlinear model of tank 1 is obtained.

$$A_1 \frac{dH}{dt} = Q_e - Q = k'\alpha - k\sqrt{H} \tag{6}$$

For tank 2, a differential equation for the nonlinear model is also found

$$A_2 \frac{dh}{dt} = Q - Q_s = k\sqrt{H} - k''\beta\sqrt{h} \tag{7}$$

Then, the proposed nonlinear model for the system is:

$$\frac{dH}{dt} = \frac{1}{A_1}[k'\alpha - k\sqrt{H}] \tag{8}$$

$$\frac{dh}{dt} = \frac{1}{A_2}[k\sqrt{H} - k''\beta\sqrt{h}]$$

Where k, k' y k'' are constants.

5   Linear Model

Modelling is applied to the case of a variable (α) opening flow in the inlet valve while a constant (β) opening flow is maintained in the outlet valve.

5.1   Linear Model for Tank 1

Using the three first terms from the Taylor series expansion [19], the nonlinear model of tank 1 is linearized, then:

$$f(\alpha, H) = A_1 \frac{dH}{dt} = k'\alpha - k\sqrt{H} \tag{9}$$

$$f(\alpha, H) \approx f(\overline{\alpha}, \overline{H}) + \frac{df}{d\alpha}(\overline{\alpha}, \overline{H})(\alpha - \overline{\alpha}) + \frac{df}{dH}(\overline{\alpha}, \overline{H})(H - \overline{H}) \tag{10}$$

The derivative and substitution in Eqs. (9) and (10) allow us to obtain:

$$f(\alpha, H) \approx f(\overline{\alpha}, \overline{H}) + k'(\alpha - \overline{\alpha}) - \frac{k}{2\sqrt{\overline{H}}}(H - \overline{H}) \tag{11}$$

$$f(\alpha, H) - f(\overline{\alpha}, \overline{H}) \approx k'(\alpha - \overline{\alpha}) - \frac{k}{2\sqrt{\overline{H}}}(H - \overline{H}) \tag{12}$$

$$f(\Delta\alpha, \Delta H) \approx k'\Delta\alpha - \frac{k}{2\sqrt{\overline{H}}}\Delta H \tag{13}$$

$$A_1 \frac{d\Delta H}{dt} \approx k'\Delta\alpha - \frac{k}{2\sqrt{\overline{H}}}\Delta H \tag{14}$$

Then, applying the Laplace transformation over Eq. (14) the transfer function, which represents the linear model of tank 1 is obtained:

$$\frac{H(s)}{\alpha(s)} = \frac{k'}{A_1 s + \frac{k}{2\sqrt{\overline{H}}}} = F(s) \tag{15}$$

5.2   Linear Model for Tank 2

Analogously, we use the Taylor series to linearize the nonlinear model from tank 2.

$$f(H, h) = A_2 \frac{dh}{dt} = k\sqrt{H} - k''\beta\sqrt{h} \tag{16}$$

$$f(H, h) \approx f(\overline{H}, \overline{h}) + \frac{df}{dH}(\overline{H}, \overline{h})(H - \overline{H}) + \frac{df}{dh}(\overline{H}, \overline{h})(h - \overline{h}) \tag{17}$$

Applying the derivative and substitution in Eqs. (16) and (17) allow us to obtain:

$$f(H,h) \approx f\left(\overline{H},\overline{h}\right) + \frac{k}{2\sqrt{\overline{H}}}(H-\overline{H}) - \frac{k''\beta}{2\sqrt{\overline{h}}}\left(h-\overline{h}\right) \tag{18}$$

$$f(H,h) - f\left(\overline{H},\overline{h}\right) \approx \frac{k}{2\sqrt{\overline{H}}}(H-\overline{H}) - \frac{k''\beta}{2\sqrt{\overline{h}}}\left(h-\overline{h}\right) \tag{19}$$

$$f(\Delta H, \Delta h) \approx \frac{k}{2\sqrt{\overline{H}}}\Delta H - \frac{k''\beta}{2\sqrt{\overline{h}}}\Delta h \tag{20}$$

$$A_2\frac{d\Delta h}{dt} \approx \frac{k}{2\sqrt{\overline{H}}}\Delta H - \frac{k''\beta}{2\sqrt{\overline{h}}}\Delta h \tag{21}$$

Applying the Laplace transformation over Eq. (21), the transfer function, which represents the linear model of tank 2 is obtained:

$$\frac{h(s)}{H(s)} = \frac{\frac{k}{2\sqrt{\overline{H}}}}{A_2 s + \frac{k''\beta}{2\sqrt{\overline{h}}}} \tag{22}$$

$$h(s) = \frac{\frac{k}{2\sqrt{\overline{H}}}}{A_2 s + \frac{k''\beta}{2\sqrt{\overline{h}}}} H(s) \tag{23}$$

$$h(s) = \left[\frac{\frac{k}{2\sqrt{\overline{H}}}}{A_2 s + \frac{k''\beta}{2\sqrt{\overline{h}}}}\right]\left[\frac{k'}{A_1 s + \frac{k}{2\sqrt{\overline{H}}}}\alpha(s)\right] \tag{24}$$

$$\frac{h(s)}{\alpha(s)} = \left[\frac{\frac{k}{2\sqrt{\overline{H}}}}{A_2 s + \frac{k''\beta}{2\sqrt{\overline{h}}}}\right]\left[\frac{k'}{A_1 s + \frac{k}{2\sqrt{\overline{H}}}}\right] = \frac{\frac{k'k}{2\sqrt{\overline{H}}}}{\left(A_1 s + \frac{k}{2\sqrt{\overline{H}}}\right)\left(A_2 s + \frac{k''\beta}{2\sqrt{\overline{h}}}\right)} = G(s) \tag{25}$$

Then, the transfer functions of the linear model in the system are defined by:

$$\frac{H(s)}{\alpha(s)} = \frac{k'}{A_1 s + \frac{k}{2\sqrt{\overline{H}}}} = F(s) \tag{26}$$

$$\frac{h(s)}{\alpha(s)} = \frac{\frac{k'k}{2\sqrt{\overline{H}}}}{\left(A_1 s + \frac{k}{2\sqrt{\overline{H}}}\right)\left(A_2 s + \frac{k''\beta}{2\sqrt{\overline{h}}}\right)} = G(s)$$

6 Stability Analysis of the Two-Tanks System

6.1 Equilibrium Points Estimation

Estimation of the equilibrium points is performed from the vectorial fields of the equations which govern the flow dynamics in tanks. These equations ensures that the inlet and outlet flow be the same for any tank. In this way, equilibrium is achieved in the tank height level.

For tank 1, the equality $Q_e = Q$ in Eq. (6) indicates the equilibrium point, such that:

$$\frac{dH}{dt} = \frac{1}{A_1}[k'\alpha - k\sqrt{H}] = 0 \tag{27}$$

Obtained H as:

$$\overline{H} = \left(\frac{k'\alpha}{k}\right)^2 \tag{28}$$

Then, the equilibrium state of the water level in tank 1 is given by $\overline{H}$.

Analogously, in Eq. (7), $Q = Q_s$ indicates the equilibrium point for tank 2, such that:

$$\frac{dh}{dt} = \frac{1}{A_2}\left[k\sqrt{H} - k''\beta\sqrt{h}\right] = 0 \tag{29}$$

From Eqs. (28) and (29) the following is obtained:

$$\overline{h} = \left(\frac{k'\alpha}{k''\beta}\right)^2 \tag{30}$$

Where $\overline{h}$ represents the equilibrium state of the water level in tank 2.

Then, the system equilibrium point is defined by:

$$P(\overline{H}, \overline{h}) = \left(\left(\frac{k'\alpha}{k}\right)^2, \left(\frac{k'\alpha}{k''\beta}\right)^2\right) \tag{31}$$

6.2 Jacobian Matrix and Eigenvalues Estimation

From the differential equations system in (8), the Jacobian Matrix evaluated on the correspondent equilibrium point is defined as:

$$J(H, h) = \begin{bmatrix} \frac{\partial F}{\partial H} & \frac{\partial F}{\partial h} \\ \frac{\partial G}{\partial H} & \frac{\partial G}{\partial h} \end{bmatrix} = \begin{bmatrix} -\dfrac{k}{2A_1\sqrt{H}} & 0 \\ \dfrac{k}{2A_2\sqrt{H}} & -\dfrac{k''\beta}{2A_2\sqrt{h}} \end{bmatrix} \tag{32}$$

To estimate the eigenvalues, the following is applied:

$$\left| J\left(\bar{H}, \bar{h}\right) - \lambda I \right| = \begin{vmatrix} -\dfrac{k}{2A_1\sqrt{\overline{H}}} - \lambda & 0 \\ \dfrac{k}{2A_2\sqrt{\overline{H}}} & -\dfrac{k''\beta}{2A_2\sqrt{\overline{h}}} - \lambda \end{vmatrix} = 0 \tag{33}$$

From Eq. (33) the following is obtained:

$$4A_1A_2\sqrt{\overline{H}\overline{h}}\lambda^2 + 2\left(kA_2\sqrt{\overline{h}} + k''\beta A_1\sqrt{\overline{H}}\right)\lambda + kk''\beta = 0 \tag{34}$$

Let:

$$n_1 = 4A_1A_2\sqrt{\overline{H}\overline{h}} \tag{35}$$

$$n_2 = 2\left(kA_2\sqrt{\overline{h}} + k''\beta A_1\sqrt{\overline{H}}\right) \tag{36}$$

$$n_3 = kk''\beta \tag{37}$$

Replacing Eqs. (35), (36) and (37) in Eq. (34) allow us to define:

$$n_1\lambda^2 + n_2\lambda + n_3 = 0 \tag{38}$$

The roots from Eq. (38) are used to build the eigenvalues:

$$\lambda_{1,2} = \frac{-n_2 \pm \sqrt{n_2^2 - 4n_1n_3}}{2n_1} \tag{39}$$

6.3 Stability in Equilibrium Points

The system stability for a certain equilibrium point is defined with the following conditions:

- If $\lambda_1 < 0$ *and* $\lambda_2 < 0$, then the equilibrium point is stable
- If $\lambda_1 < 0 \wedge \lambda_2 < 0$,, then the equilibrium point is unstable (saddle point)
- If $\lambda_1 > 0 \wedge \lambda_2 > 0$, then the equilibrium point is unstable

7 Equilibrium Points Estimation for a Study Case

For the study case the following parameters are considered:

$k' = 0.04$ m^3/s (V_e constant), $k = 0.03$ m^3/s, $k'' = 0.055$ m^3 (V_s constant), $\alpha = 0.9$ (V_e opening), $\beta = 0.6$ (V_s opening), $A_1 = 1$ m^2 (tank 1 base area), $A_2 = 1.5$ m^2 (tank 2 base area).

During experiments, opening variations in the inlet valve V_e are made accordingly with three values:

$$\alpha = \{90\%, 85\%, 89\%\}$$

While the opening percentage in the outlet valve V_s is set constant with $\beta = 60\%$. From Eq. (31) the equilibrium state for the tanks water levels is defined as:

a) For $\alpha = 90\%$, $\beta = 60\%$, the system equilibrium point $P_1\left(\overline{H}, \overline{h}\right) =$ (1.44 m, 1.1901 m) is obtained. The eigenvalues $\lambda_1 = -0.0101$, $\lambda_2 = -0.0125$ are negatives, indicating stability in the equilibrium point.

b) For $\alpha = 85\%$, $\beta = 60\%$, the system equilibrium point $P_2\left(\overline{H}, \overline{h}\right) =$ (1.284 m, 1.062 m) is obtained. The eigenvalues $\lambda_1 = -0.0107$, $\lambda_2 = -0.0132$ are negatives, indicating stability in the equilibrium point.

c) For $\alpha = 89\%$, $\beta = 60\%$, the system equilibrium point $P_3\left(\overline{H}, \overline{h}\right) =$ (1.408 m, 1.164 m) is obtained. The eigenvalues $\lambda_1 = -0.0102$, $\lambda_2 = -0.0126$ are also negatives, confirming stability in the equilibrium point.

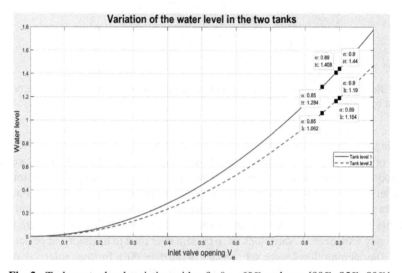

Fig. 2. Tanks water level variation with a fix $\beta = 60\%$ and $\alpha = \{90\%, 85\%, 89\%\}$.

In Fig. 2, it is observed that for an opening V_s of $\beta = 60\%$ with a fix opening V_e between $0 \leq \alpha \leq 1$, the water level of tank 1 is ascendent with $H = 1, 44$ m when $\alpha = 90\%$. This also holds true for tank 2, where water level is also ascendent with $h = 1, 1901$ m when $\alpha = 90\%$.

8 Mathematical Model Simulations for the Study Case

The models were implemented in the scientific software MATLAB R2017a. Execution was performed on a personal computer with an Intel (R) Core (TM) i5-8250U CPU @ 1.60 GHz 1.80 GHz processor with 6.00 GB of RAM.

Simulation and solution of the model were performed using the available components of Simulink in Matlab. The Runge-Kutta method was applied to the nonlinear model, whereas FCN transfer block was applied to the transfer functions of the linear model.

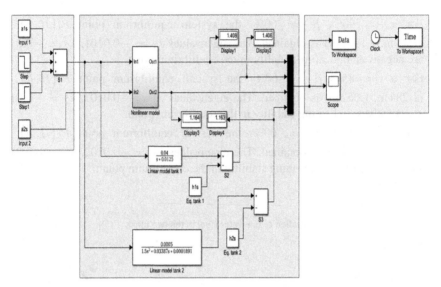

Fig. 3. Block diagram of the nonlinear and linear models of the control system in the two-tanks operated by valves.

Using the parameters of the study case (see Sect. 7), a simulation of the control system of the two-tanks in series operated by valves was performed. This allowed us to observed the behavior of the nonlinear and linear models over time (t). During simulation the outlet valve was fixed, while the inlet valve was regulated, generating variations in the tanks water levels. Figure 3, shows the simulation schematics in more detail.

Table 1, shows results of simulating the two-tank system. Variations were applied accordingly with Sect. 7. Variations were present over time in the inlet and outlet valves. At $t = 0$ s opening flows were set constant with $\beta = 60\%$ for the outlet valve, while $\alpha = 90\%$ was used in the inlet valve. At $t = 2000$ s, the opening flow was decreased by 5% with $\alpha = 85\%$ for the inlet valve. Then at $t = 3500$ s the opening flow is increased by 4% with $\alpha = 89\%$ for the inlet valve.

Figure 4, shows the dynamics of the water level in the tanks considering the nonlinear and linear models, water level changes are generated by opening variations in the inlet valve system. For an initial opening $\alpha = 90\%$ in the inlet valve both models achieved a stationary level with $H = 1,44$ m and $h = 1,19$ m for tanks 1 and 2 respectively. Simulation results of the system over time (t) are shown in Table 1. At $t = 2000$ s the opening flow is decreased a 5% ($\alpha = 85\%$) in the valve, lowering the water levels in both tanks. At $t = 2730,04461042856$ s, for the water level in tank 1, the linear and nonlinear models obtained stationary values of $H = 1,284$ m and $H = 1,28$ m respectively. Then, at $t = 2930,04421042856$ s, the water level in tank 2 achieved stationary values of $h = 1,062$ m and $h = 1,062$ m obtained by the linear and nonlinear models respectively (Figs. 3 and 4 illustrate this behavior). When the valve opening flow is increment in 4% ($\alpha = 89\%$) at $t = 3500$ s, both tanks began increase their water levels. Stationary water levels were reported for tanks 1 and 2 at $t = 4086,68774412396$ s and $t = 4286,68734412396$ s respectively. Reported values from the linear and nonlinear

Table 1. Simulation results of the two tanks system in series operated by valves.

Time	MNL Tank 1	ML Tank 1	MNL Tank 2	ML Tank 2
1755,76752188620	1,43999961546213	1,43999994814581	1,19006322338354	1,18961038075501
1855,76732188620	1,43999981673949	1,43999997161470	1,19007010100707	1,18961041827557
1955,76712188620	1,43999994148428	1,43999998796925	1,19007661192565	1,18961051856290
1999,99999999999	1,43999996527065	1,43999999125067	1,19007857220228	1,18961054610584
2000	1,43999996527065	1,43999999125067	1,19007857220228	1,18961054610584
2000,00000000001	1,43999996527062	1,43999999125064	1,19007857220228	1,18961054610584
2006,62022753665	1,42762552922792	1,42762381045521	1,18953227583623	1,18906508741603
2013,24045507329	1,41618994802250	1,41618160884516	1,18842716377639	1,18796379296354
2019,86068260992	1,40564560644850	1,40562259269335	1,18682848113206	1,18637369079115
2026,48091014656	1,39592856102461	1,39588169967422	1,18480798025361	1,18436680799015
2043,45579383099	1,37397699845610	1,37382742064916	1,17833656474925	1,17794968132097
2060,43067751541	1,35612519934906	1,35581044026832	1,17034600463791	1,17003641926653
2077,40556119984	1,34175096045615	1,34121411000183	1,16147069724134	1,16124926280197
2094,38044488427	1,33022817334516	1,32942523570303	1,15225034850163	1,15211247289125
2132,44192319865	1,31229179506012	1,31080678023652	1,13197291070193	1,13194399954401
2170,50340151303	1,30152646758370	1,29934376859632	1,11441257402178	1,11432661488807
2208,56487982741	1,29493841089593	1,29214035736583	1,10042875114882	1,10009968841506
2246,62635814179	1,29085477636207	1,28757119085586	1,08974991279288	1,08902443294862
2284,68783645618	1,28833600146111	1,28469764599414	1,08176200124191	1,08055230675180
2346,02712994927	1,28603600960128	1,28202144602056	1,07285656667335	1,07087189379413
2407,36642344237	1,28504416451864	1,28084244274790	1,06761650879413	1,06497258324618
2468,70571693546	1,28465614164376	1,28035653778811	1,06469503698353	1,06155760741119
2530,04501042856	1,28451657848358	1,28015854695064	1,06313497393943	1,05965336619567
2630,04481042856	1,28450199138253	1,28009611723224	1,06218529936912	1,05836715762164
2730,04461042856	1,28449232833391	1,28004774018748	1,06188545125269	1,05788602796148
2830,04441042856	1,28446315307823	1,27999947694808	1,06170862723026	1,05760877399812
2930,04421042856	1,28444421994232	1,27998435868292	1,06158863361176	1,05744735580074
3030,04401042856	1,28444099288541	1,27999230656735	1,06153403157967	1,05739229930755
3130,04381042856	1,28444268343003	1,27999873847668	1,06152035439243	1,05733977241296
3230,04361042856	1,28444398337189	1,28000066265365	1,06152042510593	1,05739853697128
3330,04341042856	1,28444444448303	1,28000056088689	1,06152256135333	1,05740420947107
3430,04321042856	1,28444451145156	1,28000018147776	1,06152385083300	1,05740595761033
3499,99999999997	1,28444450102677	1,28000003438405	1,06152425902254	1,05740614290177
3500	1,28444450102677	1,28000003438405	1,06152425902254	1,05740614290177
3500,00000000003	1,28444450102681	1,28000003438409	1,06152425902255	1,05740614290177
3507,40162983638	1,29538462686145	1,29098497575729	1,06208829554494	1,05794400387021
3514,80325967274	1,30536394543354	1,30105016775772	1,06321209453906	1,05902187647848
3522,20488950909	1,31444648322706	1,31025322951338	1,06481604827958	1,06056586350748
3529,60651934544	1,32271488885472	1,31866494200947	1,06681125129609	1,06249874795962
3548,11096637982	1,34067507038483	1,33702579284543	1,07286636118232	1,06842573572175
3566,61541341421	1,35497619184466	1,35176003303644	1,08008946373849	1,07558362256909
3585,11986044859	1,36626036454524	1,36347095781561	1,08787749150307	1,08338481008098
3603,62430748297	1,37513864792860	1,37274570539385	1,09575366352641	1,09135089196980
3643,13973962725	1,38808534217281	1,38640770450646	1,11169264046708	1,10767563635328
3682,65517177153	1,39582445181210	1,39465920081437	1,12501621757830	1,12153186360282
3722,17060391581	1,40058634427921	1,39976465438605	1,13538263785846	1,13245469802926
3761,68603606009	1,40355320734452	1,40295692903304	1,14320326210664	1,14078310672272
3801,20146820437	1,40537978786821	1,40493117821422	1,14902973472101	1,14703920686170
3872,57303718427	1,40719276114826	1,40690239902678	1,15615166019125	1,15473872909948
3943,94460616417	1,40786363852514	1,40763703817426	1,16004135200329	1,15898647420825
4015,31617514407	1,40807667932767	1,40787444408647	1,16201439900700	1,16116313231556
4086,68774412396	1,40813897465150	1,40794768607821	1,16296055066582	1,16221906700010
4186,68754412396	1,40816836759755	1,40798569094109	1,16354541048426	1,16288322641594
4286,68734412396	1,40817377735125	1,40799418184987	1,16371810203930	1,16308452358951
4386,68714412396	1,40817546660590	1,40799692960766	1,16376053032475	1,16313698514357
4486,68694412396	1,40817665153445	1,40799854536771	1,16377166975117	1,16315263635772
4586,68674412396	1,40817732235707	1,40799941485621	1,16377780373586	1,16316153865892
4686,68654412396	1,40817760586934	1,40799978129751	1,16378077435649	1,16316591598358
4786,68634412396	1,40817771501418	1,40799992140821	1,16378215450641	1,16316795136678
4886,68614412396	1,40817775525846	1,40799997239800	1,16378278211249	1,16316887230825
4986,68594412396	1,40817776977401	1,40799999044136	1,16378306375500	1,16316928245172
5000	1,40817777103356	1,40799999194735	1,16378309142460	1,16316932208532

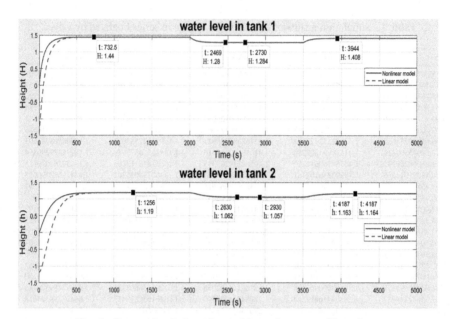

Fig. 4. System simulation of two series tanks operated by valves.

models were $H = 1,408$ m and $H = 1,408$ m for tank 1 and $h = 1,163$ m and $h = 1,164$ m for tank 2. Figures 3 and 4, show these results.

9 Conclusions and Recommendations

Using mathematical foundations such as differential equations theory, Jacobian matrix and eigenvalues allowed us to analyze the system stability of two series tanks. The stable equilibrium points were found analytically for the study case, where the opening flow was fix in the outlet valve, while opening variations occurred in the inlet valve. Thus, stable water levels were found in each tank, stabilizing the system.

The nonlinear and linear models applied to the study case were able to estimate in real time the stable water levels of the two-tank system. This ensured the demand of service, which coincided with the obtained analytical equilibrium points.

The nonlinear and linear models obtained similar results during the simulation process. However, to implement the control system, it is recommended to use the linear model, since it is composed by transfer functions which are less complex and more efficient to deploy.

It is advisable to implement the obtained results into a pilot plant. This will allow to apply the experimental phase, obtaining a utility design in the industry.

Finally, it is recommended to analyze the stability of the system with more than two tanks, especially considering flow opening variations in the inlet and outlet valves, such that: a) the inlet opening is constant while outlet flow is variable, and b) the inlet and outlet opening flows are variable.

References

1. Mahapatro, S.R., Subudhi, B., Ghosh, S., Dworak, P.: A comparative study of two decoupling control strategies for a coupled tank system. In: 2016 IEEE Region 10 Conference (TENCON). pp. 3447–3451 (2016). https://doi.org/10.1109/TENCON.2016.7848695
2. Bhamre, P.K., Kadu, C.B.: Design of a smith predictor based fractional order PID controller for a coupled tank system. In: 2016 International Conference on Automatic Control and Dynamic Optimization Techniques (ICACDOT), pp. 705–708 (2016). https://doi.org/10.1109/ICACDOT.2016.7877678
3. Younis, A.S., Moustafa, A.M., Moness, M.: Experimental benchmarking of PID empirical and heuristic tuning for networked control of double-tank system. In: 2019 15th International Computer Engineering Conference (ICENCO), pp. 162–167 (2019). https://doi.org/10.1109/ICENCO48310.2019.9027382
4. Keerthana, P.G., Gnanasoundharam, J., Latha, K.: Design of model reference adaptive controller for two tank conical interacting system. In: 2016 International Conference on Computation of Power, Energy Information and Communication (ICCPEIC), pp. 255–258 (2016). https://doi.org/10.1109/ICCPEIC.2016.7557205
5. Gurumurthy, G., Das, D.K.: A semi-analytical approach to design a fractional order proportional-integral-derivative (FOPID) controller for a TITO coupled tank system. In: 2019 IEEE Asia Pacific Conference on Circuits and Systems (APCCAS), pp. 233–236 (2019). https://doi.org/10.1109/APCCAS47518.2019.8953172
6. Naik, R. B., Kanagalakshmi, S.: Mathematical modelling and controller design for interacting hybrid two tank system (IHTTS). In: 2020 Fourth International Conference on Inventive Systems and Control (ICISC), pp. 297–303 (2020). https://doi.org/10.1109/ICISC47916.2020.9171218
7. Patel, H., Shah, V.: A fault-tolerant control strategy for non-linear system: an application to the two tank canonical noninteracting level control system. In: 2018 IEEE Distributed Computing, VLSI, Electrical Circuits and Robotics (DISCOVER), pp. 64–70 (2018). https://doi.org/10.1109/DISCOVER.2018.8674119
8. Reddy, B.A., Krishna, P.V.: Traditional and 2-SMC control strategies for coupled tank system. In: 2019 Third International Conference on Inventive Systems and Control (ICISC), pp. 682–687 (2019). https://doi.org/10.1109/ICISC44355.2019.9036440
9. Mfoumboulou, Y.D.: Design of a model reference adaptive PID control algorithm for a tank system. Int. J. Electr. Comput. Eng. 11(1), 2088–8708 (2021). https://doi.org/10.1109/ICISC44355.2019.9036440
10. Raj, R.A., Deepa, S.N.: Modeling and implementation of various controllers used for Quadruple-Tank. In: 2016 International Conference on Circuit, Power and Computing Technologies (ICCPCT), pp. 1–5 (2016). https://doi.org/10.1109/ICCPCT.2016.7530245
11. Andonovski, G., Costa, B.S.: Robust evolving control of a two-tanks pilot plant. In: 2017 Evolving and Adaptive Intelligent Systems (EAIS), pp. 1–7 (2017). https://doi.org/10.1109/EAIS.2017.7954829
12. Mendes, J., Maia, R., Araújo, R., Gouveia, G.: Evolving fuzzy controller, and application to a distributed two-tank process. In: 2018 International Conference on Electrical Engineering and Informatics (ICELTICs), pp. 55–60 (2018). https://doi.org/10.1109/ICELTICS.2018.8548831
13. Pérez, K., Vargas, H., Castro, C., Chacón, J., De la Torre, L.: Coupled tanks laboratory for teaching multivariable control concepts. In: 2018 IEEE International Conference on Automation/XXIII Congress of the Chilean Association of Automatic Control (ICA-ACCA), pp. 1–6 (2018). https://doi.org/10.1109/ICA-ACCA.2018.8609851

14. Ruiz, J.M., Giraldo, J.A., Duque, J.E., Villa, J.L.: Implementation of a two-tank multivariable process for control education. In: 2020 IX International Congress of Mechatronics Engineering and Automation (CIIMA), pp. 1–6 (2020). https://doi.org/10.1109/CIIMA50553.2020.9290318
15. Juan, Anzurez M., José, P.G., Omar, C.S.: Estabilidad de Sistemas No-lineales: Sistema de Nivel de Líquidos de Dos Tanques Interconectados. RIEE&C (2008)
16. Singh, S., Bangia, S.: LabVIEW based analysis of two input and two output interacting tank system. In: 2016 IEEE 1st International Conference on Power Electronics, Intelligent Control and Energy Systems (ICPEICES), pp. 1–5 (2016). https://doi.org/10.1109/ICPEICES.2016.7853726
17. Eredias, C.A., Pérez, M.E., Martínez, M.F.: Solución numérica y linealizada de un modelo de sistema planar autonómo no lineal, para un equipo de dos tanques interconectados. Pistas Educativas **40**(130) (2018)
18. Ogata, K.: Modern control engineering. Prentice Hall. Cuarta Edición (2002)
19. Burden, R.L., Faires, J.D., Burden, M.A.: Numerical Analysis 10th edition, Cengage Learning Editores S.A. (2016). ISBN-13 978-1305253667, ISBN-10 1305253663

Building Energy Performance: Comparison Between EnergyPlus and Other Certified Tools

Diogo Esteves[1], João Silva[1,2](✉) ⓘ, Luís Martins[2] ⓘ, José Teixeira[2] ⓘ,
and Senhorinha Teixeira[1] ⓘ

[1] ALGORITMI Centre, University of Minho, Guimarães, Portugal
st@dps.uminho.pt
[2] Mechanical Engineering and Resource Sustainability Centre (MEtRICs), University of Minho, Guimarães, Portugal

Abstract. Currently, the energy consumption study in buildings is critical in fulfilling the EU objectives to achieve carbon neutrality. There are several building energy simulation tools and other spreadsheets available on the market. One of the main software tools, the EnergyPlus software, and two Excel spreadsheets developed in Portugal to assess the thermal loads and annual energy consumption in commercial and service buildings are used and compared in this work. A simple 3D building was modeled using SketchUp2017, and the schedules, materials, and thermal loads were defined for a typical residential and service building to perform this analysis.

Concerning the thermal loads, the simulation of a building without insulation using EnergyPlus showed results for heating and cooling, respectively 10% lower and 6% higher when compared with those predicted by RECS with the transient spreadsheet. Comparing with the simplified spreadsheet for the REH, the discrepancy is more significant. These discrepancies occur because RECS and REH use a simple methodology compared with EnergyPlus. The results for heating and cooling in EnergyPlus when a building has insulation show that the heating and cooling demand is 46% lower and 25% higher compared with the results with RECS, respectively. These results show that insulation reduces the heating demand although it impairs the cooling demand when compared with the building without insulation.

Keywords: Building · Energy performance · Simulation

1 Introduction

According to the European Directive 2018/844, almost 50% of the European Union (EU) final energy consumption is used for heating and cooling, 80% are used in buildings. To meet the energy and climatic objectives proposed by the EU, it is necessary to prioritize energy efficiency and the implementation of renewable energies [1, 2]. All new buildings must be nearly zero-energy buildings (NZEB) from December 31st, 2020 [3]. The need to design buildings with energy balance near zero and improve their thermal performances

O. Gervasi et al. (Eds.): ICCSA 2021, LNCS 12949, pp. 493–503, 2021.
https://doi.org/10.1007/978-3-030-86653-2_36

requires suitable energy simulation software to estimate energy consumptions better and optimize the design and operational strategies [4].

Most thermal modeling of buildings' energy consumption in Portugal is made in three ways depending on the type of building and increasing complexity: using spreadsheets with a permanent regime model, using simplified time-dynamic software, or using detailed dynamic simulation programs.

Spreadsheets in a permanent regime model are applied to estimate the heating and cooling needs for a residential building (REH). They are built following the Regulation of Energy Performance for Residential Buildings. This regulation was approved by the Decree-Law n°118/2013, which transposed the Energy Performance of Buildings Directive (EPBD) N° 2010/31/EU of the European Parliament and the Council to the national legal system, thus establishing minimum requirements for new or subject to interventions buildings, as well as parameters and methodologies to characterize the energy performance, for all residential buildings, as a strategy to improve their thermal behavior. The envelope's thermal and energy requirements are expressed in opaque thermal transmission coefficients, glazed elements, solar factors, and ventilation requirements by imposing a minimum value for the air renovation rate. Also, maximum limits were established for heating and cooling energy needs. The Excel spreadsheet uses an almost stationary method (albeit with a thermal inertia term), which means the heat transmission takes place in a steady state featuring the building as a single zone with constant reference internal temperature. Also, it is a seasonal method, estimating heating needs for the heating season (winter) and cooling needs for the cooling season (summer).

The simplified time-dynamic software is applied to estimate the heating and cooling loads for commercial and service buildings (RECS). This software estimate heating and cooling needs by defining the building as a single zone and calculating energy balances on an hourly basis as described in EN ISO 13790 standard. The estimation of the total energy consumption is made through a simple annual calculation, based on rules and orientations presented in Regulation of Energy Performance in Commercial and Services Buildings approved by the Decree-Law n°118/2013.

In Portugal, several Excel spreadsheets have been developed by two institutes, ITeCons and PTnZEB, to facilitate various methodologies. ITeCons made one Excel spreadsheet for residential buildings and a simplified dynamic software (converted to an Excel spreadsheet) for Commercial and Services Buildings. PTnZEB developed a single spreadsheet that suits both buildings. They allow the automatic fulfillment of all information necessary for the emission of energy building certificates.

The determination of the energy consumption for a multizone building is made by computational simulations using programs accredited by the ASHRAE 140 standard. Many software tools are available, e.g., TRNSYS, TRACE 700, Energy Plus, Design Builder, et cetera [5, 6]. There are limitations or less information regarding the calculation methodologies used and detailed access to results in some situations. Hence, the specific objective is to evaluate the constraints and compare the results with the various software for the same building. The building will be commercial or residential but straightforward enough for easy implementation of different commercial software.

Regarding software tools, De Boeck et al. [6] provided a literature review about improving the energy performance of residential buildings. They highlighted that the

EnergyPlus is one of the most applied tools. In the literature, there are some research works reporting the analysis of the certification process [7], energy performance of buildings based on the national directives [8–11], and the utilization of building energy simulation tools in the analysis of different problems [12–16]. However, it is essential to understand the differences between the Excel spreadsheets developed based on the European Directives and the more sophisticated building energy simulation tools, and up to date, to the authors' knowledge, there is no information about this matter. In this way, this study intends to overcome this need by acquiring more knowledge on building thermal simulation through understanding and comparing the different methodologies used to assess the building thermal energy requirements. Furthermore, the influence of the building thermal insulation on the heat and cooling demands was analyzed.

2 Case Study

2.1 Building Description

The first step for studying thermal behavior and estimating heating and cooling needs is creating the building. The base building, presented in Fig. 1, consists of a rectangular prism, with a floor area of $6 \times 10 \, m^2$ and a ceiling height of 3 m. The south-oriented surface includes $12 \, m^2$ of glazed area. Table 2 describes the wall, roof, and floor area of the exterior envelope. For construction solutions used in the building envelope, two cases were considered: case A without thermal insulation and case B with external thermal insulation (EPS) on walls and roof with a thickness of XX mm, both expressed in Table 2 by the U- Factor (global thermal transmission coefficient). The floor was defined as an exterior surface to reduce potential differences between the Excel spreadsheets and the EnergyPlus since they calculate the ground U-factor differently. The ventilation takes place naturally by infiltration with one renovation per hour. No air conditioning has been sized as well as renewable energy systems (Table 1). The walls and roof have light color. The simple glass constitutes the glazed elements without permanent or mobile devices, and they have a solar factor of 0.85, and the glazed fraction is 1.

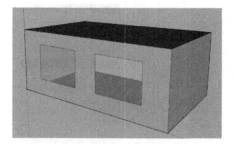

Fig. 1. The geometry of the case study.

Identification of the Climatic Zone. The building's geographic location influences the heat exchanges between the interior of the building and the outside environment due to

Table 1. Material properties of the envelope surface.

	Orientation	Area (m²)	Case A-U (W/m². K)	Case B-U (W/m². K)	Solar absorption (-)
Exterior wall	North	30.00	2.00	0,38	0.40
	South	18.00	2.00	0,38	0.40
	East	18.00	2.00	0,38	0.40
	West	18.00	2.00	0,38	0.40
Window 1	South	6.00	6.00	6.00	-
Window 2	South	6.00	6.00	6.00	-
Roof	-	60.00	Upward–1.85 Downward–1.61	Upward-0,32 Downward-0.31	0.40
Floor	-	60.00	Downward–2.06	Downward-2.06	-

factors such as the temperature and humidity of the outside air, altitude, and prevailing winds at the location [17]. The building was assumed to be located in Viana do Castelo County at an altitude of 268 m at a distance from the coast greater than 5 km.

CLIMA-SCE software version 1.05 was used to create the climate file that will later be used in the simplified and detailed dynamic simulations. The climate file used in simplified dynamic simulation has the following extension ".dat" while in EnergyPlus, it has "epw". Both climate data are written on an hourly basis and represent a Typical Meteorological Year (TMY) (Fig. 2).

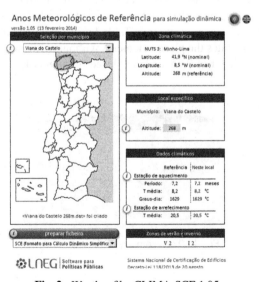

Fig. 2. Weather file: CLIMA-SCE 1.05.

Utilization Profile. In residential buildings, internal gains are considered constant for all seasons [18]. However, in the spreadsheet for simplified dynamic simulation and energy plus, internal gains vary according to the utilization profile. Internal gains are divided into internal gains by occupants, lighting, and electric equipment. For each one, three utilization profiles must be defined that depended on room type. In this study, it was defined as office space with a maximum of four persons (Fig. 3).

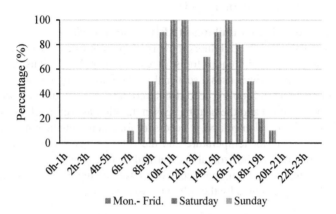

Fig. 3. Weekly occupation profile.

Thermal Loads. In residential buildings, internal heat gains assume a constant value of 4 W/m^2 [18]. still, in commercial and service buildings, they are different for each parcel and depend on the number of persons and their activity, expressed as metabolism rate. The illumination and electric equipment load depend on space type. For that case, regulation of energy performance in commercial and services buildings presented, according to space type, reference values for illumination and electric equipment. Present the occupation, illumination, and electric equipment loads used for simulations.

Table 2. Parameters defined to characterize the occupation inside the building.

Activity	Metabolism	Occupants	Thermal load (W)	Total sensible load (W)	Total latent load (W)
Sedentary	69.78	4	505.20	282.32	222.89

Table 3. Parameters defined to create the illumination and electric equipment profiles.

	Thermal load (W)	Released thermal load(%)
Space (office)	720	100
Equipment (default)	300	100

2.2 Building Energy Simulation Tools

Spreadsheet for Residential Buildings (REH)–Seasonal Method. With climate data and the characteristics of the buildings, now is it possible to estimate the heating and cooling needs for different simulation tools for residential buildings. It uses a spreadsheet in a permanent regime, which uses a seasonal method, assessing heating needs for the heating season (winter) and cooling needs for the cooling season (summer). From climate data, heating degree days and the heating season duration as well medium solar energy received in a south vertical envelope, is necessary for calculating heating energy needs, while the medium exterior temperature and the solar energy accumulated during the cooling season received on horizontal and vertical surfaces [18].

The value of the annual nominal needs of useful heating energy in heating season for building, N_{ic}, is calculated by the following expression [19]:

$$N_{ic} = \left(Q_{tr,i} + Q_{ve,i} - Q_{gu,i}\right)/A_p \tag{1}$$

where $Q_{tr,i}$, $Q_{ve,i}$, and $Q_{gu,i}$ are the transmission heat transfer through the building envelope and by ventilation and thermal increase from internal gains plus solar gains, respectively (kWh). A_p is the useful interior area of the building (m^2).

The value of the annual nominal needs of cooling useful energy in heating season for building, N_{vc}, is calculated by the following expression [19]:

$$N_{vc} = (1 - \eta_v)Q_{g,v}/A_p \tag{2}$$

where η_v is the utilization factor in cooling station, $Q_{g,v}$, gross thermal gains in cooling season (kWh) A_p is the useful interior area of the building (m^2).

The utilization factor is the function of transmission heat transfer through cooling season and air renovation and thermal gains in the cooling season.

Spreadsheet for Commerce and Services Buildings (RECS)–Simplified Dynamic Simulation. The heating and cooling demand estimation is done through an energy balance on an hourly basis according to the norm EN ISO 13790:2008. This method approximates a zone thermal to an electrical circuit establishing equivalences between electrical current and heat flows and represents in the form of an electrical scheme consisting of 5 resistors and 1 capacitance. The values of each temperature node are determined using the Crank-Nicholson iterative method, allowing the heating and cooling energy to be calculated every hour and per unit area of the thermal zone [20]. For calculation of heating and cooling energy needs is essential to choose the set-points temperature. In this case, the minimum setpoint temperature is equal to 20 °C, and the maximum temperature is 25 °C. If air indoor temperature is less than minimum setpoint temperature, then heating need is required. If air indoor temperature is higher than maximum setpoint temperature, the cooling demand is required. All equations and procedures are described in EN ISO 13790:2008 [20].

The spreadsheet given by PTnZeb and ITeCons brings a calculation motor integrated that quickly calculates the energy needs. All it needs is the climate file, created in CLIMA-SCE 1.05 software, and the introduction of all building characteristics in the spreadsheet and then run the simulation.

EnergyPlus-Detailed Dynamic Simulation. The EnergyPlus program is a collection of many program modules that work together to calculate the energy required for heating and cooling a building using various systems and energy sources. This tool was approved by ASHRAE 2004 standard, and it is based on two tools developed during the 70s after the oil crisis: BLAST and DOE-2 [21]. The software can model with excellent precision radiant and convection heat fluxes between indoor and outdoor, HVAC systems performance, heat exchanges with the ground, thermal comfort, natural, artificial, and hybrid systems. Additionally, it can simulate ventilation and airflows and allow dynamic transient simulations with a time interval between 1 h and 1 min, using an hourly representative year [21]. The simulation process started with defining the points presented in the previous subsection (Geometry, Weather zone, Schedules, Materials, Thermal loads, and setpoint temperature) [22]. After this process, the simulations are performed to obtain the heating and cooling demands. The simulation is based on the solution of the heat balance equation through the surfaces of the building [23]. It is essential to point out that the 3D model was made in SketchUp2017 and associated with the plugin OpenStudio. All building's surfaces were defined as well as their boundaries conditions. Compared with other simulations done before, it is used in the same building with the same characteristics. All building envelopes have outdoor conditions, and it was defined in only one thermal zone. After defining the 3D Model, the model was extracted to the OpenStudio plugin connected to the EnergyPlus.

3   Results and Discussion

In this section, the simulation results for the building with insulation are presented and discussed. Figure 4 shows the air temperature variation inside the building for a winter day and a summer day in RECS (simplified dynamic spreadsheet) and EnergyPlus. These days were selected because they are the warmer and cooler days of the year.

With the EnergyPlus software, air temperature reaches setpoint temperature earlier in the winter day since the setpoint should be obtained at the open time. Then, it decreases before the building closes at 20h00. On a summer day, the results are very similar.

The results of the heating and cooling demands for a building with insulation are presented in Fig. 5. It is possible to observe that EnergyPlus spend more heating and cooling time than RECS on winter and summer days.

This fact can be explained by the different methodologies used to compute the thermal losses by the building surface. With EnergyPlus, the calculation method is more detailed, and more accurate results are obtained. While in RECS, the exterior outside resistance

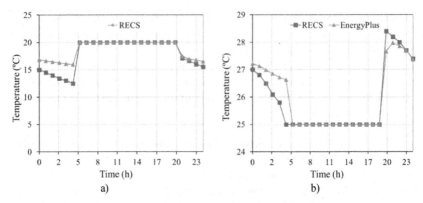

Fig. 4. Variation of the air temperature inside the building with insulation: a) winter and b) summer day.

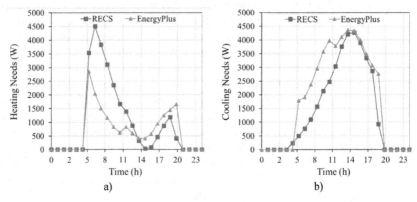

Fig. 5. Variation of the building with insulation energy demand: a) heating and b) cooling in two specific days of winter and summer season.

is constant, EnergyPlus has a comprehensive model for exterior convection by blending correlations from ASHRAE.

Furthermore, the methodology to compute the thermal inertial is different. These facts are the reasons why the results are significantly different. However, if the annual results are compared, the heating demand obtained with RECS is higher than EnergyPlus but lower in the cooling season. Regarding the REH methodology, lower results were obtained due to the simplification of the calculation of this tool where the heating and cooling demands are only computed for the summer and winter months while, with EnergyPlus and RECS tools, the heating and cooling demands are computed for all days of the year that are above or below the setpoint temperature.

Figures 6 and 7 present the air temperature and heating and cooling demand variation in the extreme conditions of the winter and summer seasons for a building without isolation. The tendency of the results is the same but, however, in the summer season, the results are not too different. Regarding the annual heating and cooling demand, the

Table 4. Comparison of the building annual heating and cooling demands between the different software tools (building with insulation).

	EnergyPlus	RECS	REH
Heating demand (kWh)	744	1,391	1,084
Cooling demand (kWh)	5,450	4,037	1,996

difference between the two tools is the same as the previous case. Since there is no insulation, the energy required in the different seasons is higher (see Table 4) (Table 5).

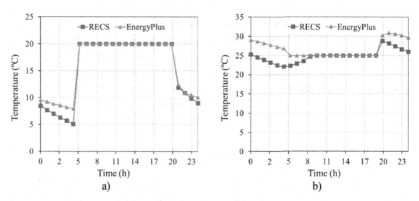

Fig. 6. Variation of the air temperature inside the building without insulation: a) winter and b) summer day.

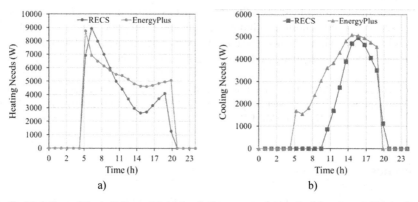

Fig. 7. Variation of the building without insulation energy demand: a) heating and b) cooling in two specific days of winter and summer season.

Table 5. Comparison of the building annual heating and cooling demands between the different software tools (building without insulation).

	EnergyPlus	RECS	REH
Heating demand (kWh)	7,750	8,580	16,494
Cooling demand (kWh)	2,639	2,481	947

4 Conclusions

There are several building energy simulation tools and other spreadsheets available on the market. One of the main software tools, the EnergyPlus software, and two Excel spreadsheets developed in Portugal to assess the thermal loads and annual energy consumption in commercial and service buildings are used and compared in this work. To perform this analysis, a simples building was modeled using Sketchup.

Concerning the thermal loads, the simulation of a building without insulation using EnergyPlus showed that heating and cooling results are 10% higher and 6% lower when compared with the values predicted by RECS. Comparing with REH, the discrepancy is more significant. This discrepancy is because RECS and REH use a simple methodology compared with Energy Plus.

The results for heating and cooling in EnergyPlus when a building has insulation show that the heating and cooling demand is lower 46% and higher 25% compared with the results from RECS, respectively. The results show that insulation reduces the heating demand but, on another side, impairs the cooling demand when compared to the building without insulation. So, it is essential to choose well the type and thickness of insulation. Another reason for high cooling demand is duo to the large size of windows are big and the non-existence of shading devices.

Despite the significant results, the model needs some improvements concerning the complexity of the building, including heating and ventilation systems and different renewable energy sources to meet the energy needs. The integration of all these aspects makes the model more realistic in determining the effective heating and cooling needs.

Acknowledgments. The authors would like to express their gratitude for the support given by FCT within the R&D Units Project Scope UIDB/00319/2020 (ALGORITMI) and R&D Units Project Scope UIDP/04077/2020 (MEtRICs).

References

1. EU Commission, Energy-European Commision 2021. https://ec.europa.eu/info/index_pt. Accessed 26 Mar 2021
2. Pérez-Lombard, L., Ortiz, J., Pout, C.: A review on buildings energy consumption information. Energy Build. **40**(3), 394–398 (2008)
3. EU Commission, Energy performance of buildings directive 2021. https://ec.europa.eu/energy/topics/energy-efficiency/energy-efficient-buildings/energy-performance-buildings-directive_en. Accessed 16 Jun 2021

4. Economidou, M., et al.: Review of 50 years of EU energy efficiency policies for buildings. Energy Build. **225**, 110322 (2020)
5. Hashempour, N., Taherkhani, R., Mahdikhani, M.: Energy performance optimization of existing buildings: a literature review. Sustain. Cities Soc. **54**, 101967 (2020)
6. De Boeck, L., Verbeke, S., Audenaert, A., De Mesmaeker, L.: Improving the energy performance of residential buildings: a literature review. Renew. Sustain. Energy Rev. **52**, 960–975 (2015)
7. Ahern, C., Norton, B.: Energy performance certification: misassessment due to assuming default heat losses. Energy Build. **224**, 110229 (2020)
8. Verichev, K., Zamorano, M., Carpio, M.: Assessing the applicability of various climatic zoning methods for building construction: case study from the extreme southern part of Chile. Build. Environ. **160**, 106165 (2019)
9. Gonzalez-Caceres, A., Lassen, A.K., Nielsen, T.R.: Barriers and challenges of the recommendation list of measures under the EPBD scheme: a critical review. Energy Build. **223**, 110065 (2020)
10. Sanhudo, L., et al.: BIM framework for the specification of information requirements in energy-related projects, Eng. Constr. Archit. Manag. (2020)
11. Ríos-Fernández, J.C., et al.: Evaluating European directives impacts on residential buildings energy performance: a case study of Spanish detached houses, Clean Technol. Environ. Policy (2021)
12. Shrivastava, R.L., Vinod, K., Untawale, S.P.: Modeling and simulation of solar water heater: a TRNSYS perspective. Renew. Sustain. Energy Rev. **67**, 126–143 (2017)
13. Sadeghifam, A.N., Zahraee, S.M., Meynagh, M.M., Kiani, I.: Combined use of design of experiment and dynamic building simulation in assessment of energy efficiency in tropical residential buildings. Energy Build. **86**, 525–533 (2015)
14. Silva, J., Brás, J., Noversa, R., Rodrigues, N., Martins, L., Teixeira, J., Teixeira, S.: Energy performance of a service building: comparison between energyplus and revit. In: Gervasi, O. (ed.) Computational Science and Its Applications – ICCSA 2020. LNCS, vol. 12254, pp. 201–213. Springer, Cham (2020). https://doi.org/10.1007/978-3-030-58817-5_16
15. Esteves, D., Silva, J., Rodrigues, N., Martins, L., Teixeira, J., Teixeira, S.: Simulation of PMV and PPD thermal comfort using energyplus. In: Misra, S. (ed.) Computational Science and Its Applications – ICCSA 2019. LNCS, vol. 11624, pp. 52–65. Springer, Cham (2019). https://doi.org/10.1007/978-3-030-24311-1_4
16. Noversa, R., Silva, J., Rodrigues, N., Martins, L., Teixeira, J., Teixeira, S.: Thermal simulation of a supermarket cold zone with integrated assessment of human thermal comfort. In: Gervasi, O. (ed.) Computational Science and Its Applications – ICCSA 2020. LNCS, vol. 12254, pp. 214–227. Springer, Cham (2020). https://doi.org/10.1007/978-3-030-58817-5_17
17. Despacho n° 15793-F, Zonamento Climático, Diário da República, **2**(234), 26–31 (2013)
18. Diário da República, Decreto-Lei n.° 118/2013, DE 20 DE agosto (desempenho energético dos edifícios), Diário da República, **159**, 4988–5005 (2013)
19. Do Emprego, M.D.E.E.: Portaria n.° 349-D/2013, de 2 de dezembro, Diário da Répuplica, **40**, 40–73 (2013)
20. ISO ISO 13790:2008-Energy performance of buildings. calculation of energy use for space heating and cooling (2016)
21. Berkeley, L., et al., EnergyPlus Essentials (2019)
22. Brackney, L., Parker, A., Macumber, D., Benne, K.: Building Energy Modeling with OpenStudio. Springer, Cham (2018). https://doi.org/10.1007/978-3-319-77809-9
23. Berkeley, L., et al.: Engineering Reference (2019)

Reconstruction of the Near-Surface Model in Complex Seismogeological Conditions of Eastern Siberia Utilizing Full Waveform Inversion with Varying Topography

Kirill Gadylshin[1][✉], Vladimir Tcheverda[1], and Danila Tverdokhlebov[2]

[1] Institute of Petroleum Geology and Geophysics SB RAS, 3 Koptug ave.,
Novosibirsk 630090, Russia
GadylshinKG@ipgg.sbras.ru
[2] LLC RN-Exploration, 8 Mozhaisky Val st., Moscow 121151, Russia

Abstract. Seismic surveys in the vast territory of Eastern Siberia are carried out in seismic and geological conditions of varying complexity. Obtaining a high-quality dynamic seismic image for the area of work is a priority task in the conditions of contrasting heterogeneities of the near-surface. For this, it is necessary to restore an effective depth-velocity model that provides compensation for velocity anomalies and calculates static corrections. However, for the most complex near-surface structure, for example, the presence of trap intrusions and tuffaceous formations, the information content of the velocity models of the near-surface area obtained on the basis of tomographic refinement turns out to be insufficient, and a search for another solution is required. The paper considers an approach based on the full waveform inversion (FWI). As the authors showed earlier, the use of multiples associated with the free surface reduces the resolution of this approach but increases the stability of the solution in the presence of uncorrelated noise. Therefore, at the first stage of FWI, the entire wavefield is used, including free surface-related multiples. The data after the suppression of multiples is then used. The obtained results demonstrate the ability of the FWI to restore complex geological structures of the near-surface area, even in the presence of high-velocity anomalies (trap intrusions).

Keywords: Near-surface · Full waveform inversion · Free surface topography

1 Introduction

This work, to some extent, acts as a development of our previous studies devoted to the consideration of full-waveform inversion for the reconstruction of deep

Supported by RSF grant 21-71-20002.

horizons (see [4,6]). Previously, we focused our attention on geological media with an effortless top of the section, typical for marine seismic observations. However, when conducting seismic observations on land, the assumption about the simplicity of the upper part of the section's structure is no longer valid, especially for Eastern Siberia. Here we are presenting our approach to the reconstruction of geological media with a complicated upper section. Really, the correct near-surface contribution is necessary when determining the structural features of the reconstruction of the geological media (shape and the location of the interfaces) and searching for its seismic parameters, like wave propagation velocities (inverse dynamic problem). Distortion of results obtained using an imprecise model of near-surface will introduce significant errors both in the depth and the shape of the target horizons, not to mention the correct restoration of their lateral variability [3].

In this regard, near-surface structure reconstruction is one of the most demanded among a wide range of seismic studies [1,2,8]. It can be argued that it is a fairly developed seismic exploration area that uses a diverse set of methods and techniques. We list only the main approaches to solve it, which we know:

– solution of the inverse kinematic problem with the subsequent introduction of the obtained static corrections. The solution here is built for near-surface layers in order to calculate the required time shifts to the original seismograms, which avoids artificial curvature of the target horizons and ensures the correct focusing of wave seismic images [9];
– application of seismic tomography methods on refracted rays for the reconstruction of the velocity structure of the near-surface and its subsequent use in calculating static corrections and constructing wave seismic images of target horizons [9,10];
– determination of the velocity structure of the near-surface using surface waves [11,12];
– full-waveform inversion methods oriented on the near-surface area [13,14].

To better understand the possibility of reconstructing the near-surface area's velocity model, it is necessary to carry out special work on seismic geological modelling. The work authors built a detailed seismogeological model based on all available geological and geophysical information for performing finite-difference modelling. As a result of modelling, high-quality synthetic data were obtained that are close to real seismic materials, which were subsequently used for full-waveform inversion.

The thickness of the sedimentary cover in the study area reaches 1600–2000 m. The upper and basal parts of the section are represented by terrigenous rocks, in the middle and lower parts - by alternating halogen and carbonate layers. The section is complicated by several levels of intrusion of magmatic intrusions. Trap intrusions and tuffaceous formations complicate the near-surface. The target horizons (terrigenous and carbonate strata) lie at the base of the section. The reservoir thickness is 10–15 m. The section's complex structure is determined by seismic and geological heterogeneities, leading to the formation of sharp and significant vertical and lateral variations in the elastic characteristics of the medium. The sizes of submerged velocity anomalies, such as intrusions, can vary from the first tens to thousands of meters, and the depth of occurrence may reach $\frac{3}{4}$ or more of the depth of the sedimentary cover. Differences in the terrain up to 200–400 m with sharp slopes of ravines and indented river valleys, exposing various types of rocks on the surface, lead not only to the complexity of restoration and taking into account the velocity characteristics of the section but also to significant variations in amplitude-frequency characteristics of the seismic signal over the area of work.

All inhomogeneities in the near-surface lead to the formation of a huge number of interference waves, which nature is not always possible to reliably identify, and which reduce the efficiency or make it useless and even impossible to use "standard" seismic processing procedures, including reconstruction of the near-surface velocity model. To improve the quality and reliability of the processing and interpretation results, it becomes necessary to carry out an in-depth analysis of the influence of changes in seismic-geological conditions on the characteristics of the seismic record and its possibility recovering a high-quality dynamic image. To assess the feasibility of algorithms and methods for reconstructing the near-surface, one needs to use synthetic seismic data modelled based on a detailed thin-layered model containing characteristic seismogeological features as close as possible to the real studied section. Based on the testing results on synthesized data based on a well-known model, one should adapt the algorithms to work on real seismic data.

2   Seismogeological Model Building

Seismic surveys in the vast territory of Eastern Siberia are carried out in seismic and geological conditions of varying complexity. The selected area to build a seismogeological model is Irkutsk region, Russia (Fig. 1). It contains anomalous and target geological objects. The original thick-layered model was built based on information about the topography and stratigraphic information from well data (Fig. 2). For detailing the near-surface model, anomalies were specified based on information about tuffs and traps' location according to a geological map. The geometry of intrusions, trap and tuff formations, as well as supply channels for the first 500m, was set according to the interpretation of non-seismic methods - electrical prospecting and gravity magnetic prospecting based on conceptual forms of volcano structures of diamondiferous regions in adjacent territories, studied in detail by drilling wells [7]. Velocity characteristics were included in the model according to well logging data. Schematically, the evolution of model building is shown in Fig. 2.

3   Method

The inverse dynamic problem is considered as a solution of a nonlinear operator equation

$$F(m) = d, \tag{1}$$

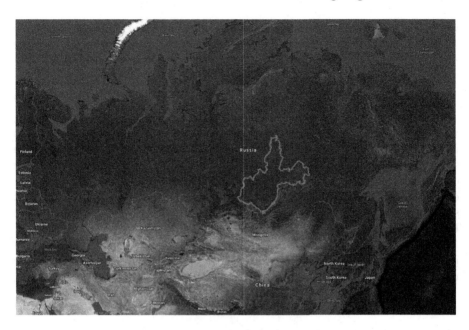

Fig. 1. The location map of the studied area. The borders of the Irkutsk region are marked in red. (Color figure online)

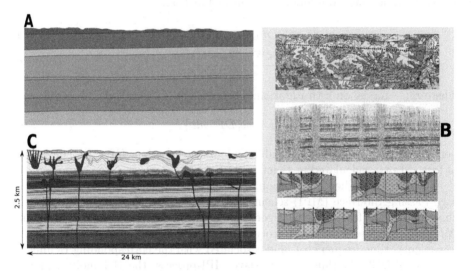

Fig. 2. The evolution of the depth model building for pressure velocities. The original thick-layered model constructed using topography and well log data (a), detailing of the model using geological maps and well information (b) and the final P-velocity model (c).

where $F : M \to D$ is a nonlinear forward operator which maps model space M into data space D. To simplify mathematical calculations, we deal with the scalar wave equation in the time domain, that is, the Helmholtz equation:

$$\triangle u(x, z) + \frac{\omega^2}{c^2(x, z)} u = f(\omega)\delta(x - x_s)\delta(z - z_s), \tag{2}$$

with absorbing boundary conditions on the base of perfectly matched layers (PML). Here (x_s, z_s) determine the coordinates of the source with the pulse shape $f(\omega)$. Calculation of the solution of the formulated problem at the location of the receivers determines the forward modelling operator F. We solve this equation using the fourth order finite difference scheme, which leads to a sparse system of linear algebraic equations (SLAE). This SLAE is solved using direct LU-based method.

The common full-waveform inversion formulation is to find the minimum point of the misfit functional, characterizing the mean square deviation of the registered data from the calculated for the current speed model:

$$m^* = argmin_{m \in M}(\|F(m) - d\|_D^2), \tag{3}$$

where $m(x, z) = c^{-2}(x, z)$ - squared slowness.

Typically, the local optimization techniques, such as conjugate gradient method, are applied to minimize the misfit function:

$$m_{k+1} = m_k + \mu_k S_k, \; S_0 = \nabla_0, \tag{4}$$

$$S_k = -\nabla_k - \frac{< \nabla_k, \nabla_k - \nabla_{k-1} >_M}{< \nabla_k, \nabla_{k-1} >_M} S_{k-1}, \tag{5}$$

where m_k – model on k-th iteration. The gradient ∇_k is calculated as follows:

$$\nabla_k = \Re\{DF^*\delta d_k\} \tag{6}$$

here $\delta d_k = F(m_k) - d$ is data residual on current iteration, DF – first derivative of forward map calculated at point m_k.

For efficient gradient calculation, we use a master/slave communication Message Passing Interface (MPI) scheme (see Fig. 3). Considering that we use Nf time-frequencies for FWI, the algorithm requires one master MPI-process and Nf slave processes. Slave MPI-process with index i calculate partial gradient for a particular time-frequency f_i and associated with the corresponding Helmholtz equation (Fig. 3). Within a single slave MPI-process, the Helmholtz equation solver is parallelized via OpenMP. On the next step, the master process gathers all partial gradients, performs the model update and send the updated model back to the slave processes. This process repeats iteratively until the convergence conditions are not satisfied.

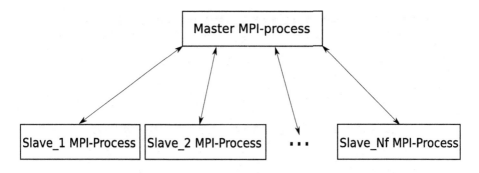

Fig. 3. The schematic master/slave communication of the FWI algorithm. Each slave MPI-process is related to the Helmholtz equation for a fixed time frequency.

4   Numerical Experiments

4.1   Model and Acquisition Geometry

We used a realistic P-velocity profile in our numerical experiments, which building process described in Sect. 2. For FWI, we choose a part of the model, designated by a red rectangle in Fig. 4. The horizontal size of the model is 24 km, while the depth reaches 2.8 km. The main difficulty in reconstructing the near-surface area caused by the presence of trap intrusions, the velocity of propagation of P-waves in which reaches 6500 m/s, thus the velocity difference when crossing the intrusion boundary reaches 4500 m/s, which significantly complicates the use of standard approaches to recover the depth velocity model. The presence of varying topography also complicates the velocity model building. It is necessary to perform full-waveform modelling on a fine mesh to mitigate numerical dispersion caused by staircase topography approximation on a rectangular grid. An example of a synthetic seismogram calculated in acoustic mode, taking into account the multiples caused by the presence of a free surface, is shown in Fig. 5. The acquisition system has 241 seismic sources and 961 receivers located at the free surface with a lateral spacing of 100 m and 25 m. The maximum source-receiver offset is 5 km. The source impulse is a Ricker wavelet with a dominant frequency 30 Hz. As an initial guess for FWI, we used a vertically heterogeneous model (see Fig. 6), which was built using the well information (x = 20 km in the true model) by Gaussian smoothing and adding a free surface topography. Thus, the starting depth velocity model contains no information about the presence of high-velocity near-surface anomalies.

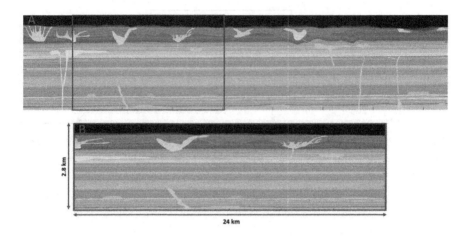

Fig. 4. The full P-velocity profile obtained after seismogeological model building (a) and the area used for performing FWI experiments (b).

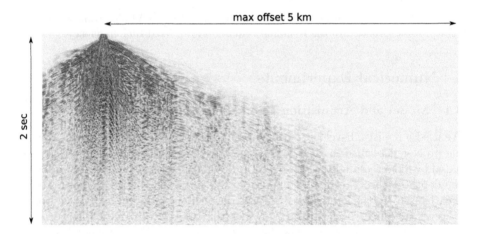

Fig. 5. The example of observed seismogram used for near-surface FWI.

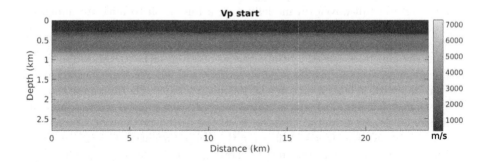

Fig. 6. The starting depth velocity model for near-surface FWI.

4.2   Inversion of Complex Near-Surface Using Data with Free Surface-Related Multiples

As shown by the authors earlier [15], taking the surface-related (SR) multiples into account during the inversion reduces the resolution and the information content of the FWI results. However, it increases their stability in the presence of uncorrelated noise. Therefore, at the first stage of FWI, we use the entire wavefield, including SR multiples. The results of the near-surface inversion are shown in Fig. 7. The inversion was performed using the temporal frequencies range from 3 10 Hz, and the target area was limited up to the depths of 1 km. As one can see, with such an acquisition system and frequency range of the source signal (presence of low frequencies), we successfully identified high-speed anomalies. However, areas directly under the traps were restored with significant distortion. Therefore, in subsequent stages of FWI (see Sect. 4.3), to improve the restored model's quality, it is proposed to use the data without SR multiples.

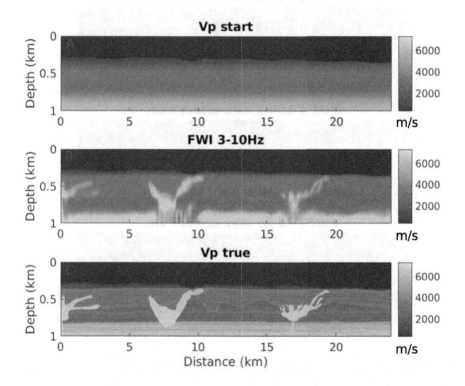

Fig. 7. The inversion of complex near surface results: starting model (a), FWI results (b), and the true model (c).

The Influence of Maximum Offset in the Data on the Near-Surface Inversion

To analyze the maximum source-receiver offsets' influence on the inversion results, we consider three different FWI scenarios. We used the same source geometry and the same receiver steps but varied the maximum offset. In the first case, we used 3 km maximum offset, in the second - 5 km (previous results), and in the last - 7 km. Typically for the real acquisition in Eastern Siberia, 5 km is the maximum available offset, and the most common maximum source-receiver offset is 3 km. The near-surface FWI results for a different scenarios are presented in Fig. 8. As one may observe, for the reasonable offset ranges (from 3 km to 7 km), there are no significant differences in the inversion results, which means that the maximum offset in the data does not affect the near-surface FWI results in our case.

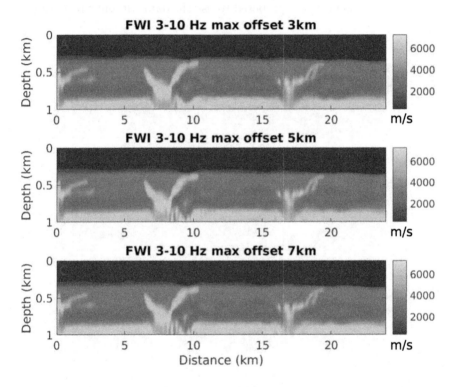

Fig. 8. The influence of the maximum offset on the inversion results: FWI using offsets up to the 3 km (a), 5 km (b), and 7 km (c).

The Influence of the Lower Available Time Frequency in the Data on the Near-Surface Inversion

As we have seen, the influence of the maximum offset on inversion results is negligible within a reasonable range. Now we are interested in the influence of the lower available temporal frequency in the observed data. We performed three FWI tests with a lower frequency 3 Hz, 5 Hz 7 Hz. The inversion results are shown in Fig. 9. We see that lower frequency affects the near-surface FWI results substantially. The bigger the lower frequency, the worse trap intrusions reconstruction we observe. The comparative analysis of relative error between the reconstructed near-surface model and true model is presented in Fig. 10. Furthermore, as we increase the lower temporal frequency, we decrease the data fitting. The corresponding misfit functional behaviour during FWI iterations for different scenarios plotted in logarithmic scale is shown in Fig. 11.

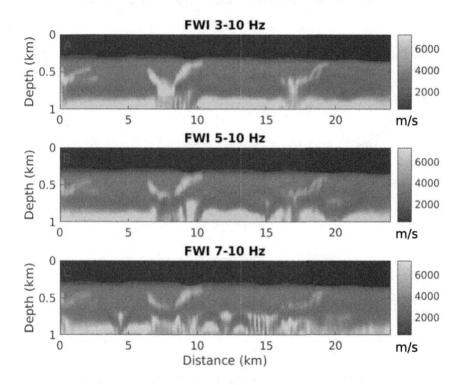

Fig. 9. The influence of the lower available time frequency in the data on the inversion results: FWI using lower 3 Hz (a), 5 Hz (b), 7 Hz (c).

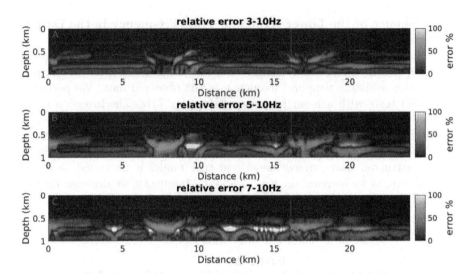

Fig. 10. The relative error between true near-surface model and the FWI model using lower 3 Hz (a), 5 Hz (b), 7 Hz (c).

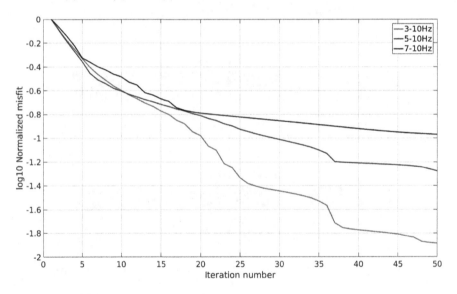

Fig. 11. The L_2 mistfits between observed and modelled data during inversion for FWI scenario using lower 3 Hz (red plot), 5 Hz (blue plot), 7 Hz (black plot). (Color figure online)

4.3   Inversion of Deep Target Horizons Using Data Without Free Surface-Related Multiples

For reconstruction of the deep target horizons to improve the restored model's quality, it is proposed to use the data without multiples associated with the

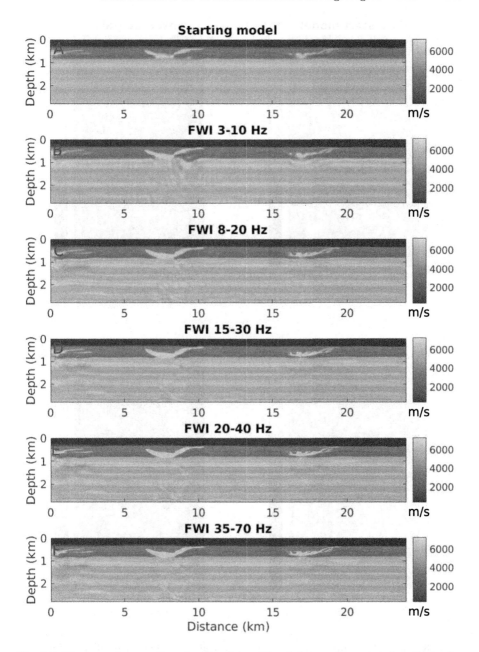

Fig. 12. The inversion results using the data without free surface related multiples for different frequencies ranges: starting model (a), FWI 3–10 Hz (b), FWI 8–20 Hz (c), FWI 15–30 Hz (d), FWI 20–40 Hz (e), and FWI 35–70 Hz (f).

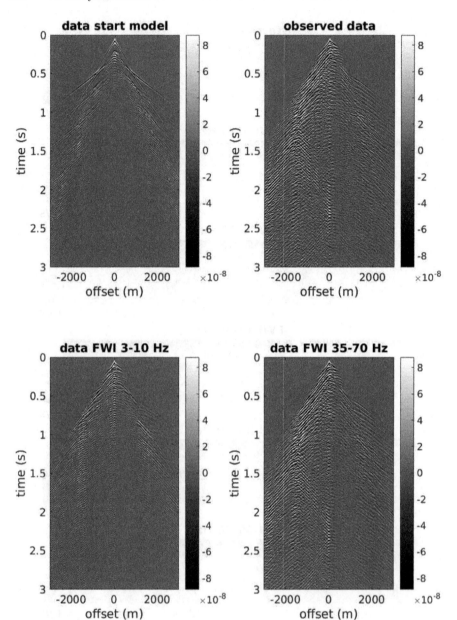

Fig. 13. The data comparison for shot located at x = 8 km on a different inversion stages: data in starting model before the near-surface FWI (a), the data modelled in a true model (b), the data after near-surface FWI (c), and the data in the final FWI model (d).

presence of a free surface. While performing FWI in the time-frequency domain, it is common practice to use overlapping frequency groups. In our case, after the near-surface reconstruction in the frequency range [3, 10] Hz with SR multiples, we used the following set of frequency groups to be used in FWI without SR multiples: [3, 10] Hz, [8, 20] Hz, [15, 30] Hz, [20, 40] Hz and [35, 70] Hz. The starting velocity model for each frequency group is the velocities from the previous FWI run (e.g. the output of FWI for a frequency range [3, 10] Hz is an input for [8, 20] Hz FWI, and so on). The evolution of depth velocity model reconstruction for a different frequency group is shown in Fig. 12. As one can see, we managed to get a detailed depth velocity model restoration in almost the entire region. To demonstrate the data fitting, we calculate the seismogram for a shot located at $x = 8$ km (see Fig. 13). As one may observe, the starting model's synthetics does not explain the observed data, while the final FWI model interprets most of the observed data.

5 Conclusions

The paper considers an approach to the near-surface P-velocity distribution inversion based on the FWI. As the authors showed earlier, the attraction of multiples associated with a free surface reduces the resolution of this approach. However, it increases the stability of the solution in the presence of uncorrelated noise. Therefore, at the first stage of FWI, one uses the entire wavefield, including free surface-related multiples. Further, the data after multiples suppression is used. The results obtained demonstrate the FWI method's ability to reconstruct complex geological structures of the near-surface area, even in the presence of high-velocity anomalies (trap intrusions).

Acknowledgements. K. Gadylshin is supported by RSF grant 21-71-20002, he performed all numerical FWI experiments. The numerical results of the work were obtained using computational resources of Peter the Great Saint-Petersburg Polytechnic University Supercomputing Center (scc.spbstu.ru).

References

1. Bloor, R., Whaler, K.: The generation of near-surface-layer models from seismic reflection traveltimes. Geophys. J. Int. **119**(3), 693–705 (1994)
2. Dou, S., et al.: Distributed acoustic sensing for seismic monitoring of the near surface: a traffic-noise interferometry case study. Sci. Rep. **7**, 11620 (2017)
3. Liu, F., Morton, S.A., Ma, X., Checkles, S.: Some key factors for the successful application of full-waveform inversion. Lead. Edge **32**(9), 1124–1129 (2013)
4. Silvestrov, I., Neklyudov, D., Kostov, C., Tcheverda, V.: Full-waveform inversion for macro velocity model reconstruction in look-ahead offset vertical seismic profile: numerical singular value decomposition-based analysis. Geophys. Prospect. **61**(6), 1099–1113 (2013)
5. Tanis, M.C., et al.: Diving-wave refraction tomography and reflection tomography for velocity model building. In: SEG 76th Annual International Meeting, Expanded Abstracts, pp. 3340–3344 (2006)

6. Tcheverda, V., Gadylshin, K.: Elastic full-waveform inversion using migration-based depth reflector representation in the data domain. Geosciences **11**, 76 (2021)
7. Tverdokhlebov, D.N., Korobkin, V.S., Danko, E.A., Kashirina, E.G., Filichev, A.V., Gaiduk, A.V.: Model based multiple waves suppression based on modeling algorithm. Geofizika **1**, 2–14 (2018)
8. Xie, H., et al.: Imaging complex geology through challenging surface terrain - a case study from West China. First Break **33**(3), 55–69 (2015)
9. Li, L.M., Luo, S.X.: The tomographic inversion static correction on complex three-dimensional surface model. Oil Geophys. Prospect. **38**(6), 636–641 (2003)
10. Zhang, W., Zhang, J.: Full waveform tomography with consideration for large topography variations. In: SEG Annual Meeting Expanded Abstracts, San Antonio, pp. 2539–2542 (2011)
11. Socco, L.V., et al.: Geophysical investigation of the Sandalp rock avalanche deposits. J. Appl. Geophys. **70**, 277–291 (2010)
12. Piatti, C., Socco, L.V., Boiero, D., Foti, S.: Constrained 1D joint inversion of seismic surface waves and P-refraction traveltimes. Geophys. Prospect. **61**, 77–93 (2013)
13. Shen, X., Tonellot, T., Luo, Y.: A new waveform inversion workflow: application to near-surface velocity estimation in Saudi Arabia. In: SEG Annual Meeting expanded abstracts, Las Vegas, USA (2012)
14. Bretaudeau, F., Brossier, R., Leparoux, D.: 2D elastic full waveform imaging of the near-surface: application to synthetic and physical modeling data sets. Near Surface Geophys. **11**(3), 307–316 (2013)
15. Gadylshin, K., Bakulin, A., Dmitriev, M., Golikov, P., Neklyudov, D., Tcheverda, V.: Effect of free-surface related multiples on near surface velocity reconstruction with acoustic frequency domain FWI. In: 76th European Association of Geoscientists and Engineers Conference and Exhibition 2014: Experience the Energy - Incorporating SPE EUROPEC 2014, Amsterdam, Netherlands, 16–19 June 2014, pp. 357–361 (2014)

Numerical Solution of Biot Equations in Quasi-static State

Sergey Solovyev, Mikhail Novikov, Alena Kopylova, and Vadim Lisitsa[(⊠)][iD]

Sobolev Institute of Mathematics SB RAS, 4 Koptug ave., Novosibirsk 630090, Russia
{solovievsa,lisitsavv}@ipgg.sbras.ru

Abstract. This paper presents a numerical algorithm to simulate low-frequency loading of fluid-filled poroelastic materials and estimate the effective frequency-dependent strain-stress relations for such media. The algorithm solves Biot equation in quasi-static state in the frequency domain. Thus, a large-scale system of linear algebraic equations have to be solved for each temporal frequency. We use the direct solver, based on the LU decomposition to resolve the system of the linear equations. According to the presented numerical examples suggested algorithm allows reconstructing the stiffness tensor within a wide range of frequencies for the realistic large-scale samples within several minutes. Thus, the estimation of the frequency-dependent stiffness tensors can be done in a routine manner and statistical data may be accumulated.

Keywords: Biot equation · Poroelasticity · Finite differences

1 Introduction

Developments in CO_2 sequestration [8,14], geothermal energy exploration [12,21] technologies raises a challenging tasks to seismic monitoring methods (4D seismic). One needs to estimate the fluid mobility and reservoir hydraulic permeability based on seismic data. Standard seismic attributes, especially the kinematic ones, are almost incentive to the changes in the reservoir structure due to the fluid substitution or partial chemical dissolution of carbonate rock matrix. However, the frequency-dependent dynamic effects can be recorded and potentially interpreted. In particular, changes in the fluid content or the pore-space geometry significantly affects the wave-induced fluid flows (WIFF). These flows are caused by the local pressure gradients within the wavefront if a seismic wave propagates in fractured-porous media [16,19]. Typically, fracture-to-background WIFF (FB-WIFF) and fracture-fracture WIFF (FF-WIFF) are considered. The FB-WIFFs appear if a low-frequency (long wavelength) wave propagates. In this case, the wave period is high enough for flow to form, even in the media

The research was supported by the Russian Science foundation grant no. 19-77-20004.

with quite low permeability. The intensity of FB-WIFF is governed by a compressibility contrast between the host rock and fracture-filling material [10,19]. High-frequency signals propagation causes FF-WIFF, defined by fracture-filling material properties and also local fracture connectivity [7,10,19]. Unfortunately, theoretical studies of this effect include consideration of relatively simple models of the media. Moreover, the fracture connectivity is considered only for the pairs of differently oriented fractures [10]. Numerical investigation of the phenomena is also restricted by such fracture connectivity criteria [10,19], except the study [9], where authors apply statistical modeling of fracture network and estimate resulting connectivity of the fractures. One of the reasons for that is the lack of efficient numerical algorithm to simulate seismic wave propagation in poroelastic media, or to solve Biot equation in quasi-static state to simulate low-frequency creeping test. Numerical upscaling experiments should satisfy two requirements. First, each considered sample should be representative [1,18], or, at least, should be much greater that the typical homogeneity size. In the case of the fractured-porous media, the characteristic size in the fracture length, typical percolation length, or connectivity index [17,25]. However, the grid step for either finite difference or finite element method is governed by the fracture width. Thus, the typical size of the discretized problem exceeds 1000 grid points in each spatial direction. Second, numerical upscaling requires a series of experiments for each statistically equivalent models of the fracture systems. Thus, an efficient numerical upscaling algorithm should be able to solve series on large-scale problems.

In this paper, we present an algorithm for numerical upscaling elastic properties of fractured-porous media in low-frequency range. To do so, we develop the algorithm to solve Biot equation in quasi-static state. The paper has the following structure. In Sect. 2, we state the problem of the poroelastic media upscaling to obtain frequency-dependent stiffness tensor, corresponding to the viscoelastic media. The finite-difference approximation of the considered boundary-value problem in considered in Sect. 3. Next, we provide the peculiarities of the numerical solution of the SLAE is provided in Sect. 4. Numerical experiments and performance analysis is presented in Sect. 5.

2   Statement of the Problem

2.1   Biot Equations in Quasistatic State

Consider the quasi-static Biot equations governing the diffusion processes in fluid-filled poroelastic media in a low-frequency regimes [2,3]. We deal with the Cartesian coordinates and restrict our considerations with 2D case, thus the equations can be written as follows:

$$\frac{\partial}{\partial x}\left[(\lambda_u + 2\mu)\frac{\partial u_x}{\partial x} + \lambda_u\frac{\partial u_z}{\partial z} + \alpha M\left(\frac{\partial w_x}{\partial x} + \frac{\partial w_z}{\partial z}\right)\right] + \frac{\partial}{\partial z}\left[\mu\left(\frac{\partial u_x}{\partial z} + \frac{\partial u_z}{\partial x}\right)\right] = 0,$$

$$\frac{\partial}{\partial x}\left[\mu\left(\frac{\partial u_x}{\partial z} + \frac{\partial u_z}{\partial x}\right)\right] + \frac{\partial}{\partial z}\left[\lambda_u\frac{\partial u_x}{\partial x} + (\lambda_u + 2\mu)\frac{\partial u_z}{\partial z} + \alpha M\left(\frac{\partial w_x}{\partial x} + \frac{\partial w_z}{\partial z}\right)\right] = 0,$$

$$\frac{\partial}{\partial x}\left[\alpha M\left(\frac{\partial u_x}{\partial x} + \frac{\partial u_z}{\partial z}\right) + M\left(\frac{\partial w_x}{\partial x} + \frac{\partial w_z}{\partial z}\right)\right] = i\omega\frac{\eta}{k_0}w_x,$$

$$\frac{\partial}{\partial z}\left[\alpha M\left(\frac{\partial u_x}{\partial x} + \frac{\partial u_z}{\partial z}\right) + M\left(\frac{\partial w_x}{\partial x} + \frac{\partial w_z}{\partial z}\right)\right] = i\omega\frac{\eta}{k_0}w_z,$$

$$(1)$$

where $\vec{u} = (u_x, u_z)^T$ is the solid matrix velocity vector, $\vec{w} = (w_x, w_z)^T$ is the vector of the relative fluid velocity with respect to the matrix, λ_u is the Lame parameter of undrained rock, μ is the shear modulus, α is Biot-Willis parameter, M is the fluid storage coefficient, η is the fluid dynamic viscosity, κ is the absolute permeability of the rock, and ω is the temporal frequency. Specific parameters λ_u, M, and α are usually estimated from the bulk moduli of drained K_d, fluid K_u, and the solid matrix K_s [13]:

$$\alpha = 1 - \frac{K_d}{K_s}, \lambda_u = K_u - \frac{2}{3}\mu, M = BK_u/\alpha,$$

$$B = \frac{1/K_d - 1/K_s}{1/K_d - 1/K_s + \phi(1/K_f - 1/K_s)}, K_u = \frac{K_d}{1 - B\alpha}.$$

2.2 Effective Visco-Elastic Media

We aim to construct the effective frequency-dependent linear elastic media (visco-elastic media) [5], so that for any stresses applied to a unit volume the average strains in the reconstructed viscoelstic and the original fluid-saturated poroelastic media should coincide. The quasi-static state of the viscoelastic wave equation is

$$\frac{\partial}{\partial x}\left[C_{11}(\omega)\frac{\partial v_x}{\partial x} + C_{13}(\omega)\frac{\partial v_z}{\partial z}\right] + \frac{\partial}{\partial z}\left[C_{55}(\omega)(\frac{\partial v_x}{\partial z} + \frac{\partial v_z}{\partial x})\right] = 0,$$

$$\frac{\partial}{\partial x}\left[C_{55}(\omega)(\frac{\partial v_x}{\partial z} + \frac{\partial v_z}{\partial x})\right] + \frac{\partial}{\partial z}\left[C_{13}(\omega)\frac{\partial v_x}{\partial x} + C_{33}(\omega)\frac{\partial v_z}{\partial z}\right] = 0,$$

$$(2)$$

where $\vec{v} = (v_x, v_z)^2$ is the displacement vector, and tensor C is the frequency-dependent stiffness tensor. Consider a rectangular domain $D = [L_1^x, L_2^x] \times [L_1^z, L_2^z]$ with the following boundary conditions applied $\sigma \cdot \vec{n} = \sigma_0$ at ∂D, where $\vec{n}$ is the outer normal.

Consider three basis loads:

- x-direction compression:

$$\sigma_{xx} = C_{11}(\omega)\frac{\partial v_x}{\partial x} + C_{13}(\omega)\frac{\partial v_z}{\partial z}|_{x=L_1^x} = \phi_x,$$

$$\sigma_{xx} = C_{11}(\omega)\frac{\partial v_x}{\partial x} + C_{13}(\omega)\frac{\partial v_z}{\partial z}|_{x=L_2^x} = \phi_x,$$

$$\sigma_{zz} = C_{13}(\omega)\frac{\partial v_x}{\partial x} + C_{11}(\omega)\frac{\partial v_z}{\partial z}|_{z=L_1^z} = 0,$$

$$\sigma_{zz} = C_{13}(\omega)\frac{\partial v_x}{\partial x} + C_{11}(\omega)\frac{\partial v_z}{\partial z}|_{z=L_2^z} = 0,$$

$$\sigma_{xz} = C_{55}(\omega)(\frac{\partial v_x}{\partial z} + \frac{\partial v_z}{\partial x})|_{\partial D} = 0,$$

$$(3)$$

Solution of the problem (2), (3) can be constructed analytically, in case of constant coefficients: $\sigma_{xx}(x, z) = \phi_x$, $\sigma_{zz}(x, z) = 0$, and $\sigma_{xz}(x, z) = 0$. Thus the components of the strain tensor can be represented as

$$
\begin{aligned}
\varepsilon_{xx} &= \frac{\partial v_x}{\partial x} = S_{11}\phi_x, \\
\varepsilon_{zz} &= \frac{\partial v_z}{\partial z} = S_{13}\phi_x, \\
\varepsilon_{xz} &= \frac{1}{2}\left(\frac{\partial v_z}{\partial x} + \frac{\partial v_x}{\partial z}\right) = S_{15}\phi_x,
\end{aligned}
\tag{4}
$$

where S_{ij} are the components of the compliance tensor (invert to stiffness tensor). If the initial loads are known and strains are computed, one may resolve the equations with respect to the components of the compliance tensor.

– z-direction compression:

$$
\begin{aligned}
\sigma_{xx} &= C_{11}(\omega)\frac{\partial v_x}{\partial x} + C_{13}(\omega)\frac{\partial v_z}{\partial z}\big|_{x=L_1^x} = 0, \\
\sigma_{xx} &= C_{11}(\omega)\frac{\partial v_x}{\partial x} + C_{13}(\omega)\frac{\partial v_z}{\partial z}\big|_{x=L_2^x} = 0, \\
\sigma_{zz} &= C_{13}(\omega)\frac{\partial v_x}{\partial x} + C_{11}(\omega)\frac{\partial v_z}{\partial z}\big|_{z=L_1^z} = \phi_z, \\
\sigma_{zz} &= C_{13}(\omega)\frac{\partial v_x}{\partial x} + C_{11}(\omega)\frac{\partial v_z}{\partial z}\big|_{z=L_2^z} = \phi_z, \\
\sigma_{xz} &= C_{55}(\omega)\left(\frac{\partial v_x}{\partial z} + \frac{\partial v_z}{\partial x}\right)\big|_{\partial D} = 0,
\end{aligned}
\tag{5}
$$

Solving problem (2), (5) one gets:

$$
\begin{aligned}
\varepsilon_{xx} &= S_{13}\phi_z, \\
\varepsilon_{zz} &= S_{33}\phi_z, \\
\varepsilon_{xz} &= S_{35}\phi_z,
\end{aligned}
\tag{6}
$$

Thus, the second column of the compliance tensor can be recovered.

– z-direction compression:

$$
\begin{aligned}
\sigma_{xx} &= C_{11}(\omega)\frac{\partial v_x}{\partial x} + C_{13}(\omega)\frac{\partial v_z}{\partial z}\big|_{x=L_1^x} = 0, \\
\sigma_{xx} &= C_{11}(\omega)\frac{\partial v_x}{\partial x} + C_{13}(\omega)\frac{\partial v_z}{\partial z}\big|_{x=L_2^x} = 0, \\
\sigma_{zz} &= C_{13}(\omega)\frac{\partial v_x}{\partial x} + C_{11}(\omega)\frac{\partial v_z}{\partial z}\big|_{z=L_1^z} = 0, \\
\sigma_{zz} &= C_{13}(\omega)\frac{\partial v_x}{\partial x} + C_{11}(\omega)\frac{\partial v_z}{\partial z}\big|_{z=L_2^z} = 0, \\
\sigma_{xz} &= C_{55}(\omega)\left(\frac{\partial v_x}{\partial z} + \frac{\partial v_z}{\partial x}\right)\big|_{x=L_1^x} = \psi, \\
\sigma_{xz} &= C_{55}(\omega)\left(\frac{\partial v_x}{\partial z} + \frac{\partial v_z}{\partial x}\right)\big|_{z=L_1^z} = \psi, \\
\sigma_{xz} &= C_{55}(\omega)\left(\frac{\partial v_x}{\partial z} + \frac{\partial v_z}{\partial x}\right)\big|_{x=L_2^x} = \psi, \\
\sigma_{xz} &= C_{55}(\omega)\left(\frac{\partial v_x}{\partial z} + \frac{\partial v_z}{\partial x}\right)\big|_{z=L_2^z} = \psi,
\end{aligned}
\tag{7}
$$

Solving problem (2), (7) one gets:

$$
\begin{aligned}
\varepsilon_{xx} &= S_{15}\psi, \\
\varepsilon_{zz} &= S_{35}\psi, \\
\varepsilon_{xz} &= S_{55}\psi,
\end{aligned}
\tag{8}
$$

Thus, the third column of the compliance tensor can be recovered.

To construct the effective anisotropic viscoelastic media one needs to solve three boundary value problems for system (1) with with boundary conditions (3), (5), and (7) respectively. After that the strains should be averaged within the domain D and systems (4), (6), and (8) should be resolved with respect to the components of the compliance tensor. However, to get a unique solution of system (1) we need to add extra no-flow boundary conditions at all boundaries:

$$\vec{w} \cdot \vec{n}|_{\partial D} = 0.$$

To simplify the interpretation of the results it is convenient to consider the seismic wave velocity and attenuation in the effective viscoelastic media [5,23, 26]. To do so, one needs to resolve the dispersion relation for the viscoelastic wave equation. However, the resulting model is anisotropic where the velocity and attenuation depend on the direction of propagation. Thus, we restrict our further considerations with the particular cases; i.e., velocity and attenuation of the quasi-longitudinal or qP-wave propagating along x and z directions and quasi-share or qS-wave propagating along x direction. In 2D orthothropic media qS-wave velocity along x and z directions coincide. We follow [4,23] to estimate the velocities and quality factors:

$$V_{px} = \Re\sqrt{\frac{C_{11}}{\rho}}, \; V_{pz} = \Re\sqrt{\frac{C_{33}}{\rho}}, \; V_s = \Re\sqrt{\frac{C_{55}}{\rho}},$$
$$Q_{px} = \frac{\Re C_{11}}{\Im C_{11}}, \; Q_{pz} = \frac{\Re C_{33}}{\Im C_{33}}, \; Q_s = \frac{\Re C_{55}}{\Im C_{55}}. \tag{9}$$

Note that the higher the quality factor the lower the attenuation.

The most time consuming part of the suggested averaging is the numerical solution of equation (1), which is described below.

3   Finite-Difference Approximation

To approximate equation (1) inside the domain $D = [L_1^x, L_2^x] \times [L_1^z, L_2^z]$ we suggest using a regular rectangular grid with steps h_x and h_z. Assume that the domain boundaries have the half-integer coordinates; moreover $L_1^x = x_{1/2}$, $L_2^x = x_{Nx-1/2}$, $L_1^z = z_{1/2}$, $L_2^z = z_{Nz-1/2}$, where N_x and N_z are the numbers of the grid points in corresponding spatial direction. A sketch of the computational domain and the grid is presented in Fig. 3.

We define the grid-functions on a staggered grids so that, u_x and w_x are placed at integer points in x direction, but half-integer points in z direction; whereas u_z and w_z are placed at integer points in z direction, but half-integer points in x direction. That is $(u_x)_{i,j+1/2} = u_x(x_i, z_{j+1/2})$, $(w_x)_{i,j+1/2} = w_x(x_i, z_{j+1/2})$, $(u_z)_{i+1/2,j} = u_x(x_{i+1/2}, z_j)$ and $(w_z)_{i+1/2,j} = w_x(x_{i+1/2}, z_j)$. Thus, the first and the third equations from (1) are approximated in the points $(x_i, z_{j+1/2})$ while the second and the fourth ones are approximated in points $(x_{i+1/2}, z_j)$. All coefficients are stored in half-integer points, assuming that they are constant within the grid cell. To preserve the second order of convergence the share modulus need to be computed at the integer points by the rule [11,15,22,24] (Fig. 1):

$$\frac{1}{\mu_{i,j}} = \frac{1}{\mu_{i-\frac{1}{2},j-\frac{1}{2}}} + \frac{1}{\mu_{i-\frac{1}{2},j+\frac{1}{2}}} + \frac{1}{\mu_{i+\frac{1}{2},j-\frac{1}{2}}} + \frac{1}{\mu_{i+\frac{1}{2},j+\frac{1}{2}}}.$$

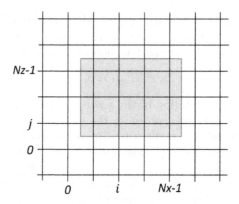

Fig. 1. A sketch of the computational domain and the mesh.

The grid cell and the positions of the solution components are provided in Fig. 2.

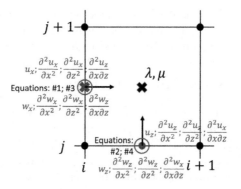

Fig. 2. The grid cell and the positions of the solution components

To approximate equation (1) we use the second order accurate finite differences using the following operators:

$$D_x[f]_{IJ} = \frac{f_{I+1/2,J} - f_{I-1/2,J}}{h_x} = \frac{\partial f}{\partial x}\bigg|_{x_I,z_J} + O(h_x^2) \qquad (10)$$

$$D_z[f]_{IJ} = \frac{f_{I,J+1/2} - f_{I,J-1/2}}{h_z} = \frac{\partial f}{\partial z}\bigg|_{x_I,z_J} + O(h_z^2) \qquad (11)$$

where f is any smooth enough function, and indices I and J can be either integer or half-integer.

Finite-difference approximation of the Eq. (1) produces a system of linear algebraic equations (SLAE) $A_0 x = b$ of the size of $N = 2(N_x-1)N_z + 2(N_z-1)N_x$ whose properties are the subject of investigation.

4   Solution of the SLAE

The properties of the constructed system of linear equations should be studied for two different cases. First is nonzero frequency, and the second one is $\omega = 0$; i.e., the static loading.

If the frequency in greater than zero; i.e., $\omega > 0$, the matrix is complex, non-symmetric. Moreover, due to the use of Neumann boundary conditions (3), or (5), or (7) the matrix is singular. The null-space of the differential operator (1), (3) is composed of three vectors:

$$(u_x, u_z, w_x, w_z) = (C_1, 0, 0, 0),$$
$$(u_x, u_z, w_x, w_z) = (0, C_2, 0, 0),$$
$$(u_x, u_z, w_x, w_z) = (C_3, -C_3, 0, 0),$$

thus the finite-difference operator has the null-space approximating the presented one.

If the frequency is equal to zero, the right-hand sides of the third and the fourth equations in (1) become trivial. Thus, one may introduce a new variable $V = \nabla \cdot \vec{w} = \frac{\partial w_x}{\partial x} + \frac{\partial w_z}{\partial z}$ and reduces system (1) to

$$\frac{\partial}{\partial x}\left[(\lambda_u + 2\mu)\frac{\partial u_x}{\partial x} + \lambda_u \frac{\partial u_z}{\partial z} + \alpha MV\right] + \frac{\partial}{\partial z}\left[\mu\left(\frac{\partial u_x}{\partial z} + \frac{\partial u_z}{\partial x}\right)\right] = 0,$$

$$\frac{\partial}{\partial x}\left[\mu\left(\frac{\partial u_x}{\partial z} + \frac{\partial u_z}{\partial x}\right)\right] + \frac{\partial}{\partial z}\left[\lambda_u \frac{\partial u_x}{\partial x} + (\lambda_u + 2\mu)\frac{\partial u_z}{\partial z} + \alpha MV\right] = 0,$$

$$\frac{\partial}{\partial x}\left[\alpha M\left(\frac{\partial u_x}{\partial x} + \frac{\partial u_z}{\partial z}\right) + MV\right] = 0,$$

$$\frac{\partial}{\partial z}\left[\alpha M\left(\frac{\partial u_x}{\partial x} + \frac{\partial u_z}{\partial z}\right) + MV\right] = 0,$$

where only three independent variables present. Thus, the approximation of the derived system will also lead to singular matrix of the rank of $3/4N$.

Therefore matrix A is ill-conditioned for either zero or non-zero frequency ω. We computed the singular value decomposition of matrix A, for small size of the problem ($N \approx 10^4$) and accounted the number of zero singular values to check the null-space dimension.

Solving system of linear equations with singular non-symmetric complex-valued matrix is a challenging task for the iterative methods [20]. Convergence of the iterative solvers is strongly affected by the choice of a preconditioner, which is not the main topic of this research. Moreover, we are dealing with the 2D problems, thus direct methods can be efficiently applied to solve SLAE. In particular, we use the sparse direct solver Intel MKL PARDISO which is very efficiently optimized for Intel architectures.

5   Numerical Experiments

5.1   Homogeneous Media

First we verified the algorithm using a homogeneous model poroelastic model with the following parameters: $\mu = 5.7 \cdot 10^9$ Pa, $\alpha = 0.87$, $\eta = 3 \cdot 10^{-3}$ Pa·s, $\kappa = 10^{-12}$ m^2, $\lambda_0 = 6.09 \cdot 10^9$ Pa, $M = 6.72 \cdot 10^9$ Pa, $\rho = 2650$ kg/m^3, $\rho = 1040$ kg/m^3, $T = 1.5$, and $\varphi = 0.3$. The last four parameters are rock density, fluid density, tortuosity, and porosity respectively. They are not used in quasi-static model, but we need them to estimate the wave propagation velocity and attenuation in the dynamic state.

To simulate the quasi-static loading of the fluid-filled porous material we considered a square domain of the size 1 m, and discretized it by the grid with steps $h_x = h_z = 0.002$ m. Thus the size of the problem was $N_x = N_z = 500$ points. We chose five frequencies $0, 0.1, 1, 10, 100, 1000$ Hz, compute the stiffness tensors and then estimate the velocity and attenuation of the fast P-wave, as presented in Fig. 3. The model was originally homogeneous and isotropic, thus the velocity is independent of the direction of propagation. The only physical factor that causes the attenuation in homogeneous poroelastic media is the Biot flow at high frequencies. However, it can not be resolved in the quasi-static state. Thus, the attenuation of the numerical solution is zero for the entire frequency range. The velocity of the numerical solution is underestimated, but the difference us about 0.1 m/s; i.e. the relative error is $3.5 \cdot 10^{-5}$ which is an acceptable level.

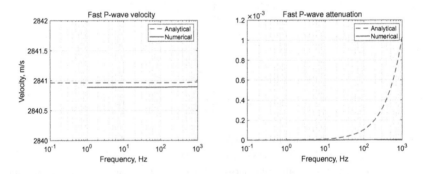

Fig. 3. Comparison of the numerical estimates (red solid line) with analytical solution (blue dashed line). Left picture presents the phase velocity, right picture illustrates the attenuation Q^{-1}. (Color figure online)

5.2   Fractured Media

To illustrate the applicability of the presented approach to estimate the dispersion and dissipation of seismic waves propagating in inhomogeneous poroelastic media, we considered fluid-filled fractured poroelastic model. We used the same models as those described in [17]. We considered two orthogonal sets of fractures. The fracture length in both sets was fixed to 50 mm and the thickness 2 mm. We generated several types of fractured models, depending on the average percolation length using the simulating annealing technique. Examples of the models are provided in Fig. 4. After that we filled in the model with the material properties. The background was unpermeable, whereas the fracture filling material was relatively hydraulically soft to support the fluid flow. The description of the model is provided in the Table 1. Due to the fracture-to-background wave-induced fluid flows [6, 16] local increase of the attenuation, thus velocity dispersion takes place.

Table 1. Material properties.

Parameter	Background	Fracture-filling material
Fluid dynamic viscosity η, Pa·s	0.001	0.001
Permeability k_0, m^2	10^{-15}	$5.5 \cdot 10^{-13}$
Density ρ, kg/m^3	2485	2458
Lame constant λ_u, Pa	$7.159 \cdot 10^9$	$2.40 \cdot 10^{10}$
Shear modulus μ, Pa	$3.0969 \cdot 10^{10}$	$1.14 \cdot 10^{10}$
Biot and Willis constant α	0.2962	0.6078
Fluid storage coefficient M, Pa	$2.01 \cdot 10^{10}$	$9.48 \cdot 10^{10}$

We used the suggested algorithm to estimate the frequency-dependent stiffness tensors for all six model types within the frequency range of $\nu \in [0, ..., 1000]$ Hz. Additionally, we directly simulated wave propagation in fluid-filled fractured-porous media using the approach, described in [17]. Wave propagation was simulated for the Ricker impulse with central frequency 1000 Hz. We provide the velocity and attenuation estimates obtained by the two methods in Figs. 5 and 6 respectively. Note, that the velocities are slightly underestimated by the suggested algorithm, whereas the attenuation matches well. In general, the obtained results show that the increase in the percolation length of the fractured system causes increase in the attenuation due to the FB-WIFF.

5.3   Performance Estimation

In all described examples we used a grid with 500 nodes in both spatial directions. We computed the effective stiffness tensors for the set of 22 frequencies from 0 1000 Hz. A single launch of the algorithm (one model, one frequency) includes the following steps:

1. Approximate problem to get matrix A and right the hand sides b;
2. PARDISO Reordering step. It's a preliminary step to decrease memory consumption and time of the next factorization step.
3. PARDISO Factorization step. It perform LU decomposition of the matrix A.

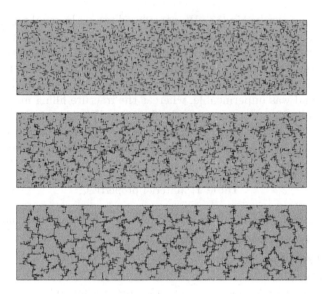

Fig. 4. Models of fractured media with different percolation length (increasing from the top to the bottom).

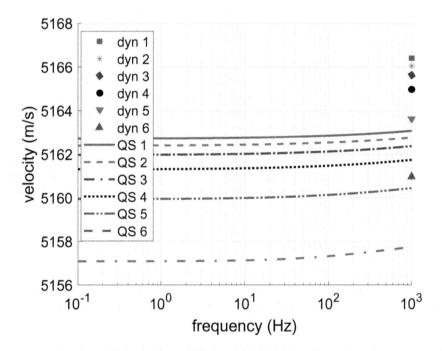

Fig. 5. Velocities of the qP wave for models with different percolation. Lines represent the quasi-static estimates, markers are used for estimates by wave propagation simulation.

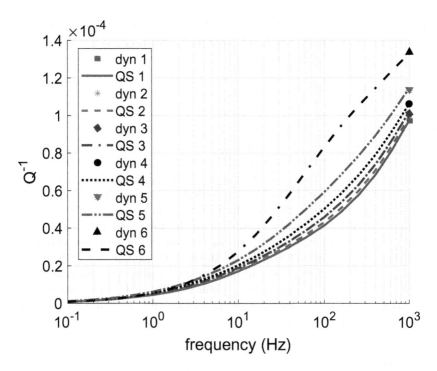

Fig. 6. Attenuation of the qP wave for models with different percolation. Lines represent the quasi-static estimates, markers are used for estimates by wave propagation simulation.

4. PARDISO Solve step. Solve the system to get solution x.
5. Postprocessing: Check the relative residual ($|b - Ax|/|b| \leq 10^{-}12$) and then construct the stiffness tensor C.

Note that if the frequency is changed only the main diagonal of matrix A should be corrected. Thus, step (1) can be done for zero-frequency but only the diagonal entries are changed at further loops of the algorithm. The reordering step (2) depends just on position non-zero elements A and can be done only once but applied for all ω. As a result, the entire algorithm (steps 1–5) is applied only to the first frequency in the raw, whereas the reduced version of the algorithm (steps 3–5) can be applied to all other frequencies.

We used Intel(R) Xeon(R) CPU E5-2690 v2 @ 3.00 GHz with 20 cores in our simulations. The computational time for one model and all 22 frequencies is ≈ 140 s. The timing profiling of the 22 runs is:

1. Approximate problem: ≈ 3 s.
2. PARDISO Reordering step: ≈ 4 s.
3. PARDISO Factorization step. $22 \times (\approx 3 - 4$ s.$)$
4. PARDISO Solve step. $22 \times (\approx 0.8$ s.$)$
5. Postprocessing: $22 \times (\approx 0.4$ s.$)$

Execution of the algorithm for the problem size of 500^2 required about 8 Gb RAM, thus it can be done on any desktop machine, or ported to a standard GP-GPU to improve performance.

6 Conclusions

We presented a numerical algorithm to simulate low-frequency loading of fluid-filled poroelastic materials and estimate the effective frequency-dependent strain-stress relations for such media. The algorithm solves Biot equation in quasi-static state. The problem is parabolic, thus, it is convenient to solve it in the frequency domain. As a result a system of linear algebraic equations have to be solved for each temporal frequency. We use the direct solver, based on the LU decomposition to resolve the SLAE. According to the presented numerical examples the suggested algorithm allows reconstructing the stiffness tensor within a wide range of frequencies $[0, ..., 1000]$ Hz, for the realistic sample size within 2–3 min. Thus, it allows to perform series of numerical experiments using simulations as the part of topological optimization techniques. Note, that our implementation is oriented on the use of CPU, whereas porting the algorithm to GPU would further improve the performance.

References

1. Bazaikin, Y., et al.: Effect of CT image size and resolution on the accuracy of rock property estimates. J. Geophys. Res. Solid Earth **122**(5), 3635–3647 (2017)
2. Biot, M.A.: Theory of propagation of elastic waves in a fluid-saturated porous solid. ii. higher frequency range. J. Acoust. Soc. Am. **28**, 179–191 (1956)
3. Biot, M.A.: Theory of propagation of elastic waves in fluid-saturated porous solid. i. low-frequency range. J. Acoust. Soc. Am. **28**, 168–178 (1956)
4. Carcione, J.M., Cavallini, F.: A rheological model for anelastic anisotropic media with applications to seismic wave propagation. Geophys. J. Int. **119**, 338–348 (1994)
5. Christensen, R.M.: Theory of Viscoelasticity, an Introduction. Academic Press, New York (1971)
6. Germán Rubino, J., Guarracino, L., Müller, T.M., Holliger, K.: Do seismic waves sense fracture connectivity? Geophys. Res. Lett. **40**(4), 692–696 (2013)
7. Guo, J., Rubino, J.G., Glubokovskikh, S., Gurevich, B.: Effects of fracture intersections on seismic dispersion: theoretical predictions versus numerical simulations. Geophys. Prospect. **65**(5), 1264–1276 (2017)
8. Huang, F., et al.: The first post-injection seismic monitor survey at the Ketzin pilot CO2 storage site: results from time-lapse analysis. Geophys. Prospect. **66**(1), 62–84 (2018)
9. Hunziker, J., et al.: Seismic attenuation and stiffness modulus dispersion in porous rocks containing stochastic fracture networks. J. Geophys. Res. Solid Earth **123**(1), 125–143 (2018)
10. Kong, L., Gurevich, B., Zhang, Y., Wang, Y.: Effect of fracture fill on frequency-dependent anisotropy of fractured porous rocks. Geophys. Prospect. **65**(6), 1649–1661 (2017)

11. Lisitsa, V., Podgornova, O., Tcheverda, V.: On the interface error analysis for finite difference wave simulation. Comput. Geosci. **14**(4), 769–778 (2010)
12. Marty, N.C.M., Hamm, V., Castillo, C., Thiéry, D., Kervévan, C.: Modelling water-rock interactions due to long-term cooled-brine reinjection in the Dogger carbonate aquifer (Paris basin) based on in-situ geothermal well data. Geothermics **88**, 101899 (2020)
13. Masson, Y.J., Pride, S.R., Nihei, K.T.: Finite difference modeling of Biot's poroelastic equations at seismic frequencies. J. Geophys. Res. Solid Earth **111**(B10), 305 (2006)
14. Menke, H.P., Reynolds, C.A., Andrew, M.G., Pereira Nunes, J.P., Bijeljic, B., Blunt, M.J.: 4D multi-scale imaging of reactive flow in carbonates: assessing the impact of heterogeneity on dissolution regimes using streamlines at multiple length scales. Chem. Geol. **481**, 27–37 (2018)
15. Moczo, P., Kristek, J., Vavrycuk, V., Archuleta, R.J., Halada, L.: 3D heterogeneous staggered-grid finite-difference modeling of seismic motion with volume harmonic and arithmetic averaging of elastic moduli and densities. Bull. Seismol. Soc. Am. **92**(8), 3042–3066 (2002)
16. Muller, T.M., Gurevich, B., Lebedev, M.: Seismic wave attenuation and dispersion resulting from wave-induced flow in porous rocks – a review. Geophysics **75**(5), 75A147-75A164 (2010)
17. Novikov, M.A., Lisitsa, V.V., Bazaikin, Y.V.: Wave propagation in fractured-porous media with different percolation length of fracture systems. Lobachevskii J. Math. **41**(8), 1533–1544 (2020)
18. Ovaysi, S., Wheeler, M., Balhoff, M.: Quantifying the representative size in porous media. Transp. Porous Media **104**(2), 349–362 (2014)
19. Rubino, J.G., Muller, T.M., Guarracino, L., Milani, M., Holliger, K.: Seismoacoustic signatures of fracture connectivity. J. Geophys. Res. Solid Earth **119**(3), 2252–2271 (2014)
20. Saad, Y.: Iterative Methods for Sparse Linear Systems. SIAM (2003)
21. Salaun, N., et al.: High-resolution 3D seismic imaging and refined velocity model building improve the image of a deep geothermal reservoir in the Upper Rhine Graben. Lead. Edge **39**(12), 857–863 (2020)
22. Samarskii, A.A.: The Theory of Difference Schemes, Pure and Applied Mathematics, vol. 240. CRC Press (2001)
23. Vavrycuk, V.: Velocity, attenuation, and quality factor in anisotropic viscoelastic media: a perturbation approach. Geophysics **73**(5), D63–D73 (2008)
24. Vishnevsky, D., Lisitsa, V., Tcheverda, V., Reshetova, G.: Numerical study of the interface errors of finite-difference simulations of seismic waves. Geophysics **79**(4), T219–T232 (2014)
25. Xu, C., Dowd, P.A., Mardia, K.V., Fowell, R.J.: A connectivity index for discrete fracture networks. Math. Geol. **38**(5), 611–634 (2006)
26. Zhu, Y., Tsvankin, I.: Plane-wave propagation in attenuative transversely isotropic media. Geophysics **71**(2), T17–T30 (2006)

Poisson Solver for Upscaling the Physical Properties of Porous Materials

Tatyana Khachkova[ID] and Vadim Lisitsa[(✉)][ID]

Sobolev Institute of Mathematics SB RAS, 4 Koptug ave, Novosibirsk 630090, Russia
lisitsavv@ipgg.sbras.ru

Abstract. In this paper, we present an iterative solver for Poisson equation. The approach is essentially oriented on upscaling the physical parameters of porous materials, such as electrical resistivity, thermal conductivity, pressure field computation. The algorithm allows solving Poisson equation for strongly heterogeneous media with small-scale high-contrast heterogeneities. The solver is based on the Krylov-type iterative method with a pseudo-spectral preconditionner, thus the convergence rate is independent on the sample size, but sensitive to the contrast of physical properties in the model. GPU-based implementation of the algorithm allows performing simulations for the samples of the size 400^3 using a single GPU.

Keywords: Pore scale · Poisson equation · GP-GPU · Numerical upscaling

1 Introduction

Numerical upscaling the physical properties of porous materials is a rapidly developing area of research, due to availability of micro-tomographic images of porous materials and computational resources. In particular, digital rock physics is getting a common tool in petrophysical analysis of rocks [4] to estimate absolute permeability [5,11], relative permeability for two- and three-phase flows [3], diffusion of chemicals for reactive transport [20,21], thermal conductivity [10], electrical resistivity [5,28]. To evaluate the effective properties of porous material one needs to simulate a specific physical process at the pore scale and then use averaging of the solution over the entire sample. The first step is the most computationally intensive and requires the numerical solution of partial differential equations. In particular, fluid flow simulation is based on the solution of Navier-Stokes or steady-state Stokes equations, where solving the Poisson equation is the principal step especially if the projection-type methods are used [8]. Numerical estimation of electrical resistivity and thermal conductivity is essentially based on the solution of Poisson equation [10,28]. Nowadays, digital rock physics laboratories are well equipped with micro-tomographic scanners but

The research was supported by the Russian Science foundation grant no. 21-71-20003.

have limited access to high performance computing facilities. Thus, an appropriate option is to use GP-GPU servers for numerical simulations in digital rock physics applications.

We present the numerical algorithm for solving Poisson equation in strongly heterogeneous media oriented on the use of graphic processor units. Strongly inhomogeneous media includes small-scale heterogeneities [6] and high amplitudes of the inhomogeneities. There are several approaches to solve the Poisson equation, but the most commonly used are the Krylov-type methods with appropriate preconditioners. Incomplete LU factorization [14], including low-rank approximations [9,17] and multi-grid [16,25] are commonly used to precondition a discretized Poisson equation. First class approaches require storing the matrix and its factors. Approaches from a different class require problems solving in a series of nested grids and solutions storage. However, if we consider strongly heterogeneous geomaterials models, a scale coarsening changes the topology and geometry of pore space, that is the main conductive model part. Thus, the problem on a coarse grid will not be closer to the true one than the problem for homogeneous media. At the same time, the spectral method [7] using GPUs can be effectively applied to solve the Poisson equation with constant conductivity.

The paper is structured as follows. We formulate the problem in Sect. 2. Finite-difference approximation is provided in Sect. 3. Description of the solver and the preconditioner are presented in Sect. 4. Numerical experiments are introduced in Sect. 5.

2   Numerical Estimation of Electrical Resistivity

The goal of this research is to simulate the electric current in rock samples using the 3D microtomographic images. Thus, two problems should be solved. First, we need to calculate the electric current in the inhomogeneous model for a given potentials difference applied to the opposite sample sides. Second, it's necessary to restore the scalar (tensor) electrical resistivity at which the resulting electric current was equal to that in the original model for a given potentials difference.

Consider the domain $\Omega = [0, X_1] \times [0, X_2] \times [0, X_3]$ with inhomogeneous media and the Poisson equation:

$$\nabla \cdot (\sigma(\boldsymbol{x})\nabla\varphi(\boldsymbol{x})) = 0, \tag{1}$$

where $\sigma = \sigma(\boldsymbol{x})$ is the electrical conductivity, $\varphi = \varphi(\boldsymbol{x})$ is the scalar electric potential. We assume that there are no charges inside the computational domain, so the right-hand sides are equal to zero. The Dirichlet boundary conditions are used at two opposite sides $x_1 = 0$ and $x_1 = X_1$, that is

$$\varphi(\boldsymbol{x})|_{x_1=0} = \Phi_0, \; \varphi(\boldsymbol{x})|_{x_1=X_1} = \Phi_1, \tag{2}$$

where Φ_0 and Φ_1 are the potentials. On the other sides of the domain, we apply Neumann boundary conditions

$$\nabla\varphi \cdot \boldsymbol{n}|_{x_2=0} = 0, \; \nabla\varphi \cdot \boldsymbol{n}|_{x_2=X_2} = 0,$$
$$\nabla\varphi \cdot \boldsymbol{n}|_{x_3=0} = 0, \; \nabla\varphi \cdot \boldsymbol{n}|_{x_3=X_3} = 0. \tag{3}$$

Neumann boundary conditions are used to simulate laboratory experiments in which a sample is placed in insulator ports.

The spatial distribution of the electric potential $\varphi(\boldsymbol{x})$ is the solution of Eq. (1), after that the electric current can be defined as

$$J_1 = \int_0^{X_2} \int_0^{X_3} \sigma(\boldsymbol{x}) \frac{\partial \varphi}{\partial x_1} dx_2 dx_3. \tag{4}$$

Note, that the current should be independent of spatial coordinates.

On the contrast, if homogeneous media with unknown conductivity $\hat{\sigma}$ is considered, Eq. (1) turns into

$$\nabla \cdot (\hat{\sigma} \nabla \hat{\varphi}(\boldsymbol{x})) = 0, \tag{5}$$

with boundary conditions (2) and (3). Solution of the this equation can be constructed analytically:

$$\varphi(\boldsymbol{x}) = \Phi_0 + \frac{\Phi_1 - \Phi_0}{X_1} x_1. \tag{6}$$

In addition, electric current in the direction x_1 satisfies the Ohm's law:

$$\hat{j}_x = \int_0^{X_2} \int_0^{X_3} \sigma(\boldsymbol{x}) \frac{\partial \varphi}{\partial x_1} dx_2 dx_3 = \hat{\sigma} \frac{\Phi_1 - \Phi_0}{X_1} S_1, \tag{7}$$

where S_1 is the surface area of the sample cross-section normal to the direction x_1.

Assuming that a given potentials difference at two opposite sides of the sample causes the same electric current either in an inhomogeneous or in a homogeneous media, that is $\hat{J}_x = J_x$, the electrical conductivity of the effective material can be reconstructed:

$$\hat{\sigma} = J_x \frac{X_1}{S_1} \frac{1}{\Phi_1 - \Phi_0}. \tag{8}$$

The most laborious computational problem in the numerical estimation of electrical resistivity or conductivity is the solution of the Poisson equation in strongly heterogeneous media, where the characteristic size of inhomogeneities is form one to ten voxels, and the conductivity amplitude varies by several orders of magnitude.

3   Finite-Difference Approximation

We propose to use a finite volume of conservative finite-difference method to approximate the Eq. (1) in an inhomogeneous domain. Let's introduce a grid with steps h_1, h_2, and h_3 in the directions x_1, x_2, and x_3 respectively, so that $(x_j)_{m_j} = h_j m_j$. Also, we introduce grid cells

$$C_{m_1, m_2, m_3} = \prod_{j=1}^3 [h_j(m_j - 1/2), h_j(m_j + 1/2)],$$

where $\prod$ denotes the direct product of sets. Let's define that the potential and conductivity are constant within the grid cell, while they can be discontinuous across the cell faces. After that, the finite-difference approximation of the Poisson equation (1) can be introduced as:

$$
\frac{1}{h_1}\left(\tilde{\sigma}_{m_1+1/2,m_2,m_3}D_1[\varphi]_{m_1+1/2,m_2,m_3} - \tilde{\sigma}_{m_1-1/2,m_2,m_3}D_1[\varphi]_{m_1-1/2,m_2,m_3}\right)
$$
$$
+\frac{1}{h_2}\left(\tilde{\sigma}_{m_1,m_2+1/2,m_3}D_2[\varphi]_{m_1,m_2+1/2,m_3} - \tilde{\sigma}_{m_1,m_2-1/2,m_3}D_2[\varphi]_{m_1,m_2-1/2,m_3}\right)
$$
$$
+\frac{1}{h_3}\left(\tilde{\sigma}_{m_1,m_2,m_3+1/2}D_3[\varphi]_{m_1,m_2,m_3+1/2} - \tilde{\sigma}_{m_1,m_2,m_3-1/2}D_3[\varphi]_{m_1,m_2,m_3+1/2}\right) = 0,
$$
$$(9)$$

where the finite-difference operators are defined as

$$
D_1[\varphi]_{m_1+1/2,m_2,m_3} = \frac{\varphi_{m_1+1,m_2,m_3} - \varphi_{m_1,m_2,m_3}}{h_1},
$$
$$
D_2[\varphi]_{m_1,m_2+1/2,m_3} = \frac{\varphi_{m_1,m_2+1,m_3} - \varphi_{m_1,m_2,m_3}}{h_2}, \qquad (10)
$$
$$
D_3[\varphi]_{m_1,m_2,m_3+1/2} = \frac{\varphi_{m_1,m_2,m_3+1} - \varphi_{m_1,m_2,m_3}}{h_3},
$$

and the electrical conductivity at the faces of grid cells is calculated as an harmonic averaging of those from the joint cells:

$$
\frac{1}{\tilde{\sigma}_{m_1+1/2,m_2,m_3}} = 2\left(\frac{1}{\sigma_{m_1+1,m_2,m_3}} + \frac{1}{\sigma_{m_1,m_2,m_3}}\right),
$$
$$
\frac{1}{\tilde{\sigma}_{m_1,m_2+1/2,m_3}} = 2\left(\frac{1}{\sigma_{m_1,m_2+1,m_3}} + \frac{1}{\sigma_{m_1,m_2,m_3}}\right), \qquad (11)
$$
$$
\frac{1}{\tilde{\sigma}_{m_1,m_2,m_3+1/2}} = 2\left(\frac{1}{\sigma_{m_1,m_2,m_3+1}} + \frac{1}{\sigma_{m_1,m_2,m_3}}\right).
$$

A detailed description of constructing the approximation can be found, for example, in [10,13,14,27]. This modification of the coefficients, preserving the second order of convergence of the finite-difference solution, is represented in [10,19,24,26].

4   Iterative Solver

The Krylov-type methods are usually used to solve a system of linear algebraic equations. However, the condition number of the system is high because the spectrum of the matrix tends to expand to the entire real axis with increasing matrix size to approximate the differential operator. In addition, in geomaterials, electrical conductivity (resistivity) can vary by several orders, while increasing the condition number. To improve convergence several possible preconditioners are used. Among them, the incomplete LU factorization [14,18], including low-rank approximations [9,17] and the multi-grid [16,25] are commonly used. The first class approaches require storing the matrix and its factors; thus, at least $4N$ of additional variables must be stored in memory, where N is the size of the problem (the number of grid points in the computational domain). Approaches from another class require to solve a set of problems (1) at a series of nested grids storing the solutions. However, if we consider strongly heterogeneous models changing the scale or a grid size changes the topology and geometry of the pore space, that is the main conductive model part. Thus, the problem on a coarse grid will not be closer to the true one than that for the homogeneous media. At the same time, the system of equations (9) with constant conductivity is efficiently solved by spectral method [7,22] using GPUs.

Let's consider system of equations (9) written in a short form:

$$L\varphi = f.$$

After preconditioning, we can get a system

$$(L_0^{-1}L)\varphi = A\varphi = L_0^{-1}f,$$

where L and L_0 are approximations of the Laplace operator for inhomogeneous and homogeneous media, respectively. The most time consuming part of the algorithm is the calculating the matrix-vector multiplication. The action of the matrix L is described by the formulas (9), while the action of L_0^{-1} requires to solve a system of equations (9) with constant conductivity for different right-hand sides. To calculate $L_0^{-1}\psi$ for an arbitrary ψ we propose to use spectral methods. Applying the Fourier transform with respect to x_2 and x_3 to the Eq. (1), we get a series of one-dimensional problems:

$$\frac{d^2\Phi}{dx_1^2} - (k_2^2 + k_3^2)\Phi = \hat{g}(x_1, k_2, k_3)/\sigma_0, \tag{12}$$

where Φ is the Fourier image of the unknown function, and $\hat{g}(x_1, k_2, k_3)$ is the image of the right-hand side. It's necessary to note, that the problem can be solved independently for each frequency. To solve one-dimensional equations, the same approximation is used as in Eq. (9), after that the Thomas algorithm for three-diagonal matrices is applied. And then we apply the inverse Fourier transform to the solution. So, the solution of the problem can be represented as:

$$\varphi_0(\boldsymbol{x}) = \frac{1}{\sigma_0}\mathcal{F}_{2,3}^{-1}\left[\hat{L}(k_2, k_3)^{-1}\mathcal{F}_{2,3}[f(\boldsymbol{x})]\right],$$

where $\mathcal{F}_{2,3}$ is the Fourier transform operator, $\hat{L}(k_2, k_3)$ is the 1D operator, that depends on the spatial frequencies k_2 and k_3. Parameter σ_0 is equal to fluid conductivity.

4.1   GPU-Based Implementation

The algorithm is implemented using GP-GPU computations, so that all main steps of the algorithm are executed using GPU with minimal data flows between the device and the host. At the preliminary step we read the model, prepro-cess it, allocate the memory to store solution and the auxiliary variables to

BCGStab and for the preconditionner. After that, all the actual computations are performed by GPU only. It includes, application of qFFT along x_2 and x_3 directions, solution of the series of 1D problems, and updating BCGStab vectors. Only after the convergence of the algorithm, the solution is transferred to the RAM.

We estimate the amount of memory needed for computations. In the RAM, we need to store the model and the solution, which have the size of $N_1 N_2 N_3$, where N_j is the number of grid points in corresponding direction x_j. For this algorithm no more RAM is allocated. Implementation of the BCGStab on the GPU requires memory to store a model, a solution, and right-hand sides, i.e., three main arrays of the same size. Moreover, the standard implementation of BICGStab still requires an additional five arrays of the specified size $N_1 N_2 N_3$ [23]. Next, we additionally store the FFT image of vector $\hat{g}(x_1, k_2, k_3)$, that is complex array of the size $N_1 N_2 N_3$, to implement the FFT-based preconditioner. Finally, the implementation of Thomas algorithm requires additional storage of the same size array. Indeed, the solution of one-dimensional problems is implemented independently for each spatial frequency $k_2^2 + k_3^2$, so that one GPU thread solves a problem only for one frequency. Please note that we solve one-dimensional problems for homogeneous media. Thus, the matrix can be described by only one parameter - the square of the spatial frequency divided by the square of the spatial step. Thomas algorithm requires a storage of matrix factors. Therefore, we need to store an array of size N_1. Moreover, such an array is required by every GPU thread, so we decided to pre-allocate memory for it; i.e., we store $N_2 N_3$ arrays of N_1 size just for the parallel Thomas algorithm implementation.

Finally, we need $11 N_1 N_2 N_3 * 8$ bytes to store all the variables on GPU in case of double-precision computations. A simple estimate shows that for numerical experiments, a 400^3 sample requires 5 Gb of memory, while a 500^3 sample requires almost 11 Gb of GPU memory.

5 Numerical Experiments

5.1 Verification of the Algorithm

Parallel Resistors. To verify the developed algorithm, we calculated the electric current for sets of simple models. First of all, a sample was considered with one conducting channel, the cross-section of which is square with a side equal to one-fifth of the sample side. At that, the sample was cubical. The electrical conductivities were for the background $\sigma_b = 0.5S$ and for the channel $\sigma_c = 5S$, the grid step was equal to 1 μm. The size of samples was varied as $5 \cdot 2^n$ points, where $n = 0, ..., 6$, i.e. from 5 to 320 points. We calculated the electric potential inside the sample and estimated the effective electrical conductivity using the formula (8). On the other hand, the model under consideration is equivalent to an electrical contour with parallel resistors, therefore, according to Ohm's law, the effective conductivity is

$$\sigma_{||} = \sigma_b S_b / S + \sigma_c S_c / S,$$

where S_b and S_c are the background and channel areas, while S is the cross-section area of the sample. For described experiments $\sigma_{||}$ was $0.68\,\mathrm{S}$ and for all considered computational domain sizes the estimated conductivity was $0,68\,\mathrm{S}$ up to machine precision. Note, that one BiCGStab iteration was required for the algorithm to converge.

Serial Resistors. The second series of experiments was carried out for layered media with layers normal to the primary direction of the potential change. This type of models is a serial resistors, and the effective conductivity can be estimated as

$$\frac{1}{\sigma_\perp} = \frac{l_b}{L}\frac{1}{\sigma_b} + \frac{l_c}{L}\frac{1}{\sigma_c},$$

where l_b is the total background thickness, and l_c is the total layers' thickness, and L is the length of the sample. We considered $\sigma_c = 5$ S, and $\sigma_b = 0.5$ S/m to provide the electric current. We changed the size of the models in the same way as before. In addition to the size of the model, we changed the model itself in two ways. First, we kept one conductive layer and increased its thickness proportional to the domain size. Second, we kept the layer thickness but increased the number of layers proportional to the size of the domain. Regardless of the size of the models, BiCGStab converged in 3 iterations with a relative residual not exceeding 10^{-13}. The numerical estimates of electrical conductivity converged to a theoretical estimate with the first order.

5.2   Performance Analysis

Dependence on the Model Size. In this section, we present experiments measuring the performance of the algorithm, depending on the samples size, contrast in the conductivity and the typical size of the heterogeneities.

We varied the domain size as $N_j \in [20, 40, 80, 160, 320]$ points. Note, that the algorithm works differently in the direction x_1 and in the directions x_2 and x_3. Thus, it is necessary to consider cases, where N_1 is fixed, while N_2 and N_3 are changed, but $N_2 = N_3$, and the opposite case, when N_1 is changed, but $N_2 = N_3$ are fixed. Thus, 25 model size combinations were considered. To create the model, a random sample was generated with the size of one homogeneous cell, equals to 4^3 voxels. A random number generator was used to fill the cells with conductivities $\sigma = 10^{R([-3,1])}$, where $R([-3,1])$ is the homogeneous distribution within the interval $[-3,1]$. To obtain stable estimates of the iterations number and the computation time, it were performed a series of 10 simulations for various statistical model realizations. We measured the number of iterations, required for convergence (Fig. 1) and time per iteration (Fig. 2). The number of iterations needed for the convergence depends on the number of grid points along x_1 direction (along the main current), but does not depend on the other two sizes of the sample. As for the time per a single iteration, it increases linearly with the size of the model in x_1 direction, since the Thomas algorithm has a linear computational complexity with respect to the size of the problem. Further, the iteration time increases quadratically with simultaneous increasing sizes along x_2 and x_3 direction, which is also reconfirmed by theoretical estimates of the FFT computational complexity.

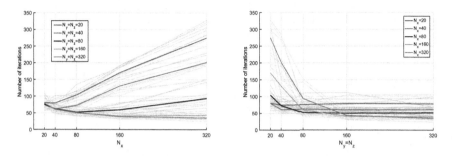

Fig. 1. Dependence of the number of iterations for convergence on the grid points number in the x_1 direction (left) and on different sizes of the model along the FFT directions (right). The thick lines correspond to the means for the statistical realizations of models, and the thin lines correspond to the separate realizations.

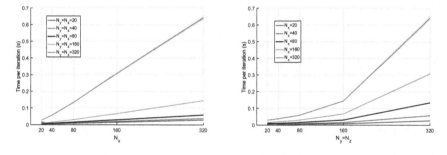

Fig. 2. Dependence of the time per iteration on the grid points number in x_1 direction (left) and on the model size along the FFT directions (right). The thick lines correspond to the means for the statistical realizations of models.

Dependence on the Resistivity Contrast. We applied the algorithm to the Bentheimer sandstone samples described in [6]. It were considered the images with a resolution of 5.58 μm and 3.44 μm per voxel, which are images B and C from [6], respectively. The size of the samples was 400^3 voxels. These two models were used to investigate the effect of the conductivity contrast on the number of iterations required for convergence. The conductivity of the fluid was equal to 5 S/m, while the conductivity of the matrix was changed $\sigma_M \in [10^{-7}, 10^4]$. We used the stopping criteria that the residual reaches 10^{-8} and ploted the number of iterations for two samples with different contrasts between the fluid and solid conductivity (Fig. 3). According to the presented results, the number of iteration is minimal if the contrast is low and increases with growing contrast.

In the Table (1) we represent the number of iterations, the time per iteration in seconds, and the wall-clock time for simulations with changed values of the matrix conductivity. Please note, that for the models with the size of 400^3, the average iteration time is about 1.22 s and the wall-clock time is proportional to the total number of iterations. However, it does not exceed 500 s for a wide range of conductivity values (Table 1).

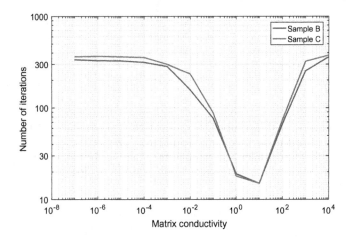

Fig. 3. The number of iterations to achieve residual 10^{-8} for changing conductivity contrasts of Bertheimer sandstone.

5.3   Statistical Models

We applied the algorithm to study the relation between the geometrical characteristics of the pore space and the formation factor. Moreover, we applied matrix reduction algorithm to study the evolution of the parameters. To simulate the interface movement, we use the level-set method [12], assuming that the dissolution rate is 0.1h per time step. We considered 100 steps of the matrix reduction process. The synthetic images were generated using truncated Gaussian field method [15] so that a statistical model is defined by two parameters: the mean porosity and the correlation length. We constructed 16 statistical models, with 100 realizations each, with porosity $\phi = \{0.05, 0.1, 0.15, 0.2\}$ and correlation length $\lambda = \{5, 10, 15, 20\}$ voxels. Thus, in total, we generated 1600 images of 250^3 voxels each. We measured the Minkowski functionals corresponding to the total porosity and specific surface area and also estimated formation factors at each step of matrix reduction process. So, 160000 simulations were done in total. Minkowski functionals for different statistical models averaged over the realizations are presented in Fig. 4. Formation factors are provided in Fig. 5. Note,

Table 1. The number of iterations, the time per iteration and the wall-clock time for the Bentheimer sandstone experiments.

σ_M	Sample B			Sample C		
	N it	Time(s)	T/it.(s)	N it	Time(s)	T/it.(s)
10^{-7}	367	454.44	1.23	363	472.43	1.3
10^{-6}	359	438.78	1.22	367	459.41	1.25
10^{-5}	357	435.28	1.22	362	439.76	1.21
10^{-4}	352	426.87	1.22	356	430.81	1.21
10^{-3}	284	346.05	1.22	300	367.0	1.24
10^{-2}	157	191.32	1.22	234	291.23	1.25
10^{-1}	77	94.58	1.32	88	110.35	1.28
10^{0}	19	25.37	1.33	18	23.08	1.33
10^{1}	15	20.07	1.24	15	20.32	1.23
10^{2}	67	83.41	1.22	74	91.57	1.23
10^{3}	254	310.2	1.21	323	396.26	1.21
10^{4}	361	438.55	1.23	367	457.5	1.24

that formation factor is well-correlated with the sample porosity. Indeed, the initial correlation length of the Gaussian random field affects the rate of the matrix reduction and changes in the specific surface. However, similar porosity values correspond to similar formation factors of the samples, regardless to the correlation length.

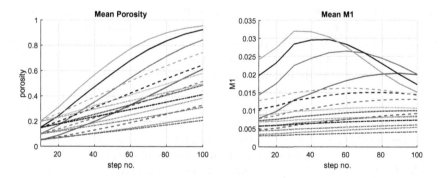

Fig. 4. Porosity and the Minkowski functional 1 for different models. Blue lines correspond to $\phi = 0.05$, red lines correspond to $\phi = 0.1$, black - $\phi = 0.15$, and green - $\phi = 0.2$, solid lines correspond to $\lambda = 5$, dashed - $\lambda = 10$, dotted - $\lambda = 15$, and dasher-dotted - $\lambda = 20$.

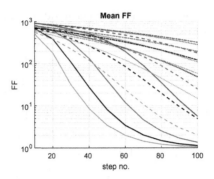

Fig. 5. Formation factors for different models. Blue lines correspond to $\phi = 0.05$, red lines correspond to $\phi = 0.1$, black - $\phi = 0.15$, and green - $\phi = 0.2$, solid lines correspond to $\lambda = 5$, dashed - $\lambda = 10$, dotted - $\lambda = 15$, and dasher-dotted - $\lambda = 20$. (Color figure online)

5.4   Application to Carbonate Rocks

We applied the developed algorithm to estimate the formation factor of the carbonate rock. The model is described in details in [2] and the CT-scans are provided in [1]. In total, four carbonate samples are available. CO_2 was injected into these samples, which resulted in partial dissolution of calcite and erosion of the pore space. On [1], ten micro-CT scans are available for each sample at different time points. The samples were grouped in two pairs depending on the flow rate used in laboratory experiments. Typically, the higher the flow rate, the faster the matrix dissolves and wormholes form. A detailed description of samples and lab experiments are provided in [2]. We use the same designations for the samples: AH and BH correspond to high flow rates, while AL and BL correspond to low flow rate. We performed simulations using subsamples of the size of 400^3 from each sample (the same for all images at different time points). After chemical dissolution of the matrix, it was formed the wide wormholes for the samples where the fluid flow rate was high, which significantly reduces the formation factors. A wormhole was formed in sample AL, where a low fluid flow rate was used due to the preexisting preferred path in the sample, which also resulted in the form-factor decrease. On the contrary, it was observed a weak dissolution in sample BL. Thus there were no changes in form-factor. We provide plots of form-factors for all four experiments as a functions of time, which clearly demonstrate the significant reduction in the rate of changes in pore space during the formation of wormholes (Fig. 6). Please note, that in all cases the iterations number before the convergence of the algorithm varies within 400–1100 iterations and depends on the geometry of the pore space and the complexity of the topology and may require additional study. In contrast, the time per iteration is stable for all experiments.

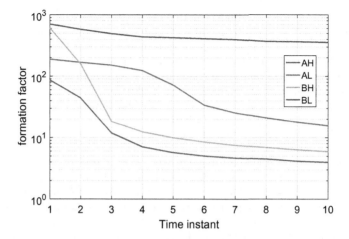

Fig. 6. Pore space of BL carbonate image at the fist (left) and last (right) instant of dissolution.

6   Conclusions

We developed the algorithm for numerical simulation of the potential electric field in porous media. The base of the algorithm is the Krylov-type solver for the discretized Poisson equation. The inverted Laplace operator for homogeneous media is used as a preconditioner, calculated using FFT in the directions, which are normal to the potential change direction. Then we solve a series of one-dimensional problems with using the Thomas algorithm. The solver is implemented on GPU, which allows solving the problems of voxel sizes up to 400^3 with a single GPU. The iterations number is practically independent of the size of the model. However, this depends on the conductivity contrast between the fluid and the rock matrix and the geometrical complexity of the pore space. We applied the algorithm to investigate the evolution of the carbonate samples form-factor during CO_2 sequestration. We have shown that due to the chemical dissolution of the carbonate matrix, the form-factor of the samples increases. However, the rate changes are slowed down when wormholes are formed.

Simulation were performed using computational resources of Peter the Great Saint-Petersburg Polytechnic University Supercomputing Center (www.spbstu.ru).

References

1. Al-Khulaifi, Y., Lin, Q., Blunt, M., Bijeljic, B.: Pore-scale dissolution by CO2 saturated brine in a multi-mineral carbonate at reservoir conditions: impact of physical and chemical heterogeneity (2019). https://doi.org/10.5285/52b08e7f-9fba-40a1-b0b5-dda9a3c83be2

2. Al-Khulaifi, Y., Lin, Q., Blunt, M.J., Bijeljic, B.: Pore-scale dissolution by co2 saturated brine in a multimineral carbonate at reservoir conditions: impact of physical and chemical heterogeneity. Water Resour. Res. **55**(4), 3171–3193 (2019)
3. Alpak, F.O., Riviere, B., Frank, F.: A phase-field method for the direct simulation of two-phase flows in pore-scale media using a non-equilibrium wetting boundary condition. Comput. Geosci. **20**(5), 881–908 (2016)
4. Andra, H., et al.: Digital rock physics benchmarks - part i: imaging and segmentation. Comput. Geosci. **50**, 25–32 (2013)
5. Andra, H., et al.: Digital rock physics benchmarks - part ii: computing effective properties. Comput. Geosci. **50**, 33–43 (2013)
6. Bazaikin, Y., et al.: Effect of CT image size and resolution on the accuracy of rock property estimates. J. Geophys. Res. Solid Earth **122**(5), 3635–3647 (2017)
7. Belonosov, M., Kostin, V., Neklyudov, D., Tcheverda, V.: 3D numerical simulation of elastic waves with a frequency-domain iterative solver. Geophysics **83**(6), T333–T344 (2018)
8. Brown, D.L., Cortez, R., Minion, M.L.: Accurate projection methods for the incompressible Navier-stokes equations. J. Comput. Phys. **168**(2), 464–499 (2001)
9. Chandrasekaran, S., Dewilde, P., Gu, M., Somasunderam, N.: On the numerical rank of the off-diagonal blocks of Schur complements of discretized elliptic PDEs. SIAM J. Matrix Anal., 2261–2290 (2010)
10. Dorn, C., Schneider, M.: Lippmann-Schwinger solvers for the explicit jump discretization for thermal computational homogenization problems. Int. J. Numer. Meth. Eng. **118**(11), 631–653 (2019)
11. Gerke, K.M., Karsanina, M.V., Katsman, R.: Calculation of Tensorial flow properties on pore level: exploring the influence of boundary conditions on the permeability of three-dimensional stochastic reconstructions. Phys. Rev. E **100**(5), 053312 (2019)
12. Gibou, F., Fedkiw, R., Osher, S.: A review of level-set methods and some recent applications. J. Comput. Phys. **353**, 82–109 (2018)
13. Haber, E., Ascher, U.M.: Fast finite volume simulation of 3D electromagnetic problems with highly discontinuous coefficients. SIAM J. Sci. Comput. **22**(6), 1943–1961 (2001)
14. Haber, E., Ascher, U.M., Aruliah, D.A., Oldenburg, D.W.: Fast simulation of 3D electromagnetic problems using potentials. J. Comput. Phys. **163**(1), 150–171 (2000)
15. Hyman, J.D., Winter, C.L.: Stochastic generation of explicit pore structures by thresholding Gaussian random fields. J. Comput. Phys. **277**, 16–31 (2014)
16. Johansen, H., Colella, P.: A cartesian grid embedded boundary method for Poisson's equation on irregular domains. J. Comput. Phys. **147**(1), 60–85 (1998)
17. Kostin, V., Solovyev, S., Bakulin, A., Dmitriev, M.: Direct frequency-domain 3D acoustic solver with intermediate data compression benchmarked against time-domain modeling for full-waveform inversion applications. Geophysics **84**(4), T207–T219 (2019)
18. Lee, B., Min, C.: Optimal preconditioners on solving the Poisson equation with Neumann boundary conditions. J. Comput. Phys. **433**, 110189 (2021)
19. Lisitsa, V., Podgornova, O., Tcheverda, V.: On the interface error analysis for finite difference wave simulation. Comput. Geosci. **14**(4), 769–778 (2010)
20. Lisitsa, V., Bazaikin, Y., Khachkova, T.: Computational topology-based characterization of pore space changes due to chemical dissolution of rocks. Appl. Math. Model. **88**, 21–37 (2020). https://doi.org/10.1016/j.apm.2020.06.037

21. Molins, S., et al.: Pore-scale controls on calcite dissolution rates from flow-through laboratory and numerical experiments. Environ. Sci. Technol. **48**(13), 7453–7460 (2014)
22. Pleshkevich, A., Vishnevskiy, D., Lisitsa, V.: Sixth-order accurate pseudo-spectral method for solving one-way wave equation. Appl. Math. Comput. **359**, 34–51 (2019)
23. Saad, Y.: Iterative Methods for Sparse Linear Systems. SIAM (2003)
24. Samarskii, A.A.: The Theory of Difference Schemes, Pure and Applied Mathematics, vol. 240. CRC Press (2001)
25. Stuben, K.: A review of algebraic multigrid. J. Comput. Appl. Math. **128**(1–2), 281–309 (2001)
26. Vishnevsky, D., Lisitsa, V., Tcheverda, V., Reshetova, G.: Numerical study of the interface errors of finite-difference simulations of seismic waves. Geophysics **79**(4), T219–T232 (2014)
27. Wiegmann, A., Zemitis, A.: EJ-HEAT: a fast explicit jump harmonic averaging solver for the effective heat conductivity of composite materials (2006)
28. Zhan, X., Schwartz, L.M., Toksöz, M.N., Smith, W.C., Morgan, F.D.: Pore-scale modeling of electrical and fluid transport in Berea sandstone. Geophysics **75**(5), F135–F142 (2010)

Properties of Multipyramidal Elements

Miroslav S. Petrov[1] and Todor D. Todorov[2]([✉])

[1] Department of Technical Mechanics, Technical University, 5300 Gabrovo, Bulgaria
[2] Department of Mathematics, Informatics and Natural Sciences, Technical
University, 5300 Gabrovo, Bulgaria

Abstract. The finite element method is based on the division of the
physical domains into a large number of small polytopes with simple geometry. Basically, the most useful finite elements can be divided
into two large groups: simplicial elements and hypercubic elements. To
keep conformity, triangulating of curved domains with complex geometry requires the usage of various kinds of transitional elements, which
are specific for any fixed Euclidean space. The paper deals with a basic
problem of the finite element method in the multidimensional spaces -
conforming coupling between hypercubic and simplicial meshes. Here we
focus on the bipyramidal elements. Some properties of such kind elements are discussed in an arbitrary Euclidean space with a dimension
greater than two.

Keywords: Transitional finite elements · Conforming coupling ·
Pyramidal elements · Bipyramids · Bihexatera

AMS Subject Classifications 65N30 · 52B11

1 Introduction

Most of the papers on the finite element method deal with numerical analysis
and error estimate in numerical approximations of the eigenvalue and boundary value problem solutions. But the successful application of the finite element
method depends on the quality of the finite element meshes. The quality of
meshes is of significant importance especially when the isoparametric multigrid
method is applied to solve elliptic problems in domains with complex geometry.
The multigrid methods require successive refinements of the initial triangulation. That is why the stability of the sequences of successive triangulations is
a crucial point for convergence of the multigrid approximations. The theory of
the finite element meshes and advanced discretization methods can be considered as a separate scientific area closely related to the finite element method.
We emphasize that the finite element meshes have been applied for computer
graphic simulations without solving any boundary or eigenvalue problems. Here,
we present some papers that deal with advanced discretization methods.

The efforts of various authors have been devoted to different refinement
strategies. Hannukainen et al. [10], Bedregal and Rivara [3], and Perdomo and

© Springer Nature Switzerland AG 2021
O. Gervasi et al. (Eds.): ICCSA 2021, LNCS 12949, pp. 546–559, 2021.
https://doi.org/10.1007/978-3-030-86653-2_40

Plaza [20] focus their research on longest-edge bisection methods. The bisection methods have been generalized to arbitrary section methods by Korotov et al. [12]. Other refinement strategies have been demonstrated by Luo et al. [13], Petrov and Todorov [17], and Verstraaten and J. Kosinka [22]. Hermosillo-Arteaga et al. [11] present recent advances in the theory of the finite element meshes. Subdivision techniques in multidimensional Euclidean space have been studied by Brandts et al. [6], Pascucci [19], and Petrov and Todorov [16].

The pyramidal elements are relatively new finite elements. In the three-dimensional case, they have been created by Bedrosian [2] in the early nineties. The pyramids play the role of transitional elements between structured and unstructured finite element meshes. The pyramidal elements have made possible the development of the hex-dominant meshes considered by several authors [14, 21, 24, 25]. The application of the pyramidal elements for constructing hybrid meshes has been described in [4, 7–9]. Mathematical models based on composite hybrid meshes have been developed in [14, 15]. In the three-dimensional domains, the conforming coupling between hexahedral and tetrahedral meshes does not require complicated transitional elements. The pyramidal elements assure conformity in a natural way, see Fig. 1. This construction has been applied by various authors, see for instance the paper by Yamakawa and Shimada [23]. The situation is completely different in the higher-dimensional spaces. A tesseract pyramid cannot be coupled directly with an arbitrary hexateron since all facets of the hexateron are pentatopes but the pyramid has not got any simplicial four-dimensional facet. Therefore, the hybrid meshes construction in higher-dimensional spaces needs a different approach. New transitional finite elements should be developed.

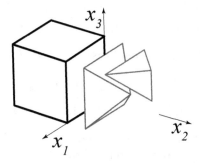

Fig. 1. A conforming coupling between hexahedron and tetrahedron finite elements.

This paper is devoted to the geometrical foundations of the finite element method and in particular the theory of meshes. One of the fundamental problems of the finite element method in an arbitrary n-dimensional Euclidean space is solved in this paper. New k-pyramidal elements are obtained and their properties are discussed. Specifically, in the five-dimensional space new cubic bipyramids and bihexateron elements are developed. A conforming coupling between penteracts and hexatera is demonstrated. Bipyramidal elements in arbitrary higher-dimensional space are uniformly refined. A plenty of new notions like k-pyramidal elements, conforming chains etc. are introduced.

Further, the paper is organized as follows. New k-pyramidal elements are defined and investigated in Sect. 2. A conforming ensemble of penteract and hexateron meshes are described in Sect. 3. Concluding remarks are presented in Sect. 4.

2   Subdivision Properties of the Bipyramidal Elements

Throughout the whole paper, the upper index stands for the dimension of the polytope.

Definition 1. *Two polytopes W_1 and W_2 are called elements from the same congruence class if one of them is obtained from the other by a rigid motion.*

In the early forties, Freudenthal has divided the n-dimensional hypercube [5] into $n!$ simplicial elements all of them from the same class. We call these simplices Freudenthal elements.

Definition 2. *We define the class $[\hat{E}^n]$ [16] of all simplices that are congruent to the quadruple Freudenthal element*

$$\hat{E}^n = [\hat{e}_1(2,0,\ldots,0),\ \hat{e}_2(2,2,0\ldots,0,0),$$

$$\hat{e}_3(1,1,1\ldots,0,0),\ldots,\hat{e}_{n-1}(1,1,\ldots,1,0),$$

$$\hat{e}_n(1,1,\ldots,1,1),\ \hat{e}_{n+1}(0,0,\ldots,0)].$$

Example 1. *In the five-dimensional case the class $[\hat{E}^5]$ is defined by the hexateron*

$$\hat{E}^5 = [\hat{e}_1(2,0,0,0,0),\hat{e}_2(2,2,0,0,0),\hat{e}_3(1,1,1,0,0),$$

$$\hat{e}_4(1,1,1,1,0),\hat{e}_5(1,1,1,1,1),\hat{e}_6(0,0,0,0,0)].$$

The elements of the class $[\hat{E}^n]$ can be obtained by coupling four Freudenthal elements. We denote the gravity center and the set of all vertices of the polytope W by $G(W)$ and $V(W)$ correspondingly.

Definition 3. *The polytope W_{-i} is obtained from $W[w_1, w_2, \ldots, w_{n+1}]$, $n \in \mathbf{N}$, by removing the node with the number i.*

Definition 4. *Let C^{n-1} be a $(n-1)$-dimensional facet of the hypercube C^n. The $(n-1)$-dimensional hypercubic pyramid $P^n[V(C^{n-1}), G(C^n)]$ is called canonical.*

The sequence $\{\nu_n\}$ is defined by $\nu_n = 2^{n-2} + 2$.

Definition 5. *The bipyramid B^n is a n-dimensional polytope with ν_n, $n \geq 3$ vertices b_i that satisfy:*

- b_i $i = 1, 2, \ldots, n - 2$ *are vertices of $(n - 2)$-dimensional hypercube C^{n-2};*
- *the line segment $b = [b_{\nu_n-1}, b_{\nu_n}]$ is not collinear to the $(n - 2)$-dimensional subspace of $\mathbf{R}^n$ determined by the hypercube C^{n-2};*
- $b \cap C^{n-2} = \emptyset$.

We consider some cases that clarify the role of the bipyramids in the lower-dimensional Euclidean spaces.

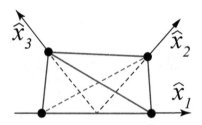

Fig. 2. The reference three-dimensional bipyramid.

The two-dimensional pyramid is the well-known triangle but a two-dimensional bipyramid does not make sense. Therefore, we begin with the three-dimensional case. The reference three-dimensional bipyramid is presented in Fig. 2. From Definition 5 with $n = 3$, we see that the three-dimensional bipyramid is actually a simplex but this is valid only in the three-dimensional case. The square pyramid is a transitional element in the interface subdomain between structured and unstructured meshes. Multigrid methods require stable refinement of all elements in the coarse triangulation. The optimal refinement strategy, see Fig. 3 for the square pyramidal elements have been described by Ainsworth and Fu [1]. All elements

$$[b_6, b_{10}, b_{11}, b_{12}], \quad [b_7, b_{10}, b_{12}, b_{13}], \quad [b_8, b_{10}, b_{13}, b_{14}], \quad [b_9, b_{10}, b_{11}, b_{14}]$$

are three-dimensional bipyramidal elements (tetrahedra). The chain

CUBE $\rightleftarrows$ PYRAMID $\rightleftarrows$ TETRAHEDRON

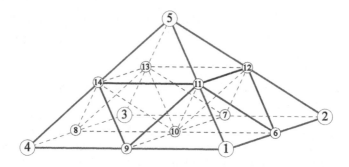

Fig. 3. The optimal partition of the square pyramid.

assure conforming coupling between structured and unstructured three-dimensional meshes, see Fig. 1.

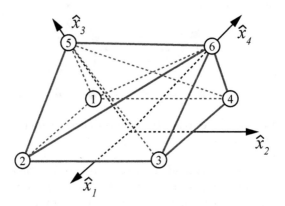

Fig. 4. The reference bipyramid.

The cubic pyramidal elements cannot guarantee conforming coupling between tesseract and pentatope meshes in the four-dimensional case. For this purpose, Petrov et al. [18] have created an additional bipyramidal finite element, Fig. 4. The four-dimensional bipyramid $B[b_i, i = 1, 2, \ldots, 6]$ can be divided into two pentatopes

$$B = S_1[b_1, b_2, b_3, b_5, b_6] \cup S_2[b_1, b_3, b_4, b_5, b_6],$$

that is why it could also be called bipentatope. The transitional four-dimensional bipyramid is related to tesseract partitions in the interface subdomains. Let us consider a tesseract element C from the interface subdomain. The transitional bipyramid is constructed as follows: the sixth and the fifth vertices are the tesseract center and the center of a three-dimensional facet C^3 of the element C; the other four nodes form a two-dimensional facet of C^3. The cubic bipyramid has also been applied to refine an arbitrary canonical cubic pyramid. Petrov et al.

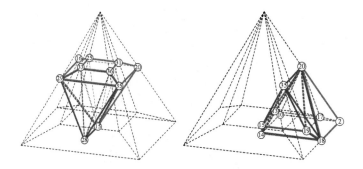

Fig. 5. Couplings between a cubic pyramid and a bipentatope.

have defined a partition operator that divides a canonical cubic pyramid into cubic pyramids and cubic bipyramids, Fig. 5. They have proved that all bipyramids from this partition are invariant and from the same congruence class. Each nondegenerated four-dimensional bipyramid B^4 (see Fig. 4) has two square pyramid facets B^4_{-5} and B^4_{-6}, and four tetrahedron facets

$$[b_1, b_2, b_5, b_6], \ [b_1, b_4, b_5, b_6], \ [b_2, b_3, b_5, b_6], \ [b_3, b_4, b_5, b_6].$$

The conformity in the four-dimensional space is assured by the chain

$$\text{TESSERACT} \rightleftarrows \text{CUBIC PYRAMID} \rightleftarrows \text{SQUARE BIPYRAMID} \rightleftarrows$$
$$\text{PENTATOPE} .$$

The bipyramids has a significant advantages since the reference bipyramid can be easily divided into a set of congruent elements. To this end let us consider the five-dimensional reference element

$$\hat{B}^5 = [\hat{b}_1(-1, -1, -1, 0, 0), \hat{b}_2(1, -1, -1, 0, 0), \hat{b}_3(1, 1, -1, 0, 0),$$

$$\hat{b}_4(-1, 1, -1, 0, 0), \hat{b}_5(-1, -1, 1, 0, 0), \hat{b}_6(1, -1, 1, 0, 0),$$

$$\hat{b}_7(1, 1, 1, 0, 0), \hat{b}_8(-1, 1, 1, 0, 0), \hat{b}_9(0, 0, 0, 1, 0), \hat{b}_{10}(0, 0, 0, 0, 1)],$$

see Fig. 6. The base of $\hat{B}^5$ can be divided into six Freudenthal elements $\hat{F}^3_i$. Then

$$\hat{B}^5 = [\hat{H}_i, \ i = 1, 2, \ldots 6, \ b_9, b_{10}],$$

$$\hat{H}_i = [V(\hat{F}^3_i), \ b_9, b_{10}].$$

On the other hand, the pyramidal elements cannot be divided only into pyramidal elements. Dividing a pyramidal element so that each edge to be partitioned into two segments we obtain gaps that are bipyramids. For a three-dimensional example we refer the reader to the paper written by Ainsworth and Fu [1], see also Fig. 3, and for the four-dimensional one to the paper by Petrov et al. [18]. Each penteract can be divided into 80 bipyramids.

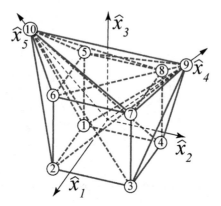

Fig. 6. The five-dimensional reference bipyramid.

Let us continue with the n-dimensional case. We introduce the ν_n-node reference element

$$\hat{B}_k^n : \begin{cases} -(1 - \hat{x}_{n-1} - \hat{x}_n) \le \hat{x}_i \le 1 - \hat{x}_{n-1} - \hat{x}_n, & i = 1, 2, \ldots, n-2 \\ 0 \le \hat{x}_{n-1} \le 1 - \hat{x}_n, \\ 0 \le \hat{x}_n \le 1 \end{cases}.$$

The largest index in the local numbering of each pyramidal element is reserved for the apex. Analogously, both apexes of the bipyramid are locally numbered by the largest indices.

Definition 6. *The set of n-dimensional hypercube, n-dimensional simplex, and all transitional elements that assure a conforming coupling between them is said to be a conforming chain.*

The whole spectre of the reference transitional elements in the n-dimensional Euclidean space with a integer $n \ge 3$ looks as follows

$$\hat{B}_k^n : \begin{cases} -\left(1 - \sum_{j=k+1}^n \hat{x}_j\right) \le \hat{x}_i \le 1 - \sum_{j=k+1}^n \hat{x}_j, & i = 1, 2, \ldots, k \\ 0 \le \hat{x}_i \le 1 - \sum_{j=i+1}^n \hat{x}_j, & i = k+1, k+2, \ldots, n-1 \\ 0 \le \hat{x}_n \le 1 \end{cases}.$$

Definition 7. *All affine equivalent elements to the element $\hat{B}_k^n$ are called transitional elements.*

The particular cases when $k = 1$, $n-2$, $n-1$ are of significant importance. The element B_1^n is a simplex, B_{n-2}^n is a bipyramid, and B_{n-1}^n is a hypercubic pyramid. Following these denotations we can say that the polytope B_{n-k}^n is a n-dimensional k-pyramid.

Further, we concentrate on the properties of the bipyramids. Describing the properties of the bipyramids we omit the index $n - 2$ and write B^n instead of B^n_{n-2} for notational simplicity. We consider two canonical pyramids P^n_1 and P^n_2 that share a common $(n - 2)$-hypercubic pyramid $P^{n-1}_{1,2}$. (Note that each five-dimensional canonical pyramid is a tesseract pyramid.) We suppose that the pyramid P^n_1 shares a common facet with a hypercubic element $\overline{C}^n$. The latter means that P^n_1 should not be refined. Additionally, the transitional elements are only located in the interface subdomain, which is canonical. Moreover, the facet $P^{n-1}_{1,2}$ should not be divided. Therefore, the transitional bipyramid B^n, in this case, should be defined by

$$B^n = [V(P^{n-1}_{1,2}), G(P^n_{2,-(2^{n-1}+1)})]. \tag{1}$$

Let

$$\mu_n = \begin{cases} 1, & \text{if } n = 3; \\ 2^{n-4}(n - 2)!, & n \geq 4. \end{cases}$$

Theorem 1. *The n-dimensional transitional bipyramid B^n $n \geq 3$ can be partitioned uniformly into μ_n quadruple Freudenthal elements.*

Proof. Let B^n be a transitional bipyramid defined by (1). Both pyramids P^n_1 and P^n_2 determine a hypercube C^n with a center the common apex of both pyramids. The bases of P^n_2 and $P^{n-1}_{1,2}$ are denoted by C^{n-1} and C^{n-2}.

(i) The three-dimensional case is trivial since the bipyramid B^3 is a single indecomposable quadruple Freudenthal element.

(ii) We consider the four-dimensional case as an exception of the general case when $n \geq 5$. The transitional bipyramid B^4 has six nodes. We define a pentatope $S[s_1, s_2, \ldots, s_5]$ by:

- the vertex s_5 is chosen to be the center of the tesseract determined by the cubic pyramids P^4_1 and P^4_2;
- the vertex s_4 is the center of $P^4_{2,-9}$;
- the vertices s_i $i = 1, 2, 3$ form a square corner of $P^3_{1,2,-5}$.

Obviously, we can create only two pentatopes following this construction. Both of them belong to $[\hat{E}]$.

Further, we suppose that $n \geq 5$. We define a simplex $S^n[s_1, s_2, \ldots, s_{n+1}]$ as follows. For the last three vertices we have $s_{n+1} = G(C^n)$, $s_n = G(C^{n-1})$ and $s_{n-1} = G(C^{n-2})$. The next $n - 5$ vertices are defined in a similar way $s_i = G(C^{i-1})$, $i = 4, 5, \ldots, n - 2$. Here, C^k is a k-dimensional facet of the hypercube C^{n-2}. The first three vertices s_i $i = 1, 2, 3$ form a square corner in a two-dimensional facet of C^{n-2}. The element $S^n \in [\hat{E}^n]$ by construction [16] (see the definition of the class $[\hat{P}^n]$ in [16] and [16, Theorem 9]).

The hypercube C^{n-1} can be partitioned into $2^{n-2}(n - 1)!$ elements of class $[\hat{E}^n]$ but only μ_n are inside the pyramid $B^n_{-\nu_n}$. Adding the vertex b_{ν_n} of B^n to each of these $(n-1)$-dimensional simplices we obtain μ_n quadruple Freudenthal elements. Thus the bipyramid B^n is divided by

$$B^n = \bigcup_{i=1}^{\mu_n} S_i^n, \quad \mathrm{vol}_n\left(S_i^n \cap S_j^n\right) = 0, \ i \neq j, \ S_i^n \in [S^n].$$

The main idea of the finite element method is to calculate all integrals in the finite element of reference. Theorem 1 guarantees that integrals in an arbitrary transitional element

$$\int_{B^n} f(x)dx = \sum_{i=1}^{\mu_n} \int_{S_i^n} f(x)dx, \quad S_i^n \in [\hat{E}^n] \ \forall i = 1, 2, \ldots, \mu_n$$

can be replaced by integrals in simplicial elements all of them from the same class. This is a significant advantage with respect to the application of the bipyramidal elements. Similar results can be obtained for the other n-dimensional k-pyramids.

Having in mind that the polytopes B_k^n and B_{k+1}^n can share B_k^{n-1} facet the conforming chain in $\mathbf{R}^n$ looks as follows

$$C^n \rightleftarrows B_{n-1}^n \rightleftarrows B_{n-2}^n \rightleftarrows \ldots \rightleftarrows B_2^n \rightleftarrows B_1^n = S^n,$$

where C^n is a n-dimensional hypercube and S^n is a n-dimensional simplex.

3   Transitional Elements in the Five-Dimensional Case

In this section, we restrict ourselves to the five-dimensional space. Let Ω be a bounded simply connected five-dimensional domain with a curved Lipschitz-continuous boundary $\Gamma = \partial\Omega$. To obtain an efficient triangulation of the domain Ω we divide it into several subdomains. We separate as large as possible canonical subdomain $\hat{\Omega} \subset \Omega$. The boundary layer Ω_B is connected to $\hat{\Omega}$ by an interface subdomain Ω_I, which is also a canonical domain. The domain $\hat{\Omega}$ is triangulated by penteract elements and the boundary layer by curved simplicial elements. The transitional elements are located in the interface subdomain Ω_I.

Let $\hat{\Pi} \subset \hat{\Omega}$ and $\Pi_I \subset \Omega_I$ be two adjacent penteract elements that share a common facet

$$T_2[\pi_1, \pi_2, \pi_4, \pi_7, \pi_9, \pi_{10}, \pi_{12}, \pi_{15}, \pi_{17}, \pi_{18}, \pi_{20}, \pi_{23}, \pi_{25}, \pi_{26}, \pi_{28}, \pi_{31}],$$

see Fig. 7. Our goal is to decompose the penteract Π_I into transitional and simplicial elements. First, we divide the element Π_I into ten tesseract pyramids P_i. The tesseract pyramid $P_2[V(T_2), \pi_{33}]$, should not be refined since P_2 and $\hat{\Pi}$ share the common facet $P_{2,-17}$. All neighbors of P_2 are refined by the same way. The opposite pyramid P_8 of the element P_2 is refined only by simplicial elements. That is why we show how one of the neighbors

$$P_1 = [T_1[\pi_i, \ i = 1, 2, \ldots, 16], \pi_{33}]$$

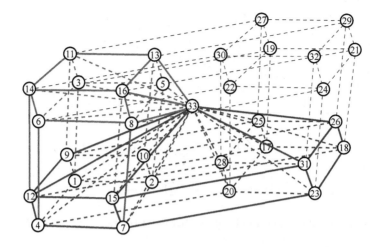

Fig. 7. The penteract Π_I.

should be refined, see Fig. 8. The pyramids P_1 and P_2 share a common cubic pyramid

$$P_{1,2}[\pi_1, \pi_2, \pi_4, \pi_7, \pi_9, \pi_{10}, \pi_{12}, \pi_{15}, \pi_{33}].$$

The pyramid $P_{1,2}$, see Fig. 7 and its facets should not be divided. Additionally, we denote the penteract center and the center of the facet T_1 by π_{33} and π_{34}, correspondingly. The point π_{35} is the center of $P_{1,2,-9}$.

We introduce a partition operator $\mathcal{L}$ that refine the bipyramid

$$B_3^5 = [b_i = \pi_i, \ i = 1, 2, \ldots, 10]$$

into twelve hexatera, see Fig. 8 and 9, as follows:

$$\mathcal{L}B = \Big\{ [b_1, b_2, b_4, b_9, b_{10}, b_{11}], \quad [b_1, b_2, b_5, b_9, b_{10}, b_{11}],$$

$$[b_1, b_4, b_5, b_9, b_{10}, b_{11}], \quad [b_2, b_3, b_4, b_9, b_{10}, b_{11}], \quad [b_2, b_3, b_6, b_9, b_{10}, b_{11}],$$

$$[b_2, b_5, b_6, b_9, b_{10}, b_{11}], \quad [b_3, b_4, b_7, b_9, b_{10}, b_{11}], \quad [b_3, b_6, b_7, b_9, b_{10}, b_{11}],$$

$$[b_4, b_5, b_8, b_9, b_{10}, b_{11}], \quad [b_4, b_7, b_8, b_9, b_{10}, b_{11}], \quad [b_5, b_6, b_8, b_9, b_{10}, b_{11}],$$

$$[b_6, b_7, b_8, b_9, b_{10}, b_{11}] \Big\}.$$

Let $H[h_i, \ i = 1, 2, \ldots, 6]$ be an arbitrary hexateron belonging to $\mathcal{L}B$. Then h_6 is the center of the penteract Π_I, h_5 is the center of the tesseract T_1, and h_4 is the center of a three-dimensional facet $P_{1,2,-9}$ of T_1. The other three vertices h_i $i = 1, 2, 3$ form a square corner in a two-dimensional facet of $P_{1,2,-9}$. This construction is valid for all elements of $\mathcal{L}B$ and guarantees that H belongs to $[\hat{E}^5]$.

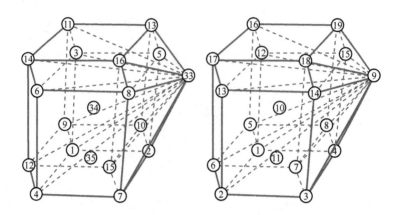

Fig. 8. The global and the local numbering of the tesseract pyramid P_1.

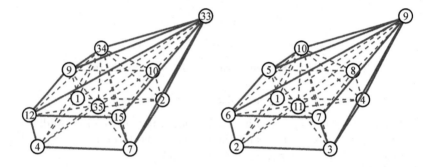

Fig. 9. The global and the local numbering of the cubic bipyramid B^5.

Any nondegenerated five-dimensional bipyramid (see Fig. 6) has two facets B_{-10} and B_{-9} that are cubic pyramids and six facets

$$[b_1, b_2, b_3, b_4, b_9, b_{10}], \quad [b_1, b_2, b_5, b_6, b_9, b_{10}], \quad [b_1, b_4, b_5, b_8, b_9, b_{10}],$$

$$[b_2, b_3, b_6, b_7, b_9, b_{10}], \quad [b_3, b_4, b_7, b_8, b_9, b_{10}], \quad [b_5, b_6, b_7, b_8, b_9, b_{10}]$$

that are four-dimensional bipyramids (see Fig. 4). The latter means that a five-dimensional bipyramid cannot be coupled conformingly with an arbitrary hexateron. To enable conforming coupling between penteract and simplicial elements in the five-dimensional space, we need another seven-node transitional element B_2^5. The new element is called bihexateron, see Fig. 10, and the reference element is defined as follows

$$\hat{B}_2^5 : \begin{cases} -\left(1 - \sum_{j=3}^5 \hat{x}_j\right) \le \hat{x}_i \le 1 - \sum_{j=3}^5 \hat{x}_j, & i = 1, 2 \\ 0 \le \hat{x}_3 \le 1 - \hat{x}_4 - \hat{x}_5, \\ 0 \le \hat{x}_4 \le 1 - \hat{x}_5, \\ 0 \le \hat{x}_5 \le 1 \end{cases}.$$

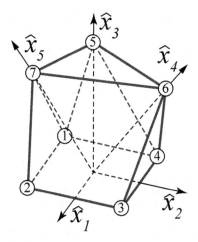

Fig. 10. The reference bihexateron $\hat{B}_2^5$.

An arbitrary nondegenerated bihexateron has seven facets, three of them

$$[\hat{b}_1, \hat{b}_2, \hat{b}_3, \hat{b}_4, \hat{b}_5, \hat{b}_6], \ [\hat{b}_1, \hat{b}_2, \hat{b}_3, \hat{b}_4, \hat{b}_5, \hat{b}_7], \ [\hat{b}_1, \hat{b}_2, \hat{b}_3, \hat{b}_4, \hat{b}_6, \hat{b}_7]$$

are four-dimensional bipyramids and other four are

$$[\hat{b}_1, \hat{b}_2, \hat{b}_5, \hat{b}_6, \hat{b}_7], \ [\hat{b}_1, \hat{b}_4, \hat{b}_5, \hat{b}_6, \hat{b}_7], \ [\hat{b}_2, \hat{b}_3, \hat{b}_5, \hat{b}_6, \hat{b}_7], \ [\hat{b}_3, \hat{b}_4, \hat{b}_5, \hat{b}_6, \hat{b}_7]$$

are pentatopes. Therefore B_3^5 and B_2^5 can share B_2^4. Obviously, a bihexateron and a hexateron can be connected conformingly since the nondegenerated hexateron has six pentatopial facets. Thus, we can assure a conforming coupling between penteracts and hexatera by applying the following chain

$$\text{PENTERACT} \rightleftarrows \text{TESSERACT PYRAMID} \rightleftarrows \text{BIPYRAMID} \rightleftarrows$$
$$\text{BIHEXATERON} \rightleftarrows \text{HEXATERON} .$$

4   Conclusion

Two kinds of new transitional finite elements are defined in the five-dimensional Euclidean space. The new seven-node bihexateron element and the new ten-node cubic bipyramid can share a common facet. A conforming coupling between structured and unstructured five-dimensional finite element meshes is obtained based on the chain penteract-tesseract pyramid-bipyramid-bihexateron-hexateron. All kinds of transitional elements are defined in the n-dimensional case, $n \geq 3$. The pyramidal elements are presented as a special kind of k-pyramidal elements. Similar presentation is found for the simplicial elements. The conforming chain is obtained in the n-dimensional case. A uniform refinement of the transitional bipyramids by quadruple Freudenthal elements is obtained. The uniform refinement essentially reduces the computational complexity of the integral computations in such elements. Additionally, the new reference elements are also triangulated uniformly.

References

1. Ainsworth, M., Fu, G.: A lowest-order composite finite element exact sequence on pyramids. Comput. Methods Appl. Mech. Eng. **324**, 110–127 (2017)
2. Bedrosian, G.: Shape functions and integration formulas for three-dimensional finite element analysis. Internat. J. Numer. Methods Engrg. **35**, 95–108 (1992)
3. Bedregal, C.: Maria-Cecilia Rivara, Longest-edge algorithms for size-optimal refinement of triangulations. Comput. Aided Des. **46**, 246–251 (2014)
4. Bergot, M., Cohen, G., Duruflé, M.: Higher-order finite elements for hybrid meshes using new nodal pyramidal elements. J. Sci. Comput. **42**, 345–381 (2010)
5. Bey, J.: Simplicial grid refinement, On Freudenthal's algorithm and the optimal number of congruence classes. Numer. Math. **85**(1), 1–29 (1998)
6. Brandts, J., Korotov, S., Křížek, M.: Simplicial finite elements in higher dimensions. Appl. Math. **52**(3), 251–265 (2007)
7. Ca, C., Qin, Q.-H., Yu, A.: A new hybrid finite element approach for three-dimensional elastic problems. Arch. Mech. **64**(3), 261–292 (2012)
8. Coulomb, J.-L., Zgainski, F.-X., Marechal, Y.: A pyramidal element to link hexahedral, prismatic and tetrahedral edge finite elements. IEEE Trans. Magn. **33**(2), 1362–1365 (1997)
9. Devloo, P., Duran, O., Gomes, S., Ainsworth, M.: High-order composite finite element exact sequences based on tetrahedral-hexahedral-prismatic-pyramidal partitions. hal-02100485 (2019)
10. Hannukainen, A., Korotov, S., Křížek, M.: On numerical regularity of the face-to-face longest-edge bisection algorithm for tetrahedral partitions. Sci. Comput. Programm. **90**(Part A) (2014) Pages 34–41
11. Hermosillo-Arteaga, A., Romo-Organista, M., Magaña del Toro, R., Carrera-Bolaños, J.: Development of a refinement algorithm for tetrahedral finite elements. Rev. Int. métodos numér. cálc. diseño ing. **37**(1), 1–21 (2021)
12. Korotov, S., Plaza, Á., Suárez, J.P.: Longest-edge n-section algorithms: properties and open problems. J. Comput. Appl. Math. **293**, 139–146 (2016)
13. Luo, X., Wang, Z., Wang, M., Shi, X.: A variable-scale refinement triangulation algorithm with faults data. J. Comput. **31**(5), 212–223 (2020)
14. O'Malley, B., Kophazi, J., Eaton, M.D., Badalassi, V., Warner, P., Copestake, A.: Pyramid finite elements for discontinuous and continuous discretizations of the neutron diffusion equation with applications to reactor physics. Prog. Nucl. Energy **105**, 175–184 (2018)
15. Martinelli, L.B., Alves, E.C.: Optimization of geometrically nonlinear truss structures under dynamic loading. REM Int. Eng. J. **73**(3), 293–301 (2020)
16. Petrov, M.S., Todorov, T.D.: Properties of the multidimensional finite elements. Appl. Math. Comput. **391**, 125695 (2021)
17. Petrov, M.S., Todorov, T.D.: Refinement strategies related to cubic tetrahedral meshes. Appl. Numer. Math. **137**, 169–183 (2019)
18. Petrov, M.S., Todorov, T.D., Walters, G., Williams, D.M., Witherden, F.D.: Enabling four-dimensional conformal hybrid meshing with cubic pyramids. Numer. Algorithms (to appear)

19. Pascucci, V.: Slow growing subdivision (SGS) in any dimension: towards removing the curse of dimensionality. EUROGRAPHICS **21**(3), 451–460 (2002)
20. Perdomo, F., Plaza, Á.: Properties of triangulations obtained by the longest-edge bisection. Central Eur. J. Math. **12**(12), 1796–1810 (2014). https://doi.org/10.2478/s11533-014-0448-4
21. Ray, N., Sokolov, D., Reberol, M., Ledoux, F., Lévy, B.: Hex-dominant meshing: mind the gap! Comput. Aided Des. **102**, 94–103 (2018)
22. Verstraaten, T.W., Kosinka, J.: Local and hierarchical refinement for subdivision gradient meshes. Pac. Graph. **37**(7), 373–383 (2018)
23. Yamakawa, S., Gentilini, I., Shimada, K.: Subdivision templates for converting a non-conformal hex-dominant mesh to a conformal hex-dominant mesh without pyramid elements. Eng. Comput. **27**, 51–65 (2011)
24. Yamakawa, S., Shimada, K.: Fully-automated hex-dominant mesh generation with directionality control via packing rectangular solid cells. Int. J. Numer. Methods Eng. **57**, 2099–2129 (2003)
25. Yamakawa, S., Shimada, K.: Increasing the number and volume of hexahedral and prism elements in a hex-dominant mesh by topological transformations. IMR, 403–413 (2003)

Multi-GPU Approach for Large-Scale Multiple Sequence Alignment

Rodrigo A. de O. Siqueira[1], Marco A. Stefanes[1], Luiz C. S. Rozante[2], David C. Martins-Jr[2], Jorge E. S. de Souza[3], and Eloi Araujo[1,2(✉)]

[1] Faculty of Computing, UFMS, Campo Grande, Brazil
`rodrigo.siqueira@ufms.br`, {`marco,feloi`}`@facom.ufms.br`
[2] Center of Mathematics, Computing and Cognition, UFABC, São Paulo, Brazil
{`luiz.rozante,david.martins`}`@ufabc.edu.br`
[3] Bioinformatics Multidisciplinary Environment (BioME),
Metrópole Digital Institute, UFRN, Natal, Brazil
`jorge@imd.ufrn.br`

Abstract. Multiple sequence alignment is an important tool to represent similarities among biological sequences and it allows obtaining relevant information such as evolutionary history, among others. Due to its importance, several methods have been proposed to the problem. However, the inherent complexity of the problem allows only nonexact solutions and further for small length sequences or few sequences. Hence, the scenario of rapid increment of the sequence databases leads to prohibitive runtimes for large-scale sequence datasets. In this work we describe a Multi-GPU approach for the three stages of the Progressive Alignment method which allow to address a large number of lengthy sequence alignments in reasonable time. We compare our results with two popular aligners ClustalW-MPI and ClustalΩ and with CUDA NW module of the Rodinia Suite. Our proposal with 8 GPUs achieved speedups ranging from 28.5 to 282.6 with regard to ClustalW-MPI with 32 CPUs considering NCBI and synthetic datasets. When compared to ClustalΩ with 32 CPUs for NCBI and synthetic datasets we had speedups between 3.3 and 32. In comparison with CUDA NW_Rodinia the speedups range from 155 to 830 considering all scenarios.

Keywords: Multiple sequence alignment · MSA · Hybrid Parallel Algorithms · Multi-GPU Algorithms · Large sequence alignment

1 Introduction

A very relevant task in bioinformatics is sequence alignment, which is routinely employed in many situations, such as comparing a query sequence with databases, comparative genome and sequence similarity searching. There are many important real world objectives which can be pursued by applying sequence alignment, such as paternity test, criminal forensics, drug discovery, personalized medicine, species evolution studies (phylogeny), just to mention a few. In

© Springer Nature Switzerland AG 2021
O. Gervasi et al. (Eds.): ICCSA 2021, LNCS 12949, pp. 560–575, 2021.
https://doi.org/10.1007/978-3-030-86653-2_41

fact, there is a myriad of techniques and tools proposed with this aim, including BLAST [1], S-W [2] and N-W [3] as the most popular ones. In particular, multiple sequence alignment (MSA) is one of the important formulations of the problem whose objective is to align multiple sequences at once. It is usually involved in phylogeny, molecular (2D and 3D) structure predictions such as proteins and RNAs, among other applications. Due to the COVID-19 pandemic, researchers are keen to reveal SARS-CoV-2 strains phylogeny hoping to understand the implications of the emerging strains in public health. Currently there are about a million strain sequences deposited (https://www.gisaid.org).

Several methods have been proposed for MSA, such as MAFFT [4], ClustalW [5], Kalign [6] and DeepMSA [7]. Yet, the assembly of optimal MSAs is highly computationally demanding considering both processing and memory requisites, since the problem is considered NP-Hard [8]. Dynamic programming based approaches retrieve an MSA with k sequences of length n in $O(2^k k^2 n^k)$. These techniques usually present important limitations regarding both length and number of input sequences to be computationally feasible [9].

Due to the aforementioned limitations and the fact that both the number of biological sequences and sequence lengths are continuously growing, finding fast solutions have led to employment of high performance computing techniques to achieve MSA as the popular aligners ClustalW_MPI [10] and ClustalΩ [11]. Hybrid parallel implementations of MSA have been recently proposed [12,13]. The former addresses only the first stage and the latter implements the three stages of the progressive alignment method, namely: i) pairwise alignment on Multi-GPUs with MPI-based communication among processes; ii) Neighbor Joining [14] implementation in a single GPU to build the guide tree; and iii) CUDA-GPU cluster implementation of the parallel progressive alignment algorithm similar to the implementation done by Truong et al. [15].

In this work we improved the three stages of the method proposed in [13] by addressing lengthy sequences during stage i), developing a scalable Neighbor Joining using Multi-GPU and paralleling Myers-Miller [16] algorithm. To the best of our knowledge this is the first Multi-GPU Neighbor Joining method in the literature. In fact, when comparing the results obtained by our method with the ClustalW-MPI and with the CUDA NW module of the Rodinia Suite [17], our proposal with 8 GPUs achieved speedups ranging from 28.5 to 282.6 with regard to ClustalW-MPI with 32 CPUs considering three NCBI datasets and three synthetic datasets. In comparison with CUDA NW_Rodinia the speedups range from 155 to 830. When compared to ClustalΩ with 32 CPUs in NCBI and synthetic datasets, we had speedups between 3.3 and 32. Regarding accuracy and quality of solution, the proposed method had a performance similar to Clustal-W and ClustalΩ, considering the benchmark BAliBASE [18].

This text is structured as follows: Sect. 2 introduces some basic concepts; Sect. 3 describes the computational model and some details about the parallel algorithms; Sect. 4 shows experimental results, including a comparative analysis. Finally, Sect. 5 presents the final remarks and future work.

2   Preliminaries

A *sequence over a finite alphabet* Σ is a finite enumerated collection of elements in Σ. The *length of a sequence* s, denoted by $|s|$, is the number of symbols of s and the *j-th element of* s is denoted by $s(j)$. Thus, $s = s(1) \ldots s(|s|)$. The set of all sequences over Σ is denoted by Σ^*. Let $S = \{s_0, s_1, \ldots, s_{k-1}\} \subseteq \Sigma^*$. An *alignment* of S is a set $A = \{s'_0, \ldots, s'_{k-1}\} \subseteq \Sigma^*_{\_} \big(= (\Sigma \cup \{\_\})^* \big)$, where: (i) $\_ \notin \Sigma$ is a new symbol called *space*; (ii) $|s'_h| = |s'_i|$; (iii) s'_i is obtained by inserting spaces in s_i; (iv) there is no j such that $s'_i(j) = \_$ for every i, $0 \le i < k$. Note that a sequence can be seen as an alignment for $k = 1$, i.e., $\{s\}$ is the single alignment of s and, hence, we sometimes refer to a sequence as an alignment. The *length of alignment* A is $|s'_i|$ and it is denoted by $|A|$. We denote by $\mathcal{A}_S$ the set of all alignments of S.

An alignment is an important tool for comparison of sequences obtained from organisms that have the same kind of relationship. It shows which part of each sequence should be compared to the other, thus suggesting how to transform one sequence into another by substitution, insertion or deletion of symbols. An alignment can be visualized placing each sequence above another as showed in the following figure with two different alignments of $(abacb, bacb, aacc)$. The left part represents the alignment $(abacb, \_bacb, a\_acc)$ and the right one represents the alignment $(aba\_cb\_, \_b\_acb\_, a\_a\_c\_c)$. Notice that the first suggests that the last c in the third sequence comes from substitution operation and the second alignment suggests that it comes from insertion operation.

$$
\begin{array}{ll}
a\ b\ a\ c\ b & a\ b\ a\ \_\ c\ b\ \_ \\
\_\ b\ a\ c\ b & \_\ b\ \_\ a\ c\ b\ \_ \\
a\ \_\ a\ c\ c & a\ \_\ a\ \_\ c\ \_\ c
\end{array}
$$

An *optimal alignment* in $\mathcal{A}_S$ is one which maximizes a given objective function whose value is also called *similarity* of S. The problem of finding an optimal alignment or even only its similarity is NP-hard for many objective functions.

This work deals with a polynomial method known as *progressive alignment* [19], which is described in 3 stages. This organization is extremely convenient because the algorithms for each stage are studied and improved independently in this work. The first stage corresponds to the *PairWise alignment* (PW) of all pairs of sequences, which builds a similarity matrix D. Using D as input, the next stage Neighbor Joining (NJ) consists in generate of a rooted binary tree T according to the similarity of each pair of sequences, which means that the more similar two nodes are, the closer they are. Each node represents an *operational taxonomic unit* (OTU): the leaves represent the sequences given in the PW step and the internal nodes represent hypothetical ancestor of theiR descendant nodes. We can refer a vertex set of a subtree of T as a set of OTUs. This tree T is known as a *guide tree* and each subtree of T corresponds to a group of closest related OTUs. The third stage, known as *Progressive Alignment* (PA), receives T as input, builds a profile for each node u of T and returns a profile of the root of T. The *profile* of the root of any subtree T' is a multiple alignment of the sequences that are leaves of T' and it is according to T' topology.

Pairwise Alignment (PW). A *scoring matrix* γ for Σ is a function $\gamma : \Sigma_{\_} \times \Sigma_{\_} \rightarrow \mathbb{R}$, such that $\gamma(\_, \_) = 0$, $\gamma(a, \_) = g$ for some constant $g \in \mathbb{R}$ and $\gamma(a, b) = \gamma(b, a)$ for each pair $a, b \in \Sigma_{\_}$. Function γ is used to attribute value for each pair of aligned sequences.

Given sequences s and t, we define a matrix H:

$$H[i, j] = \max \begin{cases} H[i-1, j-1] + \gamma(s(i), t(j)), \\ H[i-1, j] + g, H[i, j-1] + g \end{cases} \tag{1}$$

where $H[0, 0] = 0$. The value of the optimal alignment of s, t is $H[|s|, |t|]$. If $|s|$ and $|t|$ are $O(n)$, then H can be computed in $O(n^2)$ time and, once H is calculated, an optimum alignment (s', t') such that $\sum_i \gamma(s'(i), t'(i))$ is maximum can be found in $O(n)$ time. Matrix H is called *alignment matrix*.

A matrix D indexed by sequences and called *similarity matrix* is generated, where $D[s_i, s_h]$ is the value of the optimum alignment of s_i and s_h. As a consequence of γ definition, D is a symmetric matrix, which implies that we can represent D as a lower triangular matrix. Since there are $O(k^2)$ entries of D and each entry spends $O(n^2)$ time to be computed, the overall time spent in this step is $O(k^2 n^2)$. Matrix D is the input to the next step as follows.

Neighbor Joining (NJ). This stage creates *guide phylogenetic tree* T that is a binary phylogenetic tree from the similarity matrix D computed in the previous stage and it is the input of the next stage. This stage is implemented using the NJ algorithm [14]. The building begins with a star tree initially given by the set S of k sequences representing the k leaves that are the indices of D and a virtual node c in the center. In each iteration, if $|S| = 2$, it deletes the node c and connect directly the two vertices in S. Otherwise, pick u, v for which

$$(k-2)D[u, v] - \sum_{w \in S-\{u,v\}} \Big(D[u, w] + D[w, v] \Big) \tag{2}$$

is the largest. Then it deletes edges (u, c) and (c, v), creates a new vertex w (new OTU) and edges $(u, w), (v, w), (w, c)$, updates $S = (S - \{u, v\}) \cup \{w\}$ and sets

$$D[w, z] = \frac{1}{2} \Big(D[u, z] + D[v, z] - D[u, v] \Big) \tag{3}$$

for each vertex $z \neq w$ in S. It is performed in $k-2$ iterations. The expression (2) runs in time $O(k^3)$ in the first iteration and, if the computed values are stored, it spends $O(k^2)$ time in each of the next iterations. Expression (3) spends $O(1)$ time in each iteration. Thus, the total running time spent in stage 2 is $O(k^3) + (k-3) \cdot O(k^2) + (k-2) \cdot O(1) = O(k^3)$.

Progressive Alignment (PA). This stage buids an MSA by combining pairwise profiles and the guide tree described in the previous sections. For convenience, let us assume that the guide tree obtained is rooted. This assumption is not a special constraint. A rooted binary tree can be obtained from a rootless binary tree with the same set of leaves by subdividing some edge of the tree. By doing this, each internal node has exactly two children.

Given two sets S and S' of sequences, $S' \subseteq S$, and two alignments $C \in \mathcal{A}_S$ and $A \in \mathcal{A}_{S'}$, C is *compatible* with A if the elements of S' are aligned in C in the same way as in A. Feng and Doolittle [20] describe, from the alignments A and B, how to build a new alignment C that is compatible with A and B. Next, we show an example of two alignments A and B, and the respective compatible alignment C.

$$
A = \begin{array}{c} a\,c\,\llcorner\,a\,b \\ b\,b\,b\,a\,\llcorner \\ \hline a\,c\,a\,a\,a \end{array}
\qquad\qquad
C = \begin{array}{c} a\,c\,\llcorner\,\llcorner\,a\,b\,\llcorner \\ b\,b\,\llcorner\,b\,a\,\llcorner\,\llcorner \\ a\,c\,\llcorner\,a\,a\,a\,\llcorner \\ c\,a\,b\,\llcorner\,b\,a\,a \\ c\,a\,\llcorner\,\llcorner\,\llcorner\,c\,a \end{array}
$$
$$
B = \begin{array}{c} c\,a\,b\,b\,a\,a \\ \hline c\,a\,\llcorner\,\llcorner\,c\,a \end{array}
$$

The algorithm traverses the set of internal nodes of the rooted guide tree in post-order and it defines an profile for each visited OTU such that it is compatible with its children's profiles (and by transitivity, compatible with each of its descendants, including leaves). The root profile is the final MSA of the method.

Given an alignment $A = \{s'_0, \ldots, s'_{k-1}\}$, a column j of A is denoted by $A(j)$ and define $\Gamma(A(j)) = \sum_{i<h} \gamma(s'_i(j), s'_h(j))$.

Now, we describe how to get the rooted profile A of a set S. First of all, consider the corresponding profiles A_1 and A_2 of (two) root children. Suppose that $|A_1| = n_1$ and $|A_2| = n_2$, A_1 and A_2 with k_1, k_2 rows and i, j be indexes. Denote by $A_1(i) \cdot A_2(j)$ the concatenation of columns $A_1(i)$ and $A_2(j)$, i.e., the sequence with $k_1 + k_2$ elements

$$A_1[0][i], \ldots, A_1[k_1 - 1][i], A_2[0][j], \ldots, A_2[k_2 - 1][j].$$

Also, denote by $\llcorner^\ell$ the sequence of ℓ symbols equals to $\llcorner$ and suppose that $n_1, n_2 = O(N)$ for some N.

We compute the matrix also called *alignment matrix*

$$M[i, j] = \max \begin{cases} M[i-1, j-1] + \Gamma(A_1(i) \cdot A_2(j)), \\ M[i-1, j] + \Gamma(A_1(i) \cdot \llcorner^{k_2}), M[i, j-1] + \Gamma(\llcorner^{k_1} \cdot A_2(j)) \end{cases} \quad (4)$$

for each $1 \leq i \leq n_1$ and $1 \leq j \leq n_2$ with $M[0, 0] = 0$. Considering $n = n_1 + n_2$ and that each entry of M can be computed in constant time, we spend $O(n^2)$ time for compute all entries of M and since the guide tree has $O(k)$ nodes, the entire procedure spends $O(kn^2)$ time. Once M is calculated, using a similar strategy to the trace back method by Needleman-Wunsch [3], we spend $O(n)$ to obtain each profile (compatible alignment) and $O(kn)$ to obtain all profiles.

3 Parallel Algorithms

3.1 Homogeneous Hybrid Parallel Platform

In this paper, we developed an MSA algorithm based on progressive alignment strategy to execute on a hybrid parallel computing platform. That platform is based on the joint use of CPUs and GPUs in order to obtain high performance systems. Hybrid parallel platform is a two-level parallel computing model. Below we provide some features of the target hardware related to this architecture.

Suppose we have p compute nodes (**CN** for short) and each of them contains a GPU: At the upper level, we use a coarse-grained model based on Beowulf cluster, which is scalable and based on an inexpensive hardware infrastructure composed by private and dedicated interconnection network coordinated by MPI [21]. See Fig. 1. At this level a special process called master node runs in CN 0, managing the tasks of the computing nodes. On the other hand, at the lower level, we use a fine-grained model through the CUDA-enabled GPUs [22]. CUDA (Compute Unified Device Architecture) is a parallel architecture based on many-core paradigm for NVIDIA GPUs. CUDA enables programmers to write a source code and execute it on the GPU. Each GPU can have several streaming multiprocessors (SM) and each SM contains dozens or even hundreds of Single Processors (SP). All SMs access a same device global memory.

The code runs on CPU or GPU and the tasks of the computing nodes are managed by MPI. CPU creates multi-thread kernels for the GPU. GPU has its own scheduler that assigns a thread block to any SM dynamically during the execution, and the SPs within SM run the threads.

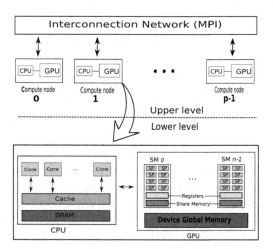

Fig. 1. Homogeneous hybrid parallel model

On CUDA programming environment we need define size of thread blocks and size of the grid, which is an abstraction for a group of thread blocks. Each thread block is assigned to an SM, and threads in a same block access a same shared memory. Moreover, the hierarchical memory consists in global memory, texture memory, shared memory and registers, where global memory is the slowest and registers are the fastest.

3.2 Parallel Algorithm

Our implementation of the MSA problem explores the two level of parallelization along the three stages of the progressive alignment. During the processing the job coordination on a CN is done by CPU which can execute local operations using CPU and GPU accordingly.

At the upper level, we use coarse-grained parallelism which manages whole process and control the distribution of the jobs to the CNs. In this level, initially the master node reads the input sequences from the disk and replicates them into the CPU memory of all CNs. Our algorithm uses a simple and efficient task allocation strategy that performs uniform load balancing between the CNs.

Let k be the number of sequences. In order to save space since we are primarily interested in computing large amount of lengthy sequences, we represent the similarity matrix D (lower triangular) by an array V of size $N = k(k-1)/2$. Each position in V represents a sequence pair to be aligned, considering the positions of the triangular matrix in lexicographic order (see Fig. 2). Assume that N is divisible by p. The array V is partitioned into p segments of size N/p, which we denote by $v_0, v_1, \ldots, v_{p-1}$. The CN i, $0 \leq i \leq p-1$, will process the elements of the segment v_i, whose positions in V are in the range $[(i)\frac{N}{p} .. (i+1)\frac{N}{p} - 1]$. Each CN i identifies the sequence pairs that will be sent to its GPU by computing (in CPU) the mapping of the elements from v_i to D, as follows:

$$l = \left\lfloor \sqrt{2(iN/p + 1)} + \tfrac{1}{2} \right\rfloor \quad \text{and} \quad c = iN/p - l(l-1)/2, \tag{5}$$

where l and c correspond, respectively, to the row and column of the element in D which is represented by the first element of v_i. Clearly, the Eq. 5 can be calculated in $O(1)$ time.

As all sequence characters belong to a small and well-defined alphabet, they can be mapped into numerical identifiers, allowing the representation of each character symbol with only 5 bits of memory (which allows to represent up to $2^5 = 32$ symbols). Hence, we can store 3 distinct sequence elements in a single 16-bit integer type using simple bit-wise operations. Effectively, this reduces by up to a third the total amount of memory required to store each sequence. Nevertheless, each sequence character can still be randomly accessed in constant time complexity with negligible overhead.

At the lower level, we use fine-grained parallelism whose jobs are computed on the CUDA-enabled GPUs. In the subsection below we detail the used approaches to perform the three stages using hybrid parallel computing.

Matrix Similarity on Multi-GPU Platform. In this stage we calculate the similarity matrix aligning pairwise sequences using a GPU version of the N-W

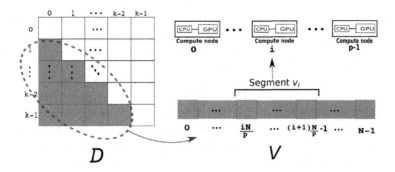

Fig. 2. Load balancing: the array V is evenly partitioned into p segments of size N/p among p computer nodes.

algorithm [3]. We align each pair of sequences in parallel on the GPU of each CN using an approach based on intra-task parallelization [23].

Each CN i receives from the master node the whole set of sequences, together with the scoring matrix γ, and identifies (through Eq. 5) the pairs v_i whose similarity value it will calculate. These data are sent to the GPU. The matrix γ is stored only once in the shared memory since it is the same for all the pairs. However, the segment v_i is sent to the GPU global memory in waves. Each element of the matrix D is computed using dynamic programming based on the Eq. 1. After aligned a set of pairs another wave of data is sent to GPU.

In this stage we save memory and gain performance by concurrently calculating (in GPUs) the similarity values using a combination of the techniques of the algorithms *DScan-mNW* and *LazyRScan-mNW* [15], as follows (see Fig. 3):

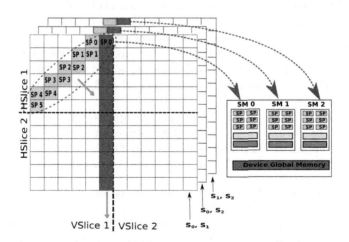

Fig. 3. Flow for calculating similarity values

Each SM stores in its shared memory its respective sequence slice and performs a diagonal scan. Each alignment matrix is iterated over the sequence in vertical slices which are computed by a thread block. Each thread acts over a row at a time. The vertical slice size is calculated so that it fits into shared memory and according to available SPs. When the thread reaches the end of the slice its value is transferred to global memory (dark green column). Then, for the next slice to resume the processing, the value of this cell is sent again to shared memory. Since our approach each thread needs only three cells at a time, instead of maintaining two contiguous rows in the shared memory we use sliding window strategy where old cells are removed from the shared memory. The similarity value is sent to the global memory as soon as they is computed.

For long sequences, we can split vertical slices in horizontal slices by forming a square mesh. As CUDA limits the number of threads per block, this improvement allows shared memory is used by more threads and hence to speedup the processing. Both combined techniques allows us to hide the access latency to global memory. However the last optimization limits the use of threads per block. On the other hand, it reduces the amount of necessary shared memory to store the similarity values being calculated. In our implementation a good equilibrium was 480 threads by block. After to compute all similarity values on GPUs these values are sent to master node where the whole matrix D is assembled.

Guide Phylogenetic Tree. Taking matrix D (previous stage) as input, we perform this stage according to the NJ algorithm [14], whose goal is to build the guide tree. Basically the NJ algorithm starts with a star tree where each leaf vertex corresponds to an OTU and iterates over the following three steps in order to joining the most similar pair of OTUs until reaches all leafs of the tree:

1. From the matrix D we have to compute each OTU pair in D by using the Formula 2. This processing generates a derived matrix in order to maintain the evolution relationship among all the OTUs.
2. By current derived matrix we choose the maximum value representing the most similar pair of OTUs. These pairs are joined to a newly created vertex which is joined the rest tree such that they form a new branch in the phylogenetic tree.
3. After joining the pair of OTUs we have to update the similarity matrix D according to Eq. 3. New row and column are created to store the similarity between the joined OTU and the remaining OTUs. Then, row and column correspondent to chosen OTUs in Step 2 are removed from the matrix D.

From parallel point of view this is a difficult task due to the data dependency among the 3 steps above, in addition there is dependency among each iteration. Despite of the large data dependency required by the method, we face the challenge to implement this stage of the alignment on Multi-GPU platform. Figure 4 illustrates the overall flow of an iteration.

Initially, the master node broadcasts the matrix D to all CNs which is copied to the GPU global memory. As soon as the CNs receive D they compute their

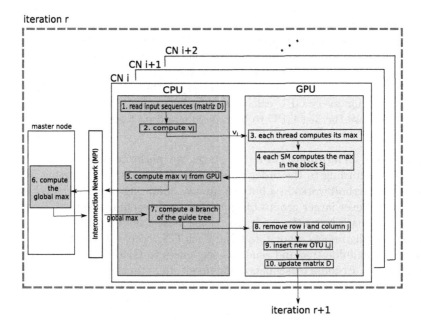

Fig. 4. MPI-CPU-GPU flow in stage two (construction of the guide tree)

respective chunks in the same scheme that of the first stage and send to GPU.
So, each GPU have to compute the range $[(i)\frac{N}{p} \dots (i+1)\frac{N}{p} - 1]$.

For the Step 1, computing Eq. 2 can be done independently within each GPU.
We avoid wasting time with redundant operations by calculating the sum of the
rows and columns attributed to GPU in advance since this values are used for
all pairs. This values are kept in the shared memory and updated at the end of
each iteration. Due to memory limit and the large amount of pairs to process we
have to launch multiple kernels to compute the derived matrix, where the exact
number of kernels depends on the number of sequences k, device global memory
and the number of SMs available. The number of pairs are evenly distributed
among the blocks.

In Step 2, choosing the largest value in D is accomplished as follows. Each
thread computes its largest value and keeps it locally. After, each thread block
applies a reduction and obtains the largest value of the block. Then, each block
sends its values to CPU and each CPU computes its largest value. Finally MPI
applies a reduction to obtain the global largest value. This value and its respective pair (i, j) are then known by all CNs. The pair (i, j) is used in each CPU
to create the new node and the tree is updated on CPU.

In Step 3, we perform the computation of the branch sizes for the new node,
the similarity value from the new node to the other sequences and update D. For
this, each GPU receives the largest value and the pair (i, j) and has to remove
row i and column j from D. In fact, these row and column are only marked as
removed. We also have to calculate the similarity of newly create OTU with the

remaining OTUs according to Eq. 3. Again, the Eq. 5 allows threads to access each OTU quickly. Hence we update matrix D and recalculate the sum for each of rest row/column to be used in next iteration. Note that a sum and a subtraction is enough to recalculate row/column sums instead of applying reduction.

An important improvement we implement is that of a threshold. At the end of each iteration every GPU calculates whether its workload is less a threshold. If so, we choose the last GPU to become idle and its task is redistributed.

Progressive Alignment. In this stage, we perform the alignment of the sequences according to the order provided by the guide tree. Starting from the leaves, the method works in a bottom-up way operating in parallel level by level and aligning ever larger groups of sequences until reaching the root (see Fig. 5).

Our approach keeps the guide tree in the master node's memory, which coordinates the alignments among its CNs. Each CN receives from the master node a node of the guide tree and sends its sequences to GPU. Thereafter, the GPU accomplishes a profile or sequence alignment and transfer it to its CPU which send it to the master node which stores it into CPU memory.

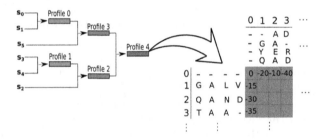

Fig. 5. Parallel progressive alignment algorithm. The guide tree lead the order of the pairwise alignments.

For all of the guide tree's nodes with height 1, several CNs run concurrently in its GPU our parallel version of the algorithm Myers and Miller [16]. Since we have to compute the alignment, not only the score, this method enables to address long sequences. In this case, we use only horizontal slices.

For the remaining nodes, note that the computation of $M[i, j]$ is based on the sum of all evolutionary distances between all possible combinations of amino acids. Hence, as we move up the tree, memory will be a bottleneck. To overcome this bottleneck we transfer only a profile and the other is sent as the slicing window technique in horizontal slices. Each GPU trhead calculates a score concurrently, for each column of the profile. These scores are projected as initialization values for the matrix M, enabling us to align the profile pair. After we join together all of their sequences into a single alignment profile.

4   Experimental Results and Discussion

We have compared our implementation, called Museqa (acronym for Multiple Sequence Aligner), with the popular aligners ClustalW-MPI and ClustalΩ. Besides, we compared the Museqa PW stage with PW_R, a CUDA N-W implementation included in the Rodinia Suite [17]. In our experiments we considered sequence datasets from NCBI and OrthoDB [24]. We have also considered three randomly generated synthetic sequence datasets called Syn20k, Syn30k and Syn40k. Table 1 presents the number of sequences and the sequence length of each dataset.

Table 1. Datasets considered in the experiments.

Datasets	Zika	Dengue	SARS-CoV2	OrthoDB	Syn20k	Syn30k	Syn40k
# of sequences	700	6,123	7,631	10,000	20,000	30,000	40,000
Average length	3,423	3,392	7,096	3,150	200	200	200

We run our experiments in two different platforms: Server A with Intel(R) Xeon(R) CPU E5-1620 3.50 GHz, 4 cores(8 threads), 64 GB RAM, 4x GeForce GTX 1080-11 GB and Server B with 2 x Intel(R) Xeon(R) CPU E5-2683 2.10 GHz, 16 cores (32 threads), 512 GB RAM, 8 x Tesla V100-16 GB.

Table 2. Comparison of Museqa run in GPUs with ClustalW-MPI run in CPUs and comparison of the stage PW of the Museqa with PW_R of the Suite Rodinia.

		a) ClustalW-MPI		b) ClustalΩ		c) R_S	d) Museqa		Speedups		
		#CPUs	#CPUs	#CPUs	#CPUs		#GPUs	#GPUs	a/d	b/d	c/d
Dataset	Stage	8	32	8	32	1	4	8			1
Zika	PW	2 h 10	52 m 03	31 s	15 s	30 m 58	14 s	3 s	1041	5	619
	NJ	15 s	2 s	14 s	9 s	–	0.11 s	0.3 s	6.67	30	–
	PA	2 m 53	2 m 20	16 m 34	27 m 27	–	1 m 15	58 s	2.41	17.1	–
Dengue	PW	196 h 07	64 h 17	8 m 43	3 m 40	14 h 28	18 m	5 m 37	687	0.65	155
	NJ	37 m 18	36 m 56	2 m 21	1 m 23	–	11 s	11 s	201	7.55	–
	PA	28 m 59	20 m 01	2 h 27	3 h 57	–	12 m 54	10 m 11	1.97	14,5	–
Sars-Cov2	PW	597 h 16	179 h 54	1 h 27	35 m 42	92 h 57	1 h 53	35 m 45	302	1	156
	NJ	1 h 08	1 h 05	40 m 41	22 m 19	–	18 s	16 s	244	84	–
	PA	28 m 58	26 m 09	14 h 24	19 h 59	–	15 m 34	14 m 18	1.83	60.8	–
OrthoDB	PW	489 h	163 h 01	44 m 45	22 m 22	108 h	24 m 27	7 m 48	1253	2.86	830
	NJ	2 h 45	2 h 33	10 m 30	7 m 30	–	29 s	7 s	1031	64.3	–
	PA	129 h 20	46 h 11	47 h 15	94 h 30	–	5 m 22	2 m 36	1065	1090	–
Syn20k	PW	2 h 51	1 h 18	32 s	14 s	3 h 20	1 m 55	32 s	146	0.44	375
	NJ	35 h 10	32 h 24	7 s	5 s	–	3 m 44	50 s	2333	0.1	–
	PA	8 h 55	3 h 11	8 h 5	9 h 4	–	2 h 17	1 h 06	2.89	7.35	–
Syn30k	PW	6 h 24	2 h 03 m	51 s	21 s	5 h 22	4 m 36	1 m 17	96	0.27	251
	NJ	165 h 35	163 h 46	11 s	7 s	–	11 m 35	2 m 38	3731	0.04	–
	PA	30 h 01	9 h 45	10 h 45	20 h 23	–	8 h 11	3 h 37	2.69	5.63	–
Syn40k	PW	11 h 28	4 h 34	1 m 12	21 s	7 h 25	8 m 54	2 m 23	115	0.15	187
	NJ	385 h	356 h	15 s	7 s	–	25 m 33	3 m 08	6817	0.04	–
	PA	90 h	27 h 12	16 h	29 h 30	–	15 h 29	10 h 56	2.49	2.70	–

Table 2 shows the runtimes of the three stages (PW, NJ and PA) using Server
A and Server B by ClustalW-MPI, ClustalΩ and our Museqa. In addition, it
shows the runtimes obtained by NW_R using Server A. For those tests we use
as input Zika Virus, Dengue Virus, SARS-CoV2, OrthoDB, Syn20k, Syn30k and
Syn40k datasets. We observe that ClustalΩ implements different strategies for
each step when compared to Museqa and ClustalW-MPI. We calculated the
average time over 3 executions for each test. Comparing with ClustalW-MPI,
the first relevant fact is that the stage PW dominates at least 95% of the time
in all NCBI dataset scenarios, while for synthetic datasets the opposite occurs:
the time spent by NJ becomes significant, surpassing the PW runtime.

The scalability of Museqa for all stages combined in terms of number of
GPUs were almost linear for all datasets considered. Note that, even though
NJ stage did not achieve linear speedups it still performed better with multiple GPUs, even for synthetic datasets where the time spent in NJ stage was
larger than in PW stage according to Table 2. Figure 6 illustrates the speedups
between ClustalW-MPI and Museqa and between ClustalΩ and Museqa regarding NJ stage only for 4 and 8 GPUs (32 CPUs for ClustalW-MPI and ClustalΩ),
highlighting the remarkable performance of Museqa compared to ClustalW-MPI
for all seven datasets, and specially for the synthetic datasets for which the time
consumption of NJ is much more significant than for NCBI datasets. Considering
the comparison between ClustalΩ NJ and Museqa NJ, Museqa NJ performed
better for real datasets (NCBI and OrthoDB), but ClustalΩ NJ had superior
performance for synthetic datasets (only marginally for Museqa with 8 GPUs).

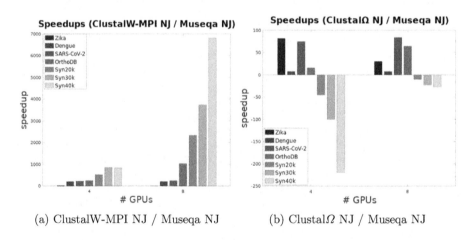

(a) ClustalW-MPI NJ / Museqa NJ   (b) ClustalΩ NJ / Museqa NJ

Fig. 6. Speedups between: a) ClustalW-MPI NJ and Museqa NJ; b) ClustalΩ NJ and
Museqa NJ; according to Table 2. Each negative value $(-X)$ in (b) means that ClustalΩ
NJ was X times faster than Museqa NJ.

Considering the total time spent of all three stages combined, as shown in
Fig. 7, Museqa with 8 GPUs was between 28.5 to 283 times faster than ClustalW-
MPI with 32 CPUs, while Museqa with 8 GPUs had speedups ranging from 2.3

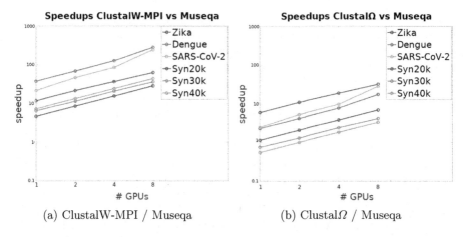

(a) ClustalW-MPI / Museqa   (b) ClustalΩ / Museqa

Fig. 7. Overall speedups (log-log scale) between ClustalW-MPI vs Museqa and ClustalΩ vs Museqa, considering 32 CPUs for ClustalW-MPI and ClustalΩ.

to 32 when compared to ClustalΩ with 32 CPUs. In addition, Fig. 7 highlights that all speedups increased almost linearly with the number of GPUs. Finally, the comparison between PW_R, which is also implemented in GPU, and Museqa PW stage performed with 8 GPUs revealed that Museqa was about two orders of magnitude faster than PW_R (speedups between 155 and 830).

Alignment Accuracy: In order to measure the accuracy of the alignments produced by Museqa, we use BALiBASE [18], which is the most widely used benchmark test sets of reference alignments. We compute BAli scores (SP and TC, which measure the alignment accuracy, ranging from 0 to 1, where 1 indicates the best possible accuracy) for Museqa, ClustalΩ and ClustalW, considering 386 alignments, which are organized in 6 BAli families covering six different situations (RV11, RV12, RV20, RV30, RV40, and RV50).

Table 3. SP and TC score average results for ClustalΩ, our proposed method (Museqa), and ClustalW, for the six considered BAli families.

	RV11		RV12		RV20		RV30		RV40		RV50	
	SP	TC	SP	TC	SP	TC	SP	TC	SP	TC	SP	TC
ClustalΩ	0.48	0.27	0.83	0.68	0.82	0.34	0.69	0.38	0.76	0.43	0.70	0.35
Museqa	0.41	0.21	0.79	0.61	0.78	0.24	0.57	0.16	0.61	0.28	0.56	0.19
ClustalW	0.48	0.24	0.80	0.64	0,79	0.26	0.62	0.25	0.65	0.30	0.62	0.27

As can be seen in Table 3, the obtained accuracy is very similar, proving the reliability of the alignments obtained by Museqa when compared to ClustalΩ and

CrustalW. However, mean values can suppress nuances regarding the differences found in each alignment. In this sense, we comparison of the measurements of scores between the three algorithms in each of the 386 alignments, and we clustered into three groups: group1 - when Museqa obtained a better score; group2 - when there was a tie between the scores; group3 - when the score obtained by Museqa was significantly lower. The group2 formed by comparisons in which the fold change between the scores is less than or equal to 1.3. And the group3 by fold change in scores greater than 1.3.

Comparing Museqa and ClaustralΩ, we notice that the SP-score possesses 18.6%, 73.8%, and 7.5% of the comparisons in group1, group2, and group3, respectively. And the TC-score holds 16.3%, 68.9%, and 14.7% of the comparisons in group1, group2, and group3, respectively. Comparing Museqa and ClustalW, the SP-score holds 36.5%, 60.3%, and 3.1% in group1, group2, and group3, respectively. And the TC-score keeps 25.3%, 70.4%, and 4.1% of the comparisons in group1, group2, and group3, respectively. These results demonstrate a high degree of congruity between the three programs.

5   Final Remarks

In this paper we described a Multi-GPU solution for the Multiple Sequence Alignment problem which implements the three stages of the progressive alignment method. All stages of the method achieved significant speedup when compared to some popular tools that are currently available for the same task.

As future work, we intend to look into three directions: i) to use a modified NJ which produces a complete binary tree, reducing the interdependence of the computations in stages 2 and 3; ii) to implement the aligments of the method using new biological information iii) to make improvements in the proposed method so that it can operate in heterogeneous hybrid parallel platforms.

Acknowledgments. We thank the High Performance Computing Center (NPAD/ UFRN) and CTEI/UFMS for providing computational resources, and grants #2018/18560-6, #2018/21934-5, São Paulo Research Foundation (FAPESP) for financial support.

References

1. Altschul, S.F., Gish, W., Miller, W., Myers, E.W., Lipman, D.J.: Basic local alignment search tool. J. Mol. Biol. **215**(3), 403–410 (1990)
2. Smith, T.F., Waterman, M.S.: Identification of common molecular subsequences. J. Mol. Biol. **147**(1), 195–197 (1981)
3. Needleman, S.B., Wunsch, C.D.: A general method applicable to the search for similarities in the amino acid sequence of two proteins. J. Mol. Biol. **48**(3), 443–453 (1970)
4. Katoh, K., Misawa, K., Kuma, K., Miyata, T.: MAFFT: a novel method for rapid multiple sequence alignment based on fast Fourier transform. Nucleic Acids Res. **30**(14), 3059–3066 (2002)

5. Larkin, M.A., et al.: Clustal W and Clustal X version 2.0. Bioinformatics **23**(21), 2947–2948 (2007)
6. Lassmann, T.: Kalign 3: multiple sequence alignment of large datasets. Bioinformatics **36**(6), 1928–1929 (2020)
7. Zhang, C., Zheng, W., Mortuza, S.M., Li, Y., Zhang, Y.: DeepMSA: constructing deep multiple sequence alignment to improve contact prediction and fold-recognition for distant-homology proteins. Bioinformatics **36**(7), 2105–2112 (2020)
8. Bonizzoni, P., Della Vedova, G.: The complexity of multiple sequence alignment with SP-score that is a metric. Theoret. Comput. Sci. **259**(1), 63–79 (2001)
9. Thompson, J.D., Linard, B., Lecompte, O., Poch, O.: A comprehensive benchmark study of multiple sequence alignment methods: current challenges and future perspectives. PloS One **6**, e18093 (2011)
10. Li, K.-B.: ClustalW-MPI: ClustalW analysis using distributed and parallel computing. Bioinformatics **19**(12), 1585–1586 (2003)
11. Sievers, F., et al.: Fast, scalable generation of high-quality protein multiple sequence alignments using Clustal Omega. Mol. Syst. Biol. **7**, 539 (2011)
12. Alawneh, L., Shehab, M.A., Al-Ayyoub, M., Jararweh, Y., Al-Sharif, A.Z.: A scalable multiple pairwise protein sequence alignment acceleration using hybrid CPU-GPU approach. Cluster Comput. **23**, 2677–2688 (2020)
13. Araujo, E., Stefanes, M.A., Ferlete, V.O., Rozante, L.C.S.: Multiple sequence alignment using hybrid parallel computing. In: 17th IEEE International Conference on Bioinformatics and Bioengineering, pp. 175–180 (2017)
14. Saitou, N., Nei, M.: The neighbor-joining method: a new method for reconstructing phylogenetic trees. Mol. Biol. Evol. **4**(4), 406–425 (1987)
15. Truong, H., Li, D., Sajjapongse, K., Conant, G., Becchi, M.: Large-scale pairwise alignments on GPU clusters: Exploring the implementation space. J. Sig. Process. Syst. **77**(1–2), 131–149 (2014)
16. Myers, E.W., Miller, W.: Optimal alignments in linear space. Comput. Appl. Biosci. CABIOS **4**(1), 11–17 (1988)
17. Che, S., et al.: Rodinia: a benchmark suite for heterogeneous computing. In: 2009 IEEE International Symposium on Workload Characterization (IISWC), pp. 44–54 (2009)
18. Thompson, J.D., Koehl, P., Ripp, R., Poch, O.: BAliBASE 3.0: latest developments of the multiple sequence alignment benchmark. Proteins Struct. Funct. Bioinf. **61**(1), 127–136 (2005)
19. Hogeweg, P., Hesper, B.: The alignment of sets of sequences and the construction of phyletic trees: an integrated method. J. Mol. Evol. **20**(2), 175–186 (1984)
20. Feng, D.-F., Doolittle, R.F.: Progressive sequence alignment as a prerequisite to correct phylogenetic trees. J. Mol. Evol. **25**(4), 351–360 (1987)
21. Gropp, W., Lusk, E., Skjellum, A.: Using MPI: Portable Parallel Programming with the Message-Passing Interface. MIT Press (1999)
22. Cook, S.: CUDA Programming: A Developer's Guide to Parallel Computing with GPUs. Elsevier (2012)
23. Liu, Y., Schmidt, B., Maskell, D.L.: MSA-CUDA: multiple sequence alignment on graphics processing units with CUDA. In: 20th IEEE ASAP, pp. 121–128 (2009)
24. Zdobnov, E.M., et al.: OrthoDB in 2020: evolutionary and functional annotations of orthologs. Nucleic Acids Res. **49**, D389–D393 (2021)

On Modeling of Interaction-Based Spread of Communicable Diseases

Arzad A. Kherani[1(✉)], Nomaan A. Kherani[1], Rishi R. Singh[1], Amit K. Dhar[1], and D. Manjunath[2]

[1] Department of Electrical Engineering and Computer Science,
Indian Institute of Technology Bhilai, Raipur, India
{arzad.alam,nomaanalam,rishi,amitkdhar}@iitbhilai.ac.in
[2] Department of Electrical Engineering,Indian Institute of Technology Bombay,
Mumbai, India
dmanju@ee.iitb.ac.in

Abstract. We use a queuing model to study the spread of an infection due to interaction among individuals in a public facility. We provide tractable results for the probability that a susceptible individual leaves the facility infected. This model is then applied to study infection spread in a closed system like a large campus, community, and model the interaction among individuals in the multiple public facilities found in such systems. These public facilities could be restaurants, shopping malls, public transportation, etc. We study the impact of relative timescales of the Close Contact Time (CCT) and the individuals' stay time in a facility on the spread of the virus. The key contribution is on using queuing theory to model time-spread of an infection in a closed population.

Keywords: M/M/∞ Queue · SEIR Model

1 Introduction

With the significant economic impact of the spread of COVID-19 in the recent times, there is a great emphasis on opening up of the various economic activities. We find an imperative need for study of dynamics of spread of such viruses in a closed facility where several individuals interact.

Several models to understand the dynamics of virus spread mechanisms exist in the literature, most of them having been derived from the SEIR models [3]. These models assume a transmission probability or a transmission rate of the infection. The determination of these transmission probabilities are usually based on empirical data available. The details of the models vary with the dominant mechanism underlying the transmission of the virus. For example, in the case of dengue fever, the models consider the mosquito bite rate [1].

In the case of COVID-19, the virus is known to spread by human-to-human contact, facilitated by droplet or aerosol transmission, leading to recommendation of social distancing norms along with mandates for wearing face masks. Models for the spread of the virus based on the dynamics of the droplets' motion

O. Gervasi et al. (Eds.): ICCSA 2021, LNCS 12949, pp. 576–591, 2021.
https://doi.org/10.1007/978-3-030-86653-2_42

in air exist [4,5]. A recent work [15] provides an interesting view of the probability of virus spread when individuals share a common physical queue, while incorporating the mechanical dynamics of aerosol (droplet) movement. The paper also provides insights into the impact of the waiting time of an individual in the queue on the transmission of the virus.

A generic abstraction of the above-mentioned key factors influencing the probability of transmission of the COVID-19 virus is the *Close Contact Time* (CCT) between an infected and a susceptible individual [2]. A representative figure used for the CCT for the COVID-19 virus spread is roughly 15 min. Such close contacts are achieved in several public facilities that one uses in daily life, for example, restaurants, shopping malls, public transportation systems, sports arena etc.

The CCT itself depends on the manner in which the individuals interact in the facility. For example, at the point of sale queues in a supermarket, the time that individuals spend in the queue depends on the extent of shopping done by the others ahead in the queue. Such systems lead to use of the single server models for the spread of the virus. Several studies exist in the literature that propose the use of such single server models to understand virus spread, not in the context of COVID-19 virus. However, for facilities such as restaurants, public transportation systems, sports arena etc., such dependency is not there, hence multi-server queuing models are also used in the literature.

Mathematical models using the queuing theory approach to capture the interaction among individuals have been considered in the case of other diseases as well. [9] uses the M/G/1 model to get the total cost of the epidemic. [10] uses the M/G/1 processor sharing queues to model an SIR (Susceptible infected Removed) epidemic with detection over time. [11] proposes an approach for SIS (Susceptible infected Susceptible) model extends to an SEIS (Susceptible Exposed infected Susceptible) model while using the M/G/N queue (multi-server) to model the system. [12] focused on Markovian queueing model as a birth-death process with emphasis on epidemiological analysis. [13] uses an M/M/1 model to study the dynamics of the Ebola virus disease. [14] provides a mix of Markovian (stochastic) and a deterministic dynamical system model for the evolution of Hepatitis C.

We use an infinite server queue to model the interaction among individuals in a public facility and provide tractable results assuming that the virus spread depends on the duration of contact between a susceptible individual and other infected individuals, thus incorporating the COVID-19-specific CCT considerations. This queuing model is then applied to study the infection spread in a closed system such as a large campus, community, to model the multiple public facilities found in such systems. We study the impact of relative timescales of the CCT and the individuals' stay time in a facility on the spread of the virus.

The paper is organized as follows: Sect. 2 describes the system model and provides the analytical expressions for probabilities of an individual getting infected in a public facility. Section 3 provides observation where an equivalent reduced load system can be identified to study the original system. Section 4 provides an approach to apply the results of Sect. 2 to a finite-population closed system. Section 5 provides simulation-based validation of the models developed in the paper.

2   Queuing Model and Analysis

Consider a facility, like a common room for a Doctor's waiting room/Beauty Salon/Restaurant/Shopping Mall/Public Transportation/sports arena, etc. Individuals arrive to such a facility, obtain service (stay in the facility for some random amount of time), and depart. Individuals can be either Susceptible or Infected.

Arrival Dynamics: Individuals arrive to the facility according to a Poisson process of rate $\lambda + \omega$.

Infected Population Arrival: The infected population arrives to the facility at a rate λ.

Susceptible Population Arrival: The susceptible population arrives to the facility at a rate ω.

Service Requirements: Each individual, susceptible or infected, requires a service worth a random time that is distributed as $Exp(\mu)$.

Service Mechanism: The service imparted to the individuals by this facility, which controls the time that an individual stays in the facility, is assumed to be that of an infinite server queue. This means that an individual's stay time in the facility is equal to its service requirement, irrespective of the number of other individuals in the facility.

Infection Spread Dynamics: A susceptible individual, while in contact with at least one infected person, gets infected depending on an independent point process of rate ρ, without depending on the number of infected individuals in the facility (as long as there is at least one infected individual). Also, a susceptible individual that gets infected while in the facility cannot start spreading the infection immediately.

Consider the scenario depicted in Fig. 1 that shows a susceptible individual (lowest rectangular block) that sees two infected (red) and two susceptible individuals (green) in the facility on its arrival. This individual is shown to become infected after some time. Another susceptible individual also makes a similar transition to infected, but these transitions are independent across the susceptible individuals, i.e., one susceptible individual becoming infected has no impact on that of other susceptible individuals.

Now, since we have assumed that the infection rate is independent of the exact number of infected individuals as long as *at least* one such individual is present in the facility, we can define the following:

p_0 The probability that a susceptible individual arriving to the facility without any infected individuals leaves **uninfected**.

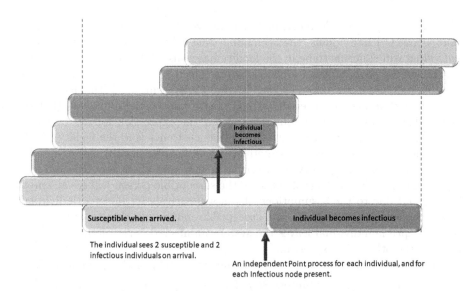

Fig. 1. The dynamics of evolution in the single facility system considered in the paper. (Color figure online)

p_1 The probability that a susceptible individual arriving to a system with exactly one infected individual leaves **uninfected**.

p_{g1} The probability that a susceptible individual arriving to a system with more than one infected individuals leaves **uninfected**.

We can observe that the dynamics of the infected individuals in the facility corresponds to the customers in an M/M/∞ queuing system, [8], with customer arrival rate λ and average service requirement of $\frac{1}{\mu}$. Let $B(\cdot)$ be the distribution of busy period length of such an M/M/∞ system. Recall that λ denotes the arrival rate of the infected individuals only, and does not include the susceptible individuals. Let $\tilde{B}(\cdot)$ denote the distribution of the *remaining* busy period length associated with $B(\cdot)$.

We can write the following equations:

Lemma 1.

$$p_{g1} = \int_{u=0}^{\infty} e^{-\rho u} p_0 e^{-\mu u} d\tilde{B}(u) + \int_{u=0}^{\infty} e^{-\rho u} \tilde{B}^c(u) \mu e^{-\mu u} du. \tag{1}$$

$$p_0 = \int_{u=0}^{\infty} \mu e^{-\mu u} e^{-\lambda u} du + \int_{u=0}^{\infty} p_1 e^{-\mu u} \lambda e^{-\lambda u} du. \tag{2}$$

$$p_1 = \int_{u=0}^{\infty} e^{-\rho u} p_0 e^{-\mu u} dB(u) + \int_{u=0}^{\infty} e^{-\rho u} B^c(u) \mu e^{-\mu u} du. \tag{3}$$

Proof: For a susceptible individual that arrives to the facility and sees more than 1 infected individuals, the remaining time till which at least one infected individual will continue to be in the facility has a distribution of $\tilde{B}(\cdot)$. Note that this is not the probability of all the *existing* infected individuals (that this susceptible individual saw on its arrival) leaving the system, but is the distribution of the first time instant, after the arrival of the susceptible individual, that the facility has no infected individuals left. Now, the expression follows by conditioning on the events where a) the susceptible customer is in the facility till the first time there is no infected individual in the facility, and b) the susceptible individual departs the facility before such the first time there is no infected individual in the facility. In case of event a), the susceptible person will see an empty system and due to the assumptions of Poisson arrivals for infected individuals and exponential stay time for the susceptible individual, the probability that this susceptible individual leaves the system uninfected is p_0. Similar reasoning leads to the other two expressions. □

Let $\hat{B}(s)$ be the Laplace-Stieltjes transform (LST) of the random variable corresponding to the distribution $B(\cdot)$. Then, the LST of $\tilde{B}(\cdot)$ is $\hat{\tilde{B}}(s) = \frac{1-\hat{B}(s)}{sE[B]}$. We thus get the following system of linear equations for p_{g1}, p_1, and p_0.

$$p_{g1} = p_0\hat{\tilde{B}}(\rho + \mu) + \frac{\mu(1 - \hat{\tilde{B}}(\rho + \mu))}{\rho + \mu}. \tag{4}$$

$$p_0 = \frac{\mu}{\lambda + \mu} + \frac{p_1\lambda}{\mu + \lambda}. \tag{5}$$

$$p_1 = p_0\hat{B}(\rho + \mu) + \frac{\mu(1 - \hat{B}(\rho + \mu))}{\rho + \mu}. \tag{6}$$

The LST of the busy period length is [8]

$$\hat{B}(s) = 1 + \frac{1}{\lambda}\left[s - \frac{1}{\int_0^\infty e^{-st-\lambda\int_0^t e^{-\mu u}du}dt}\right],$$

and

$$E[B] = \frac{e^{\frac{\lambda}{\mu}} - 1}{\lambda}.$$

Note the use of $\tilde{B}^c(u)$ in Eq. 3. The above set of linear equations can be solved and we get p_0 and p_{g1} in closed form. The probability that a randomly arriving susceptible individual then leaves the system **uninfected** is

$$p_s(\lambda, \mu, \rho) = \pi_0 p_0 + \pi_{g1} p_{g1} \tag{7}$$

where π_0 is the probability that the corresponding M/M/∞ system is empty, and $\pi_{g1} = 1 - \pi_0$. A randomly arriving susceptible individual will see the facility having no infected individuals with probability π_0.

It is known that [6,8] the distribution of number of ongoing customers in the above M/M/∞ queue is Poisson with parameter $\frac{\lambda}{\mu}$. Thus,

$$\pi_0 = e^{-\frac{\lambda}{\mu}},$$

while noting that this probability is never 0, irrespective of how large (but finite) the value of λ or $\frac{1}{\mu}$ are.

It is also known that the departure process of the individuals from an M/G/∞ facility is Poisson [7]. The susceptible individuals also see the facility as an independent M/M/∞ queue. This property will help us in the upcoming sections where we consider multiple such interconnected queues.

3   A Reduced Load Approximation

Assume that the arrival rate of the susceptible individuals is ω, so that the overall arrival rate of individuals into the facility is $\lambda + \omega$. The facility can then be broken down into two independent M/M/∞ queues where the arrivals into one facility is only of susceptible individuals, while that into another facility is only of infected individuals. The effective arrival rates and average service requirements in these two facilities are provided in the below.

The independent facility that sees only infected individuals is M/M/∞ with an arrival rate of $\lambda + \omega(1 - p_s(\lambda, \mu, \rho))$ and average service requirement of $\frac{1}{\mu}$.

The independent facility that sees only susceptible individuals is M/G/∞ with arrival rate of ω and average service requirement as determined in the following.

The susceptible individuals see an alternating renewal process corresponding to the busy-idle periods of the M/M/∞ queue of the infected individuals. Define the following:

$\Psi(x, u)$ The probability that the total time spent in busy period is less than u units in the total interval of length x, conditioned on the event that the system was in idle period at time 0.

$\Gamma(x, u)$ The probability that the total time spent in busy period is less than u units in the total interval of length x, conditioned on the event that the system was in busy period at time 0.

$Q(x)$ The probability that a susceptible individual got a service of at least x without getting infected while getting this service.

We can then write the following equation

$$Q(x) = e^{-\mu x} \left[\pi_0 \int_{u=0}^{x} e^{-\rho u} d\Psi(x, u) + (1 - \pi_0) \int_{u=0}^{x} e^{-\rho u} d\Gamma(x, u) \right]. \tag{8}$$

The distributions $\Gamma(x, u)$ and $\Psi(x, u)$ are provided in the Appendix using the results of [16].

Let $E[Q] = \int Q(x)dx$, the average service requirement of the customers in the independent queue for susceptible individuals. Then we must have the following work conservation equation satisfied

$$\frac{\lambda + \omega(1 - p_s(\lambda, \mu, \rho))}{\mu} + E[Q]\omega = \frac{\lambda + \omega}{\mu},$$

i.e., the total average work arrival rate in the original system is preserved in the new decoupled system.

4   Application to a Finite Population Model

Let there be N individuals in a system and let there be K such facilities in the system. Let the number of infected individuals at any point in time be $N_I(t)$ and the number of susceptible individuals is $N_s(t)$. Note that $N_I(t) + N_s(t) \leq N$ as there would be recovered or under-quarantine individuals at time t.

Let $\lambda_i(t)$ denote the arrival rate of infected individuals into facility i at time t, and let $\omega_i(t)$ be the rate of arrival of susceptible individuals into facility i at time t. Then, the rate of departure of susceptible individuals out of facility i is given by $\omega_i p_s(\lambda_i, \mu_i, \rho_i)$. We can see that

$$\omega_i = \sum_{j \neq i} \theta_{j,i} \omega_j p_s(\lambda_j, \mu_j, \rho_j).$$

$$\lambda_i = \sum_{j \neq i} \theta_{j,i} \left[(1 - \delta_j)\lambda_j + \omega_j(1 - p_s(\lambda_j, \mu_j, \rho_j)) \right].$$

where δ_j is the probability with which an infected individual is detected, hence quarantined, or recovered. Here $\theta_{j,i}$ is the probability that an individual coming out of facility j joins facility i.

It can be seen that

$$\delta_j = \frac{\zeta}{\mu_j + \zeta},$$

where ζ is the rate of the independent Poisson process that counts the event corresponding to detection or recovery of infection of an infected individual. Note that ζ is assumed to be independent of the facility number.

Figure 2 shows the system considered in this section. The location of a tagged individuals in the system is depicted as it moves across different facilities. The individual is initially susceptible and enters different facilities according to the routing probability $\theta_{i,j}$. While being in these facilities, the tagged individual may have come in contact with other infected individuals. The tagged individual turns infected at some point in time, but continues to visit facilities until it is detected to be infected. On detection, the tagged individual is moved to quarantine facility and comes out as a *removed* individual from such facility as it then no longer contributes to or is affected by virus spread.

Note the presence of *Private* facilities. These are the facilities where the individuals spend most of their time.

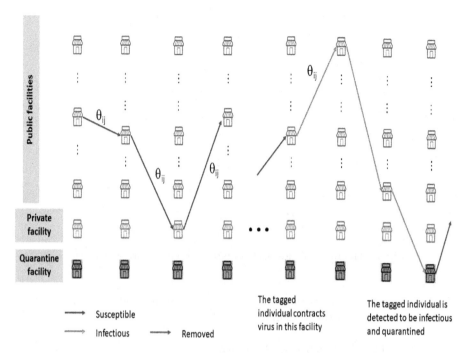

Fig. 2. Time evolution of the movement of the tagged individual across the different facilities in the multi-facility setup considered in the paper.

The rate balance proposed in this section is pictorially depicted in Fig. 3 using the pictorial representation of Fig. 2.

In the limit, $\lambda_i = 0$, $\forall i$, while the behavior of ω_i is to converge to some value, positive or zero. Note that either all $\omega_i = 0$ in the limit, or all $\omega_i > 0$.

Theorem 1. *We can show the following:*

1. $\sum_i \lambda_i > 0 \Rightarrow \Pi_i \lambda_i > 0$
2. *In the limit, when $\lambda_i = 0$, $\forall i$, ω_i are given by the unique solution to the fixed point equation*

$$\omega_i = \sum_{j \neq i} \theta_{j,i} \omega_j,$$

irrespective of the value of ρ_j.

The second result in the above theorem is very important as it shows that the limiting value of ω_i does not depend on the values of ρ_j. This indicates that in a real closed system, the steady state rates will be a constant multiple of the solution of the system of linear equations above. The constant multiple will be dependent on the total number of individuals in the system.

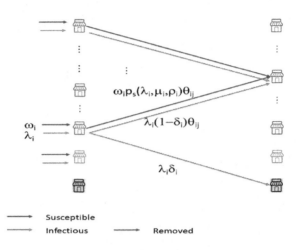

Fig. 3. The rate balance of flows into and out of various facilities.

We thus propose a discrete-time dynamical system model for the evolution of the arrival rates in the system as

$$
\left.
\begin{aligned}
\omega_i(n+1) &= \sum_{j\neq i} \theta_{j,i}\omega_j(n)p_s(\lambda_j(n),\mu_j,\rho_j), \\
\lambda_i(n+1) &= \sum_{j\neq i} \theta_{j,i}\left[(1-\delta_j)\lambda_j(n) + \omega_j(n)(1 - p_s(\lambda_j(n),\mu_j,\rho_j))\right].
\end{aligned}
\right\}
\tag{9}
$$

In this dynamical system model, the time difference between the successive updates is left unspecified.

5   Numerical Validation

5.1   Single Queue

To validate the results of the M/M/∞ queue, namely, Eq. 7, we have developed a slotted-time simulator for the M/M/∞ queue. A numerical computation of Eq. 7 is also implemented. Figure 4 provides the results for the value of $p_S(\lambda,\mu,\rho)$ as ρ is varied. It is seen from the figure that the mathematical model is accurate.

5.2   A Campus Scenario

Consider the scenario of a small closed campus or gated community. We have developed a simulator which follows the SIQR approach in order to validate the numerical results obtained from the dynamical system model of Eq. 9. It is a slot based time iterative model. In order to approximate this model to a continuous time based model and hence to a Poisson process, the slot length is set such that it is minuscule compared to the length of the smallest event in the system. The

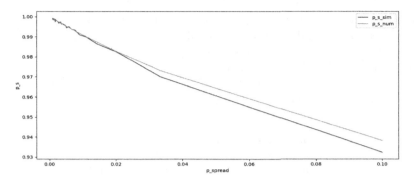

Fig. 4. Comparison of $p_S(\lambda, \mu, \rho)$ from simulation and analysis for different values of ρ.

purpose of this simulator is to simulate closed systems in which there is no inflow or outflow of individuals, rather, there is movement within the system among the various facilities. This system can be approximated to a college campus with facilities being lecture halls, stationery shops, health facilities etc. The simulator allows us to determine the scale of the simulation by allowing us to choose the population of the system. It then returns a graph which describes the time variation of the rates of arrival of susceptible and infected individuals into the public and private facilities.

We first describe the mechanism of movement. The simulator itself divides the facilities into three sets, namely Public Facilities set (ϕ), Private Facilities set (χ) and a Quarantine Facility set(ψ). Initially, ϕ and χ have a uniform distribution of people and a random array of individuals are infected. Movement of individuals between ϕ and χ is determined by a routing probability matrix and a service probability characteristic of each facility representing the average time spent by individuals in the facility. Once the service time for an individual is over, they are moved to another facility. It is assumed that the facility visited by an individual at any given time is independent of the previous facilities they have been in however the absolute probability of going to a facility may be different from that of another facility. In order to reach a steady state, we first run the simulator for 20% of the total time. During this period, there is no infected individual.

Now, we define the mechanism of transmission. Infection can only spread in ϕ. It is assumed that a susceptible individual, if there is at least one infected individual in that facility, is infected according to the next event happening in an independent Poisson process. The rate of Poisson process is the same across all facilities and isn't dependent on the number of infected individuals in a facility as long as there is at least one infected individual in that facility. Once an individual is infected, they can either recover without being detected or be detected and be sent to ψ where they spend a fixed amount of time after which it is assumed that they have recovered. Once recovered, a person can no longer contract the disease.

At the end of the simulation, a graph is returned which gives us the arrival rates of susceptible and infected individuals into ϕ and χ. As all facilities in ϕ have the same properties, it is sufficient to plot the arrival rates for any one of these facilities. By default, the simulation plots the arrival rates into the first element of ϕ. The graph so obtained is then compared with the graph obtained by the dynamical system.

Table 1. The parameters used in simulation.

Parameter	Value
Simulation time	30 days
Number of Public Facilities	20
Number of Private Facilities	1
Spread Close Contact time	15 min
Average recovery time	14 days
Average detection time	50 days
Quarantine period	14 days
Average time spent in a public facility	varied
Average time spent in a private facility	5 h

Routing matrix Θ is applicable to all the individuals, and its (i,j)-th element is given by

$$\Theta_{i,j} = \begin{cases} \theta & i \in \phi, j \in \chi, \\ \frac{1-\theta}{|\phi|-1} & i,j \in \phi, j \neq i, \\ \frac{1}{|\phi|} & i \in \chi, j \in \phi. \end{cases}$$

The slotted simulation has two phases

Settling Phase. In this phase, the system is initialized randomly, i.e., no particular consideration is given to the location of the individuals. Further, all the individuals are marked as susceptible in this phase so that there is no virus spread. The simulator is then allowed to freely run for some time so that the steady state of the arrival rates in the system is achieved.

Infection Phase. Soon after the settling phase is over, the susceptible individuals are marked as infected according to the probability set for the initial infection penetration.

The dynamical system is initialized in the following manner: starting with the number of individuals (N_{init}) in the corresponding slotted simulator, it is assumed that all the individuals are in the private facility. This then indicates that the rate of departure of individuals from the private facility is N_{init} times the service completion rate in this facility. This value is then distributed among the other facilities as per the routing matrix. The dynamical system is then allowed to freely run for some iterations so that these rates ω_i converge. Now the required percentage infection is introduced by reducing ω_i by the desired fraction (say, p_{inf}), i.e., $\omega_i \rightarrow (1 - p_{inf})\omega_i$ and letting $\lambda_i = p_{inf}\omega_i$.

Given the many parameters involved in the system, we have provided results that capture the impact of average service requirement of individuals in the system when compared with the CCT. All the numerical apparatus developed for the presented work are available online at https://github.com/Arzad-iitbh-projects/COVID-19-ICCSA.

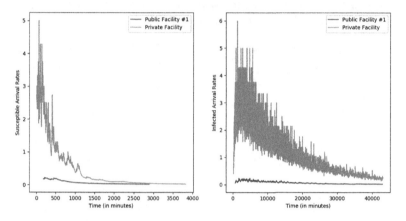

Fig. 5. The arrival rates of susceptible and infected individuals over time from the simulations for an average service time of 15 min.

Figures 5, 6, 7 and 8 show the results from our simulator and the numerical evaluation of the dynamical system. The results are for different values of average time spent in public facilities, 15 min and 30 min. Other parameters are provided in Table 1. It is seen from these figures that the dynamical system models the behaviour of the original system to a satisfactory level, and that the infection spread reduces if the average service time in a facility reduces.

Figure 9 provides the average time to extinction of susceptible population as the average service requirement increases. The results are obtained from the dynamical system model, as well as the slotted time simulator. The vertical time axis has been normalized to provide a unified view for comparison. As noted earlier, the notion of time in the dynamical system model is not accurately mapped to the real system. Note the knee-shaped behavior of this dependency, with the knee point around 15 min, which is the configured CCT in this model.

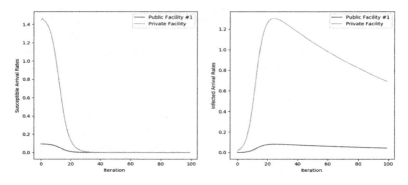

Fig. 6. The arrival rates of susceptible and infected individuals over time as per the dynamical model for an average service time of 15 min.

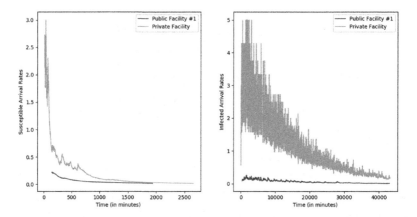

Fig. 7. The arrival rates of susceptible and infected individuals over time from the simulations for an average service time of 30 min.

The rapid decrease in the extinction time in the left-of-knee region is indicative of the very high sensitivity of $p_s(\lambda, \mu, \rho)$ when $\mu > \rho$. This indicates that keeping $\frac{1}{\mu} < \frac{1}{\rho} = CCT$ is not enough to contain the spread of a virus that spreads according to the dynamics studied in this paper. This figure also indicates that the dynamical system model is accurate in providing the qualitative analysis of the evolution of the system.

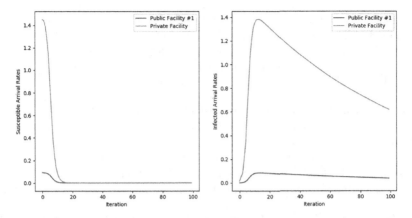

Fig. 8. The arrival rates of susceptible and infected individuals over time as per the dynamical model for an average service time of 30 min.

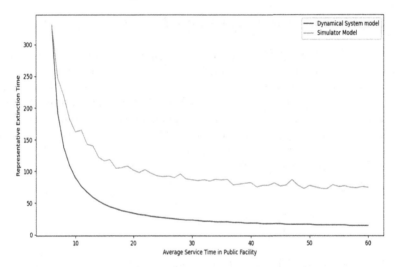

Fig. 9. Dependency of average time-to-extinction of susceptible individuals as a function of average service requirement.

6   Conclusion

This paper uses an M/M/∞ queue to model the interaction of infected and susceptible individuals at a public facility, and provides analytical derivation of the probability that a susceptible individual leaves the facility without infection. This analytical expression is validated against a slotted time M/M/∞ simulator built in-house.

We then use this model to approximate the interaction in a finite population system where the individuals interact at some of the public facilities. The interaction at these public facilities can lead to spread of the virus - this is modeled as an independent Poisson process of a constant rate. A discrete time deterministic model for such a system is provided by assume fast averaging in the related individual queues. A slotted-time simulator is developed for this system.

There are several important directions that we plan to extend the current work to:

1. When moving from the stochastic model to the deterministic one, we currently do not have a direct way of mapping the time. We think that a approach like that of [14] would help us.
2. Extending the model to incorporate M/M/1-like facilities.
3. Extending the model to incorporate an infection spread rate that increases with the number of infected individuals in the facility.

Acknowledgements. The work is supported by the grant under the SERB MATRICS Special COVID-19 Call, India.

A   Expressions for $\Psi(x, u)$ and $\Gamma(x, u)$ of Sect. 4

We will first find $\Psi(x, u)$. Since we are given that the system is Idle at time 0 and using the memoryless property of the exponential distribution, we can claim that the susceptible individual sees an alternating renewal process I_1, B_1, I_2, B_2, I_3, B_3, ... where I_i (resp. B_i is the random variable corresponding to the ith Idle period (resp. Busy period). I_i are independent and distributed as $Exp(\lambda)$, while B_i have distribution of busy period of M/M/∞ queue with arrival rate λ and service requirement $Exp(\mu)$. Let $\Psi_0(x, u) = P(\text{Total Idle Period length} > u|\text{Starting with idle period})$ so that $\Psi(x, u) = 1 - \Psi_0(x, x - u)$. Using [16, Theorem 2.1], we can show that

$$\Psi_0(x, u) = \sum_{n=0}^{\infty} (E^{(n)}(x - u) - E^{(n+1)}(x - u))B^{(n)}(u),$$

where $E^{(n)}(\cdot)$ (resp. $B^{(n)}(\cdot)$ is n-fold convolution of $Exp(\lambda)$ (resp., $B(\cdot)$).

The distribution $\Gamma(x, u)$ is obtained again using [16, Theorem 2.1]

$$1 - \Gamma(x, u) = \sum_{n=0}^{\infty} (C_n(u) - C_{n+1}(u))E^{(n)}(x - u),$$

where $C_n(\cdot) = \tilde{B} * B^{(n-1)}(\cdot)$ for $n \geq 1$, i.e., convolution of $\tilde{B}(\cdot)$ and $B^{(n-1)}(\cdot)$, with $C_0(u) = 1$, $\forall u$.

References

1. Syafruddin, S., Noorani, M.S.-M.: SEIR model for transmission of dengue fever. Int. J. Adv. Sci. Eng. Inf. Technol. **2**(5), 380–389 (2012)

2. Sayampanathan, A.-A., Heng, C.-S., Pin, P.-H., Pang, J., Leong, T.-Y., Lee, V.-J.: Infectivity of asymptomatic versus symptomatic COVID-19. The Lancet **397**(10269), 93–94 (2021)
3. Kermack, W.-O., McKendrick, A.-G.: Contributions to the mathematical theory of epidemics - I. Bull. Math. Biol. **53**(1–2), 33–55 (1991)
4. Ishii, K., Ohno, Y., Oikawa, M., Onishi, N.: Relationship between human exhalation diffusion and posture in face-to-face scenario with utterance. Phys. Fluids **33**(2), 027101 (2021)
5. Balachandar, S., Zaleski, S., Soldati, A., Bourouiba, A.-L.: Host-to-host airborne transmission as a multiphase flow problem for science-based social distance guidelines. Int. J. Multiphase Flow **132**, 103439 (2020)
6. Riordan, J.: Telephone traffic time averages. Bell Syst. Tech. J. **30**(4), 1129–1144 (1951). https://doi.org/10.1002/j.1538-7305.1951.tb03698.x
7. Mirasol, N.-M.: The output of an M/G/∞ queuing system is poisson. Oper. Res. **11**(2), 282–284 (1963)
8. Takacs, L.: Stochastic Processes. Problems and Solutions. Methuen and Co., London (1960)
9. Ball, F., Donnelly, P.: Strong approximations for epidemic models. Stoch. Process. Appl. **55**, 1–21 (1995)
10. Trapman, P., Bootsma, M.-C.-J.: A useful relationship between epidemiology and queueing theory: the distribution of the number of infectives at the moment of the first detection. Math. Biosci. **219**, 15–22 (2009)
11. Hernandez-Suarez, C.-M., Castillo-Chavez, C., Montesinos, L.-O., Hernandez-Cuevas, K.: An application of queuing theory to SIS and SEIS epidemic models. Math. Biosci. Eng. **7**(4), 809–823 (2010)
12. Okoro, O.-J.: On Markovian queueing model as birth-death process. Global J. Sci. Front. Res. Math. Decis. Sci. **13**(11), 22–38 (2013)
13. Dike, C.-O., Zainuddin, Z.-M., Dike, I.-J.: Queueing technique for ebola virus disease transmission and control analysis. Indian J. Sci. Technol. **9**(46) (2016). https://doi.org/10.17485/ijst/2016/v9i46/107077
14. Coutin, L., Decreusefond, L., Dhersin, J.-S.: A Markov model for the spread of hepatitis C virus. J. Appl. Probab. **47**(4), 976–996 (2010)
15. Mathews, S.-S.: A Computer Simulation Study on novel Corona Virus Transmission among the People in a Queue (2020). https://doi.org/10.1101/2020.05.16.20104489
16. Zacks, S.: Distribution of the total time in a mode of an alternating renewal process with applications. Seq. Anal. **31**(3), 397–408 (2012)

A Composite Function for Understanding Bin-Packing Problem and Tabu Search: Towards Self-adaptive Algorithms

V. Landero[1]([✉]), David Ríos[1], O. Joaquín Pérez[2],
and Carlos Andrés Collazos-Morales[3]

[1] Universidad Politécnica de Apodaca, Ciudad Apodaca, Nuevo León, México
{vlandero,drios}@upapnl.edu.mx
[2] Centro Nacional de Investigación y Desarrollo Tecnológico (CENIDET), Departamento de
Ciencias Computacionales, AP 5-164, 62490 Cuernavaca, México
[3] Universidad Manuela Beltrán, Bogotá, Colombia

Abstract. Different research problems (optimization, classification, ordering) have shown that some problem instances are better solved by a certain solution algorithm in comparison to any other. A literature review indicated implicitly that this phenomenon has been identified, formulated, and analyzed in understanding levels descriptive and predictive without obtaining a deep understanding. In this paper a formulation of phenomenon as problem in the explanatory understanding level and a composite function to solve it are proposed. Case studies for Tabu Search and One Dimension Bin Packing were conducted over set P. Features that describe problem instance (structure, space) and algorithm behavior (searching, operative) were proposed. Three algorithm logical areas were analyzed. Knowledge acquired by the composite function allowed designing of self-adaptive algorithms, which adapt the algorithm logic according to the problem instance description in execution time. The new, self-adaptive algorithms have a statistically significant advantage to the original algorithm in an average 91% of problem instances; other results (set P') indicate that when they obtain a best solution quality, it is significant and when they obtain the same or less solution quality, they finish significantly faster than original algorithm. The composite function can be a viable methodology toward the search of theories that permit the design of self-adaptive algorithms, solving real problems optimally.

1 Introduction

It has been seen for sorting problems, depending on the length and order of the sequence, there are algorithms that perform better than the rest [1]. Also, for NP-hard combinatorial optimization problems, the deterministic algorithms are considered adequate for smaller instances of such problems [2]. Intuition says that the difficulty of a problem instance varies with its size: large instances are usually more difficult to solve than smaller ones. However, in practice, recognizing the measuring difficulty only in terms of the instance size implies overlooking any structural property or feature of the instance, which could

© Springer Nature Switzerland AG 2021
O. Gervasi et al. (Eds.): ICCSA 2021, LNCS 12949, pp. 592–608, 2021.
https://doi.org/10.1007/978-3-030-86653-2_43

affect the problem complexity [3] and the algorithm performance. For example, classification problems show that some learning algorithms perform very well on certain problem instances depending on a set of specific features [4]. The scientific community of different disciplines, such as combinatorial optimization, machine learning, artificial intelligence and other fields of knowledge has worked for describing and analyzing the experimental relation between problem-algorithm, with the objective of solving a real problem optimally. However, its analysis has been conducted, in most cases, in descriptive and predictive understanding levels (for a major compression of this concept [5]) and it is necessary deepening more in this objective, in the explanatory understanding level, answering why the relation of certain problem instances and an algorithm produces best solutions (algorithm is very good) and why it is not good with other problem instances. Under the scope of this research (combinatorial optimization area and search algorithms), this paper contains: previous questions and formulations of reviewed literature, so too, the phenomenon is formulated as a question and a formal problem statement in the explanatory level, using and supplementing the Rice's formal nomenclature (Sect. 2); a proposed composite function to solve the stated problem (Sect. 3); a framework for the performance of the proposed composite function over the One Dimension Bin-Packing problem and Tabu Search algorithm (Sect. 4); Development of the proposed composite function, using a instances set P (Sect. 5), the discovered knowledge is used for answering how and why certain problem instances (describing structure and problem solutions space), algorithms features (describing operative and searching behavior); and the algorithm logical design contribute toward a better relation between problem-algorithm (in majority cases of reviewed literature, not all information are taken into account at the same time, problem, algorithm, logical area). A new self-adaptive search algorithm is designed for each case of study. Section 6 describes the results of self-adaptive search algorithms over instances sets P and P' ($P' \neq P$)). Conclusions and future works are drawn in Sect. 7.

2 Reviewing State of Art and Setting the Problem Statement

Using and supplementing the Rice's formal Nomenclature,

$P= \{x_1, x_2,..., x_m\}$ a set of problem instances or space for analysis.

$F=$ the problem features space generated by a description process applied to P.

$A= \{a_1, a_2,..., a_n\}$ a set of algorithms.

$Y=$ the performance space, it represents the mapping of each algorithm to a set to a set of performance metrics.

$C= \{C_1, C_2,...,C_n\}$ a partition of P, where $|A|=|C|$.

$W = \{(a_q \in A, C_q \in C) | Y_{aq,x} > Y_{\alpha,x} \cdot \forall \cdot \alpha \in (A - \{a_q\}), \forall \cdot x \in C_q\}$ is a set of ordered pairs (a_q, C_q), where each dominant algorithm $a_q \in A$ is associated with one element C_q of partition C, because this gives the best solution to partition C_q, considering a set of performance metrics mapped in set Y.

$L=$ the algorithm features space.

In descriptive traditional level, the next first research question arises from experimental relation between problem-algorithm:

1. What is the performance of algorithm $a_q \in A$ to solve the problem P?

For this, a set of algorithms A is run over a set of instances P. The general performance of each algorithm is measured by some performance metric $y \in Y$ in order to obtain a quantifiable value that could be used for performing a comparison of algorithms by means statistical analysis (statistical tests The Sign, Wilcoxon and Friedman tests, among others) or tabular analysis or graphical analysis; after that performing a results interpretation. One classic example of related work in this understanding level is [6]. Nevertheless, the results of the solution algorithms on an instance set of a problem could be incorrectly interpreted, this is, one might expect that there are pairs of search algorithms a_q and α such that a_q performs better than α on average, even if outperforms a_q at times. Such expectation could be incorrect. Wolpert describes two No Free Lunch (NFL) theorems in [7]. In general terms, it establishes that for any algorithm, a high performance over a set of problems is paid in performance over another set. The existence of instances subsets for specific problem and an algorithm for each subset is suggested by NFL theorems. For the purpose of exemplification in the reviewed literature, other classic examples of related works identified different performances of algorithms on different problem instance sets, based on the construction of an association table, where each set had one or several similar features in the context of the problem structure description [8, 9]. A few of other related works are [1, 2, 4, 10, 11]. The above indicates that one algorithm can be associated to a problem instances subset, where it is the one that solves these instances in the best way possible, for a specific problem domain. The phenomenon observed could be described by means of some set W that includes pairs in the form (a_q, C_q), where instances subset C_q better correspond to a_q for instances set P for a specific problem (see nomenclature for a major description).

In predictive level, a general research question arises:

2.   What is the best way to learn W (pairs (a_q, C_q)) in order to predict algorithm a_q that will give the best solution for a new instance from a problem?

The above question needs to consider information from the experimental relation between problem-algorithm, which is significant and has a predictive value. If this question is answered, the solution to the algorithm selection problem can be found. The algorithm selection problem (ASP) is originally formulated in [12], which is stated as:

For a given problem instance $x \in P$, with features $f(x) \in F$, find the selection mapping $S(f(x))$ into algorithm space A, such that the selected algorithm $\alpha \in A$ maximizes the performance mapping $y(\alpha(x)) \in Y$.

It can be said that the phenomenon mentioned in this paper, can be analyzed and learned in the predictive understanding level of an implicitly manner when problem ASP is being solved. ASP was generalized through different research disciplines in [13], where its solution is important. Two known approaches that are utilized by related works in solving problem ASP are algorithm portfolio and supervised learning. In the case of algorithm portfolio, the algorithm performance is characterized and adjusted to a model (model-based portfolio) either by a regression model [14] or a probability distribution model [15, 16]; other related works have also shown that some problem features are considered in the building of a model (feature-based portfolio), for example, applying supervised learning [17, 18]. In this case, many related works exhibit learning patterns (corresponding to set W in the context of this paper) from data by means of a supervised algorithm and use them for predicting the best algorithm for an unseen problem instance. Examples: Case-base reasoning [19, 20], decision trees [21, 22], Neural Networks [23] and Random Forest [24–26]. A discussion of machine learning methods applied to ASP problem can be found in [27]. Nevertheless, the principal disadvantage is that the phenomenon identified in the predictive understanding level is analyzed and studied without being fully understood. Two principal reasons: the information considered in analyses is only derived from a problem, or an algorithm, or a logical design (initializing parameter); and the built models are found to be difficult in interpreting the acquired knowledge by nature of its own structure; these are used for predictions.

In the explanatory understanding level, there are different guidelines. For example, some related works focused their analysis on problem difficulty, considering the problem structure, significant features or parameters, identifying important values from them to determine when problem is difficult and when is easy (known as Transition Phase Analysis) through the use of graphical or statistical analysis or unsupervised learning [23–25, 28, 29]. Other related works focused on algorithm performance, some deeply on the searching behavior, considering metrics that measures the trajectory, identifying when it is flat or rugged (there are fluctuations), determining whether the problem is difficult or easy for algorithm (Known as Landscape Analysis) [30–32]; others deeply focused on algorithm logical design [33, 34]; some focused on both [30, 35], for example, obtain important explanations which permit to configure the algorithm in a way that it can produce the better results. However, these explanations are not very clear, in the regard which kind of problem instances will produce better results. What are the specific features of the problem structure that help the configured algorithm better adjust to these problem instances? In [36], this information is considered important in order to adapt the algorithm logical design to the problem structure. A few related works focused on problem and algorithm [37–40], developing some visual tool or performing a graphical, or statistical, or data exploratory, or causal analysis. The reviewing of this literature indicates much effort has been taken to characterize and analyze the experimental relation of problem-algorithm under different understanding levels. However, we believe that it is necessary to pave the way in the comprehension of explanatory understanding level with a starting point, something very essential, simple, and important. Therefore, considering past efforts of the reviewed literature, the experience for previous works [41, 42], and continuing with observed in descriptive and predictive understanding levels, a simple research question arises for a domain specific:

3. Why does a problem instance subset C_q correspond better to an algorithm a_q than other instances in a specific problem domain?

In order to formulate this question as a formal problem, should be considered: as first instance, all significant information from problem (structure, solutions space) and algorithm (operative and searching behaviour), during and after execution, limitations of explanatory and predictive levels; as second instance, a methodology as guide to help obtain a formal model that can discover latent knowledge and explain the phenomenon in question for a specific problem. Following a step beyond to algorithm selection problem

(ASP), considering the above, and continuing in the improving of previous works, the phenomenon could be formulated as the next statement.

> For a set of algorithms A applied to a set of problem instances P, with problem features F, algorithm features L, the algorithms performance space Y, the algorithms performance partitions W, according to Y and an ordered pair $(a_q, C_q) \in W$; find the composite function $E(a_q, P, A, F, L, Y)$ that discovers an explanation formal model M, such that M, represents the latent knowledge from relations between features that describe: the problem features F; interest algorithm L; and provides solid foundations to explain, why certain problem instances, being the partition C_q correspond better to interest algorithm a_q, according to performance space Y, and why other partitions $(C_q)^c$ do not correspond to algorithm a_q.

3   Proposed Solution

The solution to the above problem statement, is to discover a formal model that can acquire latent knowledge, structured in some way as cause-effect relations from problem, algorithm features, which can help explain such formulation. The process is known as the discovery of causal structure of data [43] (causal model). A causal model can be defined as a causal Bayesian network [43]. It is described by expression 1.

$$M = (V, G, Z) \tag{1}$$

Specifically,

- $V = \{v_1, v_2,..,v_n\}$ is a set of observed features.
- G is a directed acyclic graph with nodes corresponding to the elements of V that represents a causal structure (V, EC); i.e.,
  $EC = \{EC_1, EC_2,..., EC_n\}$, where each $EC_i \in EC$ is a set of ordered pairs,
  $EC_i = \{(v_i, y_1), (v_i, y_2), ..., (v_i, y_n)\}$, it is
  $EC_i = \{(v_i \in V, y_k \in V)|v_i \neq y_k, y_k$ is a direct cause of v_i relative to V and there is a directed edge from y_k to v_i in $G\}$
  $Pa(v_i) = \{y_k \in V \mid (v_i, y_k) \in EC_i\}$ is a set of all direct causes of v_i.
- $Z = P(v_i = j \mid y_1 = \alpha, y_2 = \beta, ..., y_p = \gamma)$, is a function of conditional probability of v_i in the range of values j given the direct causes of v_i $\{y_1, y_2,...,y_p\} \in Pa(v_i)$, which are in the ranges of values $y_1 = \alpha, y_2 = \beta,..., y_p = \gamma$.

3.1   Composite Function E

In general terms, the composite function E (see Fig. 1) consists of analyzing the experimental relation between the problem (instances set P) – algorithm (interest algorithm $a_q, a_q \in A$) considering the space of features from problem F, the space of features L and performance Y from algorithms during execution, for discovering latent knowledge about this relation (represented by explanation model M).

The domain of proposed composite function E (Expression 2), are parameters: the interest algorithm a_q, a set of algorithms A, applied to a set of problem instances P, the

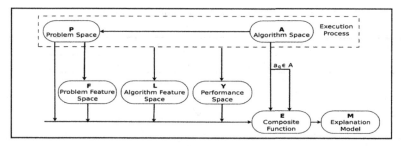

Fig. 1. General diagram of the composite function

set F obtained by f (P), the set L obtained by $l(A)$, the performance space Y, obtained by $y(A)$. The functions f and l perform a description process, before and during execution of algorithms, to obtain features that represent information about set P and A.

$$(E \cdot fz \cdot fg \cdot fv \cdot fd \cdot fc)(a_q, P, A, F, L, Y) \equiv E(fz(fg(fv(fd(fc(a_q, P, A, F, L, Y)))))) \tag{2}$$

The function y evaluates the algorithms A, by means a set of performance metrics. The acquisition of latent knowledge from experimental relation between problem-algorithm is formalized by means of the proposed composite function E, which, evaluated in each iteration, obtains the values of a set of significant features, causal relations between these, and estimations, represented by the sets V, G, Z (codomain-causal model M).

4   Framework for Proposed Composite Function E

The instance sets P and P' ($P \neq P'$) were randomly selected (324) from Beasley's OR-Library [44], the Operational Research Library [45]. The objective of each case of study is to explain the description process and the composite function of the latent relation between problem (One Dimension Bin-Packing - BPP) and algorithm (Tabu Search); an insight into the problem structure, problem solution space, the algorithm logical design, its operative behaviour, and behaviour during its searching and performance. Due to the above, four versions of Tabu Search algorithm were implemented, where each one had a specific logical design (Table 1). The methodology for initializing control parameter (PM) is applied to size of Tabu list (*nLTabu*). The static procedure is to set it as 7 [46]. The dynamic procedure is to set it as $\sqrt{n}$, where n represents the number of the objects or items of the problem instance. The methodology for the generation of an initial solution (IM) can be conducted by a random or deterministic procedure. The methodology for building the neighbourhood of a solution (NM) can achieved through one or several methods, which were proposed in [47]. The candidate list (LCANDI) size was fixed to 4*(*nLTabu* * 0.25) for all the study cases. As well as, the methodology to stop the algorithm execution (SM) was the same; after 4000 iterations or there was no improvement in the solution. Table 2 shows the study cases.

Table 1. Set of algorithms A

Variants	PM		IM		NM	
	Static	Dynamic	Random	Deterministic	One	Several
a_1	✓		✓		✓	
a_2	✓		✓			✓
a_3		✓	✓		✓	
a_4	✓			✓	✓	

Table 2. Cases of study

Case of study	Algorithms	Methodology
1	a_1, a_2	Building Neighborhood (NM)
2	a_1, a_3	Initializing Control Parameter (IM)
3	a_1, a_4	Generating Initial Solution (PM)

5 Performing Composite Function E

The function $f(P)$ performs description process for the instances set P, where the set F is obtained. After that, the set of algorithms $A(a_1, a_2, a_3, a_4)$ is applied to solve the problem instances set P. During the search and solution process, the functions $l(A)$ and $y(A)$ perform description process to obtain the algorithm features space, set L and set performance space Y. This process $(f(P), l(A), y(A))$, for all cases of study, is described in greater detail only for those features that were significant in next sections.

5.1 Problem Instances Structure Description: Function $f(P)$

There are three features, (b, d) [21] and cu proposed in this paper; b describes the proportion of the total weights of the objects that can be assigned to one container; d describes the dispersion of the quotient between the object weight and the container capacity; cu is the kurtosis of object weights (w weights and de standard deviation).

$$cu = \frac{\sum_{i=1}^{n} (w_i - \overline{w})^2}{de^4} \tag{3}$$

The problem solution space for each instance, os, is described in past works [38, 41, 42]. It is the variability of ms randomly generated solutions ($ms = 100$ produced better results). The codomain of function f is the set F (expression 4), where rows represent the problem instances and columns are the values of these features.

$$F = \{\{b_1, d_1, cu_1, os_1\}, \{b_2, d_2, cu_2, os_2\}, \ldots, \{b_m, d_m, cu_m, os_m\}\} \tag{4}$$

5.2 Algorithm Behavior Description: Function $l(A)$

The algorithm operative behaviour is described by features (nn, fn, vf), proposed in past works [38, 41, 42]. The number of neighbours built by algorithm during its search process per instance is given by nn. The number and variability of feasible solutions are given by fn, vf. The algorithm searching behaviour is described by features pn (number of inflection points), vn (number of valleys), vs (size of valleys) proposed in past works [38, 41, 42] and vd proposed in this paper. These features are obtained from the algorithm searching path. The searching inflections are the changes in the direction of fitness function from two consecutive solutions during one algorithm run; pn is the average of these for all algorithm runs (16); vn is concerned if there exists a searching pattern that refers to our concept, Valley. It is considered when there is a sequence major to sm solutions, where their fitness function values keep on decreasing ($sm = 6$ indicated be significant in past works). Then, the feature vn is the average of Valleys identified from algorithm runs. The inflection point located in an identified Valley is considered as the location point, for example one run has location points p_1, p_2, p_3 and p_4; the distance between each point is calculated, dd_1, dd_2 and dd_3. The standard deviation of these is calculated (expression 5). The average of Valleys dispersion for all algorithm runs is calculated, vd. The set L is built with the specific order as Expression, 6. Here, $L_{1,1}$ means algorithm a_1 for problem instance x_1 has the elements, $nn_{11}, fn_{11}, vf_{11}, pn_{11}$, $vn_{11}, vs_{11}, vd_{11}$ (algorithm behaviour features) and so on.

$$vd_{run} = \sqrt{\frac{\sum_{i=1}^{pn-1}\left(dd_i - \overline{dd}\right)^2}{pn - 2}} \tag{5}$$

$$L = \left\{ \begin{array}{l} \{nn_{11}, vf_{11}, pn_{11}, vn_{11}, vs_{11}, vd_{11}\}, \ldots, \{nn_{1m}, \ldots, vd_{1m}\} \\ \{nn_{21}, vf_{21}, pn_{21}, vn_{21}, vs_{21}, vd_{21}\}, \ldots, \{nn_{2m}, \ldots, vd_{2m}\} \\ \{nn_{n1}, vf_{n1}, pn_{n1}, vn_{n1}, vs_{n1}, vd_{n1}\}, \ldots, \{nn_{nm}, \ldots, vd_{nm}\} \end{array} \right\} \tag{6}$$

5.3 Performance Space Description: Function $Y(A)$

The function, y, evaluates the algorithm performance according to metrics *time* and *quality*. The metric *time* is the total of feasible and infeasible solutions built during algorithm execution. The metric *quality* is the ratio between found solution and theoretical solution [41, 42]. The codomain of the function, y, is the set, Y (performance space) which is built with the specific order as Expression 7. Here, $Y_{1,1}$ means algorithm a_1 for problem instance x_1 has the elements, $quality_{11}$ and $time_{11}$, $Y_{1,m}$ means algorithm a_1 for problem instance x_m has the elements *quality*; *time*, and so on.

$$Y = \left\{ \begin{array}{l} \{quality_{11}, time_{11}\}, \ldots, \{quality_{1m}, time_{1m}\} \\ \{quality_{21}, time_{21}\}, \ldots, \{quality_{2m}, time_{2m}\} \\ \{quality_{n1}, time_{n1}\}, \ldots, \{quality_{nm}, time_{nm}\} \end{array} \right\} \tag{7}$$

5.4 Discovering Knowledge: Functions fc, fd $(TC), fv(D), fz(G)$

The proposed composite function, E, is applied to one algorithm of interest a_q for each case of study (1, 2, 3). The algorithms, a_2, a_3, a_1 were randomly selected. The function, fc identifies the set W, considering performance space Y (based on *time* and *quality* metrics). After that, the performance scope of interest algorithm a_q is obtained, considering set W. The codomain of function fc is described by set S. Each value indicates the scope of interest algorithm for each problem instance (set P), 1 was the best, 0 otherwise. For example, $S = \{0, 1,...,1\}$ means that interest algorithm had scope: 0 for instance 1, 1 for instance 2, 1 for instance m and so on. A sets family $SF = \{F, L, Y, S\}$ is built for interest algorithm a_q and is represented by dataset TC, where $TC = \cup SF$; the tuples are instances and columns are features that describe: the structure and problem solutions space $F = f(P)$; the operative and searching behaviours of interest algorithm, obtained from set L; performance space of the interest algorithm, obtained from set Y (*time* and *quality*); performance scope from set W, value from set S. The function, fd, first normalizes the values of each by means of method min-max; values that lie within the closed interval, $[0, 1]$. After that, the method, MDL [48], is performed to discretize the values. The codomain of function fd is the discretized dataset, D. The function, fv, performs general, graphical and variance analyses of the features from dataset D for selecting the most significant. The general analysis creates bar plots, where the frequencies of the values of the features are analysed with respect to the scope (value s) of aq. The metric *quality* from dataset TC did not assume a normal distribution; therefore, for the graphical and variance analyses, it was transformed using methods of logarithm or Box-Cox, using values 2 or -2 for λ. The graphical analysis creates box plots for each feature (identified in general analysis) with respect to metric *quality*, in order to identify features that in influence it in terms of variation and locality. Finally, the function, fv, performs an analysis one variance (ANOVA), with a confidence level of 95% for each feature (identified in graphical analysis) with respect to the *quality* metric. The function, fv, did not find significant features in the study case 3; for other cases of study, 1 and 2, the codomain of function fv is the significant dataset V_1 with proposed features. For example, in case study 1, the dataset, V_1, is formed by features $b, os, nn, vf,$ pn, vn, vs, and scope of interest algorithm a_2 (S). So too, with the objective of considering metrics known and used by the scientific community (describing the algorithm searching behaviour), the auto-correlation coefficient, (ca), and the auto-correlation length, (al) ([49]), and highlight the utility of our proposed features $(pn, vn$ and $vs)$, another dataset, V_2, is built with features b, os, nn, vf, metrics ac, al, and scope of interest algorithm a_2 (S). The datasets V_1 (by means function fv) and V_2 are built in case of studies 1 and 2. The function fg performs the process of learning a causal structure (algorithm PC [43]) with a confidence level of 95% for datasets V_1 and V_2 in case studies; the causal inference software, HUGIN (Hugin Expert, www.hugin.com) was used.

Figure 2 shows the causal structures: $a) fg \rightarrow G_1$ and $b) fg \rightarrow G_2$ from datasets V_1 and V_2 for these cases of study. It is important to emphasize that the causal structure, G_1, in these cases represents clearly the direct causes (problem and algorithm significant features) of performance scope for interest algorithm a_q in performance space Y, in terms of set W. For case study 1, it is evident that the causal structure, G_2, did not yield relevant information about direct causes. In case study 2, G_2 did not yield relevant information about direct causes for algorithm performance scope with respect to algorithm behavior during its searching. These structures (G_1) were considered for the next analyses. Also, Fig. 2 shows the intervals of direct causes, obtained previously by function fd. Continuing with the composite function, E, the function, fz, referring to parameter learning algorithm Counting [43], estimates the intensity of causal relations (identified in structures G_1), see Table 3, the codomain is set Z. The codomain of composite function E for each study case are the sets V_1, G_1 and Z (causal model M). The problem instance set, P', was used as an input for each causal structure G_1 to obtain the prediction accuracy percentage, using another causal inference software NETICA (Norsys Corporation), where the obtained percentages were %78.04 and %70.37, respectively.

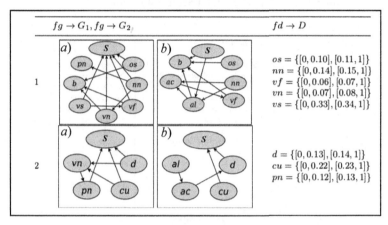

Fig. 2. Discovering knowledge

Table 3. Significant features, casual relations and estimations

	Functions	
	$fv \rightarrow V_1$	$V_1 = \{b, os, nn, vf, vn, vs, pn, S\}$
1	$fg \rightarrow G_1$	$EC = \left\{ \begin{array}{c} \{S, os\}, \{S, nn\}, \{S, vf\}, \{S, vn\}, \{vs\}, \\ \{b, os\}, \{b, fv\}, \{b, vn\}\}, \{(pn, nn)\}, \\ \{(vf, vs)\}, \{(vn, nn)\} \end{array} \right\}$
	$fz \rightarrow Z$	$P(S=1 \mid os = 2, nn = 2, vf = 2, vn = 2, vs = 2) = 92\%$ $P(S=0 \mid os = 1, nn = 1, vf = 1, vn = 1, vs = 1) = 99\%$
2	$fv \rightarrow V_1$	$V_1 = \{d, cu, pn, vn, S\}$
	$fg \rightarrow G_1$	$EC = \left\{ \begin{array}{c} \{(S, d), (S, cu), (S, pn)\}, \{(vn, d), (vn, cu)\}, \\ \{(pn, vn)\} \end{array} \right\}$
	$fz \rightarrow Z$	$P(S=1 \mid d = 2, cu = 1, pn = 2) = 99\%$ $P(S=0 \mid d = 1, cu = 2, pn = 1) = 76\%$

5.5 Analyzing Acquired Knowledge and Self-adapting Algorithms

The results from case studies 1 and 2 are reviewed deeply in Fig. 3 (Analysis 1 and 2). So too, the logical design of the new Tabu Search Self-Adapting Algorithms is shown. On the other hand, another interesting result was obtained in case study 3, where the Tabu Search algorithm was distinguished by methodology in order to generate the initial solution. Though the function, fv, of composite function E could not discover significant features to build a causal explanation model, it is important to highlight the fact that a knowledge was obtained as well. One possible interpretation of this result may be that the method used to generate the initial solution does not impact the algorithm performance, according to set Y, in solving problem instances. One similar result was observed in [31] for another optimization problem and Tabu Search algorithm.

Tabu Search Self-adaptive Algorithms (a_{e1} and a_{e2})
Begin
1 Calculates feature os from input parameters; a_{e1}
2 $nLTabu = 7$ // *Methodology for initializing input control parameter (PM)*
3 Calculates d and cu from problem parameters; a_{e2}
4 // *Methodology for initializing input control parameter (PM)*
5 **If** (value of d belongs to interval 2 and value of cu belongs to interval 1)
6 **Then** $nLTabu = \sqrt{n}$; **Else** $nLTabu = 7$; // *Size of Tabu list*
7 $tn = nLTabu$; // *Tenency of one solution in Tabu list*
8 x^{*}=Generate randomly a feasible solution initial; $x=x^{*}$; $LCANDI=\{\varnothing\}$;
9 $LTabu=\{\varnothing\}$;
10 **Repeat**
11 **Begin** a_{e1}
12 // *Methodology for building neighborhood (NM)*
13 **If** (value of os belongs to interval 1) **Then**
14 $LCANDI = N(x, LTabu)$; // Building with one method
15 **Else** $LCANDI = N(x, LTabu)$; // Building with several methods
16 $LCANDI = N(x, LTabu)$; // Building with one method a_{e2}
17 y = the best solution of $LCANDI$ and $y \notin LTabu$;
18 $LTabu = LTabu \cup \{(y, tn)\}$;
19 **For each** solution $e \in LTabu$ **do**
20 Tenency of e is decremented;
21 **If** the tenency of e has expired **Then** $LTabu = LTabu - \{(e, tn)\}$;
22 **End**
23 If $fo(y) > fo(x^{*})$ **Then** $x = y$; $y^{*}=y$;
24 **End**
25 **Until** 4000 iterations or there is no improvement
26 **return** y^{*};
27 **End**

Analysis 1. The logical design of algorithm a_2 adjusts better ($S = 1$) to the problem instances description (partition C_2) in terms of $os = 2$. Its searching behavior was flexible to find a solution close to theoretical, as it could enter and leave from an identified valleys number $vn = 2$ with a valley average size in the second interval ($vs = 2$). Algorithm a_2 wins in *quality* in the majority of instances (214 from 216) and wins in *time* 2 instances (when quality is the same). The logical design of a_2 (several methods for searching) is responsible for its high cost in its performance, in terms of *time*; it lost in 93 out of 108 problem instances, as it is not necessary to intensify the search a lot for instances with a description in terms of $os = 1$. The logical design of a_2 does not adjust ($S = 0$) to this problem description (partition C_2)^c. The algorithm, a_1, had an advantage in this situation because of its logical design, one method is considered for the searching. It is limited to finding solutions and corresponds better to this kind of problems. This knowledge gave guidelines to design a new Tabu Search self-adaptive algorithm a_{e1}.

Analysis 2. The logical design of a_3 adjusts better ($S = 1$) to problem instances description (partition C_3) in terms of $d = 2$ and $cu = 1$. This permitted to the logical design of algorithm, a_3, store more Tabu solutions. It is more restrictive in accepting a solution compared to algorithm a_1 (Tabu list size - 7); it has a searching behavior with inflection in the second interval ($pn = 2$), ranging between solutions that cannot be in the Tabu list; and it gives a solution more close to theoretical. It is found that algorithm a_3 wins in *quality* in more of half instances (164 from 278) and wins in *time* 114 instances (when *quality* is the same). However, for instances whose problem instances description (in terms $d = 1$ and $cu = 2$), the solutions that could be generated from these will be not very different; the searching behavior of algorithm a_3 had an inflection ($pn = 1$) in the first interval. Therefore, these solutions can be Tabu because the Tabu list of algorithm a_3 is long; it loses ($S = 0$) in terms of *time* in more of half instances (27 out of 46). Unlike the algorithm, a_1, which has an advantage in this situation, it is not very restrictive with the solutions generated, as it handles a small Tabu list. The acquired knowledge allowed the designing for a new self-adaptive algorithm a_{e2}.

Fig. 3. Analysis of case studies 1, 2 and Tabu Search Self-Adaptive Algorithms

6   Results Analysis

In the reviewed literature, there was no related work with the same circumstances for comparing performance of individual results. An example of this can be an improved Tabu Search algorithm for BPP. Therefore, the performance for the self-adaptive algorithms (a_{a1}, a_{a2}) was compared against that of the original analyzed algorithms (a_2, a_3), it is to say $a_{a1}-a_2$, $a_{a2}-a_3$. The comparative analysis of performance results from the algorithms was performed in two consecutive phases of analysis. The first analysis phase consisted on determining the total scope of new self-adaptive algorithms a_{a1} (287) a_{a2} (301); thereafter determining their total scope percentage, example $287/324 = 89\%$ for a_{a1}, a_{a2} (93%). Analysing the partitions $C_{a_{a1}}$ and $C_{a_{a2}}$ in more detail, a_{a1} wins 159 and a_{a2} wins 163 instances in *quality*, 128 and 138 in *time*, where the *quality* is the same for both algorithms (see Fig. 4).

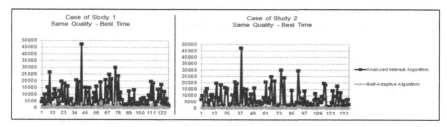

Fig. 4. Times of analyzed algorithm and self-adaptive algorithm (same quality)

The time differences are too big, it is not necessary to apply a statistical test. The self-adaptive algorithms finishing faster than the interest algorithm (a_q) analyzed in study cases. Due the above, the objective of second analysis phase is to verify the values of *quality* metric, specifically when self-adaptive algorithm has the best *quality* (159-case 1, 163-case 2). In this sense, for study cases 1 and 2, a_{a1} had best time in 80 out of 159 problem instances and a_{a2} had best time in 95 out of 163 problem instances.

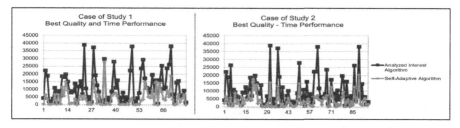

Fig. 5. Times of analyzed algorithm and self-adaptive algorithm (best quality)

Figure 5 shows these times. The metric values *time* for each one of algorithms do not assume a normal distribution. Thus, a nonparametric statistical test of two dependent samples is applied (the two sample two-side wilcoxon signed rank test) for significance levels 95% and 99%. The Dataplot statistical software (www.itl.nist.gov) was used for this test. The test statistic was 7.67 and 8.46 for study cases. The null hypothesis (equal means) is rejected (critical values 1.96, 2.57) and it means that there is a significant difference between times. The self-adaptive algorithms (μ_2) finishing faster than the original algorithm (μ_1) analyzed in study cases. Continuing with the analysis, it is necessary to verify if there is a statistically significant difference between the means of metric *quality*. The values of this metric do not assume a normal distribution. Thus, the same statistical test for significance levels was applied (test statistic 10.15 and 9.92). The null hypothesis is rejected, there is a significant difference in terms *quality*.

6.1 Other Results

The analysed original algorithm and self-adaptive algorithm from cases of study 1 and 2, were executed over set P' ($P' \neq P$). The self-adaptive algorithms had better *quality* or *time* in 166 and 156 problem instances of set P' than analysed interest algorithm, respectively. Figure 6 shows the *time* of algorithms when they have the same quality. The same wilcoxon statistical test was applied to *time* differences. The test statistic was 9.27 and 8.37 for both cases. The null hypothesis is rejected for significance levels 95% and 99% (critical values 1.96, 2.57). The self-adaptive algorithms finishing faster than the analysed interest algorithm. Also, Fig. 7 shows the times when self-adaptive algorithm has less *quality* than analysed interest algorithm. The *quality* difference average was 0.025 (very small) for 158 problem instances out of 324 and the self-adaptive algorithms finishing faster than the interest algorithm in most cases over set P'.

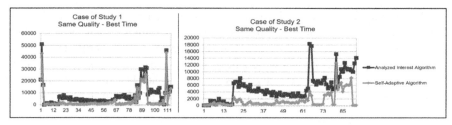

Fig. 6. Same quality of both algorithms and best times of self-adaptive algorithm

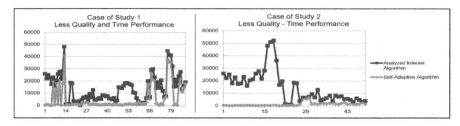

Fig. 7. Less quality and best times of self-adaptive algorithm

7 Conclusions

Most of the literature consulted from combinatorial optimization area has been focusing only on problem information or algorithm information or, rarely, on both in order to describe the experimental relation between problem-algorithm. Furthermore, in the analysis of this relation, has been identified in the descriptive understanding level that certain problem instances correspond better to a certain algorithm than other. This phenomenon has been analyzed and learned implicitly in the predictive understanding level (algorithm selection problem). However, it has not been understood at all for a specific problem. This paper goes beyond the predictive level, covering limitations identified. The phenomenon is formally formulated as question and problem at the explanatory understanding level. The composite function, E, is proposed to solve such formulation and performed for a specific domain (One Dimension Bin-Packing problem and Tabu Search algorithm) over an instance set P. A description process for problem structure and space, for algorithm performance and the operative and searching behaviors of the algorithm during execution (significant features) was proposed to build the domain of the composite function. Also, the metrics known by scientific community were used, but these were not significant in the analyses. Three important logical areas were contemplated. The knowledge acquired by the proposed composite function allowed the solving of the stated problem, understanding the phenomenon, answering the "how" and "why" for certain problem instances, algorithm significant features and the algorithm logical design contribute toward a better relation between problem-algorithm. It was applied to design self-adaptive algorithms and improve the performance considering the causal relations according to problem structure or space. On average, a 91% significant advantage of Tabu Search self-adaptive algorithms was obtained over analyzed original algorithms. Other results over set P' showed that when they obtain the same or less quality of solution, they finish significantly faster than analyzed interest algorithm and the quality difference is very small. As future work is to explore generalized features. The proposed composite function E could act as a guideline to find latent knowledge from relation problem-algorithm of other problem domains and search algorithms; this permits the adapting of logical design as well as operative and searching behaviors of an algorithm (self-adaptive algorithm) to problem structure, solutions space and behavior of algorithms (operative and searching) during execution for providing the best solution to problems.

References

1. Lagoudakis, M., Littman, M.: Learning to select branching rules in the DPLL procedure for satisfiability. Electron. Notes Discrete Math. **9**, 344–359 (2001)
2. Chr, P., Steiglitz, K.: Combinatorial Optimization: Algorithms and Complexity (1982)
3. Flum, J., Grohe, M.: Parameterized Complexity Theory. Springer, Heidelberg (2006). https://doi.org/10.1007/3-540-29953-X
4. Rendell, L., Cho, H.: Empirical learning as a function of concept character. Mach. Learn. **5**, 267–298 (1990)
5. Cohen, P.: Empirical Methods for Artificial Intelligence. The MIT Press, Cambridge (1995)

6. Barr, R., Golden, B., Kelly, J., Resende, M.: Designing and reporting on computational experiments with heuristic methods. J. Heuristics **1**(1), 9–32 (1995)
7. Wolpert, D., Macready, W.: No free lunch theorems for optimizations. IEEE Trans. Evol. Comput. **1**(1), 67–82 (1996)
8. Frost, D., Dechter, R.: In search of the best constraint satisfaction search. In: Proceedings of the National Conference on Artificial Intelligence, Seattle, vol. 94, pp. 301–306 (1994)
9. Tsang, E., Borrett, J., Kwan, A. An attempt to map the performance of a range of algorithm and heuristic combinations. In: Hallam, J., et al. (eds.) Hybrid Problems, Hybrid Solutions. Proceedings of AISB-95, vol. 27, pp. 203–216. IOS Press, Amsterdam (1995)
10. Frost, D., Rish, I., Vila, L.: Summarizing CSP hardness with continuous probability distributions. In: Proceedings of the 14th National Conference on AI, American Association for Artificial Intelligence, pp. 327–333 (1997)
11. Vanchipura, R., Sridharan, R.: Development and analysis of constructive heuristic algorithms for flow shop scheduling problems with sequence-dependent setup times. Int. J. Adv. Manufact. Technol. **67**, 1337–1353 (2013)
12. Rice, J.: The algorithm selection problem. Adv. Comput. **15**, 65–118 (1976)
13. Smith-Miles, K.: Cross-disciplinary perspectives on meta-learning for algorithm selection. ACM Comput. Surv. **41**(1), 1–25 (2009)
14. Leyton-Brown, K., Nudelman, E., Shoham, Y.: Learning the empirical hardness of optimization problems: the case of combinatorial auctions. In: Van Hentenryck, P. (ed.) CP 2002. LNCS, vol. 2470, pp. 556–572. Springer, Heidelberg (2002). https://doi.org/10.1007/3-540-46135-3_37
15. Silverthorn, B., Miikkulainen, R.: Latent class models for algorithm portfolio methods. In: Proceedings of the Twenty-Fourth AAAI Conference on Artificial Intelligence, Georgia, USA (2010)
16. Yuen, S., Zhang, X.: Multiobjective evolutionary algorithm portfolio: choosing suitable algorithm for multiobjective optimization problem. In: 2014 IEEE Congress on Evolutionary Computation (CEC), Beijing, China, pp. 1967–1973 (2014)
17. Guerri, A., Milano, M.: Learning techniques for automatic algorithm portfolio selection. In: Proceedings of the 16th Biennial European Conference on Artificial Intelligence, Valencia, Spain, pp. 475–479. IOS Press, Burke (2004)
18. Xu, L., Hoos, H., Leyton-Brown, K.: Hydra: automatically configuring algorithms for portfolio-based selection. In: Proceedings of the 25th National Conference on Artificial Intelligence (AAAI 2010), pp. 210–216 (2010)
19. Pavón, R., Díaz, F., Laza, R., Luzón, M.: Experimental evaluation of an automatic parameter setting system. Expert Syst. Appl. **37**, 5224–5238 (2010)
20. Yeguas, E., Luzón, M., Pavón, R., Laza, R., Arroyo, G., Díaz, F.: Automatic parameter tuning for evolutionary algorithms using a Bayesian case-based reasoning system. Appl. Soft Comput. **18**, 185–195 (2014)
21. Pérez, J., Pazos, R.A., Frausto, J., Rodríguez, G., Romero, D., Cruz, L.: A statistical approach for algorithm selection. In: Ribeiro, C.C., Martins, S.L. (eds.) WEA 2004. LNCS, vol. 3059, pp. 417–431. Springer, Heidelberg (2004). https://doi.org/10.1007/978-3-540-24838-5_31
22. Ries, J., Beullens, P.: A semi-automated design of instance-based fuzzy parameter tuning for metaheuristics based on decision tree induction. J. Oper. Res. Soc. **66**(5), 782–793 (2015)
23. Smith-Miles, K., van Hemert, J., Lim, X.Y.: Understanding TSP difficulty by learning from evolved instances. In: Blum, C., Battiti, R. (eds.) LION 2010. LNCS, vol. 6073, pp. 266–280. Springer, Heidelberg (2010). https://doi.org/10.1007/978-3-642-13800-3_29
24. Hutter, F., Xu, L., Hoos, H., Leyton-Brown, K.: Algorithm runtime prediction: methods & evaluation. Artif. Intell. **206**, 79–111 (2014)
25. Leyton-Brown, K., Hoos, H., Hutter, F., Xu, L.: Understanding the empirical hardness of NP-complete problems. Mag. Commun. ACM **57**(5), 98–107 (2014)

26. Munoz, M., Kirley, M., Halgamuge, S.: Exploratory landscape analysis of continuous space optimization problems using information content. IEEE Trans. Evol. Comput. **19**(1), 74–87 (2015)
27. Kottho, L., Gent, I.P., Miguel, I.: An evaluation of machine learning in algorithm selection for search problems. AI Commun. **25**(3), 257–270 (2012)
28. Lopez, T.T., Schaeer, E., Domiguez-Diaz, D., Dominguez-Carrillo, G.: Structural effects in algorithm performance: a framework and a case study on graph coloring. In: Computing Conference, 2017, pp. 101–112. IEEE (2017)
29. Fu, H., Xu, Y., Chen, S., Liu, J.: Improving WalkSAT for random 3-SAT problems. J. Univ. Comput. Sci. **26**(2), 220–243 (2020)
30. Tavares, J.: Multidimensional knapsack problem: a fitness landscape analysis. IEEE Trans. Syst. Man Cybern. Part B: Cynern. **38**(3), 604–616 (2008)
31. Watson, J., Darrell, W., Adele, E.: Linking search space structure, run-time dynamics, and problem difficulty: a step toward demystifying tabu search. J. Artif. Intell. Res. **24**, 221–261 (2005)
32. Watson, J.: An introduction to fitness landscape analysis and cost models for local search. In: Gendreau, M., Potvin, J.Y. (eds.) Handbook of Metaheuristics. International Series in Operations Research & Management Science, vol. 146, pp. 599–623. Springer, Boston (2010). https://doi.org/10.1007/978-1-4419-1665-5_20
33. Chevalier, R.: Balancing the effects of parameter settings on a genetic algorithm for multiple fault diagnosis. Artificial Intelligence, University of Georgia (2006)
34. Cayci, A., Menasalvas, E., Saygin, Y., Eibe, S.: Self-configuring data mining for ubiquitous computing. Inf. Sci. **246**, 83–99 (2013)
35. Le, M., Ong, Y., Jin, Y.: Lamarckian memetic algorithms: local optimum and connectivity structure analysis. Memetic Comput. **1**, 175–190 (2009)
36. Montero, E., Riff, M.: On-the-fly calibrating strategies for evolutionary algorithms. Inf. Sci. **181**, 552–566 (2011)
37. Pérez, J., Cruz, L., Landero, V.: Explaining performance of the threshold accepting algorithm for the bin packing problem: a causal approach. Pol. J. Environ. Stud. **16**(5B), 72–76 (2007)
38. Pérez, J., et al.: An application of causality for representing and providing formal explanations about the behavior of the threshold accepting algorithm. In: Rutkowski, L., Tadeusiewicz, R., Zadeh, L.A., Zurada, J.M. (eds.) ICAISC 2008. LNCS (LNAI), vol. 5097, pp. 1087–1098. Springer, Heidelberg (2008). https://doi.org/10.1007/978-3-540-69731-2_102
39. Quiroz-Castellanos, M., Cruz-Reyes, L., Torres-Jimenez, J., Gómez, C., Huacuja, H.J.F., Alvim, A.C.: A grouping genetic algorithm with controlled gene transmission for the bin packing problem. Comput. Oper. Res. **55**, 52–64 (2015)
40. Taghavi, T., Pimentel, A., Sabeghi, M.: VMODEX: a novel visualization tool for rapid analysis of heuristic-based multi-objective design space exploration of heterogeneous MPSoC architectures. Simul. Model. Pract. Theory **22**, 166–196 (2012)
41. Landero, V., Pérez, J., Cruz, L., Turrubiates, T., Ríos, D.: Effects in the algorithm performance from problem structure, searching behavior and temperature: a causal study case for threshold accepting and bin-packing. In: Misra, S., et al. (eds.) ICCSA 2019. LNCS, vol. 11619, pp. 152–166. Springer, Cham (2019). https://doi.org/10.1007/978-3-030-24289-3_13
42. Landero, V., Ríos, D., Pérez, J., Cruz, L., Collazos-Morales, C.: Characterizing and analyzing the relation between bin-packing problem and tabu search algorithm. In: Gervasi, O., et al. (eds.) ICCSA 2020. LNCS, vol. 12249, pp. 149–164. Springer, Cham (2020). https://doi.org/10.1007/978-3-030-58799-4_11
43. Spirtes, P., Glymour, C., Scheines, R.: Causation, Prediction, and Search, 2nd edn. The MIT Press, Cambridge (2001)
44. Beasley, J.E.: OR-Library. Brunel University (2006). http://people.brunel.ac.uk/~mastjjb/jeb/orlib/binpackinfo.html

45. Scholl, A., Klein, R. (2003). http://www.wiwi.uni-jena.de/Entscheidung/binpp/
46. Glover, F.: Tabu search - Part I, first comprehensive description of tabu search. ORSA-J. Comput. **1**(3), 190–206 (1989)
47. Fleszar, K., Hindi, K.S.: New heuristics for one-dimensional bin packing. Comput. Oper. Res. **29**, 821–839 (2002)
48. Fayyad, U.M., Irani, K.B.: Multi-interval discretization of continuous-valued attributes for classification learning. In: IJCAI, pp. 1022–1029 (1993)
49. Merz, P., Freisleben, B.: Fitness landscapes and memetic algorithm design. In: New Ideas in Optimization, pp. 245–260. McGraw-Hill Ltd., UK (1999)

General Track 2: High Performance Computing and Networks

Parallel Implementation of 3D Seismic Beam Imaging

Maxim Protasov[✉]

Institute of Petroleum Geology and Geophysics, Novosibirsk 630090, Russia
protasovmi@ipgg.sbras.ru

Abstract. The paper presents the parallel implementation of prestack beam migration of 3D seismic data. The objective is to develop the parallel imaging algorithm for processing large enough volumes of 3D data in production mode, suitable for anisotropic media, and handling 3D irregular seismic data without any preliminary regularization. The paper provides a comparative analysis of the developed migration results with the industrial version of the Kirchhoff migration on synthetic and real data.

Keywords: 3D seismic imaging · Gaussian beams · Anisotropy · Parallelization

1 Introduction

At present, in seismic data processing, prestack depth migration is one of the most computationally expensive procedures. So the migration algorithm should be optimized as much as possible and effectively parallelized. The presented algorithm for three-dimensional anisotropic depth migration in true amplitudes originates from the approach realized in a two-dimensional version [1]. But when implementing the migration of 3D seismic data, the algorithm has undergone significant changes and become closer to the method of three-dimensional Gaussian beam migration, which has become the industry standard [2]. However, this migration option is implemented for regular data in the midpoint domain for each offset [2]. The approach considered in the paper does not require a regularization, and it provides handling irregular data both in the source-receiver and in the midpoint-offset domains.

Similarly, to the two-dimensional predecessor, the algorithm is implemented by extrapolating wavefields from the observation system to a certain fixed image point along anisotropic Gaussian beams [3–5], using individual Gaussian beams instead of decomposing the wavefield into Gaussian beams. The leading algorithmic optimization consists of constructing identical Gaussian beams on the observation surface. It leads to the possibility of dividing migration into two parts: the process of summing data with weights (or decomposing data into beams) and the process of mapping the decomposed data into depth (the migration process itself). Moreover, each of these two parts uses OpenMP technology [6], providing its implementation and optimization. And, the entire migration process is effectively implemented using MPI technology [6]. In general, this

---

O. Gervasi et al. (Eds.): ICCSA 2021, LNCS 12949, pp. 611–621, 2021.
https://doi.org/10.1007/978-3-030-86653-2_44

approach made it possible to create a technological version of migration, which, in terms of computational costs, is comparable to the industrial implementation of Kirchhoff migration and superior in quality. The parallel implementation of the algorithm is verified on a representative set of synthetic and real data.

2 Seismic Migration as Asymptotic Inversion

The technique presented here provides an asymptotically correct true-amplitude image of multicomponent seismic data. In this section, the developed migration is described as an asymptotic inversion of seismic data.

2.1 Statement of the Problem

3D heterogeneous anisotropic elastic medium is considered with Lamé's parameters and density decomposing as follows:

$$c_{ijkl}(x, y, z) = c^0_{ijkl}(x, y, z) + c^1_{ijkl}(x, y, z), \rho(x, y, z) = \rho_0(x, y, z) + \rho_1(x, y, z). \quad (1)$$

The parameters $c^0_{ijkl}(x, y, z)$ and $\rho_0(x, y, z)$ describe a priori known smooth macro-velocity model/background/propagator while $c^1_{ijkl}(x, y, z)$ and $\rho_1(x, y, z)$ are responsible for its unknown rough/rapid perturbations or reflectors.

Born integral [7] describes the reflected/scattered wavefield on the surfaces:

$$\vec{u}^{obs}(x_r, y_r; x_s, y_s; \omega)$$
$$= \int G(x_r, y_r, z_r(x_r, y_r); x, y, z) \cdot L_1 \langle \vec{u}^0(x, y, z; x_s, y_s, z_s(x_s, y_s); \omega) \rangle dxdydz. \quad (2)$$

Here $(x_r, y_r, z_r(x_r, y_r))$ is the receiver coordinate, $(x_s, y_s, z_s(x_s, y_s))$ is the source coordinate, ω is the frequency and $\vec{u}^0$ is the incident wavefield propagating in a smooth background from a volumetric point source, G – Green's matrix for the smooth background and operator L_1 introducing by the rough perturbations $c^1_{ijkl}(x, y, z)$ and $\rho_1(x, y, z)$:

$$(L_1 \langle \vec{u}^0 \rangle)_j = -\sum_{i,l,l=1}^3 \frac{\partial}{\partial x_i} \left(c^1_{ijkl} \frac{\partial u^0_l}{\partial x_k} \right) - \rho_1 \omega^2 u^0_l; (x_1, x_2, x_3) \equiv (x, y, z). \quad (3)$$

The problem is to reconstruct rough perturbations of elastic parameters $c^1_{ijkl}(x, y, z)$ and density $\rho_1(x, y, z)$ or some of their combinations by resolving integral Eq. (2) with the data $\vec{u}^{obs}(x_r, y_r; x_s, y_s; \omega)$.

2.2 Asymptotic Inversion: Generalized Coordinates

On the acquisition surface, the beam center coordinates are introduced, i.e., $\bar{p}_{r0} = (x_{r0}, y_{r0})$ – receiver beam center and $\bar{p}_{s0} = (x_{s0}, y_{s0})$ – source beam center. Quasi-pressure (QP) rays are traced from every beam center to every image point $\bar{p}_i = (x_i, y_i, z_i)$ within the model. QP Gaussian beams are constructed along with these rays [3, 4], and they are denoted by $\vec{u}_{qp}^{gbr}(x_r, y_r; \bar{p}_{r0}; \bar{p}_i; \omega)$ and $\vec{u}_{qp}^{gbs}(x_s, y_s; \bar{p}_{s0}; \bar{p}_i; \omega)$. Then on the acquisition surface, integration beam weights are computed: one is the normal derivative of the Gaussian beam at the receivers, another one is the normal derivative of the scalar part of the corresponding beam in the sources:

$$\vec{T}_{qp}^{gbr}(x_r, y_r; \bar{p}_{r0}; \bar{p}_i; \omega), T_{qp}^{gbs}(x_s, y_s; \bar{p}_{s0}; \bar{p}_i; \omega). \tag{4}$$

Both parts of the Born integral (2) are multiplied by the constructed weights (4) and integrated with respect to the source and receiver coordinates. Using saddle point technique for every beam weight (it is the 3D analog of the result described in the paper [1]), one can come to the following identity:

$$\int T_{qp}^{gbs}(x_s, y_s; \bar{p}_{s0}; \bar{p}_i; \omega)\vec{T}_{qp}^{gbr}(x_r, y_r; \bar{p}_{r0}; \bar{p}_i; \omega)\vec{u}^{obs}(x_r, y_r; x_s, y_s; \omega)dx_r dy_r dx_s dy_s$$

$$= \int \vec{u}_{qp}^{gbr}(\bar{p}; \bar{p}_{r0}; \bar{p}_i; \omega) \cdot L_1 \langle \vec{u}_{qp}^{gbs}(\bar{p}; \bar{p}_{s0}; \bar{p}_i; ; \omega) \rangle d\bar{p} \tag{5}$$

Here is $\bar{p} \equiv (x, y, z)$. The representation of qP beams: $\vec{u}_{qp}^{gbr(s)} = \vec{e}_{qp}^{gbr(s)} \varphi_{qp}^{gbr(s)}$.

And computations of beam derivatives (operator L_1) retaining terms up to the first order only, together with the microlocal analysis (it is analogous to the research described in the paper [1]) of the right-hand side of (5) gives the following:

$$\int T_{qp}^{gbs}(x_s, y_s; \bar{p}_{s0}; \bar{p}_i; \omega)\vec{T}_{qp}^{gbr}(x_r, y_r; \bar{p}_{r0}; \bar{p}_i; \omega)\vec{u}^{obs}(x_r, y_r; x_s, y_s; \omega)dx_r dy_r dx_s dy_s$$

$$= \omega^2 \int \exp\{i\omega(\nabla\tau_s(\bar{p}_i; \bar{p}_{s0}) + \nabla\tau_r(\bar{p}_i; \bar{p}_{r0})) \cdot (\bar{p}_i - \bar{p})\}f(\bar{p}; \bar{p}_{r0}; \bar{p}_{s0})d\bar{p}. \tag{6}$$

Here τ_s, τ_r are travel times from the image point $\bar{p}_i$ to the corresponding acquisition point $\bar{p}_{s0}, \bar{p}_{r0}, f$ is a linearized reflection coefficient. Introducing generalized coordinates $\bar{p}_{01} = (x_{01}, y_{01}), \bar{p}_{02} = (x_{02}, y_{02})$ that depends on beam center coordinates $\bar{p}_{s0}, \bar{p}_{r0}$, the change of variables $(x_{01}, y_{01}, \omega) \rightarrow (k_x, k_y, k_z)$ in the right-hand side of (6) is used:

$$\bar{k} = (k_x, k_y, k_z) = \omega(\nabla\tau_s(\bar{p}_i; \bar{p}_{01}; \bar{p}_{02}) + \nabla\tau_r(\bar{p}_i; \bar{p}_{01}; \bar{p}_{02})). \tag{7}$$

Multiplication by generalized Beylkin's determinant [8]:

$$BelDet(\bar{p}_i; \bar{p}_{01}; \bar{p}_{02}) = \begin{vmatrix} V_x(\tau_s+\tau_r) & V_y(\tau_s+\tau_r) & V_z(\tau_s+\tau_r) \\ \frac{\partial V_x(\tau_s+\tau_r)}{\partial x_{01}} & \frac{\partial V_y(\tau_s+\tau_r)}{\partial x_{01}} & \frac{\partial V_z(\tau_s+\tau_r)}{\partial x_{01}} \\ \frac{\partial V_x(\tau_s+\tau_r)}{\partial y_{01}} & \frac{\partial V_y(\tau_s+\tau_r)}{\partial y_{01}} & \frac{\partial V_z(\tau_s+\tau_r)}{\partial y_{01}} \end{vmatrix}, \tag{8}$$

and integration of both parts of the identity (6) with respect to x_{01}, y_{01}, ω gives the final asymptotic inversion result:

$$f(\bar{p}_i; \bar{p}_{02}) = \int BelDet(\bar{p}_i; \bar{p}_{01}; \bar{p}_{02}) \cdot T_{qp}^{gbs}(x_s, y_s; \bar{p}_{01}; \bar{p}_{02}; \bar{p}_i; \omega)$$

$$\cdot \vec{T}_{qp}^{gbr}(x_r, y_r; \bar{p}_{01}; \bar{p}_{02}; \bar{p}_i; \omega) \cdot \vec{u}^{obs}(x_r, y_r; x_s, y_s; \omega)dx_r dy_r dx_s dy_s d\bar{p}_{01}d\omega. \tag{9}$$

The formula (9) is called imaging condition and provides true amplitude seismic image in generalized acquisition coordinates $\bar{p}_{01}, \bar{p}_{02}$. The last means the imaging condition provides handling any acquisition coordinates. Particularly, asymptotic inversion can utilize beam center receiver coordinates $\bar{p}_{r0} = (x_{r0}, y_{r0})$ while the final result provides averaged linearized reflection coefficient with respect to beam center source coordinates $\bar{p}_{s0} = (x_{s0}, y_{s0})$. Another helpful option uses structural dip and azimuth angle coordinates $\bar{p}_{r0} \equiv (\gamma, \theta)$ [9] in the inversion process and the final image depends on the opening angle and the azimuth of the opening angle $\bar{p}_{s0} \equiv (\beta, az)$ [9].

In the standard industrial technology of seismic processing, the data is usually processed in common midpoint and offset domains. To get the image in these terms, one should use the following representation of generalized acquisition coordinates:

$$
\begin{aligned}
\bar{p}_{01} &\equiv (x_m, y_m) = ((x_{s0} + x_{r0})/2, (y_{s0} + y_{r0})/2), \\
\bar{p}_{02} &\equiv (h_x, h_y) = (x_{r0} - x_{s0}, y_{r0} - y_{s0})
\end{aligned}
\tag{10}
$$

In this case, the inversion is done with respect to midpoints, while the final image is got with respect to offsets.

3   Implementation of Migration Operator

The software implementation utilizes the division of the whole imaging procedures by two blocks (Fig. 1). The first one computes beam attributes and stores them to the disk, and the second one provides seismic data mapping via beam attributes usage. The division into two parts makes it possible to implement offset-midpoint imaging condition (9) effectively. This way provides an opportunity to perform beam data decomposition before their mapping into the image domain. Therefore, computational cost decreases significantly because the summation of data occurs once for all image points, in contrast to the straightforward realization without beam data decomposition when there is a separate summation for each image point. Taken together, this approach provides an opportunity to implement a technological version handling industrial volumes of 3D seismic data within a reasonable computation time.

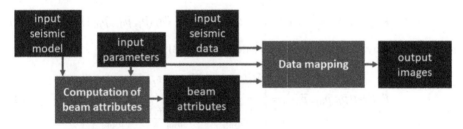

Fig. 1. General upper-level diagram for the realization of the offset-midpoint imaging procedure. In all the diagrams and schemes (Fig. 1, 2, 3, 4, and 5), dark blue color means data, and brown color implies a process (Color figure online).

3.1 Parallelization Scheme

Since the attribute computation block works separately from the migration itself, then the parallelization is performed independently for each such block also. Scheme for the attributes computation utilizes two-level parallelism. The first level uses MPI parallelization done by beam centers. Every node extracts and loads the corresponding part of the model and computes beam attributes for the beam centers portion (Fig. 2). The second level on every node provides OpenMP parallelization by the number of rays, i.e., on beams, where every CPU core provides a corresponding bunch of rays (Fig. 2).

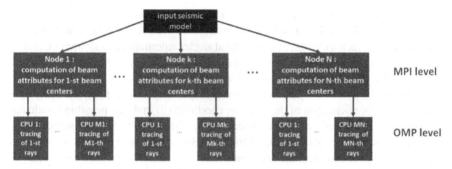

Fig. 2. Parallelization scheme of beam attributes computation procedure.

After the beam attributes computation procedure finishes, the seismic data mapping block starts processing. Its realization also utilizes MPI and OpenMP technologies. MPI realization is done for every offset with respect to the beam center midpoints (Fig. 3).

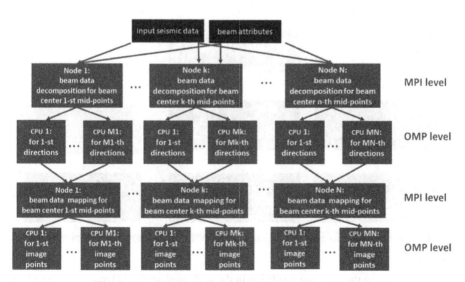

Fig. 3. Parallelization scheme of data mapping procedure.

Then the algorithm contains two steps on every MPI node. The first step provides OpenMP realization of beam decomposition with respect to directions. Every node accumulates OpenMP results, and then the second step again utilizes OpenMP technology for beam data mapping into the image domain with respect to portions of image points (Fig. 3). Every MPI process works independently, so there are no connections between MPI nodes. But the optimization implementation requires uniform data distribution between nodes which is realized based on data coordinates analysis.

3.2   The Functionality of the Algorithm

The straightforward implementation of the imaging condition (9) requires high computational costs for the migration of modern production volumes of data. Simultaneously, the realization of such a migration in structural angles automatically calculates Common Image Gathers (CIGs) depending on the opening angle and opening azimuth. Of course, this is a useful functionality required for amplitude analysis and amplitude inversion depending on the medium's parameters (for example, impedances). On the other hand, the effective implementation of the imaging condition (9) makes it possible to calculate the common image point's seismograms depending on the offsets. Such seismograms are certainly necessary, for example, for velocity analysis, but angle domain seismograms are also needed, for instance, for amplitude analysis. Therefore, within the framework of the migration algorithm based on the imaging condition (9), calculating angle domain gathers is implemented. In short, inside the migration process for each image point, a specific function provides the correspondence between the angle and the offset by searching for a stationary point. After migration, resorting and interpolation to a regular grid specified by the user takes place (Fig. 4). In this case, the additional computational time for sorting and interpolation occurs already after the migration stage.

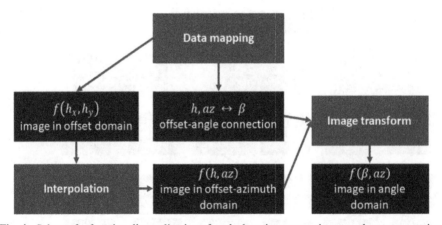

Fig. 4.   Scheme for functionality realization of angle domain common image gathers computation.

Implementing the imaging condition in the structural angles (9) makes it possible to calculate selective images or common image point seismograms depending on the angles, which are essential for the diffraction image construction. It was also required to have this option when implementing the imaging condition (9) in midpoint offset coordinates. There were two calculation options. The first option is carried out by splitting the image into angular seismograms during migration. Of course, in this case, selective images will not be as focused as in the implementation of the imaging condition (9) in structural angles. However, in many cases, they can still be quite helpful. This option requires significantly more memory (RAM) but has a very weak effect on computations' cost. On the other hand, such images are calculated by processing the total image in the Fourier domain, where the component corresponding to the calculated angles is selected (Fig. 5). A reasonable combination of these methods makes it possible to construct diffraction images of sufficiently good quality.

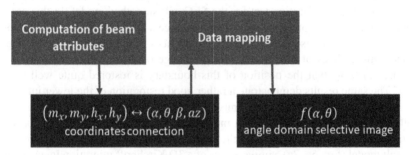

Fig. 5. Scheme for functionality realization of angle domain selective images computation.

The imaging condition (9) in terms of midpoint-offset coordinates makes it possible to implement a technological version of migration. In contrast, the migration algorithm based on the imaging condition (9) in structural angles requires much more computational resources. Conducted specific calculations for these implementations on identical data and equipment show that the computational time difference reaches two orders of magnitude (Table 1). This difference leads to the fact that structural angle realization is acceptable only for small amounts of data in a relatively small image area. On the other hand, midpoint-offset realization provides similar functionality as the structural angle version. Simultaneously, it allows handling large enough data set within reasonable computational resources, and therefore it is suitable for industrial processing.

Table 1. Computational time comparison for midpoint-offset realization and structural angle realization.

Data type	Data size	Computational time (1 CPU) structural angle realization	Computational time (1 CPU) midpoint-offset realization
2D data set	2 GB	40 h	1 h
3D data set	26 GB	4000 h	37 h

4 Numerical Examples

This section provides numerical results and a comparative analysis of the developed migration results with the industrial version of the Kirchhoff migration on 3D synthetic data and 3D marine seismic data.

4.1 Synthetic Data Set

Today, the need to test any developed algorithm on synthetics before using it on real data has become the standard. Amount of synthetic data are created for realistic models. One of the well-known 3D datasets is the free SEG-salt model data [10]. The model is 3D and contains a salt body (Fig. 6a, 7a). The dataset simulates an offshore seismic survey. For this model and data, two migration tests are performed. One of them provides Kirchhoff industrial migration test for the SEG-salt smoothed model. In this case, the model is smoothed to obtain the horizontal interface's best quality image at a depth of about 3500 m. Comparison of two-dimensional slices of the SEG-salt model (Fig. 6a, 7a) and similar slices of a three-dimensional image obtained by Kirchhoff migration (Fig. 6c, 7c) show that the position of this boundary is restored quite well in many places. The same results demonstrate a rather good restoration of the lower interface of the salt body. However, at the same time, the upper boundary is blurred and defocused. Another test provides Gaussian beam migration, using a special algorithm for a more accurate description of the ray-tracing through a salt body. Comparison of 2D slices of the SEG-salt model (Fig. 6a, 7a), similar slices of a 3D Kirchhoff migration image (Fig. 6c, 7c), and the corresponding Gaussian beam migration results (Fig. 6b, 7b) are presented. They show that the lower horizontal interface at a depth of 3500 m is reconstructed much better using Gaussian beams than with Kirchhoff migration with "optimal" model smoothing. Obviously, in the subsalt zone at a depth of between 2000 and 3000 m, the Gaussian beam imaging procedure reconstructs inclined boundary, while on the result of Kirchhoff migration, it is practically absent. Moreover, the salt boundary is more reliably determined by the Gaussian beam imaging algorithm (Fig. 6b, 7b) than by the Kirchhoff migration procedure (Fig. 6c, 7c), where migration artifacts complicate the image.

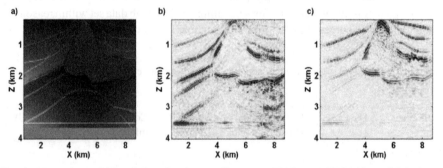

Fig. 6. 2D sections of 3D volumes for the coordinate y = 7800 m: a) SEG-salt model; b) beam imaging result; c) Kirchhoff migration image.

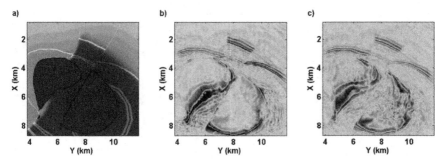

Fig. 7. 2D sections of 3D volumes at a depth of 2000 m: a) SEG-salt model; b) beam imaging result; c) Kirchhoff migration image.

4.2 Real Data Set

The developed anisotropic Gaussian beam migration is applied to the real data. For the data at the velocity model building stage, the anisotropic depth model is constructed. The industrial anisotropic Kirchhoff migration is used for the model and the data also. Gaussian beam migration results (Fig. 8a, 9a) show sufficiently focused reflected events, which indicates that the anisotropy during migration is taken into account correctly. Comparing the images obtained by beam migration (Fig. 8a, 9a) and Kirchhoff migration (Fig. 8b, 9b) provides the resolution of the Gaussian beam imaging exceeds the Kirchhoff migration resolution. The seismic horizons in the image obtained from the beam migration are traced better than those obtained by the Kirchhoff migration. The discussed images were obtained for the area 300 km^2 and the data size almost 600 Gigabytes. The computational resources contain ten nodes with 20 CPU cores on each computation node. Such data and resources provide approximately 23 h for running the beam imaging procedure (Table 2). The computation time of Kirchhoff migration is about 20 h in the same environment (Table 2). So computational times are similar, but the image quality of the beam migration is better.

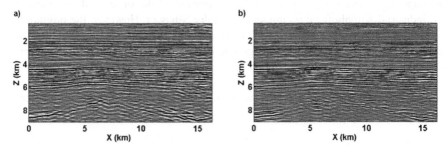

Fig. 8. 2D sections of 3D images for the coordinate y = 8.5 km: a) beam migration result; b) Kirchhoff migration result.

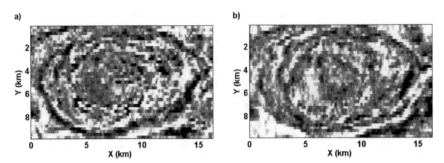

Fig. 9. 2D sections of 3D images at a depth of 6 km: a) beam migration result; b) Kirchhoff migration result.

Table 2. Computational time comparison for beam and Kirchhoff migrations computed on the real marine data set.

Migration	Data size	Computational resources	Computational time
Beam	592 GB	10 Nodes × 20 CPU cores	≈23 h
Kirchhoff	592 GB	10 Nodes × 20 CPU cores	≈20 h

5   Conclusions

The paper presents the parallel implementation of the 3D seismic beam imaging algorithm for anisotropic media. The algorithm can be applied to irregular data either in source-receiver or midpoint-offset domains. Acquisition domain realization provides similar functionality as the structural angle version, but it allows handling large enough data sets and is suitable for industrial processing. The implementation of beam imaging utilizes MPI and OpenMP technologies, effectively providing two-level parallelization with division into beam attribute and data mapping blocks with intrinsic separate data decomposition. The computational time comparison of the developed beam imaging and industrial Kirchhoff migration shows similar behavior. However, the migration of freely distributed synthetic data SEG-Salt and real data proves the developed algorithm's advantages. The obtained stacked images are cleaner and more coherent in comparison with the results of Kirchhoff migration.

Acknowledgments. The reported study was funded by RFBR and GACR, project number 20-55-26003.

References

1. Protasov, M.I.: 2-D Gaussian beam imaging of multicomponent seismic data in anisotropic media. Geophys. J. Int. **203**, 2021–2031 (2015)
2. Hill, N.R.: Prestack Gaussian-beam depth migration. Geophysics **66**, 1240–1250 (2001)

3. Nomofilov, V.E.: Asymptotic solutions of a system of equations of the second order which are concentrated in a neighborhood of a ray. Zap. Nauch. Sem. LOMI **104**, 170–179 (1981)
4. Cerveny, V.: Seismic Ray Theory. Cambridge University Press, Cambridge (2001)
5. Popov, M.M.: Ray theory and Gaussian beam for geophysicists. EDUFBA, SALVADOR-BAHIA (2002)
6. Parallel data processing homepage. https://parallel.ru/vvv/. Accessed 3 Mar 2021
7. Devaney, A.J.: Geophysical diffraction tomography. IEEE Trans. Geosci. Remote Sens. **22**, 3–13 (1984)
8. Beylkin, G.: Imaging of discontinuous in the inverse scattering problem by inversion of causual generalized Radon transform. J. Math. Phys. **26**(1), 99–108 (1985)
9. Protasov, M.I., Tcheverda, V.A., Pravduhin, A.P., Isakov, N.G.: 3D anisotropic imaging of 3D seismic data on the basis of Gaussian beams. Seismic Technol. **1**, 35–48 (2017)
10. SEG/EAGE 3D modeling Salt Model Phase-C Homepage. https://wiki.seg.org/wiki/SEG/EAGE_3D_modeling_Salt_Model_Phase-C_1996. Accessed 2 May 2016

Localization in Wireless Sensor Networks Using MDS with Inter-node Angle Estimation

Saroja Kanchi[✉]

Department of Computer Science, Kettering University, Flint, MI 48439, USA
skanchi@kettering.edu

Abstract. Localization of Wireless Sensor Networks (WSN) is an important problem that has gained a lot of attention as can be seen in the survey papers [8, 9]. Among the many techniques that have been employed to obtain accurate localization under various underlying assumptions, Multi-Dimensional Scaling (MDS) provides a centralized solution given range measurements between nodes. However, dissimilarity matrix provided to MDS is incomplete due to missing range measurements between significant numbers of nodes. Researchers have used shortest distance computations on the basis of existing range measurements to provide an estimate of the actual range measurements. This leads to significant error in localization using MDS. In this work, we introduce an improved estimate of the shortest distances between nodes based on estimated angle between nodes. We do not assume AoA (Angle of Arrival) is available, however we estimate inter-node angle to help improve distance measures. Our results show significant improvement in the performance of MDS for localization of wireless sensor networks for both range-based and range-free noisy range network models. Our simulation verifies the result on both sparse and dense networks.

Keywords: Wireless sensor networks · Localization · MDS

1 Introduction

Most everyday electronic devices currently used are equipped with sensors. Due to availability of low cost sensors that can sense various aspects of the environment such as radiation, motion, pressure, temperature etc., numerous applications have been created that rely on processing the sensor data. However sensors have limited capability in computation, energy and storage. Once deployed, sensors transmit sensed data to a centralized server or nearest sink node. Often, sensors are deployed in environments that are subject to mobility such as automotive, underwater missions, vehicles involved in combat etc., and, therefore their locations change with time. Since equipping sensors with GPS consumes tremendous energy, the locations of sensors becomes unknown over time. Localization of Wireless Sensor Networks (WSN) is the problem of finding geo-location of sensor nodes under different models of sensor communication and sensor deployment. Anchor nodes are small subset of sensor nodes in a wireless sensor network that are aware of their locations due to being equipped with GPS or stay fixed throughout

© Springer Nature Switzerland AG 2021
O. Gervasi et al. (Eds.): ICCSA 2021, LNCS 12949, pp. 622–630, 2021.
https://doi.org/10.1007/978-3-030-86653-2_45

the mission. Various deployment environments have led to numerous techniques for localization.

Techniques for localization of wireless sensor networks can be categorized using various assumptions: whether sensor nodes remain stationary or mobile during execution of the localization algorithm; whether there are anchor nodes (*anchor based*) or no anchor nodes (*anchor free*); whether distance information between the nodes is available (*range-based*) or not (*range-free*).

The *range-based* approaches assume that sensors can find the distance to neighboring sensors within their sensing radius using RSSI signal strength, or time difference of arrival (TDoA) between radio signals or time of arrival of signal (ToA). In addition, angle of arrival (AOA) of a signal [11], has be used to assist in localization of the sensors. In range-free approach, nodes are aware of the presence of neighboring nodes, but do not have the range information to them.

The node based localization based algorithm called Sweeps was developed in [12] and is related to the iterative method called trilateration. A range based localization model that considers a Bayesian approach has been proposed in [13]. Wang et al. in [14] have proposed a range free localization algorithm as an improvement to regular DV-Hop algorithm.

Among the techniques used for localization MDS (Multi-Dimensional Scaling) is powerful and works in low time complexity. MDS can be centralized or distributed. Given hybrid distance and angle of arrival measurements, [1] provide the graphical properties that guarantee unique localizability in cooperative networks. Iterative multidimensional scaling (IT-MDS) and simulated annealing multidimensional scaling (SA-MDS), which are variations classical MDS was proposed in [2]. The authors demonstrate that these perform more reliably and accurately for distributed localization. The authors in [3] present a system of tracking mobile robots and mapping an unstructured environment, using up to 25 wireless sensor nodes in an indoor setting. In [4] that authors to propose a technique relevant to networks that may not be fully localizable. The authors design a framework for two dimensional network localization with an efficient component to correctly determine which nodes are localizahle and which are not.

To obtain the distance matrix, the authors in [5] estimate the inter-tag distances using a triangular method. Then, classical MDS algorithms has been applied to determine the estimated locations of the tags. The authors in [6] propose an MDS based approach that can take advantage of additional information, such as estimated distances between neighbors or known positions for anchor nodes. The authors in [7] prove that for a network consisting of n sensors positioned randomly on a unit square and a given radio range $r = o(1)$, the resulting error is bounded, decreasing at a rate that is inversely proportional to r, when only connectivity information given. In [10], the authors address flip, rotation and translation ambiguities by exploiting the node velocities to correlate the relative maps at two consecutive instants. The authors introduce a new version of MDS, called enhanced Multidimensional Scaling (eMDS), which is able to handle these ambiguities.

In this paper we present a new way to estimate distances between those nodes for which direct range measurements are not available. We combine shortest path between nodes along with estimate of inter-node angle. Section 2 presents background of the MDS techniques, Sect. 3 describes the contribution of inter-angle estimation technique, Sect. 4 contains simulation results and Sect. 5 provides conclusion.

2 Background

We view the underlying graph of the network as a graph $G = (V, E)$, where V represent the set the nodes in the WSN, and the weight of an edge e in E between nodes u and v is the range measurement between nodes. In range-based deployment, each node is aware of the nodes within the radio range of the network. It is assumed that all nodes have the same radio range, called the radius of the network. In noise-free range measurement, each node knows the exact Euclidean distance between itself and its neighbors within the radius. In a noisy range measurement, we assume a Gaussian noise to the range measurement. In range-free deployment, each node is aware of the presence of nodes within the network radius, however, the range measurement is not known. In this case, we assume that range measurement between neighbors is the radius of the network.

The goal in all of the scenarios described above including range-based with noise-free range measurements, range-based with noisy range measurements and range-free deployments, is to find the location of the nodes each node v in V, with a mapping denoted by $m(v) = (x_v, y_v)$ such that Euclidean distance between $m(u)$ and $m(v)$ is equal to the range measurement between u and v for all nodes where range information is given.

Multidimensional Scaling (MDS) was originally created for visualizing dissimilarity of data. MDS takes as input a set of distance information between nodes assumed to be high dimensions and the distance is assumed to Euclidean distance in the dimension. In order to visualize the data, it is desirable that there be points created in lower dimensions such 2 or 3 dimensions, so that one can visualize the dissimilarity between data points. For 2 dimensional points, given $O(n^2)$ distances, i.e., distances between each pair of points is sufficient to solve n positions, however, the set of points would be subject to rotation, flip and translation and preserving the relative map. In application to localization however, we do not have all of the $O(n^2)$ distance measurements.

The dissimilarity matrix is referred to as proximity matrix P for our discussion. The goal to find the location vector, $L, = (li)$ where li represents the location of node i. Let p_{ij} represent the proximity between nodes i and j, and, Euclidean distance between location $l_i = (x_i, y_i)$ and $l_j = (x_j, y_j)$ as

$$d_{ij} = \sqrt{(x_i - x_j)^2 + (y_i - y_j)^2} \tag{1}$$

Class MDS solves for L such that $d_{ij} = f(p_{ij})$, where f is a linear transformation, given by $d_{ij} = a + b * p_{ij}$.

A common approach to classical MDS is find D, the distance matrix between coordinates, as close to given proximity matrix P as possible, such that $I(P) = D + E$, where $I(P)$ is a linear transformation of P subject to minimizing the sum of squares of the error matrix E.

This is done by singular value decomposition (SVD) of the double centered matrix of square of P, which includes the following steps

a) Square the matrix P. Double center the squared matrix and call it the matrix K. Double centering a matrix is simply subtracting the row mean and column mean from each element of the matrix and adding the grand mean of the entire matrix to each element of the matrix and then multiplying the result by $-1/2$. That is

$$k_{ij} = -\frac{1}{2}\left(p_{ij}^2 - \frac{1}{n}\sum_{j=1}^{n} p_{ij}^2 - \frac{1}{n}\sum_{i=1}^{n} p_{ij}^2 + \frac{1}{n^2}\sum_{i=1}^{n}\sum_{j=1}^{n} p_{ij}^2\right) \quad (2)$$

b) It is proved that an element in the double centered matrix is equal to

$$k_{ij} = x_i * x_j + y_i * y_j \quad (3)$$

c) Singular value decomposition is used on the double centered matrix K to obtain

$$K = UVU' \quad (4)$$

d) Now

$$L = UV^{\frac{1}{2}} \quad (5)$$

Applying classical MDS to localization involves providing the dissimilarity matrix which is equal to the distance information between nodes and using MDS to determine the location of sensor nodes. However, since range information between majority of nodes is not available, several "fillers" have been used to estimate distance between nodes that are not within the radio range, such as hop distance between nodes, or shortest distance between nodes. More accurate dissimilarity matrices, produce more accurate localization. In this paper, we present a new way to estimate distances between nodes when direct range measurements are not available. We combine shortest path between nodes along with estimate of angle as described in the next section.

3   Estimation of Euclidean Distances

While using classical MDS, given a dissimilarity matrix M of size $n \times n$ that contains the exact Euclidean distances between every pair of n nodes, MDS produces a set of n coordinate points in 2D as described in Sect. 2, where the distance between each pair of points is the corresponding weight in the dissimilarity matrix. Therefore, in order to use classical MDS for localization, one needs the distances between every pair of points that need to be localized. If ground truth is available, the result of MDS would be relative map that when rotated, reflected and translated using three anchors would result in the ground truth. In our application, in order to obtain the distance between nodes that are not adjacent in the WSN, researchers have used shortest distance using Floyd's algorithm and Dijkstra's algorithm etc. While these provide shortest distances assuming that one travels along the edges of the graph, they do not translate well for determining Eucledean distances in network graphs. However, here explore how an intermediate node can help determine the direct Eucledean distance between nodes using the fact that the nodes are uniformly distributed in a given deployment area.

Given a network $G = (V, E)$ where each edge has a weight of network radius r (range-free case), or range measured between two nodes (range-based), we consider the number of nodes that possible neighbors of a node, i.e. the nodes that are within the network radius. Since each node knows the number of nodes within its radio range, even if it does not know the exact distance between itself and its neighbor(in range-free case), we estimate that the angle between two neighbors incident at a node v is $360/|N(v)|$ where $N(v)$ is the set of neighbors of the node v. This angle information is then used at each node to estimate the Euclidean distance two nodes. Given the angle $\angle ABC$, (see Fig. 1), the distance between A and C can be found using distance between A and B, B and C and the angle $\angle BAC$. The distance is then carried over to compute the length of the shortest path between two nodes.

The algorithm used for computing shortest distances is modified using

$$D(A, C) = D(A, B) + D(B, C) + D(A, B) * D(B, C) * cos(\angle ABC) \qquad (6)$$

If nodes A and C have several nodes, then the min value of all of distances is used as an estimate of shortest between A and C.

We modify the shortest distance computation used in Floyd's algorithm using the distance computation above. It is important to note that distance between A and C is computed, it has impact on all shortest distances that use the edge (A, C) in the shortest path, thus propagating the enhance distance estimation throughout the network. See Fig. 1.

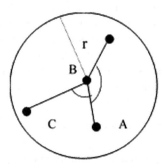

Fig. 1. Angle Estimation to computer the distance between A and C

4   Simulation Results

Simulation was performed on Matlab by creating a node layout of the Wireless Sensor Networks. The nodes were generated using a uniform distribution in a 100 by 100 unit square. For each simulation, the average of the performance of 50 networks are computed. The node density was varied between 200 and 400. The radii for sparse networks was between 10 and 14 and dense networks it was 15 to 20. Figure 2 shows sample network graphs of radius 12 (sparse) and radius 18 (dense).

Figure 3 shows that when the proposed technique of inter-angle estimation is used to compute shortest distances, the average error of the distance computed by MDS versus the ground truth is much smaller than using MDS with traditional shortest path computations. This is also true for all dense graphs with radii between 15 and 20 as shown in the same Fig. 3. It is also notable, as expected, as the network radius is higher the error is lower due to more distance being available due to larger number of node adjacencies.

Figures 4 demonstrates the results for connected networks for sparse and dense range-free graphs. In case of range free graphs, nodes can are aware of their neighbors but they are not aware of the exact distance between the node and its neighbors. We use radio range (radius) of the networks as the distance information to compute the dissimilarity matrix for MDS based localization. This leads to larger error by MDS in both traditional shortest distance computation and when inter-angle estimate is available. The error computed is error against the ground truth of the layout of the network where range measurements are available.

Figure 5 demonstrates the same performance on range based and range-free graphs with 400 nodes.

Next, we examine the effect of noise on the computation of angle estimation. The range measurements of perturbed with Gaussian noise with standard deviation of 0.1 to 0.7. Since we use 200 nodes and radius of 14, the noise in the range is from 0.7% to 5%. We compute the difference in the distance between localized nodes Figs. 5 and 6 are Note that the noise in the MDS data is much less when AoA estimation is used for computing shortest distances (Fig. 7).

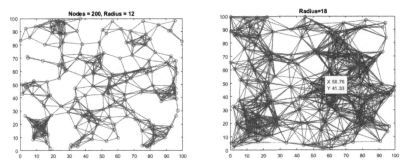

Fig. 2. Sample networks with 200 nodes and sensor radius of 12 and 18

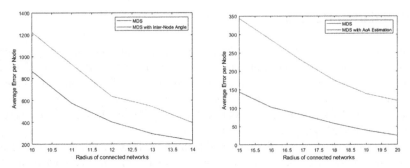

Fig. 3. Radius vs mean error in localization for range based networks

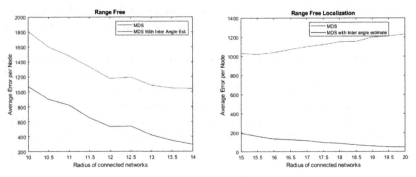

Fig. 4. Radius versus mean error in localization for range free network

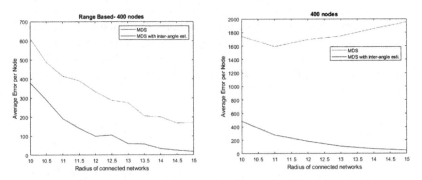

Fig. 5. Radius versus mean error in localization for 400 node networks

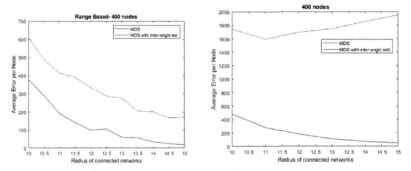

Fig. 6. Radius versus mean error in localization for 400 node networks

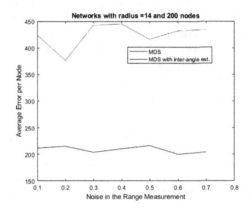

Fig. 7. Noise versus error in localization

5 Conclusion

In this paper, we present an effective technique to enhance the distance matrix provided to classical MDS localization. The enhancement is the technique in which shortest distances are computed between non-adjacent nodes. Using traditional techniques for finding shortest distances would use either Floyd's algorithm or Dijkstra's algorithm. The new technique use estimated angle between nodes assuming uniform distribution of nodes on the network. We have demonstrated that this significantly reduces localization error when using classical MDS for range free, range based computations even when there is noise in the range measurement.

References

1. Eren, T.: Cooperative localization in wireless ad hoc and sensor networks using hybrid distance and bearing (angle of arrival) measurements. EURASIP J. Wirel. Commun. Netw. **2011**(1), 1–18 (2011)

2. Biaz, S., Ji, Y.: Precise distributed localization algorithms for wireless networks. In: Sixth IEEE International Symposium on a World of Wireless Mobile and Multimedia Networks, pp. 388–394. IEEE, June 2005
3. Zhang, Y., Ackerson, L., Duff, D., Eldershaw, C., Yim, M.: Stam: a system of tracking and mapping in real environments. IEEE Wirel. Commun. 11(6), 87–96 (2004)
4. Goldenberg, D.K., et al.: Network localization in partially localizable networks. In: Proceedings IEEE 24th Annual Joint Conference of the IEEE Computer and Communications Societies, vol. 1, pp. 313–326. IEEE, March 2005
5. Shi, W., Wong, V.W.: MDS-based localization algorithm for RFID systems. In: 2011 IEEE International Conference on Communications (ICC), pp. 1–6. IEEE, June 2011
6. Shang, Y., Rumi, W., Zhang, Y., Fromherz, M.: Localization from connectivity in sensor networks. IEEE Trans. Parallel Distrib. Syst. 15(11), 961–974 (2004)
7. Oh, S., Montanari, A., Karbasi, A.: Sensor network localization from local connectivity: performance analysis for the MDS-map algorithm. In: 2010 IEEE Information Theory Workshop on Information Theory (ITW 2010, Cairo), pp. 1–5. IEEE, January 2010
8. Tian, H., Ding, Y., Yang, S.: A survey on MDS-based localization for wireless sensor network. In: Jin, D., Lin, S. (eds.) Advances in Future Computer and Control Systems, pp. 399–403. Springer, Heidelberg (2012). https://doi.org/10.1007/978-3-642-29387-0_60
9. Saeed, N., Nam, H., Al-Naffouri, T.Y., Alouini, M.S.: A state-of-the-art survey on multidimensional scaling-based localization techniques. IEEE Commun. Surv. Tutor. 21(4), 3565–3583 (2019)
10. Di Franco, C., Melani, A., Marinoni, M.: Solving ambiguities in MDS relative localization. In: 2015 International Conference on Advanced Robotics (ICAR), pp. 230–236. IEEE, July 2015
11. Niculescu, D., Nath, B.: Ad hoc positioning system (APS) using AOA. In: IEEE INFOCOM 2003. Twenty-second Annual Joint Conference of the IEEE Computer and Communications Societies (IEEE Cat. No. 03CH37428), vol. 3, pp. 1734–1743, March 2003
12. Goldenberg, D.K., et al.: Localization in sparse networks using sweeps. In: Proceedings of the 12th Annual International Conference on Mobile Computing and Networking, pp. 110–121 (2006)
13. Coluccia, A., Ricciato, F.: RSS-based localization via Bayesian ranging and iterative least squares positioning. IEEE Commun. Lett. 18(5), 873–876 (2014)
14. Wang, X., Liu, Y., Yang, Z., Lu, K., Luo, J.: Robust component-based localization in sparse networks. IEEE Trans. Parallel Distrib. Syst. 25(5), 1317–1327 (2013)

The Domination and Independent Domination Problems in Supergrid Graphs

Ruo-Wei Hung[1]([envelope]) [ORCID], Ming-Jung Chiu[1] [ORCID], and Jong-Shin Chen[2] [ORCID]

[1] Department of Computer Science and Information Engineering,
Chaoyang University of Technology, Wufeng, Taichung 413310, Taiwan
rwhung@cyut.edu.tw
[2] Department of Information and Communication Engineering,
Chaoyang University of Technology, Wufeng, Taichung 413310, Taiwan
jschen26@cyut.edu.tw

Abstract. Let G be a graph with vertex set $V(G)$ and edge set $E(G)$. A subset D of $V(G)$ is a dominating set of G if every vertex not in D is adjacent to one vertex of D. A dominating set is called independent if it is an independent set. The domination number (resp., independent domination number) of G, denoted by $\gamma(G)$ (resp., $\gamma_{\mathrm{ind}}(G)$), is the minimum cardinality of a dominating set (resp., independent dominating set) of G. The domination (resp., independent domination) problem is to compute a dominating set (resp., an independent dominating set) of G with size $\gamma(G)$ (resp., $\gamma_{\mathrm{ind}}(G)$). Supergrid graphs are a natural extension for grid graphs. The domination and independent domination problems for grid graphs were known to be NP-complete. However, their complexities on supergrid graphs are still unknown. In this paper, we will prove these two problems for supergrid graphs to be NP-complete. Then, we compute $\gamma(R_{m \times n})$ and $\gamma_{\mathrm{ind}}(R_{m \times n})$ for rectangular supergrid graphs $R_{m \times n}$ in linear time.

Keywords: Domination · Independent domination · Supergrid graph · Rectangular supergrid graph · Grid graph

1 Introduction

Let G be a graph. We will denote by $V(G)$ and $E(G)$ the vertex set and edge set of G, respectively. Let $v \in V(G)$, and let $S \subseteq V(G)$. We use $G[S]$ to represent the subgraph induced by S. Denote by $N_G(v) = \{u \in V(G) | (u, v) \in E(G)\}$ the *open neighborhood* of vertex v, while its *closed neighborhood* is represented as $N_G[v] = N_G(v) \cup \{v\}$. Generally, let $N_G(S) = \cup_{v \in S} N_G(v)$ and $N_G[S] = N_G(S) \cup S$. The *degree* of vertex v in G is the number of edges incident to v, and is denoted by $deg_G(v)$. Let D be a subset of $V(G)$. If $v \in D$ or $N_G(v) \cap D \neq \emptyset$, then we say that D *dominates* v. If D dominates every vertex of $S \subseteq V(G)$, then we

Supported by Ministry of Science and Technology, Taiwan under grant no. 110-2221-E-324-007-MY3.

O. Gervasi et al. (Eds.): ICCSA 2021, LNCS 12949, pp. 631–646, 2021.
https://doi.org/10.1007/978-3-030-86653-2_46

call that D *dominates* S. When D dominates $V(G)$, D is called a *dominating set* of G. The *domination number* of a graph G is the minimum cardinality of a dominating set of G, and it is denoted by $\gamma(G)$. A *minimum dominating set* of G is a dominating set with size $\gamma(G)$. The *domination problem* is to compute a minimum dominating set of G, and it is a well-known NP-complete problem for general graphs [16]. This problem is still NP-complete for some special graph classes, including 4-regular planar graphs [16], grid graphs [10], cubic bipartite graphs [29], etc.

Many variants of the domination problem do exist and they are to compute a minimum dominating set with some additional properties, e.g., to be independent or to induce a connected graph. These problems arise in a lots of distributed network applications in which they are requested to locate the smallest number of centers in networks so that every vertex is nearby at least one center, where the set of centers forms a minimum dominating set, and some restricted conditions are satisfied if it is the domination variant. The concept of these domination related problems have many applications and have been widely studied in literature (see [21,22]); a rough estimate says that it occurs in more than 6000 papers to date [20]. In this paper, we will study the domination problem and its one variant, namely *independent domination problem*.

A set of vertices is said to be *independent* if its any two vertices are not adjacent. An *independent dominating set* of a graph G is a dominating set I such that I is independent. The independent domination number of a graph G is the minimum size of an independent dominating set in G, and it is denoted by $\gamma_{\text{ind}}(G)$. Since an independent dominating set of G is also a dominating set of G, $\gamma(G) \leqslant \gamma_{\text{ind}}(G)$ for any graph G. The *independent domination problem* is to compute an independent dominating set of G with size $\gamma_{\text{ind}}(G)$. This problem is NP-complete for general graphs [16]. It is still NP-complete for bipartite graphs, comparability graphs [11], planar graphs [5], chordal bipartite graphs [14], cubic bipartite graphs [29], grid graphs [10], etc. However, it is polynomial solvable when the input is restricted to some special graphs, including interval graphs [6], circular-arc graphs [7], bounded clique-width graphs [12], etc. For more related works and applications on independent domination, we refer the reader to the survey on independent domination in graphs given in [17], and more results regarding this problem in [1,2,4,13].

The *two-dimensional integer grid* G^{∞} is an infinite graph in which its vertex set consists of all points of the Euclidean plane with integer coordinates, and two vertices are adjacent if and only if their (Euclidean) distance is 1. A *grid graph* is a finite and vertex-induced subgraph of G^{∞}. The *two-dimensional supergrid graph* S^{∞} is the infinite graph whose vertex set consists of all points of the plane with integer coordinates and in which two vertices are adjacent if the difference of their x or y coordinates is not larger than 1. A supergrid graph G_s is a finite and undirected graph such that $V(G_s) \subset V(S^{\infty})$ and $E(G_s) \subset E(S^{\infty})$. For a vertex $v \in V(G_s)$, it is represented as (v_x, v_y), where v_x and v_y are the x and y coordinates of v respectively. Then, $1 \leqslant deg_{G_s}(v) \leqslant 8$. Grid graphs are not subclasses of supergrid graphs, and the converse is also true: these two

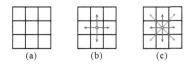

Fig. 1. (a) A set of lattices, (b) the neighbors of one lattice in a grid graph, and (c) the neighbors of one lattice in a supergrid graph, where each lattice is denoted by a vertex in a graph and arrow lines indicate the adjacent neighbors of one lattice.

graph classes contain common elements (vertices), but they are distinct since their edge sets are distinct. Clearly, grid graphs are bipartite [26], but supergrid graphs are not bipartite. A rectangular grid (or called complete grid) graph $G_{m \times n}$ has mn nodes with node $u = (u_x, u_y)$ adjacent to $v = (v_x, v_y)$ if and only if $|u_x - v_x| + |u_y - v_y| = 1$. A rectangular supergrid graph $R_{m \times n}$ is a supergrid graph with vertex set $\{(v_x, v_y)|1 \leqslant v_x \leqslant n \text{ and } 1 \leqslant v_y \leqslant m\}$ and edge set $\{(u, v)|0 \leqslant |u_x - v_x| \leqslant 1 \text{ and } 0 \leqslant |u_y - v_y| \leqslant 1\}$. Then, for $u \in V(G_{m \times n})$ and $v \in V(R_{m \times n})$, $2 \leqslant deg_{G_{m \times n}}(u) \leqslant 4$ and $3 \leqslant deg_{R_{m \times n}}(v) \leqslant 8$ if $m, n \geqslant 2$. Throughout this paper, we will denote by $(1, 1)$ the coordinates of the top-left vertex of a grid or supergrid graph.

An intuitive motivation of proposing supergrid graphs is as follows. Consider a set of lattices, depicted in Fig. 1(a), where each lattice is denoted as a vertex in a graph. For a grid graph, the neighbors of a lattice include its upper, down, left, and right lattices, see Fig. 1(b). However, in the real word and other applications, the neighbors of a lattice may also contain its upper-right, upper-left, down-right, and down-left adjacent lattices, see Fig. 1(c). For example, the sewing trace of a computerized sewing machine is such an application [23]. Thus, supergrid graphs can be used in these applications.

A brief summary of related works is given below. The domination problem on grid graphs was known to be NP-complete [10]. Many researchers have studied the domination numbers of rectangular grid graphs [8,9,19]. In [18], Gonçalves *et al.* computed $\gamma(G_{m \times n}) = \lfloor \frac{(m+2)(n+2)}{5} \rfloor - 4$ for $n \geqslant m \geqslant 16$. This result verified the conjecture in [8]. In [13], Crevals and Östergård computed $\gamma_{\text{ind}}(G_{m \times n}) = \lfloor \frac{(m+2)(n+2)}{5} \rfloor - 4$ for $n \geqslant m \geqslant 16$. In fact, $\gamma(G_{m \times n}) = \gamma_{\text{ind}}(G_{m \times n})$ for rectangular grid graph $G_{m \times n}$ by the result in [30]. In [28], the author computed $\gamma_{\text{ind}}(R_{m \times n}) = \lceil \frac{m}{3} \rceil \lceil \frac{n}{3} \rceil$. In this paper, we provide another way to compute $\gamma_{\text{ind}}(R_{m \times n})$. In [15], Gagnon *et al.* stated that $\gamma(R_{m \times n}) = \lceil \frac{m}{3} \rceil \lceil \frac{n}{3} \rceil$ is trivially known. Unfortunately, they do not provide any proof. In this paper, we will give a tight proof to verify $\gamma(R_{m \times n}) = \lceil \frac{m}{3} \rceil \lceil \frac{n}{3} \rceil$. In [23], we first introduced supergrid graphs and proved the Hamiltonian problems on supergrid graphs to be NP-complete. In [24,25,27], the Hamiltonian related properties on some special supergrid graphs have been studied. In this paper, we will prove that the domination and independent domination problems on (general) supergrid graphs are NP-complete. Then, we compute $\gamma(R_{m \times n}) = \gamma_{\text{ind}}(R_{m \times n}) = \lceil \frac{m}{3} \rceil \lceil \frac{n}{3} \rceil$ by a tight proof.

The rest of this paper is structured as follows. Section 2 introduces some notations and one related result in the literature. In Sect. 3, we prove that the domination and independent domination problems for (general) supergrid graphs are NP-complete. Section 4 computes $\gamma(R_{m \times n}) = \lceil \frac{m}{3} \rceil \lceil \frac{n}{3} \rceil$ by a formally proof. In Sect. 5, we use a known result to verify $\gamma_{\text{ind}}(R_{m \times n}) = \gamma(R_{m \times n})$. Finally, some conclusions are given in Sect. 6.

2   Preliminaries

In this section, we will introduce some notations and one result in the literature. For two sets X and Y, let $X - Y$ denote the set of elements in X that are not in Y. A path P in a graph G is a sequence of adjacent vertices starting from v_1 and ending at v_k, represented as $v_1 \rightarrow v_2 \rightarrow \cdots \rightarrow v_{k-1} \rightarrow v_k$, where all the vertices $v_1, v_2, \cdots, v_k$ are distinct except that possibly the path is a cycle when $v_1 = v_k$. A path starting from vertex v_1 and ending at vertex v_k is denoted by (v_1, v_k)-path, and a path with n vertices is denoted by P_n if no ambiguous appears.

Let G_s be a supergrid graph, and let $v \in V(G_s)$. Then, $v = (v_x, v_y)$, where v_x and v_y are the x and y coordinates of v respectively, and $1 \leqslant deg_{G_s}(v) \leqslant 8$. Rectangular supergrid graphs form a special class of supergrid graph and they first appeared in [23], in which the Hamiltonian cycle problem is linear solvable. A rectangular supergrid graph $R_{m \times n}$ is a supergrid graph with vertex set $V(R_{m \times n}) = \{v = (v_x, v_y) | 1 \leqslant v_x \leqslant n \text{ and } 1 \leqslant v_y \leqslant m\}$ and edge set $E(R_{m \times n}) = \{(u, v) | 0 \leqslant |u_x - v_x| \leqslant 1 \text{ and } 0 \leqslant |u_y - v_y| \leqslant 1\}$. Then, $R_{m \times n}$ contains m rows and n columns of vertices. In this paper, w.l.o.g. we will assume that $m \leqslant n$. Let v be a vertex in $R_{m \times n}$ with $m \geqslant 2$. The vertex v is called a *corner* of $R_{m \times n}$ if $deg_{R_{m \times n}}(v) = 3$. There are four corners of $R(m, n)$ including *upper-left*, *upper-right*, *down-left*, and *down-right* corners coordinated as $(1, 1)$, $(n, 1)$, $(1, m)$, and (n, m), respectively (see Fig. 2). The edge (u, v) is called *horizontal* (resp., *vertical*) if $u_y = v_y$ (resp., $u_x = v_x$), and is said to be *crossed* if it is neither a horizontal nor a vertical edge. For example, Fig. 2 shows a rectangular supergrid graph $R_{8 \times 10}$ and it also depicts the types of corners and edges. Note that a grid graph contains horizontal and vertical edges, but it contains no crossed edge. A sequence of consecutive horizontal (resp., vertical) edges, $e_1, e_2, \cdots, e_k$, is called a *horizontal* (resp., *vertical*) *edge-line*, denoted by L_k, where e_i and e_{i+1} contains a common vertex for $1 \leqslant i \leqslant k - 1$. Let μ, ν be two vertices in L_k such that they are only incident to one edge. Then, a horizontal or vertical edge-line is denoted by (μ, ν)-edge-line. For instance, the bold consecutive horizontal edges in Fig. 2 indicate a horizontal edge-line L_6.

In our method, we need to partition a rectangular supergrid graph into two disjoint parts. The partition is defined as follows:

Definition 1. *Let S be a rectangular supergrid graph $R_{m \times n}$. A separation operation on S is a partition of S into two vertex-disjoint supergrid subgraphs S_1 and S_2, and is called vertical (resp., horizontal) if it consists of a set of horizontal*

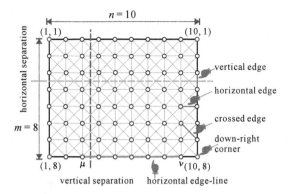

Fig. 2. A rectangular supergrid graph $R_{8 \times 10}$, where bold dashed lines indicate vertical and horizontal separations, and a set of 6 bold edges indicates a horizontal edge-line L_6.

(resp., vertical) edges. For instance, the bold dashed vertical (resp., horizontal) line in Fig. 2 depicts a vertical (resp., horizontal) separation of $R_{8 \times 10}$ which is partitioned into $R_{8 \times 3}$ and $R_{8 \times 7}$ (resp., $R_{3 \times 10}$ and $R_{5 \times 10}$).

In [9], Chang *et al.* computed the domination number $\gamma(P_n)$ of path P_n as follows:

Lemma 1. *(see [9].)* $\gamma(P_n) = \gamma(R_{1 \times n}) = \lfloor \frac{n+2}{3} \rfloor = \lceil \frac{n}{3} \rceil$.

3   NP-completeness Results

In this section, the domination and independent domination problems for (general) supergrid graphs will be proved to be NP-complete. In 1990, Clark *et al.* [10] showed that the domination problem for (general) grid graphs is NP-complete. We will reduce it to the domination and independent domination problems on supergrid graphs.

Theorem 1. *(See [10].)* *The domination problem on (general) grid graphs is NP-complete.*

We will reduce the domination problem on grid graphs to the domination problem on supergrid graphs. Given a grid graph G_g, we will construct a supergrid graph G_s to satisfy that G_g contains a dominating set D with size $|D| \leqslant k$ if and only if G_s contains a dominating set $\hat{D}$ with size $|\hat{D}| \leqslant k + 2|E(G_g)|$. The construction steps are sketched as follows. First, we enlarge the input grid graph G_g such that each edge of G_g is transformed into a horizontal or vertical edge-line L_7 with 7 edges; i.e., enlarge each edge of G_g by 7 times. Let the enlarged grid graph be G_g'. For an example, Fig. 3(b) depicts grid graph G_g' enlarged from grid graph G_g in Fig. 3(a). In the second step, each horizontal or vertical (u, v)-edge-line of graph G_g' is replaced by a (u, v)-path which is a small supergrid graph

---

Algorithm 1: The supergrid graph construction algorithm

Input: A grid graph G_g. (see Fig. 3(a))

Output: A supergrid graph G_s. (see Fig. 3(d))

Method: // an algorithm constructing a supergrid graph from a grid graph

1. enlarge G_g to a grid graph G_g' such that each edge of G_g is transformed into a horizontal or vertical edge-line L_7 with 7 edges; (see Fig. 3(b))
2. **for** each horizontal or vertical (u, v)-path in G_g', where $u, v \in V(G_g)$, replace it by a snake path to connect u and v;

   // this replaced path is a small supergrid graph and is called snake (u, v)-path, denoted by $S(u, v)$ (see Fig. 3(c)), where $u, v \in V(G_g)$ are called connectors of $S(u, v)$
3. the constructed graph is a supergrid graph G_s (see Fig. 3(d)), and **output** G_s.

---

and is an undirected path P_8, where $u, v \in V(G_g)$. The path connecting u and v is called a *snake* (u, v)-*path*, denoted by $S(u, v)$. Figure 3(c) depicts a snake (u, v)-path. Note that there exists no directionality for edges on $S(u, v)$. Finally, the constructed graph is a supergrid graph G_s. For instance, Fig. 3(d) shows the supergrid graph G_s constructed from grid graph G_g in Fig. 3(a). Note that each snake path is an undirected path P_8 with 8 vertices. Algorithm 1 presents the details of our construction.

Next, we will show that there exists an arrangement of snake paths in G_s such that they are disjoint except their connectors. The arrangement rule is sketched as follows: Consider a grid graph G_g. Since every grid graph is a subgraph of a rectangular grid graph (or called complete grid graph) with maximum number of edges, we can only consider the arrangement of snake paths for a rectangular grid graph $G_{m \times n}$, where $G_g = G_{m \times n}$. Let u and v be two adjacent vertices of $G_{m \times n}$ such that $u_x \leqslant v_x$ and $u_y \leqslant v_y$. Let P_v^u be the enlarged horizontal or vertical edge-line of edge (u, v) in G_g. We arrange snake paths of G_s by the following rule:

Case 1: $v_x = u_x + 1$. In this case, P_v^u is a horizontal edge-line. There are two subcases:

Case 1.1: u_y is odd. If u_x is odd, then the snake path $S(u, v)$ is placed above the edge-line P_v^u; otherwise, it is placed below P_v^u.

Case 1.2: u_y is even. If u_x is odd, then the snake path $S(u, v)$ is placed below the edge-line P_v^u; otherwise, it is placed above P_v^u.

Case 2: $v_y = u_y + 1$. In this case, P_v^u is a vertical edge-line. There are two subcases:

Case 2.1: u_x is odd. If u_y is odd, then the snake path $S(u, v)$ is placed to be the right of edge-line P_v^u; otherwise, it is placed to the left of P_v^u.

Case 2.2: u_x is even. If u_y is odd, then the snake path $S(u, v)$ is placed to be the left of edge-line P_v^u; otherwise, it is placed to the right of P_v^u.

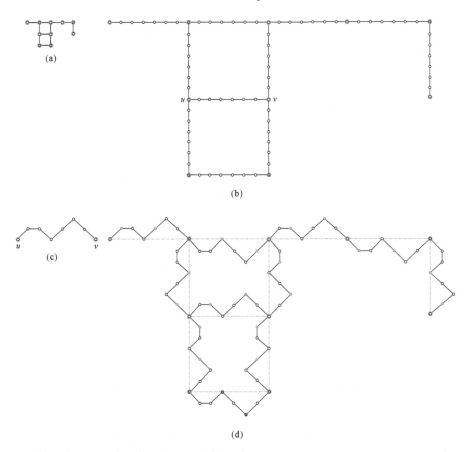

Fig. 3. (a) A grid graph G_g, (b) a grid graph G'_g by enlarging each edge of G_g 7 times, (c) a snake (u, v)-path $S(u, v)$ to replace the enlarged edge-line of G'_g, and (d) a constructed supergrid graph G_s obtained from G'_g by replacing each enlarged edge-line with a snake path in (c), where solid lines indicate the edges of G_g and G_s, double circles represent the vertices of G_g, and solid circles indicate the vertices in a dominating set of G_g or G_s.

For example, Fig. 4 depicts the arrangement of snake paths in G_s for $G_g = G_{16 \times 16}$. The above arrangement rule is called *Rule AS* (Arrange Snake paths) and it holds the following property:

Lemma 2. *Rule AS arranges the snake paths of G_s such that these paths are disjoint except their connectors.*

Proof. Consider a square with vertices (i, j), $(i + 1, j)$, $(i, j + 1)$, $(i + 1, j + 1)$ in a grid graph G_g. There are four cases:

Case 1: i, j are even or odd. By Rule AS, the snake paths with four connectors (i, j), $(i + 1, j)$, $(i, j + 1)$, $(i + 1, j + 1)$ are arranged disjoint.

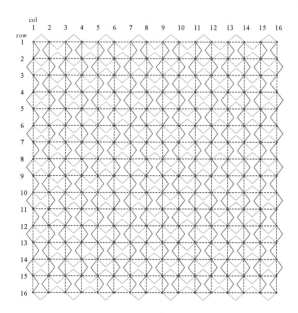

Fig. 4. An arrangement of snake paths in G_s from $G_g = G_{16 \times 16}$, where bold triangle lines indicate the snake paths of G_s.

Case 2: i is even and j is odd, or i is odd and j is even. By Rule AC, the snake paths with the four connectors (i, j), $(i+1, j)$, $(i, j+1)$, $(i+1, j+1)$ are arranged disjoint.

We have considered any case to locate snake paths. In a square of a grid graph G_g, there are at most two snake paths to be placed inside the square of paths in G_s and they are located inside the paths square across from each other. By the construction of snake paths, any two vertices of two adjacent snake paths are not adjacent except their connectors which are in G_g, and the height of each snake path equals to two. Thus, the lemma holds true.

Clearly, Algorithm 1, together with Rule AS, can be done in polynomial time. Thus, the following lemma holds true.

Lemma 3. *Given a grid graph G_g, Algorithm 1, together with Rule AS, constructs a supergrid graph G_s in polynomial time.*

To prove our NP-completeness result, we need to observe some domination properties of snake paths. Note that a snake path is a simple undirected path P_8. By Lemma 1, the following properties of snake (u, v)-path $S(u, v)$ can be easily verified:

Proposition 1. *Let $\hat{D}$ be a dominating set of G_s constructed by Algorithm 1, and let $S(u, v)$ be a snake path with connectors u and v. By Lemma 1, the following statements hold true:*

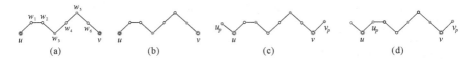

Fig. 5. The minimum dominating set of snake path $S(u, v)$ in G_s for (a) $u, v \in \hat{D}$, (b) $u \in \hat{D}$ and $v \notin \hat{D}$, (c) $u, v \notin \hat{D}$ and, $(N_{G_s}(u) - S(u, v)) \cap \hat{D} \neq \emptyset$ and $(N_{G_s}(v) - S(u, v)) \cap \hat{D} \neq \emptyset$, and (d) $u, v \notin \hat{D}$ and, $(N_{G_s}(u) - S(u, v)) \cap \hat{D} = \emptyset$ or $(N_{G_s}(v) - S(u, v)) \cap \hat{D} = \emptyset$, where $\hat{D}$ is a dominating set of G_s and solid circles indicate the vertices in $\hat{D}$.

(1) If $u, v \in \hat{D}$, then $\gamma(S(u, v) - (N_{G_s}[u] \cup N_{G_s}[v])) = 2$ *(See Fig. 5(a))*.

(2) If $u \in \hat{D}$ and $v \notin \hat{D}$, then $\gamma(S(u, v) - N_{G_s}[u]) = \gamma(S(u, v) - (N_{G_s}[u] \cup \{v\})) = 2$, and there exists a vertex v_d in $\hat{D} \cap S(u, v)$ such that v_d dominates v if $\hat{D} - S(u, v)$ does not dominate v *(See Fig. 5(b))*.

(3) If $u, v \notin \hat{D}$, then
  (3-1) if $(N_{G_s}(u) - S(u, v)) \cap \hat{D} \neq \emptyset$ and $(N_{G_s}(v) - S(u, v)) \cap \hat{D} \neq \emptyset$, then $\gamma(S(u, v) - \{u, v\}) = 2$ *(See Fig. 5(c))*;
  (3-2) if $(N_{G_s}(u) - S(u, v)) \cap \hat{D} = \emptyset$ or $(N_{G_s}(v) - S(u, v)) \cap \hat{D} = \emptyset$, then $\gamma(S(u, v) - \{u, v\}) = 3$ *(See Fig. 5(d))*.

Let $\hat{D}$ be a dominating set of G_s, and let H be a subgraph of G_s. We will denote the restriction of $\hat{D}$ to H by $\hat{D}_{|H}$. In the above proposition, we can see that for any dominating set $\hat{D}$ of supergrid graph G_s and snake (u, v)-path $S(u, v)$, $|\hat{D}_{|S(u,v) - \{u,v\}}| \geq 2$. In the following, we will prove that grid graph G_g has a dominating set D with size $|D| \leq k$ if and only if supergrid graph G_s contains a dominating set $\hat{D}$ with size $|\hat{D}| \leq k + 2|E(G_g)|$. We first prove the only if part as follows.

Lemma 4. *Assume that grid graph G_g contains a dominating set D with size $|D| \leq k$. Then, supergrid graph G_s contains a dominating set $\hat{D}$ with size $|\hat{D}| \leq k + 2|E(G_g)|$*

Proof. Consider an edge (u, v) of G_g. Let $S(u, v) = u \rightarrow w_1 \rightarrow w_2 \rightarrow w_3 \rightarrow w_4 \rightarrow w_5 \rightarrow w_6 \rightarrow v$ be the snake (u, v)-path in G_s constructed from edge (u, v) of G_g, as shown in Fig. 5(a). Initially, let $\hat{D} = D$. We then consider three cases depending on whether $u, v \in D$. These cases include (1) $u, v \in D$, (2) $u \in D$ and $v \notin D$, and (3) $u, v \notin D$. For each snake (u, v)-path $S(u, v)$ in any case, we compute two vertices w_i, w_j, $6 \geq i, j \geq 1$, of $S(u, v) - \{u, v\}$, together with u and v, to dominate $S(u, v)$, and then let $\hat{D} = \hat{D} \cup \{w_i, w_j\}$. These two vertices are depicted in Figs. 5(a)–(c). We can prove the constructed set $\hat{D}$ to be a dominating set of G_s after computing all snake paths of G_s. Because of the space limitation, we omit the details of proof.

During our construction of $\hat{D}$, $|\hat{D}| = |\hat{D}| + 2$ after computing one snake path of G_s. Since there exist $|E(G_g)|$ snake paths of G_s, we construct a dominating set $\hat{D}$ of G_s with size $|\hat{D}| = |D| + 2|E(G_g)| \leq k + 2|E(G_g)|$. For example, Fig. 3(a) shows a dominating set D of G_g with $|D| = 4$, and the dominating set $\hat{D}$ of

G_s constructed from D is shown in Fig. 3(d), where $|\hat{D}| = |D| + 2|E(G_g)| = 4 + 2 \times 11 = 26$.

Next, we will prove the if part as follows.

Lemma 5. *Assume that supergrid graph G_s contains a dominating set $\hat{D}$ with size $|\hat{D}| \leqslant k + 2|E(G_g)|$. Then, grid graph G_g has a dominating set D with size $|D| \leqslant k$.*

Proof. We will show there exists a dominating set D of G_g with size k or less. The construction steps are described as follows. First, we compute a dominating set D' of G_s from $\hat{D}$ such that $|D'| \leqslant |\hat{D}|$. Second, we remove all vertices not in G_g from D' and obtain a set D. Then, we can prove D to be a dominating set of grid graph G_g.

First, we compute D' from $\hat{D}$ as follows. Initially, let $D' = \hat{D}$. Consider a snake path $S(u, v)$ of G_s constructed from edge (u, v) of G_g, as depicted in Fig. 5(a). By Proposition 1, $|\hat{D}_{|S(u,v)-\{u,v\}}| \geqslant 2$. We will construct a set D' of G_s obtained from $\hat{D}$ such that $|D'| \leqslant |\hat{D}|$, $|D'_{|S(u,v)-\{u,v\}}| \geqslant 2$, and removing all vertices not in G_g from D' results in a dominating set D of G. We consider three cases of (1) $u, v \notin D'$, (2) $u \in D'$ and $v \notin D'$, and (3) $u, v \in D'$. First consider the case of $u, v \notin D'$. In case of Fig. 5(d), $|D'_{|S(u,v)-\{u,v\}}| \geqslant 3$ by Statement (3-2) of Proposition 1, and we then set u to be in D', i.e., $D' = D' \cup \{u\}$. Then, snake (u, v)-path $S(u, v)$ satisfies that $u \in D'$ and $v \notin D'$. It is then computed in case of $u \in D'$ and $v \notin D'$. For the other cases, we compute D' on $S(u, v)$ to satisfy $|D'_{|S(u,v)-\{u,v\}}| = 2$. We can verify that D' is a dominating set of G_s and satisfies the following properties: (due to space limitation, we omit the details of proof)

(p1) $|D'| \leqslant |\hat{D}|$,
(p2) every snake path contains exactly two vertices not in G_g, and
(p3) for each snake (u, v)-path $S(u, v)$ with $u, v \notin D'$, there exist $z_1 \in N_{G_g}(u)$ and $z_2 \in N_{G_g}(v)$ such that $z_1, z_2 \in D'$.

We finally compute a dominating set D of G_g from D' by the following steps:

(1) initially, let $D = D'$;
(2) remove all vertices of D not in D_g from D;
(3) the resultant set D will be a dominating set of G_g.

Assume by contradiction that $|D| > k$. Let $|D| = k + x$, $x > 0$. Since $D' \cap S(u, v)$ contains exactly two vertices not in G_g for each snake path $S(u, v)$, we get that $|D| = |D'| - 2|E(G_g)|$. Then,
$|D'| \leqslant |\hat{D}| = k + 2|E(G_g)| = (k + x) + 2|E(G_g)| - x = |D| + 2|E(G_g)| - x$, and hence
$|D| = |D'| - 2|E(G_g)| \leqslant (|D| + 2|E(G_g)| - x) - 2|E(G_g)| = |D| - x = (k + x) - x = k$, a contradiction. Thus, $|D| \leqslant k$.

By Lemmas 4 and 5, we summarize the following lemma:

Lemma 6. *Let G_g be a grid graph and let G_s be the supergrid graph constructed from G_g by Algorithm 1 and Rule AS. Then, G_g contains a dominating set D with size $|D| \leqslant k$ if and only if G_s contains a dominating set $\hat{D}$ with $|\hat{D}| \leqslant k + 2|E(G_g)|$.*

Obviously, the domination problem for supergrid graphs is in NP. By Theorem 1, Lemmas 2–3, and Lemma 6, we conclude the following theorem:

Theorem 2. *The domination problem on supergrid graphs is NP-complete.*

A dominating set of the supergrid graphs constructed in Lemmas 4 and 5 can be easily modified as an independent dominating set (see Figs. 5(a)–(c)). Thus, the independent domination problem on supergrid graphs is also NP-complete, and, hence, the following theorem holds true.

Theorem 3. *The independent domination problem on supergrid graphs is NP-complete.*

4   The Domination Number of Rectangular Supergrid Graphs

In this section, we will compute $\gamma(R_{m \times n})$ for rectangular supergrid graph $R_{m \times n}$ with $n \geqslant m$. To simplify notation, we will use % to denote the modulo operation. In addition, we use $\mathcal{R}_i$ to denote the set of vertices of row i in $R_{m \times n}$. By Lemma 1, $\gamma(R_{1 \times n}) = \lceil \frac{n}{3} \rceil$. Next, we consider $R_{2 \times n}$ as follows:

Lemma 7. $\gamma(R_{2 \times n}) = \lceil \frac{n}{3} \rceil$.

Proof. By the structure of $R_{2 \times n}$, a vertex of $R_{2 \times n}$ dominates at most 6 vertices including its 5 neighbors and itself. Let D be any dominating set of $R_{2 \times n}$. Then, $6 \cdot |D| \geqslant |V(R_{2 \times n})| = 2n$, and, hence, $|D| \geqslant \lceil \frac{n}{3} \rceil$. That is, $\gamma(R_{2 \times n}) \geqslant \lceil \frac{n}{3} \rceil$. Let D_1 be the minimum dominating set of $\mathcal{R}_1$. By Lemma 1, $|D_1| = \lceil \frac{n}{3} \rceil$. For each $v \in D_1$, v dominates $(v_x - 1, 2)$, $(v_x, 2)$, and $(v_x + 1, 2)$ in $\mathcal{R}_2$. In addition, v dominates at most three vertices $(v_x - 1, 1)$, $(v_x, 1)$, and $(v_x + 1, 1)$ in $\mathcal{R}_1$. Since D_1 dominates all vertices of $\mathcal{R}_1$ in $R_{2 \times n}$, it also dominates all vertices of $\mathcal{R}_2$. Thus, D_1 is a dominating set of $R_{2 \times n}$. Then, $\gamma(R_{2 \times n}) \leqslant |D_1| = \lceil \frac{n}{3} \rceil$. Therefore, $\gamma(R_{2 \times n}) = \lceil \frac{n}{3} \rceil$. For example, Fig. 6 shows the minimum dominating set of $R_{2 \times n}$ for $2 \leqslant n \leqslant 10$.

By similar arguments in proving Lemma 7, we compute $\gamma(R_{3 \times n})$ as follows:

Lemma 8. $\gamma(R_{3 \times n}) = \lceil \frac{n}{3} \rceil$.

Proof. By the structure of $R_{3 \times n}$, a vertex of $R_{3 \times n}$ dominates at most 9 vertices including its 8 neighbors and itself. Let D be any dominating set of $R_{3 \times n}$. Then, $9 \cdot |D| \geqslant |V(R_{3 \times n})| = 3n$, and, hence, $|D| \geqslant \lceil \frac{n}{3} \rceil$. That is, $\gamma(R_{3 \times n}) \geqslant \lceil \frac{n}{3} \rceil$. Let D_2 be the minimum dominating set of $\mathcal{R}_2$ in $R_{3 \times n}$. By Lemma 1, $|D_2| = \lceil \frac{n}{3} \rceil$. By the same arguments in proving Lemma 7, D_2 is a dominating set of $R_{3 \times n}$. Thus, $\gamma(R_{3 \times n}) \leqslant \lceil \frac{n}{3} \rceil$. It follows from the above upper bound and lower bound of $\gamma(R_{3 \times n})$ that $\gamma(R_{3 \times n}) = \lceil \frac{n}{3} \rceil$. For example, Fig. 7 shows the minimum dominating set of $R_{3 \times n}$ for $3 \leqslant n \leqslant 9$.

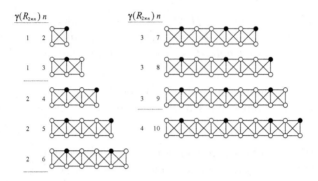

Fig. 6. $\gamma(R_{2\times n}) = \lceil \frac{n}{3} \rceil$ for $2 \leqslant n \leqslant 10$, where solid circles indicate the vertices in the minimum dominating set of $R_{2\times n}$.

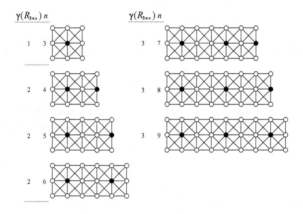

Fig. 7. $\gamma(R_{3\times n}) = \lceil \frac{n}{3} \rceil$ for $3 \leqslant n \leqslant 9$, where solid circles indicate the vertices in the minimum dominating set of $R_{3\times n}$.

Now, we consider $R_{m\times n}$ with $n \geqslant m \geqslant 3$. We will prove the following lemma by induction.

Lemma 9. *Let $R_{m\times n}$ be a rectangular supergrid graph with $n \geqslant m \geqslant 3$. Then, $\gamma(R_{m\times n}) = \lceil \frac{m}{3} \rceil \lceil \frac{n}{3} \rceil$.*

Proof. Let D_i be the minimum dominating set of $\mathcal{R}_i$, the set of vertices of row i in $R_{m\times n}$. By the construction in [9], $D_i = \{(2+3j, i)|0 \leqslant j \leqslant \lfloor \frac{n}{3} \rfloor -1\}$, and $D_i = D_i \cup \{(n, i)\}$ if $n\%3 \neq 0$. We claim that $R_{m\times n}$ contains a minimum dominating set D such that $D = \cup_{0 \leqslant i \leqslant \lfloor \frac{m}{3} \rfloor -1} D_{2+3i}$, and $D = D \cup D_m$ if $m\%3 \neq 0$.

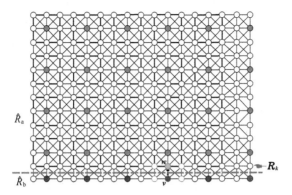

Fig. 8. The minimum dominating set of $R_{(k+1)\times n}$ when $k\%3 = 0$, where solid circles indicate the vertices in the minimum dominating set.

We will prove this claim by induction on m. Initially, let $m = 3$. By Lemma 8, the claim holds true. Assume that the claim is true when $m = k \geqslant 3$. Then, $R_{k\times n}$ has a minimum dominating set $\hat{D}$ such that $\hat{D} = \cup_{0\leqslant i\leqslant\lfloor\frac{k}{3}\rfloor-1}D_{2+3i}$, and $\hat{D} = \hat{D}\cup D_k$ if $k\%3 \neq 0$. Let $m = k+1$. Consider that $k\%3 = 0$. We first make a horizontal separation on $R_{(k+1)\times n}$ to obtain two disjoint subgraphs $\hat{R}_a = R_{k\times n}$ and $\hat{R}_b = R_{1\times n}$, where $\hat{R}_b = \mathcal{R}_{k+1}$ is the set of vertices of row $k+1$ in $R_{(k+1)\times n}$, as depicted in Fig. 8. Let D be a minimum dominating set of $R_{(k+1)\times n}$. Suppose that $D\cap\mathcal{R}_k \neq \emptyset$, where $\mathcal{R}_k$ is the set of vertices of row k as shown in Fig. 8. Then, $|D_{|\hat{R}_a}| > |\hat{D}|$. Let $w \in D\cap\mathcal{R}_k$ and $v \in \hat{R}_b$ with $w_x = v_x$, see Fig. 8. Then, $N_{R_{(k+1)\times n}}[w]\cap(\mathcal{R}_k\cup\hat{R}_b) = N_{R_{(k+1)\times n}}[v]\cap(\mathcal{R}_k\cup\hat{R}_b)$. Let $D = D-\{w\}\cup\{v\}$. Then, D is still a minimum dominating set of $R_{(k+1)\times n}$. Thus, there exists a minimum dominating set D of $R_{(k+1)\times n}$ such that $D\cap\mathcal{R}_k = \emptyset$, and hence $\hat{D} = D_{|\hat{R}_a}$. Then, $N_{R_{(k+1)\times n}}[\hat{D}]\cap\hat{R}_b = \emptyset$. So, no vertex of $D_{|\hat{R}_a} = \hat{D}$ dominates vertex of $\hat{R}_b$. To dominate $\hat{R}_b$, it needs at least $\lceil\frac{n}{3}\rceil$ vertices by Lemma 1. Let $\hat{D}_b$ be a such minimum dominating set of $\hat{R}_b$. Then, $D = \hat{D} \cup \hat{D}_b$ is a minimum dominating set of $R_{(k+1)\times n}$. By the same arguments above and Lemmas 1 and 7, the claim holds when $k\%3 = 1$ or $k\%3 = 2$. Thus, the claim holds true when $n = k + 1$. By induction, the claim holds true for $m \geqslant 3$. By the claim, $R_{m\times n}$ contains a minimum dominating set D with $|D| = \lceil\frac{m}{3}\rceil\lceil\frac{n}{3}\rceil$. Thus, $\gamma(R_{m\times n}) = \lceil\frac{m}{3}\rceil\lceil\frac{n}{3}\rceil$ for $n \geqslant m \geqslant 3$.

It immediately follows from Lemmas 1 and 7–9 that the following theorem holds.

Theorem 4. $\gamma(R_{m\times n}) = \lceil\frac{m}{3}\rceil\lceil\frac{n}{3}\rceil$ for $m, n \geqslant 1$.

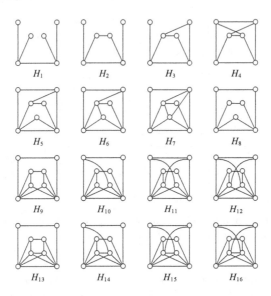

Fig. 9. A set of forbidden induced subgraphs in a graph G with $\gamma(G) = \gamma_{\text{ind}}(G)$ [30].

5   The Independent Domination Number of Rectangular Supergrid Graphs

In [28], Klobučar computed the independent domination number $\gamma_{\text{ind}}(R_{m \times n})$ of $R_{m \times n}$. In this section, we will use a different method to compute it. Since an independent dominating set of a graph G is also a dominating set, $\gamma(G) \leqslant \gamma_{\text{ind}}(G)$ for any graph G. In [3], Allan and Laskar proved that $K_{1,3}$-free graphs are graphs with equal domination and independent domination numbers. By the structure of $R_{m \times n}$, we can see that rectangular supergrid graphs are not $K_{1,3}$-free graphs, i.e., they contain induced subgraph $K_{1,3}$. However, Topp and Volkmann [30] showed that if a graph G contains no induced subgraph isomorphic to one of the graphs $H_1, H_2, \cdots, H_{16}$ in Fig. 9, then $\gamma(G) = \gamma_{\text{ind}}(G)$. The following theorem shows their result:

Theorem 5. *(see [30]). If a graph G contains no induced subgraph isomorphic to one of graphs $H_1, H_2, \cdots, H_{16}$ as shown in Fig. 9, then $\gamma(G) = \gamma_{\text{ind}}(G)$.*

By inspecting the structure of rectangular supergrid graphs, they contain H_1 and H_2 as induced subgraphs. However, rectangular supergrid graphs $R_{m \times n}$ are H_3-free graphs, i.e., they contain no induced subgraph isomorphic to H_3. Thus, $\gamma_{\text{ind}}(R_{m \times n}) = \gamma(R_{m \times n})$. In fact, we construct a minimum dominating set of $R_{m \times n}$ that is also an independent dominating set. It follows from Theorems 4 and 5 that the following theorem holds true.

Theorem 6. $\gamma_{\text{ind}}(R_{m \times n}) = \gamma(R_{m \times n}) = \lceil \frac{m}{3} \rceil \lceil \frac{n}{3} \rceil$ *for $m, n \geqslant 1$.*

6   Concluding Remarks

In this paper, we prove that the domination and independent domination problems on general supergrid graphs are NP-complete. Then, we provide a formally proof to compute the minimum dominating set of a rectangular supergrid graph. In addition, we use a simple method to verify the minimum dominating set of a rectangular supergrid graph is also a minimum independent dominating set.

References

1. Abrishami, G., Henning, M.A.: Independent domination in subcubic graphs of girth at least six. Discrete Math. **341**, 155–164 (2018)
2. Abrishami, G., Henning, M.A., Bahbarnia, F.: On independent domination in planar cubic graphs. Discuss. Math. Graph Theory **39**, 841–853 (2019)
3. Allan, R.B., Laskar, R.: On domination and independent domination numbers of a graph. Discrete Math. **23**, 73–76 (1978)
4. Bhangale, S.T., Pawar, M.M.: Isolate and independent domination number of some classes of graphs. AKCE Intern. J. Graphs Combin. **16**, 110–115 (2019)
5. Boliac, R., Lozin, V.: Independent domination in finitely defined classes of graphs. Theoret. Comput. Sci. **301**, 271–284 (2003)
6. Bui-Xuan, B.M., Telle, J.A., Vatshelle, M.: Fast dynamic programming for locally checkable vertex subset and vertex partitioning problems. Theoret. Comput. Sci. **511**, 66–76 (2013)
7. Chang, M.S.: Efficient algorithms for the domination problems on interval and circular-arc graphs. SIAM J. Comput. **27**, 1671–1694 (1998)
8. Chang, T.Y.: Domination numbers of grid graphs. Ph.D. thesis, University of South Florida, Tampa, FL (1992)
9. Chang, T.Y., Clark, W.E., Hare, E.O.: Domination numbers of complete grids I. Ars Combin. **38**, 97–112 (1994)
10. Clark, B.N., Colbourn, C.J., Johnson, D.S.: Unit disk graphs. Discrete Math. **86**, 165–177 (1990)
11. Corneil, D.G., Perl, Y.: Clustering and domination in perfect graphs. Discrete Appl. Math. **9**, 27–39 (1984)
12. Courcelle, B., Makowsky, J.A., Rotics, U.: Linear time solvable optimization problems on graphs of bounded clique-width. Theory Comput. Syst. **33**, 125–150 (2000)
13. Crevals, S., Östergård, P.R.J.: Independent domination of grids. Discrete Math. **338**, 1379–1384 (2015)
14. Damaschke, P., Muller, H., Kratsch, D.: Domination in convex and chordal bipartite graphs. Inform. Process. Lett. **36**, 231–236 (1990)
15. Gagnon, A., et al.: A method for eternally dominating strong grids. Discrete Math. Theoret. Comput. Sci. **22**(1), #8 (2020)
16. Garey, M.R., Johnson, D.S.: Computers and Intractability: A Guide to the Theory of NP-Completeness. Freeman, San Francisco (1979)
17. Goddard, W., Henning, M.A.: Independent domination in graphs: a survey and recent results. Discrete Math. **313**, 839–854 (2013)
18. Gonçalves, D., Pinlou, A., Rao, M., Thomassé, A.: The domination number of grids. SIAM J. Discrete Math. **25**, 1443–1453 (2011)
19. Guichard, D.R.: A lower bound for the domination number of complete grid graphs. J. Combin. Math. Combin. Comput. **49**, 215–220 (2004)

20. Hajian, M., Henning, M.A., Jafari Rad, N.: A new lower bound on the domination number of a graph. J. Comb. Optim. **38**, 721–738 (2019)
21. Haynes, T.W., Hedetniemi, S.T., Slater, P.J.: Fundamentals of Domination in Graphs. Marcel Dekker, New York (1998)
22. Haynes, T.W., Hedetniemi, S.T., Slater, P.J.: Domination in Graphs: Advanced Topics. Marcel Dekker, New York (1998)
23. Hung, R.W., Yao, C.C., Chan, S.J.: The Hamiltonian properties of supergrid graphs. Theoret. Comput. Sci. **602**, 132–148 (2015)
24. Hung, R.W.: Hamiltonian cycles in linear-convex supergrid graphs. Discrete Appl. Math. **211**, 99–112 (2016)
25. Hung, R.W., Li, C.F., Chen, J.S., Su, Q.S.: The Hamiltonian connectivity of rectangular supergrid graphs. Discrete Optim. **26**, 41–65 (2017)
26. Itai, A., Papadimitriou, C.H., Szwarcfiter, J.L.: Hamiltonian paths in grid graphs. SIAM J. Comput. **11**(4), 676–686 (1982)
27. Keshavarz-Kohjerdi, F., Hung, R.W.: The Hamiltonicity, Hamiltonian connectivity, and longest (s, t)-path of L-shaped supergrid graphs. IAENG Intern. J. Comput. Sci. **47**, 378–391 (2020)
28. Klobučar, A.: Independent sets and independent dominating sets in the strong product of paths and cycles. Math. Commun. **10**, 23–30 (2005)
29. Liu, C.H., Poon, S.H., Lin, J.Y.: Independent dominating set problem revisited. Theoret. Comput. Sci. **562**, 1–22 (2015)
30. Topp, J., Volkmann, L.: On graphs with equal domination and independent domination numbers. Discrete Math. **96**, 75–80 (1991)

Toward a Security IoT Platform with High Rate Transmission and Low Energy Consumption

Lam Nguyen Tran Thanh[1]([⊠]), The Anh Nguyen[2], Hong Khanh Vo[2], Hoang Huong Luong[2], Khoi Nguyen Huynh Tuan[2], Anh Tuan Dao[2], and Ha Xuan Son[3]

[1] VNPT Information Technology Company, Ho Chi Minh city, Vietnam
[2] FPT University, Can Tho city, Vietnam
[3] University of Insubria, Varese, Italy

Abstract. The Internet of Things currently is one of the most interesting technology trends. Devices in the IoT network towards mobility and compact in size, so these have a rather weak hardware configuration. One of the essential concern is energy consumption. There are many lightweights, tailor-made protocols for limited processing power and low energy consumption, of which MQTT is the typical protocol. A light and simple protocol like MQTT, however, has many problems such as security risks, reliability in transmission and reception. The current MQTT protocol supports three types of quality-of-service (QoS). The user has to trade-off between the security/privacy of the packet and the system-wide performance (e.g., transmission rate, transmission bandwidth, and energy consumption). In this paper, we present an IoT Platform Proposal to improve the security issues of the MQTT protocol and optimise the communication speed, power consumption, and transmission bandwidth, but this still responds to reliability when transmitting. We also present the effectiveness of our approach by building a prototype system. Besides, we compare our proposal with other related work as well as provide the complete code solution is publicized to engage further reproducibility and improvement.

Keywords: Internet of Things · MQTT · Quality-of-Service (QoS) · Single Sign On · Kafka

1 Introduction

In recent years, the Internet of Thing (IoT) applications have grown and applied in most fields of our life such as smart city, healthcare, supply chains, industry, agriculture. According to estimates, by 2025, the whole world will have approximately 75.44 billion IoT connected devices [1]. Timothy Chou et al. [2] claimed that the IoT system architecture consists of five layers in order from low to high: Things, Connect, Collect, Learn, and Do.

The Things layer contains actuators that control physical devices or sensors to collect environmental parameters. The Connect layer connects devices with

© Springer Nature Switzerland AG 2021
O. Gervasi et al. (Eds.): ICCSA 2021, LNCS 12949, pp. 647–662, 2021.
https://doi.org/10.1007/978-3-030-86653-2_47

applications and users. The Collect layer is responsible for aggregating the data returned by the Things layer devices. Finally, Learn layer is used to analyze data to give suggestions to the Do layer in response to received data. The Things and Connect layers can be considered as the two most important since they provide input data for the upper layers.

The Things layer usually consists of devices in the IoT which have limitations in network connectivity, power, and processing capabilities [3]. Therefore, how can we optimize the processing capacity and energy consumption of the devices while still having to meet some basic requirements on communication speed and the confidentiality of information for the whole system? This issue is determined by the protocols themselves within the Collect layer.

There are 5 popular protocols used for IoT platform, namely Hypertext Transfer Protocol (HTTP), Constrained Application Protocol (CoAP), Extensible Messaging and Presence Protocol (XMPP), Advanced Message Queuing Protocol (AMQP), and Message Queuing Telemetry Protocol (MQTT) [4]. For communications in limited networks (constrained networks), MQTT and CoAP are proposed to use [5]. Corak et al. found that the MQTT protocol has faster Packet creation time and Packet transmission time is twice as fast as with the CoAP protocol [6]. For developers of low bandwidth and memory devices, MQTT is also the most preferred protocol [7]. In addition, a comparison of the energy consumption level shows that the MQTT protocol consumes less energy than CoAP [8]. Therefore, we assess the energy consumption as well as the existing security risks of the MQTT protocol.

The MQTT protocol has three levels of QoS, ranging from 0 to 2^1 and these QoS levels are related to the level of confidence in the transmission of the packet (i.e., QoS has the lowest confidence level and QoS-2 has the confidence level highest reliability). Shinho et al. [9] demonstrate that the packet loss rate and the packet transmission rate of the QoS-0 level are the highest. Besides, Jevgenijus et al. [10] found that the energy consumption of the QoS-0 level is only about 50% of the QoS-2 level and the communication bandwidth of QoS-0 is also lower than that of QoS-2. Therefore, in this paper, we present a system design that takes advantage of the fast speed, low energy consumption and bandwidth while still meeting the reliability of the QoS-2 level.

The current MQTT protocol only provides identity, authentication and authorization for the security mechanism [11]. L. Lundgren et al. [12] indicate that data can be obtained by subscribing to any topic of the MQTT broker that is public on the internet. This is a serious limitation in terms of security. If the attacker subscribes to MQTT topics with the client ID, the victim will experience a denial of service (DOS) state and all information sent to the victim will be forwarded to the attacker [13].

MQTT supports authentication by username and password pair for authorization mechanism, but this authentication mechanism is not encrypted [11]. According to a survey by Zaidi et al. [14] about Shodan, the world's first IoT search engine for Internet-connected devices, points out that there are 67,000

---

[1] https://mqtt.org/.

MQTT servers on the public Internet with most of them without authentication. For authorization, access to a specific topic based on the access list (ACL). This mechanism must be defined in advance in the MQTT broker configuration file and must restart the MQTT broker service if the users want to apply the new access-list configuration. This is inconvenient and difficult to expand especially for systems with billions of devices [15].

Besides MQTT's security flaws, IoT systems are vulnerable to user behavior. Users often tend to ignore security issues especially privacy until the loss of critical data [16]. Subahi et al. [17] found that a significant proportion of IoT users were not fully aware of their behavior. Therefore, with IoT systems having billions of users and devices, it is difficult to manage the behavior of all users.

In this paper, we propose the model of IoT Platform using message queue, which is able to take advantage of the reliable transmission capability of MQTT QoS-2 but still satisfies the fast message transmission rate. Besides, this paper shows how to combine the MQTT and Oauth protocols to improve the Authentication and Authorization mechanism of the current MQTT protocol. We also offer a management model of users, things and channels information to prevent DOS attacks and attack the availability of systems using MQTT protocol. In addition, we offer a method of strict management of the communication process to ensure that users control their information-sharing channels to limit the user's careless behavior. To engage further reproducibility or improvement in this topic, finally, we share the completely code solution which is publicized on the our Github[2].

The remainder of the paper is organized as follows. The next two sections present the background and related work. In Sects. 4 and 5, we introduce the IoT Platform to build a prototype system and its implementation, respectively. In the evaluation section, we discuss our results. Finally, we summarize the main idea of this work and discuss potential directions for future work in the conclusion.

2   Background

2.1   MQTT Protocol

MQTT (Message Queue Telemetry Transport) is a messaging protocol in a publish/subscribe model, using low bandwidth and high reliability. MQTT architecture consists of two main components: Broker and Clients. In which, MQTT Broker is the central server, it is the intersection point of all the connections coming from the client. The main task of the broker is to receive messages from all clients, and then forward them to a specific address. Clients are divided into two groups: publisher and subscriber. The former is the user that publishes messages on a specific topic. Whereas, the latter is the user that subscribe to one or more topics to receive messages going to these topics. In the MQTT protocol, there are 3 levels of QoS as follows:

---

[2] https://github.com/thanhlam2110/mqtt-sso-kafka.

- QoS-0 (at-most-once): each packet is transmitted to the destination up to once.
- Qos-1 (at-least-once): each packet is passed to the destination at least once, meaning packet iteration can occur.
- QoS-2 (exactly-once): each packet is sent to its destination only once.

2.2   Oauth Protocol and Single Sign-On

Oauth is an authentication mechanism that helps a third party applications to be authorized by the user. The main purpose of the Oauth protocol is to access user resources located on another application. Oauth version 2 is an upgrade version of Oauth version 1, an authentication protocol that allows applications to share a portion of resources with each other without authentication via username and password the traditional way. Thereby, it helps to limit the hassle of having to enter username, password in many places for many applications that are difficult for users to manage.

In Oauth, there are four basic concepts[3]:

- Resource owners: are the users who have the ability to grant access, the owner of the resource that the application wants to get.
- Resource servers: are the places which store resources, capable of handling access requests to protected resources
- Clients: are third-party applications that want to access the shared resource of the owner (i.e., prior to access, the application needs to receive the user's Authorization).
- Authorization servers: are the authentications that check the information the user sent from there, grants access to the application by generating access tokens. Sometimes the same Authorization server is the resource server.

Token is a random code generated by the Authorization server when a request comes from the client. There are two types of tokens, namely the access token and the refresh token. The former is a piece of code used to authenticate access, allowing third-party applications to access user data. This token is sent by the client as a parameter in the request when it is necessary to access the resource in the Resource server. The access token has a valid time (e.g., 30 min, 1 h), when it expired, the client had to send a request to the Authorization server to get the new access token. Whereas, the latter is also generated by Authorization server at the same time with accessed token but with different function. Refresh token is used to get the new access token when it expires, so the validity period is longer than the access token.

Single Sign-On (SSO) is a mechanism that allows users to access multiple applications with just one authentication. SSO simplifies administration by managing user information on a single system instead of multiple separate authentication systems. It makes it easier to manage users when they join or leave an organization [18]. SSO supports many authentication methods such as Oauth, OpenID, SAML, and so on.

---

[3] https://oauth.net/2/.

2.3 Kafka

Kafka[4] is a distributed messaging system. Kafka is capable of transmitting a large amount of messages in real-time, in case the receiver has not received the message, the message is still stored on the message queue and on the disk to ensure safety. The Kafka architecture includes the main components: producer, consumer, topic, and partition. Kafka producer is a client to publish messages to topics. Data is sent to the partition of the topic stored on the broker. Kafka consumers are clients that subscribe and receive messages from topic, consumers are identified by group names. Many consumers can subscribe to the same topic. Data is transmitted in Kafka by topic, when it is necessary to transmit data for different applications, it is possible to create many different topics. Partition is where to store data of a topic. Each topic can have one or more partitions. On each partition, the data is stored permanently and assigned an ID called offset. In addition, a set of Kafka server is also called a broker and the zookeeper is a service to manage the brokers.

3 Related Work

3.1 Oauth and MQTT

Paul Fremantle et al. [15] used Oauth to enable access control in the MQTT protocol. The paper results show that IoT clients can fully use OAuth token to authenticate with an MQTT broker. The paper demonstrates how to deploy the Web Authorization Tool to create the access token and then embed it in the MQTT client. However, the paper does not cover the control of communication channels, so when the properly authenticated MQTT client is able to subscribe to any topic on the MQTT broker, this creates the risk of data disclosure. The paper presents the combined implementation of Oauth and MQTT for internal communication between MQTT broker and MQTT client in the same organization, but not the possibility of applying for inter-organization communication. Therefore, in our paper, we implement a strict management mechanism for users, devices, and communication channels.

Benjamin Aziz et al. [19] invested Oauth to manage the registration of users and IoT devices. These papers also introduce the concept of Personal Cloud Middleware (PCM) to perform internal communication between the device and a third-party application on behalf of the user. PCM is an MQTT broker that isolates and operates on a Docker or operating system. Each user has their PCM, and this can help limit data loss. However, Benjamin Aziz et al. also said that they do not have a mechanism for revoking PCM when users are no longer using IoT services.

Lam et al. [20] tested the ability to combine MQTT and Oauth through Single Sign On to enhance security. The author proposes a users and things

---

[4] https://Kafka.apache.org/.

management model - tree model. This architecture has a single user representative for the organization, thus allowing to quickly isolate all users and devices of an organization when the organization is attacked.

3.2 Kafka and MQTT

A.S. Rozik et al. [21] found that the MQTT broker does not provide any buffering mechanism and cannot be extended. When large amounts of data come from a variety of sources, both of these features are essential. In the Sense Egypt IoT platform, A.S. Rozik et al. have used Kafka as an intermediary system to transport messages between the MQTT broker and the rest of the IoT system, which improves the overall performance of the system as well as provides easy scalability.

Moreover, [22] presented Kafka Message Queue and MQTT broker's combined possibilities in Intelligent Transportation System. The deployment model demonstrates the ability to apply to bridge MQTT with Kafka for low latency and handle messages generated by millions of vehicles. They used MQTT Source Connector to move messages from MQTT topic to Kafka Topic and MQTT Sink connector to move messages from Kafka topic to MQTT topic.

In the implementation of the IoT Platform, we also adopt and extend this technique by building APIs that allow users to map their topics.

3.3 Reducing MQTT Energy Consumption

A new approach to the MQTT-SN protocol [23], publishes sensor data (a.k.a the smart gateway selection method) that estimates end-to-end content delay and message loss, during the transmission of content in all levels of QoS. Al-Ali et al. [24] argued that MQTT QoS-2 excluded from test cases due to its large overhead the smart energy management for smart homes and cities. Similarly, Toldinas et al. [10] estimated the energy consumption in transferring data using lightweight MQTT protocol over it different QoS levels (i.e., QoS = 0, QoS = 1, and QoS = 2). However, these works ignore the security and privacy aspects, that is highlighted in [25]. The author investigated the side effects of reliability on IoT communication protocols but they considered MQTT as reliable protocol.

Lam et al. [26] demonstrated an architecture that combines MQTT broker, Single Sign On, and kafka message queue. This combination allows no need to trade-off speed and reliability when communicating with power consumption (this is related to QoS-0 and QoS-2 levels) while still ensuring security. of the system.

4   IoT Platform Proposal

The IoT Platform is a set of APIs combined with system architecture such as Single Sign-On system, Kafka Message Queue, and MQTT broker that provide as following:

- Authenticate information believes the user, thereby granting Oauth access token and refresh token for the user.
- Create and manage user information in the tree model and the unlimited number of user levels created.
- Allowing users to participate in an IoT system capable of creating logical information management of physical devices/ applications and communication channels.
- Allows sending and receiving messages locally when correctly defining a user, a thing, and a specific communication channel.
- Allow users to send and receive messages between two various organizations through the Kafka message queue.

4.1   System Architecture

Figure 1 presents an architectural proposal model of the IoT framework.

- MQTT Broker Cluster is a set of MQTT brokers connected together to distribute the system. MQTT Broker Cluster plays the role of collecting data from IoT devices and transporting control commands from the user to the device according to the MQTT protocol.
- Server Single Sign-On authenticates the user, creates access and refreshes the token according to the Oauth protocol.
- SSO database: contains management information of users, channels and things.
- The Kafka Message Queue stores messages in partitions. This prevents the message from being lost when the receiver has problems.
- MQTT Proxy API provides an interface for users - things to communicate with the MQTT broker.
- API SSO Proxy: provides an interface for users and authenticated parties to communicate with SSO server.

4.2   Software Architecture

To meet the goals set out by the IoT Framework, we provide a number of definitions of the components involved in the system, the design of the database, and the interactions among these components.

4.2.1   Users

Who use IoT services. By constructing the user hierarchy in the model tree with the child's user_parent_id value equal to the parent user's username. Our model allows the creation and management of multiple levels of users and undepending on the organisation's characteristics. This tree-modelled user hierarchy makes suitable for companies, especially when it comes to decentralizing a specific user or changing the state of operation for a series of child users at the same time w.r.t crashes (only the ACTIVE user can request the access token). User information

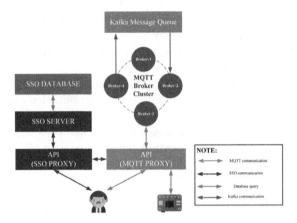

Fig. 1. IoT platform proposal

is generated when registering/using an IoT service or is created by the parent user and provided to their child users.

Each user has a unique user_id value, which conforms to the UUID standard[5] and is managed by the IoT Platform (user is not aware of this value). When publishing or subscribing, the user must pass the access token obtained from the Single Sign-On server, which contains the user_id information and is used as the `clientID` value in the MQTT protocol. In this way, we have enhanced the authentication and authorization mechanisms for the user as well as minimize the risk of a denial-of-service attack when hackers subscribe to a topic with the clientID of the user affecting the accessibility use of the system-wide.

4.2.2   Things

The information about physical devices or applications owned by the user. To create things, the user needs to call the API provided by the IoT Platform, passing in his valid Oauth token. Things information includes two values thing_id and thing_key, conform to the UUID standard and use the equivalent of a username and password values. Only things possess a valid pair of thing_id and thing_key provided by the IoT Platform, that can communicate with the MQTT broker. Since things are not allowed to publish or subscribe directly to the MQTT broker. This layer API validates thing_id and thing_key submitted by things. Similar to the user, this way to create device management information, helps increase security and reduce the risk of denial-of-service attacks.

4.2.3   Channels and Map Things to Channels

In our proposed IoT Platform, channels are the logical concept of managing topics (Kafka and MQTT) that publishes and subscribe to messages by users and things. Users who want to create a channel must call the IoT Platform's API

---

[5] https://tools.ietf.org/html/rfc4122.

and pass in their valid token Oauth. The channels information has a channel_id value that is unique and follows the standard UUID.

To communicate through the IoT Platform Proposal, the user has to assign things to the channel by calling the API and passing in the existing access token, thing_id, thing_key and channel_id information. The purpose of this process is to allow only one thing with a valid thing_id and thing_key to publish and subscribe to messages on a predefined channel (mapped). From there, avoid the client can subscribe to any topic. This strongly increases the authorization mechanism, which is a flexible way that the original MQTT protocol did not support. The mechanism of assigning things to a channel also enhances security since only things mapped to the channel is able to publish and subscribe to messages on this channel. By implementing this process, users are also able to master their own communication channels.

4.2.4 Publish and Subscribe Message

The process of publishing and subscribing messages via IoT Platform Proposal is described as follows: To perform communication in the IoT Platform, users create two channels with the following roles:

- The channel used to send messages is called "send-channel". The "send-channel" is an MQTT topic and the device will publish the message to this channel through the MQTT Proxy API. Actually, the API uses channel_id but for brevity, we show it by the name of the channel.
- The channel used to receive messages is called a "receive-channel". The "receive-channel" acts as both the Kafka topic and the MQTT Topic.

After creating two channels, we call the API that creates the MQTT source connector, to map the MQTT Topic and the Kafka topic. In this case, we map the "send-channel" and the "receive" channel. This process allows messages from Thing to be sent to "send-channel" to be automatically forwarded to "receive-channel". Right now, the message is being stored in the Kafka message queue, namely, the topic "receive-channel" partition. In order for users to receive messages on this "receive-channel" topic over the MQTT protocol, we implement an MQTT sink connector. This process allows, messages to Kafka topic "receive-channel" are received by users who subscribe to the MQTT topic "receive-channel".

During Publish and Subscribe message, the MQTT Proxy API also calls the authentication service from the SSO Proxy API to validate the token, check for thing_id, thing_key, and assign things to the channel. The process of publishing and subscribing to the message is shown in Fig. 2.

5 Implement

5.1 Database

As explained in Sect. 4.2, the IoT Platform allows to manage users, things, channels information and implement map things to channel. In practical implementation, we use MongoDB as a NoSQL database management system, and to

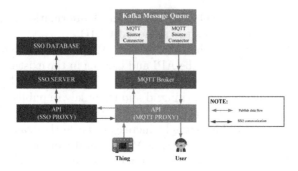

Fig. 2. Process publish a message to the public

implement the model tree (select, update, delete) outlined in Sect. 4.2.1, we used Aggregation techniques are provided by MongoDB[6].

5.2 Single Sign-On

In the prototype system, we use the open-source CAS Apereo[7] to provide the Single Sign-On service. CAS Apereo supports many protocols for implementing single sign-on services such as Oauth, SAML, OpenID, and so on. The IoT Platform protocol used to communicate with the Single Sign-On server is Oauth. In our implementation, the clients do not interact directly with the CAS server but instead, we provide the set of APIs through the SSO Proxy layer (API SSO Proxy layer).

5.3 Mosquitto MQTT Broker

Mosquitto is an open-source to implement an MQTT broker that allows to transmit and receive data according to MQTT protocol. Mosquitto is also part of the Eclipse Foundation[8]. Mosquitto is very light and has the advantages of fast data transfer and processing speed, high stability.

5.4 Prototype Model

The Prototype system we deployed on Amazon EC2[9] infrastructure consists of three servers as shown in Table 1.

According to the MQTT protocol, the first server deploys the Mosquitto MQTT Broker service in the Prototype model to collect and forward messages. The second server, deploying MQTT Broker 1b service for public communication.

---

[6] https://docs.mongodb.com/manual/reference/operator/aggregation/graphLookup/.

[7] https://apereo.github.io/cas/6.3.x/index.html.

[8] The project iot.eclipse.org (https://iot.eclipse.org/).

[9] https://aws.amazon.com/ec2/pricing/.

Table 1. The list of servers in the prototype

Server	Role	Server configuration
MQTT Broker	Deploy service MQTT broker Deploy API MQTT Proxy	CPU 1 RAM 1GB
Kafka server	Deploy the Kafka service	CPU 1 RAM 1GB
SSO server	Deploy Single Sign-On service Deploy the SSO database service Deploy the service API SSO Proxy	CPU 2 RAM 1GB

Also, on the first server, we deploy the MQTT Proxy API service to provide interfaces for users and things that communicate with MQTT brokers through the API. Whereas, the second server implements the Single Sign-On service, which creates access to tokens and refreshes tokens for users according to the Oauth protocol. Besides, the SSO Proxy API service provides interfaces for users, things, and the MQTT Proxy API performs SSO server communication through the API. In addition, the second server also has MongoDB installed to store user information, thing, and channel, presented in Sect. 4.2.

The Kafka service, deployed on a third server, acts as a message queue to store messages exchanged through the MQTT broker. This ensures that when things or users experience interruptions, notifications are not lost as well as limit infinite timeouts or packet repetitions.

6 Evaluation

Our proposed model has two goals. The first is to strengthen the security mechanism for MQTT Protocol. The second is to build a mechanism to take advantage of the advantages of MQTT QoS-0 rather than QoS-2, such as transmission speed, bandwidth and low energy consumption while ensuring reliability when transmitting packets. To test these goals, we execute the following scenarios.

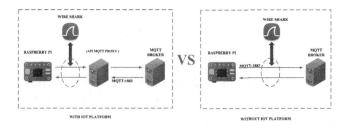

Fig. 3. The first scenario test model

The first scenario uses Wireshark software[10] to capture packets when things send messages to the MQTT broker in two cases. In the first case, things do connect and publish the message directly to the MQTT broker. In the second case, things do connect and publish the message through the MQTT Proxy API class provided by the IoT Platform Proposal. The first scenario model is shown in Fig. 3. The outcomes captured of the first scenario from the Wireshark software are shown in Fig. 4

Fig. 4. Capture message of the two cases (with and without IoT Platform)

The resulting snapshot shows that publish the message directly to the MQTT broker, it takes two steps: "Connect" and "Publish" in the case of things. Performing packet capture can obtain all `ClientID`, username, and password in the "Connect" packet and MQTT topic along with the message content in the "Publish" packet. When things publish a message via IoT Platform Proposal, it only takes one step of "Publish". The content of the message is encrypted because the IoT Platform supports TLS. After the IoT Platform receives the message, it will use the information in the message such as access token, channel_id, things_id and things_key to authenticate the request, make a connection to the MQTT broker, and transport the packet according to the received topic. From here, the use of the IoT Platform Proposal significantly improves the security risks of the MQTT protocol.

Whereas, the second scenario examines the packet transfer rate, throughput, and the number of things that can publish messages simultaneously in three test-cases:

- **Test 1**: things connect directly to the MQTT broker and do publishing messages according to QoS-2.
- **Test 2**: things connect to the MQTT broker through the IoT Platform Proposal and do publish the message according to QoS-0. In this case, do not apply security checks such as: validity of access token, check assigning things to channel.
- **Test 3**: things connect to MQTT broker via IoT Platform Proposal, publish messages according to QoS-0 and apply security check function.

---

[10] Wireshark (https://www.wireshark.org/) is a network packet analyzer software (a network packet analyzer) capable of monitoring and monitoring packets in real-time.

The test we use is Apache Jmeter[11] that is fully written in Java language, and usable for performance testing on static resources, dynamic resources and Web applications. This is used to simulate a large number of virtual users, large requests on a server, a group of server/network, an object to test tolerance, load test or analyze the response time. Jmeter firstly makes the requests and sends them to the server according to the predefined method. Then, it receives responses from the server, collects them and displays information in the report (report). Jmeter has many report parameters, but we consider in two parameters: throughput and error, with throughput (request/s), which is the number of requests processed by the server per second and error (%) is the percentage of requests that fail to total requests. Test results are shown in Table 2.

Table 2. Test results of the second scenario

	Result	100 CCU	200 CCU	300 CCU	400 CCU	500 CCU
Test 1	Throughput	48.8	97.4	139.1	193.4	–
	Error	0	0	0	47%	–
Test 2	Throughput	95.3	189.6	312.5	226.6	208.7
	Error	0	0	0	0	0
Test 3	Throughput	48.8	50.1	51.4	50	38.8
	Error	0	0	0	0	0

Table 2 shown that the use of QoS-2 helps ensure the packet is transmitted to the destination and not duplicated in Test 1. However, to achieve this, we have to trade system performance such as lower throughput and higher error rate since the server has to handle more when sending messages in QoS-2. We noted the effectiveness of MQTT and Kafka Message Queue's combination w.r.t in Test 2. Using QoS-0 for faster processing speed and throughput was much higher than using QoS-2, the packets, in this case, have still send the guaranteed and did not duplication thanks to Kafka's capabilities. The third test is completely acceptable as our IoT Platform proposal has added the test and validation mechanisms as outlined in Sect. 4.2.

To develop a larger scenario and increase the number of devices/users authorized quickly, other security issues such as security, privacy, availability for objects are still the challenges. For the security aspect, further works will be deployed in different scenarios like healthcare environment [27–29], cash on delivery [30,31]. For the privacy aspect, we will exploit attribute-based access control (ABAC) [32,33] to manage the authorization process of the IoT Platform via the dynamic policy approach [34–36]. Besides, we will apply the blockchain benefit to improve the availability issues [37–39]. Finally, we eliminate reliance on MQTT Broker for the data collection process and routing and building a more proactive method of data collection for the low-energy devices.

---

[11] Apache Jmeter https://jmeter.apache.org/.

7   Conclusion

In this paper, we propose a combination method of Oauth protocol, Single Sign-On, user management model, thing and channel to improve the security of MQTT protocol. The method of assigning things to the channel helps the system strictly manage the communication channels to minimize the careless behavior of users when sharing data. Besides, MQTT and Kafka Message Queue's combination allows our approach to accepting low-resource devices such as low communication speeds, low energy consumption, and low transmission bandwidths while still providing reliability in transmitting. The evaluation section also shows the effectiveness of the proposal IoT Platform.

References

1. Alam, T.: A reliable communication framework and its use in internet of things (IoT). CSEIT1835111/Received **10**, 450–456 (2018)
2. Chou, T.: Precision-Principles, Practices and Solutions for the Internet of Things. McGraw-Hill Education, New York (2017)
3. Karagiannis, V., et al.: A survey on application layer protocols for the internet of things. Trans. IoT Cloud Comput. **3**(1), 11–17 (2015)
4. Niruntasukrat, A., et al.: Authorization mechanism for MQTT-based internet of things. In: 2016 IEEE International Conference on Communications Workshops (ICC), pp. 290–295. IEEE (2016)
5. Jaikar, S.P., Iyer, K.R.: A survey of messaging protocols for IOT systems. Int. J. Adv. Manag. Technol. Eng. Sci. **8**(II), 510–514 (2018)
6. Çorak, B.H., et al.: Comparative analysis of IoT communication protocols. In: 2018 International Symposium on Networks, Computers and Communications (ISNCC), pp. 1–6. IEEE (2018)
7. Hillar, G.C.: MQTT Essentials-A Lightweight IoT Protocol. Packt Publishing Ltd, Birmingham (2017)
8. Martı, M., Garcia-Rubio, C., Campo, C.: Performance evaluation of CoAP and MQTT_SN in an IoT environment. In: Multidisciplinary Digital Publishing Institute Proceedings, vol. 31, no. 1, p. 49 (2019)
9. Lee, S., et al.: Correlation analysis of MQTT loss and delay according to QoS level. In: The International Conference on Information Networking 2013 (ICOIN), pp. 714–717. IEEE (2013)
10. Toldinas, J., et al.: MQTT quality of service versus energy consumption. In: 2019 23rd International Conference Electronics, pp. 1–4. IEEE (2019)
11. Mendez Mena, D., Papapanagiotou, I., Yang, B.: Internet of things: survey on security. Inf. Secur. J. Glob. Perspect. **27**(3), 162–182 (2018)
12. Lundgren, L.: Light-Weight Protocol! Serious Equipment! Critical Implications! In: Defcon 24 (2016)
13. Anthraper, J.J., Kotak, J.: Security, privacy and forensic concern of MQTT protocol. In: Proceedings of International Conference on Sustainable Computing in Science, Technology and Management (SUSCOM), Amity University Rajasthan, Jaipur-India (2019)

14. Zaidi, N., Kaushik, H., Bablani, D., Bansal, R., Kumar, P.: A study of exposure of IoT devices in India: using Shodan search engine. In: Bhateja, V., Nguyen, B.L., Nguyen, N.G., Satapathy, S.C., Le, D.-N. (eds.) Information Systems Design and Intelligent Applications. AISC, vol. 672, pp. 1044–1053. Springer, Singapore (2018). https://doi.org/10.1007/978-981-10-7512-4_105

15. Fremantle, P., et al.: Federated identity and access management for the internet of things. In: 2014 International Workshop on Secure Internet of Things, pp. 10–17. IEEE (2014)

16. Tawalbeh, L., et al.: IoT Privacy and security: challenges and solutions. Appl. Sci. 10(12), 4102 (2020)

17. Subahi, A., Theodorakopoulos, G.: Detecting IoT user behavior and sensitive information in encrypted IoT-app traffic. Sensors 19(21), 4777 (2019)

18. Radha, V., Reddy, D.H.: A survey on single sign-on techniques. Procedia Technol. 4, 134–139 (2012)

19. Fremantle, P., Aziz, B.: Oauthing: privacy-enhancing federation for the internet of things. In: 2016 Cloudification of the Internet of Things (CIoT), pp. 1–6. IEEE (2016)

20. Nguyen, L., Thanh, T., et al.: Toward a unique IoT network via single sign-on protocol and message queue. In: International Conference on Computer Information Systems and Industrial Management (2021)

21. Rozik, A.S., Tolba, A.S., El-Dosuky, M.A.: Design and implementation of the sense Egypt platform for real-time analysis of IoT data streams. Adv. Internet Things 6(4), 65–91 (2016)

22. Hugo, Å., et al.: Bridging MQTT and Kafka to support C-ITS: a feasibility study. In: 2020 21st IEEE International Conference on Mobile Data Management (MDM), pp. 371–376. IEEE (2020)

23. Roy, D.G., et al.: Application-aware end-to-end delay and message loss estimation in Internet of Things (IoT)-MQTT-SN protocols. Future Gener. Comput. Syst. 89, 300–316 (2018)

24. Al-Ali, A.-R., et al.: A smart home energy management system using IoT and big data analytics approach. IEEE Trans. Consum. Electron. 63(4), 426–434 (2017)

25. Safaei, B., et al.: Reliability side-effects in Internet of Things application layer protocols. In: 2017 2nd International Conference on System Reliability and Safety (ICSRS), pp. 207–212. IEEE (2017)

26. Thanh, L.N.T., et al.: UIP2SOP: a unique IoT network applying single sign-on and message queue protocol. IJACSA 12(6) (2021)

27. Son, H.X., Chen, E.: Towards a fine-grained access control mechanism for privacy protection and policy conflict resolution. Int. J. Adv. Comput. Sci. Appl. 10(2), 507–516 (2019)

28. Duong-Trung, N., et al.: Smart care: integrating blockchain technology into the design of patient-centered healthcare systems. In: International Conference on Cryptography, Security and Privacy, pp. 105–109 (2020)

29. Duong-Trung, N., et al.: On components of a patient-centered healthcare system using smart contract. In: Proceedings of the International Conference on Cryptography, Security and Privacy, pp. 31–35 (2020)

30. Le, H.T., et al.: Introducing multi shippers mechanism for decentralized cash on delivery system. Money 10(6) (2019)

31. Le, N.T.T., et al.: Assuring non-fraudulent transactions in cash on delivery by introducing double smart contracts. IJACSA 10(5), 677–684 (2019)

32. Hoang, N.M., Son, H.X.: A dynamic solution for finegrained policy conflict resolution. In: The International Conference on Cryptography, Security and Privacy, pp. 116–120 (2019)

33. Son, H.X., Hoang, N.M.: A novel attribute-based access control system for fine-grained privacy protection. In: The International Conference on Cryptography, Security and Privacy, pp. 76–80 (2019)

34. Xuan, S.H., et al.: Rew-XAC: an approach to rewriting request for elastic ABAC enforcement with dynamic policies In: International Conference on Advanced Computing and Applications, pp. 25–31. IEEE (2016)

35. Thi, Q.N.T., Dang, T.K., Van, H.L., Son, H.X.: Using JSON to specify privacy preserving-enabled attribute-based access control policies. In: Wang, G., Atiquzzaman, M., Yan, Z., Choo, K.-K.R. (eds.) SpaCCS 2017. LNCS, vol. 10656, pp. 561–570. Springer, Cham (2017). https://doi.org/10.1007/978-3-319-72389-1_44

36. Son, H.X., Dang, T.K., Massacci, F.: REW-SMT: a new approach for rewriting XACML request with dynamic big data security policies. In: Wang, G., Atiquzzaman, M., Yan, Z., Choo, K.-K.R. (eds.) SpaCCS 2017. LNCS, vol. 10656, pp. 501–515. Springer, Cham (2017). https://doi.org/10.1007/978-3-319-72389-1_40

37. Ha, X.S., et al.: DeM-COD: novel access-control-based cash on delivery mechanism for decentralized marketplace. In: International Conference on Trust, Security and Privacy in Computing and Communications, pp. 71–78. IEEE (2020)

38. Ha, X.S., Le, T.H., Phan, T.T., Nguyen, H.H.D., Vo, H.K., Duong-Trung, N.: Scrutinizing trust and transparency in cash on delivery systems. In: Wang, G., Chen, B., Li, W., Di Pietro, R., Yan, X., Han, H. (eds.) SpaCCS 2020. LNCS, vol. 12382, pp. 214–227. Springer, Cham (2021). https://doi.org/10.1007/978-3-030-68851-6_15

39. Son, H.X., Le, T.H., Quynh, N.T.T., Huy, H.N.D., Duong-Trung, N., Luong, H.H.: Toward a blockchain-based technology in dealing with emergencies in patient-centered healthcare systems. In: Bouzefrane, S., Laurent, M., Boumerdassi, S., Renault, E. (eds.) MSPN 2020. LNCS, vol. 12605, pp. 44–56. Springer, Cham (2021). https://doi.org/10.1007/978-3-030-67550-9_4

Towards a Novel Vehicular Ad Hoc Networks Based on SDN

Houda Guesmi$^{(\boxtimes)}$, Anwar Kalghoum, Ramzi Guesmi, and Leïla Azouz Saïdane

CRISTAL LAB, National School of Computer Science, Manouba, Tunisia
{houda.guesmi,anwar.kalghoum,leila.saidane}@ensi-uma.tn,
ramzi.guesmi@gmail.com

Abstract. Driven by long traffic jams and numerous road accidents, vehicle networks (Vehicular Ad hoc NETwork, VANET) have emerged to make the journey more pleasant, the road safer and the transport system more efficient. Today's vehicle network architectures suffer from scalability issues as it is challenging to deploy services on a large scale. These architectures are rigid, difficult to manage and suffer from a lack of flexibility and adaptability due to vehicular technologies' heterogeneity.

Over the past few years, the emerging paradigm of Software-Defined Networking (SDN) network architecture has become one of the most important technologies for managing large-scale networks such as vehicle networks. By the first vision and under the SDN paradigm umbrella, we propose a new VANET network architecture based on the SDN paradigm named "SDN-based vehicular ad hoc networks" (SDN-VANET). Through our simulation, we show that in addition to the flexibility and fine programmability brought by the SDN paradigm, the latter opens the way to the development of efficient network control functions.

Keywords: VANET · Software Defined Networking · SDN · openFlow · Distributed controller

1 Introduction

Nowadays, we spend more and more time in transport, whether in personal vehicles or public transport. In addition, our uses of the means of communication have become more and more nomadic, especially with the significant advances in information and communication technologies (ICT). Therefore, the concept of ad hoc vehicle networks emerged as a result of this development to offer a wide variety of services, ranging from improving road safety to optimizing traffic, including entertainment. Driver and passengers. Indeed, this type of network is mainly characterized by the nodes' high mobility, the incredibly dynamic topology, and the network's enormous scale. Vehicle networks represent a projection of Intelligent Transportation Systems (ITS), in which vehicles can communicate with each other and with infrastructure along roads. In this chapter, we introduce the basic concepts of vehicle networks before describing the state of the art of the software-defined networking paradigm applied to vehicle networks.

© Springer Nature Switzerland AG 2021
O. Gervasi et al. (Eds.): ICCSA 2021, LNCS 12949, pp. 663–675, 2021.
https://doi.org/10.1007/978-3-030-86653-2_48

Despite the development and rapid emergence of vehicle networks with futuristic capabilities, their architectures suffer from several shortcomings. In addition to the problem of the heterogeneity of network equipment, which causes great difficulties in management and integration [1], we can cite, for example, (i) the lack of scalability in the deployment of services on a large scale in such a dense and dynamic topology such as that of vehicle networks [1,2], (ii) lack of intelligence, mainly due to the closed aspect of vehicular equipment and their inherent characteristics such as the absence of programmability and their development dependence on suppliers. This implies severe and challenging to manage architectures, (iii) the lack of flexibility and adaptability, induced by the great diversity of deployment environments and the vast heterogeneity of wireless communication technologies (4G/5G, WiFi, etc.).

Vehicular Ad-hoc Networks (VANETs) [1,3,4] architectures today suffer from scalability issues as it is tough to deploy solutions on a large scale. These architectures are rigid, difficult to manage, and suffer from a lack of flexibility and adaptability in control [1]. Therefore, it isn't easy to choose the right solution, given the current context, due to the diversity of deployment environments and the wide variety of solutions. These constraints limit the functionality of the system and often lead to the under-exploitation of network resources. Thus, the need for new, more flexible, and scalable architectures becomes an absolute requirement to face the new needs of the next generations of vehicle networks.

In recent years, the Software-Defined Networking (SDN) paradigm [5] has been proposed as an innovative solution to manage large-scale networks such as VANET. Indeed, several works [1] have shown that by separating the control plane from that of the data and that by adopting a centralized control mode, SDN can provide flexibility, scalability, and programmability to the architectures of VANET. SDN also makes more efficient use of network resources and introduces new services.

In line with the first vision and under the SDN paradigm umbrella, we provide a new network architecture that allows joint control of different vehicle access network technologies. We also explore the possibilities offered by such an architecture combining both the advantages of hybridization of various technologies and the SDN paradigm properties. We show through a few use cases that in addition to the flexibility and fine programmability brought by the SDN paradigm, the latter paves the way for efficient network control functions.

This article will combine the VANET paradigm and the SDN architecture and propose a VANET architecture based on the SDN paradigm. Our proposal called SDN-based vehicular ad hoc networks strategy (SDN-VANET), relies on SDN controllers. The experimental results show that SDN-VANET significantly reduces the network load while improving the VANET network's performance.

This article is organized as follows. Section 2 presents some preliminary concepts such as Software-defined networking and vehicle networks. We present a synthesis of existing work in the scientific literature in Sect. 3. Section 4 details the proposed approach. Section 5 describes the simulation results of a use case. Finally, the last Sect. 6 concludes this article.

2 Software-Defined Networking

Initially designed for wired networks and, in particular, data centers, SDN has had tremendous success in industry and academia. In 2012, Google engineers announced that they had switched to using SDN to connect their WANs (Wide Area Networks) to data centers [6]. There are currently even network devices compatible with SDN and available on the market, such as the OpenStack [7] or OpenDaylight [6] controller. Since then, researchers have explored all the possibilities to take advantage of SDN's advantages to improve performance and facilitate the management of today's vehicle network architectures.

2.1 Principle and Characteristic of SDN

The concept of software-defined network, widely known as Software-Defined Networking (SDN) [6], is an emerging new paradigm of network architecture primarily based on (i) a physical separation between the control plane (i.e., the functionalities which ensure the management of the network) and the data plane (i.e., the functionalities which assure the transfer of the data), and (ii) control and a logically centralized intelligence in one or more software controllers. In SDN, controllers have a holistic view of the entire network state and manage other network data plane equipment. These become simple transmitters/receivers of data with minimal intelligence. SDN promises to bring flexibility, scalability, and programmability to vehicle network architectures today. They also facilitate network management and introduce new services [6].

Architecture. SDN is based on a hierarchical three-layer architecture, see Fig. 1:

- Data plane layer: The data plane represents all of the equipment on the network, often called broadcast equipment, which only sends and receives data with minimal intelligence. The broadcast equipment also performs the actions of the controller.
- Control plane layer: The control plane represents all network equipment, often called SDN controllers, which centralize network intelligence and manage other data plane broadcast equipment.
- Application layer: it groups all the services and applications of the systems installed on the SDN controller.

2.2 Interface de Communication

To allow communication between the three planes, SDN defines several unified communication interfaces [6], see Fig. 1:

- North-bound API: it allows communication and data exchange between the SDN controller on the control plane and network applications. The type of information exchanged and its forms and frequencies, depending on each network application. There is no standardization for this interface.

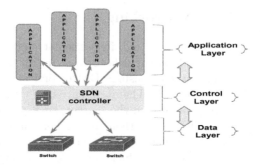

Fig. 1. Software-defined network (SDN) architecture

– South-bound API: it refers to the different APIs that allows communication between control plane equipment and data plane equipment. OpenFlow [6] is the most widely used standard for this interface.

2.3 OpenFlow

OpenFlow is a communication standard for SDN managed by the ONF (Open Networking Foundation) [6], a foundation of industrialists whose primary mission is to design and promote network equipment compatible with to facilitate its marketing. OpenFlow defines two types of network equipment: OpenFlow controllers, software that centralizes all network control functions, and OpenFlow vStwichs, which are virtual switches that only perform packet data transfer functions. Each vSwitch has a flow table containing controller flow entries. Indeed, the controller manages the vStwichs by installing flow rules in the flow table.

3   Related Work

We mainly focus on work involving the adoption of SDN as an architecture for vehicular networks. This paradigm mainly aims to separate the data plane from the control plane.

Based on the architecture proposed in [8,9] take advantage of the global view provided by the SDN Controller to calculate routing paths in a centralized manner. The simulation compares these approaches to traditional VANET routing protocols. The performance results show that these approaches outperform traditional approaches in terms of reliability and latency. This further demonstrates the value of integrating the SDN paradigm into VANET networks and seeing vehicles as nodes programmable via SDN.

To respond to scaling issues, especially in very dense environments, the authors [10] propose a decentralized SD-VANET architecture in which the control plane is distributed. As expected, the results show that the distribution of the control plane improves the scalability of the network (measured in the number of requests of the data plane processed according to the density of the vehicles) while ensuring an acceptable delay of data delivery.

We cite the architecture proposed by Xiaohu Ge et al. [11] in which they consider a heterogeneous data plane following a hierarchical control plane. They propose using a fog cell based on multi-hop links where a vehicle functions as a gateway to minimize the frequent changes of RSU (handover) attachment points.

Jianqi Liu et al. [12] propose a heterogeneous vehicular network architecture based on control via SDN assisted by MEC (Mobile Edge Computing) to support the most demanding services in terms of latency. Two safety and non-safety use cases are simulated and the performance shows that the proposed architecture supports the expected prerequisites in terms of latency, reliability, and throughput.

Data to guide the control of the vehicular network has also been mentioned recently in the literature. We cite the positioning work [13] listing all the opportunities offered by the use of data for network control. They notably advocate the use of machine learning-based approaches to learn the dynamics of the vehicular network and perform effective network control.

In the same vein and the context of a vehicular network controlled via SDN, the authors focus on developing a specific network control function [14]. It is about the allocation of resources. They propose an approach that takes advantage of the global vision of the network. Their mechanism is based on machine learning techniques.

4   SDN-VANET: SDN-Based Vehicular Ad Hoc Networks

In line with the vision conveyed by our work, namely the proposal for an architecture of vehicular networks, we propose the adoption of the SDN paradigm as the main base of our architecture. We are taking advantage of the contributions of the SDN paradigm to facilitate management and improve VANET control. This paves the way for the development of new network control algorithms that take advantage of 1) insight into the state of various communication networks; 2) the ability to jointly control these networks dynamically and with scaling up (Hierarchical control); and 3) knowledge of the environment in which vehicles operate from the data that orbit this ITS system. This data can come from the various players in the system, for example, ITS operators, road authorities, etc. (Data-driven control).

Figure 2 illustrates the overall view of the proposed architecture. Indeed, integrating the SDN paradigm into the architecture consists of separating the control plane and the data plane. As shown in Fig. 2, SDN controllers hold the network's intelligence and enforce various network policies via specific protocols (e.g. an extension of the OpenFlow standard). The data plane nodes ensure the routing of data according to the instructions provided by the controllers.

We present in what follows the three fundamental design principles of our architecture, namely: 1) Heterogeneous data plane (vehicle, BS, sensor, Object IoT); 2) Hierarchical control plane and 3) data-driven plane.

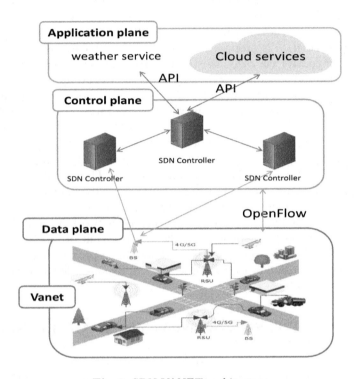

Fig. 2. SDN-VANET architecture

4.1   Key Principles of the Proposed Architecture

Heterogeneous Data Plane. As shown in Fig. 2, the data plane is made up of vehicles, RSUs, base stations (BS), temperature sensors, and IoT objects, all of which can be programmed via SDN.

However, we have chosen to consider the vehicles as programmable nodes via SDN. This choice is initially motivated by the first studies applying the SDN paradigm to VANET networks [2,3]. These studies show that routing is more efficient when calculated centrally in SDN controllers, compared to be distributed computing using conventional routing protocols (OLSR, AODV). We are also convinced that the global vision of the network offered by SDN reduces the interference and risk of collisions through topology control (e.g. transmission parameters, etc.). Therefore, we assume that the interface implemented by the nodes allows controllers to perform control beyond routing, for example, power control, choice of wireless channels, etc.

The data plane nodes are controlled in a transparent manner following a unified model such as OpenFlow (or a southern interface specific to the mobile wireless context), regardless of the technology or the manufacturer of the equipment. In fact, from an SDN controller point of view, the difference between the nodes of the data plane lies in their characteristics and the functionalities supported by each node.

The Data Plane Layer comprises all the network equipment that is solely responsible for collecting and transmitting information, RSUs, and all network equipment located between the cellular base station (BS) and the SDN controller. In our architecture, the data plane layer is divided according to the mobility of its components into two sublayers:

- Fixed data plane: it is said to be fixed because it is made up of static RSUs and all the fixed components of the network (sensors and IoT objects), ensuring the transmission of data between the base station and the SDN controller.
- Mobile data plan: it is said to be mobile because it comprises mobile vehicles, mobile sensors, mobile IoTs objects and Drones.

4.2   Hierarchical Control Plan

SDN controllers are the central part of the architecture. They host all the network control functions to define the various rules to be communicated to the data plane's nodes. They are connected to the various nodes via wired or wireless links, depending on their placement and the type of nodes (Vehicle, RSU, BS).

We consider two main types of controllers:

- Global SDN Controller: The Global SDN Controller has compelling storage and compute capacity and has a holistic view of the entire network topology. It only intervenes to process requests that require a global view of the entire network and/or very significant resources in terms of computation and storage. The global SDN controller is considered the master of all network SDN controllers.
- Core SDN Controllers: They have robust storage and compute capabilities. Central SDN Controllers handle specific requests requiring a central view of the network's state; large compute and storage resources, operations that are not time-sensitive, or local, lower level SDN controllers cannot serve that. A central SDN controller acts in terms of control as a slave to the overall SDN controller. The choice of organizing the control plane hierarchically is motivated by the vision conveyed by our work, namely joint control of these networks. In effect, the central controller builds a global view of the communication infrastructure using each network's information controller. It defines and sends to each controller the global rules that describe the network's general behavior and define the specific rules to be implemented by each node of the network.

4.3   External Data-Driven Control

SDN controllers can call on external data to perform more efficient network control. This data can come from external actors (e.g. weather service, road manager, etc.) and can be used to enrich the overall view of the network and derive a potential view of the state of the network. This allows for proactive/anticipatory network control. Figure 3 shows an example of data exchange between a few actors in the system and their use for network control and the design of ITS services.

The data plane nodes regularly send information to the control plane concerning their states and the characteristics of the links that connect them. This information can be coupled with external data to enrich the vision built by SDN controllers. We refer to the example of predicting potential network quality to perform anticipatory network monitoring.

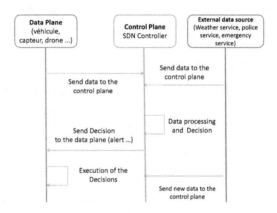

Fig. 3. External data-driven control

5   Simulation

The experiment aims to show how the global view of the network established at the controller level, combined and enriched with the external actors' data, allows a more informed and efficient network control to support ITS services effectively.

We show through evaluations how the SDN Controller can take advantage of its holistic view of current and potential network loads to guide the node in selecting the point of network attachment with the best expected performance.

We first present the simulation tools supporting the SDN-VANET architecture simulation prerequisites. We describe the tools chosen as part of our studies. Finally, we present the simulation results obtained.

5.1 Simulation Tools

To study, evaluate, compare our approach with the VANET without SDN, the simulation environment must combine both:

- Support for SDN programmable networks (e.g. via OpenFlow),
- A wireless communication medium (e.g. 802.11p/LTE vehicular communication standard),
- A vehicle mobility support.

To perform simulations requiring the use of a real SDN controller and/or the OpenFlow protocol, we opt for the choice of MiniNet-WiFi [15,16]. This tool supports the main prerequisite for simulating a vehicle network programmable via SDN. This is the implementation of the OpenFlow protocol and its integration with various nodes (RSU, vehicles, etc.), as well as the possibility of including real SDN controllers.

5.2 Description de la Simulation

The simulation is based on a simple scenario of a Vanet network. Indeed, we are simulating an SDN-VANET network with a topology deployed over an area of $12,000 \times 12,000$ m, which contains an area of the road not covered by the fixed infrastructure. A cellular base station (BS) is placed at the side of the road 1000 m from vehicles. The route is divided into virtual segments of size equal to 200 m each. The node density is 220 vehicles. Each vehicle has two wireless communication interfaces: a DSRC (IEEE 802.11p) interface with a transmission range of up to 300 m and a 4G cellular interface. The vehicles travel at speeds between 15 and 25 m/s. The simulation time is 300 s and the packet generation rate is 10 packets/s.

The performance metrics we used are:

- Flow rule installation delay: represents the time elapsed since a vehicle requests a new flow rule from the SDN controller and the moment when the flow rule is installed in the vehicle flow table.
- Average central processing unit (CPU) usage: Measured at the level of each vehicle.

5.3 Flow Rule Installation Delay

In this experiment, we study the impact of the distance to the SDN controller, which represents the physical distance between the SDN controller and vehicles, on the installation time of flow rules. So, we simulate a scenario where a set of vehicles send 120 packets to request new flow rules from an SDN controller, and we vary the distance between the vehicle and the SDN controller. We repeat this experiment for different package sizes and different numbers of vehicles.

The results in Fig. 4 and Fig. 5 clearly show that the average installation time of flow rules increases with the number of vehicles. This, perhaps justified by the fact that as the number of vehicles increases, the interference increases, and consequently, the waiting time increases. Also, as the distance and the packets' size increase, it is clear that the time required to transmit the end-to-end packets increases. The simulation clearly shows that our solution's application makes it possible to reduce the flow rule installation delay, unlike VANET without SDN.

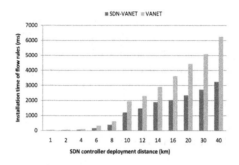

Fig. 4. Flow rule installation delay 80 vehicles

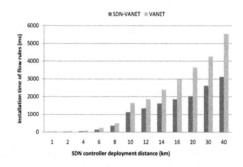

Fig. 5. Flow rule installation delay 120 vehicles

5.4 Average CPU Usage

Figure 6 shows the CPU usage, depending on the number of vehicles. We can see that CPU usage increases in proportion to the number of vehicles. The comparison shows that our method made it possible to considerably reduce CPU resource consumption compared to the VANET without SDN.

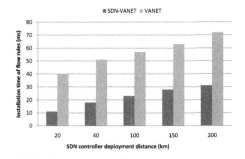

Fig. 6. Average CPU usage

5.5 SDN-VANET Contributions for Vehicle Networks

The main advantages of our SDN-VANET solution in-vehicle networks are:

- Facilitate efficient network management: thanks to the separation of control and data dissemination functionalities and the logically centralized control mode, the integration of SDN in-vehicle networks makes it possible to simplify the network architecture and offer more flexible management and faster configuration of the network. Indeed, instead of configuring and managing each vehicle individually, all network management is centralized in the SDN controller.
- More efficient use of network resources: With the real-time global view of the SDN controller over the entire network topology and the centralized control mode, it becomes easy to identify the current traffic state, which allocates all types of network resources more efficient (i.e., bandwidth, spectrum, transmission power, etc.).
- Facilitate scaling and reduce costs: SDN's logically centralized control mode makes it easy to deploy services and offer new features at scale with minimal cost. This, by only intervening at the SDN controller level instead of acting at each vehicle's level separately. For example, the SDN controller can easily handle a large-scale software update by simply installing entries in remote vehicle flow tables.

6 Conclusion

In this article, we have proposed a new VANET architecture based on SDN to benefit from the SDN approach and improve the VANET network's performance.

The simulations' results clearly showed the positive effect of using the SDN paradigm on the VANET network. This integration considerably reduces CPU consumption and Flow rule installation delay in the VANET network. Our solution represents an important step for the deployment of the VANET architecture in real Internet networks.

In future work, we will use the controller's controllability to enhance other features of VANET, such as intelligent control of network resources.

References

1. Bhatia, J., Modi, Y., Tanwar, S., Bhavsar, M.: Software defined vehicular networks: a comprehensive review. Int. J. Commun Syst **32**(12), e4005 (2019). https://doi.org/10.1002/dac.4005
2. Mishra, R., Singh, A., Kumar, R.: VANET security: issues, challenges and solutions. In: 2016 International Conference on Electrical, Electronics, and Optimization Techniques (ICEEOT). IEEE (2016). https://doi.org/10.1109/iceeot.2016.7754846
3. Al-Heety, O.S., Zakaria, Z., Ismail, M., Shakir, M.M., Alani, S., Alsariera, H.: A comprehensive survey: benefits, services, recent works, challenges, security, and use cases for SDN-VANET. IEEE Access **8**, 91028–91047 (2020). https://doi.org/10.1109/access.2020.2992580
4. Islam, M.M., Khan, M.T.R., Saad, M.M., Kim, D.: Software-defined vehicular network (SDVN): a survey on architecture and routing. J. Syst. Archit. 101961 (2020). https://doi.org/10.1016/j.sysarc.2020.101961
5. Zhang, Y., Cui, L., Wang, W., Zhang, Y.: A survey on software defined networking with multiple controllers. J. Netw. Comput. Appl. **103**, 101–118 (2018). https://doi.org/10.1016/j.jnca.2017.11.015
6. Singh, S., Jha, R.K.: A survey on software defined networking: architecture for next generation network. J. Netw. Syst. Manag. **25**(2), 321–374 (2016). https://doi.org/10.1007/s10922-016-9393-9
7. Kristiani, E., Yang, C.-T., Huang, C.-Y., Wang, Y.-T., Ko, P.-C.: The implementation of a cloud-edge computing architecture using openstack and kubernetes for air quality monitoring application. Mob. Netw. Appl. **26**(3), 1070–1092 (2020). https://doi.org/10.1007/s11036-020-01620-5
8. Ku, I., Lu, Y., Gerla, M., Gomes, R.L., Ongaro, F., Cerqueira, E.: Towards software-defined VANET: architecture and services. In: 2014 13th Annual Mediterranean Ad Hoc Networking Workshop (MED-HOC-NET). IEEE (2014). https://doi.org/10.1109/medhocnet.2014.6849111
9. Ji, X., Yu, H., Fan, G., Fu, W.: SDGR: an SDN-based geographic routing protocol for VANET. In: 2016 IEEE International Conference on Internet of Things (iThings) and IEEE Green Computing and Communications (GreenCom) and IEEE Cyber, Physical and Social Computing (CPSCom) and IEEE Smart Data (SmartData). IEEE (2016). https://doi.org/10.1109/ithings-greencom-cpscom-smartdata.2016.70
10. Kazmi, A., Khan, M.A., Akram, M.U.: DeVANET: decentralized software-defined VANET architecture. In: 2016 IEEE International Conference on Cloud Engineering Workshop (IC2EW). IEEE (2016). https://doi.org/10.1109/ic2ew.2016.12
11. Ge, X., Li, Z., Li, S.: 5G software defined vehicular networks. IEEE Commun. Mag. **55**(7), 87–93 (2017). https://doi.org/10.1109/mcom.2017.1601144
12. Liu, J., Wan, J., Zeng, B., Wang, Q., Song, H., Qiu, M.: A scalable and quick-response software defined vehicular network assisted by mobile edge computing. IEEE Commun. Mag. **55**(7), 94–100 (2017). https://doi.org/10.1109/mcom.2017.1601150
13. Liang, L., Ye, H., Li, G.Y.: Toward intelligent vehicular networks: a machine learning framework. IEEE Internet Things J. **6**(1), 124–135 (2019). https://doi.org/10.1109/jiot.2018.2872122
14. Khan, S., et al.: 5G vehicular network resource management for improving radio access through machine learning. IEEE Access **8**, 6792–6800 (2020). https://doi.org/10.1109/access.2020.2964697

15. dos Reis Fontes, R., Rothenberg, C.E.: Mininet-WiFi. In: Proceedings of the 2016 ACM SIGCOMM Conference. ACM (2016). https://doi.org/10.1145/2934872. 2959070

16. Ghosh, S., et al.: SDN-sim: integrating a system-level simulator with a software defined network. IEEE Commun. Stand. Mag. **4**(1), 18–25 (2020). https://doi.org/10.1109/mcomstd.001.1900035

Implementing a Scalable and Elastic Computing Environment Based on Cloud Containers

Damiano Perri[1,2]([✉])(iD), Marco Simonetti[1,2](iD), Sergio Tasso[2](iD),
Federico Ragni[3], and Osvaldo Gervasi[2](iD)

[1] Department of Mathematics and Computer Science, University of Florence,
Florence, Italy
damiano.perri@unifi.it
[2] Department of Mathematics and Computer Science, University of Perugia,
Perugia, Italy
[3] Information Systems Division, University of Perugia, Perugia, Italy

Abstract. In this article we look at the potential of cloud containers and we provide some guidelines for companies and organisations that are starting to look at how to migrate their legacy infrastructure to something modern, reliable and scalable. We propose an architecture that has an excellent relationship between the cost of implementation and the benefits it can bring, based on the "Pilot Light" topology. The services are reconfigured inside small docker containers and the workload is balanced using load balancers that allow horizontal autoscaling techniques to be exploited in the future. By generating additional containers and utilizing the possibilities given by load balancers, companies and network systems experts may model and calibrate infrastructures based on the projected number of users. Containers offer the opportunity to expand the infrastructure and increase processing capacity in a very short time. The proposed approach results in an easily maintainable and fault-tolerant system that could help and simplify the work in particular of small and medium-sized organisations.

Keywords: High availability · Docker · Load balancing · Elastic computing · Disaster recovery · High performance computing · Public cloud · Private cloud · Hybrid cloud

1 Introduction

The cloud is one of the biggest technological innovations of recent years [1,2]. In fact, it enables the creation of complex, reliable and available technological infrastructures that provide services of various kinds: from calculation services to storage ones, to servers for the contents' distribution via web pages. There are various forms of cloud: for example we can use a public cloud or set up a private cloud [3–5]. In the case of a public cloud, hardware provided by third-party companies is used, such as Amazon AWS, Microsoft Azure, Google Cloud,

© Springer Nature Switzerland AG 2021
O. Gervasi et al. (Eds.): ICCSA 2021, LNCS 12949, pp. 676–689, 2021.
https://doi.org/10.1007/978-3-030-86653-2_49

and so on. Through small initial investments, only what is actually used is paid for, and the hardware which is made available to the users, will be maintained by the companies.

In contrast to the public cloud, there is the private cloud. This term refers to infrastructures that are typically corporate, providing virtualised and reduntant hardware and highly reliable services for employees, associated staff, or a external users who consume the content produced by the company. In this case, the company will have the responsibility of maintaining and configuring the hardware needed to set up and build the infrastructure. This initial disadvantage, however, is counterbalanced by a huge strategic advantage: the data (either sensitive either ultra-sensitive data) is completely entrusted to the management of the company itself and does not have to transit to third-party servers, so this solution may be preferable for certain applications [6]. The private cloud is a type of development architecture in which the computational resources are reserved and dedicated for the organisation managing the system.

In the public cloud, on the other hand, the services offered by a provider use pools of machines that also accommodate other users. There is a third alternative, called hybrid cloud, which represents a combination of the two architectural strategies [7]. In the hybrid cloud, part of the services are given by a third-party provider, and some subsystems are allocated within the corporate platform. For example, it is reasonable to imagine a situation where the provider is demanded to maintain a certain number of virtual machines and a certain number of disks for data storage while a second pool of hard disks containing encrypted sensitive data or databases for permanent information storage, is located inside the private organisation.

The Covid-19 pandemic has highlighted a feature common to many countries around the world: many organisations and companies have inadequate infrastructure to meet the pressing needs arising as a result of the digitisation processes that have become extremely urgent. Our intent is to provide a model that can speed up the technological evolution of companies and organisations or improve their current computing infrastructure. The solution we propose, supported by practical experimentation, allows a transition from an obsolete infrastructure of an SME (small and medium-sized enterprise), representing a typical case, to a system based on a cluster whose services are deployed by docker containers, and delivered through a cluster of nodes configured according to the best practices of an High Availability (HA) approach. The idea sprang from a report of the Italian National Institute of Statistics (STAT), which examined active companies with at least 10 employees, finding out that the Information Technology (IT) infrastructure are still inadequate in many cases to their needs.[1]

2   Related Works

The cloud is being studied and analysed by many researchers around the world because of its strong capabilities. In the course of this decade, it is expected that

---

[1] http://dati.istat.it/Index.aspx?DataSetCode=DCSP_ICT.

the cost-benefit ratio will continue to increase and there will be new patterns of developments focused on the Internet of Things (IOT) [8], and other emerging technologies. The impacts that this type of architecture have on the environment are also studied. The data centres which are required by the Cloud, consume large amounts of energy and it is necessary to make accurate estimates of the pollution that will be produced in the next few years [9]. The cloud can also be used to make sites and content available for use in the world of education [10,11]. A number of researchers are tackling extremely topical and interesting subjects, e.g. techniques for implementing and exploiting Function as a Service (FaaS) [12]. FaaS are serverless systems, where programmers can insert their own snippets of code (Python, PHP, etc.) and call them up via APIs. The result is that virtual machines automatically execute the code, without the necessity of setting up a server. The cloud can also be used to do complex computational calculations that require expensive GPUs and CPUs to perform machine learning calculations [13–16].

Docker containers are lightweight cloud technologies that are dominating among IT solutions because they allow applications to be released faster and more efficiently than in the case of virtual machines. The adoption of Docker containers in dynamic, heterogenous environments and the ability to deploy and effectively manage containers across multiple clouds and data centers has made these technologies dominant and fundamental [17]. The improvements in terms of increased performance and reduced overhead have made the cloud container approach indispensable for building cloud environments that keep pace with the demands emerging from various application domains [18,19].

3   Towards Scalable and Reliable Services

Legacy service delivery architectures are based on a single, centralised server. The main disadvantage of this solution is its low maintainability and the lack of fault tolerance. If a hardware problem or a hacker attack occurred, service delivery might be compromised. Monolithic architectures also suffer from another disadvantage: they cannot effectively scale in case of peak demands.

Other types of architectures support scaling. Scaling can be of two types: vertical or horizontal [20–22]. Vertical scaling is defined as an operation that increases the machine hardware resources, for example, rising the number of vCPUs or the amount of GiB of RAM. Horizontal scaling is defined as the operation that creates replicas of the server (node) that provides the service. The new node is identical to the original one and will help it to respond to client requests. In the case of monolithic architectures, horizontal scaling is not possible.

Vertical scaling, on the other hand, involves shutting down the machine and upgrading the hardware. This is unsuitable and not acceptable by modern standards. The problem of fault tolerance must also be taken into account.

3.1   Disaster Recovery

Legacy architectures often do not have particularly complex plans for fault management or data loss [23,24]. Disaster recovery (DR) is the procedure implemented to restore the functionality of a system, suffering a disaster, a damage: for example, the loss of system data, the loss of user data, the compromise of security following a hacker attack or hardware damage due to a natural disaster. When planning internal disaster recovery policies, three objectives must be defined: the Service Level Agreement (SLA), the Recovery Time Objective (RTO) and the Recovery Point Objective (RPO) [25,26].

The SLA is a percentage value that indicates the minimum uptime that the system guarantees during the year: for example, if a system has an SLA value of 99%, it means that in one year it could be unavailable for 3 d 15 h 39 m 29 s. An SLA of 99.99% instead could accept an annual downtime of 8 h 45 m 56 s. The higher the desired SLA value, the higher the costs to build an architecture to meet our demand.

The RTO is the maximum acceptable time between the interruption of the service and the moment when it is restored. In the case of legacy architectures the RTO is generally about 48 h; for example, in case of hardware problems, spare parts must be found, repairs or replacements must be made, and the system must be restored.

The RPO indicates the maximum acceptable time between the last data backup and time data loss because of a disaster. This indicates how much data time, in terms of hours, we can accept to lose. Legacy architectures often rely only on RAID 1 storage systems (mirroring) and do not perform daily incremental and offsite backups. Unlike the SLA, we have that the lesser RTO and RPO time is required by the system, the higher the cost to achieve this requirement is. The types of disaster recovery plans [27] that can be implemented are summarised in Fig. 1.

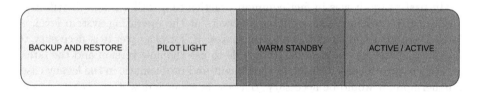

Fig. 1. Disaster recovery plans

Generally, legacy architectures implement a "Backup and Restore" type of architecture. This means that in case of critical problems, the only thing that can be done is to shut down the machine (if it is still working), restore the backup on the repaired machine or on a new machine, and restart the system with the data updated to the last backup. The "backup and restore" mode has an RPO in hours and an RTO in the 24 to 48 h range.

"Pilot Light" mode creates hourly or daily backups and maintains a complete copy of the architecture in a separate place, away from the system. In the event of a disaster, IT technicians will initiate backup and services will be restored. The "pilot light" mode guarantees an RPO in minutes and an RTO in hours.

Alternatively, there is the "Warm Standby" mode. In this mode, a backup system is replicated on a different location with respect to the main system and is synchronised in near real-time. This mode provides for a synchronisation in seconds (RPO) and an RTO time in minutes.

On the other hand, the "Active/Active" mode is the last possible architecture according to the diagram shown, and it is the safest and the most expensive from the point of view of service availability. It requires that two or more service delivery systems are always synchronised and active at the same time. A load balancing service is required to sort requests between the two systems. For example, an active/active system of the type 70/30 could be configured. A 70% of the requests will go to the main system and a 30% of the requests will go to the secondary backup system.

If the main system experiences a problem, the secondary backup system will be active and the user will not experience any particular problem. This type of architecture provides for an RPO around milliseconds' time, or even 0, and an RTO potentially equal to 0. It should be noted that, as we move towards the right-hand side of the Fig. 1, the protection and ability of the system to resist faults increases, so too the costs of implementing the architecture.

There are several problems with Legacy architectures: first, we can be exposed to ransomware hacking attacks that can irretrievably destroy our data store; second, there could be a software problem or a badly crafted query by a software developer that would irrevocably wipe out the database; therefore, other kinds of software problems can exist: minor portability issues among operating systems.

An example of legacy architecture with no virtual machines is shown in the Fig. 2A.

In past years, it was in fact common practice to install a Linux distribution and directly configure the application servers at the operating system level. A further problem is related to system updates. If, for example, it is necessary to update the PHP server in order to be able to use the new version and the latest security improvements, this might be difficult and problematic in the legacy case. Furthermore, it would be necessary to assess how much downtime we can accept in order to carry out the update; but there would be no guarantee that a rollback could be carried out quickly in the event of incompatibility problems.

3.2   Legacy with Virtual Machines

A slightly better solution, falling however under legacy configurations, is to use several Virtual Machines (VMs), one for each service-providing application. An example is shown in the Fig. 2B. In the second scenario, using Virtual Machines, programs may be moved from one system to another, because virtual machines are kept in one or more files readily transferred across devices. The side effect is that they have a significant hardware impact. Since a complete system needs to be executed, a lot of hardware (especially RAM) resources are used even when not required by the applications. In addition, further clock cycles are used to perform and maintain the secondary operating system activities, which also waste energy. The installation of the operating system also requires that virtual machines occupy a considerable amount of disk space, each time that a new one is generated. There are consequently two benefits of this architecture: the ability to make backups quickly, for example by duplicating the Virtual Machine-related files and the independence from the hardware subsystem and operating system.

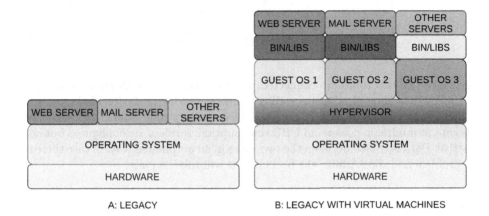

Fig. 2. Legacy architecture

4   The Proposed Architecture

In this section we describe the main techniques we used to implement a modern, reliable and highly available system (HA). The technology behind the proposed architecture is based on the use of Docker containers, which have the enormous advantage of being much leaner and more efficient than a virtual machine. Containers also make it possible to isolate an application at the highest level, making it a completely separate entity.

The first step is to design and configure the services and applications using Docker containers, as described in Subsect. 4.1 [28, 29]. Docker also guarantees high security thanks to its container architecture with separate storage spaces and access permissions [30]. In the Subsect. 4.3 we explain the way we can set up the databases in a Master/Slave configuration.

Once the services are up and running inside the containers, a distributed, redundant network environment must be prepared using load balancers, as described in Sect. 4.2. A networked, redundant and available file system must then be set up as described in Sect. 4.4. The proposed architecture is shown in the Fig. 4. As it is shown, two availability zones are configured according to the Pilot Light

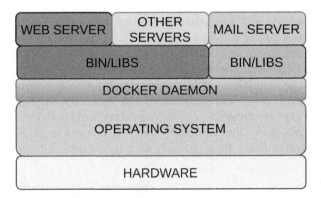

Fig. 3. Docker architecture

scheme. The two zones are connected to each other via a VPN connection. The primary zone allows horizontal scaling thanks to the use of the load balancer and small servers inside docker containers. The secondary zone is on IDLE state and is kept at minimum power and CPU consumption; services are configured but not active. Data is copied between the two zones in an automated way. In case there is a problem and zone A fails, the backup zone will take its place.

4.1 Docker

Docker allows to build an architecture as shown in the Fig. 3. Each service, such as the Web Server or the Mail Server, is encapsulated within its own container. Containers are defined by "yml" files, and an example of them is given in the code shown in Listing 1.1, where a Web Server Apache container is defined, exposing the HTTP and HTTPS ports. Each "yml" file may contain the definition of one or more containers. Containers are extremely light from a computational point of view and do not instantiate a real operating system; the applications that run inside them only need to allocate the libraries and binaries necessary for the application to work. Containers are stateless by definition, which means they have no true running state. To ensure data consistency, we should mount the folders of the filesystems where we want to execute input/output operations in a permanent way within them.

4.2 The Load Balancing Service

Load balancers are fundamental to the implementation of an HA architecture. They are customizable and can be adapted to a wide range of applications. The standard task of a load balancer is to distribute incoming requests to a pool of worker nodes. The nodes will process the requested information and if necessary provide the output to the users. The requests load can be managed in two different ways: balanced and unbalanced.

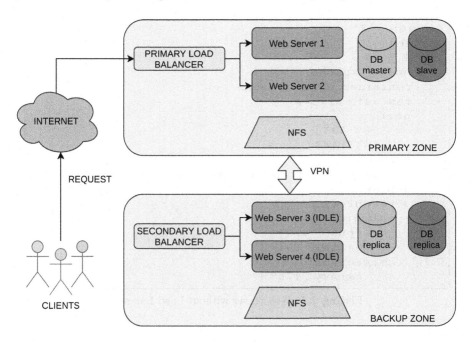

Fig. 4. Pilot light architecture

In the balanced mode each node receives a quantity of requests equal to $1/N$, with N equal to the number of nodes. This type of balancing can be implemented in the case of an Active/Active architecture.

In unbalanced mode, percentage values can be defined to indicate the amount of requests each node will receive. For example, a 60/40 configuration allows 60% of requests to be sent to Node1 and 40% of requests to Node2.

Moreover, there are various request scheduling algorithms [31,32]. The first algorithm is called round-robin. Requests are sorted cyclically across nodes using the round-robin algorithm. This method ensures that nodes receive an equal amount of requests regardless of their CPU use or complexity. A second algorithm is Least Outstanding Requests (LOR) which sorts requests across nodes trying to balance the number of "unprocessed requests". In our architecture, two load balancers are configured. The first is HAproxy placed in a docker container. HAproxy only exposes ports 80 HTTP and 443 HTTPS [31]. Its role is to obtain requests from clients and sort them within the primary zone, evenly distributing the workload across nodes. So, The file `haporxy.cfg` must be configured. The Listing 1.2 defines a new docker container with the image haproxy. The container will expose the correct ports and the custom configuration file "myHaproxy.cfg" will be mounted into the file system.

```
 1  version: '3'
 2  services:
 3    web:
 4      image: apache
 5      container_name: apache_web
 6      restart: always
 7      ports:
 8        - "80:80"
 9        - "443:443"
10      volumes:
11        - "/home/user/myWebsite:/var/www/html/"
12      deploy:
13        resources:
14          limits:
15            cpus: '4.0'
16            memory: 2048M
17          reservations:
18            memory: 1768M
```

Listing 1.1. Web Server without Load Balancer

4.3   The Database Service

A MariaDB RDBMS cluster is configured by defining 'yml' files in the Master/Slave configuration [33,34]. In a "yml" file it is in fact possible to define more containers by adding more elements in the "services" branch.

We have provided a single master database whose data is directly saved to the system disk. As this is a master slave mode, it is important to enable logging on the file system so that debugging can be carried out and configuration errors can be analysed. The Slave database containers have a "yml" definition similar to that of the Master: differences only concern the names and IP addresses. We will now describe how the load balancer for Databases is configured. The docker image used is mariadb/maxscale [35]. First we need to publish the MariaDB MaxScale REST API, an HTTP interface, which generates data in JSON format, offering visual management tools. MariaDB MaxScale splits requests in

```
$ docker-compose exec maxscale maxctrl list servers
```

Server	Address	Port	Connections	State	GTID
server1	master	3306	0	Master, Running	0-3000-5
server2	slave1	3306	0	Slave, Running	0-3000-5

Fig. 5. Database: master and slave

```
1  version: '3'
2  services:
3   haproxy:
4    image: haproxy:2.3.5
5    hostname: haproxy
6    ports:
7     - 80:80
8     - 443:443
9    volumes:
10    - /myHaproxy.cfg:/usr/local/etc/haproxy/haproxy.
        cfg
11   deploy:
12    placement:
13     constraints: [node.role == worker]
```

Listing 1.2. "Definition of the haproxy"

such a way that write instructions are sent to the Master container and read instructions are balanced among the Master and Slave containers. A specific user for maxscale, with **GRANT ALL** privileges, must then be defined, acting in the Master database. MariaDB MaxScale is distributed with a BSL (Business Source License) and is capable of doing much more than just load balancing: it also has the ability to perform failover and switchover. The failover mode allows to monitor and operate even if one of the nodes in the database is in an unhelpful state. If, for example, the Master database were to crash, the Max Scale load balancer would be able to promote the Slave database to the role of new Master. The configuration involves a few steps: definition of the servers, creation of the monitoring service, definition of the traffic routing using **readwritesplit** mode and finally the configuration of the listner using the mariadbclient protocol and its TCP port. The resulting configuration is shown in Fig. 5.

4.4 Network File System

The file system must be highly available, reliable and distributed. Nodes must be able to access the file system where they will safely store data, regardless of where it is located within the infrastructure. To do this, we have chosen to implement the GlusterFS network file system [36,37]. GlusterFS is distributed, open source and highly scalable.

The file system of the individual nodes has been configured to use XFS [38]. GlusterFS includes commands for defining the list of trustworthy servers that comprise the trusted pool for sharing disk space for replication. They are defined by executing the command **gluster peer probe <hostname>** command, specifying the various nodes. In order to create the gluster volume, it is necessary to specify in sequence: the volume name, the type (e.g. Replica), and the nodes involved with their brick paths.

You need to authorise the four nodes running GlusterFS to connect to the created volumes. To do this, we need to specify the IP addresses for each node that we want to connect to the gluster volume. The parameter to use is `auth.allow` and then start the volume. Finally, the common folder where the data will be stored must be created, for example `/var/gvolume`. This folder must be created in each node that will use it. To mount it, simply use the specific script `mount.glusterfs`. In order to speed up the start-up time of the glusterd daemon and ensure that the reboot process is automated, we recommend automount. GlusterFS is one of the easiest persistent storage solutions to implement, combined with the use of SSD disks, which are now widely used.

4.5   Scaling

Scaling is simple to implement adopting our recommended design. Our architecture enables to expand the computational power available in the system without requiring substantial structural modifications or shutting it down. Assume, for example, that we expect to have five times as many users in November as we do the rest of the year. In this situation, we might install more nodes, for example, by adding containers containing a Web Server instance. The HAproxy load balancer will take care of sorting requests across nodes as described in Sect. 4.2. Because our design involves operating in a private cloud environment, it is possible that we may need to employ more hardware and processors. The docker setup, on the other hand, comes to our rescue. Since all services are described by "yml" files, it will not be necessary to reconfigure the machines from scratch. All is needed is to launch new containers within the Linux distribution and tell HAproxy which new IPs are to be used in the pool of web servers. These procedures may be carried out without ever shutting down or disrupting the infrastructure.

5   Conclusion and Future Developments

Until recently, the term "cluster" was associated with huge corporations and data centres. Thanks to the Open Source software available, everyone has the opportunity to deploy a Docker Cluster. Containers allow for the expansion of infrastructure and the increase of computing capability that may be provided in a relatively short period of time. Companies and network system engineers may model and calibrate infrastructures based on the estimated number of users by introducing additional containers and utilising the capabilities of load balancers. Autoscaling refers to more sophisticated approaches that can generate or delete containers based on the number of users currently present. We have achieved long-term dependability and a low RPO and RTO time using the Pilot Light model, which will ensure that we do not lose data and keep our services available to consumers in the case of various problems occur (such as hacker attacks or natural catastrophes). In the future, we want to offer pre-configured Docker images that consumers may freely utilise. Furthermore, we want to establish a pre-configured architecture utilising the Infrastructure as a Code (IAAC) paradigm, which allows the entire virtual structure to be described in a text file and then automated reproduced in the many organisations where it may be required.

References

1. Hayes, B.: Cloud computing. Commun. ACM **51**(7), 9–11 (2008). https://doi.org/10.1145/1364782.1364786. ISSN 0001-0782
2. Antonopoulos, N., Gillam, L.: Cloud Computing. Springer, London (2010). https://doi.org/10.1007/978-1-84996-241-4. ISBN 978-1-4471-2580-8
3. Basmadjian, R., De Meer, H., Lent, R., Giuliani, G.: Cloud computing and its interest in saving energy: the use case of a private cloud. JoCCASA **1**(1), 1–25 (2012). https://doi.org/10.1186/2192-113X-1-5
4. Doelitzscher, F., Sulistio, A., Reich, C., Kuijs, H., Wolf, D.: Private cloud for collaboration and e-Learning services: from IaaS to SaaS. Computing **91**(1), 23–42 (2011). https://doi.org/10.1007/s00607-010-0106-z
5. Li, A., Yang, X., Kandula, S., Zhang, M.: Comparing public-cloud providers. IEEE Internet Comput. **15**(2), 50–53 (2011)
6. Ren, K., Wang, C., Wang, Q.: Security challenges for the public cloud. IEEE Internet Comput. **16**(1), 69–73 (2012)
7. Li, J., Li, Y.K., Chen, X., Lee, P.P., Lou, W.: A hybrid cloud approach for secure authorized deduplication. IEEE Trans. Parallel Distrib. Syst. **26**(5), 1206–1216 (2014)
8. Buyya, R.: A manifesto for future generation cloud computing: research directions for the next decade. ACM Comput. Surv. **51**(5), 1–38 (2018). https://doi.org/10.1145/3241737. ISSN 0360-0300
9. Gill, S.S., Buyya, R.: A taxonomy and future directions for sustainable cloud computing: 360 degree view. ACM Comput. Surv. (CSUR) **51**(5), 1–33 (2018). https://doi.org/10.1145/3241038. ISSN 0360-0300
10. Santucci, F., Frenguelli, F., De Angelis, A., Cuccaro, I., Perri, D., Simonetti, M.: An immersive open source environment using godot. In: Gervasi, O., et al. (eds.) ICCSA 2020. LNCS, vol. 12255, pp. 784–798. Springer, Cham (2020). https://doi.org/10.1007/978-3-030-58820-5_56
11. Simonetti, M., Perri, D., Amato, N., Gervasi, O.: Teaching math with the help of virtual reality. In: Gervasi, O., et al. (eds.) ICCSA 2020. LNCS, vol. 12255, pp. 799–809. Springer, Cham (2020). https://doi.org/10.1007/978-3-030-58820-5_57
12. Sreekanti, V., et al.: Cloudburst: stateful functions-as-a-service. Proc. VLDB Endow. **13**(12), 2438–2452 (2020). https://doi.org/10.14778/3407790.3407836. ISSN 2150-8097
13. Biondi, G., Franzoni, V., Gervasi, O., Perri, D.: An approach for improving automatic mouth emotion recognition. In: Misra, S., et al. (eds.) ICCSA 2019. LNCS, vol. 11619, pp. 649–664. Springer, Cham (2019). https://doi.org/10.1007/978-3-030-24289-3_48
14. Perri, D., Simonetti, M., Lombardi, A., Faginas-Lago, N., Gervasi, O.: Binary classification of proteins by a machine learning approach. In: Gervasi, O., et al. (eds.) ICCSA 2020. LNCS, vol. 12255, pp. 549–558. Springer, Cham (2020). https://doi.org/10.1007/978-3-030-58820-5_41

15. Benedetti, P., Perri, D., Simonetti, M., Gervasi, O., Reali, G., Femminella, M.: Skin cancer classification using inception network and transfer learning. In: Gervasi, O., et al. (eds.) ICCSA 2020. LNCS, vol. 12249, pp. 536–545. Springer, Cham (2020). https://doi.org/10.1007/978-3-030-58799-4_39

16. Perri, D., Sylos Labini, P., Gervasi, O., Tasso, S., Vella, F.: Towards a learning-based performance modeling for accelerating deep neural networks. In: Misra, S., et al. (eds.) ICCSA 2019. LNCS, vol. 11619, pp. 665–676. Springer, Cham (2019). https://doi.org/10.1007/978-3-030-24289-3_49

17. Abdelbaky, M., Diaz-Montes, J., Parashar, M., Unuvar, M., Steinder, M.: Docker containers across multiple clouds and data centers. In: 2015 IEEE/ACM 8th International Conference on Utility and Cloud Computing (UCC), pp. 368–371 (2015). https://doi.org/10.1109/UCC.2015.58

18. Maliszewski, A.M., Vogel, A., Griebler, D., Roloff, E., Fernandes, L.G., Oa, N.P.: Minimizing communication overheads in container-based clouds for HPC applications. In: 2019 IEEE Symposium on Computers and Communications (ISCC), pp. 1–6 (2019). https://doi.org/10.1109/ISCC47284.2019.8969716

19. Zhang, W.Z., Holland, D.H.: Using containers to execute SQL queries in a cloud. In: 2018 IEEE/ACM International Conference on Utility and Cloud Computing Companion (UCC Companion), pp. 26–27 (2018). https://doi.org/10.1109/UCC-Companion.2018.00028

20. Gong, Z., Gu, X., Wilkes, J.: Press: predictive elastic resource scaling for cloud systems. In: 2010 International Conference on Network and Service Management, pp. 9–16. IEEE (2010)

21. Vaquero, L.M., Rodero-Merino, L., Buyya, R.: Dynamically scaling applications in the cloud. ACM SIGCOMM Comput. Commun. Rev. 41(1), 45–52 (2011)

22. Mao, M., Li, J., Humphrey, M.: Cloud auto-scaling with deadline and budget constraints. In: 2010 11th IEEE/ACM International Conference on Grid Computing, pp. 41–48. IEEE (2010)

23. Alhazmi, O.H., Malaiya, Y.K.: Assessing disaster recovery alternatives: on-site, colocation or cloud. In: 2012 IEEE 23rd International Symposium on Software Reliability Engineering Workshops, pp. 19–20. IEEE (2012)

24. Alhazmi, O.H., Malaiya, Y.K.: Evaluating disaster recovery plans using the cloud. In: 2013 Proceedings Annual Reliability and Maintainability Symposium (RAMS), pp. 1–6. IEEE (2013)

25. Chang, V.: Towards a big data system disaster recovery in a private cloud. Ad Hoc Netw. 35, 65–82 (2015)

26. Khoshkholghi, M.A., Abdullah, A., Latip, R., Subramaniam, S.: Disaster recovery in cloud computing: a survey (2014)

27. Hamadah, S.: Cloud-based disaster recovery and planning models: an overview. ICIC Express Lett. 13(7), 593–599 (2019)

28. Boettiger, C.: An introduction to Docker for reproducible research. ACM SIGOPS Oper. Syst. Rev. 49(1), 71–79 (2015)

29. Turnbull, J.: The Docker Book: Containerization is the new virtualization. James Turnbull (2014)

30. Combe, T., Martin, A., Di Pietro, R.: To docker or not to docker: a security perspective. IEEE Cloud Comput. 3(5), 54–62 (2016)

31. Prasetijo, A.B., Widianto, E.D., Hidayatullah, E.T.: Performance comparisons of web server load balancing algorithms on HAProxy and Heartbeat. In: 2016 3rd International Conference on Information Technology, Computer, and Electrical Engineering (ICITACEE), pp. 393–396. IEEE (2016)

32. Pramono, L.H., Buwono, R.C., Waskito, Y.G.: Round-robin algorithm in HAProxy and Nginx load balancing performance evaluation: a review. In: 2018 International Seminar on Research of Information Technology and Intelligent Systems (ISRITI), pp. 367–372. IEEE (2018)
33. Bartholomew, D.: Getting started with MariaDB. Packt Publishing Ltd (2013)
34. Wood, W.: MariaDB solution. In: Migrating to MariaDB, pp. 59–71. Springer, Berkeley (2019). https://doi.org/10.1007/978-1-4842-3997-1_5
35. Zaslavskiy, M., Kaluzhniy, A., Berlenko, T., Kinyaev, I., Krinkin, K., Turenko, T.: Full automated continuous integration and testing infrastructure for MaxScale and MariaDB. In: 2016 19th Conference of Open Innovations Association (FRUCT), pp. 273–278. IEEE (2016)
36. Boyer, E.B., Broomfield, M.C., Perrotti, T.A.: Glusterfs one storage server to rule them all. Technical report, Los Alamos National Lab. (LANL), Los Alamos, NM (United States) (2012)
37. Selvaganesan, M., Liazudeen, M.A.: An insight about GlusterFS and its enforcement techniques. In: 2016 International Conference on Cloud Computing Research and Innovations (ICCCRI), pp. 120–127. IEEE (2016)
38. Pawlowski, B., Juszczak, C., Staubach, P., Smith, C., Lebel, D., Hitz, D.: NFS version 3: design and implementation. In: USENIX Summer, Boston, MA, pp. 137–152 (1994)

Author Index

Printed in the United States
by Baker & Taylor Publisher Services